QUANTUM HALL EFFECT

QUANTUM HALL EFFECT

Editor

MICHAEL STONE

Department of Physics
University of Illinois at Urbana
USA

Published by

World Scientific Publishing Co. Pte. Ltd.
P O Box 128, Farrer Road, Singapore 912805
USA office: Suite 1B, 1060 Main Street, River Edge, NJ 07661
UK office: 57 Shelton Street, Covent Garden, London WC2H 9HE

British Library Cataloguing-in-Publication Data
A catalogue record for this book is available from the British Library.

First published 1992
Reprinted 2001

QUANTUM HALL EFFECT

ISBN 981-02-0883-9
ISBN 981-02-0884-7 (pbk)

Printed in Singapore.

CONTENTS

INTRODUCTION

This book is a compilation of reprinted papers on the theory of the quantum Hall effect. Although most of the "classic" papers are included, the collection is not complete. I have selected the papers with a view to illustrating some of the topological aspects of the quantum Hall effect and so making connection with other topics of contemporary interest in theoretical physics. Because of this I hope the book will be of value to researchers in many areas and not just to the student of condensed matter physics seeking an introduction to the Hall effect.

It is worth advertising the Hall effect to a larger readership. Good ideas are often invented independently in condensed matter and particle physics, and much effort would be saved if the interdisciplinary analogies were widely known. An early case of this occurred in 1929–31, when the anomalous Hall coefficients of *Al*, *Be* and other metals led to the discovery, by Peierls, Wilson, and Heisenberg, of the importance of "holes", as mobile charge carriers. During the same period, Dirac came up with the notion of anti-electrons as holes in an infinite sea of negative-energy electrons. The analogy seems obvious in retrospect, and the experimental support for holes should have led to an earlier acceptance of Dirac's exotic positive electron — but there is no evidence that either concept influenced the development of the other, and Dirac's hole does not seem to have been taken seriously until the experimental discovery of the positron in 1932. Sometimes a concrete model provided by a condensed matter phenomenon can lead to a better understanding of its more abstract relatives. In 1963–4, for example, the discoverers of the Higgs mechanism for mass generation in spontaneously broken gauge theories were encouraged by the analogy with the Ginzburg–Landau theory of the Meissner effect. Later, the renaissance of quantum field theory in the 1970s was aided by Wilson's appreciation of the equivalence between the continuum limit and second-order phase transitions, and by the consequent realization that perturbation-theory infinities were simply manifestations of the fractal self-similarity of the field configurations in the scaling region. Today, string theorists and others studying the roles of conformal invariance, analyticity, and topology in two-dimensional systems will find that the quantum Hall effect provides them with simple and concrete exemplars of such abstract concepts as holomorphic bundles, vertex operators, chiral bosons and Wess–Zumino–Witten models.

The reprinted papers are grouped, roughly by topic, into five chapters. Each chapter begins with a short introduction where I try to put the papers into a larger context. I then either give a more detailed overview of the topic, or explain some technical points not covered in the reprints themselves. These introductions and overviews are not meant to be comprehensive reviews. They do not contain an exhaustive set of references, and the works I do refer to are, by and large, those that I have found most useful for learning the subject. The same principle has governed my selection of the reprinted papers: they are not always the earliest works on a topic, but are those papers that I have read with profit, and so would like to have close at hand in a more convenient form than a disorganized heap of preprints or battered xerox copies.

Certain important topics have been omitted. For example, I have not discussed the important Landauer–Buttiker transmission-coefficient approach to the Hall effect. This formalism is indispensable in mesocopic systems where the topological approach cannot deal with the anomalous effects produced by selective populated edge-states, but including it would have led the book away from its principal thrust. For similar reasons, I have also omitted the topological instanton formalism, where localization field-theory is extended to include magnetic field effects. The second of these topics has been well reviewed in the book *The Quantum Hall Effect* edited by E. Prange and S. M. Girvin (Springer 1987). Other books that the reader may wish to consult are the reprint volume, *Quantum Hall Effect: A Perspective*, edited by A. H. MacDonald (Kluwer Academic Publishers 1989), which takes a broader historical approach than this present volume, and the more specialized book by T. Chakraborty and P. Pietilinen, *The Fractional Quantum Hall Effect* (Springer Series in Solid State Science vol. 85, Springer 1988).

There are many people I must thank for making this book possible. Some I thank for sharing their insights with me. They may not remember the casual remarks that cast sudden light in the murky shadows of my understanding, but I would yet be lost without such help from, among others: Duncan Haldane, Jason Ho, Steve Kivelson, Dung-Hai Lee, Qian Niu, Nick Read, Xiao-Gang Wen, Yong-Shi Wu, and Shou-Cheng Zhang. Especial thanks must go to my colleague Eduardo Fradkin, both for infecting me with his enthusiasm, and for his patient explanation of things that I found obscure but were obvious to him. I must credit my past and present graduate students, Frank Gaitan, Bill Goff, and especially Juan Martinez, for keeping me supplied with their ideas, and for correcting those of mine when they (all too often) needed it. I also thank Kyre Mithrandir, Dorothy Day, and Nan Hyland for keeping my spirits up, feeding me, and helping correct my English. Finally, this book would not exist without the gentle persuasion of Miss P. H. Tham of World Scientific, who convinced me not only that it was a worthwhile project, but also that it would involve only a finite amount of work on my part.

Michael Stone
Urbana 1992

Chapter 1

INTEGER EFFECT

1.1 Introduction

The integer quantum Hall effect (IQHE) was discovered in early 1980 by Klauss von Klitzing. He was investigating the charge-transport properties of high mobility two-dimensional electron gases at low temperatures and high magnetic fields, using samples provided by his collaborators, G. Dorda and M. Pepper[1]. Von Klitzing found that, for certain ranges of the magnetic field, the longitudinal resistance of his samples became very small. At the same values of the field, plots of the Hall conductance, the ratio of the transverse potential difference to the total current, against B showed flat regions, or ***plateaux***. He soon realized that the Hall conductance on the plateaux was constant from sample to sample, and approximately ne^2/h where n was an integer. As the accuracy of the measurements improved, it became clear that n was an integer to a very high precision. There was clearly something fundamental at work.

This project had been motivated by earlier suggestions [2,3] that interesting physics would be found in the Hall conductivity at low temperatures, but no one seems to have foreseen the almost exact quantization of the Hall conductivity. It is now possible to fabricate devices with n an integer to 1 part in 10^7, and in January 1990 the quantum Hall effect became the basis for the US standard of resistance. Long before this, the quantum Hall effect had bred much beautiful theory, a small part of which is reprinted in this volume.

For his discovery von Klitzing won the 1985 Nobel Prize in Physics.

1.2 Landau Levels

We can make no better beginning in our study of two-dimensional electron gases than by reviewing Landau's classic analysis of the quantum mechanics of an electron in a uniform magnetic field. His ideas underlie every subsequent development. We choose the ***Landau gauge*** where

$$A_x = By, \qquad A_y = 0, \tag{1.2.1}$$

and the single-particle Schrödinger equation takes the form

$$\left(-\frac{1}{2m}\partial_x^2 - \frac{1}{2m}(\partial_y - ieBx)^2\right)\psi(x,y) = E\psi(x,y). \tag{1.2.2}$$

Translational invariance in the y direction suggests writing

$$\psi(x,y) = e^{iky} f_n(x), \tag{1.2.3}$$

and substituting this in (1.2.2) reveals that $f_n(x)$ is an eigenfunction of the harmonic oscillator equation

$$\left(-\frac{1}{2m}\partial_x^2 + \frac{1}{2m}(k - eBx)^2\right) f_n(x) = E_n f_n(x). \tag{1.2.4}$$

From the harmonic oscillator ground-state, we generate a family of wavefunctions

$$\psi_k(x,y) = e^{iky} e^{-\frac{eB}{2}(x-k/eB)^2} \tag{1.2.5}$$

which are extended in the y direction but localized in x. These states all have the same energy

$$\epsilon_0 = \frac{eB}{m}\frac{1}{2}, \tag{1.2.6}$$

and, by taking suitable linear combinations of them, we can produce eigenstates that are localized in y and extended in x, or indeed states localized in both directions simultaneously.

There are also families of wavefunctions derived from the excited states of the oscillator. They have energies

$$\epsilon_n = \frac{eB}{m}(n + \frac{1}{2}). \tag{1.2.7}$$

The quantity $\omega_c = eB/m$ has dimension T^{-1} and is the frequency of the classical cyclotron orbit. The corresponding harmonic oscillator energy $\hbar\omega_c$ sets the scale for the physics. Taking m to be the free electron mass, we find, for $B = 1$ Tesla, that $\omega_c = .17 \times 10^{12}$ Hz, corresponding to an energy of about 10^{-4} eV, or ≈ 1.2 degrees Kelvin. The system should be at lower temperature than this for us to observe any effects of the energy quantization. Actually, in a solid, m is replaced by the *effective mass*, m^*, which depends on the band structure of the device being used, and may be about 30 times smaller than the free electron mass; the energy scale is then larger than 1.2 K.

Since the k in (1.2.5) is at our disposal, these *Landau levels* are highly degenerate. The simplest way to calculate the density of states is to assume periodic boundary conditions in the y direction. This restricts k to the set

$$k_n = \frac{2\pi n}{L_y}. \tag{1.2.8}$$

As noted earlier, our chosen gauge has the feature that the wavefunctions are extended in the y direction but sharply peaked about the abscissa, $x = k/eB$. When the system has a finite extent, L_x, in the x direction, then k must be restricted to a range $\Delta k = eBL_x$ to ensure that the states lie in the box. Ignoring boundary effects from when the wavefunctions begin to overlap the edges, we can estimate the number of available states as

$$N = \frac{eB}{2\pi} L_x L_y. \tag{1.2.9}$$

This degeneracy *per unit area* may be viewed as each state occupying an irreducible region of area $2\pi/eB$. For $B = 1$ Tesla this would be a square with sides of about $.6 \times 10^{-7}$ m, and there are

2.7×10^{14} states/m^2 — a typical electron density in a quantum Hall device. This density of states does not depend on the effective mass of the electron or any other material properties.

1.3 Elementary Theory of the Quantum Hall Effect

There is always current density

$$j_y = \frac{1}{2mi}(\psi^\dagger \partial_y \psi - (\partial_y \psi^\dagger)\psi) \tag{1.3.1}$$

associated with the state $\psi(x, y)$. For

$$\psi_k(x, y) = e^{iky} e^{-\frac{eB}{2}(x - k/eB)^2} \tag{1.3.2}$$

and after normalizing the wavefunction, the current is

$$j_y(x) = \sqrt{\frac{eB}{\pi}} \frac{1}{L_y} \frac{k - eBx}{m} e^{-eB(x-k/eB)^2}. \tag{1.3.3}$$

The integral of (1.3.3) over x would give the current, I_ψ, carried by the state, but this integral is zero because the wavefunction is symmetric about $x = k/eB$. Now an electric field breaks this symmetry. This field requires us to include a potential $V(x) = -eEx$ in the Schrödinger equation, but the solutions are still harmonic oscillator wavefunctions, although now centered about $x = k/eB + mE/eB^2$. The formula for the current retains the original $(x - k/eB)$ factor, so the asymmetry in the integrand results in a nonzero value of $I_\psi = eE/BL_y$. The current is in the $-y$ direction.

To obtain the *total* current, we must add the contributions from all occupied states. If the Fermi energy, ϵ_f, lies between Landau levels, the levels below ϵ_f will be full and those above, empty. Multiplying by the number of states in the lowest Landau level, we find it provides a total current of

$$I_0 = \frac{e^2 E}{2\pi} L_x. \tag{1.3.4}$$

We repeat this for a total of n filled Landau levels, and write the potential difference across the conductor as $\mathcal{E} = EL_x$. We find that

$$I_{tot} = \frac{ne^2}{2\pi} \times \mathcal{E}. \tag{1.3.5}$$

Up to now we have been indulging ourselves by using "theorists units", in which $\hbar = 1$, or $h = 2\pi$. Restoring Planck's constant to its proper place leads to

$$I_{tot} = \frac{ne^2}{h}\mathcal{E}. \tag{1.3.6}$$

This gives the Hall conductance.

In two dimensions *conductivity* is equal to *conductance* so the off-diagonal part of the Hall conductivity tensor is

$$\sigma_{xy} = \frac{ne^2}{h}. \tag{1.3.7}$$

There is no current in the direction of the applied field so the longitudinal conductivity σ_{xx} vanishes. Equivalently, there is no potential drop in the direction of the current so the longitudinal resistivity ρ_{xx} also vanishes. We have found that the Hall conductivity tensor is purely transverse, and depends only on the number of filled levels, and on a simple combination of fundamental constants.

A purely transverse conductivity tensor has the experimentally important consequence that the current flows along the equipotentials, and that the total current across a line joining a pair of equipotentials depends only on the potential difference between them. Armed with this observation, we may measure the Hall conductivity without requiring detailed knowledge of the geometry of the Hall device, or of the exact distribution of currents and electric fields within it. All we need is a relatively simple measurement of the total current and the transverse voltage.

When precise measurements are made, it is found that, as long as the sample is good enough to observe the quantum Hall effect, the ratio of the longitudinal current to the Hall voltage is an integer times e^2/h to a precision of a few parts in 10^8. The theory presented above seems to be very good. Far *too* good in fact. It provides no explanation as to why the result is unaffected by the inevitable impurities and disorder, nor does it explain the plateaux. How can there be a *finite range* of magnetic field where the sample contains precisely the right number of electrons to fill an integral number of Landau levels? Remember that the number needed depends on the field.

1.4 Scattering Theory

The first reprinted paper [rep.1] contains a demonstration that elastic scattering from a single delta-function impurity does not affect the quantization of the Hall conductance. Prange later [4] gave an attractive and simplified generalization of this result, which I recount here.

We begin by looking at the effects of scattering centers on the current carried by a full Landau level. We will worry later about how the level came to be exactly filled. With the electric field included, the unnormalized wavefunctions are

$$\psi_k(x,y) = Ne^{iky}e^{-\frac{eB}{2}(x-k/eB-mE/eB^2)^2}, \tag{1.4.1}$$

and have a k dependent energy

$$\epsilon(k) = -eE\langle x\rangle = k\frac{E}{B} + em\frac{E^2}{B^2}. \tag{1.4.2}$$

From the energy we find a group velocity

$$v_y = \frac{\partial\epsilon}{\partial k} = -\frac{E}{B}, \tag{1.4.3}$$

which is just the *Hall drift velocity* at which the Lorentz force balances the electric field. In classical language, this is the velocity at which the guiding centers of the cyclotron orbits move.

Consider now the effect of an *elastic* scatterer, centered on the point x_0, y_0, on a state of given energy ϵ. Upstream of the scatterer (the electron is drifting in the $-y$ direction so this is $y > y_0$) the states have the form of (1.4.1). Downstream, $y < y_0$, the electron still has its initial energy and is therefore in a linear combination of states with this energy. There is only one such state — the initial one — and since the amplitude must be unaltered, the scattered wavefunction can differ from the incoming one only by a phase-shift factor, $\psi_k \to e^{-i\delta(k)}\psi_k$.

The scattering will be small for those ψ_k whose x range is such that the wavefunction does not overlap with the scattering center. This means that $e^{-i\delta(k)} \approx 1$, but does not necessarily mean that δ itself is small; it may be some multiple of 2π. To see the significance of the number of 2π's, we take periodic boundary conditions in the y direction by demanding

$$\psi(x, L_y) = e^{-i\phi}\psi(x, 0). \tag{1.4.4}$$

After including the phase-shift, (1.4.4) selects the allowed states to be at

$$k_n L_y + \delta(k) + \phi = 2\pi n. \tag{1.4.5}$$

When L_y is large, sums over states become an integral over k weighted by a density of states $\rho(k)$ derived from (1.4.5)

$$\rho(k) = \frac{\partial n}{\partial k} = \frac{1}{2\pi}(L_y + \frac{\partial \delta}{\partial k}). \tag{1.4.6}$$

If we start with no impurities and gradually switch on the scattering centers, the total number of states in the system must be unchanged, but some continuum states will become bound and no longer included among the states counted by (1.4.6). Their absence is detectable however: as we sweep k, and hence x, through a scatterer, the phase shift will decrease by $2\pi n_b$ where n_b is the number of states sucked out from that part of the continuum and bound to the potential. This is the Landau-level version of Levinson's theorem [5].

The Levinson theorem is rather pretty in itself, but, when combined with another simple argument, it becomes powerful. We use the phase ϕ that was quietly slipped into the periodic boundary condition in (1.4.4). Including ϕ is gauge equivalent to including a constant $eA_y = \phi/L_y$ in the Schrödinger equation. Since the current operator j_y is the functional derivative of the Hamiltonian with respect to A_y, it is not hard to see that the current carried by a state of fixed n is given by

$$I(k) = -e\frac{\partial \epsilon}{\partial \phi} = e\frac{\partial \epsilon}{\partial k}\left(L_y + \frac{\partial \delta}{\partial k}\right)^{-1}. \tag{1.4.7}$$

We recognize the inverse of the density of states in this expression. When we sum over all scattering states (the bound states are insensitive to ϕ and carry no current) in some interval of the x axis, the density of states factor $\rho(k)$ cancels, and we find

$$I_0 = e\int_{k_1}^{k_2} dk\rho(k)\frac{\partial \epsilon}{\partial k}(2\pi\rho(k))^{-1} = e\frac{(E(k_2) - E(k_1))}{2\pi}. \tag{1.4.8}$$

The total current carried by the Landau level is seen to depend only on the difference in energy of the states at the left- and right-hand edges of the strip of conductor! This energy difference is the electro-chemical potential difference $\mathcal{E}$ between the edges, and is what would be measured by a voltmeter as the Hall voltage.

Equation (1.4.8) shows that scatterers do not affect the current carried even if they have bound some of the continuum states. It must be that the remaining states "speed up" and carry proportionally more of the current.

To develop some intuition about the bound states, we can use a semiclassical approximation [rep.2]. In this approximation, the electrons drift along the equipotentials and the allowed states are governed by a quantization principle which selects orbits enclosing an integer number of flux quanta

$$\int_{orbit} Bd^2x = 2\pi N. \tag{1.4.9}$$

For any sufficiently strong potential there will be closed equipotentials satisfying this condition, even if the potential is repulsive. Indeed, if there is no electric field, nearly all the impurity equipotentials form closed loops around local maxima or minima of the potential. These loops are "localized" states. There will be at most one energy at which an infinitely long extended equipotential may exist. This is the unique energy corresponding to the equipotential "shoreline" at which the potential maxima cease to be the peaks of islands in an infinite sea, and instead the minima become the deep points of isolated lakes in an infinite continent. (This percolation localization is rather different from Anderson localization, which is due to coherent backscattering.)

Now, a bonus! Since the trapped states carry *no* net current, it cannot matter whether they are occupied or not. It is the irrelevence of electrons missing from these states that provides the explanation as to why we do not need exactly enough electrons to fill an integer number of Landau levels. It also explains why the Hall plateaux are larger for low currents: at low potential gradients there are more irrelevent bound states. The details of this argument are in [rep.2], and are also discussed in chapter III of [4].

1.5 Laughlin's Gauge Principle

The arguments of the last section have a gone a long way towards explaining the physics of the integer Hall effect. The remaining puzzle lies in (1.4.7), where the vital cancellation of the density of states factor, the only part of the expression to depend on the scattering processes, seems miraculous. Now, at about the same time as Prange was showing that scatterers did not affect the current, Laughlin made the first of his incisive contributions to the theory of the quantum Hall effect by relating the Hall conductance to a fundamental physical principle — Gauge covariance [rep.3]. This principle explains the miracle.

Laughlin's argument, and its subsequent exegesis by Halperin [rep.4], is in some sense a reformulation of the the scattering formalism presented in **1.4**. But, more significantly for future developments, it is also intimately related to a concept from the theory of anomalies in Quantum Field Theory. These anomalies are in their turn physical applications of the Index theory of elliptic operators (see [6] for an account for physicists), some of the deepest mathematical results of recent years. I am fairly confident, though, that Laughlin's argument was conceived via physical insight, and without the benefit of these latter connections.

The underlying idea is to make an round-trip excursion in the space of gauge-equivalent hamiltonians, *i.e.*, a trip that starts and finishes at hamiltonians which are equivalent up to a gauge transformation. Such hamiltonians must have identical spectra. This identity applies, however, only to the spectrum regarded as a set. Even if we chose to label eigenstates so that they vary continuously with the hamiltonian, a round trip in gauge-equivalent hamiltonian space does not guarantee that *individual eigenstates* return to their original places. When the space of gauge transformations is disconnected, an eigenstate may well end up being gauge equivalent to a different eigenstate of the starting hamiltonian. This permutation of the spectrum is called *spectral flow*.

Laughlin's realization of this idea was to continously vary the phase ϕ of (1.4.4) from 0 to 2π. After a gauge transformation

$$\psi(x,y) \to \psi(x,y)e^{i\phi y/L_y}, \tag{1.5.1}$$

this is equivalent to varying

$$A_y \to A_y + 2\pi/L_y. \tag{1.5.2}$$

The first representation shows that we have a path in hamiltonian space which returns the set of energy levels to its starting point. Intermediate points, where ϕ is not equal to 0 modulo 2π, give different energy spectra, and are not gauge equivalent. In the absence of impurities, the state with quantum number n marches in the $-x$ direction through a distance equal to $2\pi/eBL_y$ and ends up being gauge equivalent (if we vary A_y) or identical (if we vary ϕ) to the state with $n \to n-1$. Even if there are scatterers in the region $0 < d < x < L_x - d < L_y$, this "register shift" must occur for states in the intervals $[0,d]$ and $[L_y - d, L_y]$. If, in addition, there are no unoccupied states to be redistributed inside the disordered region, the net result is to transfer one electron from side of the system to the other. The eEx potential then gives a calculable change in the energy of the system. From this Laughlin obtained $d\epsilon/d\phi$, and so by similar arguments to (1.4.7), the current (for details see [rep.3]). The result is a formula identical in content to (1.4.8).

1.6 Fiber Bundles

The next set of reprinted papers [rep.5] – [rep.11] reveal some of the topological mysteries adumbrated in my discussion of [rep.3]. These papers explore detailed consequences of round-trip excursions in Hamiltonian space by assuming the electron gas is confined on manifolds, such as spheres and tori, that have no boundary. This absence of boundaries seems a technical complication, but it is one with surprising ramifications.

Begin by considering the Schrödinger problem for a particle moving on a flat torus. Typically we represent a torus as a rectangle, with the understanding that as a particle disappears through the right-hand boundary it immediately re-appears at a corresponding point on the left-hand boundary. In quantum mechanics, we implement this rule by taking periodic boundary conditions for the wave function

$$\psi(L_x,y) = \psi(0,y) \qquad \psi(x,L_y) = \psi(x,0). \tag{1.6.1}$$

These conditions make the wavefunction a well-defined continuous *function* on the torus in the sense that after pasting the edges of the rectangle together to make a real toroidal surface the function has no jumps, and each point on the surface assigns a *unique* value to ψ.

When we try to carry out the same program for a particle moving in a uniform magnetic field we at once meet a problem: while the magnetic field itself is constant there is no constant, or even periodic, choice for the gauge potentials. Earlier, for example, we chose the Landau gauge, $A_y = Bx$, $A_x = 0$; clearly the x in A_y is not a periodic function and cannot define A_y as a *function* on the torus. A_y has to be something other than a function. To make mathematical sense of the toroidal periodicity we must combine translations with gauge transformation. Specifically, as the particle moves out of the right-hand edge of the rectangle representing the torus, we must perform the gauge transformation that prepares it for motion in the A_μ field it will encounter when it reappears at the left. We therefore modify the boundary conditions to

$$\psi(L_x,y) = e^{ieBL_xy}\psi(0,y), \qquad \psi(x,L_y) = \psi(x,0). \tag{1.6.2}$$

The y-dependent phase-factor resets the gauge potential A_y at the right-hand boundary back to its value at the left-hand boundary. This modification seems innocent but there is an important consistency constraint lurking in it: we can find the value of $\psi(L_x, L_y)$ from that of $\psi(0,0)$ by using either the first *or* the second equation of the pair in (1.6.2). Since we must find the same $\psi(L_x, L_y)$ whichever equation we use, we need to satisfy the condition

$$e^{ieBL_xL_y} = 1, \tag{1.6.3}$$

and the magnetic flux, BL_xL_y, through the torus must be quantized to be some whole number, N, of "flux units". That is

$$eBL_xL_y = 2\pi N. \tag{1.6.4}$$

The condition (1.6.4) is not unreasonable. If we had confined the particles in a box with rigid walls, the individual states in a Landau level will no longer be exactly degenerate in energy because those at the edges will have their energy pushed up. The formula (1.2.9) would then be only approximate. With periodic boundary conditions there are no potentials to lift the degeneracy, and there should be an integer number of exactly degenerate states in each level. This number is N.

The quantization condition has the consequence that we cannot continuously alter the flux through the finite torus. It also means that if we introduce a torus as an intermediate tool in an argument, as we do when we combine Bloch waves with a magnetic field [7], then physical effects may depend discontinuously on the field. This is illustrated by the fractal pattern of Hofstadter energy levels, and by the appearance of the diophantine equations in [rep.5] whose solutions determine the Hall conductance.

The wavefunction in (1.6.2) is no longer a function.† After rolling up the $L_x \times L_y$ rectangle to make a torus there is no longer a unique way to assign a value to ψ at any point. Instead it is some kind of *twisted* function. Such creatures are known in mathematic as *sections*. To be more precise, ψ is a section of a one-dimensional *fiber-bundle* over the torus.

In fiber-bundle language, the torus becomes the *base space*, and the gauge field A_μ a *connexion* on the bundle. (Being an former Mathematical Tripos student, and so brought up on Whittaker and Watson's "*Shew*" and "*Point of inflexion*", I have come to like the archaic spelling of the word for this technical sense; then we can use the modern spelling "connection" elsewhere without worrying about being misunderstood.) The B field is the *curvature* of the connexion.

The integral over the torus of $1/2\pi$ times the curvature of the connexion is the magnetic flux, which as we have seen is quantized. Inspection of (1.6.4) and (1.6.2) reveals that N measures the number of times the phase of the wavefunction is twisted as we go from $x = L_x, y = 0$ to $x = L_x, y = L_y$, gluing the right-hand edge wavefunction to back to the left-hand edge wavefunction. This twisting number is a *topological invariant*, and any continuous deformation in the problem must keep the flux and the phase winding number constant. This particular topological invariant is called the *first Chern character* of the bundle.

1.7 Adiabatic Transport and the Kubo Formula

Reprints [rep.5] – [rep.11] combine topology with linear response theory. When the conductivity tensor is purely transverse there is no dissipation, and the Kubo formula reduces to an application

†This should not be too surprising; the wave "function", ψ, in quantum mechanics is never really a *function* of space-time. In a moving frame ψ acquires factor of $\exp(-ikmx - mv^2t/2)$ [8], and this is no way for a self-respecting function of x and t to behave.

of the adiabatic theorem. The idea of the adiabatic theorem is to seek the solution to the time dependent Schrödinger equation

$$i\partial_t|\psi_0(t)\rangle = \hat{H}(t)|\psi_0(t)\rangle \tag{1.7.1}$$

by writing

$$|\psi_0(t)\rangle = \sum_{n=0}^{\infty} a_n|n,t\rangle \exp\{-i\int^t E_0 dt\}. \tag{1.7.2}$$

We choose the complete sets of states $|n,t\rangle$ to be eigenstates of the "snapshot" hamiltonian, $\hat{H}(t)$,

$$H(t)|n,t\rangle = E_n(t)|n,t\rangle. \tag{1.7.3}$$

Insert (1.7.2) into (1.7.1), take components and assume that $|\psi_0(t)\rangle$ stays close to $|0,t\rangle$. This leads to

$$\dot{a}_0 + a_0\langle 0|\partial_t|0\rangle \approx 0, \tag{1.7.4}$$

$$a_m \approx ia_0\langle m|\partial_t|0\rangle \frac{1}{(E_m - E_0)}. \tag{1.7.5}$$

Up to first order in time-derivatives of the states, we find

$$|\psi_0(t)\rangle = e^{-i\int^t E_0(t)dt + i\gamma_{Berry}} \left\{ |0\rangle + i\sum_{m\neq 0} \frac{|m\rangle\langle m|\partial_t|0\rangle}{E_m - E_0} + \ldots \right\}. \tag{1.7.6}$$

Berry's phase, $a_0 = \exp i\gamma_{Berry}$, is the solution of the differential equation (1.7.4). This phase-factor is needed to take up the slack between the arbitrary phase choice made when defining the $|n,t\rangle$, and the specific phase selected by the Schrödinger equation as it evolves the state. [9]

Equation (1.7.6) is only useful if the energy denominator $(E_m - E_0)$ is bounded away from zero. This requires that the state which we are trying to follow adibatically must never become degenerate with any other state. We will assume this to be true in the rest of this section — but bear in mind that it is an important physical assumption.

To compute the Hall conductivity, we apply (1.7.6) to a many-body hamiltonian $\hat{H}(A_x(t), A_y(t))$ on a torus of size $L_x \times L_y$. $\hat{H}$ depends on the vector potential of the fixed external magnetic field A^{mag}, and on the slowly varying perturbations θ and φ

$$\begin{aligned} A_x(t) &= A_x^{Mag} + \varphi/L_x, \\ A_y(t) &= A_y^{Mag} + \theta/L_y. \end{aligned} \tag{1.7.7}$$

The current operator $\hat{I}_x$ is given by

$$\hat{I}_x = \frac{\partial \hat{H}}{\partial \phi}, \tag{1.7.8}$$

and an external Hall voltage, $\mathcal{E}$, is imposed by setting $\dot{A}_y = E_y$ and $-L_y E_y = \mathcal{E}$. Then $\dot{\theta} = \mathcal{E}$.

From (1.7.6) we get

$$\langle\psi_0|\hat{I}_x|\psi_0\rangle = i\sum_{m\neq 0} \left(\frac{\langle 0|\hat{I}_x|m\rangle\langle m|\partial_\theta|0\rangle}{E_m - E_0} - \frac{\partial_\theta(\langle 0|)|m\rangle\langle m|I_x|0\rangle}{E_m - E_0} \right) \dot{\theta}. \tag{1.7.9}$$

We can simplify this by using (1.7.8) in conjunction with the identity

$$\langle m|\partial_\varphi|n\rangle = -\partial_\varphi(\langle m|)|n\rangle = \frac{\langle m|\partial_\varphi \hat{H}|n\rangle}{E_m - E_n}, \tag{1.7.10}$$

and find

$$\langle\psi_0|\hat{I}_x|\psi_0\rangle = -i\{\langle\partial_\varphi 0|\partial_\theta 0\rangle - \langle\partial_\theta 0|\partial_\varphi 0\rangle\}\mathcal{E}. \tag{1.7.11}$$

Here the notation $|\partial_\varphi 0\rangle$ is shorthand for $\partial_\varphi|0\rangle$, and similarly for φ.

The expression in parenthesis in (1.7.11) is nothing but the curvature associated with the Berry phase of the family of states $|0,\theta,\varphi\rangle$. From (1.7.4), the phase itself is given by integrating the one-form

$$\mathcal{A} = \langle 0|\partial_\theta 0\rangle d\theta + \langle 0|\partial_\varphi 0\rangle d\varphi. \tag{1.7.12}$$

Its curvature is the exterior derivative, or curl, of this,

$$\mathcal{B} = d\mathcal{A} = \{\langle\partial_\varphi 0|\partial_\theta 0\rangle - \langle\partial_\theta 0|\partial_\varphi 0\rangle\}\, d\varphi\wedge d\theta. \tag{1.7.13}$$

The set of rays $|0,\theta,\varphi\rangle$ constitutes a fiber bundle with the manifold of boundary angles as its base space. If we average (1.7.11) over the boundary angles θ,φ we are evaluating the Chern character of the bundle, so it is not surprising that there will be a quantization condition. Before we derive this condition, let us take time out to consider more general Hilbert space bundles.

1.8 Hilbert Space Bundles

Suppose we have a Hilbert space H and a collection of orthonormal sets of states $|n,x\rangle \in H$, $n = 1, N$, there being one set for each point, x, in a manifold M. For example, M may parametrize a set of hamiltonians $\hat{H}(x)$, and the $|n,x\rangle$ could be the N lowest energy eigenstates for $\hat{H}(x)$. These $|n,x\rangle$ will span an x-dependent subspace V_x of the Hilbert space H. We say that we have a fiber-bundle, or more specifically a *vector-bundle*, E_V, over the base space M. The fiber above x is the vector space V_x.

The topologically interesting properties of the bundle come from the way in which the subspaces V_x are twisted and glued to each other as we patch together the parameter space. We wish to be able to characterize this twisting. To do this we first introduce a *connexion* on the bundle, *i.e.*, a notion of parallel transport and a covariant derivative. Using the notation $|dn\rangle = d|n\rangle$ for the (vector valued) exterior derivative of $|n\rangle$, we define the connexion as a matrix valued 1-form

$$\omega_{nm} = \langle n|dm\rangle. \tag{1.8.1}$$

Since $d\langle n|m\rangle = 0 = \langle dn|m\rangle + \langle n|dm\rangle$, we have $\omega_{nm} = -\omega^*_{mn}$ and the matrix is an element of the Lie algebra of $U(N)$.

If the states $|n,x\rangle$ span the whole of H this would be the natural connexion because we would define parallel transport on M by keeping the vector fixed in H, only the choice of basis varying with x^μ. The covariant derivative of an x-dependent state $\sum a(n)|n\rangle$ would then be

$$\nabla_\mu\left(\sum_n a(n)|n\rangle\right) = \sum_n (|n\rangle\partial_\mu a(n) + a(n)\partial_\mu|n\rangle) \tag{1.8.2}$$

$$= \sum_n |n\rangle \left(\partial_\mu a(n) + \sum_m \omega_{nm} a(m) \right). \tag{1.8.3}$$

In this case we have (with the wedge product of forms being understood)

$$d\omega_{nm} = \langle dn|dm\rangle = \sum_p \langle dn|p\rangle\langle p|dm\rangle = -\sum_p \langle n|dp\rangle\langle p|dm\rangle = -\omega_{np}\omega_{pm} \tag{1.8.4}$$

or

$$\Omega = d\omega + \omega^2 = 0. \tag{1.8.5}$$

The 2-form Ω is the *curvature* of the connexion. Being zero in this case, the connexion is said to be *flat*.

When V is not the entire space H, we extend the set $|n\rangle$ to a basis of H by adding states that span the orthogonal complement subspace, W. Then ω may be decomposed into blocks

$$\omega = \begin{matrix} & V & W \\ \begin{matrix} V \\ W \end{matrix} & \multicolumn{2}{c}{\begin{pmatrix} \omega_{VV} & \omega_{VW} \\ \omega_{WV} & \omega_{WW} \end{pmatrix}} \end{matrix}. \tag{1.8.6}$$

The natural connexion on the bundle E_V is ω_{VV}, *i.e.*, the $N \times N$ submatrix ω_{nm}, $1 \le n, m, \le N$. This connexion transports the vector by attempting to keep it as parallel to itself as possible, but it must keep projecting the vector back into the chosen subspace, V_x. In this manner, for example, the Riemann connexion transports tangent vectors on a sphere or other surface. In quantum mechanics, with V a subspace of degenerate states for some family of hamiltonians, slowly varying the hamiltonian of the time-dependant Schrödinger equation evolves states within the subspace by parallel transporting them with this connexion. In the simplest case, where the space V is a bundle of one-dimensional rays, the transport provides Berry's phase and $\omega = \mathcal{A}$ is the Berry-phase "vector potential". For higher dimensional subspaces we have the non-abelian generalization of Berry transport introduced by Wilczek and Zee [10].

Because of the projections, the curvature two-form

$$\begin{aligned} \Omega_V &= d\omega_{VV} + \omega_{VV}\omega_{VV} \\ &= -\omega_{VW}\omega_{WV} \end{aligned} \tag{1.8.7}$$

is no longer zero, but depends on the extent to which the states try to poke out of the subspace.

It is sometimes useful to express the curvature in terms of $P = \sum |n\rangle\langle n|$, the projector onto the subspace V. An easy exercise [rep.9] [rep.10] shows that

$$\sum |n\rangle \Omega_{nm} \langle m| = P(dPdP)P. \tag{1.8.8}$$

The degree of topological twisting of the bundle is measured by *characteristic classes*. The ones relevent here are called the *Chern characters*. They are defined by, [6]

$$ch_n(E_V) = \mathrm{Tr}\left[\frac{1}{n!}(\frac{i}{2\pi}\Omega_V)^n\right]. \tag{1.8.9}$$

Like the ordinary magnetic field strength, the ch_n are closed differential $2n$-forms, and their integrals over $2n$-dimensional submanifolds of M are integer-valued topological invariants of the bundle. They do not change if one alters the connexion continuously.

Using $\Omega_V = -\omega_{VW}\omega_{WV}$ and the cyclic symmetry of traces, we find that

$$ch_n(E_V) + ch_n(E_W) = 0. \tag{1.8.10}$$

This result is an example of a general theorem: the bundle obtained by forming the direct sum of two vector spaces over a common base space, called the *Whitney sum*, and written $E_V \oplus E_W$, has

$$ch_n(E_V \oplus E_W) = ch_n(E_V) + ch_n(E_W). \tag{1.8.11}$$

In (1.8.10), $ch_n(E_V \oplus E_W)$ is zero, since $H = V \oplus W$ and the bundle with the whole Hilbert space as its fiber is *trivial*. "Trivial" is here being used in the technical sense that the bundle is an untwisted product $E_H = H \times M$.

For the first Chern character the theorem is obvious, since

$$ch_1(E_V \oplus E_W) = \sum_{|n\rangle \in V \oplus W} \langle dn|dn\rangle = \sum_{|n\rangle \in V} \langle dn|dn\rangle + \sum_{|n\rangle \in W} \langle dn|dn\rangle. \tag{1.8.12}$$

Of particular interest is the bundle of one-dimensional bundle of N-fermion states $|\Psi\rangle_x$, each a Slater determinant of N single-particle states $|\psi_n\rangle$. The Berry connexion, $\mathcal{A}$, on the bundle of one-dimensional spaces E_Ψ is easily found by differentiating the Slater determinant row-by-row and taking the inner product. All the unaltered single-particle wavefunctions give unity, so

$$\mathcal{A} = \langle \Psi | d\Psi \rangle = \sum_n \langle \psi_n | d\psi_n \rangle \tag{1.8.13}$$

and

$$B = d\mathcal{A} = \sum_n \langle d\psi_n | d\psi_n \rangle. \tag{1.8.14}$$

The first Chern character of this line bundle is therefore equal to the first Chern character of the bundle having the N-dimensional spaces of single particle states as its fibers.

1.9 Hall Conductance as a Topological Invariant

Let us return to (1.7.11), and the bundle of one-dimensional rays of many-body states for the quantum Hall problem. We will show that the Hall conductance for any particular value of θ, φ is not quantized, but, when the ground state is non-degenerate, the *average* Hall conductance is an integer multiple of e^2/h.

For the moment, ignore electron-electron interactions and assume that we may take the states $|0, \theta, \varphi\rangle$ to be a Slater determinant $|\Psi_{\theta,\varphi}\rangle$. We need to evaluate

$$ch_1(\psi) = \frac{1}{2\pi i} \int_{T^2} d\theta d\phi \left(\langle \partial_\theta \psi | \partial_\phi \psi \rangle - \langle \partial_\phi \psi | \partial_\theta \psi \rangle \right) \tag{1.9.1}$$

$$= \frac{1}{2\pi i} \int_{\partial T^2} \left(\langle \psi | \partial_\phi \psi \rangle d\phi + \langle \psi | \partial_\theta \psi \rangle d\theta \right). \tag{1.9.2}$$

We select phases for the states by choosing arbitrarily at some point θ_0, ϕ_0, and then smoothly continuing the phase choice over the rectangle, $0 \geq \theta \leq 2\pi$, $0 \geq \phi \leq 2pi$. When the hamiltonian is periodic in the boundary angles, the states at corresponding boundary points must be equal up to a phase, so

$$|\psi_{2\pi,\phi}\rangle = e^{i\chi_1(\phi)}|\psi_{0,\phi}\rangle \tag{1.9.3}$$

and

$$|\psi_{\theta,2\pi}\rangle = e^{i\chi_2(\theta)}|\psi_{\theta,0}\rangle. \tag{1.9.4}$$

Comparing the phase of $|\psi_{0,0}\rangle$ with $|\psi_{2\pi,2\pi}\rangle$ shows that

$$\chi_1(2\pi) + \chi_2(0) - \chi_2(2\pi) - \chi_1(0) = 2\pi M \tag{1.9.5}$$

for some integer M. Inserting (1.9.3) and (1.9.4) into (1.9.2) shows that $M = ch_1\Psi$.

There is a slight technicality that should be discussed before we can apply this quantization condition directly to (1.7.11). We described two distinct formulations for introducing the boundary twists θ, φ into the problem. In the first they appear as phases in the boundary conditions, and in the second as gauge potentials. The hamiltonian is only manifestly periodic in θ, φ in the *first* formulation. In the second formulation we cannot write down equations like (1.9.3) and (1.9.4) since the equalities are true only up to gauge transformations. Furthermore, we are compelled to use the second formulation, since otherwise we have a different Hilbert space for each value of θ, φ and would have to do more work to define what is meant by inner products of wavefunctions satisfying different boundary conditions. Fortunately, we only need to use (1.9.3),(1.9.4) in the inner products in (1.9.2). Equality up to gauge transformations implies equality of inner products, and this is quite adequate for deducing that, when the ground state of the system is non-degenerate, the average Hall conductance is quantized in multiples of e^2/h.

We must now evaluate the number M and try to relate it to the number of filled Landau levels. To do this, we remember that we began by switching off electron-electron interactions. The resultant many-fermion Slater determinant states have a Berry-phase curvature that is a sum over the curvatures of *any* set of single-particle states spanning the same space as the states in the Fermi sea. There is no requirement that the sea states themselves be non-degenerate — only that they never become linearly dependent. To compute the sum over the occupied states, it suffices to use our theorem that the Chern character is additive under direct (Whitney) sums of bundles and is a topological invariant. In particular, we can compute for the unperturbed system. Here, although the individual states are degenerate, do not have one dimensional Berry phases, or even return to themselves as periodic functions of the boundary angles, we know the Hall conductance from section (1.2), and hence without calculating it we know that the Chern character of the N dimensional subspace spanned by each Landau level contributes unity to the sum.

When we introduce substrate interactions, the degeneracy of the states is generically lifted, the Landau level broadens into a band of states, and the individual states span rays that now return to themselves as the θ or φ wind through 2π. The Chern character of the level is now the sum of the Chern characters of the individual states, but it still totals unity for each level. Unless there is cancellation between different states, there will be only one state that contributes from each level. We declare any state that has zero Chern character to be "localized". As in section (1.3), we see that the band of states deriving from the Landau level is composed mostly of localized states, but each Landau level has at least one, and possibly only one, state that is extended, and they alone contribute to the linear response formula for the Hall current.† When the fermi surface lies among the localized states, *i.e., in a mobility gap*, its exact location is unimportant. These concepts are illustrated with considerable graphical beauty in [rep.11].

When electron-electron interactions are included, the states will no longer be simple Slater determinants, but, provided we can adiabatically switch the electron-electron interactions without the ground state ever becoming degenerate, the Chern character of the bundle of ground states cannot change — so neither can the Hall conductivity.

Appendix 1. Working with the Total Space

This appendix is designed to highlight a more sophisticated approach to fiber bundles than that used in the main text. I will demonstrate the power of fiber bundle concepts in organizing one's thoughts by using them to solve the quantum mechanics of a charged particle moving on a sphere surrounding a magnetic monopole [11][12]. The conventional approach to this problem involves selecting a gauge for the A field, which, because of the monopole, is necessarily singular at a Dirac string located somewhere on the sphere, and then delving into properties of Gegenbauer polynomials. Eventually one finds the gauge-dependent wavefunction. By working with the *total space* of the fiber bundles we can solve the problem in all gauges at once, and the problem becomes a simple exercise in group theory.

First we need some general definitions. The fiber bundles that arise in a general gauge theory with group G are called *Principal G-Bundles*, and the wavefunction are sections of *associated* bundles. A principal G-bundle comprises a manifold P called the *total space* together with a projection, π, to the *base space*, B, (in our case B will be the sphere S^2) such that the *fiber*, the inverse-image of a point $x \in B$, is a copy of the group G

$$\pi : P \to B, \qquad \pi^{-1}(x) = G. \tag{1A.1}$$

Locally P is required to be a product $P \equiv B \times G$, but globally it may be twisted. We locally parameterize P by ordered pairs (x, g) where $x \in B$ and $g \in G$. To define a bundle *associated* with P it is simplest to define the *sections* of the associated bundle. Let $\varphi_i(x, g)$ be a function on the total space P with a set of indices i carrying some representation $\rho(g)$ of G. We say that $\varphi_i(x, g)$ is a section of an associated bundle if it varies in a particular way as we run up and down the fibers by acting on them from the right with elements of G. We require

$$\varphi_i(x, gh) = \rho_{ij}^{-1}(h)\varphi_j(x, g). \tag{1A.2}$$

These functions are going to be the wavefunctions of the particle moving on the base space. The representation ρ plays the role of the charge, and (1A.2) are the gauge transformations.

The gauge field in which the particle moves is the connexion on the bundle. The formal definition of a connexion is a decomposition of the tangent space $T_p(P)$ of P at $p \in P$ into a *horizontal subspace*, $H_p(P)$, and a *vertical subspace*, $V_p(P)$. We require that $V_p(P)$ be the tangent space to the fibers and $H_p(P)$ to be a complementary subspace, *i.e.*, the direct sum should be the whole tangent space

$$T_p(P) = H_p(P) \oplus V_p(P). \tag{1A.3}$$

† Only one state may contribute to the Kubo formula, but this should not be taken to mean that in an actual device all the current is carried by one state. When we apply an external potential some of the localized states "ionize" and carry current.

The horizontal subspaces must all be images of each other under the action of G on the fibers from the right.

Given a curve $x(t)$ in the base space we can, by solving the equation

$$\dot{g} + \frac{\partial x^\mu}{\partial t} A_\mu(x) g = 0, \tag{1A.4}$$

lift it to a curve $(x(t), g(t))$ in the total space, whose tangent is everywhere horizontal. The A_μ are a set of Lie algebra valued functions, depending only on the choice of horizontal subspace, such that the vector $(dx, -A_\mu dx^\mu g)$ is horizontal for each small displacement dx in the tangent space to B. The directional derivative along the lifted curve is

$$\dot{x}^\mu D_\mu = \dot{x}^\mu \left(\left(\frac{\partial}{\partial x^\mu} \right)_g - A^a_\mu \bar{L}_a \right). \tag{1A.5}$$

In (1A.5), $\bar{L}_a$ is a right-invariant vector field on G, *i.e.*, a differential operator on functions on G. The D_μ are a set of vector fields, the *covariant derivatives*, which span the horizontal subspace at each point. They satisfy the relation

$$[D_\mu, D_\nu] = -F^a_{\mu\nu} \bar{L}_a. \tag{1A.6}$$

The curvature, $F_{\mu\nu}$, is given in terms of the structure constants of the Lie algebra of G $[\lambda_a, \lambda_b] = f^c_{ab}\lambda_c$ by

$$F^c_{\mu\nu} = \partial_\mu A^c_\nu - \partial_\nu A^c_\mu + f^c_{ab} A^a_\mu A^b_\nu, \tag{1A.7}$$

and the commutator of the covariant derivatives is proportional to $\bar{L}$ and therefore lies entirely in the vertical subspace.

We can make contact with the more familiar definitions of covariant derivatives by remembering that right invariant vector fields are derivatives that involve infinitesimal multiplication from the *left*. The usual definition is

$$\bar{L}_a \varphi_i(x, g) = \lim_{\epsilon \to 0} \frac{1}{\epsilon} \left(\varphi_i(x, (1 + \epsilon\lambda_a) g) - \varphi_i(x, g) \right). \tag{1A.8}$$

So, for a section of the associated bundle,

$$\begin{aligned} \bar{L}_a \varphi_i(x, g) &= \lim_{\epsilon \to 0} \left(\rho_{ij}(g^{-1}(1 - \epsilon\lambda_a) g) \varphi_j(x, g) - \varphi_i(x, g) \right) / \epsilon \\ &= - \rho_{ij}(g^{-1})(\lambda_a)_{jk} \rho_{kl}(g) \varphi_l(x, g) = \left((-g^{-1}\lambda_a g) \varphi \right)_i, \end{aligned} \tag{1A.9}$$

and

$$D_\mu \varphi = (\partial_\mu \varphi)_g + g^{-1} A_\mu g \varphi. \tag{1A.10}$$

This still does not look too familiar because the derivatives with respect to x_μ are being taken at *fixed* g. We normally *fix a gauge* by making a choice of g for each x_μ. The conventional wavefunction $\varphi(x)$ is then $\varphi(x, g(x))$. So, using $\varphi(x, g(x)) = g^{-1}(x)\varphi(x, 1)$, we have

$$\partial_\mu \varphi = (\partial_\mu \varphi)_g + \left(\partial_\mu g^{-1} \right) g \varphi = (\partial_\mu \varphi)_g - \left(g^{-1} \partial_\mu g \right) \varphi. \tag{1A.11}$$

From this we get a derivative on functions $\varphi(x)$ defined (locally) on the base space, B. This is the usual covariant derivative, now containing gauge fields which are gauge transformations of a fixed A

$$\nabla_\mu = \partial_\mu + (g^{-1}A_\mu g + g^{-1}\partial_\mu g) = \partial_\mu + \omega_\mu. \tag{1A.12}$$

These covariant derivatives on the base space obey

$$[\nabla_\mu, \nabla_\nu] = g^{-1}F_{\mu\nu}g = \Omega_{\mu\nu}. \tag{1A.13}$$

We usually regard the curvature, $\Omega_{\mu\nu}$, as a 2-form and write $\Omega \equiv \Omega_{\mu\nu}dx^\mu dx^\nu$. The connexion $\omega \equiv \omega_\mu dx^\mu$ is a 1-form. Then, using the exterior derivative, d, we can write compactly

$$\Omega = d\omega + \omega \wedge \omega. \tag{1A.14}$$

These operations and definitions still seem rather abstract. We see their power when we utilize them to construct the monopole bundle and its associated wavefunctions. Consider the representation matrices $D^{(J)}_{mn}(g)$. Now, $g \in G$ could be an element of any compact Lie group, but for concreteness think of $G = SU(2)$. The total space is then the manifold of $SU(2)$ which is the three-sphere S^3. The $D^{(J)}_{mn}(g)$ are thus functions on S^3 parameterized by the Euler angles. We will take a subgroup H as the gauge group, and the coset space G/H as the base space of a principal H-bundle. The projection π is just the usual projection $G \to G/H$, and the fibers are copies of H. The functions $D^{(J)}(g)$ are not in general *functions* on the coset space G/H since they depend on the choice of representative. Instead, because of the representation property, they vary with the choice of representative in a well-defined way,

$$D^{(J)}_{mn}(gh) = D^{(J)}_{mn'}(g)D^{(J)}_{n'n}(h). \tag{1A.15}$$

Since we are dealing with compact groups the representations can be taken to be unitary and

$$\begin{aligned}[D^{(J)}_{mn}(gh)]^* &= [D^{(J)}_{mn'}(g)]^*[D^{(J)}_{n'n}(h)]^* \\ &= D^{(J)}_{nn'}(h^{-1})[D^{(J)}_{mn'}(g)]^*.\end{aligned} \tag{1A.16}$$

This is the correct variation under the right action of the group H for the set of functions $[D^{(J)}_{mn}(gh)]^*$ to be sections of a bundle associated with the principal fiber bundle $G \to G/H$. The representation ρ is not necessarily that defined by the label (J) because irreducible representations of G may be reducible under H; ρ depends on what representation of H the index n belongs. If ρ is the identity representation, then the functions are functions on G/H in the ordinary sense. For $G = SU(2)$ and H the $U(1)$ subgroup generated by J_3, the quotient space is just S^2 and projection is the ***Hopf map***: $S^3 \to S^2$. Parameterizing $SU(2)$ with Euler angles

$$D^{(J)}_{mn}(\theta, \phi, \psi) = \langle J, m| e^{-i\phi J_3} e^{-i\theta J_2} e^{-i\psi J_3} |J, n\rangle \tag{1A.17}$$

shows that the Hopf map consists of simply forgetting about ψ, so Hopf:$[(\theta,\phi,\psi) \in S^3] \mapsto [(\theta,\phi) \in S^2]$. The bundle is twisted because S^3 is not a product $S^2 \times S^1$. Taking $n = 0$ in (1A.16) gives us functions independent of ψ, and we obtain the well-known identification of the spherical harmonics with representation matrices

$$Y_m^L(\theta,\phi) = \sqrt{\frac{2L+1}{4\pi}}[D_{m,0}^{(L)}(\theta,\phi,0)]^*. \quad (1A.18)$$

For $n = \Lambda \neq 0$ we get sections of the bundle. These are the monopole harmonics:

$$\mathcal{Y}_{m,\Lambda}^{(J)}(\theta,\phi,\psi) = \sqrt{\frac{2J+1}{4\pi}}[D_{m,\Lambda}^{(J)}(\theta,\phi,\psi)]^*. \quad (1A.19)$$

The monopole harmonics have a non-trivial dependence on the choice we make for ψ at each point on S^2, and we cannot make a globally smooth choice; we always encounter a point where there is a string singularity. Apart from being useful for the study of monopoles, these functions occur in molecular physics as the wavefunctions for the rotational degrees of freedom of excited diatomic molecules [13].

We must show that these harmonics are eigenfunctions of the Schrödinger operator, $-\nabla^2$, containing the gauge field connexion, just as the spherical harmonics are eigenfunctions of the laplacian on the sphere. This is a simple geometrical exercise: because they are irreducible representations, the $D^{(J)}(g)$ are automatically eigenfunctions of the quadratic Casimir operator

$$(J_1^2 + J_2^2 + J_3^2)D^{(J)}(g) = J(J+1)D^{(J)}(g). \quad (1A.20)$$

The J_i can be either right or left-invariant vector fields on G; the quadratic Casimir is the same second-order differential operator in either case, and it is a good guess that it is proportional to the laplacian on the group manifold. Taking a locally geodesic co-ordinate system (in which the connexion vanishes) confirms this: $J^2 = -\nabla^2$ on the three-sphere. The operator in (1A.19) is not the laplacian we want, however. What we need is the ∇^2 on the two-sphere, G/H, including the the connexion. This ∇^2 operator differs from the one on the total space since it must contain only differential operators lying in the horizontal subspaces. There is a natural notion of orthogonality in the Lie group, deriving from the Killing form, and it is natural to choose the horizontal subspaces to be orthogonal to the fibers of G/H. Since multiplication on the right by the subgroup generated by J_3 moves one up and down the fibers, the orthogonal displacements are obtained by multiplication on the right by infinitesimal elements made by exponentiating J_1 and J_2. The desired ∇^2 is thus made out of the left-invariant vector fields (which act by multiplication on the right), J_1 and J_2 only. The wave operator must be

$$-\nabla^2 = J_1^2 + J_2^2 = J^2 - J_3^2. \quad (1A.21)$$

Applying this to the $\mathcal{Y}_{m,\Lambda}^{(J)}$ we see that they are eigenfunctions of $-\nabla^2$ on S^2 with eigenvalues $J(J+1) - \Lambda^2$. The energy levels for our monopole problem therefore have energies

$$E_{J,m} = \frac{1}{2I}(J(J+1) - \Lambda^2). \quad (1A.22)$$

For general G/H there will be a similar construction although we will, in general, have to adapt the basis of the representation of G to the required representations of H, else we will not automatically find irreducible representations of H, and thus not get eigenvectors of ∇^2.

Reprints for Chapter 1

[rep.1] *Quantized Hall resistance and the measurement of the fine-structure constant*, R. E. Prange, Phys. Rev. B23 (1981) 4802-4805.

[rep.2] *Localization, percolation and the quantum Hall effect*, S. A. Trugman, Phys. Rev. B27 (1983) 7539-7546.

[rep.3] *Quantized Hall conductivity in two-dimensions*, R. B. Laughlin, Phys. Rev. B23 (1981) 5632-5633.

[rep.4] *Quantized Hall conductance, current-carrying edge states and the existence of extended states in a two-dimensional disordered potential*, B. I. Halperin, Phys. Rev. B25 (1982) 2185-2190.

[rep.5] *Quantized Hall conductance in a two-dimensional periodic potential*, D. J. Thouless, M. Kohmoto, M. P. Nightingale, M. den Nijs, Phys. Rev. Lett. 49 (1982) 405-408.

[rep.6] *Quantized Hall conductance as a topological invariant*, Q. Niu, D. J. Thouless, Y-S. Wu, Phys. Rev. B31 (1985) 3372-3377.

[rep.7] *Quantum Hall effect with realistic boundary conditions*, Q. Niu, D. J. Thouless, Phys. Rev. B35 (1987) 2188-2197.

[rep.8] *Topological invariant and the quantization of the Hall conductance*, M. Kohmoto, Ann. Phys. (NY) 160 (1985) 343-354.

[rep.9] *Quantization of the Hall conductance for general multiparticle Schrödinger Hamiltonians*, J. E. Avron, R. Seiler, Phys. Rev. Lett. 54 (1985) 259-262.

[rep.10] *Adiabatic quantum transport in multiply connected systems*, J. E. Avron, A. Raveh, B. Zur, Rev. Mod. Phys. 60 (1988) 873-915.

[rep.11] *Localization, wave-function topology, and the integer quantum Hall effect*, D. P. Arovas, R. N. Bhatt, F. D. M. Haldane, P. B. Littlewood, R. Rammal, Phys. Rev. Lett. 60 (1988) 619-622.

Other References for Chapter 1

[1] K. von Klitzing, G. Dorda, M. Pepper, Phys. Rev. Lett. 45 (1980) 494.

[2] T. Ando, Y. Matsumoto, Y. Uemura, J. Phys. Soc. Japan 39 (1975) 279.

[3] S. Kawaji, T. Igarishi, J. Wakabayashi, Prog. Theor. Phys. Suppl. 57 (1975) 176.

[4] *The Quantum Hall Effect* edited by E. Prange and S. M. Girvin (Springer, 1987).

[5] R. Joynt, R. E. Prange, Phys. Rev. B 29 (1984) 3303.

[6] T. Eguchi, P. B Gilkey, A. J Hanson, Phys. Rep 66 (1980) 215.

[7] D. G. Hofstadter, Phys. Rev. B14 (1976) 2239.

[8] G. Baym, *Lectures on Quantum Mechanics* (Benjamin/Cummings, 1973) Prob. 1, p 533.

[9] M. V. Berry, Proc. Roy. Soc London, Ser A 392 (1984) 45.

[10] F. Wilczek, A. Zee, Phys, Rev. Lett. 52 (1984) 2111.

[11] Y-S. Wu, Scientia Sinica 21 (1978) 193.

[12] M. Stone, Nuc. Phys. B 314 (1989) 557.

[13] see for example the footnote on p317 of L. D. Landau and E. M. Lifshitz *Quantum Mechanics (non relativistic theory)*, Course of theoretical physics v.3, Third edition (Pergamon Press, 1977).

Quantized Hall resistance and the measurement of the fine-structure constant

R. E. Prange
Department of Physics and Astronomy and Center for Theoretical Physics, University of Maryland, College Park, Maryland 20742
(Received 10 October 1980; revised manuscript received 17 February 1981)

An elementary, exact calculation of two-dimensional electrons in crossed electric and magnetic fields with a δ-function impurity is carried out in the quantum limit. A state localized on the impurity exists and carries no current. However, the remaining mobile electrons passing near the impurity carry an extra dissipationless Hall current exactly compensating the loss of current by the localized electron. The Hall resistance should thus be precisely h/e^2, as found experimentally by Klitzing *et al.* Other possible sources of deviation from this result are briefly examined.

In a recent Letter, v. Klitzing *et al.*[1] have reported a high-accuracy measurement of e^2/h (to one part in 10^5) which after improvements and together with the known value of the speed of light, c, potentially[2] would provide a measurement of the fine-structure constant of precision greater than that currently available (one part in 10^7). Their result is based on a measurement of the quantized Hall resistance in a two-dimensional electron gas, as realized in the inversion layer of a metal-oxide-semiconductor field-effect transistor. We here provide an elementary calculation which has a bearing on their result, and is a step in the direction of estimating theoretically the accuracy to which e^2/h is determined by the experiment.

Since free electrons which fill an integral number of Landau levels give a Hall resistance precisely an integral fraction of h/e^2, the problem is one of treating the imperfections, which might give rise to ordinary resistance, and/or to localized states which could cause the Hall resistance to deviate from its ideal value. (We do not treat the electrons as interacting, an omission which future work must remedy.) We here work out the instructive, elementary, and essentially exactly solvable case of two-dimensional electrons with a single δ-function impurity.

The main result is that (a) *a localized state exists,* which (b) carries *no current,* but (c) *the remaining nonlocalized states* carry an *extra Hall current* which *exactly compensates for that not carried* by the localized state. Thus, provided *all the nonlocalized states* of the appropriate Landau level are filled, the *total Hall current carried by the level is precisely the same as in the absence of impurities and localized states.*

We consider then free electrons of mass m ($=0.2m_e$ as appropriate for silicon) in the xy plane, subjected to an electric field E in the negative x direction, and a magnetic field B in the z direction. (The geometry is given in Fig. 1.) Spin and valley degeneracies are ignored, and attention is confined to the case in which only the ground Landau level is occupied. We choose m as the unit of mass, mc/eB as the unit of time, and $(hc/2\pi eB)^{1/2} \equiv l_B$ as the unit of length. (The cyclotron radius, $k_f hc/2\pi eB$, does not enter the problem.) The drift velocity, $cE/B = -e\Phi_H/W$ is denoted by v, where Φ_H is the Hall vol-

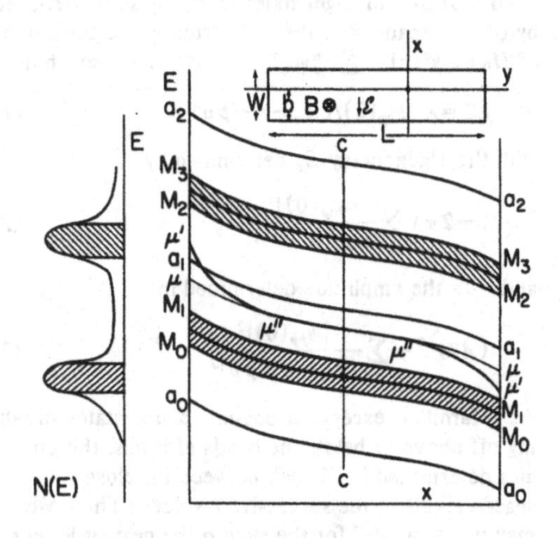

FIG. 1. The inset gives the geometry of the single-impurity problem. To the right below is the energy diagram of the lowest two Landau levels, as banded by impurities, and bent by the electric potential. The lines $a_l - a_l$ bound regions of energy states deriving from distinct levels. The lines $M_l - M_l$ are mobility edges separating localized and delocalized states. The hatched regions are those of delocalized states. The electrochemical potential is given by the line $\mu - \mu'' - \mu'' - \mu$. The position of this line depends on the total density of electrons, i.e., on the gate voltage. The states are occupied below the line $\mu' - \mu'' - \mu'' - \mu'$. The excess of electrons (holes) in the regions $\mu - \mu' - \mu''$ is the source of the Hall voltage. At the left is a schematic diagram of the density of states along the cut $c - c$.

tage, and W is the width of the device in the x direction. The Hamiltonian is $H = H_0 + 2\pi\lambda\delta^2(r)$, with $H_0 = \frac{1}{2}[-\partial^2/\partial x^2 + (p_y + x)^2] - vx$. The eigenstates of H_0 are

$$\psi_{np}(r) = \exp(ip_y y) H_n(x)\phi(x)/(2^n n! L)^{1/2} \ .$$

(The Landau gauge has been used.) The length of the sample in the y direction is L, H_n is the Hermite polynomial, $x = x + p$, $p = p_y - v$, and $\phi(x) = e^{-x^2/2}/\pi^{1/4}$. The eigenenergy corresponding to state ψ_{np} is $n + vp$. The values of p are $2\pi k/L$ with k integer and $-p$ ranges from $-b$ to $W-b$, where the impurity is at the origin of coordinates which is located a distance b above the lower edge of the sample. The total number of p values is thus $WL/2\pi$. This gives the degeneracy of the Landau level. For a sample as described in Ref. 1, we have in our units (with $B = 18$ T), $L = 7\times10^4$, $W = 8\times10^3$, and $v \sim 1/W \sim 10^{-4}$. We shall treat L and W as macroscopic and v as small, but it will be seen that $1/L$ is by no means the smallest number in the problem. The potential strength λ is of order unity, i.e., $\lambda \gg 1/L$.

To find the full eigenstates of H, the state, denoted by ψ^α, is expanded in the "unperturbed" eigenstates of H_0 as $\psi^\alpha(r) = \sum c^\alpha_{np}\psi_{np}(r)$. It is easily seen that

$$c^\alpha_{np} = A^\alpha \psi_{np}(0)/(E_\alpha - n - pv) \qquad (1)$$

with the eigenenergy E_α determined by

$$1 = 2\pi\lambda \sum_{np} \frac{|\psi_{np}(0)|^2}{E_\alpha - n - pv} \qquad (2)$$

and with the amplitude determined by

$$(A^\alpha)^{-2} = \sum_{np} \frac{|\psi_{np}(0)|^2}{(E_\alpha - n - pv)^2} \ . \qquad (3)$$

As is familiar, except for possible bound states breaking off above or below the bands of levels, the energies determined by (2) fall between the closely spaced levels of the successive p values. Thus, we may use as a label for the state α the nearest level of the system unperturbed by the impurity. For simplicity only levels belonging to the zeroth Landau level are considered.

We begin by making the approximation of retaining only the term $n = 0$ in the sums. This is suggested by the "strong magnetic field limit" which is usually taken in the literature.[3] The idea is that it is adequate to diagonalize the subspace consisting of the (nearly) degenerate states of one Landau level, if the levels are sufficiently separated. The current in this case may be obtained without the necessity of finding the explicit eigenfunctions and is completely independent of the form of the scattering potential. The (number) current operator in the x direction is $j_x = (1/i)(\partial/\partial x)$, and it is immediately seen that none of the states carries current along the electric field. In fact, for the exact solution to be found, the same is true. The current operator in the y direction is $j_y = p_y + x = p + x + v$. When this operator acts on a state ψ^α it gives $v\psi^\alpha$ plus a state orthogonal to ψ^α, because the operator $p + x$ changes the Landau level, and by assumption, there is only one Landau level in the sum defining ψ^α. Thus, all states belonging to the Landau level carry the same Hall current.

Since a localized state can certainly carry no current, we conclude that whenever the approximation is valid, there are no localized states. We shall see that for the case of the δ-function potential, there is a localized state and the approximation is not valid.

We therefore must solve the complete problem, keeping the admixture into the wave function of all Landau levels. An immediate difficulty arises because the sum on n in (2) does not converge if performed after the sum over p, at least if it is assumed that the latter sum may be approximated by an integral from $-\infty$ to ∞, using the largeness of L and W, and that, for large n, the term pv in the denominators may be neglected. However, the finiteness of the integration range will become a factor when the spatial extent of the wave functions ψ_{np} starts to become equal to the sample width, and there will be an effective cutoff at $n = M \sim W^2$. (The large magnitude of this cutoff is special to the δ-function potential. Finite-range potentials will have much smaller effective cutoffs.)

To evaluate (2) and (3), the sum over p is replaced by a principal value integral plus a contribution coming from the p values in the immediate neighborhood of the singular point. By introducing $k_\alpha - \delta_\alpha = LE_\alpha/2\pi v$, where k_α is an integer and $2|\delta_\alpha| < 1$, as well as $p_\alpha = 2\pi k_\alpha/L$ Eq. (2) may be rewritten as

$$1 = \lambda G(E_\alpha) + \lambda\phi(p_\alpha)^2\sigma_\alpha/v \ . \qquad (2a)$$

Here $G(E)$ is the principal value integral, equal to $\sum_0^M (E-n)^{-1}$ for $|E-n| \gg v$. (Since G is of order $\ln M$, the large but finite cutoff does not lead to an intolerably large G.) The discrete sum is convergent and is given by $\sigma_\alpha = -\pi\cot(\pi\delta_\alpha)$. In the same way the amplitudes are expressed as

$$(A^\alpha)^{-2} = -\left|\frac{\partial G}{\partial E_\alpha}\right| \Big/ 2\pi + \phi(p_\alpha)^2 \frac{L(\sigma_\alpha^2 + \pi^2)}{(2\pi v)^2} \ . \qquad (3a)$$

Consider first states for which $p_\alpha \gg 1$, that is, states which hardly overlap the impurity and for which $\phi(p_\alpha)^2 \ll 1/L$ is very small indeed. Except for the case that $G(E_\alpha) = 1/\lambda$, σ_α will necessarily be enormous, the second term on the right of (3a) will dominate, and the state ψ^α will differ insignificantly from the corresponding unperturbed state.

For the special energy E_R satisfying $G(E_R) = 1/\lambda$, however, $\sigma_\alpha \sim 1$, and the first term on the right-hand side of (3a) dominates. Under the circumstances, there is exactly one level for which this is true. This also subsumes the case for impurities near the sample edges, which may have bound states lying outside the quasicontinuum. (One may also take the thermodynamic limit, $L\to\infty$, in which case the state in question becomes an exceedingly narrow resonance, without changing the result.) This is a localized state, whose wave function is $\psi^R=\exp[-(x^2+y^2-2ixy)/4]/\sqrt{2\pi}$ (for $v=0$, and neglecting the interlevel mixing). It is a peculiarity of this system that the spatial extent of the localized state is controlled by the magnetic field when the potential fluctuation is short ranged and does not become large even when λ and E_R become small. (This may modify the theory of Anderson localization for weakly localized states.) The current of the localized state, j^R, of course vanishes as can be verified by direct computation which gives

$$j_y^R=v\left\{1-\frac{(A^R)^2}{2\pi}\sum(n+1)[(E_R-n)^{-2}-(E_R-n-1)^{-2}]\right\}=0\ . \tag{4}$$

Next, the remaining class of states is considered, namely, those with energies E_α whose corresponding p_α is of order unity, that is, states made up of unperturbed eigenstates which have a significant overlap with the impurity. The eigenstates ψ^α do *not* overlap the impurity, of course, since they must be orthogonal to ψ^R which is localized at the site. The energies E_α are of order v, and thus $G(E_\alpha)\cong \mathrm{Re}G_0(p_\alpha)/v$ which is given by

$$G_0(p_\alpha)=\int dp\,\frac{\phi(p)^2}{p_\alpha-p+i\eta}\ . \tag{5}$$

Then, $G\sim 1/v$, $\phi(p_\alpha)\sim 1$, $\sigma_\alpha\sim 1$, and the second term on the right-hand side of (3a) dominates because L is large. The Hall current carried by such a state may be evaluated to leading order in v as

$$j_y^\alpha=v+(A^\alpha)^2/\pi v\ . \tag{6}$$

The sum over states α is smooth and may be replaced by an integral over p_α. It is found that

$$j^0\equiv\sum_\alpha(j_y^\alpha-v) = v\left(\frac{2}{\pi}\right)\int dp\,\mathrm{Im}\left(\frac{1}{G_0(p)}\right)\ . \tag{7}$$

The integral is evaluated as $\pi\int dp\,p^2\phi(p)^2=\pi/2$, by recognizing its analytic properties. Thus, j^0, which is the excess current carried by the electrons which pass near the impurity, is just $j^0=v$, exactly enough to compensate for the failure of the localized state to carry a current.

The case of two δ-function potentials may be studied as well. In this case, if the potential sites are separated by a distance $l\gg 1$, they do not interact, and the single-impurity problem gives the answer. (This may be extended to many well separated impurities.) On the other hand, if the separation l is small compared with unity, the potential acts as a single site. Only when $l\sim 1$ do the impurity levels interact and start to form an impurity band.

The preceding results, as well as the lore of Anderson localization due to Mott and others,[1,4,5] support the conjecture[1,3] that the following picture holds. In the absence of an electric field, and in the presence of a considerable number of impurities, defects or other potential fluctuations, and in a sufficiently strong magnetic field, a given Landau level will be broadened into a band. The central region of this band will be delocalized states. Beyond a mobility edge there will be localized states, which can only conduct current by hopping processes. The delocalized states of different bands will thus be separated by a region of localized states. The delocalized states of each Landau "band," however, collectively carry a total Hall current I, in the presence of an electric field or potential gradient, which is

$$I=(-ev/L)LW/2\pi=e^2\Phi_H/2\pi(=e^2\Phi_H/h)\ . \tag{8}$$

Our calculation thus tends to confirm the expectation that the quantum Hall resistance is h/e^2.

This result applies to a dilute system of δ-function impurities. Ando *et al.*[3] also obtained a quantized Hall current within the framework of their approximations, which are the assumption of "high magnetic field," the single-site approximation with the effects of scattering taken into account self-consistently, and the assumption that there are gaps between the resulting "impurity bands." The spirit of the "high-field" approximation seems to be the same as discussed earlier, but in detail it is different since the elegant formalism of Kubo *et al.*[6] is used. (This method breaks the current correlation expression for the conductivity into two parts, one of which is treated exactly, and the other approximately.) Although the conditions for the validity of the high-field approximation have not been spelled out, it seems likely that it will be valid when the field is so great that the potential hardly varies on the scale of l_B. The δ-function results on the other hand ought to be qualitatively valid when the potential is confined to a small region compared with l_B. Given the value of l_B appropriate for the experiment (70 Å), it is unlikely that either approximation is *a priori* very good. The experiment, however, is evidence that the results which have been obtained in these limiting cases must be valid under very general conditions.

Aside from these questions which somehow must

receive a favorable answer if the experiment is to be explained at all, there are several other considerations which might lead to effects at the part in 10^6 level. In particular, W is not so large that corrections of order $1/W$ can be tolerated. Thus, the edge effects must be carefully investigated. It is known from the theory of the Landau diamagnetism, for example, that the surface Landau levels tend to carry current in a direction counter to that of the bulk. Perhaps this kind of correction can be avoided along the lines suggested by the foregoing, namely, that any surface states or other anomalous states will be compensated for by an increase in the Hall current of the delocalized states. Another small effect needing investigation is the nonparabolicity of the energy bands.

It is interesting to ask whether the quantum Hall current is a supercurrent in the sense of the theory of superconductivity, and whether a persistent current can be set up. There is no dissipation connected with the Hall current *per se,* since it is perpendicular to the field. It is a supercurrent in the sense that the wave function is locked into place by an energy gap, and it is because of the vector potential that the current exists.

Thus the more interesting question is whether there will be a small current parallel to the electric field, that is, whether in this direction, the system is a perfect insulator. In our model, such a current can come only by taking into account inelastic processes so far neglected, which give rise to a change in occupation of the states. When an entire level of current-carrying states are filled there is no possibility of changing their occupation without large energy cost. The localized states also are activated so they too are perfectly insulating at sufficiently low temperature. Thus we expect that the Hall potential can be maintained without dissipation and that a persistent current can be set up.

This raises the possibility of photoinducing a potential drop in the y direction. If light falls on the junction, it can excite electrons into the unfilled delocalized states of a higher Landau level, and these electrons will provide a current in the x direction which will in turn cause a shift in the direction of the Hall current and lead to a potential drop along the length of the sample. By controlling the frequency of the radiation, something about the position of the mobility edges may be inferred.

The approximation of constant electric field must also be examined. It is not known where the charge which gives rise to this field resides, at least in the case that the diagonal component of the conductivity, σ_{xx}, vanishes, and the actual electric field configuration may depend on how the Hall current is set up. If the charge is localized toward the edges of the sample, it will attract an equal and opposite image charge in the facing metal a few hundred nm away. The potential of such a line of dipoles will vary most pronouncedly in the first few hundred nm away from the edge. If this is the case, most of the Hall current will flow along the edges of the sample, and the inner part will carry practically no current. This will increase the effective value of the local v and corrections of order v^2 could start to play a role. There is, fortunately, no evidence thus far that the electric field is far from constant.[7] The situation might arise, however, in a Corbino (disk with center hole) geometry where a Hall voltage could be applied by moving up external charges.

In the actual experimental configuration,[1] the primary charge giving rise to the Hall potential presumably resides in the localized states. We thus envisage a situation which is schematically shown in the figure. The charges in these localized states will be unable to relax toward equilibrium at low temperature since they require thermally activated inelastic processes to change their state. The Landau band will then bend to follow the potential, and a picture as in the figure will result.

ACKNOWLEDGMENTS

I wish to thank Dennis Drew for introducing me to this problem and to express gratitude to him and Victor Korenman for fruitful discussions on the subject. This work was supported in part by NSF Grant No. DMR 7908819.

[1]K. v. Klitzing, G. Dorda, and M. Pepper, Phys. Rev. Lett. 45, 494 (1980).

[2]Several laboratories have already apparently achieved 1-ppm precision by this method. E. Braun, E. Staben, and K. v. Klitzing, PTB-Mitteilungen 90, 350 (1980); B. Taylor (private communication).

[3]T. Ando, Y. Matsumoto, and Y. Uemura, J. Phys. Soc. Jpn. 39, 279 (1975), and preceding papers.

[4]R. J. Nicholas, R. A. Stradling, and R. J. Tidley, Solid State Commun. 23, 341 (1977).

[5]H. Aoki and H. Kanimura, Solid State Commun. 21, 45 (1977).

[6]R. Kubo, S. J. Miyake, and N. Hashitsume, in *Solid State Physics,* edited by F. Seitz and D. Turnbull (Academic, New York, 1965), Vol. 17, p. 269.

[7]K. v. Klitzing (private communication).

PHYSICAL REVIEW B VOLUME 27, NUMBER 12 15 JUNE 1983

Localization, percolation, and the quantum Hall effect

S. A. Trugman
*Laboratory of Atomic and Solid State Physics, Cornell University, Ithaca, New York 14853**
and Department of Applied Physics, Stanford University, Stanford, California 94305
(Received 19 January 1983)

Noninteracting electrons in a smooth two-dimensional random potential are localized in the large magnetic field limit. In contrast to Anderson localization, eigenstates with large localization lengths occur with a probability proportional to a universal power of their size, with the power given in terms of percolation critical exponents. Adding a parallel electric field $\vec{\mathscr{E}}$ causes extended states to appear in numbers proportional to a power of $\mathscr{E}$. This implies a nonlinear broadening of steps in the quantized Hall conductivity. The results for a parallel electric field are obtained by considering a graded percolation problem, in which the probability that a site is occupied varies with position.

I. INTRODUCTION

It is believed that in two dimensions without a magnetic field any amount of disorder will localize all of the electronic wave functions.[1] Experiments have been performed in which an intense perpendicular magnetic field is applied to electrons that are confined to two dimensions in metal-oxide semiconductors and heterojunctions.[2,3] These experiments have demonstrated that the Hall conductivity is quantized in units of e^2/h.

The following question arises: Do the electronic states remain localized in the presence of a magnetic field? The simplest point of view[4] is that there must be extended (or arbitrarily large) states since localized states carry no current and nonzero Hall conductivity is observed. Purely theoretical arguments reaching this conclusion for sufficiently weak potential have been given by Aoki and Ando,[5] Laughlin,[6] and Halperin.[4]

It is of interest to know whether a finite fraction of the eigenstates is extended. If so, the steps in the quantum Hall conductivity will never become sharp, even in the limit where the magnetic field is large and the temperature and electric field go to zero. There have been speculations[7] that this may be the case (the assumption is also implied by Refs. 5 and 6). A further agrument in favor of a nonzero fraction of extended states is that is difficult to imagine a single delocalized electron carrying sufficient current to be observed in a macroscopic experiment.

Contrary to these expectations and in agreement with remarks in Refs. 3 and 8, we find that for a smooth random potential in the limit of an arbitrarily large magnetic field B, the fraction of extended states is zero. (Because the number of states per unit area in a given Landau level is proportional to B, it is possible that the number of extended states per unit area does not vanish as $B\rightarrow\infty$, even though the fraction of states that are extended goes to zero.) Our main result is to demonstrate a *quantitative* relation between the localization problem and two-dimensional percolation. [A qualitative correspondence has been previously noted; see Refs. 3 and 8–10(a).] The distribution of eigenstates of large spatial extent is given by a universal relation involving percolation critical exponents. A second relation is derived that describes the fraction of extended eigenstates as a function of the tangential electric field, which implies a nonlinear broadening of the Hall conductivity steps.

The contents of the remainder of this paper are as follows: Section II describes the electron eigenfunctions and their relation to continuum percolation. The distribution law for large eigenstates is derived. Section III treats the effect of adding a tangential electric field. A lower bound for the number of extended states in a general potential is derived, as well as a relation for a random potential. The model is shown to imply a linear relation between Hall voltage and current, but a nonlinear broadening of the steps in Hall conductivity versus electron density. Section IV includes a description of the electron Green's function and conclusions. The appendix is a rough calculation of the magnitude of the nonlinear step broadening, which appears to be in an experimentally accessible regime.

II. EIGENFUNCTIONS AND PERCOLATION

We consider the eigenfunctions of the two-dimensional Hamiltonian

$$H=\frac{1}{2m_{\rm eff}}\left[\vec{p}-\frac{e\vec{A}}{c}\right]^2+V(x,y)\,, \tag{1}$$

where $V(x,y)$ is a smooth position-dependent potential that arises from inhomogeneities. Electron-electron interactions are neglected. The magnetic field $\vec{B}=\vec{\nabla}\times\vec{A}$ is of constant magnitude B in the $\hat{z}$ direction. Let $l\equiv(\hbar c/eB)^{1/2}$ be the radius of the ground Landau orbit, and $\omega_c\equiv eB/m_{\rm eff}c$ be the cyclotron frequency.

In the limit in which the magnetic field becomes arbitrarily large (V slowly varying on the scale l), the eigenfunctions ψ become quite simple: $|\psi|$ is large only in the vicinity of a constant-energy surface of the potential V. Perpendicular to the line of constant V, the wave function extends a distance of order l, which vanishes in the limit of $B\to\infty$. Other authors, including Tsukada[9] and Prange and Joynt,[8] have noted that the eigenfunctions of energy $E+\hbar\omega_c(n+\frac{1}{2})$ are localized about the classical orbit $V(\vec{x})=E$ in the large-B limit. The latter work contains an illuminating derivation based on the path integral representation of the electron propagator. The motion of electrons along equipotential contours is analogous to that of superconducting or superfluid vortices in an inhomogeneous film. Percolation theory implies universal power-law behavior for vortices as well.[10(b)]

Although we will not need the explicit form of the electron eigenfunctions, they can be approximated by

$$\tilde{\psi}(u,v)=C(u)\chi_n(v)e^{i\phi(u,v)} \tag{2}$$

in the limit $B\to\infty$. The variable u parametrizes distance along the constant energy surface $V(\vec{x})=\text{const}$, and v parametrizes distance orthogonal to this line. The index $n=0,1,2,\ldots$ is the Landau level and the energy eigenvalue is $E=V+\hbar\omega_c(n+\frac{1}{2})$. The function χ_n is the nth harmonic oscillator function,

$$\chi_n=H_n\left[\frac{v}{l}\right]\exp\left[-\frac{v^2}{2l^2}\right],$$

$$C^2(u)\sim\frac{1}{|\vec{\nabla}V(u,v)|_{v=0}}\,,$$

and $\phi(u,v)$ is a gauge-dependent phase.

One can show that the wave function $\tilde{\psi}(u,v)$ given by Eq. (2) satisfies the Schroedinger equation $H\tilde{\psi}=E\tilde{\psi}$ in the regions where $\tilde{\psi}$ is large, with fractional errors of order $r_c^{-1}(\hbar c/eB)^{1/2}$ and also of order

$$\hbar m_{\rm eff}|\vec{\nabla}V|(c/\hbar eB)^{3/2}$$

(r_c is the local radius of curvature of the constant-energy surface). These errors vanish on smooth constant-energy surfaces in the limit $B\to\infty$.

Because $\phi(u,v)$ must change by an integral multiple of 2π around a circuit, only a discrete set of constant-energy surfaces corresponds to electron eigenfunctions. In the high-B limit, each allowed surface encloses an additional area of hc/Be. The electron density is then everywhere equal to nBe/hc if n Landau levels are filled. (These points are discussed in more detail in Ref. 9.)

Consider two approximate eigenfunctions given by Eq. (2), ψ_1 and ψ_2, that are on opposite sides of a saddle and whose constant-energy surfaces are a distance d apart at their closest approach. The eigenfunctions are approximate because they were obtained by retaining terms in H only to second order in v. For a large but finite B, there is a nonvanishing overlap integral η between ψ_1 and ψ_2. A true eigenstate will then be an admixture of ψ_1 and ψ_2. Note, however, that

$$\eta\sim\exp(-d^2eB/4\hbar c)\,,$$

whereas the minimum energy mismatch ΔE of ψ_1 and ψ_2 goes like $1/B$, barring an accidental degeneracy.[11] The ratio $\eta/\Delta E$ vanishes as $B\to\infty$. This implies that as $B\to\infty$, an electronic eigenstate has a large amplitude only in the neighborhood of a single (connected) constant-energy surface.

We now consider a random potential $V(\vec{r})$ which is assumed to have the following properties: $V(\vec{r})$ is a random variable whose distribution is given by $\text{Prob}[V(\vec{r})=\epsilon]=\rho(\epsilon)$. The distribution function $\rho(\epsilon)$ is continuous, bounded, and independent of $\vec{r}$ with $\int_{-\infty}^{\infty}\rho(\epsilon)d\epsilon=1$. We choose the zero of energy such that $\langle V(\vec{r})\rangle=0$, where enclosure by angular brackets denotes an average over $\vec{r}$. The correlation function $\langle V(\vec{r})V(\vec{r}+\vec{x})\rangle$ goes rapidly to zero for $|\vec{x}|\gg b$, where b is the correlation length of the potential.[12] Finally, we assume that V is slowly varying on a sufficiently small length scale. Many functions satisfy the above requirements, such as Gaussian white noise with a smooth high-frequency cutoff at $k\approx b^{-1}$.

As $B\to\infty$, the density of states is proportional to $\rho(E)$. If one seeks eigenfunctions of energy E,[13] a convenient construction is to color in all points $\vec{r}$ such that $V(\vec{r})\le E$. Then as $B\to\infty$, the eigenfunctions will be localized on the perimeters of the colored areas (see Fig. 1). For a random V, the

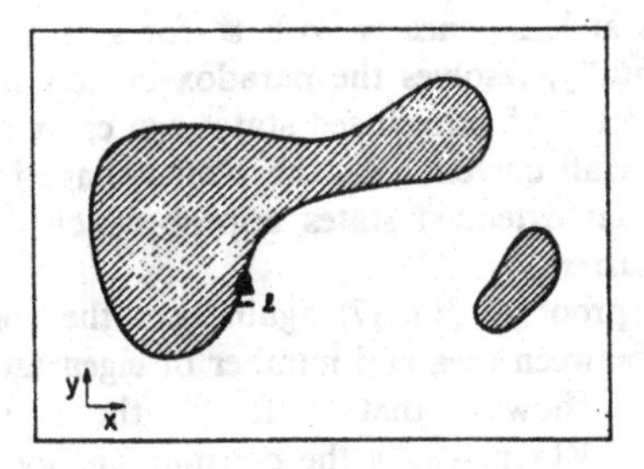

FIG. 1. Points $\vec{r}$ such that $V(\vec{r})\leq E$ are shaded. Eigenfunctions of energy E are localized within a distance l of the perimeter of a shaded region.

properties of the connected colored regions are those of a continuum percolation problem.[14] In the usual site-percolation problem, a lattice point $\vec{r}_j$ is considered to be occupied (colored) with a probability p. One is interested in the properties of connected clusters as a function of p. In the continuum problem described above, $\vec{r}_j$ is replaced by the continuous variable $\vec{r}$ and p is replaced by E through

$$p(E)=\int_{-\infty}^{E}\rho(\epsilon)d\epsilon\ . \quad (3)$$

Percolation theory immediately implies a number of universal results for the case in which the potential V is random. There is a unique energy E_c (the percolation threshold) where E_c is the smallest energy for which an infinite connected colored region exists. For $E<E_c$ there is a length

$$\xi(E)\sim|E-E_c|^{-\nu} \text{ as } E\to E_c \quad (4)$$

such that connected clusters of spatial extent L greater than ξ are exponentially rare. (The spatial extent L could be loosely defined as the "diameter" of a connected region or quantitatively as the radius of gyration.) In two dimensions the percolation critical exponent ν is $\frac{4}{3}$.[15] Thus for $B\to\infty$ the electronic wave functions of energy E will typically have a spatial extent of no larger than the order of $\xi(E)=c_1|E-E_c|^{-\nu}$. For $E>E_c$ an infinite connected colored region is present. One might suppose that this would imply eigenfunctions of infinite spatial extent. Such is not the case, however. For $E>E_c$ it is convenient to identify the wave functions with the perimeters of the *uncolored* regions.[16] Above E_c the connected uncolored regions are of finite extent, with maximum typical size again given by Eq. (4). Because of this symmetry, we will need to consider only the finite connected regions for $E<E_c$.

There are no states of infinite extent except perhaps at $E=E_c$, which is a set of measure zero. Thus we have that *for a random potential as $B\to\infty$, the fraction of eigenstates of infinite extent is zero.* Note that we have used the assumption that the distribution function $\rho(\epsilon)$ is continuous, so that in the vicinity of E_c Eq. (3) implies that $E-E_c$ is proportional to $p-p_c$. In contrast, Prange and Joynt[8] have considered a model potential in which $V=0$ except in disconnected regions in which V is arbitrary. That potential has a δ function in $\rho(\epsilon)$ exactly at the percolation threshold, and in fact has a finite fraction of states of infinite extent for $B\to\infty$.

The energy dependence of the localization length ξ given in Eq. (4) disagrees with that given by Ono,[16(a)] who uses the self-consistent Born approximation to find $\xi(E)\sim l\exp(\gamma^2/E^2)$. One could attribute the difference to the fact that Ono considered a random potential that is a sum of δ functions, so that his $V(\vec{r})$ is never smooth on the scale of l, no matter how large the magnetic field. The localization of an eigenfunction on the perimeter of a percolation cluster may also be too subtle an effect to be seen in the self-consistent Born approximation.

In the absence of infinite eigenstates, we investigate those of large spatial extent. Let $f(R)$ be the fraction of eigenstates whose extent $L\geq R$. (The number of eigenstates per unit area is given by the fraction f multiplied by eB/hc.) We make the following assumptions, the first two of which are discussed by Stauffer[14]:

(1) The scaling hypothesis: The number density n_s of clusters of large area s is given by

$$n_s\sim s^{-\tau}g[(p_c-p)s^{\sigma}]\ ,$$

where $g(z)$ is analytic with $g(0)=1$, and $g(z)$ goes rapidly to zero for large z.

(2) The ramification hypothesis: Large clusters of area s have a typical perimeter t with $s\sim t^{p'}$ and $p'=1$.

(3) We further assume that as p is increased to $p+dp$, a cluster of area s and perimeter t will be coated with an additional area dA which is proportional to tdp. The fraction of the total area occupied by constant-energy surfaces of perimeter t or greater is then

$$f\sim\int_0^{1/2}d\epsilon\int_t^{\infty}ds\,s^{1-\tau}g(\epsilon s^{\sigma})\to f\sim t^{2-\tau-\sigma}\ . \quad (5)$$

Using the relation between the perimeter and the diameter of a connected region $t\sim R^{1/\sigma\nu}$, we obtain the relation that for large R

$$f(R)\sim R^{-p_1}\ . \quad (6)$$

The exponent

$$p_1=(\tau-2+\sigma)/\sigma\nu=(1+\beta)/\nu ,$$

where the last equality follows from scaling relations among exponents. With the use of $\beta=\frac{5}{36}$ (Ref. 15), $p_1=\frac{41}{48}$. As expected, the fraction of states of extent larger than R vanishes as $R\rightarrow\infty$.

Equation (6) states that large eigenstates appear with a probability proportional to a power of their diameter. This is in sharp contrast to Anderson localization in the absence of a magnetic field, where $f(R)$ goes to zero exponentially.[1]

One subtlety should be mentioned with respect to the above derivation. Equation (5) is most naively obtained by assuming that each cluster of area s has a single (external) perimeter $t\sim s$. Clusters, however, have internal perimeter as well, and it is the total perimeter (internal plus external) that is proportional to s. To incorporate this fact, define the average perimeter $\bar{t}$ of a cluster as $\bar{t}\equiv\sum_j t_j^2/\sum_j t_j$, where the sum is over the external and all internal perimeters. If $\bar{t}\sim s$, one can show that $p_1=(1+\beta)/\nu$ still obtains. A sufficient condition for $\bar{t}\sim s$ is that the external perimeter remains a nonzero fraction of the total perimeter. Computer simulations by Leath and Reich[16(b)] show that as s increases, the external perimeter is a *very* slowly decreasing fraction of the total perimeter. The largest simulated clusters (2000 sites) still have over 75% external perimeter. This issue is not yet sufficiently well understood to know whether $p_1=(1+\beta)/\nu$ or whether this is a slight underestimate of the correct exponent.

III. TANGENTIAL ELECTRIC FIELDS

A. Lower bound for general $V(x,y)$

In experiments on quantized Hall conductivity, electrons move in the presence not only of a magnetic field normal to the surface, but also of an electric field tangent to the surface. When σ_{xx} vanishes, this electric field is simply the Hall voltage divided by the sample width. We will show that such an electric field has a drastic effect on the localization of electronic states. We assume a smooth potential $V(x,y)$ whose magnitude and slope are bounded: $|V(x,y)|\leq M$, $|\vec{\nabla}V(x,y)|\leq M'$. The potential need not be random (it could, for example, have long-range correlations). Then, in the presence of a tangential electric field $\vec{\mathscr{E}}$ in the limit $B\rightarrow\infty$, a nonzero fraction of the electronic eigenstates is delocalized. The fraction f_∞ is bounded below by f_b:

$$f_\infty(\mathscr{E})\geq e\mathscr{E}/(M'+e\mathscr{E})\equiv f_b(\mathscr{E}) . \qquad (7)$$

Thus the fraction of states of infinite extent increases at least linearly with $\mathscr{E}$ for small $\mathscr{E}$. This, incidentally, resolves the paradox of how a vanishing fraction of delocalized states can carry a macroscopic Hall current. As the Hall voltage increases, additional extended states appear which carry the extra current.

The proof of Eq. (7) again uses the correspondence between area and number of eigenstates. One must show that for the potential $U(x,y)=V(x,y)-e\mathscr{E}y$ the constant-energy surfaces that are infinite in extent comprise a fraction of the total area no smaller than f_b. The sample is assumed infinite in the $\hat{y}$ direction, which is the direction of the electric field. It is convenient to define $\tilde{V}(x,y)$ as the function $V(x,y)$ multiplied by a factor $G(x)$, where $G(x)=1$ for $0<x<L_1$, $G(x)=0$ for $x\leq-\Delta$ and $L_1+\Delta\leq x$, and $G(x)$ interpolates smoothly between zero and one in between. Define

$$\tilde{V}(x,y)\equiv V(x,y)G(x) ,$$
$$\tilde{U}(x,y)\equiv\tilde{V}(x,y)-e\mathscr{E}y . \qquad (8)$$

Thus $\tilde{V}=V$ in the region $0\leq x\leq L_1$, and $\tilde{V}$ vanishes outside the region $-\Delta<x<L_1+\Delta$.

Consider a constant-energy surface of $\tilde{U}$ containing the point $(\bar{x},\bar{y})$ with $\bar{x}<-\Delta$. This constant-energy surface is simply the line $y=\bar{y}$ for $x\leq-\Delta$ [see Fig. 2(a)]. As the line enters the region $-\Delta\leq x\leq L_1+\Delta$ it must remain in the band $|y-\bar{y}|\leq M/e\mathscr{E}$ since it is a constant-energy surface. In particular it cannot escape to $y=\pm\infty$ for a fixed nonzero electric field. The line cannot reemerge at $x<-\Delta$ and $y\neq\bar{y}$ because then again it would not be a constant-energy surface. Thus the

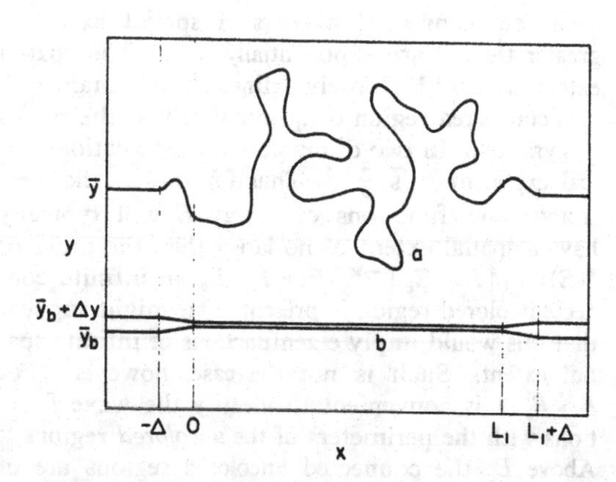

FIG. 2. Electric field is in the $\hat{y}$ direction. *a*, a constant-energy surface of $\tilde{U}$ for a random potential; *b*, a band of constant-energy surfaces that occupies the minimum possible area for $0\leq x\leq L$ (general potential).

line must reemerge at $x > L + \Delta$ where its equation is $y = \bar{y}$.[17] This is an extended constant-energy surface; it has an arc length of at least L_1 as the sample width $L_1 \to \infty$.

What fraction of the area is occupied by lines that extend across the sample? A band of states in the region $\bar{y}_b \le y \le \bar{y}_b + \Delta y$ for $x < -\Delta$ will occupy the smallest possible area in the region $0 \le x \le L_1$ if it moves straight across the sample in a region in which $|\vec{\nabla} U|$ has its maximum possible value of $M' + e\mathscr{E}$ [see Fig. 2(b)].[18] The minimum fraction of the area occupied by such states is then $f_b = e\mathscr{E}/(M' + e\mathscr{E})$, which proves Eq. (7).

Corollary. It is straightforward to show that if $e\mathscr{E} > M'$ all constant-energy surfaces are extended, so that $f_\infty = 1$.

B. Random $V(x,y)$

If V is a random potential then the constant-energy surfaces in the presence of a small electric field are tortuous and resemble the upper line in Fig. (2) more than the lower. The actual fraction $f_\infty(\mathscr{E})$ of extended states will then increase more rapidly with electric field than does the lower bound, Eq. (7). We propose that

$$f_\infty(\mathscr{E}) \sim \mathscr{E}^{p_2} \tag{9}$$

may be exact for $\mathscr{E} \to 0$, with

$$p_2 = (\tau - 2 + \sigma)/\sigma(1+\nu) = (1+\beta)/(1+\nu) = \tfrac{41}{84} .$$

Equation (9) is motivated as follows: One is interested in the connected regions of the plane such that

$$U(x,y) = V(x,y) - e\mathscr{E}y \le E .$$

This is a type of graded percolation problem in which the probability p that a point is occupied (colored) is not simply a constant, but rather a function of position. As a result, the locally defined correlation length $\xi(\vec{r})$ will vary from point to point in space. For fixed $\mathscr{E}$ and V, there will be a length which we denote by $L(\mathscr{E})$, such that finite clusters that extend a distance greater than L are exponentially rare. To estimate L, consider a cluster of extent l_y in the y direction. Such a cluster will be strongly suppressed if the local correlation length $\xi(\vec{r})$ is much less than l_y in large parts of the cluster. For nonzero $\mathscr{E}$, the local correlation length $\xi = c_1 |E - E_c|^{-\nu}$ can be no larger than the order of $c_1 |e\mathscr{E} l_y|^{-\nu}$ in large parts of the cluster. If the cluster is to occur with substantial probability, one must have

$$l_y \lesssim c_1 |e\mathscr{E} l_y|^{-\nu} ,$$

or

$$l_y \lesssim c' \mathscr{E}^{-\nu/(1+\nu)} = L(\mathscr{E}) . \tag{10}$$

The application of a small electric field $\mathscr{E}$ is not expected to substantially modify the statistics of clusters of extent $R \ll L(\mathscr{E})$, whereas it will strongly affect those with $R \gg L(\mathscr{E})$. It is reasonable to assume that the area that was occupied by constant-energy surfaces of extent $R \gtrsim L(\mathscr{E})$ in the absence of an electric field will be converted to f_∞ when the field is applied. Equations (6) and (10) then imply the desired result, Eq. (9).

C. Quantum Hall effect

The Hall current is carried by extended states. Since the number of such states is a singular function of the parallel electric field [Eq. (9)], one might expect a nonlinear relation between the current I_x and the Hall voltage V_H. Such is not, however, the case, and in fact this model predicts a linear relation $I_x = \sigma_{xy} V_H$. This relation has been previously obtained by Laughlin[6] as a consequence of gauge invariance. We include the following brief derivation, which obtains the same result from a different point of view.

Assume that the first Landau level is filled and for simplicity neglect spin and valley degeneracies. In the high-B limit, the electron number density $\bar{n}$ is everywhere equal to eB/hc.[19] The expectation value of the local electron velocity operator is

$$\langle \vec{v}(\vec{r}) \rangle = c \frac{\vec{\mathscr{E}}(\vec{r}) \times \vec{B}}{B^2} . \tag{11}$$

$\vec{\mathscr{E}}(\vec{r})$ is the electric field in the x-y plane, which has contributions from the local random potential and from an external field,

$$\vec{\mathscr{E}}(r) = \vec{\mathscr{E}}_{\text{ext}} - (1/e)\vec{\nabla} V(\vec{r}) .$$

Equation (11) is correct both classically and quantum mechanically.[20(a)] The current I_x that passes through the line $x = $const for a sample width L_2 in the $\hat{y}$ direction is

$$I_x = \int_0^{L_2} dy\, \bar{n} e (\vec{v} \cdot \hat{x}) \to I_x = \frac{e^2}{h} \left[\left(\int_0^{L_2} \mathscr{E}_{\text{ext}} \cdot \hat{y} \right) - \frac{1}{e} [V(x, L_2) - V(x,0)] \right] . \tag{12}$$

In an experiment, the measured Hall potential V_H is the sum of the electrostatic potential and the difference in Fermi levels[4] [respectively, the first and second terms inside the braces of Eq. (12)]. One recovers the result $I_x = \sigma_{xy} V_H$, with $\sigma_{xy} = e^2/h$. A simple generalization gives $\sigma_{xy} = ne^2/h$ for n occupied Landau levels.

Equation (9) implies a nonlinear broadening of the steps in σ_{xy} versus electron density. Consider a metal-oxide-semiconductor field-effect transistor (MOSFET) at $T=0$ in a nonzero parallel electric field $\vec{\mathscr{E}}$, with electrons added to the lowest Landau level by varying the gate voltage. The first electrons will occupy localized states with $E \ll E_c$. Then they will fill extended states at $E \approx E_c$, and finally the localized states with $E \gg E_c$. The current will increase only as the extended states are filled. By Eq. (9), the number of extended states is proportional to $\mathscr{E}^{(1+\beta)/(1+\nu)}$, with $\mathscr{E}$ proportional to the Hall voltage. One then has the following nonlinear effect: *The width of the steps in the Hall conductivity as a function of electron density is proportional to* $V_H^{(1+\beta)/(1+\nu)}$. The Appendix discusses the observability of this effect.

IV. DISCUSSION AND CONCLUSIONS

Percolation theory implies several additional properties of the electron eigenstates and Green's functions for very high magnetic fields. For $\mathscr{E}=0$ the eigenstates are closed curves of vanishing thickness $l \sim B^{-1/2}$. The statistics of the small closed curves are *not* universal, and depend in detail on the short-wavelength properties of the random potential V. In contrast, the large eigenstates have universal properties, including a density given by Eq. (6). They are rough objects (fractals), with a perimeter that grows faster than their diameter ($t \sim R^{1/\sigma\nu}$, with $1/\sigma\nu = \frac{91}{48}$).

We now consider some properties of the (retarded) electron propagator defined by

$$G(\vec{x}_2, \vec{x}_1, \omega) = \sum_m \frac{\psi_m^*(\vec{x}_2)\psi_m(\vec{x}_1)}{\omega - E_m/\hbar + i\eta} . \qquad (13)$$

For fixed $\vec{x}_1$ there is a (connected) constant-energy surface S that contains $\vec{x}_1$. Let R_m be the distance to the point on S that is farthest from $\vec{x}_1$. $G(\vec{x}_2)$ will have a substantial amplitude for those points $\vec{x}_2$ that are near S. This amplitude tends on the average neither to grow nor diminish as $R_{12} \equiv |\vec{x}_2 - \vec{x}_1|$ increases. Then suddenly as R_{12} increases past R_m, $\vec{G}(x_2)$ decreases very rapidly with increasing R_{12} for all $\vec{x}_2$. Even for the most favorable $\vec{x}_2$, G decreases like

$$\exp[-eB(R_{12} - R_m)^2/2\hbar c]$$

for R_{12} near R_m. This arises because increasing R_{12} past R_m forces $\vec{r}_2$ off of the constant-energy surface and into the Gaussian tail of the wave function. For random starting points $\vec{x}_1$, R_m will occur with a probability given by Eq. (6).

Conclusions. We have partially characterized the eigenstates of noninteracting electrons in a two-dimensional random potential and strong magnetic field. As $B \to \infty$, a fraction 1 of the eigenstates are localized. (As mentioned in the Introduction, this does not imply that no extended states exist, but rather that they comprise a vanishing fraction of the total.) Large eigenstates occur in numbers proportional to a power of their spatial extent, with the power given in terms of percolation critical exponents. The result is quite unlike Anderson localization that occurs for $B=0$, where large eigenstates are exponentially rare.

For $B \to \infty$, a tangential electric field $\mathscr{E}$ destroys localization, and creates extended states in numbers proportional to a power of $\mathscr{E}$. This implies a nonlinear broadening of the steps in quantum Hall conductivity, again given in terms of percolation critical exponents.

We have considered only the limiting distribution functions as $B \to \infty$. There is as yet no systematic way to perturb away from this limit, such as an expansion in powers of B^{-1}. In particular, the possibility remains that the fraction of extended states goes smoothly to zero as $B \to \infty$, but does not actually vanish at any finite magnetic field. One might investigate finite magnetic fields by considering a kind of Anderson localization problem in which the sites are constant-energy surfaces, and are connected by hopping terms. One feature of this model is that the sites themselves can become arbitrarily large near E_c, so that an eigenfunction near the band center could be localized when measured in terms of sites but extended in terms of distance.

Note added. After this work was completed, we learned of a study by Luryi and Kazarinov.[20(b)] These authors have adopted a similar point of view and have discussed (among other topics) finite-size effects, in contrast to the results for extended systems that are obtained here.

ACKNOWLEDGMENTS

The author has benefited from conversations with numerous colleagues, including Seb Doniach, James Sethna, Elihu Abrahams, Mustansir Barma, Sudip Chakravarty, Patrick Lee, Richard Prange, John Samson, Sara Solla, and John Wilkins. This work was supported at Cornell University by National Science Foundation Grant No. DMR-80-20429.

Part of this research was conducted at the Department of Applied Physics, Stanford University, supported by an IBM Fellowship and by the U.S. Office of Naval Research Contract No. N00014-82-K-0524.

APPENDIX: EXPERIMENTAL ASPECTS OF NONLINEAR BROADENING

The appendix contains a rough argument that the nonlinear broadening of the steps in the Hall conductivity may be experimentally observable. Let w be the width of the Hall step divided by the distance between Hall steps at successive Landau levels. (The width can be defined as the inverse of the maximum slope of $d\sigma_{xy}/d\bar{n}$.) Equation (9) implies

$$w=c_2\mathscr{E}^{P_2}\,. \tag{A1}$$

It is of interest to determine c_2 to know whether nonlinear broadening should be observable at reasonable Hall voltages.[21] An accurate estimate is difficult because c_2 is not universal and depends on details of the random potential V, including its correlation length b and average magnitude $\bar{V}\equiv\langle V^2\rangle$. We will give what is at best an order-of-magnitude estimte.[22]

The discussion preceding Eq. (9) implies that $f(\mathscr{E})$ is O(1) for an electric field the order of an average $|\vec{\nabla}V|$, where $\langle|\vec{\nabla}V|\rangle\approx\bar{V}/b$. Equation (A1) becomes

$$w=\alpha(e\mathscr{E}b/\bar{V})^{P_2}\,, \tag{A2}$$

with α a dimensionless constant of order unity. Both b and $\bar{V}$ are highly sample dependent. We use $\bar{V}\approx k_BT$ at the temperature T at which the steps experimentally get fairly broad due to thermal excitation of electrons. $T\approx5$ K might be typical for some samples. At $B=20$ T, the Landau length is 60 Å. The percolation arguments used are valid only for $b\gg l$; we assume a sample with $b\approx500$ Å. For a sample of width 50 μm, Eq. (A2) implies that a Hall voltage of ≈1 mV will give a step of width $w=0.05$. This is an experimentally accessible regime. In fact, the steps published by Klitzing *et al.*[2] may be nonlinearly broadened by the Hall voltage.

We are unaware of any published studies of $w(V_H)$. (Nonlinear studies of σ_{xx} do, however, exist.[23])

---

*Present address.

[1]E. Abrahams, P. W. Anderson, D. C. Licciardello, and T. V. Ramakrishnan, Phys. Rev. Lett. 42, 673 (1979).

[2]K. v. Klitzing, G. Dorda, and M. Pepper, Phys. Rev. Lett. 45, 494 (1980).

[3]M. A. Paalanen, D. C. Tsui, and A. C. Gossard, Phys. Rev. B 25, 5566 (1982).

[4]B. I. Halperin, Phys. Rev. B 25, 2185 (1982).

[5]H. Aoki and T. Ando, Solid State Commun. 38, 1079 (1981).

[6]R. B. Laughlin, Phys. Rev. B 23, 5632 (1981).

[7]D. J. Thouless, J. Phys. C 14, 3475 (1981).

[8]R. E. Prange and R. Joynt, Phys. Rev. B 25, 2943 (1982).

[9]M. Tsukada, J. Phys. Soc. Jpn. 41, 1466 (1976).

[10](a) R. F. Kazarinov and Serge Luryi, Phys. Rev. B 25, 7626 (1982); S. V. Iordansky, Solid State Commun. 43, 1 (1982); (b) S. A. Trugman and S. Doniach, Phys. Rev. B 26, 3682 (1982).

[11]Such accidental degeneracies would occur systematically for a periodic potential, but can be neglected for the random potential that we will consider.

[12]More specifically, the assumption is that the correlations between $V(\vec{r}_1)$ and $V(\vec{r}_2)$ are sufficiently short ranged that the continuum percolation problem defined by V is in the universality class of the uncorrelated lattice percolation problem.

[13]The zero-point energy has been subtracted off, so that the eigenvalue of H is $E+\hbar\omega_c(n+\frac{1}{2})$.

[14]Review articles on percolation have been written by D. Stauffer, Phys. Rep. 54, 2 (1979) and by J. W. Essam, Rep. Prog. Phys. 43. 833 (1980).

[15]This presumably exact result follows from the equivalence of percolation and the $q=1$ state Potts model; see the review by E. K. Riedel, Physica (Utrecht) 106A, 110 (1981), and references therein.

[16]Equivalently, the eigenfunctions lie on the boundaries of holes in the infinite colored cluster. (a) Y. Ono, J. Phys. Soc. Jpn. 51, 2055 (1982), and unpublished; (b) P. L. Leath and G. R. Reich, J. Phys. C 11, 4017 (1978).

[17]We have assumed that V is such that with probability 1, a constant-energy surface will not simply terminate or intersect itself at other than a simple saddle point.

[18]Note that trajectories that begin with different y for $x<-\Delta$ have different energies and cannot intersect, so that none of the area is double counted.

[19]See Ref. 9; corrections are of order $(m_{\rm eff}\omega_c^2)^{-1}\nabla^2V$.

[20](a) R. Kubo, S. J. Miyake, and N. Hashitsume, in *Solid State Physics*, edited by F. Seitz and D. Turnbull (Academic, London, 1965), Vol. 17, p. 269; (b) S. Luryi and R. F. Kazarinov, Phys. Rev. B 27, 1386 (1983), and unpublished. We also thank Dr. Luryi for an informative conversation.

[21]In an actual experiment, Eq. (A1) will not be obeyed down to $\mathscr{E}=0$. For sufficiently small $\mathscr{E}$, the nonlinear

broadening will become smaller than the broadening due to nonzero temperature and an insufficiently large magnetic field; $w(\mathscr{E})$ will then cross over and approach a constant as $\mathscr{E}\rightarrow 0$.

[22]We implicitly assume that neither $\rho(\epsilon)$ nor $\langle V(\vec{r}_1)V(\vec{r}_2)\rangle$ have very long tails.

[23]M. A. Paalanen (private communication).

Quantized Hall conductivity in two dimensions

R. B. Laughlin
Bell Laboratories, Murray Hill, New Jersey 07974

(Received 20 January 1981)

It is shown that the quantization of the Hall conductivity of two-dimensional metals which has been observed recently by Klitzing, Dorda, and Pepper and by Tsui and Gossard is a consequence of gauge invariance and the existence of a mobility gap. Edge effects are shown to have no influence on the accuracy of quantization. An estimate of the error based on thermal activation of carriers to the mobility edge is suggested.

There has been considerable interest in the remarkable observation made recently by von Klitzing, Dorda, and Pepper[1,2] and by Tsui and Gossard[2] that, under suitable conditions, the Hall conductivity of an inversion layer is quantized to better than one part in 10^5 to integral multiples of e^2/h. The singularity of the result lies in the apparent total absence of the usual dependence of this quantity on the density of mobile electrons, a sample-dependent parameter. As it has been proposed[1] to use this effect to define a new resistance standard or to refine the known value of the fine-structure constant, an important issue at present is to what accuracy the quantization is exact, particularly in the regime of high impurity density. Some light has been shed on this question by the renormalized weak-scattering calculations of Ando,[3] who has shown that the presence of an isolated impurity does not affect the Hall current. A similar result has been obtained recently by Prange,[4] who has shown that an isolated δ-function impurity does not affect the Hall conductivity to lowest order in the drift velocity $v=cE/H$, even though it binds a localized state, because the remaining delocalized states carry exactly enough extra current to compensate for its loss. The exactness of these results and their apparent insensitivity to the type or location of the impurity suggest that the effect is due, ultimately, to a fundamental principle. In this communication, we point out that it is, in fact, due to the long-range phase rigidity characteristic of a supercurrent, and that quantization can be derived from gauge invariance and the existence of a mobility gap.

We consider the situation illustrated in Fig. 1, of a ribbon of two-dimensional metal bent into a loop of circumference L, and pierced everywhere by a magnetic field H_0 normal to its surface. The density of states of this system, also illustrated in Fig. 1, consists, in the absence of disorder, of a sequence of δ functions, one for each Landau level. These broaden, in the presence of disorder, into bands of extended states separated by tails of localized ones. We consider the disordered case with the Fermi level in a mobility gap, as shown.

We wish to relate the total current I carried around the loop to the potential drop V from one edge to another. This current is equal to the adiabatic derivative of the total electronic energy U of the system with respect to the magnetic flux ϕ through the loop. This may be obtained by differentiating with respect to a uniform vector potential A pointing around the loop, in the manner

$$I=c\frac{\partial U}{\partial \phi}=\frac{c}{L}\frac{\partial U}{\partial A} \quad . \tag{1}$$

This derivative is nonzero only by virtue of the phase coherence of the wave functions around the loop. If, for example, all the states are localized then the only effect of A is to multiply each wave function by $\exp(ieAx/\hbar c)$, where x is the coordinate around the loop, and the energy change and current are zero. If a state is extended, on the other hand, such a gauge transformation is illegal unless

$$A=n\frac{hc}{eL} \quad . \tag{2}$$

In the case on noninteracting electrons, phase coherence enables a vector potential increment to

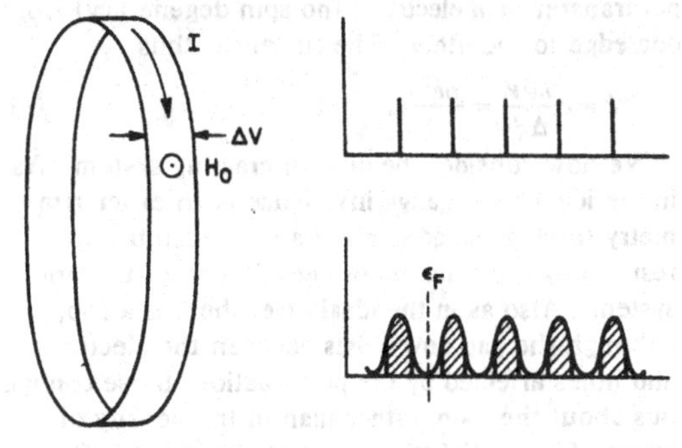

FIG. 1. Left: Diagram of metallic loop. Right: Density of states without (top) and with (bottom) disorder. Regions of delocalized states are shaded. The dashed line indicates the Fermi level.

change the total energy by forcing the filled states toward one edge of the ribbon. Specifically, if one adopts the usual isotropic effective-mass Hamiltonian,

$$H = \frac{1}{2m^*}\left(\vec{p} - \frac{e}{c}\vec{A}\right)^2 + eE_0 y \ , \tag{3}$$

where E_0 is the electric field across the ribbon, and adopts Landau gauge

$$\vec{A} = H_0 y \hat{x} \ , \tag{4}$$

then the wave functions, given by

$$\psi_{k,n} = e^{ikx}\phi_n(y - y_0) \ , \tag{5}$$

where ϕ_n is the solution to the harmonic-oscillator equation

$$\left[\frac{1}{2m^*}p_y^2 + \frac{1}{2m^*}\left(\frac{eH_0}{c}\right)^2 y^2\right]\phi_n = (n + \tfrac{1}{2})\hbar\omega_c\phi_n \ , \tag{6}$$

and y_0 is related to k by

$$y_0 = \frac{1}{\omega_c}\left(\frac{\hbar k}{m^*} - \frac{cE_0}{H_0}\right) \ , \tag{7}$$

are affected by a vector potential increment $\Delta A\hat{x}$ only through the location of their centers, in the manner

$$y_0 \to y_0 - \frac{\Delta A}{H_0} \ . \tag{8}$$

The energy of the state, still given by

$$\epsilon_{n,k} = (n + \tfrac{1}{2})\hbar\omega_c + eE_0 y_0 + \tfrac{1}{2}m^*(cE_0/H_0)^2 \tag{9}$$

thus changes linearly with ΔA. This gives rise to the derivative in Eq. (1), which may be conveniently evaluated via the substitution

$$\frac{\partial U}{\partial \phi} \to \frac{\Delta U}{\Delta \phi} \tag{10}$$

with $\Delta\phi = hc/e$ a flux quantum. Since, by gauge invariance (2), adding $\Delta\phi$ maps the system back into itself, the energy increase due to it results from the net transfer of n electrons (no spin degeneracy) from one edge to the other. The current is thus

$$I = c\frac{neV}{\Delta\phi} = \frac{ne^2V}{h} . \tag{11}$$

We now consider the dirty interacting system. As in the ideal case, gauge invariance is an exact symmetry forcing the addition of a flux quantum to result only in excitation or deexcitation of the original system. Also as in the ideal case, there is a gap, although the gap now exists between the electrons and holes affected by the perturbation, those contiguous about the loop, rather than in the density of states. Since adiabatic change of the many-body Hamiltonian cannot excite quasiparticles across this gap, it can only produce an excitation of the charge-transfer variety discussed in the ideal case, although the number of electrons transferred need not be the ideal number, and can be zero, as is the case for most systems with gaps. Therefore, Eq. (11) is always true, as a bulk property, for some integer n whenever the local Fermi level lies everywhere in a gap in the extended-state spectrum.

At the edges of the ribbon, the effective gap collapses and communication between the extended states and the local Fermi level is reestablished. Particles injected into this region rapidly thermalize to the Fermi level, in the process losing all memory of having been mapped adiabatically. This would be a significant source of error in Eq. (11) were it not for the fact that *isothermal* differentiation with respect to ϕ, the thermodynamically correct procedure for obtaining I, is equivalent to adiabatic differentiation in the sample interior and is reversible. Thus, slow addition of $\Delta\phi$ physically removes a particle from the local Fermi level at one edge of the ribbon and injects it at the local Fermi level of the other, acting as a pump. Since the Fermi energy is defined as the change in V resulting from the injection of a particle, and since eV is defined to be the Fermi-level difference, edge effects are not a source of error in Eq. (11).

Several other sources remain to be investigated, including possible ϕ dependence, the effect of substituting the ring geometry of Fig. 1 for the usual strip geometry, and effects of tunneling. However, we find it intuitively appealing that the quantum effect should go hand in hand with the persistence of currents, and thus that the physically significant source of error should be thermal activation of carriers to the mobility edge. These carriers produce a large, but finite, normal resistance per square R, which in the steady-state strip geometry, results in a Hall resistance too small in the amount

$$\left|\frac{\Delta R_H}{R_H}\right| = \left(\frac{R_H}{R}\right)^2 \ . \tag{12}$$

In summary, we have shown that the quantum Hall effect is intimately related to the extended nature of the states near the center of the disorder-broadened Landau level, and that edge effects do not influence the accuracy of the quantization. We speculate that the only significant source of error is thermal activation of carriers to the mobility edge.

ACKNOWLEDGMENTS

I am grateful to P. A. Lee, D. C. Tsui, R. E. Prange, and H. Störmer for helpful discussions.

[1]K. V. Klitzing, G. Dorda and M. Pepper, Phys. Rev. Lett. 45, 494 (1980).

[2]Identical behavior has been seen in GaAs-$Al_xGa_{1-x}As$ heterostructures. D. C. Tsui and A. C. Gossard (unpublished).

[3]T. Ando, J. Phys. Soc. Jpn. 37, 622 (1974).

[4]R. E. Prange (unpublished).

PHYSICAL REVIEW B VOLUME 25, NUMBER 4 15 FEBRUARY 1982

Quantized Hall conductance, current-carrying edge states, and the existence of extended states in a two-dimensional disordered potential

B. I. Halperin
Lyman Laboratory of Physics, Harvard University, Cambridge, Massachusetts 02138
(Received 21 August 1981)

When a conducting layer is placed in a strong perpendicular magnetic field, there exist current-carrying electron states which are localized within approximately a cyclotron radius of the sample boundary but are extended around the perimeter of the sample. It is shown that these quasi-one-dimensional states remain extended and carry a current even in the presence of a moderate amount of disorder. The role of the edge states in the quantized Hall conductance is discussed in the context of the general explanation of Laughlin. An extension of Laughlin's analysis is also used to investigate the existence of extended states in a weakly disordered two-dimensional system, when a strong magnetic field is present.

I. INTRODUCTION

In a recent paper Laughlin has given a very elegant and general explanation of the phenomenon that under appropriate conditions, for a two-dimensional sample in a strong magnetic field, at $T=0$, the Hall conductance is quantized in *exact* multiples of the unit e^2/h.[1-4] The purpose of the present paper is to discuss some curious properties of electronic states in a magnetic field that are implied by Laughlin's analysis, and, incidentally, to clarify some details of Laughlin's argument. In particular, it is shown in Secs. II and III below that states at the perimeter of the sample are quasi-one-dimensional states which carry a current, and which do not become localized in the presence of a disordered potential of moderate strength. The perimeter states play an important role in the Hall measurement, if the Fermi levels are different at two edges of the sample.

Following the method of Laughlin,[1] we consider a film of annular geometry, in a magnetic field perpendicular to the plane of the film. In this case, the currents at the inner and outer edge are in opposite directions, and they contribute no net current around the annulus if the Fermi levels are the same at the two edges. If the two Fermi levels differ by an amount $e\Delta$, however, we find that the edge states contribute a net current δI around the ring given by

$$\delta I = ne^2\Delta/h \ , \tag{1}$$

where n is an integer. This contribution is consistent with the quantized Hall conductance, as the chemical potential difference Δ is included, along with any electrostatic potential present, in the potential difference that would be measured by a voltmeter connected between the inner and outer edges of the ring. Of course, the edge current and the quantity Δ are taken into account automatically in the general analysis of Laughlin.[1]

In Sec. IV below, we use an extension of Laughlin's analysis to investigate the question of whether extended states can exist in principle in the interior of a two-dimensional disordered system. We conclude that there must exist a band of extended states in the vicinity of the Landau energy, or at least an energy at which the localization length diverges, if the random potential is weak compared to the cyclotron energy $\hbar\omega_c$.

II. IDEAL SAMPLE

Let us first consider a collection of noninteracting electrons, confined in an ideal uniform film of annular geometry, with a uniform magnetic field $\vec{B}_0$ perpendicular to the plane of the sample. (See Fig. 1.) We assume in addition that there is a magnetic flux Φ, confined to the interior of a solenoid magnet threading the hole in the annulus, and we shall be able to vary the flux Φ without changing the magnetic field in the region where the electrons are confined. (This is a slight modification of the cylinder geometry considered by Laughlin.) We shall assume that no electric field is present so that the electrostatic potential seen by the electrons is constant in the interior of the film, and we assume that the dimensions of the annulus are very large compared to the cyclotron radius r_c for electrons in the magnetic field. We adopt the gauge where the vector potential $\vec{A}$ points in the

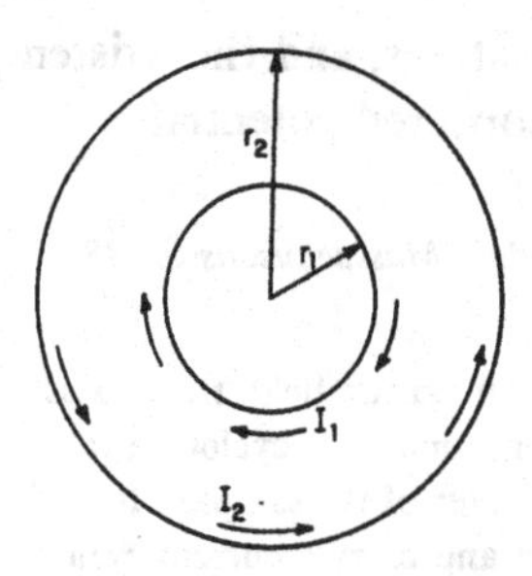

FIG. 1. Geometry of sample. Annular film, in region $r_1<r<r_2$ is placed in uniform magnetic field B_0, pointing out of the page. Additional magnetic flux Φ is confined to region $r<r_1$. Curved arrows show direction of currents I_1 and I_2 at the boundaries of film.

azimuthal (θ) direction, and the magnitude of $\vec{A}$ depends only on the distance from the center of the annulus:

$$A=\tfrac{1}{2}B_0 r+\Phi/2\pi r\ . \tag{2}$$

Away from the edges of the film, the electronic states in this geometry have the form

$$\psi_{m,\nu}(\vec{r})\simeq \text{const}\times e^{im\theta}f_\nu(r-r_m)\ , \tag{3}$$

where m and ν are integers, with $\nu\geq 0$, f_ν is the $\nu+1$ eigenstate of a one-dimensional harmonic oscillator, and the radius r_m is determined by

$$B_0\pi r_m^2=m\Phi_0-\Phi\ . \tag{4}$$

Here Φ_0 is the flux quantum, hc/e. The width of f is of order r_c, where r_c is the cyclotron radius. Of course, Eq. (3) is only applicable if r_m is in the range $r_1<r_m<r_2$, with r_m-r_1 and r_2-r_m large compared to r_c. We shall assume throughout that r_c is small compared to r_1 and r_2-r_1. The energies of the states (3) are given by the Landau formula

$$E_{m,\nu}=\hbar\omega_c(\nu+\tfrac{1}{2})\ , \tag{5}$$

where ω_c is the cyclotron frequency determined by the field B_0 and the carrier effective mass m^*:

$$\omega_c=|eB_0|/m^*c\ . \tag{6}$$

The electron density $|\psi_{m,\nu}(r)|^2$ associated with Eq. (3) is symmetric about the radius r_m, and decays rapidly for $|r-r_m|/r_0>>1$. The current carried by the state is given by

$$I_{m,\nu}=\frac{e}{m^*}\int_0^\infty dr\,|\psi_{m,\nu}(\vec{r})|^2\left[\frac{m\hbar}{r}-\frac{eA(r)}{c}\right]$$

$$\simeq\frac{e^2B_0}{m^*c}\int_0^\infty dr\,|\psi_{m,\nu}|^2(r_m-r)\ . \tag{7}$$

The integral may be taken over the radial coordinate r, at any fixed value of θ. The net current vanishes for states in the interior of the annulus, since the probability densities of the harmonic oscillator states are symmetric about the point $r=r_m$.

The situation is very different when r_m is closer than a few times r_c to an edge of the sample. Then the condition that the wave function vanish at the edges of the sample will shift the energies of the eigenstates away from the Landau energies (5).

Let us focus our attention on the behavior near the outer edge of the annulus, and let us continue to use the index ν to label the number of nodes in the radial wave function. We may then write the electronic wave functions as

$$\psi_{m,\nu}(\vec{r})=\text{const}\times e^{im\theta}g_\nu(r-r_m,r_2-r_m)\ , \tag{8}$$

where $g_\nu(x,s)$ is a wave function which is defined in the region $-\infty<x<s$ and has ν nodes, which vanishes for $x\to s$ and $x\to-\infty$, and which obeys the eigenvalue equation

$$\left[-\frac{\hbar^2}{2m^*}\frac{d^2}{dx^2}+\frac{B_0^2e^2x^2}{2m^*c^2}\right]g_\nu=Eg_\nu\ . \tag{9}$$

Now it is clear that the eigenvalue $E_{m,\nu}$ will approach the value $E_\nu=\hbar\omega_c(\nu+\frac{1}{2})$, for $r_2-r_m>>r_c$. The energy $E_{m,\nu}$ will increase monotonically as r_m increases, passing through the value $E_{m,\nu}=\hbar\omega_c(2\nu+\frac{3}{2})$, when $r_m=r_2$, and increasing eventually as $(r_m-r_2)^2e^2B_0^2/2m^*c^2$ for $r_m-r_2>r_c$. The energy curve is sketched in Fig. 2.

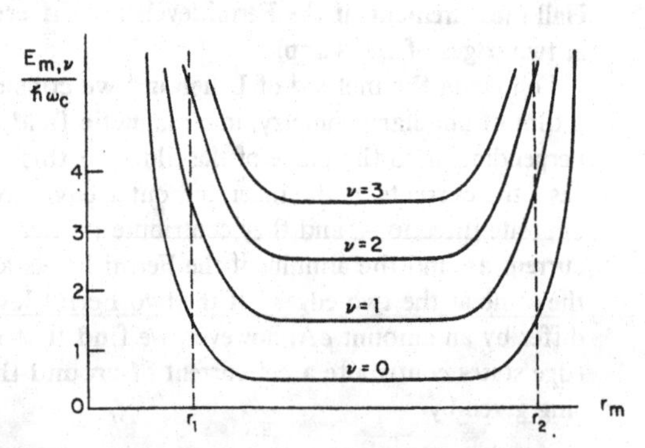

FIG. 2. Energy levels of nonrandom system, in units of $\hbar\omega_c$, as a function of the parameter r_m. The latter quantity is determined by the azimuthal quantum number m, according to Eq. (4), and it is the radius at which the azimuthal current density vanishes for quantum number m. The radius r_m is the center of the wave function $\psi_{m\nu}$ provided that r_m is not too close to the boundary r_1 or r_2.

Since the density $|\psi_{m,\nu}(\vec{r})|^2$ is no longer symmetric about $r=r_m$, we no longer expect that $I_{m,\nu}=0$. In fact, it is readily established that

$$I_{m,\nu}=-c\frac{\partial E_{m,\nu}}{\partial \Phi}=\frac{e}{h}\frac{\partial E_{m,\nu}}{\partial m} . \tag{10}$$

For $B_0>0$, we find that $I_{m,\nu}>0$, for $r_m \simeq r_2$, while $I_{m,\nu}<0$, near the inner edge $r_m \simeq r_1$.

Note that the quantity $|\partial E_{m,\nu}/\partial m|$ is just the energy separation between adjacent energy levels for a given quantum number ν. Thus the total current carried by states of a given ν in a small-energy interval δE is equal to $(e/h)\delta E$ at the outer edge of the sample, and $-(e/h)\delta E$ at the inner edge. (We neglect here any spin or valley degeneracy of the carriers.)

Let us suppose that the Fermi level lies in between the energies E_ν of two Landau levels $\nu=n-1$ and $\nu=n$, in the interior of the sample. Suppose also that near r_2 and r_1 there are Fermi levels $E_F^{(2)}$ and $E_F^{(1)}$, respectively, which differ from each other, but still lie in the interval between E_{n-1} and E_n. Then the total current carried by the edge states between $E_F^{(2)}$ and $E_F^{(1)}$ is clearly given by $neh^{-1}(E_F^{(2)}-E_F^{(1)})$, in agreement with Eq. (1).

In a real experiment, the measured Hall potential eV is the sum of an electrostatic potential eV_0 and the difference in Fermi levels $E_F^{(2)}-E_F^{(1)}$. The edge current is then only a *fraction* of the total Hall current, given by $(E_F^{(2)}-E_F^{(1)})/eV \approx \alpha n r_c \hbar\omega_c C/e^2$ where C is the capacitance per unit length of the edge states, and α is a number of order unity.

III. DISORDERED SAMPLE

Now we must show that the edge currents are not destroyed by a moderate amount of disorder in the sample. Let us consider the effect of a weak random potential $V(\vec{r})$, with $|V(\vec{r})| << \hbar\omega_c$. Let us consider for simplicity a situation where the Fermi level E_F lies midway between the unperturbed Landau energies E_0 and E_1. It is then clear that there will be no energy eigenstates with E near E_F in the interior of the sample, but there will remain two bands of states with E near E_F which are radially localized near r_2 and r_1, respectively.

Consider an energy eigenstate ψ from the band at r_2, and write the state as superposition of the eigenstates $\psi_{m\nu}$ of the nonrandom system:

$$\psi(\vec{r})=\sum_{m\nu} c_{m\nu}\psi_{m\nu}(\vec{r}) . \tag{11}$$

The expansion coefficient $c_{m\nu}$ will be relatively large for $\nu=0$ and r_m near to, but slightly smaller than r_2. The coefficient $c_{m\nu}$ will be smaller by a factor of order $V(\vec{r})/\hbar\omega_c$ for $\nu\geq 1$, and $c_{m\nu}$ will be "exponentially small" for $|r_2-r_m| >> r_c$.

The azimuthal current carried by the state ψ is given by

$$\langle I\rangle=\sum_{m\nu\nu'} c_{m\nu}^* c_{m\nu'} I_{m\nu\nu'} , \tag{12}$$

where

$$I_{m\nu\nu'}\equiv\frac{e}{2\pi m^*}\int\int dr\, d\theta\, \psi_{m\nu}^*(\vec{r})\psi_{m\nu'}(\vec{r}) \times\left[\frac{m\hbar}{r}-\frac{eA(r)}{c}\right] . \tag{13}$$

Note that azimuthal current must be independent of θ, since current conservation requires $\vec{\nabla}\cdot\langle\vec{j}(\vec{r})\rangle=0$, where $\langle\vec{j}(\vec{r})\rangle$ is the current density carried by any exact eigenstate of the Hamiltonian. We see that $I_{m\nu\nu'}$ is identical to $I_{m\nu}$ when $\nu=\nu'$. Furthermore, for r_m near r_2, I_{m01} is of the same order as I_{m0}, namely of order $e\omega_c r_c/r_2$. It follows then that the off-diagonal contribution ($\nu\neq\nu'$) to Eq. (12) cannot cancel the positive diagonal contribution ($\nu=\nu'=0$), when $V(\vec{r})/\hbar\omega_c$ is small; hence the current $\langle I\rangle$ is nonzero. It follows also from current conservation that the eigenstate ψ is not localized azimuthally in any region of θ, but must be spread more or less uniformly around the annulus.

It is clear, physically, that the situation is unaltered if there are a few isolated regions with $V(\vec{r}) >> \hbar\omega_c$. Although there may be localized bound states or resonances in the regions of strong potential, the current-carrying edge states will simply be displaced, locally, to go around these regions. Of course, if the random potential becomes sufficiently strong that electron scattering rate is large compared to ω_c, it is no longer useful to employ the Landau levels as starting points and the arguments given here breakdown.

The arguments given above can be extended, with little difficulty, to the case where E is midway between the $\nu=1$ and $\nu=2$ Landau levels, etc. In this case there will be several values of ν for which the expansion coefficients $c_{m\nu}$ can be large. The contributions of the off-diagonal terms ($\nu\neq\nu'$) in Eq. (12) to the current carried by the state ψ are nevertheless small for $V(\vec{r})/\hbar\omega_c << 1$, because the matrix element $I_{m\nu\nu'}$ is diagonal in m, while the largest values of $c_{m\nu}$ occur at different values of m

for different oscillator levels ν. The reasoning clearly breaks down, on the other hand, if E_F is too close to an unperturbed energy E_ν.

Our argument that $\langle I \rangle \neq 0$ for an edge state in a weakly disordered system did not show that the current carried satisfies Eq. (1) exactly in this case. The validity of this equation may be most easily established by considering what happens as one adiabatically increases the threading flux Φ by one flux quantum, in the manner described by Laughlin, in Ref. 1. We shall not repeat that analysis here in detail, but we shall mention some essential features in the following section.

IV. DO EXTENDED STATES EXIST IN TWO DIMENTIONS?

A starting point of Laughlin's analysis of the quantized Hall conductance is the assumption that for a collection of noninteracting electrons in an infinite two-dimensional sample with weak disorder, in a strong perpendicular magnetic field, there exist energy bands of extended states ("Landau levels") separated by energy regions of localized states and/or energy gaps where there are no states at all. Laughlin shows that if the Fermi energy occurs at a position outside the bands of extended Landau states, and if the flux Φ threading the hole of an annular sample is increased adiabatically by one flux quantum Φ_0, then the net effect will be to transfer an integral number n electrons from the Fermi level at the outer edge to states at the Fermi level of the inner edge of the sample. Since the net change in the energy of the sample is $-neV$, where V is the voltage difference between the outer and inner edge of the sample, and since the work done in the flux change is equal to $-c^{-1}I\Phi_0$, where I is the current around the loop, Laughlin has established that $I/V = nec/\Phi_0$.

It is natural to identify the integer n with the number of bands of extended states below the Fermi energy (multiplited by the spin and valley degeneracy of the carriers), and it is natural to suppose that for weak disorder this number will be the same as the number of Landau levels that would occur below E_F in the absence of disorder. If the disorder is sufficiently strong, however, so that all states below E_F are localized, then we would obtain the integer $n = 0$, and the quantized Hall conductance would not be observed.

It is now generally believed that in the absence of a magnetic field or other mechanisms to break the time-reversal invariance of the Schrödinger equation, the electronic states in a two-dimensional random potential are *always* localized, in principle.[5] When time-reversal symmetry is broken, the leading term responsible for localization in the renormalization-group equations is known to be absent; nevertheless, it has remained an open question whether extended states can exist in a two-dimensional system under these conditions.[5,6]

If two-dimensional states were actually always localized there would seem to be a serious problem, in principle, with the starting point of Laughlin's theory. One could take the point of view that the experimental existence of a nonzero, quantized Hall conductance is sufficient evidence for the existence of extended states, and that further discussion of this point is unnecessary.[7] For the sake of intellectual completeness, however, it seems worthwhile to note that the existence of extended states and of nonzero Hall conductance can actually be demonstrated theoretically, at least in the case of a weakly disordered sample in a strong magnetic field, by an extension of Laughlin's arguments, which will be given below. (Actually, we cannot rule out the possibility that the energy regions of extended states have vanishing width in the limit of an infinite sample, but this possibility would still be compatible with a nonzero quantized Hall conductance.) In addition, the theoretical argument can be applied directly to the theoretically important case of noninteracting electrons, whereas the electron-electron interactions could *a priori* be important in the experimental systems.[8] In the discussion below, we shall in fact confine ourselves to the noninteracting case, although a small modification of the arguments also confirms that the electron-electron interaction, if it is not too strong, will not destroy a nonzero, quantized Hall conductance.[9]

We begin by generalizing the annular geometry of Fig. 1 as follows. We divide the sample into three concentric regions, bounded by radii $r_1 < r_1' < r_2' < r_2$. For $r_1 < r < r_1'$ and $r_2' < r < r_2$, we assume the potential $V(\vec{r}) = 0$. For $r_1' < r < r_2'$, we assume a weak random potential $V(\vec{r}) << \hbar\omega_c$. There is no macroscopic electrostatic field present, and we assume infinite reflecting walls at r_1 and r_2 as before. We shall assume the dimensions of the sample to be arbitrarily large compared to any microscopic length.

The electronic energy levels in this geometry are indicated in Fig. 3. In the border regions $r_1 < r < r_1'$ and $r_2' < r < r_2$, the analysis of Sec. II applies, and the electronic states are well under-

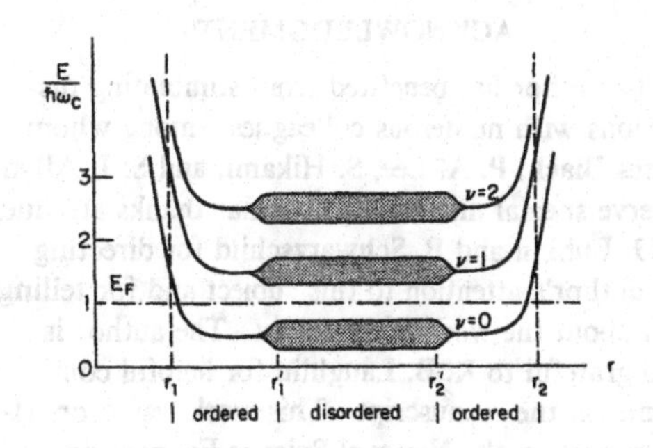

FIG. 3. Energy bands, as a function of position for the inhomogeneous geometry described in Sec. IV. Regions $r_1 < r < r_1'$ and $r_2' < r < r_2$ contain ordered "ideal" conductor, while region $r_1' < r < r_2'$ contains a weak random potential $V(\vec{r})$.

stood. The states have energies $E_{m\nu}$ which are given by the Landau formula $E_\nu = \hbar\omega_c(\nu + \frac{1}{2})$, except at the boundaries r_1 and r_2, where they are pushed upward in energy as in Fig. 2. Now, in the interior disordered region $r_1' < r < r_2'$, we expect that the states will occur in a series of energy bands of finite width, centered about the energies E_ν. If the potential $V(\vec{r})$ is sufficiently weak there should be no states in the region midway between two Landau levels. (Alternatively, if there are a small number of strong impurities, there may be a small density of isolated impurity levels in the mid-gap region; these states will be localized on a scale of order r_c, however, and will not be important for our argument.)

We may now choose one of two hypotheses:

(a) The states in the disordered region are *localized at all energies* with a finite energy-dependent localization length $\lambda(E)$.

(b) The states near the center of each magnetic energy band are *delocalized*, or at least $\lambda(E) \to \infty$ for some energy E in the band.

We shall adopt hypothesis (a), and see that this leads to a contradiction.

Assume that, initially, all electron states in the sample are filled up to a Fermi level E_F, which we choose to lie at the energy $\hbar\omega_c$, midway between the $\nu=0$ and $\nu=1$ Landau levels. Let $\lambda_{\max}$ be the maximum value of $\lambda(E)$, for $E < E_F$, and choose $r_2' - r_1' >> \lambda_{\max}$. Let us now increase adiabatically the flux Φ through the hole in the annulus, by one flux quantum Φ_0. Since, initially, there was no net current flowing in the sample, there is no work done in this process, or, more accurately, the work $-c^{-1}\int I\,d\Phi$ is inversely proportional to the size of the system, as the induced current will be small for large r. We also know that the electronic wave functions in the ordered regions will contract slightly during the flux change so that at the end there is one state unoccupied just below E_F, at $r \approx r_2$, and one new state occupied just above E_F, at $r \approx r_1$.

This change in occupation costs no energy in the limit of a large sample. If, however, in the disordered region the states below the Fermi surface are all localized, there will be no way to transport an electron across this region, since, as discussed by Laughlin, localized states remain unchanged during the flux increase, except for an uninteresting phase factor $e^{i\theta(\vec{r})}$. Then the electron removed from $r \approx r_2$ must be "transferred" to a new occupied state at $r \approx r_2'$, and the new electron at $r \approx r_1$ must be associated with a hole near $r \approx r_1'$. However, by construction, there are no states in the interior of the sample with energy near E_F (except perhaps for some strongly localized impurity states, whose occupation cannot change during the flux increase). It follows that the required change of occupation must cost an energy of order $\hbar\omega_c$, which would be a violation of conservation of energy. Therefore, there must be at least some delocalized states below the Fermi level, even in the disordered region of the sample.

It is interesting to ask what happens to the above argument when the random potential is sufficiently strong that all states below the Fermi energy are localized in the disordered region. It seems that the bands of extended states do not disappear, but rather are pushed upwards in energy as the disorder is increased, and that the Hall conductance ceases when the lowest extended band rises above the Fermi energy.[10] In an inhomogeneous geometry such as that considered above, there will be current-carrying states at the Fermi level near the boundaries of the disordered region (radii r_1' and r_2') analogous to the current-carrying states at the edge of the sample. Under these circumstances, the addition of a flux quantum will transfer an electron from a state at r_2 to a state at r_2' and another electron from a state at r_1' to a state at r_1, so that the Laughlin argument cannot be applied to the sample as a whole. The Hall current will then be determined by the voltage drops across the nondisordered regions only.

As a final remark, we note that by using the geometry described above, we have put Laughlin's derivation of the exact quantization of the Hall conductance in a form which does not require any *a priori* assumption about the behavior of extended states in the disordered region, during the adiabatic change of Φ. We have only made use of the known behavior of the wave functions in the ordered boundary regions, and the relatively trivial behavior of any localized states at the Fermi level during the change of Φ. The transfer of charge through the disordered region, and the quantized relation for I/V, are then implied by conservation of energy and particle number.

ACKNOWLEDGMENTS

The author has benefited from stimulating discussions with numerous colleagues, among whom James Black, P. A. Lee, S. Hikami, and S. J. Allen deserve special mention. Particular thanks are due to G. Lubkin and B. Schwarzschild for directing the author's attention to this subject and for telling him about the work of Laughlin. The author is also grateful to R. B. Laughlin for helpful comments on the manuscript. This work was supported in part by the National Science Foundation through Grant No. DMR77-10210.

---

[1]R. B. Laughlin, Phys. Rev. B 23, 5632 (1981).

[2]The quantized Hall conductance was predicted originally, on the basis of an approximate calculation of a simple model, by T. Ando, Y. Matsumoto, and Y. Uemura, J. Phys. Soc. Jpn. 39, 279 (1975). Subsequently, the quantization was observed to hold with great precision experimentally by K. V. Klitzing, G. Dorda, and M. Pepper, Phys. Rev. Lett. 45, 494 (1980), and by others. Recent experimental results from a number of laboratories are reported in the Proceedings of the Fourth International Conference on Electronic Properties of Two-Dimensional Systems, New London, New Hampshire, 1981 [Surf. Sci. (in press)].

[3]See also the discussion of R. E. Prange, Phys. Rev. B 23, 4802 (1981), and references therein, and H. Aoki and T. Ando, Solid State Commun. 38, 1079 (1981).

[4]D. J. Thouless [J. Phys. C 14, 3475 (1981)] has taken an approach quite different from Laughlin's, and has presented an argument, based on the convergence of perturbation theory, that the quantization of Hall conductance remains exact in the presence of a random potential smaller than $\frac{1}{2}\hbar\omega_c$.

[5]See P. A. Lee and D. S. Fisher, Phys. Rev. Lett. 47, 882 (1981), and references therein.

[6]P. A. Lee (private communication); S. Hikami (private communication).

[7]Ando and Aoki, Ref. 3, and also Thouless, Ref. 4, have also emphasized that a nonzero Hall conductance at $T=0$ implies the existence of some extended states below the Fermi level.

[8]It has been suggested by H. Fukuyama, P. M. Platzman, P. A. Lee, and P. W. Anderson that the electron-electron interaction must be taken into account explicitly if one wishes to understand experiments in which the carrier density is varied and the Fermi level passes through a region of extended states [H. Fukuyama (private communication); H. Fukuyama and P. M. Platzman (unpublished)].

[9]One way to consider electron-electron interactions is via a *Gedanken* experiment, where the electron-electron interaction applies only when the electrons are inside the disordered region $r_1' < r < r_2'$ of Fig. 3. If the interaction is not too strong, there must remain an energy gap between the first and second Landau levels. If necessary, we may add a constant background potential in the disordered region to keep the Fermi level in this gap. The requirements of conservation of energy and of particle number then imply that a nonzero integral number of electrons is transferred through the disordered region, when the flux Φ is increased by one flux quantum, just as in the noninteracting case, analyzed in Sec. IV. If there is a finite density of localized states at the Fermi level in the noninteracting case, then we must make the additional reasonable assumption that these states remain localized (i.e., nonconducting) in the presence of the electron-electron interaction.

[10]This suggestion was also made by R. B. Laughlin (private communication).

Quantized Hall Conductance in a Two-Dimensional Periodic Potential

D. J. Thouless, M. Kohmoto,[a] M. P. Nightingale, and M. den Nijs
Department of Physics, University of Washington, Seattle, Washington 98195
(Received 30 April 1982)

The Hall conductance of a two-dimensional electron gas has been studied in a uniform magnetic field and a periodic substrate potential U. The Kubo formula is written in a form that makes apparent the quantization when the Fermi energy lies in a gap. Explicit expressions have been obtained for the Hall conductance for both large and small $U/\hbar\omega_c$.

PACS numbers: 72.15.Gd, 72.20. Mg, 73.90.+b

The experimental discovery by von Klitzing, Dorda, and Pepper[1] of the quantization of the Hall conductance of a two-dimensional electron gas in a strong magnetic field has led to a number of theoretical studies of the problem.[2-6] It has been concluded that a noninteracting electron gas has a Hall conductance which is a multiple of e^2/h if the Fermi energy lies in a gap between Landau levels, or even if there are tails of localized states from the adjacent Landau levels at the Fermi energy. However, it can be concluded from Laughlin's[2] argument that the Hall conductance is quantized whenever the Fermi energy lies in an energy gap, even if the gap lies within a Landau level. For example, it is known that if the electrons are subject to a weak sinusoidal perturbation as well as to the uniform magnetic field, with $\varphi=p/q$ magnetic-flux quanta per unit cell of the perturbing potential, each Landau level is split into p subbands of equal weight.[7] One might expect each of these subbands to give a Hall conductance equal to e^2/ph, and that is what the classical theory of the Hall current suggests, but according to Laughlin each subband must carry an integer multiple of the Hall current carried by the entire Landau level. This result appears even more paradoxical when it is realized that p, the number of subbands, can become arbitrarily large by an arbitrarily small change of the flux density. This paper contains a calculation of the Hall conductance for such a system, both in the limit of a weak periodic potential and in the tight-binding limit of a strong periodic potential. We have derived explicit expressions for the Hall currents carried by the various subbands, and show how the paradox is resolved.

We consider electrons in a potential $U(x,y)$ which is periodic in x,y with periods a, b, and in a uniform magnetic induction B perpendicular to the plane of the electrons. The band structure of such a system depends critically on $\varphi = abeB/h$, which is the number of flux quanta per unit cell. We take φ to be a rational number p/q; the behavior for irrational values of φ can be deduced by taking an appropriate limit. We use the Landau gauge in which the vector potential has components $(0, eBx)$. In this gauge the eigenfunctions of the Schrödinger equation can be chosen to satisfy the generalized Bloch condition

$$\psi_{k_1k_2}(x+qa,y)\exp(-2\pi ipy/b - ik_1qa) = \psi_{k_1k_2}(x,y+b)\exp(-ik_2b) = \psi_{k_1k_2}(x,y), \tag{1}$$

where k_1 (modulo $2\pi/aq$) and k_2 (modulo $2\pi/b$) are good quantum numbers.[8] We can now define functions $u_{k_1k_2} = \psi_{k_1k_2}\exp(-ik_1x - ik_2y)$ which satisfy the generalized periodic boundary conditions

$$u_{k_1k_2}(x+qa,y)e^{-2\pi ipy/b} = u_{k_1k_2}(x,y+b) = u_{k_1k_2}(x,y), \tag{2}$$

and are eigenfunctions of a Hamiltonian

$$\hat{H}(k_1,k_2) = \frac{1}{2m}\left(-i\hbar\frac{\partial}{\partial x} + \hbar k_1\right)^2 + \frac{1}{2m}\left(-i\hbar\frac{\partial}{\partial y} + \hbar k_2 - eBx\right)^2 + U(x,y). \tag{3}$$

The components of the velocity operator are then given by $\hbar^{-1}$ times the partial derivatives of $\hat{H}$ with respect to k_1, k_2.

There are two quite different approaches to the problem of calculating the Hall conductance σ_H. Laughlin[2] and Halperin[6] have studied the effects produced by changes in the vector potential on the states at the edges of a finite system. By this technique the quantization of the conductance is made explicit, but it is not obvious that the result is insensitive to boundary conditions. An alternative approach is to use the Kubo formula for a bulk two-dimensional conductor. In previous work using this method[3-5] it has not been made obvious that an integer value for the conductance must be obtained.

Because of the relation between the velocity operator and the derivatives of $\hat{H}$, the Kubo formula can be written as

$$\sigma_H = \frac{ie^2}{A_0\hbar}\sum_{\epsilon_\alpha < E_F}\sum_{\epsilon_\beta > E_F}\frac{(\partial\hat{H}/\partial k_1)_{\alpha\beta}(\partial\hat{H}/\partial k_2)_{\beta\alpha} - (\partial\hat{H}/\partial k_2)_{\alpha\beta}(\partial\hat{H}/\partial k_1)_{\beta\alpha}}{(\epsilon_\alpha - \epsilon_\beta)^2}, \tag{4}$$

where A_0 is the area of the system and $\epsilon_\alpha, \epsilon_\beta$ are eigenvalues of the Hamiltonian. This can be related to the partial derivatives of the wave functions u, and gives

$$\sigma_H = \frac{ie^2}{2\pi h}\sum\int d^2k\int d^2r\left(\frac{\partial u^*}{\partial k_1}\frac{\partial u}{\partial k_2} - \frac{\partial u^*}{\partial k_2}\frac{\partial u}{\partial k_1}\right)$$
$$= \frac{ie^2}{4\pi h}\sum\oint dk_j\int d^2r\left(u^*\frac{\partial u}{\partial k_j} - \frac{\partial u^*}{\partial k_j}u\right), \tag{5}$$

where the sum is over the occupied electron subbands and the integrations are over the unit cells in r and k space. The integral over the k-space unit cell has been converted to an integral around the unit cell by Stokes's theorem. For nonoverlapping subbands ψ is a single-valued analytic function everywhere in the unit cell, which can only change by an r-independent phase factor θ when k_1 is changed by $2\pi/aq$ or k_2 by $2\pi/b$. The integrand reduces to $\partial\theta/\partial k_j$. The integral is $2i$ times the change in phase around the unit cell and must be an integer multiple of $4\pi i$.

The problem of evaluating this quantum number remains. We have considered the potential

$$U(x,y) = U_1\cos(2\pi x/a) + U_2\cos(2\pi y/b), \tag{6}$$

both in the limit of a weak periodic potential ($|U| \ll \hbar\omega_c$) and in the tight-binding limit of a strong periodic potential. In the weak-potential limit the wave function can be written as a superposition of the nearly degenerate Landau functions in the same Landau level N:

$$u_{k_1k_2}=\sum_{n=1}^{p} d_n \sum_{l=-\infty}^{\infty} \chi_N\left(x-\frac{\hbar k_2}{eB}-lqa-\frac{nqa}{p}\right)\exp\left[-ik_1\left(x-lqa-\frac{nqa}{p}\right)+2\pi iy\frac{(lp+n)}{b}\right], \tag{7}$$

where χ_N is the appropriate oscillator wave function. Since the term $U_1\cos(2\pi x/a)$ is diagonal in the Landau functions and $U_2\cos(2\pi y/b)$ changes the value of n by unity, the amplitudes d_n satisfy the secular equation of the form

$$V\exp(-iqak_1/p)d_{n-1}+2V'\cos(qbk_2/p+2\pi nq/p)d_n+V\exp(iqak_1/p)d_{n+1}=Ed_n, \tag{8}$$

with V and V' proportional to U_2 and U_1,[7] and with $d_{n+p}=d_n$. This is known as Harper's equation.[9] Its spectrum has been studied in detail by Hofstadter[10] for the iostropic case $V=V'$. We have made considerable use of his results.

Numerical solution of the eigenvalue problem (8) and direct substitution in Eqs. (7) and (5) are possible but laborious. We have done this for a number of cases. The results obtained in this way are entirely concordant with the general results we have obtained by examining the limit $V \ll V'$. The quantum number of a subband cannot change unless the gaps close up; we believe that none of the gaps closes when V/V' is varied, but we have no proof of this. For small V/V' only one component d_n is appreciably different from zero, except for qbk_2 close to a multiple of π, where there is a changeover in the dominant component. These values of k_2 are the locations of the energy gaps which have opened for small V/V'. In the interval $0 \leq k_2 < 2\pi/b$, the rth gap repeats itself q times, at values of qbk_2 equal to an odd (even) multiple of π, if $p-r$ is odd (even). Here the rth and $(r+1)$th band change their dominant component d_n according to $n_r \to n_r+s_r$, $n_{r+1} \to n_{r+1}-s_r$. The value of s_r is independent of k_2 and determined by the Diophantine equation

$$r=s_rq+t_rp, \tag{9}$$

where $|s|\leq p/2$. The rth gap is of order $(V/V')^{|s_r|}$. Each time, the wave function of band r picks up a phase from the off-diagonal terms in Eq. (8): q times qak_1s_r/p and q times $-qak_1s_{r-1}/p$. The total phase change in d_n is

$$d_n(k_1,k_2)=d_n(k_1+2\pi/aq,k_2)$$
$$=d_n(k_1,k_2+2\pi/b)\exp[-i\theta_d(k_1)] \tag{10a}$$

$$\theta_d(k_1)=qak_1(t_{r-1}-t_r)+qak_1/p. \tag{10b}$$

The term qak_1/p in θ_d cancels against the phase change which is already explicit in Eq. (7). It represents the classical Hall current. The total Hall current carried by the rth band is quantized according to $\sigma_H=(e^2/h)(t_r-t_{r-1})$. If the Fermi surface is located in the rth gap of the Nth Landau level, the total Hall conductance is equal to

$$\sigma_H=(e^2/h)(t_r+N-1), \tag{11}$$

with t_r the solution of Eq. (9). This has an unambiguous solution, expect for the case of p even, $r=\frac{1}{2}p$, where there is no gap.

For $q=1$ this gives $t=0$ in the first half of the Landau level and $t=1$ in the second half, so that only the central subband of each Landau level carries the Hall current. For $q=2$, the values of t are alternately 1 and 0, so that alternate subbands carry ± 1 times the Hall current of the Landau level. In general each subband carries one of two possible Hall currents which differ from one another by q units. For example, in the case $q/p=\frac{11}{7}$, the first 11 values of t are -3, 5, 2, -1, -4, 4, 1, -2, 6, 3, 0, so that the Hall current is proportional to -3 or 8 in each subband.

Figure 1 gives an intuitive explanation of the results. The abscissa represents the variable k_2, or equivalently the position of the center coordinate of the state in the x direction, while the ordinate gives the energy as a function of this variable. Application of an electric field in the y direc-

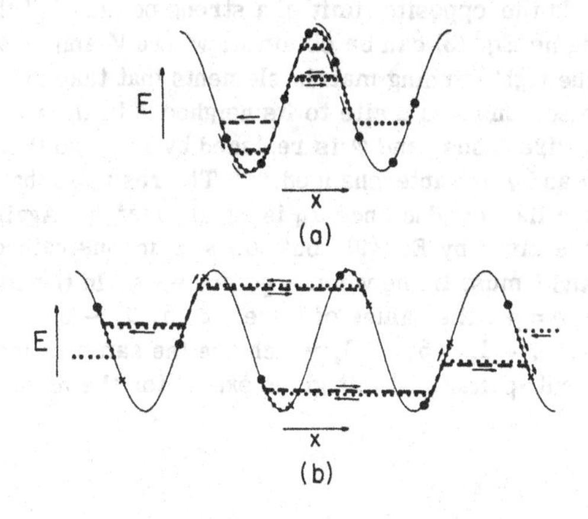

FIG. 1. Motion of electrons in the x direction under the influence of an electric field in the y direction for $V \ll V'$, for the two cases (a) $\varphi=5$ and (b) $\varphi=\frac{5}{7}$.

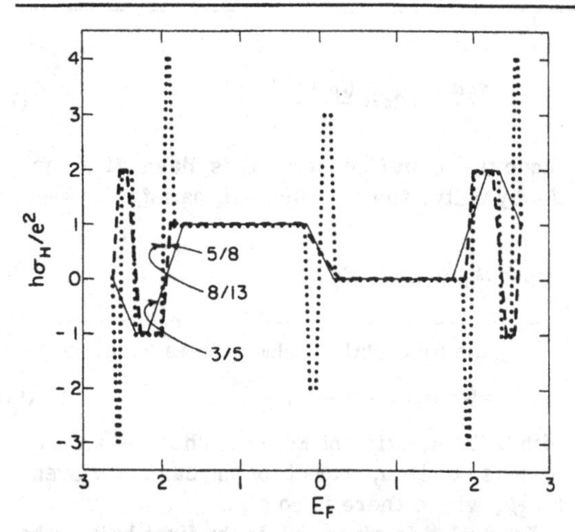

FIG. 2. Hall current as a function of Fermi energy for a weak isotropic substrate. Truncating the continued-fraction expansion of $\frac{1}{2}\sqrt{5}-\frac{1}{2}$, we obtained $q/p=\frac{3}{5}$, $\frac{5}{8}$, $\frac{8}{13}$,.... Within the energy bands, the Hall current was not calculated; the interpolation merely serves as a guide to the eye.

tion gives a steady change of k_2, and therefore a steady motion of a representative point along the $E(k_2)$ curve, until the energy gets very close to the energy of a state whose k_2 value differs by a multiple of $2\pi/b$. Then the two states exchange places rather than crossing in energy. Figure 1(a) shows the case $p=5$, $q=1$. Four of the orbits are closed and give no motion in the x direction, while a Hall current of +1 is given by the middle orbit. Figure 1(b) shows $p=5$, $q=3$, where three orbits give a negative Hall current equal to -1, and the other two give a positive current +2.

In the opposite limit of a strong potential U the same Eq. (8) can be obtained, where V and V' are the tight-binding matrix elements that take an electron from a site to its neighbors in the x and y directions, and φ is replaced by $1/\varphi$, so that p and q are interchanged.[9,10] The result is that the Hall conductance σ_H is equal to te^2/h. Again, t is given by Eq. (9), but now s is unconstrained and t must lie between $-\frac{1}{2}q$ and $+\frac{1}{2}q$. In the case $p/q=\frac{7}{11}$, the values of t are -3, 5, 2, -1, -4, 4, 1, -2, -5, 3, 0, which are the same as the weak-potential limit gives except for the ninth gap, where + 6 has been replaced by -5. This pattern is easy to understand, as p subbands constitute one Landau level, so we get a Hall current t if we count up tp (modulo q) levels from the bottom of the band, and $-t$ if we count down tp (modulo q) levels from the top of the band.

It is generally true that the larger the integer part of tp/q, the smaller will be the corresponding energy gap. The complexity of the structure is associated with the smaller energy gaps as illustrated in Fig. 2; for these smaller energy gaps, the electric field has to be very small for linear response theory to be valid. Stronger fields will cause tunneling across the gap. It is this restriction, and corresponding restrictions on disorder and deviations from the simple sinusoidal potential, that resolves the paradox of the sensitivity of the Hall conductance to the precise value of φ. In fact, as Hofstadter[10] and Wannier[11] have shown, each energy gap persists over a continuous range of φ. The Hall conductance is constant in a particular energy gap, even for irrational values of φ.

Streda[12] has recently obtained a result for the Hall conductance that is in agreement with our results.

This work was supported by the National Science Foundation under Grant No. DMR7920785.

(a)Present address: Department of Physics, University of Illinois, Urbana, Ill. 61801.

[1]K. von Klitzing, G. Dorda, and M. Pepper, Phys. Rev. Lett. 45, 494 (1980).
[2]R. B. Laughlin, Phys. Rev. B 23, 5632 (1981).
[3]H. Aoki and T. Ando, Solid State Commun. 38, 1079 (1981).
[4]R. E. Prange, Phys. Rev. B 23, 4802 (1981).
[5]D. J. Thouless, J. Phys. C 14, 3475 (1981).
[6]B. I. Halperin, to be published.
[7]A. Rauh, G. H. Wannier, and G. Obermair, Phys. Status Solidi (b) 63, 215 (1974).
[8]J. Zak, Phys. Rev. 134, A1607 (1964).
[9]P. G. Harper, Proc. Phys. Soc., London, Sect. A 68, 874 (1955).
[10]D. Hofstadter, Phys. Rev. B 14, 2239 (1976).
[11]G. H. Wannier, Phys. Status Solidi (b) 88, 757 (1978).
[12]P. Streda, to be published.

PHYSICAL REVIEW B VOLUME 31, NUMBER 6 15 MARCH 1985

Quantized Hall conductance as a topological invariant

Qian Niu, D. J. Thouless,* and Yong-Shi Wu†
Department of Physics FM-15, University of Washington, Seattle, Washington 98195
(Received 21 September 1984)

Whenever the Fermi level lies in a gap (or mobility gap) the bulk Hall conductance can be expressed in a topologically invariant form showing the quantization explicitly. The new formulation generalizes the earlier result by Thouless, Kohmoto, Nightingale, and den Nijs to the situation where many-body interaction and substrate disorder are also present. When applying to the fractional quantized Hall effect, we draw the conclusion that there must be a symmetry breaking in the many-body ground state. The possibility of writing the fractionally quantized Hall conductance as a topological invariant is also discussed.

I. INTRODUCTION

In the experiments on both integral[1] and fractional[2] quantized Hall effect it is found that the appearance of a plateau in the Hall conductance is always accompanied by a dip in the longitudinal conductance. This well-observed fact suggests that the existence of the Fermi gap (the energy gap or mobility gap in which the Fermi energy of the system lies) is a necessary condition for the quantization of the Hall conductors. On the other hand, since the phenomenon is quite independent of the details of the devices used in the experiments, this condition must also be sufficient (of course, at zero temperature and in weak electric field).

By now, in the integral case, this relationship has been quite established by perturbation theory[3,4] or by gauge-invariance argument.[5] The latter is more profound for it only uses global properties of the electron system in the external fields. But the solenoid device typically employed in this theory seem to be artificial to most of the known experiments.

Another nonperturbative approach was proposed by Thouless *et al.*[6] (henceforth referred to as TKNdN), who considered an infinite two-dimensional electron gas in a periodic substrate potential commensurate to the perpendicular magnetic field. The Hall conductance calculated from the Kubo formula was rewritten into an integral which shows quantization explicitly. This expression has the advantage that it is independent of the detailed structure of the periodic potential. Later this integral was recognized as the first Chern class of a U(1) principal fiber bundle on a torus.[7,8] The fibers are the magnetic Bloch waves and the torus corresponds to the magnetic Brillouin zone.

Unfortunately, this theory cannot allow either impurity disorder or many-body interactions, because the use of Bloch waves is quite essential to their derivations. In this paper we generalize TKNdN's idea so that an invariant expression can still be constructed in the general case.

The method we are going to use is quite parallel to the generalized formulation of Niu and Thouless[9] for the quantization of particle transport induced by a potential varying slowly and periodically. We use the same geometry as used by TKNdN, consequently we share with them the same deficiency of ignoring the edge effect.[10] The many-body wave functions are required to satisfy a particular boundary condition described by two fixed phase parameters. We then prove that the Hall conductance becomes independent of the phase parameters in the thermodynamic limit, so that it can be averaged over all the phases that prescribe different boundary conditions. The averaged quantity which equals the Hall conductance possesses an expression whose value is quantized explicitly. This expression is of the same form as TKNdN's integral, except that the roles played by the Bloch wave numbers are now played by the phase parameters in the boundary conditions. Thus the same topological identification can be made in the new theory.

In our generalized formulation it is found that the Hall conductance is quantized in an integer times e^2/h as long as the Fermi gap is finite and the many-body ground-state energy is nondegenerate. This quantized value is topological in the sense that it is unchanged under a variation of the potentials so long as the Fermi gap is kept open. Also, the result does not depend on how this gap is generated. It could be generated by the action of the magnetic field alone (Landau gap), together with a periodic substrate potential (gaps between the subbands), or with the many-body interactions.

To obtain a fractional quantization, we have to require, in addition to a finite Fermi gap, that the ground-state energy is degenerate and the ground states have a discrete symmetry breaking. In this case, the Fermi gap must be generated by the many-body interactions, since otherwise the degeneracy cannot be obtained, nor can the symmetry breaking. Recently, Tao and Wu[11] generalized Laughlin's gauge-invariance argument; our result agrees with theirs.

The degeneracy in the ground-state energy at fractional fillings has been clearly demonstrated by the numerical calculation of Su[12] for a few small systems with torus geometry. On the other hand, such degeneracy was not found in Haldane's[13] numerical calculations with spherical geometry. At the present time we cannot conclude (although we suspect) whether the Haldane system will eventually present a degeneracy in the thermodynamic limit, because it is not clear how the spherical geometry could be fitted into our formulation.

Recently, the topological nature of the quantized Hall effect has been revealed in a different approach by Levine, Libby, and Pruisken.[14] They showed that the effective Lagrangian in a replica treatment of a two-dimensional (2D) disordered electron system in a transverse magnetic field contains a nonperturbative topological term which gives rise to the quantization of the Hall conductance. Unfortunately, this theory seems unable to deal with the many-body interaction and therefore the fractionally quantized Hall effect.

In Sec. II we present the formulation generalizing TKNdN's theory in the integral quantized Hall effect. In Sec. III we illustrate the idea of how the fractional quantized Hall conductance could be expressed as a topological invariant. Finally, in Sec. IV we give a brief discussion of the relation between our formulation and Laughlin's gauge-invariance argument.

II. THE INTEGER CASE

The method we will use relies on the fact that for an energy in the Fermi gap the Green function falls off exponentially with distance, and so the current induced by an electric field is a local function of the field. Therefore the response to a field in the interior (away from the edge) of a two-dimensional system can be calculated with only exponentially small errors by replacing the realistic boundary conditions with convenient artificial ones.

Edge currents are a separate problem and cannot be calculated in this way, since they depend on a delicate balance between the diamagnetic currents at the two edges. However, it is possible to devise geometrical conditions under which there are no edge currents. For example, we could consider a Corbino disc or a cylinder with the emf applied in the azimuthal direction.

We now consider a two-dimensional interacting electron system in both a magnetic field $B\hat{z}$ perpendicular to the plane and an electric field $E\hat{x}$ in the plane. The substrate potential may or may not be periodic in space. The Hall current which flows in the y direction can be calculated by the Kubo formula[15] derived from a linear-response theory as

$$\sigma=\frac{ie^2\hbar}{A}\sum_{n\,(>0)}\frac{(v_1)_{0n}(v_2)_{n0}-(v_2)_{0n}(v_1)_{n0}}{(E_0-E_n)^2}\,, \tag{2.1}$$

where $A=L_1L_2$ is the area of the system; the subscripts 0 and n label the ground state and the excited states of the N-body Hamiltonian in the absence of the external electric field:

$$H=\sum_{i=1}^{N}\left[\frac{1}{2m_i}\left[-i\hbar\frac{\partial}{\partial x_i}\right]^2+\frac{1}{2m_i}\left[-i\hbar\frac{\partial}{\partial y_i}-eBx_i\right]^2\right]+\sum_{i=1}^{N}U(x_i,y_i)+\sum_{j=1}^{N}\sum_{\substack{i=1\\ i<j}}^{N}V(|\mathbf{r}_i-\mathbf{r}_j|)\,. \tag{2.2}$$

Correspondingly, E_0 and E_n are the eigenenergies. The velocity operators appearing in the Kubo formula are given by

$$v_1=\sum_{i=1}^{N}\frac{1}{m_i}\left[-i\hbar\frac{\partial}{\partial x_i}\right],\qquad v_2=\sum_{i=1}^{N}\frac{1}{m_i}\left[-i\hbar\frac{\partial}{\partial y_i}-eBx_i\right]. \tag{2.3}$$

A realistic boundary condition is

$$\psi(x_i=L_1)=\psi(x_i=0)=0\,, \tag{2.4a}$$

$$\psi(y_i+L_2)=e^{i\beta L_2}\psi(y_i)\,, \tag{2.4b}$$

$$i=1,2,\ldots,N$$

where the phase parameter β is independent of the particle indices, as is required by the total antisymmetry. But we are only interested in the bulk contribution to the Hall conductance; the condition (2.4a) can then be relaxed to the following form:

$$\psi(x_i+L_1)=e^{i\alpha L_1}e^{i(eB/\hbar)y_iL_1}\psi(x_i)\,, \tag{2.4c}$$

where the y-dependent phase factor is necessary for the Hamiltonian to be Hermitian. At this moment one should note that the boundary conditions (2.4b) and (2.4c) are appropriate only when the particular Landau gauge (2.2) is chosen, but in general we can use instead the gauge-covariant boundary conditions of the form

$$\mathscr{T}_i(L_1,\hat{x})\psi(x_i)=e^{i\alpha L_1}\psi(x_i)\,, \tag{2.4d}$$

$$\mathscr{T}_i(L_2,\hat{y})\psi(y_i)=e^{i\beta L_2}\psi(y_i)\,, \tag{2.4e}$$

$$i=1,2,\ldots,N$$

where $\mathscr{T}_i(L_1,\hat{x})$ and $\mathscr{T}_i(L_2,\hat{y})$ are the single-particle magnetic translation operators in the x and y directions.[16] With this generalization the argument will follow the same line, so we would rather stay with the special case (2.4b) and (2.4c).

Now we make the unitary transformation

$$\phi_n=\exp[-i\alpha(x_1+\cdots x_N)]\times\exp[-i\beta(y_1+\cdots+y_N)]\psi_n\,. \tag{2.5}$$

Then (2.1) becomes

$$\sigma=\frac{ie^2}{A\hbar}\sum_{n\,(>0)}\frac{\left\langle\phi_0\left|\frac{\partial\tilde{H}}{\partial\alpha}\right|\phi_n\right\rangle\left\langle\phi_n\left|\frac{\partial\tilde{H}}{\partial\beta}\right|\phi_0\right\rangle-\left\langle\phi_0\left|\frac{\partial\tilde{H}}{\partial\beta}\right|\phi_n\right\rangle\left\langle\phi_n\left|\frac{\partial\tilde{H}}{\partial\alpha}\right|\phi_0\right\rangle}{(E_0-E_n)^2}\,, \tag{2.6}$$

where $\tilde{H}$ is the transformed Hamiltonian. Equivalently, $\tilde{H}$ can be obtained from H by the following replacement:

$$-i\frac{\partial}{\partial x_i}\to -i\frac{\partial}{\partial x_i}+\alpha,\quad -i\frac{\partial}{\partial y_i}\to -i\frac{\partial}{\partial y_i}+\beta\,. \tag{2.7}$$

It is clear that $(1/\hbar)\partial\tilde{H}/\partial\alpha$ and $(1/\hbar)\partial\tilde{H}/\partial\beta$ are just the transformed velocity operators. By a simple manipulation we can express (2.6) in terms of the partial derivatives of the transformed wave function for the ground state of the many-body system:

$$\sigma=\frac{ie^2}{\hbar A}\left[\left\langle\frac{\partial\phi_0}{\partial\alpha}\middle|\frac{\partial\phi_0}{\partial\beta}\right\rangle-\left\langle\frac{\partial\phi_0}{\partial\beta}\middle|\frac{\partial\phi_0}{\partial\alpha}\right\rangle\right]$$

$$=\frac{ie^2}{\hbar}\left[\left\langle\frac{\partial\phi_0}{\partial\theta}\middle|\frac{\partial\phi_0}{\partial\varphi}\right\rangle-\left\langle\frac{\partial\phi_0}{\partial\varphi}\middle|\frac{\partial\phi_0}{\partial\alpha}\right\rangle\right], \tag{2.8}$$

where $\theta=\alpha L_1$ and $\varphi=\beta L_2$.

So far the derivatives are formal and we still cannot see why the Hall conductance should be quantized. To proceed further, we need to assume that there is always a finite energy gap between the ground state and the excitations under any given boundary conditions of the form in (2.4b) and (2.4c). Also, it is plausible to say that the bulk conductance as given by the Kubo formula should be insensitive to the boundary conditions if the particles do not have long-range correlations in the ground state. We leave the justification of this point to the Appendix. In fact, in the special case of zero interaction and flat substrate potential, one can explicitly show that the above expression is indeed independent of the parameters, even without taking the thermodynamic limit.

Consequently we can equate σ with its average over all the phases $(0\le\theta<2\pi,\ 0<\varphi\le 2\pi)$ that specify different boundary conditions, i.e.,

$$\sigma=\bar{\sigma}=\frac{e^2}{h}\int_0^{2\pi}\int_0^{2\pi}d\theta\,d\varphi\frac{1}{2\pi i}\left[\left\langle\frac{\partial\phi_0}{\partial\varphi}\middle|\frac{\partial\phi_0}{\partial\theta}\right\rangle-\left\langle\frac{\partial\phi_0}{\partial\theta}\middle|\frac{\partial\phi_0}{\partial\varphi}\right\rangle\right]. \tag{2.9}$$

This is of the same form as the integral that appeared in TKNdN's original theory, except that positions of the Bloch wave numbers are now taken by the phase parameters θ and φ. Because of the energy gap, the ground state must go back to itself (up to an overall phase factor) as θ or φ changes by 2π, unless the ground state is not uniquely determined by the boundary condition. Thus the Hall conductance is quantized into an integer times e^2/h whenever the ground state is nondegenerate and is separated from the excited states by a finite energy gap.

The integral in (2.9) is actually a topological invariant. It is the first Chern class of a U(1) principal fiber bundle of the ground-state wave functions on the base manifold of a torus T^2 parametrized by the phases θ and φ. Originally such a recognition was made by Avron, Seiler, and Simon[7] in the context of TKNdN's original theory. The base manifold was the magnetic Brillouin zone, and the fibers were the single-particle Bloch waves. Recently, Kohmoto[8] finished a detailed analysis showing how the abstract topological idea is applied to their expression for the quantized Hall conductance. But since the use of single-particle Bloch waves is essential in TKNdN's theory, this topological idea loses sense as soon as the many-body interaction and the substrate disorder are taken into account.

Fortunately, all these can be recovered by manipulating the phases describing the boundary conditions. Apparently, the Hall conductance should be calculated under a fixed boundary condition as given in (2.8). But the insensitivity of the physical quantity to the boundary conditions allows us to make an average over the phases. In this way an expression for the Hall conductance similar to that of TKNdN's is obtained, so the same topological words can apply to the problem in the rather general situation.

In fact, the recognition of the Hall conductance as a topological invariant is not only of mathematical formality, but also of physical content, because we can explain the stability of the quantization of the Hall conductance against various kinds of perturbations. The reason lies in the fact that the existence of a finite Fermi gap above the ground state is a discrete property which does not depend upon the potentials continuously. Also, in the presence of slight disorder in the substrate, the Fermi level can be locked into the impurity spectra and allows the mobility gap to open in a finite range of the magnetic field. This explains the plateaus of the Hall conductance at the quantized values.

III. THE FRACTIONAL CASE

As mentioned in the introduction, the observation of fractional quantized Hall effect[2] (FQHE) is also accompanied with a vanishing of the longitudinal conductance in the zero-temperature limit. This suggests that a Fermi gap must also exist at the fractional fillings near which FQHE is observed. Several theoretical calculations[17–19] have already justified this point. Furthermore, the ground state is shown to be liquidlike, so correlation between electrons decays rapidly as their separation becomes large. Thus we can continue to use the method employed in the preceding section to equate the Hall conductance to its average over all different boundary conditions as in (2.9). Since a nondegenerate ground state always leads to an integral quantization, we must require a degeneracy in order to explain the fractional quantization. In this case, Eq. (2.9) should be written as

$$\sigma=\bar{\sigma}$$

$$=\frac{e^2}{hd}\sum_{K=1}^{d}\int_0^{2\pi}\int_0^{2\pi}d\theta\,d\varphi\frac{1}{2\pi i}\left[\left\langle\frac{\partial\phi_K}{\partial\varphi}\middle|\frac{\partial\phi_K}{\partial\theta}\right\rangle-\left\langle\frac{\partial\phi_K}{\partial\theta}\middle|\frac{\partial\phi_K}{\partial\varphi}\right\rangle\right], \tag{3.1}$$

where d is the degree of the degeneracy, and $\{\phi_K\}$ is an orthogonal basis spanning the ground-state Hilbert space. In the above expression we have also used the fact that there is no coupling between different ground states, because they are macroscopically separated in the sense that they cannot be obtained from one another by a few number of single-particle excitations.

Unlike the nondegenerate case, the integral over θ and φ is no longer a topological invariant, since the variation of θ or φ by 2π does not necessarily lead each ground state back to itself. But the summation over the integrals may still be a topological invariant. We consider this possibility in the following.

Consider the $1/p$ fillings first. We start from the parent states similar to those proposed by Tao and Thouless.[17] The single-particle states in the ground Landau level are

$$e^{i\lambda y}u(x-\lambda b^2),\ b^2=\frac{\hbar}{eB}\ . \tag{3.2}$$

In order to satisfy the boundary conditions (2.4b) and (2.4c), we make the following linear combinations:

$$W_m(\alpha,\beta)=\sum_{n=-\infty}^{\infty} e^{i\delta_m\lambda_n b^2}e^{i\lambda_n y}u(x-\lambda_n b^2)\ , \tag{3.3}$$

where

$$\lambda_n=\beta+\frac{2\pi}{L_2}pn,\ \delta_n=\alpha+\frac{2\pi}{L_1}m \tag{3.4}$$

and the linear dimensions of the system is chosen to give $1/p$ filling of the ground Landau level:

$$\frac{N}{L_1L_2}=\frac{1}{2\pi pb^2}\ . \tag{3.5}$$

Then one can show that these satisfy the following boundary conditions:

$$W_m(x+L_1)=e^{i\alpha L_1}e^{iL_1y/b^2}W_m(x)\ , \tag{3.6}$$

$$W_m(y+L_2)=e^{i\beta L_2}W_m(y)\ . \tag{3.7}$$

Also one should notice that there are N different states of the form (3.3) (for fixed α and β), since W_{m+N} differs from W_m by only an overall phase factor. This is not surprising because we only used those states in (3.2) whose centers (in the x direction) are separated by a multiple of $(2\pi/L_2)pb^2$ to construct the linear combinations. The parent state of the many-body system is then made by taking the determinant of these single-particle states. We claim that this parent state is equivalent to that proposed by Tao and Thouless,[17] because different linear combinations of the single-particle states give the same determinantal wave function. One can check explicitly that the same boundary conditions as in (3.6) and (3.7) are satisfied by the parent state just constructed, i.e.,

$$\psi(x_l+L_1)=e^{i\alpha L_1}e^{iL_1y_l/b^2}\psi(x_l)\ , \tag{3.8}$$

$$\psi(y_l+L_2)=e^{i\beta L_2}\psi(y_l)\ , \tag{3.9}$$

$$l=1,2,\ldots,N\ .$$

Now we come to the interesting point. Clearly the above boundary condition is specified by $\theta=\alpha L_1$ (mod 2π) and $\varphi=\beta L_2$ (mod 2π). But the parent state cannot be uniquely determined in the same way, since a variation of αL_1 by 2π does not lead the parent state back to itself. In other words, under a given boundary condition we can construct many different parent states from which the true ground states would be generated by the adiabatic turning on of the many-body interactions. Since a variation of βL_2 by 2π and/or of αL_1 by $p2\pi$ do lead the state back to itself, p different parent states are obtained:

$$\psi(\theta,\varphi),\psi(\theta+2\pi,\varphi),\ldots,\psi(\theta+(p-1)2\pi,\varphi)\ . \tag{3.10}$$

These states are orthogonal since they are constructed from different sets of single-particle states. In fact, the true ground states generated from them are also orthogonal to one another. The reason is that these parent states belong to different eigenstates (with different eigenvalues) of the magnetic translation operator $\mathcal{T}((2\pi/L_1)b^2,\hat{\mathbf{y}})$ which translates all the particles by $(2\pi/L_1)b^2$ along the y direction, while the operator commutes with the total Hamiltonian and its interaction part from which the S matrix is constructed.[18] In other words the true ground states obtained by acting the S matrix on the parent states must also belong to the eigenstates of $\mathcal{T}((2\pi/L_1)b^2,\hat{\mathbf{y}})$ with different eigenvalues, therefore they must be orthogonal to one another.

On the other hand, different parent states are connected by the total magnetic translation $\mathcal{T}((2\pi/L_2)b^2,\hat{\mathbf{x}})$ and its powers. Since this operator also commutes with the S matrix, the states generated from different parent states must have the same energy.

Let us now look back to formula (3.1), where ϕ_K are just the true ground states transformed by relation (2.5). If there are no other ground states other than those generated from our parent states, then (3.1) can be written as

$$\sigma=\bar{\sigma}$$

$$=\frac{e^2}{hp}\int_0^{2\pi p}d\theta\int_0^{2\pi}d\varphi\frac{1}{2\pi i}\left[\left\langle\frac{\partial\phi_1}{\partial\varphi}\middle|\frac{\partial\phi_1}{\partial\theta}\right\rangle-\left\langle\frac{\partial\phi_1}{\partial\theta}\middle|\frac{\partial\phi_1}{\partial\varphi}\right\rangle\right], \tag{3.11}$$

where we have absorbed the summation into the integration over the extended range $[0\le\theta\le p2\pi]$, because different ground states can be obtained from ϕ_1 by continuous variation of θ by 2π, 4π, etc. We can regard the extended zone $[0\le\theta<p2\pi,\ 0\le\varphi<2\pi]$ as a torus, because ϕ_1 is led back to itself (up to an overall factor) as θ change by $2\pi p$ or φ by 2π. Thus

$$\sigma=\frac{e^2}{h}\frac{c}{p}\ , \tag{3.12}$$

where c is the integer given by the integral. As in the nondegenerate case we can continue to attach a topological meaning to this integer if the extended zone is regarded as the base manifold.

To determine the integer c we may turn off the varying part of the substrate potential; the integer is unchanged assuming the Fermi gap is unclosed by this process. We then transform to the moving frame in which the external electric field becomes zero, we will find no electron current in this frame. Consequently the Hall conductance is directly related to the filling factor $1/p$, so c is unity.

In the above argument we used the concept of parent states to illustrate the idea how the fractional quantization can be obtained within our frame. We do not mean to give a rigorous proof, but we suggest that the following scenario might correspond to the physical reality:

(1) At a $1/p$ filling with p odd and small, the ground-state energy of the system has a p-fold degeneracy. As the phase parameter θ changes by 2π, 4π, etc., the system goes from one ground state to another, and it comes back to the original ground state after θ varies by $2\pi p$.

(2) At a q/p filling, again with p odd and small, the degeneracy is jp-fold with $j \leq q$. But these states fall into j groups, and the states in each group transform in the same way as in (1).

(3) As the density of electrons deviates a small but finite amount from one of the fillings considered above, the ground states may seek a similar structure as in the neighboring case in order to gain the commensurate energy (which is negative) achieved at the neighboring small denominator fillings. If this is right, then the finite plateaus of the Hall conductance observed near the small and odd denominator fillings can be understood.

(4) Since p is even, conjectures (1) and (2) might still be true, but (3) may break down by the fractional statistics recently suggested by a number of people.[20]

Before closing this section we would like to add one more comment. The argument about the degeneracies of the ground state has been presented in terms of the wave function proposed by Tao and Thouless,[17] but it can also be presented in terms of Laughlin's wave function.[19] With the boundary condition (2.4b) (periodicity in the y direction) a wave function confined to the region $0 < x_1 < hNp/eBL_2$ can be written in the form

$$\phi_1 = \exp\left[-\frac{1}{2b^2}\sum_{i=1}^{N} x_i^2 + \beta\sum_{i=1}^{N} z_i\right] \times \prod_{j=1}^{N}\prod_{\substack{i=1 \\ i<j}}^{N} (e^{2\pi z_i/L_2} e^{2\pi z_j/L_2})^p \tag{3.13}$$

where $z_j = x_j + iy_j$. This has the same local properties as the wave function Laughlin writes in cylindrical geometry or Haldane[21] in spherical geometry. This state can be modified to satisfy the condition (2.4c) by expanding it in terms of the single-particle Landau states (3.2), and replacing each of these by its periodic continuation in the x direction (3.3). This state is also degenerate with the $p-1$ orthogonal states which can be constructed by the action of the magnetic translation operator on it.

IV. RELATION TO LAUGHLIN'S GAUGE ARGUMENT

According to Laughlin's theory,[5] the Hall conductance can be expressed in terms of the number of electrons transported from one edge of the sample to the other after an adiabatic change of the gauge flux by one quantum. We now try to relate this charge transport to the Kubo formula we have been using. The Hamiltonian with a time-dependent gauge parameter is given by (2.2) with the replacement

$$-i\frac{\partial}{\partial y_i} \to -i\frac{\partial}{\partial y_i} + \beta(t) . \tag{4.1}$$

The current in the x direction induced by the adiabatic variation of $\beta(t)$ can be calculated from the formula[9,22]

$$J_x = \frac{i\hbar^2\dot{\beta}(t)}{L_1} \sum_{n(>0)} \frac{(v_2)_{0n}(v_1)_{n0} - (v_1)_{0n}(v_2)_{n0}}{(E_0 - E_n)^2} , \tag{4.2}$$

where the indices 0 and n indicate the instantaneous eigenstates of the Hamiltonian. The charge transport in a period T during which $\beta(t)$ changes by $2\pi/L_2$ (corresponding to a flux quantum) is thus

$$\begin{aligned} C &= \int_0^T dt\, J_x \\ &= \int_0^{2\pi/L_2} d\beta \frac{i\hbar^2}{L_1} \sum_{n(>0)} \frac{(v_2)_{0n}(v_1)_{n0} - (v_1)_{0n}(v_2)_{n0}}{(E_0 - E_n)^2} \\ &= \frac{h}{e^2}\int_0^{2\pi} \frac{d\varphi}{2\pi}\sigma(\varphi) , \end{aligned} \tag{4.3}$$

or

$$\bar{\sigma} = \frac{e^2}{h} C , \tag{4.4}$$

where $\sigma(\varphi)$ is the Hall conductance calculated from the Kubo formula under a fixed gauge $\beta \equiv \varphi L_2$.

Note that it is the averaged Hall conductance that corresponds to the charge transport, so Laughlin's argument really involved an approximation. In fact, as he pointed out in his paper, when he made use of the Faraday's theorem, he actually replaced the adiabatic derivative $dU/d\Phi$ by the fraction $\Delta U/\Phi_0 \propto C$. Here U is the total energy of the electron system, Φ the flux, and Φ_0 the flux quantum. In other words, his approximation is of the same nature as ours.

One final comment. When Laughlin[5] tried to establish the quantization of the Hall conductance, he actually based his argument upon the belief that as the electronic states in the bulk go back to themselves, the particle transport from one edge to the other must be an integral number. Although this idea is physically intuitive, it is still not obvious because of the wave nature of the electrons. Our formulation presents a rigorous proof of this idea.

ACKNOWLEDGMENTS

This work was supported by the National Science Foundation under Grant No. DMR-83-19301, by the Royal Society of London, and by the U.S. Department of Energy under Contract No. DE-AC06-81ER40048.

APPENDIX

The insensitivity of the Hall conductance to the boundary conditions is most easily understood in the noninteracting case. Then the Kubo formula (2.6) can be written in terms of the single-particle quantities as

$$\sigma=\frac{ie^2}{A\hbar}\sum_{\epsilon_m(<\epsilon_f)}\sum_{\epsilon_n(>\epsilon_f)}\frac{\left\langle\varphi_m\left|\frac{\partial\tilde{h}}{\partial\alpha}\right|\varphi_n\right\rangle\left\langle\varphi_n\left|\frac{\partial\tilde{h}}{\partial\beta}\right|\varphi_m\right\rangle-\left\langle\varphi_m\left|\frac{\partial\tilde{h}}{\partial\beta}\right|\varphi_n\right\rangle\left\langle\varphi_n\left|\frac{\partial\tilde{h}}{\partial\alpha}\right|\varphi_m\right\rangle}{(\epsilon_m-\epsilon_n)^2} \tag{A1}$$

where $\tilde{h}$ is the transformed single-particle Hamiltonian and φ_n are the single-particle wave functions. With some manipulations, this can in turn be written in terms of the Green functions as

$$\sigma=\frac{ie^2}{A\hbar}\oint\frac{dz}{2\pi i}\mathrm{Tr}\left[g\frac{\partial\tilde{h}}{\partial\alpha}g\frac{\partial\tilde{h}}{\partial\beta}g\right], \tag{A2}$$

where the integral contour surrounds the filled state energies. The derivative of the Hall conductance σ with respect to θ is

$$\frac{\partial\sigma}{\partial\theta}=\frac{ie^2}{A\hbar L_1}\oint\frac{dz}{2\pi i}\mathrm{Tr}\left[2g\frac{\partial\tilde{h}}{\partial\alpha}g\frac{\partial\tilde{h}}{\partial\alpha}g\frac{\partial\tilde{h}}{\partial\beta}g+g\frac{\partial\tilde{h}}{\partial\alpha}g\frac{\partial\tilde{h}}{\partial\beta}g\frac{\partial\tilde{h}}{\partial\alpha}g+g\frac{\partial^2\tilde{h}}{\partial\alpha^2}g\frac{\partial\tilde{h}}{\partial\beta}g\right]. \tag{A3}$$

Now, because of the existence of the Fermi gap the energy parameter z can be chosen away from the spectrum of the extended states, hence the Green function $g(\mathbf{r},\mathbf{r}')$ is exponentially bounded as $|\mathbf{r}-\mathbf{r}'|$ becomes large.[4] Thus $\partial\sigma/\partial\theta$ is of the order of $(l/L_1)e^2/h$, with l being the localization length of the Green function. Similarly we have

$$\frac{\partial\sigma}{\partial\varphi}\sim\frac{l}{L_2}\frac{e^2}{h}. \tag{A4}$$

When the many-body interactions are taken into account, we can use the method used in Ref. 9 to estimate $\partial\sigma/\partial\theta$ and $\partial\sigma/\partial\varphi$. Although the manipulations are complicated, the same result can be obtained under the assumption that the electrons do not have long-range correlations in the ground state.

---

*Also at Cavendish Laboratory, University of Cambridge, Madingley Road, Cambridge CB3 0HE, United Kingdom.

†On leave from Department of Physics, University of Utah, Salt Lake City, UT 84112 (address after April 1, 1985).

[1]K. V. Klitzing, G. Dorda, and M. Pepper, Phys. Rev. Lett. 45, 494 (1980).

[2]D. C. Tsui, H. L. Störmer, and A. C. Gossard, Phys. Rev. Lett. 48, 1559 (1982); H. L. Störmer, D. C. Tsui, A. C. Gossard, and J. C. M. Hwang, in Proceedings of the 16th International Conference on Physics of Semiconductors (unpublished).

[3]H. Aoki and T. Ando, Solid State Commun. 38, 1079 (1981); R. E. Prange, Phys. Rev. B 23, 4802 (1981); R. E. Prange and R. Joynt, *ibid.* 25, 2943 (1982).

[4]D. J. Thouless, J. Phys. C 14, 3475 (1981).

[5]R. B. Laughlin, Phys. Rev. B 23, 5632 (1981).

[6]D. J. Thouless, M. Kohmoto, M. P. Nightingale, and M. den Nijs, Phys. Rev. Lett. 49, 405 (1982).

[7]J. Avron, R. Seiler, and B. Simon, Phys. Rev. Lett. 51, 51 (1983); B. Simon, Phys. Rev. Lett. 51, 2167 (1983).

[8]M. Kohmoto (unpublished).

[9]Q. Niu and D. J. Thouless, J. Phys. A 17, 2453 (1984).

[10]B. I. Halperin, Phys. Rev. B 25, 2185 (1982).

[11]R. Tao and Y. S. Wu, Phys. Rev. B 30, 1097 (1984).

[12]W. P. Su, Phys. Rev. B 30, 1069 (1984).

[13]F. D. M. Haldane (private communication).

[14]H. Levine, S. B. Libby, and A. M. M. Pruisken, Nucl. Phys. B 240, 30 (1984); 240, 49 (1984); 240, 71 (1984).

[15]At zero temperature this form of the Kubo formula is most easily derived by first-order static perturbation expansion.

[16]J. Zak, Phys. Rev. 134, A1607 (1964).

[17]R. Tao and D. J. Thouless, Phys. Rev. B 28, 1142 (1983).

[18]R. Tao, Phys. Rev. B 29, 636 (1984).

[19]R. B. Laughlin, Phys. Rev. Lett. 50, 1395 (1983); D. Yoshioka, B. I. Halperin, and P. A. Lee, Surf. Sci. 142, 155 (1984).

[20]B. I. Halperin, Phys. Rev. Lett. 52, 1583 (1984); D. Arovas, J. R. Schrieffer, and F. Wilczek, Phys. Rev. Lett. 53, 722 (1984); R. Tao and Y. S. Wu (unpublished).

[21]F. D. M. Haldane, Phys. Rev. Lett. 51, 605 (1983).

[22]D. J. Thouless, Phys. Rev. B 27, 6083 (1983).

PHYSICAL REVIEW B VOLUME 35, NUMBER 5 15 FEBRUARY 1987-I

Quantum Hall effect with realistic boundary conditions

Qian Niu
Department of Physics, University of Illinois at Urbana-Champaign, 1110 West Green Street, Urbana, Illinois 61801

D. J. Thouless
Department of Physics, University of Washington, Seattle, Washington 98195
(Received 29 September 1986)

Earlier theories of the quantum Hall effect depend on boundary conditions (cylindrical, toroidal, etc.) which are very different from that of the experimental devices (essentially striplike). To remove this discrepancy, we show that the Hall conductivity has an exponential locality property whenever the Fermi energy lies between the levels of the bulk extended states, and that this is true in spite of the edge states at the Fermi energy. We describe in detail how this locality property can be used to adapt the gauge-symmetry argument of Laughlin and the topological-invariant approach of Niu, Thouless, and Wu to conditions that are much closer to real experimental ones. The resulting conclusion is that the boundary correction to the quantization of the Hall conductance is exponentially small when the system size is large compared with a microscopic length (typically the magnetic length).

I. INTRODUCTION

For the most part the theory of the integer quantum Hall effect[1] (QHE) is well understood, and there is no mystery about why the Hall conductance is a multiple of e^2/h with such high accuracy. On the other hand there seems to be a gap between the theoretical derivations of the QHE and the experimental measurements, in that the theories seem, for the most part, to rely on special forms of boundary conditions which do not correspond to the actual conditions used in experiments. In this paper we examine these boundary conditions more closely and show how the gaps between theory and experiment can be closed.

Figure 1 illustrates a typical device used to make a four-terminal measurement of the Hall voltage for an inversion layer in a strong magnetic field. A strip of inversion layer [essentially a two-dimensional (2D) electron layer] is connected to a source and a drain (three dimensional) at the two ends. A strong and nearly uniform magnetic field is applied in the direction normal to the inversion layer. A current I is passed through the strip. A pair of voltage probes on opposite sides of the strip, also part of the inversion layer, allow the voltage V_H across the strip to be measured with a voltmeter attached to the probes. A second pair of probes further along the side enables the longitudinal component of the resistance to be measured simultaneously. The Hall conductance is

$$\sigma_H = I/V_H \ . \tag{1}$$

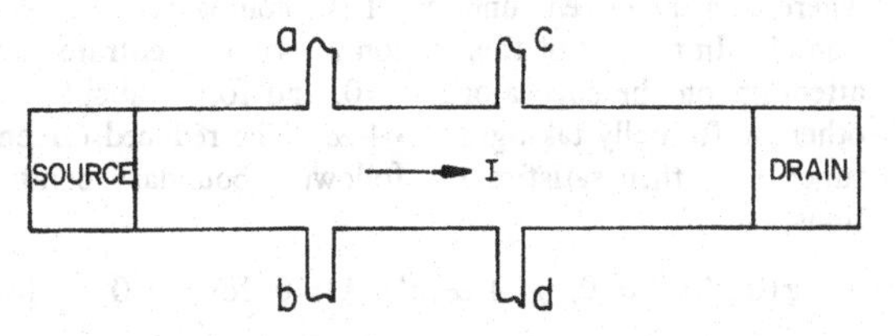

FIG. 1. Illustration of a typical device (strip geometry) used to make a four-terminal measurement of the QHE. The magnetic field is in the normal direction. A current I is passed through the strip. Between a and b the Hall voltage is measured and between a and c the longitudinal voltage is measured.

This is found to be quantized in multiples (integer for the integer QHE, exact fractions with quite small denominators for the fractional QHE) of e^2/h whenever the longitudinal component of the resistance vanishes. Energy dissipation does occur near the ends of the strip, and, for the plateaus of the QHE, the voltage drop measured in a two-terminal measurement is very close to V_H, as can be seen from the arguments of Kawaji.[2] It is, however, an experimental fact that the disturbance of the two ends has little influence on the high precision of the measured quantization of σ_H, so long as the strip is long enough.[3] This immediately suggests that the Hall conductivity has a locality property.

Theoretical analyses are all agreed that there should be a precise quantization of the Hall conductance at sufficiently low temperatures whenever the Fermi energy lies in a gap of the density of states, or in an energy region where there are electron states, but these are all localized. Such an energy gap or mobility gap will be signaled by the vanishing of the longitudinal component of resistance (or, equivalently, of the longitudinal component of conductance). Derivations of this quantization are, however, based on geometries very different from typical experimental situations. For example, the very important argument of Laughlin[4] is based on a cylindrical geometry for the device, with the current flowing round the cylinder. Niu *et al.*[5] assumed that the device was a torus. Halperin[6] gave a modified version of Laughlin's argument in which an annulus was used, and particular attention was paid to the currents flowing round the circles bounding the annulus. There is also a set of arguments based on perturbation theory[7–9] which show that in the bulk two-dimensional system the two-dimensional conductivity, the ratio of the current per unit length to the transverse electric field, is quantized, and is undisturbed by perturbations due to disorder or interactions; integration of this relation leads to the conclusion that the Hall conductance is indeed quantized away from the edges. All these arguments use boundary conditions that are very different from those used in real systems.

The key observation is that, when the Fermi energy is in a mobility gap, the conductivity, which is purely transverse, depends only on the local environment, up to exponentially small terms. The locality property of the transverse conductivity σ_{xy} can best be analyzed through the Kubo formula expressed in terms of Green functions.[5,9] The conductivity is expressed as an integral over the energy parameter of the Green function taken round a contour in the complex plane which cuts the real axis only at the Fermi energy and at some energy below the spectrum of the system. If the Fermi energy is in a mobility gap then the Green function is an exponentially decreasing function of distance for all values of the energy on the contour of integration. We have used this property in earlier work both to show that there are no perturbative corrections to the Hall conductance to any power of the electric field,[10] and to justify averaging over generalized periodic boundary conditions.[5]

On a strip geometry, there are states which damp exponentially away from the edges, but are extended along the edges. The energies of these edge states fill the gap or mobility gap between the levels of the extended states in the bulk. We are thus forced to consider the effects of the edge states on the long-distance behavior of the Green function. Our main result is that the Green function behaves in a semilocal manner when its two position variables are confined in a narrow region along an edge, that is, it is only extended in one direction along the edge, but is exponentially localized in the other direction. Since the Kubo formula involves a product of two or three (depending on the specific expression) Green functions, a full exponential locality of the Hall conductivity is then ensured.

As we will show in the text, the locality of the Hall conductivity, together with a current conservation law, leads to an important conclusion about the finite-size correction to the quantization of the Hall conductance. Under quite general conditions, the correction will be exponentially small when the linear size of the system is large compared with a microscopic length which is typically the magnetic length. This estimate gives a much smaller bound to the correction than those given earlier.[11]

In Sec. II we present a detailed calculation of the Green function for a free-electron system in a long strip bounded by two edges. In Sec. III we derive a version of the Kubo formula which is particularly suitable for the locality analysis. At the end of this section, a primary argument to justify the use of the cylinder geometry is also given. In Sec. IV we use an argument of adiabatic charge transport to justify the use of the torus geometry. In Sec. V, we describe in detail how Laughlin's argument can be adapted to conditions that are much closer to real experimental conditions. In Sec. VI we examine the expression for the Hall conductance as a topological invariant[5,12,13] and show how this is also applicable under conditions more realistic than those used in earlier discussions. For the most part we restrict the discussion to the integer QHE, but where the argument can be extended to cover the fractional QHE we mention the fact.

II. THE SEMILOCALITY OF THE GREEN FUNCTION

Consider a 2D free-electron gas on a strip under a uniform magnetic field $\mathbf{B}=B\hat{\mathbf{z}}$ in the normal direction of the surface. The strip is assumed to be infinitely long in the y direction and to have edges along $x=0$ and $x=L_x$. It is convenient to use the Landau gauge $\mathbf{A}=(0,Bx,0)$ so that the Schrödinger equation has the separable form

$$\left[-\frac{\hbar^2}{2m}\frac{\partial^2}{\partial x^2}+\frac{1}{2m}\left[-i\hbar\frac{\partial}{\partial y}+eBx\right]^2\right]\psi(x,y)=E\psi(x,y)\ , \quad (2)$$

which is to be solved under the boundary conditions

$$\psi(x=0)=\psi(x=L_x)=0\ . \quad (3)$$

The wave functions can be written in the product form $e^{-iky}\psi_n(x,k)$, where $\psi_n(x,k)$ is an eigenstate of the one-dimensional (1D) problem

$$\left[\frac{d^2}{dx^2}+\nu+\tfrac{1}{2}-\tfrac{1}{4}(k-x)^2\right]\psi(x,k)=0\ . \quad (4)$$

In the last equation we have used $\sqrt{\hbar/2eB}$ as the length scale in the x direction and $\sqrt{2\hbar/eB}$ in the y direction, and have expressed E as $(\nu+\frac{1}{2})(\hbar eB/m)$.

The Green function in the spectral representation is

$$G(\mathbf{r},\mathbf{r}';z)=\int_{-\infty}^{\infty}\frac{dk}{2\pi}e^{ik(y'-y)}\sum_{n=0}^{\infty}\frac{\psi_n(x,k)\psi_n(x',k)}{z-E_n(k)}=\int_{-\infty}^{\infty}\frac{dk}{2\pi}e^{ik(y'-y)}g(x,x';\nu,k)\ , \quad (5)$$

where g is the Green function of the homogeneous equation (4). In the rest of this section we will concentrate our attention on the edge along $x=0$, and forget about the other by formally taking $L_x\to+\infty$. The reduced Green function g then satisfies the following boundary conditions:

$$g(0,x';\nu,k)=0,\quad g(+\infty,x';\nu,k)=0\ \text{ for } x'>0\ . \quad (6)$$

As a 1D problem, we can write for $\nu\neq 0,1,2,3\ldots$ (away from the bulk Landau levels)

$$g(x,x';\nu,k)=\frac{1}{W}[D_\nu(x-k)f(x',k)\theta(x-x')+f(x,k)D_\nu(x'-k)\theta(x'-x)]\ , \quad (7)$$

where $D_\nu(x)$ is the solution of Eq. (4) (with $k=0$) that vanishes as $x\to+\infty$. The function $\theta(x)$ in Eq. (7) is just the unit step function, while the other functions are given in terms of $D_\nu(x)$ by

$$f(x,k)=D_\nu(k)D_\nu(x-k)-D_\nu(-k)D_\nu(k-x)\,,$$
$$W=-2D_\nu(-k)D_\nu(0)D'_\nu(0)\,. \tag{8}$$

The function $D_\nu(x)$ is called the parabolic cylinder function in the literature of mathematical physics, and some of its important properties are listed in the Appendix.

To be explicit, let us assume $x>x'>0$, then we have

$$g(x,x';\nu,k)=\frac{D_\nu(x-k)D_\nu(k-x')}{2D_\nu(0)D'_\nu(0)}-\frac{1}{2D_\nu(0)D'_\nu(0)}\frac{D_\nu(x-k)D_\nu(x'-k)D_\nu(k)}{D_\nu(-k)} \tag{9}$$

The first term on the right-hand side of the above equation is the reduced Green function in the absence of the boundary, and when inserted in the expression (5), produces a Gaussian-like localized behavior of $G(\mathbf{r},\mathbf{r}';z)$ in the separation $(\mathbf{r}-\mathbf{r}')$ in all directions. Thus, if there is an extended behavior of $G(\mathbf{r},\mathbf{r}';t)$ it must come from the contribution of the second term which is induced by the boundary condition at $x=0$. It is therefore sufficient to consider the following object:

$$\Delta(\mathbf{r},\mathbf{r}';\nu)\equiv\int_{-\infty}^{\infty}\frac{dk}{2\pi}e^{ik(y'-y)}\frac{D_\nu(x-k)D_\nu(x'-k)D_\nu(k)}{D_\nu(-k)}\,. \tag{10}$$

First, consider the case that ν is real negative or ν is away from the real axis of the energy plane. Then $D_\nu(-k)$ has no zero in a finite strip, I_k, parallel to the real axis in the complex k plane. Since $D_\nu(k)$ is analytic in the k plane (excluding ∞), the integrand in Eq. (10) is analytic in the strip I_k. Within I_k, the function $D_\nu(k)$ has the following asymptotic behavior:

$$k^\nu e^{-(1/4)k^2}\quad\text{as } \mathrm{Re}(k)\to+\infty\,,$$
$$k^{-\nu-1}e^{(1/4)k^2}\quad\text{as } \mathrm{Re}(k)\to-\infty\,, \tag{11}$$

so that the integrand in (10) behaves (in absolute value) as

$$\frac{1}{|k|}e^{-(1/2)|k|(x+x')},\quad |k|\to\infty \tag{12}$$

for fixed x and x' (which are positive). Thus, we can replace the contour of integration of (10) by a line below or above the real axis to make $\Delta(\mathbf{r},\mathbf{r}';\nu)$ an apparent localized function of $y-y'$. The localization of Δ in $(y-y')$ is therefore exponential.

Now suppose ν is real positive and is between two bulk Landau levels n and $n+1$. In this case, the function $D_\nu(-k)$ has $n+1$ zeros on the real axis of the k plane, corresponding to the $n+1$ edge states at energy ν. We denote these zeros as k_j, $j=0,1,2,\ldots,n$. If we add to ν an infinitesimal imaginary part $i\epsilon$, then these zeros become

$$k_j+\frac{\partial k_j}{\partial\nu}i\epsilon\,, \tag{13}$$

Since the energy levels decrease as k increases [see Eq. (4)], the derivative is real negative. Thus, when ϵ is positive the zeros move just below the real axis. From the analytic property of $D_\nu(-k)$ there is no other zero in a finite strip I_k parallel to the real axis. The asymptotic behavior of $\Delta(\mathbf{r},\mathbf{r}';\nu+i\epsilon)$ as a function of $(y-y')$ can be readily found as follows.

Let us fix $\epsilon>0$. For $y-y'<0$ we replace the contour of integration in (10) by a line which is a finite distance above the real axis (but still within I_k). The new integral equals the original one because there are no poles of the integrand in the region between the new and the old contour, and because the asymptotic behavior of the integrand still holds as in Eq. (12). It is therefore clear that $\Delta(\mathbf{r},\mathbf{r}';\nu+i\epsilon)$ is exponentially small for $y\ll y'$, for k has a finite positive imaginary part along the new contour. In the other case, $y-y'>0$, we replace the contour of integration in (10) by a line below the real axis. The integral along the new contour gives an exponentially small contribution to $\Delta(\mathbf{r},\mathbf{r}'\nu+i\epsilon)$ for $y\gg y'$. In addition to this, we must consider now the contributions from the poles of the integrand [the zeros of $D_\nu(-k)$]. The contributions from the poles are plane waves in $(y-y')$, which make $\Delta(\mathbf{r},\mathbf{r}';\nu+i\epsilon)$ extended for $y\gg y'$.

Thus, the retarded Green function $G^+(\mathbf{r},\mathbf{r}';z)$ for z between two bulk Landau levels behaves quite differently for $y\ll y'$ and $y\gg y'$. In the former case it is exponentially small, while in the latter case it is plane-wave-like. This particular behavior of the Green function can in fact be intuitively understood from the following classical picture. A classical electron circles around anticlockwise in the $x-y$ plane under a magnetic field in the z direction. When its orbit hits the wall at $x=0$, the electron bounces and travels to the positive y direction along the edge. It is this one way traveling behavior of the electron that makes the Green function a semilocalized function of $y-y'$.

The advanced Green function G^- also has the semilocality property but in the opposite sense. It is localized for $y\gg y'$ and extended for $y\ll y'$. This is not surprising, for changing ϵ from positive to negative effectively reverses the time.

So far we have only considered the behavior of the Green function in the y direction (along the edge). The behavior in the x direction can also be studied through the integral in Eq. (10), with the properties of the parabolic cylinder function $D_\nu(k)$. The results are quite simple. For z between two bulk Landau levels, the Green function (either retarded or advanced) is an exponentially localized function of the separation $|x-x'|$.

Before closing this section, we would like to make some remarks about the results obtained so far. First, the localization lengths in the behavior of the Green functions should be of order unity in the dimensionless variables, if the energy z is not very close to the bulk Landau levels. When we put the length scales back into the equations, we should find that the localization lengths are of order of the magnetic length $\sqrt{\hbar/eB}$. Second, we expect that the general features of the Green functions should hold in the presence of a disordered potential. One primary reason for this is that the nature of the spectrum for the extended states including the extended edge states are quite the same with or without disorder.[6,14] With disorder, the localization length may however be different from the magnetic length.

III. LOCALITY OF THE HALL CONDUCTIVITY

The Kubo formula for the Hall conductivity has been the basis for many theories of the quantum Hall effect. In the following we give a brief derivation of this formula in a form which is particularly suitable for the analysis of its locality. The arguments used in the derivation can be largely found from a paper of Streda and Smrcka.[15]

In the limit of linear response, a uniform electric field E in the x direction produces a perturbation ρ_1 to the density matrix, given by

$$\rho_1 = -ieE \int_0^\infty dt\, e^{-\epsilon t} e^{-iHt}[x,\rho_0]e^{iHt} , \quad (14)$$

where ρ_0 is the unperturbed density matrix (namely the Fermi-Dirac distribution operator), and ϵ is a positive infinitesimal. The y component of the electric current density at $\mathbf{r}=\mathbf{r}_0$ is then given by

$$J_y(\mathbf{r}_0) = ie^2E \int_0^\infty dt\, e^{-\epsilon t}\,\mathrm{Tr}(j_y e^{-iHt}[x,\rho_0]e^{iHt}) , \quad (15)$$

where j_y is the current density operator given in terms of the velocity operator V_y by

$$j_y = \tfrac{1}{2}[V_y\delta(\mathbf{r}-\mathbf{r}_0) + \delta(\mathbf{r}-\mathbf{r}_0)V_y] . \quad (16)$$

The Hall conductivity at $\mathbf{r}_0$ is simply

$$\sigma_H(\mathbf{r}_0) = ie^2 \int_0^\infty dt\, e^{-\epsilon t}\,\mathrm{Tr}(j_y e^{-iHt}[x,\rho_0]e^{iHt}) . \quad (17)$$

Now we can expand the trace operation in the energy eigenspace and carry out the time integration to yield

$$\begin{aligned}\sigma_H(\mathbf{r}_0) &= e^2 \sum_{m,n} \rho_0(E_m) \left[\frac{(j_y)_{mn}X_{nm}}{E_m-E_n+i\epsilon} + \text{c.c.} \right] \\ &= e^2 \int_{-\infty}^{\infty} d\eta\, \rho_0(\eta) \\ &\quad \times \sum_{m,n} \delta(\eta - E_m) \left[\frac{(j_y)_{mn}X_{nm}}{\eta+i\epsilon-E_n} + \text{c.c.} \right] ,\end{aligned} \quad (18)$$

where $\rho_0(\eta)$ is now the Fermi-Dirac distribution function. Using the definition of the Green functions

$$G^{\pm}(\eta) = \frac{1}{\eta \pm i\epsilon - H} \quad (19)$$

and the formal relation

$$\delta(\eta - H) = \frac{i}{2\pi}(G^+ - G^-) , \quad (20)$$

we can rewrite Eq. (18) as

$$\begin{aligned}\sigma_H(\mathbf{r}_0) &= \frac{ie^2}{2\pi} \int_{-\infty}^{\infty} d\eta\, \rho_0(\eta)\,\mathrm{Tr}[(G^+ - G^-)(j_yG^+x + xG^-j_y)] \\ &= \frac{ie^2}{2\pi} \int_{-\infty}^{\infty} d\eta\, \rho_0(\eta)\,\mathrm{Tr}(G^+j_yG^+x - G^-xG^-j_y) \\ &= \frac{-e^2}{2\pi} \int_{-\infty}^{\infty} d\eta\, \rho_0'(\eta) \int_{-\infty}^{\eta} d\eta'\,\mathrm{Tr}[G^+(\eta')j_yG^+(\eta')x - G^-(\eta')j_yG^-(\eta')x] .\end{aligned} \quad (21)$$

The integral over η' in the last expression can be turned into a contour integral in the complex energy plane with the result

$$\begin{aligned}\sigma_H(\mathbf{r}_0) &= \frac{ie^2}{2\pi} \int_{-\infty}^{\infty} d\eta\, \rho_0'(\eta) \int_{C(\eta)} dz\,\mathrm{Tr}[G(z)j_yG(z)x] \\ &= \frac{ie^2}{2\pi} \int_{-\infty}^{\infty} d\eta\, \rho_0'(\eta) \int_{C(\eta)} dz \int\int d\mathbf{r}\,d\mathbf{r}'\,G(\mathbf{r},\mathbf{r}',z)j_yG(\mathbf{r}',\mathbf{r};z)x ,\end{aligned} \quad (22)$$

where in the last step the double coordinate integration represents the trace operation and the current density operator j_y acts on the $\mathbf{r}'$ variable of the second Green function. The contour $C(\eta)$ in the energy integration of (22) is now in the complex energy plane surrounding the energy spectrum below η, with the infinitesimal segment $(\eta - i\epsilon, \eta + i\epsilon)$ being omitted.

Equation (22) is the Kubo formula that has the desired form for the locality analysis. At low enough temperature ($k_BT \ll \hbar\omega_c$), the quantity $\rho_0'(\eta)$ is highly peaked about the chemical potential which we assume to lie between two bulk Landau levels or the levels of extended states in the bulk. In the absence of boundaries, the contour $C(\eta)$ is away from the spectrum of extended states, so that the Green functions are exponentially localized in the separation $|\mathbf{r}-\mathbf{r}'|$. Because of the δ function $\delta(\mathbf{r}'-\mathbf{r}_0)$ contained in the operator j_y, the contribution to $\sigma_H(\mathbf{r}_0)$ only comes from a neighborhood of $\mathbf{r}_0$. This neighborhood has a typical linear scale of a magnetic length, corresponding to the localization length of the Green functions. Now, suppose our system is in a strip bounded by two parallel edges. In this case, the contour $C(\eta)$ can come close to the spectrum of the edge states near the Fermi energy. These edge states are extended along the edges, which make the Green functions also extended along the edges. According to the analysis of the previous sections, the extendedness of the Green functions is, however, only in one way near each edge. Together with the fact that the

Green functions are still localized in the direction perpendicular to the edges, it is now easy to see that the product $G(\mathbf{r},\mathbf{r}',z)j_yG(\mathbf{r}',\mathbf{r},z)$ is in fact localized in all directions in the separation $(\mathbf{r}-\mathbf{r}')$. Again, because of the δ function contained in the operator j_y, the contribution to $\sigma_H(\mathbf{r}_0)$ only comes from a small neighborhood of $\mathbf{r}_0$. This completes our locality analysis of the Hall conductivity.

In the usual experiments of the quantum Hall effect, the Hall voltage is measured across a section (say, at $y=y_0$) in the middle of the strip (see Fig. 1). The Hall conductance can thus be expressed by

$$\sigma_H(y_0)=\frac{1}{L_x}\int_0^{L_x} dx_0\sigma_H(\mathbf{r}_0) , \tag{23}$$

where L_x is the width of the strip at y_0. All the complications due to the two ends, where the strip is connected to the source and the drain, die out exponentially at y_0 if the ends are sufficiently far away from the voltage probes. This explains why the Hall conductance can be so accurately determined in spite of the shorting effect at the two ends.

Owing to the exponential locality of the Hall conductivity, we can use whatever boundary conditions at the two ends which is convenient for a theoretical analysis. In particular, we can use a periodic boundary condition which joins the two ends making the strip into a cylinder. If this is done, the Hall conductance can be related to the coherent response of the extended states with respect to the change of a magnetic flux through the center of the cylinder as was done by Laughlin.[4] The gauge symmetry of the system then leads to the quantization of the Hall conductance averaged over a flux quantum. At this step, one can see very clearly the important role played by the locality of the Hall conductivity in establishing the relevance of Laughlin's theory to the experiments.

The average procedure over the flux quantum was not justified in Laughlin's paper. In a paper of Niu *et al.*,[5] an argument is given that explains why the Hall conductance can be replaced by its average. The theoretical basis of this argument is again the locality of the Hall conductivity, although the effect of the edge states (which is now shown to be harmless) was not taken into account by us.

In Sec. V we will give a more detailed analysis of Laughlin's theory.

IV. THE TORUS GEOMETRY

It is convenient to start with the cylinder geometry, the use of which has primarily been justified in Sec. III. Let us continue to use (x,y) as the coordinates for the cylinder surface, where x is confined into the interval $[0,L_x]$ while y and $y+L_y$ are identified as the same. Suppose we have a magnetic flux through the center of the cylinder. We denote $C(x_0,\phi)$ as the induced charge transport through the circle at $x=x_0$ when this flux is increased adiabatically from zero to ϕ. Following Laughlin's idea we equate the Hall conductance with the flux derivative of the adiabatic charge transport,

$$\sigma_H(y_0)=\tilde{C}(x_0)\equiv\left.\frac{\partial C(x_0,\phi)}{\partial\phi}\right|_{\phi=0} , \tag{24}$$

where $\phi_H(y_0)$ is defined in Eq. (23), and x_0 may be taken as $L_x/2$. Later we will give an expression for $\tilde{C}(x_0)$, from which we can show that $\tilde{C}(x_0)$ has exponentially small contributions from the region $|x-x_0|\gg l$, where l is the magnetic length. Thus, if $L_x\gg l$, then we can change the boundary conditions at $x=0$ and $x=L_x$ with exponentially small error introduced into $\tilde{C}(x_0)$ and hence into $\sigma_H(y_0)$ by Eq. (24). In particular, we can use a periodic boundary condition which joins the edges of the cylinder together to form a torus. Thus, if we can prove Eq. (24) and establish the exponential locality of $\tilde{C}(x_0)$, then we can justify the use of the torus geometry.

Let us first derive a formula for $\tilde{C}(x_0)$ to show its locality. Through the dependence on the flux, the Hamiltonian varies adiabatically in time. To the first order in $\dot\phi$, the perturbation ρ_1 to the density matrix is determined by the equation of motion

$$i\hbar\dot\rho_0+[\rho_1,H]+i\epsilon\rho_1=0 , \tag{25}$$

where $\dot\rho_0$ is the time derivative of the instantaneous density matrix corresponding to the Fermi-Dirac distribution at fixed flux, and ϵ is an positive infinitesimal. Taking the matrix element of Eq. (25) between two orthonormal instantaneous eigenstates of the Hamiltonian, we obtain

$$(\rho_1)_{mn}=i\hbar\frac{\langle\dot m|n\rangle\rho_0(E_n)+\langle m|\dot n\rangle\rho_0(E_m)}{E_n-E_m+i\epsilon} , \tag{26}$$

where we have used the identity

$$\langle m|\dot\rho_0|n\rangle=-\langle\dot m|n\rangle\rho_0(E_n)-\langle m|\dot n\rangle\rho_0(E_m) , \tag{27}$$

because ρ_0 is diagonal in the basis of the eigenstates. The induced current through the circle at $x=x_0$ is given by the trace of the product of ρ_1 and the current operator $i_x\equiv(-e/2)[V_x\delta(x-x_0)+\delta(x-x_0)V_x]$, that is

$$\begin{aligned}I_x(x_0)&=\sum_{m,n}\frac{i\hbar[\langle\dot m|n\rangle\rho_0(E_n)+\langle m|\dot n\rangle\rho_0(E_m)]}{E_n-E_m+i\epsilon}(i_x)_{nm}\\&=i\hbar\int d\eta\,\rho_0(\eta)\sum_n\delta(\eta-E_n)(\langle\dot n|G^-i_x|n\rangle\\&\qquad-\langle n|i_xG^+|\dot n\rangle) .\end{aligned} \tag{28}$$

The adiabatic time derivative of an eigenstate can be written as

$$\begin{aligned}P_n|\dot n\rangle&=\frac{1}{E_n-H}P_n\dot H|n\rangle\\&=\tfrac{1}{2}[G^+(E_n)+G^-(E_n)]\dot H|n\rangle ,\end{aligned} \tag{29}$$

where P_n is the operator that projects off the state $|n\rangle$. Using the above result and the formal relation (20), Eq. (28) can be rewritten as

$$I_x(x_0)=\frac{\hbar}{4\pi}\int d\eta\,\rho_0(\eta)\,\mathrm{Tr}[i_x G^+(G^+ + G^-)\dot{H}(G^+ - G^-)+\text{c.c.}]$$

$$=\frac{\hbar}{4\pi}\int d\eta\,\rho_0(\eta)\,\mathrm{Tr}\left[i_x\left[G^+G^+\dot{H}G^+ + G^-\dot{H}G^-G^- + \frac{\partial}{\partial\eta}(G^+\dot{H}G^-)\right]\right], \tag{30}$$

where in the last step we have used the following identities:

$$[G^+,G^-]=0,\quad G\dot{H}G=\dot{G},\quad GG=-\frac{\partial}{\partial\eta}G\,. \tag{31}$$

The Hamiltonian depends on time through the y component of the canonical momentum which contains an additional term of $e\phi/L_y$. As a result of this, we have $\dot{H}=e\dot{\phi}V_y/L_y$. The charge transport $C(x_0,\phi)$ is just the time integral of $I_x(x_0)$, so that the quantity $\tilde{C}(x_0)$ as defined in Eq. (24) is given by

$$\tilde{C}(x_0)=\frac{\hbar e}{4\pi L_y}\int d\eta\,\rho_0(\eta)\,\mathrm{Tr}\left[i_x\left[G^+G^+V_yG^+ + G^-V_yG^-G^- + \frac{\partial}{\partial\eta}(G^+V_yG^-)\right]\right], \tag{32}$$

where everything is now evaluated at $\phi=0$. To reveal the locality of $\tilde{C}(x_0)$, we integrate the right-hand side of the above equation by parts to get the result

$$\tilde{C}(x_0)=\frac{-\hbar e}{4\pi L_y}\int d\eta\,\rho_0'(\eta)\left[\mathrm{Tr}(i_xG^+V_yG^-)+\int_{-\infty}^{\eta}d\eta'\,\mathrm{Tr}[i_x(G^+G^+V_yG^+ + G^-V_yG^-G^-)]\right]. \tag{33}$$

At low enough temperature ($k_BT \ll \hbar\omega_c$), $\rho'(\eta)$ is highly peaked about the chemical potential which lies between the levels of the extended states in the bulk. For each term in the integral over η', the contour of integration can be deformed into the complex energy plane, so that the only place that η' is close to the spectrum is at the Fermi energy. The Green functions are then exponentially local functions of the separation in the x directions as were shown in Sec. II. As a result of the locality of the Green functions and the δ functions $\delta(x-x_0)$ contained in the operator i_x, the quantity $\tilde{C}(x_0)$ has exponentially small contributions from the region $|x-x_0| \gg l$, where l is the magnetic length.

Having shown the locality of $\tilde{C}(x_0)$, we now proceed to establish the identity

$$\sigma_H(y_0)=\tilde{C}(x_0)\,, \tag{34}$$

which was stated in Eq. (24) and is given here for convenience. Since the two sides of the above equation depend on different variables, we must show that $\tilde{C}(x_0)$ is independent of x_0 and that $\sigma_H(y_0)$ is independent of y_0. The reason for the constancy of $\tilde{C}(x_0)$ can be found from the current conservation law. It is important to notice that we only need the adiabatic current $I_x(x_0)$ in a vanishingly small neighborhood of the flux to evaluate the quantity $\tilde{C}(x_0)$. In this neighborhood the effect of charge accumulation or decumulation due to the edges can be neglected. The reason for the constancy of $\sigma_H(y_0)$ can be found in a similar way. We are thus left to show the identity (34) when both sides are averaged over the position variables.

From a simple electrodynamic argument, the adiabatic current $I_x(x_0)$ is in fact the Hall current under the electromotive force generated by the changing flux. The averaged quantity $\langle\tilde{C}\rangle\equiv(1/L_x)\int dx_0\tilde{C}(x_0)$ is just the Hall conductivity σ_{xy} (with a minus sign) over the whole surface. On the other hand, $\langle\sigma_H\rangle\equiv(1/L_y)\int dy_0\sigma_H(y_0)$ is the average of the Hall conductivity σ_{yx} over the surface. The Onsager relation $\sigma_{xy}=-\sigma_{yx}$ immediately leads to the identity $\langle\sigma_H\rangle=\langle\tilde{C}\rangle$, which is what we wanted to prove.

To summarize, we started with a cylinder geometry for the Hall system, and reexpressed the Hall conductance as the flux derivative of the adiabatic charge transport from one edge to the other. The new expression can be justified by an electrodynamic argument originally used by Laughlin, or by mathematical manipulations on the Kubo formula. Since the charge transport can be evaluated from a circle in the middle of the cylinder, we can use the locality property of the Green function to show its insensitivity to the boundary conditions at the two edges. The flux derivative of the charge transport is calculated in a vanishing neighborhood of the flux, so it is a property of the ground state of the system at zero flux. Then we can join the two edges to turn the cylinder into a torus without worrying about the inability of using a flux in the new geometry. On the torus we can restore the new expression for the Hall conductance back to the Kubo formula, from which a topological invariant expression can be derived.

In Sec. VI we will study a Hall system which has a topology of a torus with a hole in it. As will be pointed out there, such a system could be realized in real experiments. For such a system, a topological invariant expression for the Hall conductance will be derived to show its quantization.

V. LOCALITY AND LAUGHLIN'S ARGUMENT

Laughlin's[4] argument involves consideration of a device with the geometry of a finite cylinder, although it can equally well be applied to an annulus or any other figure with the same topology. A uniform magnetic field is applied normal to the cylindrical surface, and there is a solenoid along the axis of the cylinder through which a magnetic flux Φ passes. It is supposed that this flux Φ can be varied without changing the magnetic field which acts on the surface. This arrangement is shown in Fig. 2. It is supposed, in the simplest form of this argument, that the electrons are noninteracting and at zero temperature,

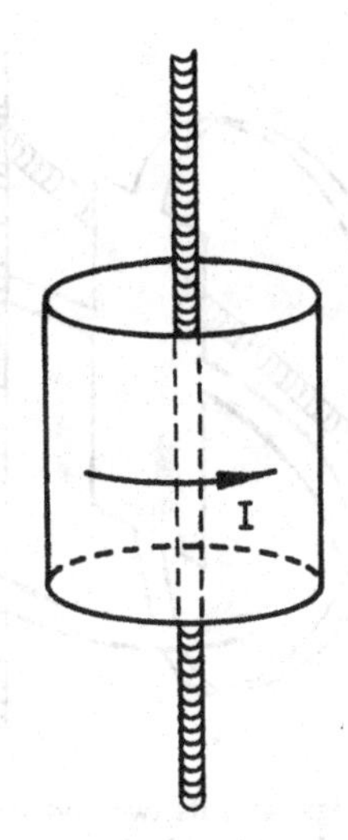

FIG. 2. Cylinder geometry used in Laughlin's theory. Through the center of the cylinder a magnetic flux is passed. The flux derivative of the induced adiabatic charge transport from one edge to the other determines the ratio of the Hall current I flowing around the cylinder to the voltage drop between the two edges.

and that the Fermi energy is such that there are no bulk extended states near that energy; there may be localized states at that energy, of a size much less than the circumference of the system, and there will in general be extended states at the edges of the cylinder which are confined to a region much smaller than the height of the cylinder.[6]

The effect of changing the flux Φ passing through the solenoid by one quantum unit h/e is now considered. For the localized states this merely maps each one into the equivalent state related by a gauge transformation. For each extended state in flux Φ there is an equivalent gauge-transformed state for flux $\Phi+h/e$, but the continuous change of flux can map the set of occupied extended states into a different set of states. Occupied bulk states must remain occupied, since their energies are well away from the Fermi energy, but it is possible for an integer number n of electrons to be transferred from one edge to the other. By Faraday's law the changing flux produces an electromotive force (emf) round the cylinder, and so the ratio of the integral over time of the current from one edge to the other to the time integral of the emf in ne^2/h. If this relation for the integrated current produced by a discrete change in the flux can be changed to a differential relation, the quantization of the ratio between the voltage down the cylinder and the current round it follows. This has the same topology as is usual in the Corbino disk arrangement.

There are two very important conclusions that can be drawn from this argument. The first is that it is only the existence of a mobility gap between the regions of bulk extended states that is necessary to produce the integer QHE, and the gap in which the Fermi energy lies does not have to be obtained by perturbation from an ideal system of electrons in a uniform potential. The second as pointed out by Tao and Wu,[16] is that a fractional QHE, with a conductance pe^2/qh, where p and q are integers with no common factors, implies that in flux Φ there are q equivalent ground states, and the system is mapped through all q of them when the flux is changed by q quanta.

There are a number of points in this argument which have been criticized. The first is that the replacement of the ratio of a charge transfer to a flux change by the conductance, which is its limiting value for an infinitesimal change in flux, is not justified. The second is that a special sort of geometry is assumed, and this is not obviously related to the usual geometry of experimental devices. We show how the locality properties of the Green functions can be used to overcome both of these objections. There is a further criticism of the theory, which is that it is assumed that the rate of change of flux is so slow, or the electric field so small, that adiabatic theory can be used for the charge transfer. We have addressed this problem in an earlier paper,[10] but we had to use perturbative arguments rather than the Laughlin argument to show that the restriction to vanishingly weak electric fields is not essential.

Now it is possible to ask how the Laughlin argument[4] could be applied to the kind of experimental arrangement sketched in Fig. 1. The quantity in which we are interested is the current flowing across a line such as the one shown in the figure going from the end of one voltage probe to the end of the opposite probe divided by the voltage difference between the two ends of the line. The Kubo formula allows the current density at a point to be expressed in terms of the integral of the two-particle Green function of the many-electron system multiplied by the electric field. In a mobility gap the relevant Green function falls off exponentially when the separation between the two points (the point where the current density is calculated, and the point in the integral where the electric field is measured) is increased. This is true despite the existence of extended edge states, as is shown in Sec. II. This exponential localization of the Green function is the key property that allows us to estimate the effect of modifying the boundary conditions. We can expect the localization length to be the order of magnitude of the magnetic length, which is about 10 nm under typical conditions.

Two changes can be made to bring the experimental arrangement shown in Fig. 1 into the correspondence with the geometry of Laughlin's argument shown in Fig. 2, and the locality of the Green function allows us to argue that each of these changes will have a negligible effect on the result. The first change is to replace the current driven round a circuit connected to the source and drain by two passive electron reservoirs at the source and drain. The same current will still flow, but it is to be driven by an emf applied to the voltage probes. Conditions are exactly the same except in the neighborhood of the source and drain, and these are very far from the line in which we are interested, so only exponentially small changes in the ratio of current to voltage occur.

The other change is to replace the emf supplied by the

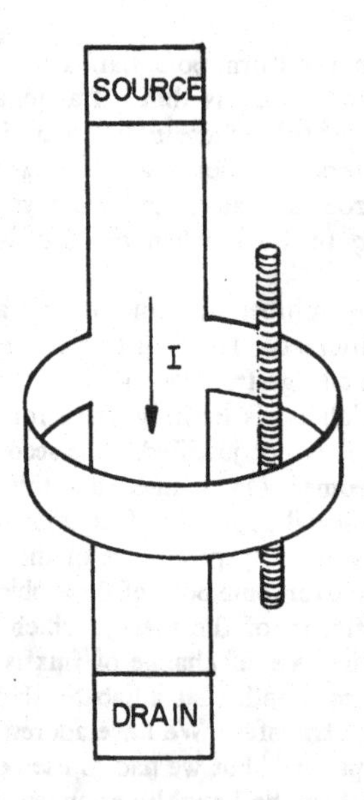

FIG. 3. A pair of opposite voltage leads in Fig. 1 is connected by a ribbon. Through the hole between the ribbon and the main part of the device passes a changing magnetic flux which drives the Hall current.

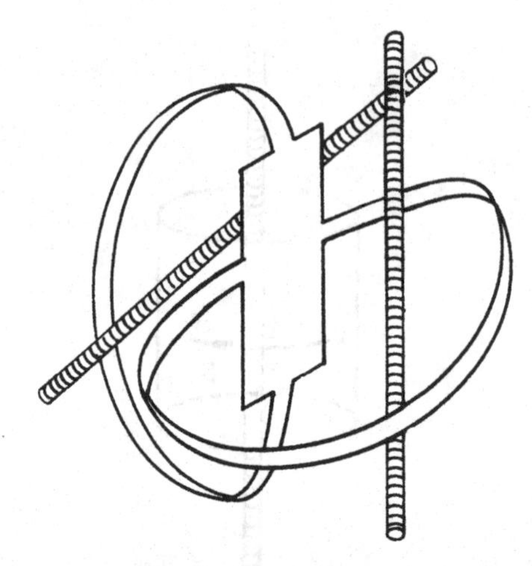

FIG. 4. A rectangle with its two ends connected by a strip (the current lead) and its two sides by another strip (the voltage lead). Through the holes are the magnetic flux solenoids.

voltmeter by a voltage source similar to that assumed in Laughlin's argument. The two ends of the pair of opposed voltage probes are connected to one another by a strip of two-dimensional material, and a solenoid is inserted between this strip and the main part of the device as shown in Fig. 3. The emf is produced by a uniform rate of change of the flux through the solenoid. The nature of the emf should be irrelevant to the physical effects, and any changes in the local conditions in the voltage probes due to replacement of the voltmeter leads by more two-dimensional material can be compensated by imposing local irrotational electrostatic fields. The Laughlin argument can be applied directly to his device shown in Fig. 3. Change of flux by one quantum must drive an integer number of electrons from one side of the system to the other, unless there is a ground state with q-fold degeneracy, in which case the fractional QHE is obtained.

VI. QUANTUM HALL CONDUCTANCE AS A TOPOLOGICAL INVARIANT

Niu *et al.*[5] showed that the quantum Hall conductance of a torus could be written as a topological invariant which is stable against the presence of disorder in the substrate, interactions between the electrons, etc. Although a torus geometry can be mathematically transformed from a strip geometry up to exponentially small errors in the Hall conductance, such a geometry is nevertheless impossible to be realized without using a magnetic monopole source for a uniform magnetic field, so we are now constructing a more realistic version of the theory.

We consider the system shown in Fig. 4 which consists of a rectangle, with its two ends connected by a strip which we call the "current lead," and its two sides connected by a strip which we call the "voltage lead." All of this, the rectangle and the two sets of leads, are made of the same two-dimensional material and are in the same approximately uniform magnetic field. This is evidently a system that could be realized in practice. It has the topology of a torus with a simple hole in it, and its edge is a single simply connected curve. We also suppose that there are two solenoids with flux ϕ_I and ϕ_V which pass through each of the two sets of leads. The boundary conditions round each set of leads are periodic functions (up to a gauge transformation) of ϕ_I and ϕ_V, while an emf can be induced around either set of leads by making a uniform change in the appropriate flux.

Now we use the Kubo formula[17,5,9] to determine the total current I which flows round the current leads in response to a weak electric field $\mathbf{E}_V$ induced by a steady change of the flux ϕ_V. We write the emf induced round the voltage leads as V_V. It is convenient to introduce a fictional electric field $\mathbf{E}_I$ which is locally irrotational, has equipotentials that go all the way round the voltage leads, and has a line integral V_I round the current leads; such a field could be induced by a steady change of ϕ_I if the solenoid were appropriately placed in relation to the voltage leads to make the equipotentials go all round the leads. The current can be written as

$$I=\sum_{\alpha(\neq 0)}\frac{i\hbar}{V_I(\varepsilon_0-\varepsilon_\alpha)^2}\left[\left\langle\Psi_0\right|\int \mathbf{j}\cdot\mathbf{E}_V\left|\Psi_\alpha\right\rangle\left\langle\Psi_\alpha\right|\int \mathbf{j}\cdot\mathbf{E}_I\left|\Psi_0\right\rangle-\left\langle\Psi_0\right|\int \mathbf{j}\cdot\mathbf{E}_I\left|\Psi_\alpha\right\rangle\left\langle\Psi_\alpha\right|\int \mathbf{j}\cdot\mathbf{E}_V\left|\Psi_0\right\rangle\right], \quad (35)$$

where Ψ_0 is the many-body ground state of energy ε_0, the Ψ_α are excited states of energy ε_α. The operator $\mathbf{j}$ is the sum over all electrons of

$$\mathbf{j}=(e/m)(-i\hbar\nabla-e\mathbf{A}) . \tag{36}$$

The integrals involving $\mathbf{j}\cdot\mathbf{E}_I$, which give the current that flows in response to the perturbation $\mathbf{j}\cdot\mathbf{E}_V$, are written as integrals over the entire space and divided by V_I, but, because of the continuity of current flow, they could equally well be written as integrals between two close equipotentials of $\mathbf{E}_I$ and divided by the difference in potential δ. If these equipotentials go round the system inside the voltage leads it can be argued that the Green functions implied by Eq. (35) are insensitive to the boundary conditions applied round the current leads, and so the resulting current I is independent of the flux ϕ_I. This insensitivity to boundary conditions will also be used at a later stage in the argument.

The operators in Eq. (35) can be expressed in terms of the partial derivatives of the Hamiltonian H with respect to the flux through the solenoids. We have

$$\int \mathbf{j}\cdot\mathbf{E}_I=-V_I\frac{\partial H}{\partial \phi_I}=-\frac{V_I e}{\hbar}\frac{\partial H}{\partial \eta_I} , \tag{37}$$

and a similar equation for $\mathbf{j}\cdot\mathbf{E}_V$. This enables us to rewrite Eq. (35) as

$$I=\sum_{\alpha\,(\neq 0)}\frac{ie^2V_V}{\hbar(\varepsilon_0-\varepsilon_\alpha)^2}\left[\left\langle\Psi_0\left|\frac{\partial H}{\partial\eta_V}\right|\Psi_\alpha\right\rangle\left\langle\Psi_\alpha\left|\frac{\partial H}{\partial\eta_I}\right|\Psi_0\right\rangle-\left\langle\Psi_0\left|\frac{\partial H}{\partial\eta_I}\right|\Psi_\alpha\right\rangle\left\langle\Psi_\alpha\left|\frac{\partial H}{\partial\eta_V}\right|\Psi_0\right\rangle\right]$$
$$=\frac{ie^2V_V}{\hbar}\left[\left\langle\frac{\partial\Psi_0}{\partial\eta_V}\middle|\frac{\partial\Psi_0}{\partial\eta_I}\right\rangle-\left\langle\frac{\partial\Psi_0}{\partial\eta_I}\middle|\frac{\partial\Psi_0}{\partial\eta_V}\right\rangle\right] . \tag{38}$$

Now it can be argued that the phase η_V is averaged over because its steady change with time provides the emf V_V, and we have already argued that the current is independent of η_I, so we can certainly average over that to get the Hall conductance as

$$\frac{I}{V_V}=\frac{ie^2}{2\pi h}\int_0^{2\pi}d\eta_V\int_0^{2\pi}d\eta_I\left[\left\langle\frac{\partial\Psi_0}{\partial\eta_V}\middle|\frac{\partial\Psi_0}{\partial\eta_I}\right\rangle-\left\langle\frac{\partial\Psi_0}{\partial\eta_I}\middle|\frac{\partial\Psi_0}{\partial\eta_V}\right\rangle\right] . \tag{39}$$

The double integral in this equation now has the form of the topological invariant that defines the first Chern class of the mapping of the torus onto the complex projective space of many-particle wave functions.[12,13] Provided that continuous changes in η_V,η_I by multiples of 2π map the ground state $|\Psi_0\rangle$ into itself, apart from phase factors, this integral is an integer multiple of 2π, and we get the integer QHE. Edge states are unimportant in this argument, as there is a single edge in the system, and so no current can result from transfer of electrons from one edge to the other. The detailed geometrical shape of the sample, substrate disorder, slight inhomogeneity of the magnetic field, and electron-electron interactions are unimportant, provided that the ground state remains isolated from other states, so that there is not a continuum of other states into which the continuous gauge transformations can map the ground state.

The fractional QHE will occur if there is a discrete set of q equivalent states into which $|\Psi_0\rangle$ can be mapped. In such a case η_V must be changed by $2\pi q$ before the state returns to its initial form, so that the integral on the right side of Eq. (39) must be replaced by an integral equal to $1/q$ times an integer multiple of 2π.

The relation between this system shown in Fig. 4 and the experimental situation shown in Fig. 1 is very similar to the relation described for the Laughlin geometry in Sec. V. We argued in Sec. V that the voltmeter leads can be replaced by a loop of two-dimensional material with an emf applied around it, so we have already justified the special form of the voltage leads in Fig. 4. The fact that the current density depends only on the local environment of the line across which it is measured allows us to replace the source and drain by the current leads of Fig. 4, provided that we also apply some local electric fields (with zero circulation) to maintain the pattern of current flow in the system.

ACKNOWLEDGMENT

This work was supported in part by the National Science Foundation under Grants No. DMR-84-15063 and No. DMR-83-19301.

APPENDIX: PROPERTIES OF $D_\nu(k)$

In this Appendix we list some of the important properties of the function $D_\nu(k)$, following the reference book by Bateman.[18] The solutions of the equation

$$\frac{d^2f}{dk^2}+(\nu+\tfrac{1}{2}-\tfrac{1}{4}k^2)f=0 ,$$

are called parabolic cylinder functions. For any value of ν there is a solution $D_\nu(k)$, which damps to zero as $k\rightarrow+\infty$. From the reflectional symmetry of the equation, $D_\nu(-k)$ is also a valid solution. If ν is zero or a positive integer corresponding to the harmonic oscillator levels, the two solutions $D_\nu(k)$ and $D_\nu(-k)$ become linearly dependent, and

$$D_\nu(k)=2^{-(1/2)\nu}e^{-k^2/4}H_\nu(k/\sqrt{2})$$

under an appropriate normalization for $D_\nu(0)$, where H_ν is the Hermite polynomial of degree ν. In the following, we are interested in the case with energies off the bulk Landau levels, so we will only consider those values of ν which are off the oscillator levels mentioned above.

(i) The functions $D_\nu(k)$ and $D_\nu(-k)$ are linearly independent of each other, and are analytical functions of k in the whole complex plane excluding ∞.

(ii) The function $D_\nu(k)$ has the following asymptotic expansions when $|k|\to+\infty$:

$$k^\nu e^{-(1/4)k^2} \quad \text{for } -\tfrac{3}{4}\pi<\arg k<\tfrac{3}{4}\pi\,,$$

$$-\frac{(2\pi)^{1/2}}{\Gamma(-\nu)}e^{\nu\pi i}k^{-\nu-1}e^{(1/4)k^2} \quad \text{for } \tfrac{3}{4}\pi<\arg k<\tfrac{5}{4}\pi\,,$$

$$-\frac{(2n)^{1/2}}{\Gamma(-\nu)}e^{-\nu\pi i}k^{-\nu-1}e^{(1/4)k^2} \quad \text{for } -\tfrac{5}{4}\pi<\arg k<-\tfrac{3}{4}\pi\,.$$

(iii) For ν real and positive, $D_\nu(k)$ has $[\nu+1]$ real zeros, where $[\nu+1]$ denotes the largest integer less than $\nu+1$. For ν real and negative, there are no real zeros. In the above cases, there may be zeros off the real axis, but the smallest distance of these zeros to the real axis should be at least of order 1. The complex zeros with large magnitudes should lie about the lines of $|\arg k|=\frac{3}{4}\pi$, following from the asymptotic expansions of $D_\nu(k)$. For non-real values of ν, there are no zeros on the real axis. In any case, the positions of the zeros should depend continuously on ν (except at $\nu=0,1,2,\ldots$), and the complex zeros are not close to the real axis if ν is not close to positive portion of the real axis of the energy plane.

(iv) For $\nu\neq 0,1,2,\ldots$, $D_\nu(k)$ has infinite number of zeros on the complex plane. This statement can be proved easily by using the growth theory of entire functions.[19] Suppose $D_\nu(k)$ has n zeros ($n<\infty$), then $D_\nu(k)$ can be written as $P_n(k)Q(k)$, where $P_n(k)$ is an nth degree polynomial having the zeros of $D_\nu(k)$, and $Q(k)$ is an entire function without zeros. From the asymptotic behavior of $D_\nu(k)$ we know that $D_\nu(k)$ and hence $Q(k)$ has growth of order 2. A simple theorem says that $Q(k)$ must be of the from $\exp(ak^2+bk+c)$, where a, b, and c are constants. But this form of $Q(k)$ cannot give the correct asymptotic behavior of $D_\nu(k)$ in all directions, unless ν is zero or an positive integer. This contradiction proves our statement.

[1] K. von Klitzing, G. Dorda, and M. Pepper, Phys. Rev. Lett **45**, 494 (1980).
[2] S. Kawaji, Surf. Sci. **73**, 46 (1978).
[3] K. von Klitzing, *Festkörperprobleme, Advances in Solid State Physics*, edited by J. Treusch (Vieweg, Braunschweig, 1981), Vol. XXI.
[4] R. B. Laughlin, Phys. Rev. B **23**, 5632 (1981).
[5] Q. Niu, D. J. Thouless, and Y. S. Wu, Phys. Rev. B **31**, 3372 (1985).
[6] B. I. Halperin, Phys. Rev. B **25**, 2185 (1982).
[7] H. Aoki and T. Ando, Solid State Commun. **38**, 1079 (1981).
[8] R. E. Prange, Phys. Rev. B **23**, 4802 (1981).
[9] D. J. Thouless, J. Phys. C **14**, 3475 (1981).
[10] Q. Niu and D. J. Thouless, Phys. Rev. B **30**, 3561 (1984).
[11] H. Levine, S. B. Libby, and A. M. M. Pruisken, Nucl. Phys. B **240**, 30 (1984); B. Shapiro, J. Phys. C 19, 4709 (1986).
[12] D. J. Thouless, M. Kohmoto, P. Nightingale, and M. den Nijs, Phys. Rev. Lett. **49**, 405 (1982).
[13] J. E. Avron, R. Seiler, and B. Simon, Phys. Rev. Lett. **51**, 51 (1983).
[14] T. Ando, J. Phys. Soc. Jpn. **52**, 1740 (1983); H. Aoki, J. Phys. C **16**, 1893 (1983); H. Levine, S. B. Libby, and A. M. M. Pruisken, Phys. Rev. Lett. **51**, 1915 (1984); G. C. Aers and A. H. MacDonald, J. Phys. C **17**, 5491 (1984); H. Aoki and T. Ando, Phys. Rev. Lett. **54**, 831 (1985).
[15] P. Streda and L. Smrcka, Phys. Status. Solidi. B **70**, 537 (1975).
[16] R. Tao and Y. S. Wu. Phys. Rev. B **30**, 1097 (1984).
[17] R. Kubo, J. Phys. Soc. Jpn. **12**, 570 (1957).
[18] H. Bateman, *Higher Transcendental Functions* (McGraw-Hill, New York, 1953), Vol. II.
[19] A. S. B. Holland, *Introduction to the Theory of Entire Functions* (Academic, New York, 1973).

Reprinted from ANNALS OF PHYSICS
All Rights Reserved by Academic Press, New York and London

Vol. 160, No. 2, April 1, 1985
Printed in Belgium

Topological Invariant and the Quantization of the Hall Conductance

MAHITO KOHMOTO*

Department of Physics and the Materials Research Laboratory, University of Illinois at Urbana-Champaign, Urbana, Illinois 61801

Received March 27, 1984

The topological aspects of wavefunctions for electrons in a two dimensional periodic potential with a magnetic field are discussed. Special attention is paid to the linear response formula for the Hall conductance σ_{xy}. It is shown that the quantized value of σ_{xy} is related to the number of zeros of wavefunctions in the magnetic Brillouin zone. A phase of wavefunctions cannot be determined in a unique and smooth way over the entire magnetic Brillouin zone unless the magnetic subband carries no Hall current. © 1985 Academic Press, Inc.

1. INTRODUCTION

The discovery of the quantization of the Hall conductance by von Klitzing, Dorda, and Pepper [1] is one of the most important results in condensed matter physics in recent years. Since there had been no prediction of such a precise quantization, the subject has been under intensive theoretical investigation [2]. The ideal two dimensional electron system has integral values of the Hall conductance in unit of e^2/h if the Fermi energy lies in a gap between Landau levels. The experimental result is astonishing because, in a real system, we would expect corrections of various sorts, due to, for example, electron–electron interactions, impurities, quasi-two-dimensionality, substrate potentials, finite size of samples, etc.

The effects of periodic potentials have been discussed by several authors [3–6]. Spectrum of electrons in a crystal with a magnetic field can display an amazing complexity including various kinds of scaling and a Cantor set structure [7–9]. However, it was shown that the Hall conductance is still an integral multiple of e^2/h as long as the Fermi energy lies in a gap.

This paper discusses topological aspects of a two-dimensional periodic systems in a magnetic field. The Hall conductance is shown to be represented by a topological invariant which is naturally an integer. This is a sequel to Ref. [3] in a spirit similar to Laughlin's arguments [10] that an exact quantization must be a consequence of a general principle which is determined by the geometrical nature of the problem.

* Current address: Department of Physics, University of Utah, Salt Lake City, Utah 84112.

0003-4916/85 $7.50

This subject has been first discussed by Avron, Seiler and Simon [6] using homotopy theory. More recently, Simon [11] made a connection between this topological invariant and Berry's geometrical phase factor in the quantum adiabatic theorem [12]. The purpose of the present paper is to illustrate and explain the topological invariant of Thouless *et al.* and Avron *et al.* in a rather simple physical picture.

In Section II, some relevant features of Bloch electrons in a magnetic field are reviewed. The implications of the topological features for wavefunctions are discussed. In Section III, the linear response (Nakano–Kubo) formula for the Hall conductance is introduced. It is shown that a contribution of a single filled band to σ_{xy} is given by an integer associated with a topological invariant. In Section IV, the problem is further analyzed by the theory of fiber bundles. It is shown that the quantized Hall conductance is related to the first Chern number which characterizes a principal $U(1)$ bundle.

II. Bloch Electrons in a Uniform Magnetic Field

The Schrödinger equation for a 2-D non-interacting electron system in a uniform magnetic field perpendicular to the plane is written as

$$H\Psi = \left[\frac{1}{2m}(\mathbf{p} + e\mathbf{A}))^2 + U(x, y)\right] \Psi = E\Psi, \tag{2.1}$$

where the momentum $\mathbf{p} = -i\hbar\nabla$ and the gauge potential $\mathbf{A}$ are in the x–y plane. We consider the case where $U(x, y)$ is periodic in both the x- and y-directions, i.e.,

$$U(x + a, y) = U(x, y + b) = U(x, y). \tag{2.2}$$

The system is invariant under a translation by a along the x-direction or a translation by b along the y-direction. However, the Hamiltonian is not invariant under these translations. The reason for this is that the gauge potential $\mathbf{A}$ is not constant in spite of the fact that the magnetic field is uniform. An appropriate gauge transformation is required to make the Hamiltonian invariant. Let us introduce some formalism in order to better describe the point discussed above and also to clarify the topological aspects of the problem. Let $\mathbf{R}$ be a Bravais lattice vector, i.e.,

$$\mathbf{R} = n\mathbf{a} + m\mathbf{b},$$

where n and m are integers. For each Bravais lattice vector $\mathbf{R}$ we define a translation operator T_R which, when operating on any smooth function $f(\mathbf{r})$, shifts the argument by $\mathbf{R}$:

$$T_R f(\mathbf{r}) = f(\mathbf{r} + \mathbf{R}). \tag{2.3}$$

This operator is explicitly written as

$$T_R = \exp\{(i/\hbar)\,\mathbf{R}\cdot\mathbf{p}\}. \tag{2.4}$$

If T_R is applied to the Hamiltonian (2.1), the potential $U(\mathbf{r})$ is left invariant. However, the gauge potential is transformed to $\mathbf{A}(\mathbf{r}+\mathbf{R})$ which is not generally equal to $\mathbf{A}(\mathbf{r})$. Instead, $\mathbf{A}(\mathbf{r})$ and $\mathbf{A}(\mathbf{r}+\mathbf{R})$ differ by a gradient of a scalar function since the magnetic field is uniform:

$$\mathbf{A}(\mathbf{r}) = \mathbf{A}(\mathbf{r}+\mathbf{R}) + \nabla g(\mathbf{r}). \tag{2.5}$$

Let us consider the magnetic translation operators [13–15]

$$\begin{aligned}\hat{T}_R &= \exp\{(i/\hbar)\,\mathbf{R}\cdot[\mathbf{p}+e(\mathbf{r}\times\mathbf{B})/2]\}\\ &= T_R \exp\{(ie/\hbar)(\mathbf{B}\times \boldsymbol{R})\cdot\mathbf{r}/2\}.\end{aligned} \tag{2.6}$$

If the symmetric gauge $\mathbf{A} = (\mathbf{B}\times\mathbf{r})/2$ is taken, $\hat{T}_R$ leaves the Hamiltonian invariant, i.e.,

$$[\hat{T}_R, H] = 0.$$

Now, we look for eigenstates which simultaneously diagonalize $\hat{T}_R$ and H. However, note that the magnetic translations do not commute with each other in general since

$$\hat{T}_a\hat{T}_b = \exp(2\pi i\phi)\,\hat{T}_b\hat{T}_a, \tag{2.7}$$

where $\phi = (eB/h)\,ab$ is a number of magnetic flux in the unit cell. When ϕ is a rational number, $\phi = p/q$ (p and q are integers which are relatively prime), we have a subset of translations which commute with each other. We take an enlarged unit cell (magnetic unit cell) which an integral multiple of magnetic flux goes through. For example, if the Bravais latice vectors of the form

$$\mathbf{R}' = n(q\mathbf{a}) + m\mathbf{b} \tag{2.8}$$

are taken, then p magnetic flux quanta are in the magnetic unit cell which is formed by the vectors $q\mathbf{a}$ and $\mathbf{b}$. The magnetic translation operators $\hat{T}_{R'}$ which correspond to these new Bravais lattice vectors commute with each other.

Let ψ be an eigenfunction which diagonalizes H and $\hat{T}_{R'}$ simultaneously, then it is easy to show that the eigenvalues of $\hat{T}_{qa}$ and $\hat{T}_b$ are given by

$$\begin{aligned}\hat{T}_{qa}\psi &= e^{ik_1 qa}\psi,\\ \hat{T}_b\psi &= e^{ik_2 b}\psi,\end{aligned} \tag{2.9}$$

where k_1 and k_2 are generalized crystal momenta and can be restricted in the magnetic Brillouin zone: $0 \leqslant k_1 \leqslant 2\pi/qa$, $0 \leqslant k_2 \leqslant 2\pi/b$. The eigenfunctions are labeled by k_1 and k_2 in addition to a band index α and are written in a Bloch form as

$$\psi_{k_1k_2}^{(\alpha)}(x, y) = e^{ik_1x+ik_2y}u_{k_1k_2}^{(\alpha)}(x, y). \tag{2.10}$$

Equations (2.9) with (2.6) give the property of $u_{k_1k_2}^{(\alpha)}(x, y)$

$$\begin{aligned} u_{k_1k_2}^{(\alpha)}(x+qa, y) &= e^{-i\pi py/b}u_{k_1k_2}^{(\alpha)}(x, y), \\ u_{k_1k_2}^{(\alpha)}(x, y+b) &= e^{i\pi px/qa}u_{k_1k_2}^{(\alpha)}(x, y), \end{aligned} \tag{2.11}$$

These are the generalized Bloch conditions. Note that a gauge transformation $\mathbf{A}' = \mathbf{A} + \nabla f$ changes the phase of a wavefunction, $\psi' = e^{-ie/\hbar f}\psi$. A gauge invariant, hence meaningful quantity, is the phase change around the boundary of the magnetic unit cell. From Eq. (2.11), the phase change is given by $2\pi p$. Writing the wavefunctions as

$$u_{k_1k_2}(x, y) = |u_{k_1k_2}(x, y)| \exp[i\theta_{k_1k_2}(x, y)], \tag{2.12}$$

then one has

$$p = \frac{-1}{2\pi}\int dl \cdot \frac{\partial\theta_{k_1k_2}(x, y)}{dl}, \tag{2.13}$$

where $\int dl$ represents a counterclockwise line integral around the boundary of the magnetic unit cell. As we will see below, Eq. (2.13) is gauge-invariant although $\theta_{k_1k_2}(x, y)$ depends on a gauge. This equation represents an important topological feature of the system. Consider an arrow whose directional angles are given by the phase $\theta_{k_1k_2}(x, y)$ of the wavefunction. The arrow rotates p times as we go around the boundary. This gives a topological constraint to the wavefunction. Consider a zero of the wavefunction. If we go around clockwise a small circle which contains the zero, the corresponding arrow rotates once either clockwise or counterclockwise. Therefore we can regard a zero of a wavefunction as a vortex which has a vorticity either 1 or -1 corresponding to clockwise or counterclockwise rotation of the arrow, respectively. Cases where we have a multiple rotation are considered to be special ones of having several vortices at the same point. The magnetic field forces a wavefunction to have $-p$ vorticity in the magnetic unit cell. This is a topological constraint because the total vorticity in the magnetic unit cell is independent of a particular potential chosen.

III. Linear Response Formula for the Hall Conductance

It is useful to write the Schrödinger equation (2.1) in a form

$$\hat{H}(k_1, k_2)\, u_{k_1k_2}^{\alpha} = E^{\alpha}u_{k_1k_2}^{\alpha}, \tag{3.1}$$

with

$$\hat{H}(k_1, k_2) = \frac{1}{2m}(-i\hbar\nabla + \hbar\mathbf{k} + e\mathbf{A})^2 + U(x, y), \tag{3.2}$$

where $\mathbf{k}$ is a vector whose x- and y-components are k_1 and k_2, respectively. Note that the eigenvalue E^α depends on $\mathbf{k}$ continuously. For a fixed band index α, a set of possible values of E^α with $\mathbf{k}$ varying in the magnetic Brillouin zone forms a band (magnetic subband).

When a small electric field is applied, a resulting current may be given by the linear response (Nakano–Kubo) formula. A linear response of current in the perpendicular direction to the applied electric field is represented by the Hall conductance

$$\sigma_{xy} = \frac{e^2\hbar}{i} \sum_{E^\alpha < E_F < E^\beta} \frac{(v_y)_{\alpha\beta}(v_x)_{\beta\alpha} - (v_x)_{\alpha\beta}(v_y)_{\beta\alpha}}{(E^\alpha - E^\beta)^2}, \tag{3.3}$$

where E_F is a Fermi energy and the summation implies the sum over all the states below and above the Fermi energy. The indices α and β label bands. One needs $\mathbf{k}$ to specify a state in addition to the band index. The existence of the index $\mathbf{k}$ must implicitly be understood wherever the band index appears. To obtain the matrix elements of the velocity operator $\mathbf{v} = (-i\hbar\nabla + e\mathbf{A})/m$, it is sufficient to integrate over one magnetic unit cell

$$(\mathbf{v})_{\alpha\beta} = \delta_{k_1 k_1'} \delta_{k_2 k_2'} \int_0^{qa} dx \int_0^b dy\, u^{\alpha *}_{k_1 k_2} \mathbf{v} u^\beta_{k_1 k_2}, \tag{3.4}$$

where the states are normalized as $\int_0^{qa} dx \int_0^b dy\, |u|^2 = 1$. In Eq. (3.3) the velocity operators can be replaced by partial derivatives of the $\mathbf{k}$-dependent Hamiltonian (3.2), since only off-diagonal matrix elements are considered

$$\begin{aligned} (v_x)_{\alpha\beta} &= \frac{1}{\hbar}\left\langle \alpha \left| \frac{\partial \hat{H}}{\partial k_1} \right| \beta \right\rangle, \\ (v_y)_{\alpha\beta} &= \frac{1}{\hbar}\left\langle \alpha \left| \frac{\partial \hat{H}}{\partial k_2} \right| \beta \right\rangle. \end{aligned} \tag{3.5}$$

Furthermore the matrix elements of the partial derivatives of $\hat{H}$ are written as

$$\begin{aligned} \left\langle \alpha \left| \frac{\partial \hat{H}}{\partial k_j} \right| \beta \right\rangle &= (E^\beta - E^\alpha)\left\langle \alpha \left| \frac{\partial u^\beta}{\partial k_j} \right.\right\rangle \\ &= -(E^\beta - E^\alpha)\left\langle \left. \frac{\partial u^\alpha}{\partial k_j} \right| \beta \right\rangle, \qquad j = 1 \text{ or } 2. \end{aligned} \tag{3.6}$$

From Eqs. (3.5) and (3.6), Eq. (3.3) is written as

$$\sigma_{xy} = \frac{e^2}{i\hbar} \sum_{E^\alpha < E_F < E^\beta} \left(\left\langle \frac{\partial u^\alpha}{\partial k_2} \middle| \beta \right\rangle \left\langle \beta \middle| \frac{\partial u^\alpha}{\partial k_2} \right\rangle - \left\langle \frac{\partial u^\alpha}{\partial k_1} \middle| \beta \right\rangle \left\langle \beta \middle| \frac{\partial u^\alpha}{\partial k_1} \right\rangle \right).$$

Using the identity $\sum_{E^\alpha < E_F < E^\beta} (|\alpha\rangle\langle\alpha| + |\beta\rangle\langle\beta|) = 1$, we have

$$\sigma_{xy}^{(\alpha)} = \frac{e^2}{h} \frac{1}{2\pi i} \int d^2k \int d^2r \left(\frac{\partial u_{k_1k_2}^{\alpha *}}{\partial k_2} \frac{\partial u_{k_1k_2}^{\alpha}}{\partial k_1} - \frac{\partial u_{k_1k_2}^{\alpha *}}{\partial k_1} \frac{\partial u_{k_1k_2}^{\alpha}}{\partial k_2} \right), \tag{3.7}$$

where $\sigma_{xy}^{(\alpha)}$ is a contribution of the Hall conductance from a completely filled band α. Let us define a vector field in the magnetic Brillouin zone by

$$\hat{\mathbf{A}}(k_1, k_2) = \int d^2r u_{k_1k_2}^* \nabla_k u_{k_1k_2} = \langle u_{k_1k_2} | \nabla_k | u_{k_1k_2} \rangle, \tag{3.8}$$

where ∇_k is a vector operator whose components are $\partial/\partial k_1$ and $\partial/\partial k_2$. The band index α is omitted from the wavefunction, since we will consider only a contribution from a single band only. The contribution is written from Eqs. (3.7) and (3.8) as

$$\sigma_{xy}^{(\alpha)} = \frac{e^2}{h} \frac{1}{2\pi i} \int d^2k [\nabla_k \times \hat{\mathbf{A}}(k_1, k_2)]_3, \tag{3.9}$$

where, $[\ ]_3$ represents the third component of the vector.

The integration is over the magnetic Brillouin zone: $0 \leqslant k_1 \leqslant 2\pi/qa$, $0 \leqslant k_2 \leqslant 2\pi/b$. An important observation here is that the magnetic Brillouin zone is topologically a torus T^2 rather than a rectangular in **k**-space. The two points in **k**-space $k_1 = 0$ and $2\pi/qa$ (or, $k_2 = 0$ and $2\pi/b$) must be identified as the same point. Since a torus does not have a boundary, the application of Stokes' theorem to Eq. (3.9) would give $\sigma_{xy}^{(\alpha)} = 0$ if $\hat{\mathbf{A}}(k_1, k_2)$ is uniquely defined on the entire torus T^2. A possible non-zero value of $\sigma_{xy}^{(\alpha)}$ is a consequence of a non-trivial topology of $\hat{\mathbf{A}}(k_1, k_2)$. Note that the identification of the magnetic Brillouin zone as a torus T^2 is essential here. Non-trivial $\hat{\mathbf{A}}(k_1, k_2)$ can only be constructed when the global topology of the base space is non-contractible.

In order to understand the topology of $\hat{\mathbf{A}}(k_1, k_2)$, let us first discuss a "gauge transformation" of a special kind. Suppose $u_{k_1k_2}(x, y)$ satisfies the Schrödinger equation (3.1), then so does $u_{k_1k_2}(x, y)\, e^{if(k_1,k_2)}$, where $f(k_1, k_2)$ is an arbitrary smooth function of k_1 and k_2 and is independent of x and y.

This introduces a transformation

$$u'_{k_1k_2}(x, y) = u_{k_1k_2}(x, y) \exp[if(k_1, k_2)]. \tag{3.10}$$

Since this is a change of the overall phase of the wavefunction, any physical quantity remains invariant under this transformation. From Eq. (3.8) the corresponding transformation of $\hat{\mathbf{A}}(k_1, k_2)$ is given by

$$\hat{\mathbf{A}}'(k_1, k_2) = \hat{\mathbf{A}}(k_1, k_2) + i\nabla_k f(k_1, k_2). \tag{3.11}$$

It is easy to see from Eqs. (3.9) and (3.11) that $\sigma_{xy}^{(\alpha)}$ is invariant under the transformation (3.10).

The non-trivial topology arises when the phase of the wavefunction cannot be determined uniquely and smoothly in the entire magnetic Brillouin zone. The transformation (3.10) implies that the overall phase factor for each state vector $|u_{k_1k_2}\rangle$ can be chosen arbitrary. This phase can be determined, for example, by demanding that a component of the state vector $u_{k_1k_2}(x^{(0)}, y^{(0)}) = \langle x^{(0)}, y^{(0)} | u_{k_1k_2}\rangle$ is real. However, this convention is not enough to fix the phase on the entire magnetic Brillouin zone, since $u_{k_1k_2}(x^{(0)}, y^{(0)})$ vanishes for some (k_1, k_2). The existence of zeros of $u_{k_1k_2}(x, y)$ has been shown in Section II. For the sake of simplicity, consider the case where $u_{k_1k_2}(x^{(0)}, y^{(0)})$ vanishes only at one point $(k_1^{(0)}, k_2^{(0)})$ in the magnetic Brillouin zone. See Fig. 1. Divide T^2 into two pieces H_{I} and H_{II} such that H_{I} contains $(k_1^{(0)}, k_2^{(0)})$. We adopt a different convention in H_{I} so that another component of the state vector $u_{k_1k_2}(x^{(1)}, y^{(1)}) = \langle x^{(1)}, y^{(1)} | u_{k_1k_2}\rangle$ is real, where $(x^{(1)}, y^{(1)})$ and H_{I} are chosen such that $u_{k_1k_2}(x^{(1)}, y^{(1)})$ does not vanish in H_{I}. Thus the overall phase is uniquely determined on the entrie T^2. In Fig. 1, a phase of one component of the state vector $u_{k_1k_2}(x^{(0)}, y^{(0)}) = \langle x^{(0)} y^{(0)} | u_{k_1k_2}\rangle$ is schematically drawn.

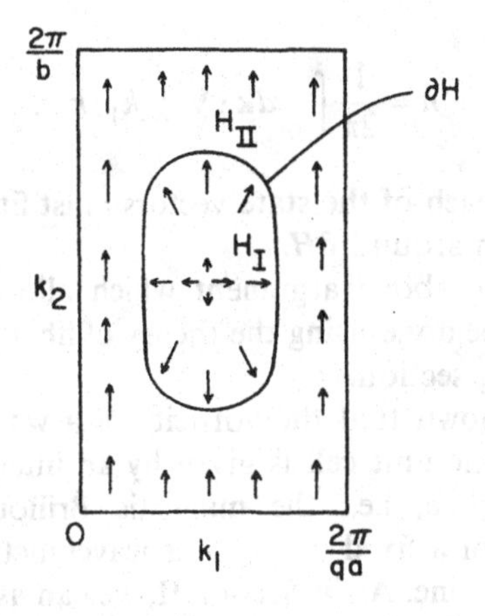

FIG. 1. Schematic diagram of a phase of a wavefunction in the magnetic Brillouin zone. The Brillouin zone is actually a torus, so the edges $(k_1, k_2) = (0, k_2)$ and $(2\pi/qa, k_2)$; and also the edges $(k_1, 0)$ and $(k_1, 2\pi/b)$ must be identified.

Note that the overall phase of the state vector is well defined at $(k_1^{(0)}, k_2^{(0)})$ even though a phase of a single component $u_{k_1k_2}(x^{(0)}, y^{(0)}) = \langle x^{(0)}, y^{(0)} | u_{k_1k_2}\rangle$ cannot be defined there. At the boundary ∂H of H_{I} and H_{II}, we have a phase mismatch

$$|u_{k_1k_2}^{\text{II}}\rangle = \exp[i\chi(k_1, k_2)]|u_{k_1k_2}^{\text{I}}\rangle, \tag{3.12}$$

where $\chi(k_1, k_2)$ is a smooth function of (k_1, k_2) on ∂H.

This non-trivial topology of $|u_{k_1k_2}\rangle$ is simply carried over to that of $\hat{\mathbf{A}}(k_1, k_2)$. Smooth vector fields $\hat{\mathbf{A}}_{\text{I}}(k_1, k_2)$ and $\hat{\mathbf{A}}_{\text{II}}(k_1, k_2)$ are defined on H_{I} and H_{II}, respectively, by Eq. (3.8). The phase mismatch of the state vector given by Eq. (3.12) induces the following relation between $\hat{\mathbf{A}}_{\text{I}}(k_1, k_2)$ and $\hat{\mathbf{A}}_{\text{II}}(k_1, k_2)$ on ∂H:

$$\hat{\mathbf{A}}_{\text{II}}(k_1, k_2) = \hat{\mathbf{A}}_{\text{I}}(k_1, k_2) + i\nabla_k\chi(k_1, k_2). \tag{3.13}$$

Now, in Eq. (3.9) we can apply Stokes' theorem to H_{I} and H_{II} separately

$$\sigma_{xy}^{(\alpha)} = \frac{e^2}{h}\frac{1}{2\pi i}\left\{\int_{H_I} d^2k[\nabla_k \times \hat{\mathbf{A}}_{\text{I}}(k_1, k_2)]_3 + \int_{H_{\text{II}}} d^2k[\nabla_k \times \hat{\mathbf{A}}_{\text{II}}(k_1 k_2)]_3\right\}$$

$$= \frac{e^2}{h}\frac{1}{2\pi i}\int_{\partial H} d\mathbf{k}\cdot[\hat{\mathbf{A}}_{\text{I}}(k_1, k_2) - \hat{\mathbf{A}}_{\text{II}}(k_1, k_2)], \tag{3.13}$$

where $\int_{\partial H} d\mathbf{k}$ represents a line integral on ∂H and the sign change occurs because ∂H has the opposite orientation for H_{I} and H_{II}. Using the relation between $\hat{\mathbf{A}}_{\text{I}}$ and $\hat{\mathbf{A}}_{\text{II}}$, Eq. (3.12), we find

$$\sigma_{xy}^{(\alpha)} = \frac{e^2}{h}n, \tag{3.14}$$

with

$$n = \frac{1}{2\pi}\int_{\partial H} d\mathbf{k}\cdot\nabla_k\chi(k_1, k_2). \tag{3.15}$$

n must be an integer for each of the state vectors must fit together exactly when we complete a full revolution around ∂H.

A generalization of the above argument which allows $u_{k_1k_2}(x^{(0)}, y^{(0)})$ to have more than one zero can be done using the theory of fiber bundles [16]. This will be discussed in the following section.

In Section II, it was shown that the vorticity of a wavefunction $u_{k_1k_2}(x, y)$ for a fixed (k_1, k_2) in a magnetic unit cell is given by an integer $-p$. There is a similar structure in the dual space, i.e., the magnetic Brillouin zone T^2. Consider a wavefunction $u_{k_1k_2}(x, y)$ for a fixed (x, y). This wavefunction has a number of zeros in the magnetic Brillouin zone. As in Section II, we can assign vorticities 1 or -1 to each zero by considering the phase in the neighborhood of a zero. The quantized value of $\sigma_{xy}^{(\alpha)}$ is given by the total vorticity of the wavefunction in the magnetic Brillouin zone.

IV. Fiber Bundle and Chern Class

The topological structure discussed in Section III has a close resemblance to that of the Dirac magnetic monopole bundle which is a principal $U(1)$ bundle over a sphere S^2 [16].

We have been considering the normalized eigenstate $|u_{k_1k_2}\rangle$. Since $\exp[if(k_1, k_2)]|u_{k_1k_2}\rangle$ is also a normalized eigenstate, it is natural to consider a principal $U(1)$ bundle over T^2. A torus is covered by four neighborhoods $\{H_i\}$, $i=1,..., 4$ where each H_i is a subspace of R^2. For example, if we define four regions by

$$
\begin{aligned}
H'_1 &= \{(k_1, k_2)|0<k_1<\pi/qa, 0<k_2<\pi/b\},\\
H'_2 &= \{(k_1, k_2)|\pi/qa<k_1<2\pi/qa, 0<k_2<\pi/b\},\\
H'_3 &= \{(k_1, k_2)|\ \pi/qa<k_1<2\pi/qa, \pi/b<k_2<2\pi/b\},\\
H'_4 &= \{(k_1, k_2)|\ 0<k_1<\pi/qa, \pi/b<k_2<2\pi/b\},
\end{aligned}
\tag{4.1}
$$

then we can choose $H_i\,(i=1,..., 4)$ to be slightly larger regions each of which completely includes $H'_i\,(i=1,..., 4)$. A principal $U(1)$ bundle is a topological space which is locally isomorphic to $H_i \times U(1)$ in each neighborhood H_i. We consider a specific fiber whose global topology is determined by the eigenstate $|u_{k_1k_2}\rangle$. A construction of a fiber bundle is as follows: Take a component of the state vector $u_{k_1k_2}(x^{(0)}, y^{(0)}) = \langle x^{(0)}y^{(0)}|u_{k_1k_2}\rangle$ which does not vanish in the overlaps of H_i. Since H_i is contractible, it is possible to choose a phase convention such that the phase factor $\exp[i\theta^{(i)}(k_1, k_2)] = u^{(i)}_{k_1k_2}(x^{(0)}, y^{(0)})/|u^{(i)}_{k_1k_2}(x^{(0)}, y^{(0)})|$ is smooth in each neighborhood H_i except at zeros of $u_{k_1k_2}(x^{(0)}, y^{(0)})$. As exemplified in Section III, it is not possible in general to have a global phase convention which is good to all the neighborhoods. As a result, we have a transition function Φ_{ij} in the overlap between two neighborhoods, $H_i \cap H_j$:

$$
\Phi_{ij} = \exp i[\theta^{(i)}(k_1 k_2) - \theta^{(j)}(k_1, k_2)] = \exp[if^{(ij)}(k_1 k_2)]. \tag{4.2}
$$

If we regard Φ_{ij} to be a map Φ_{ij}: $U(1) \to U(1)$, a principal $U(1)$ bundle over T^2 is completely specified by this transition function.

Thus, we have constructed a nontrivial fiber bundle. Fiber bundles are classified by certain integers characterizing the transition functions. These integers also correspond to integrals involving a bundle curvature when we put connections on the bundles. We may write a connection 1-form as

$$
\begin{aligned}
\omega &= g^{-1}Ag + g^{-1}dg\\
&= A + id\chi,
\end{aligned}
\tag{4.3}
$$

where $g = e^{i\chi} \in U(1)$ is a fiber. We choose a 1-form A by

$$
A(k_1, k_2) = \hat{A}_\mu(k_1, k_2)\, dk_\mu = \langle u_{k_1k_2}|\frac{\partial}{\partial k_\mu}|u_{k_1k_2}\rangle\, dk_\mu. \tag{4.4}
$$

The transition functions of the form (4.2) act on fibers by left multiplication. In an overlap of two neighborhoods H_i and H_j, a transition function $\Phi = \Phi_{ij}$ relates the local fiber coordinated g and g' in H_j and H_i as

$$g' = \Phi g. \tag{4.5}$$

This is equivalent to the "gauge transformation" Eq. (3.10). From Eqs. (3.11), (4.2) and (4.5), a transformation of $A(k_1, k_2)$ is given by

$$A' = \Phi A \Phi^{-1} + \Phi d\Phi^{-1} = A - i\frac{\partial f}{\partial k_\mu} dk_\mu. \tag{4.6}$$

It can be shown that ω is invariant under the transformations (4.5) and (4.6). So, ω is indeed a legitimate connection 1-form with a choice of A in Eq. (4.4).

Since a connection is given, we have a differential geometry on the topological space. The curvature is given by

$$F = dA = \frac{\partial \hat{A}_\mu}{\partial k_\nu} dk_\nu \wedge dk_\mu. \tag{4.7}$$

Since $c_1 = (i/2\pi) F$ is the first Chern form, an integral of c_1 over the entire manifold T^2,

$$C_1 = \frac{i}{2\pi}\int F = \frac{i}{2\pi}\int dA = \frac{i}{2\pi}\int \frac{\partial \hat{A}_\mu}{\partial k_\nu} dk_\nu \wedge dk_\mu, \tag{4.8}$$

is the first Chern number. This number is an integer which is independent of a particular connection chosen. It is only given by the topology of the principal $U(1)$ bundle which is constructed from the state vector $|u_{k_1k_2}\rangle$.

A comparison of Eqs. (3.9) and (4.8) gives

$$\sigma_{xy}^{(\alpha)} = -\frac{e^2}{h} C_1, \tag{4.9}$$

i.e., a contribution to the Hall conductance from a single filled band in unit of e^2/h is given by minus the first Chern number.

V. Concluding Remarks

The methods of differential geometry have been essential in modern gauge theory as well as in Einstein's theory of gravity. Maxwell's electromagnetism is nothing but an Abelian gauge theory. The differential geometry in the real physical space is given by a gauge potential.

In the present problem of two dimensional periodic potentials with a uniform magnetic field, we have a differential geometry on the reciprocal k-space (magnetic

Brillouin zone) which is a torus T^2. A connection is given by $\hat{\mathbf{A}}(k_1, k_2)$ (see Eqs. (3.8) and (4.3)). The phase factor of the state vector in the magnetic Brillouin zone is a section of a principal $U(1)$ bundle. The first Chern number characterizing this fiber bundle is an integer and is given by a certain integral involving $\hat{\mathbf{A}}(k_1, k_2)$. It is remarkable that this integral can be identified as the linear response formula for $\sigma_{xy}^{(\alpha)}$: a contribution to the Hall conductance from a single band. There is no higher order Chern number associated with the $U(1)$ bundle since the base space is two dimensional. Therefore $\alpha_{xy}^{(\alpha)}$ is the only quantized quantity of this kind.

The quantized value of $\alpha_{xy}^{(\alpha)}$ is related to zeros of wavefunction in the magnetic Brillouin zone. An integer vorticity can be assigned to each of the zeros from a local structure around it. The total vorticity in the magnetic Brillouin zone gives $\alpha_{xy}^{(\alpha)}$. Therefore if the topological structures of wavefunctions are known, we do not have to perform the four dimensional integration in Eq. (3.7) to calculate $\alpha_{xy}^{(\alpha)}$.

When $\sigma_{xy}^{(\alpha)}$ takes a non-zero value, wavefunctions have a non-trivial topological structure in the magnetic Brillouin zone. This means that a phase of wavefunctions cannot be determined globally. We need several different phase conventions to cover the entire magnetic Brillouin zone.

The topological nature of the formula for $\alpha_{xy}^{(\alpha)}$ implies that it will be unchanged when the potential is varied in a way that the band α remains non-degenerate. This gives a convenient calculational method since we can take the most convenient form of a potential to obtain $\alpha_{xy}^{(\alpha)}$ [3].

In case two bands touch each other while a potential is varied, we can still expect that the sum of the Hall conductances remains unchanged. However, each conductance may not be conserved when the bands split again. This has been proved by Avron *et al.* [6].

The linear response (Nakano–Kubo) formula for the Hall conductance has played a crucial role in relating an abstract mathematical quantity (the Chern number) to the quantity of physical interests. There are serious foundational questions regarding this formula [17]. The physical system we have considered is dissipationless. There is no longitudinal current [18] when the Fermi energy lies in a gap. In this case, it is possible to have an alternative derivation of the formula which is free from the criticisms of van Kampen and others.

The topological aspects of the quantized Hall effect in periodic potentials have been discussed. It will be of great interest to generalize the above arguments to more realistic models of the experimental systems which should include the effects of disorder and/or electron–electron interactions.

ACKNOWLEDGMENTS

I have had useful discussions with K. Aomoto, T. Eguchi, Mathew Fisher, M. den Nijs, Y. Oono, Michael Stone, D. J. Thouless and A. Tsuchiya. It is a pleasure to thank J. Avron and B. Simon for illuminating discussions about their work. This work is supported by the NSF Grant DMR-80-20250.

References

1. K. von Klitzing, G. Dorda, and M. Pepper, *Phys. Rev. Lett.* **45** (1980), 494.
2. D. J. Thouless, "Proceedings, 5th International Conference on Electronic Properties of Two-Dimensional Systems," Oxford, England, 1983 and references therein.
3. D. J. Thouless, M. Kohmoto, M. P. Nightingale, and M. den Nijs, *Phys. Rev. Lett.* **49** (1982), 405.
4. P. Streda, *J. Phys. C* **15** (1982), L717.
5. D. Yoshioka, *Phys. Rev. B* **27** (1983), 3637.
6. J. Avron, R. Seiler, and B. Simon, *Phys. Rev. Lett.* **51** (1983), 51.
7. M. Azbel, *Soviet Phys. JETP* **19** (1964), 634.
8. D. R. Hofstadter, *Phys. Rev. B* **14** (1976), 2239.
9. M. Kohmoto, *Phys. Rev. Lett.* **51** (1983), 1198.
10. R. B. Laughlin, *Phys. Rev. B* **23** (1981), 5632.
11. B. Simon, *Phys. Rev. Lett.* **51** (1983), 2167.
12. M. V. Berry, *Proc. Roy. Soc. London* **392** (1984), 45.
13. G. A. Peterson, thesis, Cornell University, Ithaca, New York, 1960, unpublished.
14. E. Brown, *Phys. Rev.* **133** (1964), A1038; *Solid State Phys.* **22** (1968), 313.
15. J. Zak, *Phys. Rev.* **134** (1964), A1602.
16. T. Eguchi, P. B. Gilkey, and A. J. Hanson, *Phys. Rep.* **66** (1980), 213; C. Nash and S. Sen, "Topology and Geometry for Physicists," Academic Press, New York, 1983.
17. J. F. C. van Velsen, *Phys. Rep.* **41** (1978), 135.
18. A. Widom, *Phys. Rev. B* **28** (1983), 4858.

Printed by the St. Catherine Press Ltd., Tempelhof 41, Bruges, Belgium

Quantization of the Hall Conductance for General, Multiparticle Schrödinger Hamiltonians

Joseph E. Avron
Department of Physics, Technion–Israel Insitute of Technology, Haifa 32000, Israel

and

Ruedi Seiler
Fachbereich 3, Mathematik Technische Universität Berlin, D-1000 Berlin, Germany
(Received 15 October 1984)

We describe a precise mathematical theory of the Laughlin argument for the quantization of the Hall conductance for general multiparticle Schrödinger operators with general background potentials. The quantization is a consequence of the geometric content of the conductance, namely, that it can be identified with an integral over the first Chern class. This generalizes ideas of Thouless *et al.*, for noninteracting Bloch Hamiltonians to general (interacting and nonperiodic) ones.

PACS numbers: 03.65.Bz, 02.40.+m, 72.20.My

The integer quantization of the Hall conductance has been explained by Laughlin[1] making clever use of a nontrivial geometry: a ring threaded by a flux tube, combined with a gauge argument. The impact of this work on the development of the subject cannot be overestimated.

Our purpose here is to describe a precise mathematical theory of this argument. The two key issues are, first, Laughlin's identification of a physical quantity as the Hall conductance averaged over one unit of quantum flux (of the flux tube that threads the ring). Following Laughlin we shall slightly abuse the terminology and call it the Hall conductance.

The second main theme will be the identification of the geometric content of the Hall conductance: Roughly speaking, there is a natural notion of curvature describing how the state of the system is parallel transported in the Hilbert space of states. The Hall conductance is a suitable integral of the corresponding curvature. More precisely, it is an integral over the first Chern class.[2] This was first recognized by Thouless *et al.* in the special case of noninteracting Bloch Hamiltonians.[3] What is shown here is that this holds generally, with electron-electron interaction and general background potential. (No flux averaging is necessary for Bloch Hamiltonians.) Bellissard generalized the result of Thouless *et al.* from rational to real magnetic flux.[4]

We shall replace Laughlin's condition that the Fermi energy lies in the region of the localized states (which is not an appropriate condition for multiparticle Hamiltonians) by a condition of nondegeneracy of the multiparticle ground states (for all fluxes).

Our work has been independent of, but is nevertheless closely related to, a recent published paper of Niu and Thouless.[5] The general framework is similar, although there are some differences in the details and in the approach. In both works one needs the ground state to be separated from the rest of the spectrum by a finite gap (the nondegeneracy condition). In both approaches one considers time-periodic Hamiltonians (in Niu-Thouless only up to unitary equivalence). In Niu and Thouless the time dependence resides in the substrate potential and it comes from a Galilean transformation that removes the electric field. In our case, the time dependence comes from generating the electromotive force by a flux tube, and so resides in the minimal coupling term in the Hamiltonian. In both, strict quantization is obtained only after a suitable averaging: In Niu-Thouless the averaging is over boundary conditions and here, averaging is over the fluxes in Laughlin's flux tube.

We feel that avoiding the Galilean transformation is a distinct advantage of the present approach. Also, essentially all the structure we shall use of the Hamiltonian is minimal coupling. This makes it clear, for example, that there are no relativistic corrections, nor spin effects, nor finite-volume corrections, not even exponentially small ones, something which is less obvious in Ref. 5. We also note that our approach is more geometric and in our opinion, simpler.

We stress that strict quantization is proven for the Hall conductance averaged over the fluxes in Laughlin flux tubes. It is interesting to investigate under what conditions the Hall conductance for most flux values is close to its average. This problem has been investigated by Thouless[6] and Niu-Thouless,[5] who showed that under suitable additional conditions this is indeed the case in the thermodynamic limit of infinitely large systems. This is reasonable as it is easy to see that variations in the conductance are the same as those caused by changing the lengths of the connecting leads. The results in Ref. 5 are essentially perturbative, and are based on the assumption that correlations decay fast.

The nondegeneracy condition turns out to be a sufficient condition for integer quantization at zero temperature and finite volume (we shall study the infinite-volume limit elsewhere). From this one can also learn about fractions: Tao and Wu[7] and Tao[8] have extended Laughlin's argument to degenerate ground state and have shown that the ground-state degeneracy is related to the denominator of the fractional conductance. The present analysis extends to this case as well and agrees with these results. However, since degeneracies are nongeneric it is not clear how one can get plateaus at fractions. We speculate that this occurs via diamagnetism as the relation between the external and internal fields is singular near degeneracies because the ground-state energy is not smooth there. We have nothing to say about the odd-denominator rule.

We find it convenient to formulate the problem in the following setting. Consider a domain Λ in the plane, with two holes. The holes are threaded with two flux tubes with fluxes $\phi = (\phi_1, \phi_2)$. On the boundary of Λ we impose Dirichlet boundary conditions. Now consider the multielectron Schrödinger Hamiltonian for this system, $H(\phi)$, depending on ϕ through minimal coupling. The geometry, shown in Fig. 1(a), is motivated by that of the physical Hall effect shown in Fig. 1(b). The motivation is clear once the leads that connect to the sample in Fig. 1(b) are considered as part of the system. [One may remove the leads in Fig. 1(b) at the expense of imposing periodic boundary conditions.] We take one of the fluxes, say ϕ_1, linearly increasing with time thereby replacing the battery in Fig. 1(b). The second flux is the flux tube introduced by Laughlin in his original argument and is shown also in Fig. 1(b). The translation between the two geometries will always be evident and henceforth we stick with that of Fig. 1(a).

Adding one unit of quantum flux to ϕ_1 or ϕ_2 is equivalent to a gauge transformation. $H(\phi)$ is therefore a continuous function of two variables with a natural period of 2π in each (in units where $\hbar = e = 1$). There is now a formal analogy with Bloch Hamiltonians in two dimensions where $H(k)$ is a function of the

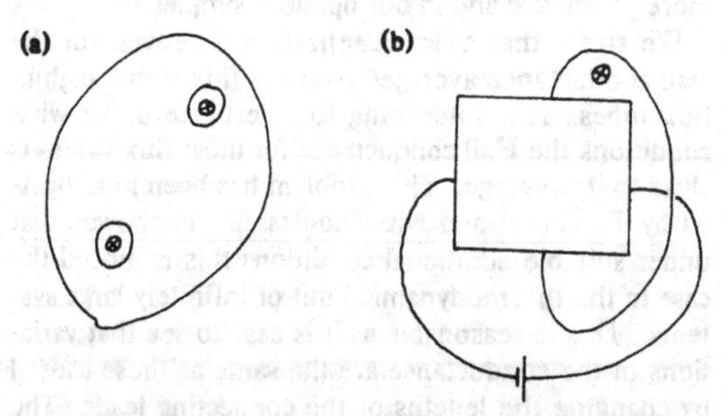

FIG. 1. (a) The domain Λ with the two holes threaded by the two flux tubes; (b) the standard Hall settings with the connecting leads considered as part of the system. Both geometries have two independent closed cycles. The battery in (b) is replaced by a time-dependent flux tube in (a). Both have an extra flux tube which is put there "by hand" in the Laughlin argument.

two Bloch momenta with natural period given by the Brillouin zone. This suggests that the ideas developed by Thouless and co-workers[3,5] and by Thouless[6] for Bloch Hamiltonians could be applied in the present circumstances as well.

For our purposes it is convenient to consider Hamiltonians that are smooth and periodic functions of ϕ. To achieve this, introduce two cuts in Λ so that the resulting set, $\tilde{\Lambda}$, is simply connected. The vector potential associated with ϕ is the gradient of the function $F(\phi)$ which is regular in $\tilde{\Lambda}$ and with discontinuities ϕ_1 and ϕ_2 across the cuts. $\tilde{H}(\phi) = \exp[-iF(\phi)]H(\phi) \times \exp[iF(\phi)]$ is formally ϕ independent. The ϕ dependence enters through the $\exp(i\phi_{1,2})$ boundary conditions that the wave function has to satisfy across the cuts. $\tilde{H}(\phi)$ is manifestly periodic in ϕ. $H(\phi)$ is periodic up to unitary equivalence.

All we shall need in order to prove the quantization is to assume that $H(\phi)$ has a nondegenerate ground state for all values of ϕ. Let us examine this condition. Recall that according to the Wigner-von Neumann no-crossing theorem, eigenvalue crossing has codimension three.[9] This says that Hamiltonians that depend on two parameters will not have any degeneracy (generically) while those that depend on three parameters have points of degeneracy. Thus, with ϕ as parameters there are no crossings, but if one varies the magnetic field as well, there will be crossings at special values of the magnetic field. These are the special values of B where the Hall conductance jumps. The proof of integer conductance we give holds for generic Hamiltonians at zero temperature. Had there been a way to prepare the system at an excited and isolated eigenstate, the Hall conductance would be quantized then too. It follows that at finite temperature it is, generically, a thermal average of integers.[10]

Let $|\Omega\rangle$ be the ground state of $H(\phi)$ with energy $E(\phi)$. By the assumption of nondegeneracy, it can be chosen normalized and smooth in ϕ (for all ϕ). The same holds for $|\tilde{\Omega}\rangle$. However, it is in general impossible to require periodicity in ϕ as well. In contrast, the spectral projection, $\tilde{P}(\phi) = |\tilde{\Omega}\rangle\langle\tilde{\Omega}|$, is both smooth and periodic in ϕ.

We shall establish the following set of formulas for the (flux-averaged) Hall conductance $\langle\sigma_H\rangle$:

$$2\pi\langle\sigma_H\rangle = \frac{i}{2\pi}\int_{\partial T}\langle\Omega|d\Omega\rangle \tag{1a}$$

$$= \frac{i}{2\pi}\int_T \langle d\tilde{\Omega}|d\tilde{\Omega}\rangle \tag{1b}$$

$$= \frac{i}{2\pi}\int_T \mathrm{Tr}(dPPdP) \tag{1c}$$

$$= \frac{i}{2\pi}\int_T \mathrm{Tr}(d\tilde{P}\tilde{P}d\tilde{P}) \tag{1d}$$

$$= \frac{i}{2\pi}\int_T \langle\Omega|dH\hat{R}^2 dH|\Omega\rangle. \tag{1e}$$

T is the square of length 2π and ∂T is its boundary.

The physics lie in showing that (1a) follows from a reasonable definition of the Hall conductance. The equivalence of the various formulas turns out to be simple. Equations (1a) and (1b) are analogs of the basic equations in Thouless *et al.* and therefore express an integer by their argument. This is a standard mathematical fact and we shall sketch a simple proof of it towards the end of this note. Equations (1c) and (1d) were used before in Ref. 11. Equation (1e) is the standard Kubo formula (this is explained below). Here $\hat{R}=(2\pi i)^{-1}\oint R(z)/(z-E)dz=R(1-P)$ is the reduced resolvent.

If $\phi_1=-Vt$, $\phi_2=$const, there is an electromotive force V around hole 1 and "Hall current" $I_2=\sigma_{21}V$ around hole 2. We shall establish a formula for σ_{21}, in the limit $V\to 0$ assuming that at time $t=0$ the system is in a state Ψ which is an eigenstate of $H(\phi_1=0,\phi_2)$. We may choose $\Psi(t=0,\phi_2)=\Omega(\phi_1=0,\phi_2)$. Ψ evolves in time according to the Schrödinger equation $-iV\partial_1\Psi=H(\phi)\Psi$. In the limit $V\to 0$, Ψ evolves adiabatically in the ϕ_1 variable. Therefore we shall have to use in the following the adiabatic wave function

$$\Psi^{\text{ad}}(\phi)=\lim_{V\to 0}[\exp(i/V)\int_0^{\phi_1}d\phi_1' E(\phi_1',\phi_2)]\Psi(\phi)$$

which satisfies the evolution equation[12]

$$\partial_1\Psi^{\text{ad}}=[\partial_1 P,P]\Psi^{\text{ad}}. \tag{2}$$

As a result of the fundamental expression for the current[1] $I_2=\langle\Psi|\partial_2 H|\Psi\rangle$, one finds for the conductivity

$$\sigma_{21}=-i(\partial_2\langle\Psi,\partial_1\Psi\rangle+\langle\partial_1\Psi,\partial_2\Psi\rangle-\langle\partial_2\Psi,\partial_1\Psi\rangle). \tag{3}$$

The adiabatic wave function does not represent $\Psi(\phi)$ well enough to compute σ_{21}, but it is sufficient to compute $\langle\sigma_{21}\rangle$.[13] The reason for that is that $\langle\sigma_{21}\rangle$ can be written as a line integral (rather than a surface integral):

$$2\pi\langle\sigma_{21}\rangle=-\frac{i}{2\pi}\{\int_0^{2\pi}d\phi_1\langle\Psi,\partial_1\Psi\rangle|_{\phi_2=0}^{2\pi}+\int_T\langle d\Psi,d\Psi\rangle\} \tag{4}$$

$$=-\frac{i}{2\pi}\int_0^{2\pi}d\phi_2\langle\Psi,\partial_2\Psi\rangle|_{\phi_1=0}^{2\pi} \tag{5}$$

$$=-\frac{i}{2\pi}\int_0^{2\pi}d\phi_2\langle\Psi^{\text{ad}},\partial_2\Psi^{\text{ad}}\rangle|_{\phi_1=0}^{2\pi}. \tag{6}$$

In going from (4) to (5) we used Stokes theorem; (6) follows from (5) due to periodicity of the ground-state energy $E(\phi)$. Since $\Psi^{\text{ad}}(\phi)$ is a ground-state wave function it follows from our spectral assumption that $\Psi^{\text{ad}}(\phi)$ equals $\Omega(\phi)$ up to a phase $\exp i\gamma(\phi)$ (Berry's phase[14,15]). Hence we get for the Hall conductivity

$$2\pi\langle\sigma_{21}\rangle=-\frac{i}{2\pi}\int_0^{2\pi}d\phi_2\{i\partial_2\gamma+\langle\Omega,\partial_2\Omega\rangle\}|_{\phi_1=0}^{2\pi}=-\frac{i}{2\pi}[\Gamma(2\pi)-\Gamma(0)]-\frac{i}{2\pi}\int_0^{2\pi}d\phi_2\langle\Omega,\partial_2\Omega\rangle|_{\phi_1=0}^{2\pi}, \tag{7}$$

where we used the notation $\Gamma(\phi_2)=\gamma(\phi_1=2\pi,\phi_2)$ and the fact $\gamma(\phi_1=0,\phi_2)=0$. To compute the Γ's we use Eq. (2)

$$0=\langle\Psi^{\text{ad}},\partial_1\Psi^{\text{ad}}\rangle=i\partial_1\gamma+\langle\Omega,\partial_1\Omega\rangle.$$

Integrating yields

$$\Gamma(2\pi)-\Gamma(0)=i\int_0^{2\pi}d\phi_1\langle\Omega,\partial_1\Omega\rangle|_{\phi_2=0}^{2\pi}. \tag{8}$$

Inserting (8) into (7) gives (1a).

A simple computation gives $d\langle\Omega|d\Omega\rangle=\langle d\Omega|d\Omega\rangle=\text{Tr}(dPPdP)$. This establishes $a=c$ and $b=d$. The equivalence $c=d$ follows from

$$\text{Tr}(dPPdP)=\text{Tr}(d\tilde{P}\tilde{P}d\tilde{P})+id\,\text{Tr}(\tilde{P}dF). \tag{9}$$

$\tilde{P}dF$ is continuous in ϕ on T so the boundary term vanishes upon integration. $c=e$ follows from the operator identity:

$$dPPdP=\hat{R}(dH)P(dH)\hat{R}, \tag{10}$$

and the cyclicity of the trace.

Equation (1e) is Kubo's formula. To see that write $\mathbf{A}_i=\phi_i\mathbf{a}_i$ for the two vector potentials with $\mathbf{a}_i$ normalized by $\int_{c_j}\mathbf{a}_i\cdot\mathbf{d}_1=2\pi\delta_{ij}$; c_j is a loop around hole j. $dH=-\frac{1}{2}\{\mathbf{v}\cdot\mathbf{a}_i\}d\phi_i$ with $\mathbf{v}=\sum_{i=1}^{N}\mathbf{v}_i$ the cannonical velocity and N is the number of particles. Since in (1e) only the antisymmetric part enters, one of the terms in dH must be identified with the perturbation and the second with the observable. If we take $\phi_1(t)$ and ϕ_2 fixed it follows from minimal coupling that the $d\phi_1$ term in dH is indeed the perturbation. To see that the $d\phi_2$ term is the response we note that without loss of generality we can choose $\mathbf{a}_2=(2\pi)^{-1}(\text{grad}\theta)$ where the polar coordinates (r,θ) have their origin in hole 2. The $d\phi_2$ term in dH is now naturally interpreted as the quantum mechanical observable associated with the rotation frequency $\theta/2\pi$ around hole 2. After some algebra this is seen to reduce (1e) to standard forms of the Kubo formula (with a ϕ average). It is interesting that this version is free from the uncontrolled interchange of the $V\to 0$, $t\to\infty$ limits, or the adiabatic switching which makes the derivation of the dissipative part of Kubo formulas mathematically formal.

We conclude with a proof that (1b) is an integer when integrated on any two-dimensional manifold without a boundary (in our case a 2-torus).

Take $|\tilde{\Omega}(\phi)\rangle$ as our fixed basis and $|\tilde{\Psi}_1(\phi)\rangle$ satisfying the adiabatic dynamics along the loop 1. $|\tilde{\Omega}\rangle$ and $|\tilde{\Psi}_1\rangle$ are related by the holonomy[14] (Berry's) phase: $|\tilde{\Psi}_1\rangle = (\exp i\gamma_1)|\tilde{\Omega}(\phi)\rangle$. Now if 1 is the boundary of S, $\gamma_1 = i\int_1 \langle\tilde{\Omega}|d\tilde{\Omega}\rangle = i\int_S \mathrm{Tr}(d\tilde{P}\tilde{P}d\tilde{P})$.

On a closed orientable manifold such a 1 is in fact the boundary of both $S_{\rm out}$ and $S_{\rm in}$. Therefore, from the uniqueness $|\tilde{\Psi}\rangle$:

$$\left(\int_{S_{\rm out}} - \int_{S_{\rm in}}\right)\mathrm{Tr}(d\tilde{P}\tilde{P}d\tilde{P}) = 2\pi(\text{integer}). \qquad (11)$$

The integrand is smooth. Hence in the limit that $S_{\rm in}$ shrinks to a point and $S_{\rm out}$ is the entire manifold we obtain the requisite result $\exp i\int_T \mathrm{Tr}(d\tilde{P}\tilde{P}d\tilde{P}) = 1$.

In conclusion, we have shown that Laughlin's argument and its extension to fractions by Tao and Wu can be understood in terms of the first Chern class of the line bundle over the torus associated to the ground state of $\tilde{H}(\phi)$.

We thank H. Cycon, C. Erdmann, M. Klein, A. Raphaelian, and in particular Boris Shapiro, Barry Simon, and Rainer Wüst for useful discussions. One of us (J.A.) thanks the Technische Universität Berlin for its kind hospitality.

---

[1]R. B. Laughlin, Phys. Rev. B 23, 5632–5633 (1981).

[2]J. W. Milnor and J. D. Stasheff, *Characteristic Classes,* Annals of Mathematics Studies, Vol. 76 (Princeton Univ. Press, Princeton, 1974).

[3]D. J. Thouless, M. Kohmoto, M. P. Nightingale, and M. den Nijs, Phys. Rev. Lett. 49, 405–408 (1982).

[4]J. Bellissard, private communication. This most elegant version makes use of K theory.

[5]Q. Niu and D. J. Thouless, J. Phys. A 17, 2453–2462 (1984).

[6]D. J. Thouless, Phys. Rev. B 27, 6083–6087 (1983).

[7]R. Tao and Y. S. Wu, Phys. Rev. B 30, 1097–1098 (1984).

[8]R. Tao, to be published.

[9]J. von Neumann and E. P. Wigner, Phys. Z 30, 467–470 (1929).

[10]In the nongeneric case, where bands can cross, the ideas developed here together with a generalization of the adiabatic theorem can be used to show that the conductance is a fraction with a denominator that is a divisor of the number of bands that cross. This will be discussed in greater detail elsewhere: J. E. Avron, R. Seiler, and B. Simon, to be published; also see H. L. Stormer, A. Chang, D. C. Tsui, J. C. M. Hwang, and W. Wiegmann, Phys. Rev. Lett. 50, 1953 (1983); R. Laughlin, Phys. Rev. Lett. 50, 1395 (1983); F. D. M. Haldane, Phys. Rev. Lett. 51, 1142 (1983); B. Halperin, Phys. Rev. Lett. 52, 1583 (1984).

[11]J. E. Avron, R. Seiler, and B. Simon, Phys. Rev. Lett. 51, 51–53 (1983).

[12]T. Kato, J. Phys. Soc. Jpn. 5, 435–439 (1950).

[13]Ψ can be expanded in powers of V (adiabatic expansion). The zero-order term is up to a phase $\Psi^{\rm ad}$; to get σ_{21} one has to put the first-order term $\hat{R}^2\partial_1 H\Psi^{\rm ad}$ into the formula for the Hall current I_2. This gives σ_{21} in the form of a Kubo formula.

[14]M. V. Berry, Proc. Roy. Soc. London, Ser. A 392, 45–57 (1984).

[15]B. Simon, Phys. Rev. Lett. 51, 2167–2170 (1983).

Adiabatic quantum transport in multiply connected systems

J. E. Avron

Department of Physics, Technion, Haifa 32000, Israel
and Division of Physics, Mathematics, and Astronomy, California Institute of Technology, Pasadena, California 91125

A. Raveh and B. Zur

Department of Physics, Technion, Haifa 32000, Israel

The adiabatic quantum transport in multiply connected systems is examined. The systems considered have several holes, usually three or more, threaded by independent flux tubes, the transport properties of which are described by matrix-valued functions of the fluxes. The main theme is the differential-geometric interpretation of Kubo's formulas as curvatures. Because of this interpretation, and because flux space can be identified with the multitorus, the adiabatic conductances have topological significance, related to the first Chern character. In particular, they have quantized averages. The authors describe various classes of quantum Hamiltonians that describe multiply connected systems and investigate their basic properties. They concentrate on models that reduce to the study of finite-dimensional matrices. In particular, the reduction of the "free-electron" Schrödinger operator, on a network of thin wires, to a matrix problem is described in detail. The authors define "loop currents" and investigate their properties and their dependence on the choice of flux tubes. They introduce a method of topological classification of networks according to their transport. This leads to the analysis of level crossings and to the association of "charges" with crossing points. Networks made with three equilateral triangles are investigated and classified, both numerically and analytically. Many of these networks turn out to have nontrivial topological transport properties for both the free-electron and the tight-binding models. The authors conclude with some open problems and questions.

CONTENTS

LIST OF SYMBOLS

$\mathbf{A}$	vector potential
$\mathbf{A}(x,e)$	vector potential associated with an edge, Eq. (4.8)
C	closed two-surfaces, the set of cells of a graph
$\mathbb{C}$	the complex numbers
C_j	cuts in Ω
$D(P)$	the singularity set for the projection P
$D(a)$	the U(1) incidence matrix of a graph, Eq. (3.9)
E	the set of edges in a graph, energies
$E_j(\mathbf{A})$	the jth eigenvalue of $H(\mathbf{A})$
F	the faces of a graph, the vector in Eq. (4.36)
$G(V)$	point group acting on vertices of a graph
$G(E)$	point group acting on edges of a graph
$G(F)$	point group acting on faces of a graph
$H(\mathbf{A})$	Hamiltonian associated with the gauge field $\mathbf{A}$
H_{ad}	adiabatic Hamiltonian, Eq. (6.1)
$H_2(S)$	the homology groups of S
I	inversion, Eq. (9.5)
$\lvert \mathbf{1}\rangle$	the vector $(1,1,1,\ldots,1)$
$I(m,\psi)$	current around the mth hole at the state ψ, Eq. (5.4)
$\langle I(m,\psi)\rangle$	flux-averaged current around the mth hole at the state ψ, Eq. (5.5)
$I(e),I(f)$	edge and loop currents
$I(\psi)$	current one-form at the state ψ, Eq. (5.6a)
$\langle I(\psi)\rangle$	flux-averaged current one-form at the state ψ, Eq. (5.6b)
$I_{\text{Ad}}(m,\psi)$	adiabatic current around the mth hole at the state ψ
L_{ij}	the inductance matrix of a network
$M(k)$	metric in $\mathbb{C}^{\lvert V\rvert}$, Eq. (4.40), induced from that in the Hilbert space
P	projector on a spectral subspace, Eq. (7.1)
P_j	projector on a spectrum below the jth gap
Q	$1-P$
$\mathbb{R}$	the reals
$S(k,v)$	scattering matrix for vertex v and wave number k
S_ϕ	a small two-sphere centered at ϕ
$\mathbb{T}^h$	the h-dimensional torus
$T(k)$	a linear map from $\mathbb{C}^2$ to the Hilbert space,

	Eq. (11.1)
$T_m(\Phi)$	a line in the torus passing through Φ
$T_{lm}(\Phi)$	a planar slice of the torus passing through Φ
U	unitary operator
U_{Ad}	the adiabatic evolution operator, Eq. (6.2)
$U(n)$	the space of unitary $n\times n$ matrices
V	the set of vertices in a graph
$\mathbf{V}, V_t, V_\Phi$	scalar potentials, Eq. (2.5)
$W(x)$	eigenvalue function of fast variables, (4.5)
$\mathbf{W}(x)$	renormalized eigenvalue function of fast variables
$\mathbf{Z}$	the integers
$\mathbf{a}(x,e)$	indefinite integral of vector potential associated with the edge e
$\mathbf{a}(e)$	definite integral of vector potential associated with the edge e
c	a cell in a graph
$\mathrm{ch}(P,C)$	first Chern number, Eq. (8.7)
d	exterior derivative with respect to Φ
$\mathbf{d}$	exterior derivative with respect to x
e	an edge in a graph, the charge of the electron
$\|e\|$	the length of an edge
f	face of a graph
$\|f\|$	the number of edges of f
g, g_{Ad}	the physical and adiabatic conductance
$g_{\mathrm{Ad}}(P)$	the adiabatic conductance at the state P
$[g_{\mathrm{Ad}}(P)]_{jm}$	the jm coefficient of the adiabatic conductance matrix at the state P
$\langle g_{\mathrm{Ad}}(P)\rangle$	the average conductance at the state P
$h(\Phi)$	a 2×2 Hermitian matrix function, Eq. (7.2)
$h(k,a)$	the de Gennes–Alexander matrix, Eq. (4.28)
$h_0(k,a)$	the de Gennes–Alexander matrix for a graph with unit edges, Eq. (4.30)
h	the number of holes
k	wave number
v	a vertex in a graph
$\|v\|$	the valence (coordination number) of a vertex
(v,u)	an edge directed from v to u
$[v,e]$	the vertex-edge incidence matrix, Eq. (3.2)
z	complex number
z_3	the cubic root of unity
$z_j^{(2)}$	an element of the basis for the second homology
α,β	complex numbers
Δ	the Laplacian of a graph Eq. (3.5)
∂	the boundary operator
∂_j	$\equiv\partial_{\Phi_j}$
$\hat{\epsilon}$	the unit vector for ϵ
ϵ_0,ϵ	a real four-vector, Eq. (7.2)
η	wave function of the fast variable, Eq. (4.5)
Φ	the vector of fluxes
Φ_j	the jth entry in Φ
ϕ	flux values for level crossings
$\boldsymbol{\phi}$	a three-vector of fluxes where levels cross
γ_j	a closed contour
$\Gamma(A),\Gamma_t$	the adiabatic connection, Eq. (6.3)
κ	a vector in $\mathbb{C}^{\|V\|}$, Eq. (4.18)
$\lambda(v)$	vertex potential, Eq. (4.2)
Λ	real valued functions, gauge transformations
ψ_t	a time-dependent solution of the Schrödinger equation
ψ_{Ad}	a time-dependent solution of the adiabatic evolution equation
$\psi_j(A)$	the jth eigenfunction of $H(A)$
$\psi(v)$	wave function on the vertices
$\psi_{\mathrm{in/out}}$	incoming and outgoing waves toward a vertex, Eq. (4.14)
σ	the triplet of Pauli matrices
(θ,ρ)	cylindrical coordinates
$\theta(k,v)$	the angle associated with unitary point scatterers, Eq. (4.23)
$\omega,\omega[P]$	curvature two-form associated with the projection P, Eq. (6.7)
Ω	a multiply connected domain in coordinate space
χ	characters of groups, Eq. (7.12)
$\|\psi\rangle, \|\varphi\rangle$	a pair of degenerate eigenvectors
$\langle\cdot\|\cdot\rangle$	scalar product of fast variables, Eq. (3.9)
$\langle\cdot,\cdot\rangle$	scalar product in $\mathbb{C}^{\|V\|}$ induced from the Hilbert space
$(\cdot,\cdot)$	the ordinary scalar product in $\mathbb{C}^{\|V\|}$
$\|\cdot\|$	cardinality
$\#(i,j)$	number of common vertices of triangles i and j
$\otimes$	magnetic flux going into the plane
$\odot$	magnetic flux coming out of the plane

I. INTRODUCTION

In this work we study adiabatic transport in multiply connected systems. A typical system might be a mesoscopic piece of normal metal with several holes, shown schematically in Fig. 1(a), or the network shown in Fig. 1(b), which is made of thin, mesoscopic, normal-metal wires. Yet another setting is the array of Josephson junctions. Figure 1(b) is a degenerate version of Fig. 1(a) having the same topology.[1] In either case, the holes are threaded by flux tubes, carrying independent fluxes, which serve as drives and controls. The electrons (Cooper pairs) do not feel the magnetic fields associated with the fluxes; they feel only the vector potentials. This generalizes the setting in the original Bohm-Aharonov effect (Aharonov and Bohm, 1959, 1961) in two ways. First, we allow several, possibly many, holes. Second, we allow for time-dependent fluxes. In the Bohm-Aharonov effect there is one hole and the flux is fixed. The insistence on several holes turns out to be important in that the theory trivializes in the one-hole and two-hole situations. So,

[1]The two figures are of the same homotopy type.

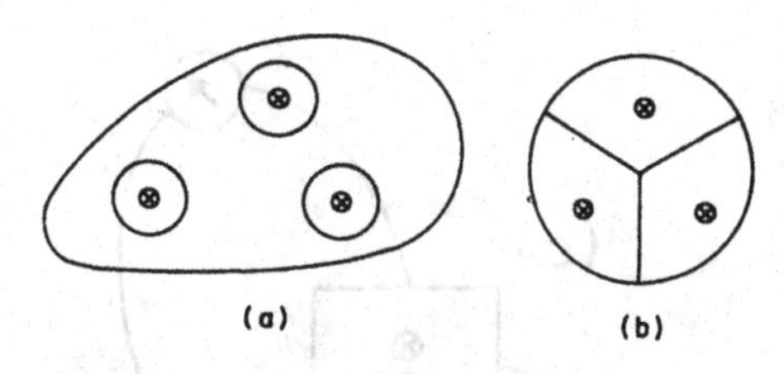

FIG. 1. A multiply connected domain Ω in $\mathbf{R}^2$: (a) with three-holes threaded by three flux tubes $\Phi_{1,2,3}$; (b) a degenerate version of (a) made of thin connecting wires.

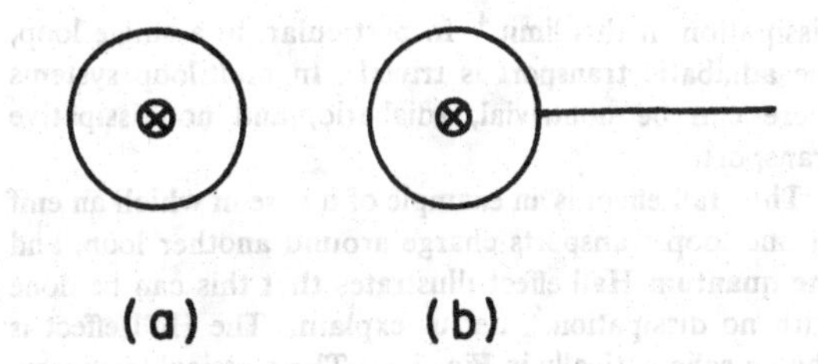

FIG. 2. Multiply connected domains with a single hole: (a) a ring; (b) a lasso.

for the most part, we shall consider systems with at least three holes. Time-dependent fluxes are used to drive the system by generating electromotive forces (emf's) around the respective holes.

By a general result (Byers and Yang, 1961), observables, in the time-independent case, depend on the flux periodically with period of one flux quantum. In emu and atomic units the flux quantum unit is 2π. Due to this periodicity, it turns out that for certain purposes the flux space may be identified with the multidimensional torus. This may be viewed as the basic reason for many of the interesting features of the Bohm-Aharonov effect and its multiflux generalizations. It turns out that, for certain purposes, flux space can also be viewed as a multitorus in time-dependent situations where the time dependence is adiabatic. This, more than the topology of the system in coordinate space, is the basic reason for the topological aspects of the adiabatic transport that we discuss.

As in the Bohm-Aharonov effect, the phenomena we study are a result of quantum coherence and so require that the wave function be coherent over the sample and "know" about the holes. Such rigidity is present in superconductors over macroscopic lengths and in normal metals under more stringent conditions on length and temperature scales. We shall return later to the setting of superconductivity. Let us first briefly review the normal-metal situation.

When the length scale of the system is mesoscopic (i.e., a few hundred angstroms) and the temperature is in the sub-Kelvin range, the electronic wave function is coherent over the entire system and quantum effects are important. Quantum conductance in mesoscopic systems at low temperatures is a rapidly developing subject, partly because of its obvious technological significance and partly because of the interest in quantum coherence. The reader may wish to consult the review of Imry (1986) on this burgeoning subject. A considerable amount of work has been devoted to multiply connected systems with a single hole, such as the ring and the lasso in Fig. 2 (see, for example, Umbach *et al.*, 1984). Periodicity of the *dissipative conductance* in the flux threading the hole, with the period 2π, has been observed in the thin rings (Chandrasekhar *et al.*, 1985; Webb *et al.*, 1985). In thick, dirty rings the period is halved to π (Alt'shuler *et al.*, 1981; Sharvin and Sharvin, 1981; Alt'shuler, 1985). It is also expected that rings have persistent currents that depend (periodically) on the fluxes and that flow even in the absence of driving emf's (Büttiker *et al.*, 1983). Another quantum aspect is nonlocality. For example, the (dissipative) conductance of the tail of the lasso in Fig. 2(b) depends (periodically) on the flux threading the ring. The conductances in mesoscopic systems can be sample specific (Gefen *et al.* 1984; Büttiker *et al.*, 1985), and Onsager relations can be subtle (Benoit *et al.*, 1986; Büttiker, 1986b). Gefen and Thouless (1987) looked at mechanisms that lead to dissipation in small rings and, in particular, at the role of quantum interference and localization.

When the geometry becomes more complicated, as in the case of multihole systems, the various issues discussed above become issues in the more complicated settings. However, new issues, that have no analog in the one-loop setting, also arise. Nondissipative quantum conductances that have topological significance is one. This is the subject of this work.[2]

In multiply connected systems with several holes, a matrix of transport coefficients relates the current around one hole to the emf around another. Although somewhat pedantic, it is useful to distinguish between charge transport and conductance: conductances are defined as the linear coefficients that relate currents to emf's in situations where the emf's are (asymptotically, in the distant future) constants or harmonic functions of time. Charge transports relate the charges transported around holes to the increase in some of the fluxes by a single quantum, where the fluxes are asymptotically (both in the distant past and distant future) time independent.[3] The adiabatic transport coefficients are defined by the limit where the fluxes change adiabatically (the time scale is determined by the minimal gap in the spectrum), so the emf's are all weak. By general principles, *for finite systems*, there is no

[2]In Yurke and Denker (1984), quantum network means the study of capacitors and inductors quantized by imposing the canonical commutation relations. This is a different problem.

[3]Due to persistent currents, the transport we consider is actually associated with the *excess* charge transported by the increase in flux.

dissipation in this limit.[4] In particular, in a single loop, the adiabatic transport is trivial. In multiloop systems there can be nontrivial, adiabatic, and nondissipative transport.

The Hall effect is an example of a case in which an emf in one loop transports charge around another loop, and the quantum Hall effect illustrates that this can be done with no dissipation.[5] Let us explain. The Hall effect is shown schematically in Fig. 3(a). The classical (ordinary) Hall effect (Hall, 1879) is basically a room-temperature phenomenon, and the (integer) quantum Hall effect (von Klitzing *et al.*, 1980; see also Prange and Girvin, 1987 for a collection of review articles) is a low-temperature phenomenon. Quantum mechanics turns out to be important already in the ordinary Hall effect, even though it is observable in macroscopic systems at elevated temperatures. Indeed, there is no classical explanation of the anomalous (holelike) Hall coefficients that occur in certain materials. The classical Hall transport is accompanied by some dissipation. In the quantum Hall effect, the transport is nondissipative (at least in the region of the plateaus).

The quantum Hall effect may be viewed as a quantum phenomenon associated with multiply connected systems. The multiple connectivity is, in fact, central in some theories.[6] Indeed, the original argument of Laughlin (1981) is a gauge argument applied to a geometry of a single loop. Subsequent theories (see below) that rely on the identification of the Hall conductance with a standard topological object actually need two loops (Avron and Seiler, 1985; Niu and Thouless, 1987). Since the topological view will be central to much of what follows, we recall the motivation for the two loops. It is presented in Fig. 3(b), where the battery in Fig. 3(a) has been replaced by a time-dependent flux tube, and the ammeter in Fig. 3(a) by a second, independent, flux tube. (We shall explain later precisely in what sense flux tubes play the role of batteries and ammeters.) This structure leads to the identification of the Hall conductance with the first Chern number. (We shall discuss below the differential geometric and topological significance of the first Chern numbers.)

The study of nondissipative conductances in networks is closely related to the topological-geometric view of the quantum Hall effect. In fact, we have chosen to study networks partly because it is a setting that is tailored to the theory and does not have some of the difficulties that the Hall effect presents. In the Hall effect, the multiconnectivity is not believed to be an essential feature of the actual systems; the wave function is probably not coherent in the leads and the electronic circuitry. In networks the wave function is assumed to be coherent throughout.

Networks differ from the Hall effect not only in setting but also in some of the basic properties of the transport. The Hall conductance is actually *a property of the sample*, and the connecting leads in Fig. 3 are not believed to be important for the actual value of the Hall conductance. (This is believed to be related to the fact that the samples are macroscopic.) In networks the conductance matrix

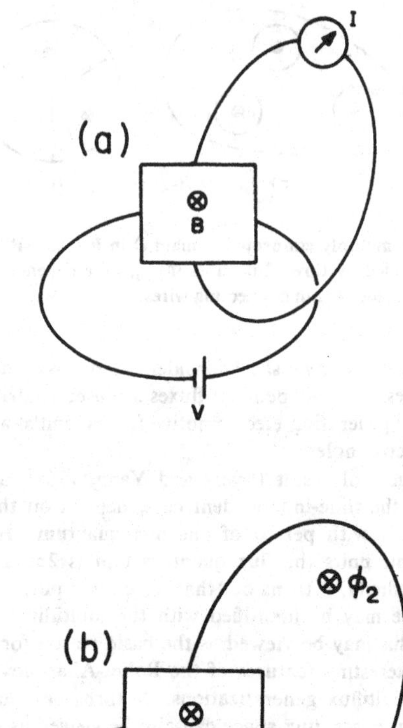

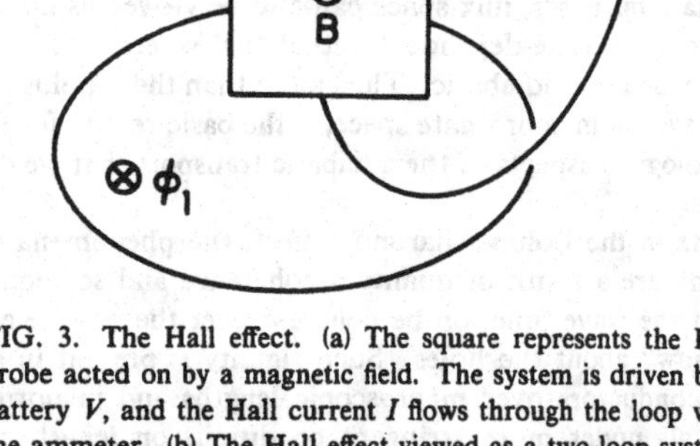

FIG. 3. The Hall effect. (a) The square represents the Hall probe acted on by a magnetic field. The system is driven by a battery V, and the Hall current I flows through the loop with the ammeter. (b) The Hall effect viewed as a two-loop system. The battery is replaced by a time-dependent flux tube, and the ampermeter by a time-independent flux tube.

[4]The convention we follow here is that the adiabatic limit is *not defined* when eigenvalues cross.

[5]Classical networks are another example in which emf's in one-loop transport charges around another loop. This phenomenon is, however, intrinsically related to dissipation: the associated transport matrix is symmetric. The nondissipative transport that we study below has an antisymmetric transport matrix. We thank S. Ruschin for a discussion on this point.

[6]The Born–von Karman periodic boundary conditions that are common in much of solid-state physics may also be viewed as a way of effectively making the system multiply connected.

reflects the multiconnectivity. In a sense, networks are like a quantum Hall effect without a magnetic field acting directly on the system and without a Hall probe.

There is actually a further and deeper connection with the integer quantum Hall effect. One of the interesting theoretical developments that followed the experimental work of von Klitzing has been the recognition that nondissipative Hall coefficients have geometric significance. The geometry that enters is related to the way (a family of) spectral subspaces curve inside the (infinite-dimensional, flat) Hilbert space. Kubo's formula for the Hall conductance is an expression for this curvature. The curvature is related to the Berry or holonomy phase (Simon, 1983; Berry, 1984). This fact[7] was noted by Thouless, Kohmoto, Nightingale, and den Nijs (1982) in a special case, and was later recognized to hold in great generality (Niu and Thouless, 1984; Avron and Seiler, 1985; Tao and Haldane, 1986). In particular, as we shall see, the geometric interpretation is not special to the Hall conductance and holds also for networks.

The classical Chern-Gauss-Bonnet-type formulas say that integrals of curvatures over closed surfaces are integers. The integers associated with integrals of the aforementioned curvature over closed two-dimensional manifolds are known as first Chern numbers (Chern, 1979; Choquet-Bruhat *et al.*, 1982). In the topological-geometrical view, the quantization of the Hall conductances observed by von Klitzing can be interpreted as a combination of the basic facts that nondissipative transports are curvatures and that flux space is a torus with Gauss-Bonnet-Chern theorem (Niu and Thouless, 1984; Avron and Seiler, 1985; Kohmoto, 1985; Niu, Thouless, and Wu, 1985; Avron, Seiler, and Shapiro, 1986; Tao and Haldane, 1986; Avron, Seiler, and Yaffe, 1987; Niu and Thouless, 1987). The precise statement is, in fact, that for general multiparticle Hamiltonians describing *finite systems*, the (adiabatic) Hall conductance, averaged over the threading flux in the current loop *at zero temperature*, is generically quantized to be integer multiples of $e^2/2\pi\hbar$ ($=1/2\pi$, in the present units). [Why the Hall experiments measure averages, and why the quantization is stable for finite temperatures, finite emf's, and macroscopic systems are questions outside the scope of this work and not yet fully understood in a general framework (Laughlin, 1981; Niu and Thouless, 1984; Avron, Seiler, and Shapiro, 1986; Kunz, 1987; Niu and Thouless, 1987).]

Multiply connected systems with many holes offer a rich setting, as there are many conductances related to curvatures. Such conductances have quantized averages. The averaging, even when the number of fluxes is large, is

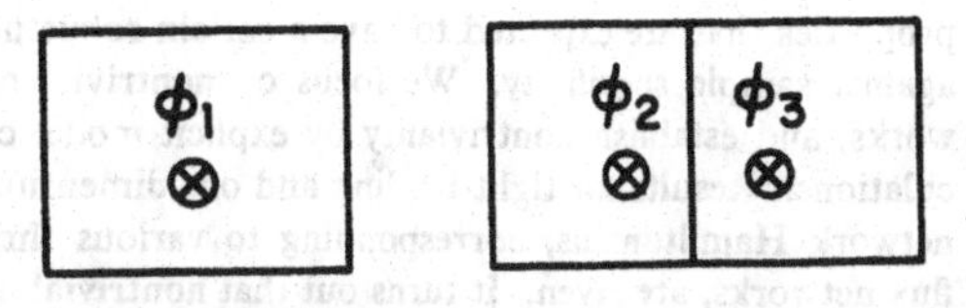

FIG. 4. A disconnected network with three loops that has trivial topological conductances.

only over the *single* flux, distinguished by the current loop. The integers obtained are therefore functions of the remaining fluxes and are periodic with period of a flux unit. (The setting is also richer because it allows, in principle, for higher Chern numbers.)

As in the theory of the Berry phase, degeneracies (points of level crossings) play the role of sources for the Chern numbers and can be assigned integer "charges" (Herzberg and Longuet-Higgins, 1963; Longuet-Higgins, 1975; Alden Mead and Truhlar, 1979; Simon, 1983; Berry, 1984). Flux space, with points of charges removed, has an interesting second homology, which, together with a basic curvature (two-form), determines the Chern numbers. The case in which the charges are discrete is particularly simple, for then the information can be organized in a table. This is the generic situation for three-flux networks.

The adiabatic transport coefficients relate three distinct topological spaces: the physical network in three-dimensional space; the multidimensional punctured torus in flux space; and, finally, the bundle of spectral subspaces in the Hilbert space. The Chern numbers describe the twisting of these bundles and are related to the geometry of the network. For example, in a disconnected network, like Fig. 4, the transport coefficients associated with loops in distinct components are naturally expected to vanish. An outstanding problem is a deeper understanding of the ways the three topological spaces are related. It is interesting to recall that formulas that relate electric properties and topological properties of networks in ordinary three space have a distinguished predecessor, Ampere's law is related to the linking number and plays a role in knot theory (Flanders, 1963).

We say that a network is trivial if *all* its Chern numbers vanish (or are ill defined because of nongeneric crossings). We shall see that networks with one or two fluxes are trivial in this sense. This is not to say that the transport properties of one- and two-loop networks are trivial. In a trivial network there can still be a current flowing in one loop due to an emf in another. The (adiabatic) current is trivial only in the sense that its average over the flux in the current loop vanishes and that it lacks the topological significance of Chern character.[8] Nontriviality guarantees interesting *topological* transport

[7]The nontriviality of the bundle that arises in the study of Schrödinger operators with periodic structure and magnetic fields was first noted by several Soviet authors as early as 1980 (see Dubrovin and Novikov, 1980; Novikov, 1981; Lyskova, 1985, and references therein).

[8]In the case of nongeneric crossings, the adiabatic limit may become empty.

properties that are expected to have a certain robustness against sample specificity. We focus on nontrivial networks, and establish nontriviality by explicit model calculations. Results for tight-binding and one-dimensional network Hamiltonians, corresponding to various three-flux networks, are given. It turns out that nontrivial networks abound, and as a rule of thumb networks are nontrivial except for a reason. Because of the topological nature of the problem, at least the qualitative features of the results, and in particular the nontriviality, should survive for more realistic Hamiltonians.

The nontrivial three-flux networks that we shall describe below are a three-way quantum switch. By this we mean that depending on the value of the controlling flux in loop 1, say, the average current in loop 2, due to an emf in loop 3, will either flow, not flow, or flow in reverse. The switch is periodic in the controlling flux with period 2π. It is an honest three-state switch in the sense that the average current in loop 2 is a 1, 0, or -1 multiple of the emf. The switch is stable in the sense that each state is determined by an interval in the controlling flux. In fact, the controlling flux has to be varied on the order of a (fraction of *a*) flux unit to alter the state of the switch. Multistate switches, such as the five- and seven-state switches, where the currents are multiples of the emf's that are larger than unity, presumably arise in the study of multiloop networks. The nine networks that we have analyzed have only a few loops, and four of them turn out to be three-way switches for most states. Some of the networks turn out to be two-way switches in some states, that is, the averaged current can be made to change direction when the flux is reversed, but cannot be stopped. The other five networks are trivial.

II. SCHRÖDINGER OPERATORS FOR MULTIPLY CONNECTED DOMAINS

Consider a multiply connected domain Ω [see Fig. 1(a)] embedded in three- or two-dimensional Euclidean space. Suppose that Ω is a finite, smooth manifold and has smooth boundary $\partial\Omega$. Ω is threaded by h independent flux tubes. Φ_j is the flux in the jth tube, and Φ is the vector $(\Phi_1,\ldots,\Phi_h)$. Let $\mathbf{A}$ be the associated vector potential. We think of $\mathbf{A}$ as a one-form, or a vector field, as is convenient. $\mathbf{A}$ is closed on Ω, that is, the associated magnetic field $\mathbf{d}\,\mathbf{A}$ vanishes on Ω, expressing the fact that all the fluxes are outside Ω. $\mathbf{d}$ is the exterior derivative, and we use boldface to denote that the differentials are taken with respect to the coordinates in configuration space. We allow time-dependent fluxes so $\mathbf{A}$ may generate an electric field $-\partial_t\mathbf{A}$. The electric field need not vanish on Ω, since the emf's around the holes are $-\dot{\Phi}$. Φ are the periods of $\mathbf{A}$, that is,

$$\Phi_j=\int_{\gamma_j}\mathbf{A}\,, \tag{2.1}$$

where γ_j is a loop in Ω around the jth flux tube. Unless otherwise stated we shall always assume that a gauge has been chosen so that the fluxes are represented by a pure vector potential, that is, there is no Φ-dependent scalar potential. Also, it is convenient to choose the vector potential so that $\Phi=0$ corresponds to $\mathbf{A}=0$. Finally, the situations we have in mind are those in which the time dependence resides in the fluxes alone: we do not consider cases in which the flux tubes are, say, jiggled inside the holes. This means that we take $\mathbf{A}$ to be linear in Φ.

Given $\mathbf{A}$, we associate with it a Schrödinger operator $H(\mathbf{A})$ for electrons moving in Ω. In the single-particle case, with no background fields, the operator is (up to a factor $\frac{1}{2}$)

$$H(\mathbf{A})=\tfrac{1}{2}(-i\nabla-\mathbf{A})^2\,, \tag{2.2}$$

where $(-i\nabla-\mathbf{A})$ is the canonical velocity. Dirichlet boundary conditions are imposed on $\partial\Omega$. The general multiparticle case is more complicated only in that the operator is decorated by particle indices and interaction terms *that are* Φ *independent*. These complications do not affect the analysis below in any essential way, except for a messier notation. We stick with Eq. (2.2) for the sake of clarity.

$H(\mathbf{A})$ determines the dynamics according to the Schrödinger equation,

$$i\partial_t\psi_t=H(\mathbf{A})\psi_t\,. \tag{2.3}$$

In much of the following we shall be interested in the limit where the flux tubes, and therefore $\mathbf{A}$, change adiabatically. As is well known, the analysis of this limit reduces to the spectral study of the family of operators $H(\mathbf{A})$, i.e., time may be regarded as a parameter. We denote by $E_j(\mathbf{A})$ and $\psi_j(\mathbf{A})$ the eigenvalues and eigenfunctions of $H(\mathbf{A})$ for fixed fluxes, i.e.,

$$H(\mathbf{A})\psi_j(\mathbf{A})=E_j(\mathbf{A})\psi_j(\mathbf{A})\,. \tag{2.4}$$

By general principles (Kato, 1966; Reed and Simon, 1972–1978), $H(\mathbf{A})$ has a discrete spectrum in $[0,\infty)$, something borne out by the notation. Equation (2.4) does not determine $\psi_j(\mathbf{A})$ uniquely, as there is a phase ambiguity. A convenient choice will be singled out when needed.

Gauge transformations play an important, and occasionally confusing, role. A general gauge transformation may depend on the fluxes, and so is time dependent if the fluxes are, and may have an additional time dependence not coming from the fluxes. Let U be a multiplication unitary, that is, locally, $U=\exp(-i\Lambda)$, which is smooth in Φ and t and $x\in\Omega$ (x may be a local coordinate). The primed and unprimed systems, related (locally) by

$$\psi'=U\psi\,,\quad \mathbf{A}'=\mathbf{A}-\mathbf{d}\Lambda\,,$$
$$H'(\mathbf{A}')=UH(\mathbf{A})U^{\dagger}+V_t+\dot{\Phi}\cdot V_{\Phi}\,, \tag{2.5}$$
$$V_t\equiv -iU\partial_tU^{\dagger}=-\partial_t\Lambda,\quad V_{\Phi}\equiv -iU\,dU^{\dagger}\,,$$

have the same electric and magnetic fields acting *on* Ω and describe equivalent dynamics. Nonboldface d denotes the exterior derivative with respect to the fluxes.

$\dot{\Phi}\cdot V_{\Phi}$ is the canonical pairing between the vector $\dot{\Phi}$, describing the flow in flux space, and the one-form V_{Φ}.

$d\Lambda$ has periods that are integral multiples of 2π:

$$\int_{\gamma_j} d\Lambda = 2\pi n_j, \quad n_j \in \mathbf{Z}, \tag{2.6}$$

where, as before, γ_j is a loop in Ω around the jth hole. Such U's do not have a smooth continuation to the holes if the periods are nonzero, and this is why the fields in the holes can be different. Indeed, from Eqs. (2.1) and (2.5),

$$\Phi' = \Phi - 2\pi n, \tag{2.7}$$

where n is a vector with integer components n_j. It follows from Eq. (2.5) that if the fluxes are *time independent*, Hamiltonians with distinct fluxes that are related by Eq. (2.7) are unitarily equivalent and describe equivalent dynamics. So, for time-independent questions, flux space may be thought of as a multitorus. We shall refer to this as the Bohm-Aharonov periodicity. It is also known as the Byers-Yang theorem (Byers and Yang, 1961). As we shall see, this will also carry over to some time-dependent questions in the adiabatic limit.

The next question we want to consider is the relation between dynamics that have the same Ω, same fluxes, and same emf's, but different flux tubes. For example, consider two systems, identical except for the fact that some of the flux tubes have been moved about in their respective holes (see Fig. 5). In the time-dependent case, such systems have inequivalent dynamics. This is easy to understand in physical terms: different flux tubes, or flux tubes that are positioned in different places in the holes, have different A's acting on Ω, and induce different electric fields $-\partial_t \mathbf{A}$ on Ω. These electric fields are related only through the same emf's on each loop, but their local behavior is different. There is no reason for the dynamics in such cases to be equivalent. The corresponding Hamiltonians are not related by a gauge transformation.

Let us consider some concrete examples of flux tubes that illustrate the different fields and dynamics that can arise. One of the examples is a choice of flux tubes that make the Schrödinger operator periodic in the fluxes. The existence of such a choice will play a role in later sections.

Consider a planar Ω and let $x=(\rho_j,\theta_j)$ be cylindrical coordinates with origin in the jth hole. Let

$$\mathbf{A}(x) = \sum_{j=1}^{h} \Phi_j d(\theta_j)2\pi \tag{2.8}$$

(away from $\theta=0$ and $\theta=2\pi$). This $\mathbf{A}$ is manifestly closed on Ω (it is locally exact) and satisfies Eq. (2.1). Suppose now that Φ is time dependent. Then there is a nonvanishing electric field acting on Ω given by

$$-\sum_{j=1}^{h} \dot{\Phi}_j \frac{\hat{\theta}_j}{2\pi\rho_j}. \tag{2.9}$$

The field clearly depends on the choice of origin of the coordinates, which is the position of the tube in the holes. For example, if Ω is a thin ring, the electric field is uniform only if the tube is placed at the center of the ring. There is no reason for the dynamics to be determined by Φ and $\dot{\Phi}$ alone; further details about the flux tubes matter.

As a second example for a choice of flux tubes or vector potentials, consider the *singular* gauge field

$$\mathbf{A}(x) = \sum_{j=1}^{h} \Phi_j \hat{n}(x)\delta(x-C_j), \tag{2.10}$$

where C_j are cuts in Ω that make it simply connected (see Fig. 6) and $\hat{n}$ is a unit vector orthogonal to the cut. This $\mathbf{A}$ also satisfies Eq. (2.1). For fixed Φ, the Hamiltonians corresponding to various choices for the cuts C_j in Eq. (2.10) [or the tubes in Eq. (2.8)] are unitarily equivalent. However, if the Φ are time dependent, the dynamics are distinct. In fact, the electric field associated with Eq. (2.10) is zero everywhere except on the cuts and is given by

$$-\sum_{j=1}^{h} \dot{\Phi}_j \hat{n}(x)\delta(x-C_j). \tag{2.11}$$

The choice of C_j clearly matters.

The Schrödinger operator associated with the flux tubes of Eq. (2.10) is $-\Delta$ on the cut domain, Fig. 6, with $\exp(i\Phi_j)$ boundary conditions across the cut C_j. The differential operator is thus independent of Φ, and the Φ dependence comes solely from the boundary condition, which is manifestly periodic in Φ. The nice thing about this choice of flux tubes is that they give a Hamiltonian that is manifestly periodic in Φ as well. For this choice, the time evolution is as if flux space were the h torus $\mathbf{T}^h$. And this holds even for time-dependent fluxes. For other choices of the flux tubes, say Eq. (2.8), this is not the case,

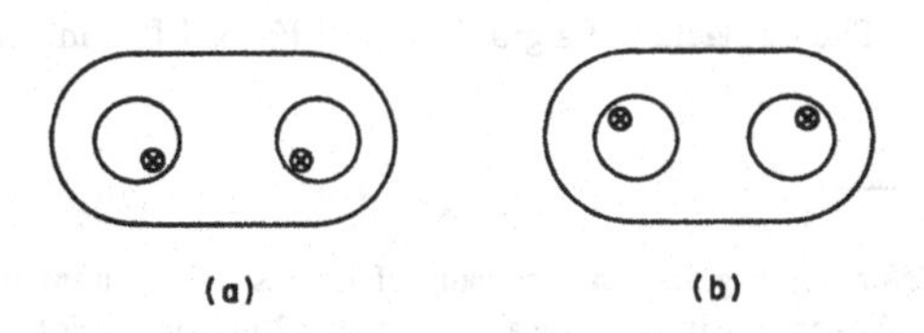

FIG. 5. Two identical systems, except that the flux tubes have been placed in different positions in the holes.

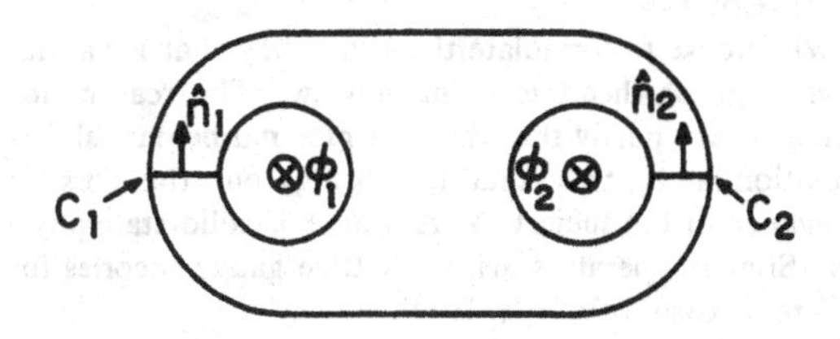

FIG. 6. Ω made simply connected by drawing appropriate cuts C_j. $\hat{n}$ is normal to the cut.

and flux space is better thought of as $\mathbb{R}^h$.

It is, of course, natural to ask what dynamical properties are independent of the choice of flux tubes and are functions of Φ and $-\dot{\Phi}$ alone. A related question is under what condition does the Bohm-Aharanov periodicity extend to time-dependent situations. It is, of course, reasonable to expect that in the adiabatic situation, some form of the Bohm-Aharonov periodicity should survive. However, it may survive for certain observables but not for others. Note that once certain properties are known to be functions of the fluxes and the emf's alone, one may choose the flux tubes to be those that make the Hamiltonian periodic in the fluxes. We shall use this to establish the Bohm-Aharonov periodicity of the averaged adiabatic transport coefficients.

A reader interested in the general structure and theory may want, at this point, to skip the next two sections and proceed with Sec. V on loop currents. The following two sections are devoted to the formulation of special classes of Hamiltonians for which the analysis reduces to the study of finite matrices. This simplification has no bearing on the general theory of adiabatic transport that we describe in Secs. V–XI, but it is of considerable use in the actual computation of the transport properties for specific networks, something we return to in Secs. XII and XIII.

III. TIGHT-BINDING HAMILTONIANS

From formal point of view, the Hamiltonians of the preceding section were distinguished by periodic dependence (up to unitary equivalence) on a set of parameters, i.e., fluxes. The simplest operators of this kind are, of course, periodic matrix functions. Finite matrices are especially useful when one is interested in concrete examples and in actual computations and not just in the general structure of the theory. Operators of the tight-binding type retain much of the structure of the original Schrödinger operator and lead to the study of finite matrices.

The tight-binding model arises from the Schrödinger equation in the limit of strongly attractive atomic potentials, hence the name. For a system made of N atoms, each contributing n atomic levels, the relevant Hilbert space, in the one-particle case, is $\mathbb{C}^{nN}$. The tight-binding Hamiltonian is the restriction of the appropriate Schrödinger operator to this subspace. We shall consider the case of $n=1$.

We choose to formulate them in a way that is natural from a graph-theoretic point of view. The reasons for doing so are partly that this is a nice mathematical formulation and partly that this brings out the possible relevance of the subject to areas outside solid-state physics. Similar operators arise in lattice gauge theories for different reasons (Wilson, 1974).

With the network Ω we associate a directed graph, Fig. 7, with vertices V, edges E, faces F, the cells C. There are, of course, no cells in planar graphs. A cautionary word about terminology in warranted. We used the term graph in a more restrictive sense than is usual in graph theory, as the graphs we consider have metric properties. For example, in Secs. XII and XIII we focus on graphs made of equilateral triangles. In graph theory edges have no lengths, so there is no notion of equilateral triangles. Consequently the notion of planarity is also distinct from that in (nonmetric) graph theory.

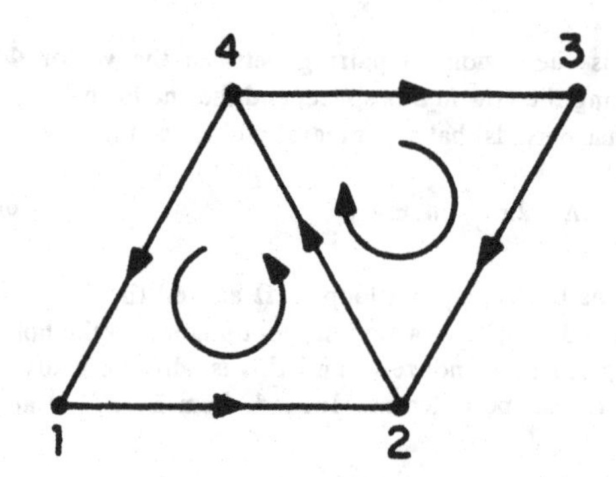

FIG. 7. A directed graph with four vertices, five edges, and two faces. The valence of vertex 2 is 3 and of vertex 1 is 2.

We denote by a lower-case letter an element of a set denoted by an upper-case letter, so v is a vertex in V. Moreover, $|\cdot|$ assigns to $\cdot$ a number. Thus $|V|$ and $|E|$ are the number of vertices and edges, etc. $|e|$ is the length of e and $|v|$ is the valence (coordination number) of the vertex v. We consider only simple graphs in which edges are uniquely specified by their vertices. $e=(v,u)$ is an edge directed from v to u. Figure 8 shows a graph that is not simple.[9]

We consider physical networks, embedded in Euclidean space. For connected networks the vertices, edges, etc., are related by the Euler characteristic,

$$|C|-|F|+|E|-|V|=-1\ . \tag{3.1}$$

The incidence matrix gives an algebraic description of the graph (Wilson, 1972; Biggs, 1974). $[v,e]$ denotes the vertex-edge incidence matrix, defined by

$$[v,(v_1,v_2)]=\delta(v,v_2)-\delta(v,v_1)\ . \tag{3.2}$$

The edge-face $[e,f]$ and face-cell $[f,c]$ incidence matrices are similarly defined. They obey the "boundary of the boundary is zero" rule (Patterson, 1969):

$$\sum_E [v,e][e,f]=0,\quad \sum_F [e,f][f,c]=0\ . \tag{3.3}$$

The Laplacian of a graph is the $|V|\times|V|$ matrix

[9]Nonsimple graphs are actually of interest, since nontrivial bundles appear in this case already for 2×2 matrices. For simple graphs, the lowest rank of matrices with nontrivial bundles is 4.

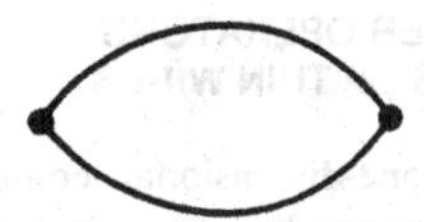

FIG. 8. A graph that is not simple. The two vertices do not determine a single edge.

$$[\Delta]_{uv} \equiv -\sum_{E} [u,e][v,e] . \tag{3.4}$$

For simple graphs,

$$[\Delta]_{uv} = \begin{cases} -|v| & \text{for } u = v , \\ +1 & \text{for } (u,v) \in E , \\ 0 & \text{otherwise} . \end{cases} \tag{3.5}$$

On a graph of a regular lattice, Δ is the discrete Laplacian. In particular, if ψ is a function on the vertices such that $\Delta\psi=0$, then ψ is harmonic in the usual sense that its value at a given vertex is the mean of its neighboring values. It is known that the spectral properties of Δ are related to the topology of the graph (Wilson, 1972; Biggs, 1974).

The operator in Eq. (3.5) describes a tight-binding model for a "molecule" made of atoms placed at the vertices of the graph, in the absence of magnetic fields. There is unit hopping between atoms connected by an edge, and the electrons experience an on-site interaction $|v|$. From a solid-state physics point of view, a simpler and more natural operator to study is one in which the diagonal in Eq. (3.5) is replaced by zero. (This is the case if the atoms at the vertices of the graph are identical.) We stick with the graph-theoretic choice (but the solid-state terminology).

The Laplacian carries no information on the fluxes threading the graph. From a graph-theoretic point of view, fluxes lead to the consideration of a "gauge incidence matrix" that is a U(1) generalization of Eq. (3.2).

Fluxes and gauge fields can be defined intrinsically on the graph. It is not necessary to think of them as embedded in Euclidean space. Fluxes $\Phi(f)$ are defined on the faces, and gauge fields $a(e)$ are defined on the edges. The fluxes are constrained by the zero divergence of the magnetic fields

$$\sum_{F} [f,c]\Phi(f)=0 , \tag{3.6}$$

and are related to the gauge fields by the discrete version of Eq. (2.1),

$$\Phi(f)=\sum_{E} [e,f]\mathbf{a}(e) . \tag{3.7}$$

By Eq. (3.3), such $\Phi(f)$'s automatically satisfy Eq. (3.6). Pure gauge fields are "gradients" of functions on the vertices,

$$\mathbf{a}_0(e)=\sum_{V} [v,e]\Lambda(v) , \tag{3.8}$$

with $\Phi(f)\equiv 0$ by Eq. (3.3). The space of nontrivial pure gauges is $|V|-1$ dimensional [since Λ=const has $\mathbf{a}_0(e)\equiv 0$]. The space of all gauge fields is $|E|$ dimensional and the space of admissible fluxes is $|F|-|C|$ dimensional, due to the constraint of Eq. (3.1). The number of flux tubes, or holes, is therefore $h=|F|-|C|$. From the Euler characteristic, Eq. (3.1), it follows that the flux tubes determine the gauge fields modulo pure gauges.

The gauge incidence matrix $D(\mathbf{a})$ is the linear map from $\mathbb{C}^{|V|}$ to $\mathbb{C}^{|E|}$ defined by

$$[D(\mathbf{a})]_{e,v} \equiv [v,e]\exp\{i[v,e]\mathbf{a}(e)/2\} . \tag{3.9}$$

The tight-binding Hamiltonian is, by analogy with Eq. (3.4),

$$H(\mathbf{a}) \equiv D^{\dagger}(\mathbf{a})D(\mathbf{a}) . \tag{3.10a}$$

Explicitly

$$[H(\mathbf{a})]_{uv} = \begin{cases} |v| & \text{for } u = v , \\ -\exp[i\mathbf{a}(e)] & \text{for } e=(u,v) , \\ 0 & \text{otherwise} . \end{cases} \tag{3.10b}$$

As a tight-binding model, $H(\mathbf{a})$ describes a "molecule" in a magnetic field so that the fluxes through the various faces are given by Φ. The magnetic fields modify the hopping terms to $\exp[i\mathbf{a}(e)]$. The graph-theoretic formulation leads to the on-site potential $|v|$, and, as discussed above, this is not a particularly natural choice in the tight-binding model. In tight binding it is more natural to let the on-site potential be a fixed constant, if all the atoms are identical. In the general case of distinct atoms, the on-site potential depends on the binding energy of the atom. Moreover, in the general case, the hopping need not have identical magnitudes.

We shall stick with Eq. (3.10) for the sake of concreteness. This choice does not affect the analysis, for we focus on stable (topological) properties. In particular, setting the diagonal to zero would not change the overall picture, but would modify details. For regular graphs, that is, graphs in which $|v|$ is the same for all vertices, like the tetrahedron, the two notions essentially coincide.

$H(\mathbf{a})$ is a $|V|\times|V|$ self-adjoint matrix that reduces to $-\Delta$ for $\mathbf{a}(e)=0$. We recall some of its basic properties. First, one has the (sharp) bounds

$$2\max_{v\in V} |v| \geq H(\mathbf{a}) \geq 0 . \tag{3.11}$$

The right-hand side of Eq. (3.11) follows directly from Eq. (3.10a). The left-hand side follows from consideration of the eigenvalue equation for the largest component of the eigenvector. Actually, $H(\mathbf{a})$ is strictly positive definite if $\Phi \neq 2\pi n$, $n\in\mathbb{Z}^h$, and has a one-dimensional kernel if $\Phi=2\pi n$. To see this, note that from

$$\langle\psi|H(\mathbf{a})|\psi\rangle = \|D(\mathbf{a})|\psi\rangle\|^2 \tag{3.12}$$

it follows that the kernel of $H(\mathbf{a})$ is the kernel of $D(\mathbf{a})$. The equation $D(\mathbf{a})\psi=0$ says that

$$\psi(u)=\exp[i\mathbf{a}(e)]\psi(v), \quad e=(u,v) . \tag{3.13}$$

This is consistent on the graph provided the periods of $\mathbf{a}(e)$ on the loops are integer multiples of 2π, that is, provided $\Phi=2\pi n$. In particular, for $\mathbf{a}(e)=0$, the ground state 0 has the eigenvector $(1,1,1,\ldots)$.[10]

Tight-binding Hamiltonians gauge transform in the usual way:

$$H'(\mathbf{a})\equiv H(\mathbf{a}')+\partial_t\Lambda ,$$
$$\mathbf{a}'(e)\equiv\mathbf{a}(e)+\sum_V [v,e]\Lambda(v) . \tag{3.14}$$

$H(\mathbf{a})$ and $H(\mathbf{a}')$ are unitarily equivalent and so have the same spectrum.

$H(\mathbf{a})$ is strictly periodic in the gauge fields:

$$H(\mathbf{a})=H(\mathbf{a}+2\pi n), \quad n\in\mathbf{Z}^{|E|} . \tag{3.15}$$

This relates Hamiltonians with different fluxes:

$$\Phi'(f)=\Phi(f)+2\pi\sum_E [e,f]n(e) . \tag{3.16}$$

For the applications that we consider in Secs. XII and XIII, this relation is enough to guarantee the existence of gauges in which the Hamiltonian is periodic in all the fluxes.

The space of gauge fields is $\mathbf{R}^{|E|}$, but because of Eq. (3.15) it may be thought of as $\mathbf{T}^{|E|}$. In $\mathbf{R}^{|E|}$ sits an $\mathbf{R}^{|V|-1}$ subspace of pure gauge transformations. The space of distinct gauge fields is therefore $\mathbf{R}^h$, where h is the number of holes:

$$h\equiv|F|-|C|=|E|-|V|+1 . \tag{3.17}$$

In view of Eq. (3.15), the gauge-distinct Hamiltonians are naturally defined on $\mathbf{T}^h$.

In conclusion, the simplest set of models with the structure of the general case discussed in Sec. II are periodic matrices of the tight-binding type. As we shall see, even for quite small matrices interesting things happen. An example with 4×4 matrices will be analyzed in detail in Sec. XII. This elementary aspect is one of the appeals of the theory.

[10] $H(\mathbf{a}=\pi)$ has top state with energy $2|v|$ for the same eigenvector if the graph is regular.

IV. SCHRÖDINGER OPERATORS FOR NETWORKS OF THIN WIRES

Networks of one-dimensional connecting wires are idealizations corresponding to physical networks in the limit that the widths of the wires are small relative to all other length scales in the problem. The corresponding Schrödinger operators have, of course, the basic features of the general case discussed in Sec. II, and like the tight-binding model offer some simplifications. The first and obvious simplification is that the partial differential operator of Sec. II is replaced by an ordinary differential operator. The second simplification is that, at least for the case of "free electrons," the problem can be further reduced to matrices. In fact, the matrices turn out to be close relatives of the tight-binding Hamiltonians.

There are several reasons for considering this subclass. First, it is a natural class to consider and it is a useful description of various physical settings. Second, to examine the stability of the transport properties of networks, it is useful to compare how sensitive they are to the dynamics. Free electrons are in some sense on the other end of the spectrum from those that are tightly bound, and so offer an interesting alternative dynamics.

The formulation of wave equations on networks of one-dimensional wires has a long history, partly because the setting arises in many areas of physics: single-mode acoustic and electromagnetic waveguide networks (Mittra and Lee, 1977; Ramo, Whinery, and van Duzer, 1984); organic molecules (Ruedenberg and Scherr, 1953; Platt, 1964); superconductivity in granular and artificial materials (deGennes, 1981; Alexander, 1983); and mesoscopic quantum systems (Imry, 1986). The construction of wave equations for such networks is a topic in its own right. Ruedenberg and Scherr, who were apparently among the first to address the problem, based their formulation on the analysis of the limit of wires of finite thickness. Alexander generalized this to networks in external magnetic fields. Recently, the problem came of age in a series of mathematical works by Exner and Šeba (1987), whose formulation is based on the von Neumann theory of self-adjoint extensions of formal differential operators.

Our aims in this section are, first, to formulate Schrödinger operators for one-dimensional networks; second, to motivate this formulation by showing the relation to the original partial differential operator; and finally, to describe the reduction to a matrix problem for "free electrons." This section contains a fair amount of known material and is of an expository nature. Readers familiar with Alexander's work may want to skip it.

In the limit of narrow wires, part of the geometric information in Ω is lost. Some of it translates to dynamical information in the one-dimensional wave operator in the form of various potentials. We shall first formulate, in an *ad hoc* way, the operator, and later make the connection with the limiting procedure. With the network we associate the following potentials: (1) the "vector" potentials $\mathbf{A}(x,e)$, which are roughly the tangential component of

A at the point x on the edge e; (2) scalar potentials $V(x,e)$; and (3) vertex potentials $\lambda(v)$, which are associated with the vertices. All these potentials are real.

The wave function ψ is a vector in $\oplus_E L^2(e)$, with components $\psi(x_i,e_i)$. The Schrödinger operator, acting on the edge e, is the ordinary differential operator

$$(H(\mathbf{A})\psi)(x,e)=([-i\partial_x-\mathbf{A}(x,e)]^2+V(x,e))\psi(x,e) . \tag{4.1}$$

ψ has a unique value at the vertices and satisfies the Sturm-Liouville type of boundary conditions,

$$\sum_E [v,e][(-i\partial_x-\mathbf{A}(x,e))\psi](x,e)\,|_v=-i\lambda(v)\psi(v) , \tag{4.2}$$

where $\psi(v)\equiv\psi(x,e)\,|_v$.

We first discuss Eq. (4.1) and examine the relation of the potentials to the magnetic and geometric information in the original multidimensional problem, and then we discuss Eq. (4.2).

Consider the eigenvalue problem

$$(-i\nabla-\mathbf{A})^2\Psi=E\Psi \tag{4.3}$$

in a striplike domain, Fig. 9, with Dirichlet boundary conditions. Note the absence of a scalar potential in Eq. (4.3). Ω is essentially straight, with x the coordinate along it and y that in the transverse direction. In the limit of small width, the y coordinate becomes[11] a "fast variable" relative to the "slow variable" x, so Ψ admits a Born-Oppenheimer decomposition (Born and Oppenheimer, 1927)

$$\Psi(x,y)\simeq\psi(x)\eta_x(y) , \tag{4.4}$$

where $\eta_x(y)$ is a normalized eigenstate of the transverse (fast) motion, i.e.,

$$(-i\partial_y-\mathbf{A}_y)^2\eta_x(y)=W(x)\eta_x(y) . \tag{4.5}$$

Write $\langle\mu\,|\,\eta\rangle$ for the scalar product in the y coordinates, i.e.,

$$\langle\mu\,|\,\eta\rangle(x)\equiv\int dy\,\mu_x^*(y)\eta_x(y) . \tag{4.6}$$

$\langle\eta\,|\,\eta\rangle(x)=1$ by normalization. With the ansatz Eq. (4.4), the slow variable $\psi(x)$ solves the ordinary differential equation

$$\{[-i\partial_x-\mathbf{A}_x(x)-i\langle\eta\,|\,\partial_x\eta\rangle(x)]^2+W(x)$$
$$+\langle\eta\,|\,\partial_x\eta\rangle^2(x)+\langle\partial_x\eta\,|\,\partial_x\eta\rangle\}\psi(x)=E\psi(x) . \tag{4.7}$$

By the normalization of η, $i\langle\eta\,|\,\partial_x\eta\rangle(x)$ is a real-valued function. $\mathbf{A}_x(x)\equiv\langle\eta\,|\,\mathbf{A}_x\,|\,\eta\rangle(x)$. We see that the vector potential in Eq. (4.1) has the tangential component of the vector potential $\mathbf{A}$, corrected by gauge fields arising from the transverse motion:

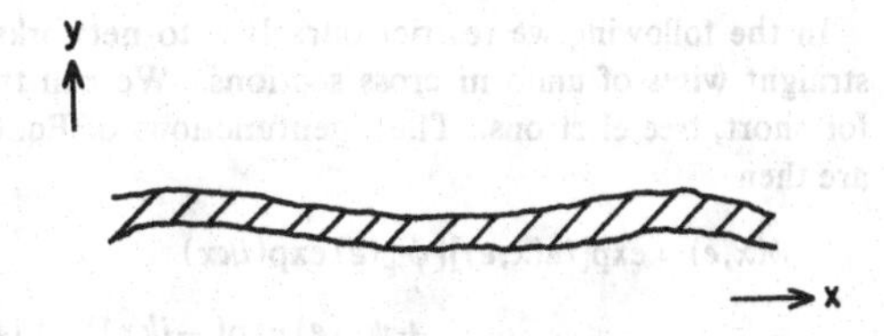

FIG. 9. A narrow striplike domain for which the study of the Laplacian reduces to the study of an ordinary differential equation in the Born-Oppenheimer approximation.

$$\mathbf{A}(x,e)\equiv\langle\eta\,|\,\mathbf{A}_x\,|\,\eta\rangle(x)+i\langle\eta\,|\,\partial_x\eta\rangle(x) . \tag{4.8}$$

The emergence of gauge potentials in the slow dynamics whose origin is in the fast dynamics is a feature of the Born-Oppenheimer theory (Alden Mead and Truhlar, 1979; Combes, Duclos, and Seiler, 1981; Wilczeck and Zee, 1984; Alden Mead, 1987).

We now examine the scalar potentials. Since the original operator, Eq. (4.6), had no scalar potential, the scalar potentials in Eq. (4.7) have their origins in the geometry of Ω and the dynamics of the fast variables. As the width of Ω shrinks to zero, $W(x)$ is dominant and shoots to $+\infty$ as $(\text{width})^{-2}$ by the uncertainty principle. This makes E in Eq. (4.7) shoot to $+\infty$ as well. To obtain a finite limit, a simple renormalization is required, which is familiar from the study of points interactions (Albeverio *et al.*, 1988). Suppose that as the width shrinks to zero

$$W(x)-E\to W(x)-k^2 , \tag{4.9}$$

where $\mathbf{W}(x)$ and k^2 are both finite. That is, we remove from $W(x)$ and E the same large constant (which may be identified with the constant attractive potential that keeps the electrons confined to Ω). This, together with the other two (subdominant) potential terms in Eq. (4.7), defines $V(x,e)$ in Eq. (4.1). It is instructive to note that for the limit Eq. (4.9) to exist, the wires must be of almost uniform width. Indeed, differentiating Eq. (4.9) gives

$$\partial_x W(x)=O(1) , \tag{4.10}$$

so $\partial_x(\text{width})$ is of order $(\text{width})^3$.

In the special case of uniform and straight wires we may choose

$$\eta_x(y)=\left[\exp\left[i\int^y \mathbf{A}_y(x,y')dy'\right]\right]\mu(y)\to\mu(y) \tag{4.11}$$

independent of x and $\mathbf{A}$. In this case $V(x,e)$ is constant (which one can take to be zero) and $\mathbf{A}(x,e)$ is the tangential component of the vector potential.[12]

---

[11]By scaling the x and y coordinates to order unity, say, one finds that the x coordinate has a heavy mass associated with it while the y coordinate has a light mass.

[12]Twisting wires have additional geometric information that translates to dynamical information. In the case of the wave equation (without vector potential) this has been studied in Berry (1987), Haldane (1986), and Kugler and Shtrikman (1987).

In the following we restrict ourselves to networks of straight wires of uniform cross sections. We call these, for short, free electrons. The eigenfunctions of Eq. (4.1) are then

$$\psi(x,e)=\exp[ia(x,e)][\psi_+(e)\exp(ikx) + \psi_-(e)\exp(-ikx)] , \quad (4.12)$$

where

$$a(x,e)\equiv \int_0^x \mathbf{A}(y,e)dy , \quad (4.13)$$

and ψ_+ and ψ_- are the amplitudes of the forward- and backward-moving waves on e.

We now want to "explain" the boundary condition (4.2). It was originally derived by Ruedenberg and Scherr (1953) by considering the zero-width limit of the multidimensional junction. Exner and Šeba show that such boundary conditions describe all the self-adjoint extensions of the operator (4.1). The scattering theory of junctions, an approach that has gained popularity in the quantum theory of complex systems (see, for example, Anderson *et al.*, 1980; Shapiro, 1983), gives some insight into Eq. (4.2).

Let $\psi_{in/out}(v)$ be the $|v|$-vector of incoming/outgoing amplitudes toward the junction v, with components $\psi_{in/out}(e,v)$ (where the edges e are incident on v):

$$\begin{aligned}\psi_{in}(e,v)&=\delta(1,[v,e])\psi_+(e)\exp\{i(k\,|e|+a(e))+\delta(-1,[v,e])\psi_-(e)\} ,\\ \psi_{out}(e,v)&=\delta(1,[v,e])\psi_-(e)\exp\{i(-k\,|e|+a(e))+\delta(-1,[v,e])\psi_+(e)\} ,\end{aligned} \quad (4.14)$$

where

$$\mathbf{a}(e)\equiv\mathbf{a}(|e|,e), \quad \mathbf{a}(-e)\equiv-\mathbf{a}(e) , \quad (4.15)$$

satisfy Eq. (3.7).

With each vertex v associate a scattering matrix $S(k,v)$, which is a unitary $|v|\times|v|$ matrix relating the incoming and outgoing amplitudes:

$$S(k,v)\psi_{in}(v)=\psi_{out}(v) . \quad (4.16)$$

The current flowing toward v on e is $k[\,|\psi_{in}(v,e)|^2-|\psi_{out}(v,e)|^2]$, so the total current flowing toward the vertex v satisfies Kirchoff's first law,

$$k\sum_e[\,|\psi_{in}(v,e)|^2-|\psi_{out}(v,e)|^2]=0 , \quad (4.17)$$

by the unitarity of S. Current is conserved at the vertices. Conversely, current conservation implies the unitarity of S.[13]

Define point junctions by the requirement that ψ has a unique continuation to the vertices, that is,

$$\psi_{in}(v,e)+\psi_{out}(v,e)=\langle\kappa(v)|\psi_{in}(v)\rangle , \quad (4.18)$$

where $\langle\kappa(v)|$ is a $|v|$-vector characterizing the vertex and the right-hand side is e independent. In vector notation, using Eq. (4.16),

$$[1+S(k,v)]|\psi_{in}\rangle=\langle\kappa(v)|\psi_{in}(v)\rangle|\mathbb{I}\rangle , \quad (4.19)$$

where $|\mathbb{I}\rangle$ is the $|v|$-vector

$$|\mathbb{I}\rangle\equiv(1,1,1,\ldots,) . \quad (4.20)$$

Equation (4.19) gives the operator relation

$$S(v)=-1+1|\mathbb{I}\rangle\langle\kappa(v)| . \quad (4.21)$$

Since S is unitary, the two equations, $S^\dagger S=1$ and $SS^\dagger=1$, lead to

$$\begin{aligned}&-|\mathbb{I}\rangle\langle\kappa(v)|-|\kappa(v)\rangle\langle\mathbb{I}|+|\kappa(v)\rangle\,|v|\,\langle\kappa(v)|=0 ,\\ &|\kappa(v)\rangle\,|v|\,\langle\kappa(v)|=|\mathbb{I}\rangle\langle\kappa(v)|\kappa(v)\rangle\langle\mathbb{I}| .\end{aligned} \quad (4.22)$$

The second equation says that $|\kappa(v)\rangle$ is proportional to $|\mathbb{I}\rangle$. The first constrains the proportionality constant to lie on a circle with radius $1/|v|$ in the complex plane, which is tangent to the imaginary axis:

$$S(k,v)=-1+(2/|v|)\cos[\theta(k,v)] \times\exp[i\theta(k,v)]|\mathbb{I}\rangle\langle\mathbb{I}| , \quad (4.23)$$

with $0\le\theta(k,v)\le\pi$. This singles out a circle in the $|v|^2$-dimensional space of unitary matrices.[14] Note that (in the spinless case considered here) $S(k,v)$ of a point junction is automatically time-reversal invariant: time reversal says that $(\psi_{in}(v))^*=\psi_{out}(v)$, from which it follows that $S=S^t$. Equation (4.23) is of this form.

To relate $\theta(k,v)$ of Eq. (4.23) and $\lambda(v)$ of Eq. (4.2), rewrite Eq. (4.2) as

$$k\langle\mathbb{I}|(\psi_{in}(v)-\psi_{out}(v))\rangle =-i\lambda(v)\langle\mathbb{I}|(\psi_{in}(v)+\psi_{out}(v))\rangle/|v| . \quad (4.24)$$

Substituting Eq. (4.23) in Eq. (4.16) and then in Eq. (4.24)

[13]Current conservation, Eq. (4.17), implies that $S^\dagger S$ must be a diagonal unitary and positivity implies that it is the identity. We are indebted to L. Sadun for pointing this out to us.

[14]After the completion of this work, we received JINR preprints (Exner and Šeba, 1987) in which Eq. (4.23) is derived. We thank F. Gesztesy for drawing our attention to these works.

gives

$$\lambda(v)=-|v|\,k\tan[\theta(k,v)]\ . \tag{4.25}$$

If $\lambda(v)\neq 0$, then, as the energy k^2 varies from 0 to ∞, the scattering matrix traverses the full circle of point scatterers in the space of unitaries.[15]

We conclude this section by describing a method, following Alexander (1983), that further reduces the study of the operators associated with one-dimensional networks to the study of finite matrices.[16] The wave function on the edge $\mathbf{e}=(v,u)$ can be written in terms of its values on the vertices, u and v:

$$\psi(x,e)=\exp[ia(x,e)]\{\psi(u)\sin[k(|e|-x)]+\psi(v)\sin(kx)\exp[-ia(e)]\}/\sin(k|e|)\ , \tag{4.26}$$

where $k\geq 0$. $\psi(x,e)$ has, by construction, a unique value on the vertices and satisfies the eigenvalue equation (4.1) with zero scalar potential. Substitution in the boundary conditions, Eq. (4.2), gives the matrix equation

$$h(k,\mathbf{a})\psi=0\ , \tag{4.27}$$

with ψ the $|V|$-vector on the vertices and $h(k,\mathbf{a})$ the $|V|\times|V|$ Alexander-de Gennes matrix:

$$[h(k,\Phi)]_{u,v}=\delta(u,v)\left[\sum_{E}|[e,v]|\cot(k|e|)-\lambda(v)/k\right]-\sum_{E}\delta([u,v],e)\frac{\exp[ia(e)]}{\sin(k|e|)}\ . \tag{4.28}$$

The equation $\det[h(k,a)]=0$ determines the spectrum $\{k_j^2(\Phi)\,|\,j=0,1,\ldots,\Phi\in T^h\}$ of the Schrödinger operator. This is an implicit-eigenvalue problem for the matrix $h(k,\Phi)$.

In the special case of no scattering potential at vertices, $\lambda(v)\equiv 0$, and edges of equal lengths, which one may then take to be unity with no loss of generality, Eq. (4.28) simplifies to

$$h_0(k,a)\psi=0\ , \tag{4.29}$$

where

$$[h_0(k,a)]_{uv}=\begin{cases}|v|\cos(k) & \text{for } u=v\ ,\\ -\exp[ia(e)] & \text{for } e=(u,v)\ ,\\ 0 & \text{otherwise}\ .\end{cases} \tag{4.30}$$

Here $h_0(k,\mathbf{a})$ is periodic in $\mathbf{k}$ and $\mathbf{a}$ with period 2π. It is remarkable how close it is to the tight-binding Hamiltonians $H(\mathbf{a})$ of Sec. III. In fact, for a given graph,

$$h_0(0,\mathbf{a})=H(\mathbf{a})\ . \tag{4.31}$$

The Bohm-Aharnov periodicity and the gauge properties discussed in Sec. III transfer to this case as well. An interesting new ingredient is that the parameter space can now be thought of as the $(h+1)$-dimensional torus. The h torus is as in the tight-binding case. An extra period comes from k.

The map $k^2\rightarrow\cos(k)$ maps the unbounded spectrum of the Schrödinger operator on a finite set. More explicitly, for fixed flux, the eigenstates are naturally paired (because of the symmetry of the cosine function), and each has two quantum numbers:

$$k_{+jn}(\Phi)=k_j(\Phi)+2\pi n\ , \tag{4.32a}$$

$$k_{-jn}(\Phi)=[2\pi-k_j(\Phi)]+2\pi n\ . \tag{4.32b}$$

Here $k_j(\Phi)$ is in $[0,\pi]$, and n natural. As we shall show, j runs over the finite set $1,\ldots,|V|$. [$k_{\pm jn}(\Phi)$ is constrained to be positive.]

In contrast with the situation in the tight-binding case, the Hilbert space can accommodate an infinite number of fermions. The n dependence is simple and explicit, and it is natural to expect that the problem can be fully analyzed by thinking of k as an angle and considering one period of $\cos(k)$.

Consider the set $\{(k,\Phi)\,|\,\Phi\in[-\pi,\pi]^h,\ k\in[0,2\pi],\ \det[h_0(k,a)]=0\}$. It may be thought of as the graph of "energy bands" over flux space restricted to its basic periods. Some of its basic properties are listed below.

(a) The bands are periodic in Φ (by Bohm-Aharonov periodicity); they are invariant under the reflection $\Phi\rightarrow-\Phi$ (by complex conjugating), and also under $k\rightarrow 2\pi-k$ [by the symmetry of $\cos(k)$].

(b) For fixed Φ there are $|V|$ bands, counting multiplicity, in the interval $k\in[0,\pi]$. [This follows from the self-adjointness of $h_0(k,a)$ and property (e) below.]

(c) The set of bands has a property that we shall call "π-shift" invariance, which is reminiscent of an electron-hole symmetry. By this we mean that the kernels of $h_0(k,a)$ and $h_0(k+\pi,\mathbf{a}+\pi)$ coincide. π is the vector $\pi(1,1,\ldots,1)$ in $\mathbb{R}^{|E|}$. This follows from

$$h_0(k,\mathbf{a})=-h_0(k+\pi,\mathbf{a}+\pi)\ . \tag{4.33}$$

[15]This does not explain why λ is k independent. Formally, this can be related to the question of the choice of self-adjoint extension; see Exner and Šeba (1987).

[16]Some of the formulas that we shall write below do not make sense if $\sin(k|e|)=0$. These should be interpreted as the limit when k has a small imaginary part.

A shift of π of the gauge fields translates to the shift of the fluxes:

$$\Phi(f)\to\Phi(f)+\pi\sum_E[e,f]$$
$$=\Phi(f)+\pi|f|\,\mathrm{mod}(2\pi)\;. \tag{4.34}$$

$|f|$ is the number of edges of f. If $|f|$ is even, then Eq. (4.34) is the identity on the flux torus. If, however, $|f|$ is odd, Eq. (4.34) is a shift of π in the corresponding flux. We conclude that the spectrum is invariant under

$$(k,\Phi)\to(k+\pi,\Phi+\pi F)\;, \tag{4.35}$$

where F is the $|F|$-vector

$$F(f)=|f|\;. \tag{4.36}$$

This is a global property of the spectrum, not necessarily a property of any given band. This is illustrated in Fig. 10.

(d) The bands touch the plane $\cos(k)=1$ at the single point $\Phi=0$. This follows from Eq. (4.31) and the strict positivity of the tight-binding Hamiltonians, proven in Sec. III. When combined with Eq. (4.35) this also gives the result that the spectrum touches $\cos(k)=-1$ at the single point $F\pi$. Note that these are the points where Eq. (4.26) is ill defined.

(e) $h_0(k,\mathbf{a})>0$, as an operator identity, for $\cos(k)>1$, and similarly $h_0(k,\mathbf{a})<0$ for $\cos(k)<-1$. For $\cos(k)>1$, this is seen from the fact that $h_0(k,\mathbf{a})$ is an increasing function of $\cos(k)$, and from (d) above. Combining this with (c) above gives the result for $\cos(k)<1$.

For a given graph, the implicit eigenvalue problem associated with free electrons, Eq. (4.29), and the explicit eigenvalue problem associated with the tight-binding Hamiltonian describe rather different physics. Consequently it is remarkable that both lead to the reduction of a partial differential operator to related matrix problems. In fact, more is true. For *regular graphs* the eigenvalue problem (and, as we shall see below, also the topological conductances) turn out to be simply related:

$$\cos(k)=1-E/|v|\;, \tag{4.37}$$

where E is the eigenvalue for the graph-theoretic tight-binding model, and k^2 is the corresponding eigenvalue for free electrons. (A similar relation holds for the standard tight-binding model with zero on-site interaction, but again, only for regular graphs.)

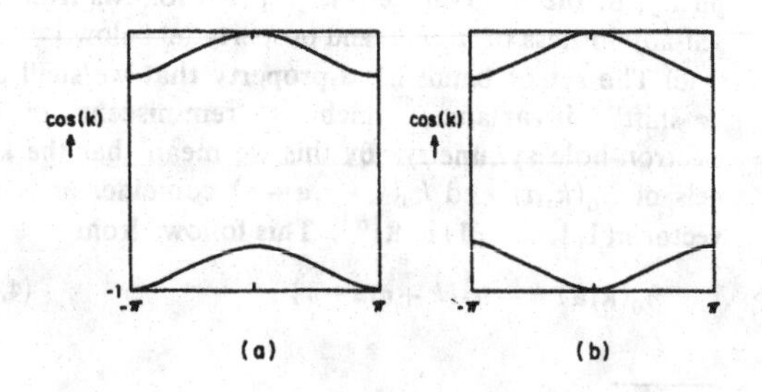

FIG. 10. Possible (schematic) spectra for $h_0(k,\Phi)$: (a) when the flux is through a face with an odd number of edges; (b) when the flux is through a face with an even number of edges.

The last point we discuss in this section is the normalization of the wave function. The length in the original Hilbert space induces a metric in the vector space $\mathbb{C}^{|V|}$. Writing $|\psi\rangle$ for the two-vector $(\psi(u),\psi(v))$ associated with the edge $e=(u,v)$ we find for the eigenstates

$$\|\psi(x,e)\|^2=\langle\psi|M(k;e)|\psi\rangle\;, \tag{4.38a}$$

where $M(k;e)$ is the 2×2 positive, Hermitian matrix

$$M(k;e)=\frac{1}{2\sin^2(k|e|)}\times\begin{bmatrix}\xi(k,e) & \exp[i\mathbf{a}(e)]\eta(k,e)\\ \exp[-i\mathbf{a}(e)]\eta(k,e) & \xi(k,e)\end{bmatrix},$$

with

$$\xi(k,e)\equiv|e|-\frac{\sin(2k|e|)}{2k} \tag{4.39}$$

and

$$\eta(k,e)\equiv\frac{\sin(2k|e|)}{2k}-|e|\cos(k|e|)\;.$$

As a consequence, eigenstates are normalized according to

$$\|\psi\|^2=\sum_E\|\psi(e,x)\|^2=\sum_{u,v\in V}\psi^*(u)[M(k)]_{uv}\psi(v)\;, \tag{4.38b}$$

where $M(k)$ is the $|V|\times|V|$ (positive) matrix

$$[M(k)]_{uv}=\begin{cases}\displaystyle\sum_E\frac{\xi(e,k)|[e,v]|}{2\sin^2(k|e|)} & \text{for } u=v\;,\\[2ex] \displaystyle\frac{\exp[i\mathbf{a}(e)]\eta(e,k)}{2\sin^2(k|e|)} & \text{for } e=(u,v)\;,\\[2ex] 0 \text{ otherwise}\;. \end{cases} \tag{4.40}$$

For determining the spectrum, the correct normalization is not an issue. For most observables, however, it is. In particular, it is important for the calculation of the transport. We return to this in Secs. X and XI. As we shall see there, for the *averaged* transport coefficients, it turns out that because of the topological interpretation the normalization is, in fact, not an issue.

V. LOOP CURRENTS

In circuit theory, it is convenient to introduce loop currents so that Kirchhoff's first law of current conservation at each vertex is automatically satisfied. Let $I(f)$ be the loop current associated with the elementary face f and $I(e)$ be the current in the edge e. The two are related by

$$I(e)=\sum_F[e,f]I(f)\;. \tag{5.1}$$

Loop currents are the basic objects in quantum networks. In this section we define and examine the observables associated with them.

We motivate the definition by the following consideration: to "measure" the current flowing around the mth flux, consider a virtual change $\delta\Phi_m$. This creates a virtual emf around the mth loop, which does not affect the state of the system (this is what we mean by virtual), except that the current now goes through a potential drop. The virtual change in energy of the system is

$$\delta E = -I(m)\delta\Phi_m \ . \tag{5.2}$$

This suggests that $-\partial_m H(A)$ is the observable associated with the mth loop current. Here, and throughout, we use the convention $\partial_m \equiv \partial_{\Phi_m}$. For the Schrödinger operator of Sec. II,

$$-2\partial_m H(A) = (-i\nabla - A)\cdot(\partial_m A) + (\partial_m A)(-i\nabla - A) \ . \tag{5.3}$$

It is convenient to introduce form notation such that

$$I(\psi) \equiv \sum_{m=1}^{h} I(m,\psi)d\Phi_m \ , \tag{5.4a}$$

$$\langle I(\psi)\rangle \equiv \sum_{m=1}^{h} \langle I(m,\psi)\rangle d\Phi_m \ , \tag{5.4b}$$

$$dH \equiv \sum_{m=1}^{h} (\partial_m H)d\Phi_m \ , \tag{5.4c}$$

with the flux-averaged current defined by

$$\langle I(m,\psi)\rangle \equiv \frac{1}{2\pi}\int_0^{2\pi} d\Phi_m I(m,\psi) \ . \tag{5.5}$$

Equation (5.4) enables us to write some of the formulas without excessive indexing.

The loop current for a system at the state ψ is[17]

$$I(m,\psi) = -\langle\psi|\partial_m H|\psi\rangle \tag{5.6a}$$

in components and

$$I(\psi) = -\langle\psi|dH|\psi\rangle \tag{5.6b}$$

as a one-form.

Equation (2.5) can be used to define loop currents for different choices of gauge.

It is instructive to examine the relation of the loop-current operator with the conventional current-density operator, given by (up to factor $\frac{1}{2}$)[18]

$$(-i\nabla - \mathbf{A})\delta(x-y) + \delta(x-y)(-i\nabla - \mathbf{A}) \ . \tag{5.7}$$

In a single ring, loop and edge currents coincide, so we consider this case. Let $x = (\rho,\theta)$ be the canonical cylindrical coordinates with the z axis threading the ring. The current operator associated with a section of the ring with fixed azimuthal angle θ_0 is

$$\left[\frac{\delta(\theta-\theta_0)}{\rho(\theta_0)}v_\theta + v_\theta\frac{\delta(\theta-\theta_0)}{\rho(\theta_0)}\right] , \tag{5.8a}$$

where v_θ is the velocity operator in the θ direction. Equation (5.8a) follows from integrating the current density over the θ_0 sections of the ring. The average current (over θ) is therefore

$$\frac{1}{2\pi}\left[\frac{1}{\rho(\theta)}v_\theta + v_\theta\frac{1}{\rho(\theta)}\right] . \tag{5.8b}$$

Suppose now that the ring is threaded by a flux tube, so that the electric field associated with it is azimuthally oriented. The vector potential describing the most general flux tube is

$$\mathbf{A}(x) = \Phi\, d\Lambda(x) \ , \tag{5.9}$$

where

$$\Lambda(\theta+2\pi) = \Lambda(\theta)+1 \ . \tag{5.10}$$

The loop-current operator is

$$\frac{(\partial_\theta\Lambda)}{\rho(\theta)}v_\theta + v_\theta\frac{(\partial_\theta\Lambda)}{\rho(\theta)} \ . \tag{5.11}$$

Comparing Eqs. (5.8) and (5.11), we see that the normalized weight $\delta(\theta-\theta_0)$ in Eq. (5.8a) is replaced by the normalized weight $1/(2\pi)$ in Eq. (5.8b), and by a general normalized weight $(\partial_\theta\Lambda)$. The weight (which need not be positive) in Eq. (5.11) characterizes the flux tube. For example, the flux tube of Eq. (2.10), associated with a singular gauge field, that is supported on a half-line along the radial direction, has all the potential drop at θ_0. The associated Λ is a staircase function of the azimuthal angle, i.e., the integral part of $\Phi(\theta-\theta_0)/2\pi$. The corresponding loop current coincides with the current (5.8a). If, on the other hand, the tube is that of Eq. (2.8) and generates uniform fields $\Lambda(\theta) = (\Phi/2\pi)\theta$, the loop coincides with the average edge current, Eq. (5.8b). In steady state, current is conserved and is independent of the choice of Λ. The loop current (5.11) and the edge currents (5.8) coincide.

The basic definition of loop currents says that we may interpret the fluxes threading the loops as "ammeters" of loop currents.[19] The examples discussed above show that

---

[17]For the corresponding thermodynamic identity, see, for example, Byers and Yang (1961).

[18]The following is meaningful for the Schrödinger operators of Secs. II and IV. For tight-binding Hamiltonians, one has to use the virtual work argument of Eq. (5.2).

[19]One way to actually measure the loop currents is to observe the change in the fluxes resulting from induction. This is distinct from what is meant here by the statement that fluxtubes are ammeters. The theoretical framework neglects all induction effects. Induction is briefly discussed in Sec. XIV.

different ammeters, i.e., different flux tubes, measure distinct, although related, currents that coincide in steady states. The situation is like that in Sec. II, where different driving flux tubes gave distinct dynamics even though the fluxes and emf's were the same. Clearly, a particularly natural and convenient set of observables to focus on are those that are independent of the flux tubes and depend only on the fluxes Φ and emf's $-\dot{\Phi}$. This corresponds to our desire to define the transport coefficients as ratios of currents to voltages, paying no regard to how the electric field and currents are distributed. As we shall see in Secs. VI and VIII, if the transport coefficients are defined via flux averaging, they have this property: the same (averaged) loop currents are measured by all "ammeters" and "batteries."

Suppose now that $H(\mathbf{A})$ is time dependent, because some of the fluxes are. Let $|\psi_t\rangle$ solve the time-dependent Schrödinger equation. Then, using the chain rule, Eq. (5.4), we can rewrite the mth loop current as

$$I(m,\psi_t)=-i\partial_t\langle\psi_t|\partial_m\psi_t\rangle\ , \tag{5.12a}$$

which, in form notation, is

$$I(\psi_t)=-i\partial_t\langle\psi_t|d\psi_t\rangle\ . \tag{5.12b}$$

Equations (5.4) and (5.12) are the basic equations of "loop transport." In the next section we shall see that, in the adiabatic limit, the loop currents are related to geometrical properties of the bundle of spectral subspaces of the Hamiltonians.

VI. ADIABATIC EVOLUTION AND TRANSPORT

When the fluxes generating $\mathbf{A}$ vary slowly in time, the evolution generated by $H(\mathbf{A})$ respects the spectral structure. This is the content of the adiabatic theorem. Namely, let $P(\mathbf{A})$ be a spectral projection for $H(\mathbf{A})$. Then states in $P(\mathbf{A})$ at time 0 evolve to states in $P(\mathbf{A})$ at time t, up to a small error term. Technical conditions aside, the theorem holds provided $P(\mathbf{A})$ is separated by gaps from the rest of the spectrum. We make this assumption throughout.

A convenient way to study the evolution in the adiabatic limit is to introduce an *adiabatic Hamiltonian*, an idea that goes back to Kato (1950). It generates an evolution that respects the adiabatic theorem with no error. As we shall see below, there is a choice to be made. One choice is to require that the evolution approximate the physical evolution the best it can. This leads to the following choice of generator (Avron, Seiler, and Yaffe, 1987):

$$H_{\rm Ad}(\mathbf{A},P)\equiv H(\mathbf{A})+\Gamma_t(\mathbf{A})\ ,\qquad \Gamma_t(\mathbf{A})\equiv i[\partial_t P,P]\ . \tag{6.1a}$$

Another choice, the one originally picked by Kato, is to take as generator

$$H_{\rm K}(\mathbf{A},P)\equiv\Gamma_t(\mathbf{A})\ . \tag{6.1b}$$

Although Eq. (6.1b) appears to be less motivated, it is, in certain ways, a more convenient choice for our purposes. In either case, the evolutions $U_{\rm Ad}$ have the property

$$U_{\rm Ad}(t)P(\mathbf{A}(0))=P(\mathbf{A}(t))U_{\rm Ad}(t)\ . \tag{6.2}$$

This is the precise meaning of the statement that the evolution respects the spectral structure of the Hamiltonian. Because of this, $U_{\rm Ad}$ has geometric significance as it defines transport of states in the bundle of spectral subspaces associated with the projection $P(\mathbf{A})$.

It is convenient to use form notation:

$$\Gamma(\mathbf{A})\equiv i[dP,P],\quad dP\equiv\sum_k(\partial_k P)d\Phi_k\ . \tag{6.3}$$

Equation (6.1b) can be rewritten as

$$H_{\rm K}(\mathbf{A},P)=\partial_t\Phi\cdot\Gamma(\mathbf{A})\ , \tag{6.4}$$

where the vector $\partial_t\Phi$ gives the flow in flux space. A centered dot denotes the pairing of vectors and forms, so, for example, $\partial_j\cdot d\Phi_k=\delta_{jk}$. $\Gamma(\mathbf{A})$, being a one-form, has a natural interpretation as the connection on the bundle of projections.

$U_{\rm Ad}$ parallel transports vectors in $P(\mathbf{A})$, and so is useful in the study of the geometry of the spectral subspace. Curvature is related to the noncommutativity of transport in different directions (Arnol'd, 1978), so parallel transport around a closed loop in flux space need not be the identity but rather a general unitary in $U(n)$, $n=\dim(P)$. The jk component of the curvature is

$$\omega_{jk}=\frac{U(\text{infinitesimal loop in } jk \text{ plane})-1}{|\text{area of loop}|}\ . \tag{6.5}$$

From the evolution equation, the curvature associated with the projection P, for a small square loop in flux space, is given by[20]

$$\omega_{jk}=P[\partial_j P,\partial_k P]P\ . \tag{6.6}$$

In a component-free notation, ω is the curvature two-form:

$$\omega=P(dP)\wedge(dP)P\ . \tag{6.7}$$

It is imaginary as

$$\omega^\dagger=-\omega\ . \tag{6.8}$$

We now proceed to show that the components of ω are the matrix elements of the conductance.

The adiabatic current in the mth loop, $I_{\rm Ad}(m,\psi)$, is the obvious analog of Eq. (5.12), where ψ now solves the adi-

---

[20]Up to factors of i this is essentially the standard formula that says that the curvature is $dA-iAA$, where A is the connection one-form. For the generator in Eq. (6.1a) the curvature is $-iTP(\partial_j H-\partial_k H)P+P[\partial_j P,\partial_k P]P$, where T is the period associated with the loop. The first term in this expression is the "dynamical phase" and has vanishing periods. The second is the same as in Eq. (6.7).

abatic evolution. Let $P_j(\mathbf{A})$ be the one-dimensional projection $P_j(\mathbf{A})\equiv|\psi_j(\mathbf{A})\rangle\langle\psi_j(\mathbf{A})|$ associated with an isolated eigenvalue $E_j(\Phi)$. We may, and do, choose $\psi_j(\mathbf{A})$ according to the adiabatic evolution along the path shown in Fig. 11. This determines $\psi_j(\mathbf{A})$ with no ambiguity up to an overall constant phase on $\Phi\in\mathbf{R}^h$. Undoing the calculation leading to Eq. (5.12) [which turns out to be the same as replacing H in Eq. (5.6) by the adiabatic generator Eq. (6.1a)], one finds

$$I_{\mathrm{Ad}}(m,\psi_j)=-\partial_m E_j(\Phi) -i\{\langle\partial_m\psi_j(\mathbf{A})|\partial_t\psi_j(\mathbf{A})\rangle -\langle\partial_t\psi_j(\mathbf{A})|\partial_m\psi_j(\mathbf{A})\rangle\}\ . \tag{6.9}$$

By assumption, no levels cross, so the derivatives are well defined (see Sec. VII). Using the Schrödinger equation for the adiabatic evolution, Eq. (6.1a),

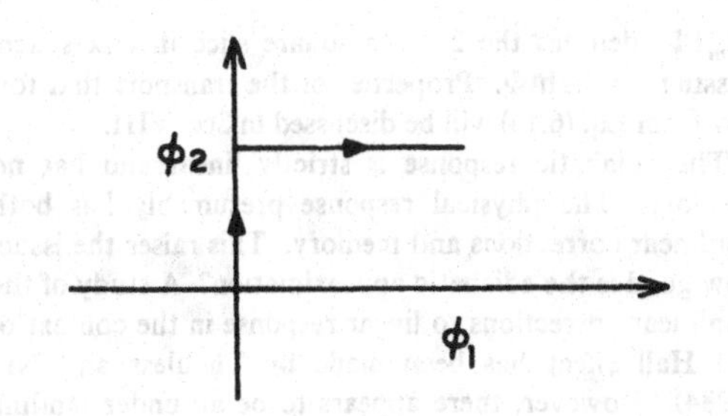

FIG. 11. A choice of paths for the adiabatic phase factors.

$$i\partial_t\psi_j=i\dot{\Phi}\cdot d\psi_j=H_{\mathrm{Ad}}(\mathbf{A})\psi_j\ , \tag{6.10}$$

and the fact that

$$\langle\psi_j|H_{\mathrm{Ad}}(\mathbf{A})|\psi_j\rangle=E_j(\Phi) \tag{6.11}$$

gives

$$I_{\mathrm{Ad}}(m,\psi_j)=-\partial_m E_j(\Phi)-i\sum_k\dot{\Phi}_k(t)(\langle\partial_m\psi_j(\mathbf{A})|\partial_k\psi_j(\mathbf{A})\rangle-\langle\partial_k\psi_j(\mathbf{A})|\partial_m\psi_j(\mathbf{A})\rangle)\ . \tag{6.12}$$

In index-free notation,

$$I_{\mathrm{Ad}}(P_j)=-dE_j-i\langle d\psi_j(\mathbf{A})|d\psi_j(\mathbf{A})\rangle\cdot\dot{\Phi}(t)=-dE_j-i\,\mathrm{Tr}[\omega(P_j)]\cdot\dot{\Phi}(t)\ . \tag{6.13}$$

This equation is the basic equation of adiabatic transport and plays a central role in all that follows. The transport matrix $i\,\mathrm{Tr}[\omega(P_j)]$ in Eq. (6.13) is, in view of Eq. (6.7), the curvature two-form. Written in longhand, $i\,\mathrm{Tr}[\omega(P)]$ is an $h\times h$ matrix with loop indices.

The current in Eq. (6.12) or Eq. (6.13) is affine in the emf's: in the absence of emf's there are persistent currents given by $-dE_j$. Persistent currents are common in atomic physics and manifest themselves in diamagnetism. Persistent currents also occur in macroscopic systems with macroscopic coherence, such as superconducting rings. Büttiker, Imry, and Landauer (1983) have suggested that persistent currents also occur in mesoscopic normal-metal rings, but this has not yet been observed. Another remarkable feature of Eq. (6.13) is that the currents at time t are determined by the emf's at the same time: the adiabatic transport has no memory. This implies that the adiabatic ac conductance is frequency independent (being the Fourier transform of a delta function in time).[21] Finally, there are no power-law correction terms in the adiabatic transport (Klein and Seiler, 1988).

The adiabatic transport coefficients $i\,\mathrm{Tr}[\omega_j(P)]$ depend on the flux tubes, i.e., depend on a choice of vector potential $\mathbf{A}$, for P depends on $\mathbf{A}$. This is as one expects from the discussion in the previous sections. Distinct flux tubes drive the system differently and measure its response differently. When we consider the transport of current *averages*, two nice things happen. First, the transport coefficients become independent of the flux tubes and become a property of the fluxes alone. Second, the persistent currents disappear, and one gets the usual linear response.

The average mth loop current is, by Eq. (5.5),

$$\langle I_{\mathrm{Ad}}(m,P_j)\rangle=-\frac{i}{2\pi}\int_0^{2\pi}d\Phi_m\{\langle\partial_m\psi_j(\mathbf{A})|\partial_t\psi_j(\mathbf{A})\rangle-\langle\partial_t\psi_j(\mathbf{A})|\partial_m\psi_j(\mathbf{A})\rangle\}=\frac{i}{2\pi}\int_{T_m(\Phi)}\mathrm{Tr}[\omega(P_j)]\cdot\dot{\Phi}\ . \tag{6.14a}$$

$T_m(\Phi)$ is the line $\Phi+\lambda\partial_m$, λ going from 0 to 2π. The persistent currents, being complete derivatives, have vanishing averages and drop from (6.14). The average transport coefficients, defined as the Fourier transform of the kernel in Eq. (6.13), written out as a matrix, are

$$\begin{aligned}2\pi[\langle g_{\mathrm{Ad}}\rangle(P_j)]_{km}&=\frac{i}{2\pi}\int_0^{2\pi}d\Phi_k\int_0^{2\pi}d\Phi_m\{\langle\partial_m\psi_j(\mathbf{A})|\partial_k\psi_j(\mathbf{A})\rangle-\langle\partial_k\psi_j(\mathbf{A})|\partial_m\psi_j(\mathbf{A})\rangle\}\\&=\frac{i}{2\pi}\int_{T_{km}(\Phi)}\langle d\psi_j(\mathbf{A})|d\psi_j(\mathbf{A})\rangle=\frac{i}{2\pi}\int_{T_{km}(\Phi)}\mathrm{Tr}[\omega(P_j)]\ .\end{aligned} \tag{6.14b}$$

[21]For the quantum Hall effect there is experimental support for the quantization of the ac conductance (Kuchar *et al.*, 1987).

$T_{km}(\Phi)$ denotes the $2\pi\times 2\pi$ square slice in flux space, passing through Φ. Properties of the transport that follow from Eq. (6.14) will be discussed in Sec. VIII.

The adiabatic response is strictly linear and has no memory. The physical response presumably has both nonlinear corrections and memory. This raises the issue, how good is the adiabatic approximation? A study of the nonlinear corrections to linear response in the context of the Hall effect has been made by Thouless and Niu (1984). However, there appears to be no understanding of these issues from a general mathematical point of view. For questions about tunneling, the adiabatic limit is known to be very good in the sense that corrections are often exponentially small in the time scale. Tunneling is, however, only one of the issues. And, although it is natural to expect that $g_{\rm Ad}$ and $\langle g_{\rm Ad}\rangle$ approximate the physical transport g and $\langle g\rangle$, this has not been shown in any great generality. Avron, Seiler, and Yaffe (1987) have shown that $\langle g_{\rm Ad}\rangle$ and $\langle g\rangle$ are close if interpreted as charge transport (with error that is polynomial in the inverse time scale), and suggest that $g_{\rm Ad}$ and g, without averaging, may actually *not* be close to each other. The case of constant and harmonic emf's appears to be largely open, at least from a rigorous mathematical point of view.

VII. LEVEL CROSSINGS

The assumption that $E_j(\Phi)$ is an isolated eigenvalue entered in several places in the previous section. First, the adiabatic theorem requires no level crossings; second, dP_j and dE_j may not exist at crossing; and, finally, the zero average of the persistent current relies on $E_j(\Phi)$'s being smooth and periodic. So points in flux space where levels cross are where the theory in the previous section breaks down. In an almost dialectic fashion, these points are also the source of nontrivial transport. If no levels cross anywhere in flux space, $\langle g_{\rm Ad}\rangle(P)$ vanishes identically. If levels do cross, the adiabatic transport is not defined at crossings, but is defined away from crossings and may be nontrivial. This says that crossings are where the "sources" of the transport are located. For this reason, getting a handle on crossings is a key issue in understanding the transport.

There are three questions that we address in this section: (1) What is the local behavior of P near level crossings? (2) How big is the set of crossing points in the generic case? and (3) What can one say about crossings when the network has symmetries? This is a "service section," where relevant information that is needed later is collected. Many of the results are standard. [For a general overview on level crossings see Berry (1983).]

The Hamiltonians introduced in Secs. II, III, and IV are all entire functions of the fluxes. The projection $P_j(\Phi)$ is given by

$$P_j(\Phi)=\frac{1}{2\pi i}\int_{\gamma_j}\frac{dz}{H(\Phi)-z}\ , \tag{7.1}$$

where γ_j is a contour surrounding the jth piece in the spectrum, Fig. 12. For convenience of notation we suppress henceforth the index j. $P(\Phi)$ inherits the smoothness of $H(\Phi)$ as long as γ stays outside the spectrum. If $H(\Phi)$ is periodic, or periodic up to unitary equivalence, then $P(\Phi)$ inherits that too. Finally, if $P(\Phi)$ is a finite-dimensional projection, then $\mathrm{Tr}(H^k(\Phi)P(\Phi))$ is smooth in Φ for all k. In particular, isolated energy bands are smooth. However, when gaps in the spectrum close, so that γ is pinched, smoothness may be lost. Let $D(P)$ be the set of points in flux space where P is not smooth. Points in $D(P)$ must be points of level crossings. Because of Eqs. (2.5)–(2.7), if $\phi\in D(P)$ then $\phi+2\pi k\in D(P)$. So if we think of flux space as $\mathbf{R}^h$, $D(P)$ is periodic there. It is, however, better to think of flux space as $\mathbf{T}^h$.

Consider the local behavior of $E_j(\Phi)$ and $P_j(\Phi)$ near a two-level crossing at ϕ. Restricting the Hamiltonian to the degenerate subspace at ϕ gives a 2×2 Hermitian matrix function

$$h(\Phi)\equiv\begin{bmatrix}\langle\psi|H(\mathbf{A}|\psi\rangle\langle\psi|H(\mathbf{A})|\varphi\rangle\\ \langle\varphi|H(\mathbf{A})|\psi\rangle\langle\varphi|H(\mathbf{A})|\varphi\rangle\end{bmatrix}$$

$$\equiv\epsilon_0(\Phi)1+\epsilon(\Phi)\cdot\sigma\ , \tag{7.2a}$$

where $|\psi\rangle$ and $|\varphi\rangle$ are the two independent eigenvectors of $H(\mathbf{A})$ at ϕ. Here $\epsilon(\Phi)$ is a *real* three-vector valued function, and σ is the triplet of Pauli matrices. The two eigenvalues of Eq. (7.2a) are

$$E_\pm(\Phi)=\epsilon_0(\Phi)\pm|\epsilon(\Phi)|\ , \tag{7.3a}$$

from which it follows that $\epsilon(\phi)=0$. The eigenprojections are

$$P_\pm(\Phi)=[1\pm\hat{\epsilon}(\Phi)\cdot\sigma]/2\ . \tag{7.4}$$

$\hat{\epsilon}$ is the unit vector associated with ϵ. Because of the absolute value in Eq. (7.3a) and the normalization to unit vectors in Eq. (7.4), neither $E_\pm(\Phi)$ nor $P_\pm(\Phi)$ need be smooth at ϕ. In ϵ space, Eq. (7.3a) describes a *conic*, Fig. 13. Berry and Wilkinson (1984) call such points *diabolic*. The eigenvalues are continuous in ϵ but not smooth, and the projections are not even continuous near $\epsilon=0$. [The fact, as well as the example, are classical results due to Rellich (1969).]

The behavior in Φ space can be more complicated, and things depend on the way Φ space is mapped on ϵ space. The simplest case is when the map is characterized by its linear piece. Because we are interested mostly in three-flux networks, we restrict ourselves to the case in which

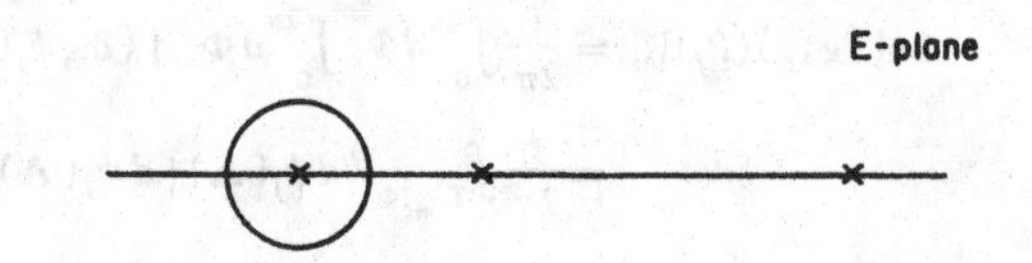

FIG. 12. A contour in the complex plane associated with a spectral projection.

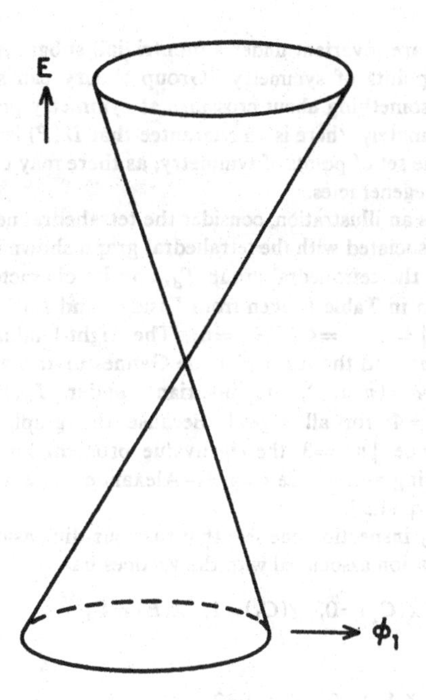

FIG. 13. A conic singularity.

Φ space is three dimensional. This is a particularly simple situation. Since ϵ is entire in Φ,

$$\epsilon(\Phi)\sim\sum_{m=1}^{3}(\partial_m\epsilon)\,|_{\phi}(d\Phi_m)\ . \tag{7.5}$$

We denote the linearized map from Φ to ϵ by $\nabla\otimes\epsilon$, that is,

$$[\nabla\otimes\epsilon]_{jk}\equiv\partial_j\epsilon_k\ . \tag{7.6a}$$

Now, if the linearized map is of rank 3, then the description of the singularity as conic holds in the Φ variables as well. In particular, in three-flux networks, diabolic points are those where

$$\det[\nabla\otimes\epsilon]\,|_{\phi}\neq0\ . \tag{7.6b}$$

For the tight-binding Hamiltonians, the linearized map, Eq. (7.6b), can be written down by inspecting the graph. This is especially useful if one wants to compute the charges "by hand." Let us briefly describe this.

Let $e=(u,v)$ be an edge and $\mathbf{a}(e)$ the associated vector potential. For the tight-binding Hamiltonian $H(\mathbf{a})$ one has

$$\left\langle\psi\left|\frac{\partial H}{\partial\mathbf{a}(e)}\right|\varphi\right\rangle=i[-\psi^*(u)\varphi(v)e^{i\mathbf{a}(e)}+\psi^*(v)\varphi(u)e^{-i\mathbf{a}(e)}]\ . \tag{7.7}$$

For the sake of simplicity suppose that a gauge has been chosen so that $\mathbf{a}(e)=\Phi(e)$ for h of the edges, and $\mathbf{a}(e)=0$ for all other edges. (All the examples we consider in Secs. XII and XIII are of this form.) Suppose that ψ and φ span the degenerate subspace. Then from Eqs. (7.7) and (7 2a) one finds for the derivative of the map

$$\frac{\partial(\epsilon_2+i\epsilon_1)}{\partial\Phi(e)}=\psi^*(u)\varphi(v)e^{i\Phi(e)}-\psi^*(v)\varphi(u)e^{-i\Phi(e)}\ ,$$
$$\frac{\partial\epsilon_3}{\partial\Phi(e)}=\mathrm{Im}[\psi^*(u)\psi(v)e^{i\Phi(e)}-\varphi^*(u)\varphi(v)e^{i\Phi(e)}]\ . \tag{7.8}$$

Similar analysis can be made for two-level crossings in the de Gennes–Alexander problem, Eq. (4.29). Let $|\psi\rangle$ and $|\varphi\rangle$ be two degenerate eigenvectors of $h_0(k_0,\phi)$, with eigenvalue zero. Construct the 2×2 Hermitian matrix function of Φ and k,

$$\epsilon_0(k,\Phi)+\epsilon(k,\Phi)\cdot\sigma$$
$$\equiv\begin{bmatrix}\langle\psi|h_0(k,\Phi)|\psi\rangle & \langle\psi|h_0(k,\Phi)|\varphi\rangle\\ \langle\varphi|h_0(k,\Phi)|\psi\rangle & \langle\varphi|h_0(k,\Phi)|\psi\rangle\end{bmatrix}. \tag{7.2b}$$

This defines a map from (k,Φ) space to the (ϵ_0,ϵ) space. The energy bands near (k_0,ϕ) are given by the vanishing of the determinant,

$$\epsilon_0^2(k,\Phi)-\epsilon^2(k,\Phi)=0\ . \tag{7.3b}$$

This is a conic in (ϵ_0,ϵ). If the Jacobian of the map is nonvanishing, it is also a conic in (k,Φ). The assumption that the degeneracy is isolated translates to $\epsilon(k,\Phi)\neq0$ for (k,Φ) on a small three-sphere centered at (k_0,ϕ).

We now turn to the second question, how big is $D(P)$? This question was first posed by von Neumann and Wigner (1929), who also proposed a counting rule that gives the answer. A somewhat more precise formulation of the question is, in the space of "all" Hamiltonians what is the dimension of the space of Hamiltonians with degenerate eigenvalues? This, of course, depends on what one means by "all." The convention in the statistical theory of spectra (Dyson, 1964; Porter, 1965) is that for problems with magnetic fields, with or without spin, "all" means complex Hermitian matrices.

The space of Hermitian $n\times n$ matrices is an n^2-dimensional vector space. To illustrate the von Neumann–Wigner strategy we start with a warmup and show that the space of nondegenerate Hermitian matrices is of full dimension.

The unitary that diagonalizes a given Hermitian matrix with fixed nondegenerate spectrum,

$$E_1<E_2<\cdots<E_n\ , \tag{7.9a}$$

is determined up to a diagonal unitary matrix. So there is a one-to-one correspondence between nondegenerate Hermitian matrices with fixed spectrum and elements of

$$\mathrm{U}(n)/[\mathrm{U}(1)]^n\ . \tag{7.10a}$$

Since $\dim[\mathrm{U}(n)]=n^2$, the space in Eq. (7.10a) is $n(n-1)$ dimensional, which together with the n dimensions associated with varying E_j gives n^2, the full dimension.

Now consider the Hermitian matrices with, say, a degenerate ground state. Equation (7.7) is replaced by

$$E_1=E_2<\cdots<E_n\ . \tag{7.9b}$$

The corresponding diagonalizing unitaries are identified with elements of

$$U(n)/[U(1)]^{n-2}\times U(2)]\ . \tag{7.10b}$$

This space is $n(n-1)-2$ dimensional. The dimension of the space associated with varying E_j is $n-1$ so, altogether, the space with a twofold ground-state degeneracy is of dimension n^2-3. The codimension is 3 and is independent of n. [For an alternative derivation, see Avron and Simon (1978).] It has $n-1$ components. More generally, the space of Hermitian matrices with m-fold degeneracy has $n-m+1$ components with codimension given by[22]

$$(m-1)+\dim U(m)-m\dim U(1)=m^2-1\ . \tag{7.11}$$

Thus the $n-1$ components with twofold degeneracy are connected by "filaments" of codimension 8, with triplet degeneracies. The codimensions are independent of the size of the matrices and hold for operators that are limits of matrices and have *discrete* spectra. [For recent interesting mathematical developments on the crossing rule see Friedland *et al.* (1984).]

The von Neumann–Wigner theorem suggests that a family of operators depending on n parameters has eigenvalue crossings on a set of codimension 3 in parameter space. This is an ansatz. It is not a theorem, because the n-parameter family may be embedded in a special way in the space of all Hermitian matrices. In fact, when taken too literally, the ansatz has easy counterexamples.[23] The ansatz says that $D(P)$ is of codimension 3 and is a set of points in a three-flux network, lines in four-flux network, etc. We find that the ansatz holds for many networks, and when it fails, it does so for an identifiable and often interesting reason.

We have already mentioned the fact that $D(P)$ acts as a source of transport and that getting a handle on $D(P)$ is the major step in the calculation of $\langle g_{\rm Ad}\rangle(P)$. It is therefore natural to consider symmetric networks in which group-theoretic methods can be used to give information on $D(P)$ and P. Symmetric networks have a point symmetry group G associated with the graph.. G induces representations on the vertices, edges, and faces, which we denote by $G(V)$, $G(E)$, and $G(F)$, respectively. $D(P)$ is invariant under $G(F)$. Points in the flux space that are invariant under a (nontrivial) subgroup of $G(F)$ are points of symmetry. Group theory can sometimes say something about crossings at symmetry points. Unfortunately, there is no guarantee that $D(P)$ is contained in the set of points of symmetry, as there may be accidental degeneracies.

As an illustration, consider the tetrahedral network. It is associated with the tetrahedral graph shown in Fig. 14. G is the tetrahedral group T_d, and its character table is given in Table I taken from Landau and Lifshitz (1977). $|V|=|F|=4$, $|E|=6$. The tight-binding Hamiltonian and the Alexander–de Gennes matrix with $\Phi\equiv0$, or $\Phi=(\pi,\pi,\pi)$, are invariant under $T_d(V)$. [Take $a(e)=\Phi$ for all edges.] Because the graph has fixed valence $|v|=3$, the eigenvalue problem for the tight-binding and the de Gennes–Alexander cases are related by Eq. (4.37).

By inspection one sees that the four-dimensional representation associated with the vertices has

$$\chi(C_2)=0,\quad \chi(C_3)=1,\quad \chi(E)=4\ ,$$
$$\chi(S_4)=0,\quad \chi(\sigma_d)=2\ . \tag{7.12}$$

So the four-dimensional Hilbert space decomposes according to

$$A_1+T_2\ . \tag{7.13}$$

This says that $\Phi=0$ and $\Phi=(\pi,\pi,\pi)$ have points of triple degeneracy. For $\Phi=0$ the nondegenerate subspace is $\lambda(1,1,1,1)$, $\lambda\in\mathbb{C}$ and is associated with the eigenvalue zero for the tight-binding Hamiltonian, and $\cos(k\,|e|)=1$ for the de Gennes–Alexander matrix. The triple degeneracy lies in the orthogonal complement to $(1,1,1,1)$, with eigenvalue 4 in the tight-binding case and $\cos(k\,|e|)=-\frac{1}{3}$ for the de Gennes–Alexander case. At $\Phi=(\pi,\pi,\pi)$, in the gauge where all bonds are -1, the same vector is the top state, with eigenvalue 6 for the tight-binding case and $\cos(k\,|e|)=-1$ in the de Gennes–Alexander case. The triply degenerate state/has eigenenergy 2 in the tight-binding case and $\cos(k\,|e|)=\frac{1}{3}$ in the de Gennes–Alexander case [cf. Eq. (4.35)].

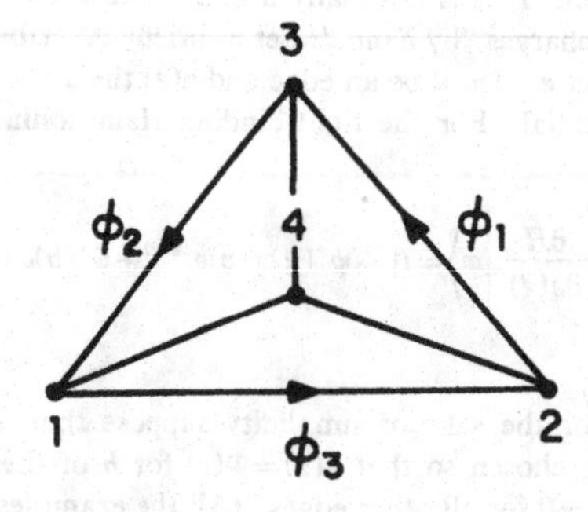

FIG. 14. The tetrahedral graph.

[22]In the case of real symmetric matrices the $U(m)$ and $U(1)$ in Eq. (7.11) are replaced by $O(m)$ and $O(1)$, respectively. This corresponds to the spin-zero, time-reversal-invariant situation. In the spin-$\frac{1}{2}$ time-reversal-invariant situation, $Sp(m)$ and $Sp(1)$ replace $U(m)$ and $U(1)$. This will be discussed in a forthcoming work of one of us (J.E.A.) with R. Seiler and B. Simon.

[23]Consider the Schrödinger equation on the line with potential $V(x;\phi)$, depending on n parameters ϕ so $V\to\infty$ at $|x|\to\infty$. The von Neumann–Wigner ansatz gives codimension 2 in this (real) case and clearly fails arbitrarily badly as the spectrum is simple for all ϕ due to a Wronskian identity.

TABLE I. Character table for the tetrahedral group.

T_d	E	$8C_3$	$3C_2$	$6\sigma_d$	$6S_4$
A_1	1	1	1	1	1
A_2	1	1	1	-1	-1
E	2	-1	2	0	0
T_1	3	0	-1	-1	1
T_2	3	0	-1	1	-1

Methods of group theory can also be used to reduce the matrix problem to smaller invariant spaces along symmetry lines. This can sometimes be used to compute explicitly the band functions along such symmetry directions. We shall see several applications of this in Secs. XII and XIII that permit analytic calculations of the crossing points as solutions to quadratic equations.

Finally, we want to recall that methods originally developed in the context of the Bloch theory of solids (Herring, 1937) can be used to tell when time reversal together with other symmetries implies level crossings.

Points of level crossings contain global spectral information. It turns out that for the tetrahedron they occur at symmetry points, so the global character of the problem is taken care of by symmetry. The global aspect of the problem is also the hard part of the analysis. Networks with no symmetry can therefore only be analyzed numerically. Even for networks with symmetry, where group-theoretic methods give some crossing points, there appears to be no way, except brute (numerical) force, to show that there are no other "accidental" points of level crossings. It would be useful to have "sum rules" that would indicate whether a set of crossings is complete or not. One set of such rules will be described in Sec. X.

VIII. GEOMETRY AND THE FORM CALCULUS FOR PROJECTIONS

We want to examine properties of the adiabatic transport $i\,\mathrm{Tr}[\omega(P)]$ of Eq. (6.13). The calculus involved is that of forms of projections. As pointed out by Bellissard (1986a), there is a relation to noncommutative geometry (Connes, 1969; Witten, 1986).

One important question that we have already raised in Sec. V, and that we address here is what transport properties depend only on the emf's and fluxes but not on other details about the flux tubes. From a mathematical point of view the answer turns out to be quite simple: transport is described by a closed two-form. A natural equivalence is to identify closed two-forms that differ by an exact form. This is the cohomology associated with the two-form, and, as we shall explain, it has the desired properties. In particular, it implies that the flux-averaged transport has this property. These, as well as related issues, are the subject of this section.

$i\,\mathrm{Tr}[\omega(P)]$ is a closed two-form. To see this, note that $P^2=P$ gives

$$P(dP)+(dP)P=dP,\quad P(dP)P=Q(dP)Q=0\,, \tag{8.1}$$

where $Q\equiv 1-P$. It follows that

$$P(dP)^j=P(dP)^jQ,\ j \text{ odd}\,,\qquad P(dP)^j=P(dP)^jP,\ j \text{ even}\,. \tag{8.2}$$

Now, since $P^2=P$ and $d^2=0$, forms that are constructed from P and (dP) have the property that odd forms map P to Q and Q to P. In particular, the trace of any odd form vanishes. Since ω is an even form, $d\omega$ is an odd form, and so $d\,\mathrm{Tr}[\omega(P)]=0$.

Another interesting property of $\mathrm{Tr}[\omega(P)]$ is linearity. That is, if P_1 and P_2, as functions of the fluxes Φ, are mutually orthogonal, then

$$\mathrm{Tr}(\omega(P_1+P_2))=\mathrm{Tr}(\omega(P_1))+\mathrm{Tr}(\omega(P_2))\,. \tag{8.3}$$

This is remarkable because $\omega(P)$ is cubic in P. In view of Eq. (6.13), Eq. (8.3) may be interpreted as the additivity of the conductance for noninteracting fermions.

We want to consider different flux tubes, that carry the same fluxes, in the same basket. The Hamiltonians $H(\mathbf{A})$ and $H(\mathbf{A}')$ are unitarily equivalent, and so the corresponding projections are related by unitaries $U(\Phi)$. For the applications in Sec. XI we do not wish to exploit the full unitarity of U, and instead consider the more general setting where $P'\equiv UPU^\dagger$ and

$$U^\dagger U=1\,. \tag{8.4}$$

(We do not need to assume $UU^\dagger=1$ for the following calculation.) P' is a projection if P is. By explicit, tedious calculation,

$$\omega(P')=UP(\omega(P)+id(PVP)-(PVP)^2)PU^\dagger\,, \tag{8.5}$$

where, as before [cf. Eq. (2.5)], $V\equiv -iU^\dagger dU$. As a consequence

$$\mathrm{Tr}(\omega(P'))=\mathrm{Tr}(\omega(P))+id\,\mathrm{Tr}(PVP)\,. \tag{8.6}$$

$\mathrm{Tr}(\omega(P'))$ and $\mathrm{Tr}(\omega(P))$ differ by an exact form: a cohomology class is singled out. So, if we consider the transport two-form, $i\,\mathrm{Tr}(\omega(P))$, modulo exact forms, we get a transport property that is common to all flux tubes and depends only on the fluxes. We shall return presently to the question of what it means in practice to look at transport in cohomology. Before doing that, however, let us consider another important consequence of Eq. (8.6).

The gauge fields $\mathbf{A}$ and $\mathbf{A}'$ are linear in Φ. Therefore (locally) $U(\Phi)=\exp(i\Phi\Lambda)$, and $V=\Lambda\,d\Phi$ is a one-form that is independent of Φ. So, the Φ dependence of $(PVP)(\Phi)$ is determined by P. Now, from Sec. II we know that there are choices of flux tubes that make P periodic in flux space with a period of one flux unit. Any other choice of flux tubes is related to this one by an appropriate $U(\Phi)$. For this choice, each of the two terms on the right-hand side of Eq. (8.6) is periodic in Φ and so is the left-hand side. We conclude that the adiabatic transport $i\,\mathrm{Tr}(\omega(P))$ is periodic for all flux tubes [even those for which the Hamiltonian is not periodic, like Eq. (2.7)]. This establishes the fact that, for the adiabatic transport, flux space may be identified with the torus T^h.

This may be viewed as a generalization of the Bohm-Aharonov periodicity, i.e., the Byers-Yang theorem, to a class of time-dependent problems.

How does one measure a cohomology class? The answer to this is suggested by de Rham theory (Flanders, 1963), which says that the study of periods is a way to study the cohomology. Consider a closed, two-dimensional surface C in that portion of flux space where P is smooth, i.e., in $T^h/D(P)$. The Chern number associated with the projection P and the surface C is

$$\mathrm{ch}(P,C) \equiv \frac{1}{2\pi} \int_C i\, \mathrm{Tr}(\omega(P)) . \tag{8.7}$$

By Stokes-Poincaré theorem (Arnol'd, 1978) this is an invariant in cohomology, i.e.,

$$\mathrm{ch}(P,C) = \mathrm{ch}(P',C) , \tag{8.8}$$

where $P' = UPU^\dagger$ and is real by Eq. (6.8). (It is actually an integer by a more complicated argument.) Also, if C is homologous to C' in $T^h/D(P)$, then, by the closedness of ω (and Stokes-Poincaré theorem again)

$$\mathrm{ch}(P,C) = \mathrm{ch}(P,C') . \tag{8.9}$$

Of particular interest are two-dimensional sections of the torus $T_{kl}(\Phi)$ that do not intersect $D(P)$. Combining the definition of Eq. (8.7) with the basic formulas for the average transport, Eq. (6.15), we have

$$[\langle g_{\mathrm{Ad}} \rangle (P_j)]_{kl} = \mathrm{ch}(P_j, T_{kl}(\Phi)) . \tag{8.10}$$

We see that the *average adiabatic transports* are properties of the cohomology class. They depend on the fluxes and emf's and are independent of the detailed fields generating them. We note, also, that from the periodicity in flux space the kl transport coefficient in Eq. (8.10) is a function of all Φ but Φ_k and Φ_l. So, in a two-flux network, the averaged transport coefficient is an antisymmetric 2×2 matrix of numbers (it is actually the zero matrix, as we shall see later), and in a three-flux network it is a 3×3 antisymmetric matrix whose entries are functions of one variable, etc.

$\mathrm{ch}(P,C)$ are topological invariants in the sense that they are independent of the details of the flux tubes and depend only on the fluxes. Actually, they are invariants in a stronger sense of deformations of the network Hamiltonian H. To see this, note that a general (self-adjoint) deformation of H can be decomposed into three "transversal" pieces: (1) deformations of the set of level crossings $D(P)$; (2) deformations of the spectrum that keep the spectral subspaces invariant; and (3) isospectral deformations, i.e., deformations of the spectral subspaces that keep the spectrum invariant. $\mathrm{ch}(P,C)$ is invariant under (1), for if C does not intersect $D(P)$, it also does not intersect a deformation of $D(P)$. $\mathrm{ch}(P,C)$ is invariant under (2), for $\omega(P)$ is invariant (it depends only on the projection P, not the energies E). Finally, isospectral deformations are given by unitaries and the invariance of $\mathrm{ch}(P,C)$ under these follows from Eq. (8.6).

We now recall the proof that $\mathrm{ch}(P,C)$ is an integer if P

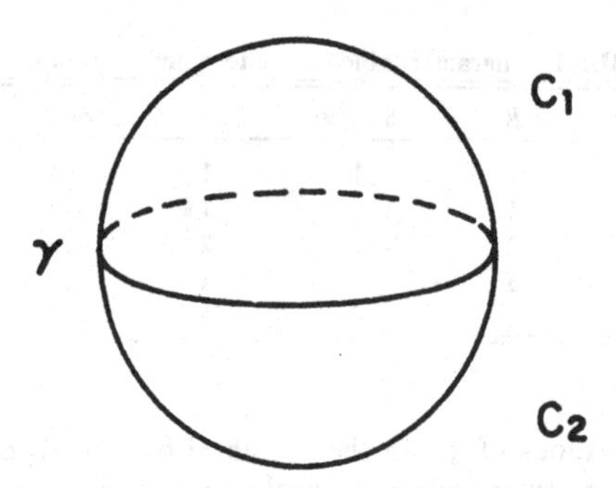

FIG. 15. A covering of a closed surface S^2 by two contractible surfaces with a common boundary γ.

is one dimensional. [By Eq. (8.3) this extends to finite-dimensional P's that are deformations of sums of one-dimensional ones.] Let $C = C_1 + C_2$, with $C_{1,2}$ contractible and $\gamma = \partial C_1 = -\partial C_2$ (see Fig. 15). Since C_1 is contractible there is a choice of smooth phase for ψ_1 so that $P = |\psi_1\rangle\langle\psi_1|$. Then

$$i \int_{C_1} \mathrm{Tr}(\omega(P)) = i \int_\gamma \langle \psi_1 | d\psi_1 \rangle . \tag{8.11}$$

The left-hand side is independent of the choice of (phase for) ψ_1 [see Eq. (7.1)]. The right-hand side is Berry's phase (Berry, 1984), associated with the parallel transport along the curve γ. Since the adiabatic evolution along γ is unique and is independent of whether we think of γ as the boundary of C_1 or of C_2, it must be that

$$\int_\gamma \langle \psi_1 | d\psi_1 \rangle = \int_\gamma \langle \psi_2 | d\psi_2 \rangle + 2\pi i n , \tag{8.12}$$

with n integer. It follows that

$$\int_C \mathrm{Tr}(\omega(P)) = \int_{C_1} \mathrm{Tr}(\omega(P)) + \int_{C_2} \mathrm{Tr}(\omega(P)) = 2\pi i n , \tag{8.13}$$

proving the integrality.

We conclude by noting the physically obvious fact that there is no transport between the disconnected pieces of a network. Namely, if the loops l and m belong to disconnected pieces in the network, then $[\langle g_{\mathrm{Ad}} \rangle (P)]_{lm} = 0$. To see this, choose a representation in which the Hamiltonian factorizes as

$$H_1(\Phi_1) \otimes 1 + 1 \otimes H_2(\Phi_2), \quad \Phi \equiv (\Phi_1, \Phi_2) . \tag{8.14}$$

This relation carries over to the projections and it makes $\omega(P(\Phi))$ vanish identically on any two-surface made of fluxes in disconnected components.

IX. TIME REVERSAL

Time reversal is the single most powerful tool in the analysis of networks. For Schrödinger operators, time reversal is the statement

$$H(\Phi) = U_0 H^*(-\Phi) U_0^\dagger , \tag{9.1}$$

where U_0 is a fixed (Φ-independent) unitary. For spinless electrons, in a gauge where $H(\Phi=0)$ is real, and in the "coordinate representation," $U_0=1$. For spin-$\frac{1}{2}$ electrons in the same gauge and representation, $U_0=i\sigma_y$. It follows that

$$P(\Phi)=U_0P^*(-\Phi)U_0^\dagger \,, \tag{9.2}$$

and so

$$i\,\mathrm{Tr}(\omega(\Phi))=-i\,\mathrm{Tr}(\omega(-\Phi)) \,. \tag{9.3}$$

We write $\omega(\Phi)$ for $\omega(P(\Phi))$. Equation (9.3) is Onsager's relation (Onsager, 1931).[24] It also follows from Eq. (9.2) that $D(P)$ is invariant under inversion:

$$D(P)=I[D(P)] \,, \tag{9.4}$$

where

$$I(\Phi)=-\Phi \,, \tag{9.5}$$

if one thinks of flux space as $\mathbb{R}^h$. If one thinks of flux space as $\mathbb{T}^h$, then Eq. (9.5) is interpreted modulo 2π.

From Eq. (9.3) we get

$$\mathrm{ch}(P_j,C)=-\mathrm{ch}(P_j,I[C]) \,. \tag{9.6}$$

Let $T_{lm}(\Phi)$ be the lm slice of the torus in flux space passing through Φ. Then

$$I[T_{lm}(\Phi)]=T_{lm}(-\Phi) \,; \tag{9.7}$$

see Fig. 16. Combining Eq. (8.10) with Eqs. (9.6) and (9.7) gives the antisymmetry of transport in the fluxes:

$$\langle g_{\mathrm{Ad}}\rangle(\Phi)=-\langle g_{\mathrm{Ad}}\rangle(-\Phi) \,. \tag{9.8}$$

This is a weaker version of Onsager's relation. (It is a statement about averages only.) In particular, in a network with only two fluxes, $\langle g_{\mathrm{Ad}}\rangle$ is a 2×2 matrix of numbers, and so zero. (Zero is the only antisymmetric number.)

Equation (9.8) makes certain periods vanish. Suppose that T_{lm} is a section that is invariant under inversion, e.g., the planes $\Phi_k=0$ ($\Phi_k=\pi$), $k\neq l,m$. If T_{lm} does not intersect $D(P)$ then, from Eq. (9.8),

$$\mathrm{ch}(P,T_{lm})=0, \quad T_{lm}=I[T_{lm}] \,. \tag{9.9}$$

If the two-torus T_{lm} intersects $D(P)$, the period is not defined. However, as we shall see, something can be said about this case as well.

We have already mentioned the fact that $D(P)$ acts as the source of charge transport. In a three-flux network $D(P)$ is, generically, a discrete set, and the "charge" of the point $\phi\in D(P)$ is defined by $\mathrm{ch}(P,S_\phi^2)$ where S_ϕ^2 is a small two-sphere centered at ϕ (with outward drawn normal). Now, if $\phi\in D(P)$, $-\phi\in D(P)$ by Eq. (9.4), and[25]

$$I[S_\phi^2]=-S_{-\phi}^2 \tag{9.10}$$

(see Fig. 17). Note the difference in the way inversion acts on the orientation of two-tori and two-spheres. Combining Eqs. (9.10) and (9.6),

$$\mathrm{ch}(P,S_\phi^2)=\mathrm{ch}(P,S_{-\phi}^2) \,. \tag{9.11}$$

Degeneracies that are images of each other under inversion have *the same* charge.

We now return to sections of the torus that are invariant under inversion, $T_{lm}(\Phi)=I[T_{lm}(\Phi)]$, and intersect $D(P)$. Let $\hat{n}$ be the normal to the section. Then $I[T_{lm}(\Phi+\hat{n}\epsilon)]=T_{lm}(\Phi-\hat{n}\epsilon)$, and the translated sections will not intersect $D(P)$ for ϵ small. Therefore, their periods are well defined. From the closedness of $\omega(P)$ and Eq. (9.8) we obtain

$$\begin{aligned}\mathrm{charge}(T_{lm}(\Phi)) &= (\mathrm{ch}(P,T_{lm}(\Phi+\hat{n}\epsilon)) \\ &\quad -\mathrm{ch}(P,T_{lm}(\Phi-\hat{n}\epsilon)) \\ &=2\,\mathrm{ch}(P,T_{lm}(\Phi+\hat{n}\epsilon)) \,. \end{aligned}\tag{9.12}$$

Thus such planes are evenly charged. We learn that in either case, whether the invariant planes are charged or not, the periods of their ϵ translates are determined by the charges on the invariant planes. We shall make use

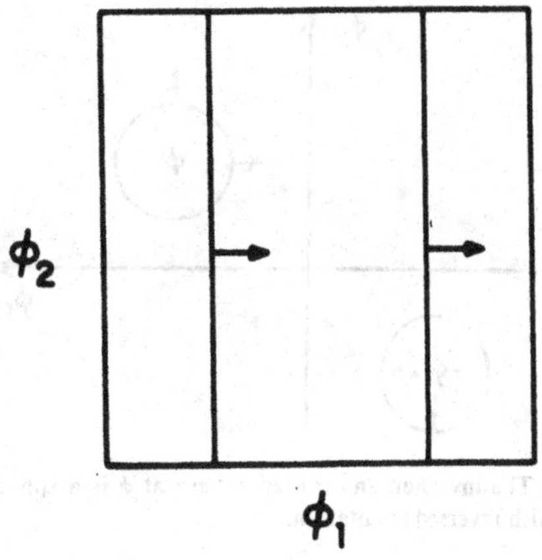

Φ_2

Φ_1

FIG. 16. The inversion of the section $T_{lm}(\Phi)$, with, say, $l=2$, $m=3$, is the lm section though $-\Phi$ with the same orientation, i.e., $T_{lm}(-\Phi)$.

[24]Strictly speaking, Eq. (9.3) is only a statement about the *adiabatic transport*, defined in Sec. VI. Whether it also applies to the physical transport depends on how well the two approximate each other. See the discussion in Sec. VI.

[25]To see why Eq. (9.9) holds, recall that the area form of the n-sphere is

$$\sum_j (-)^{j+1}\frac{x_j}{|x|}dx_0\wedge\cdots\wedge dx_{j-1}\wedge dx_{j+1}\wedge\cdots\wedge dx_n \,.$$

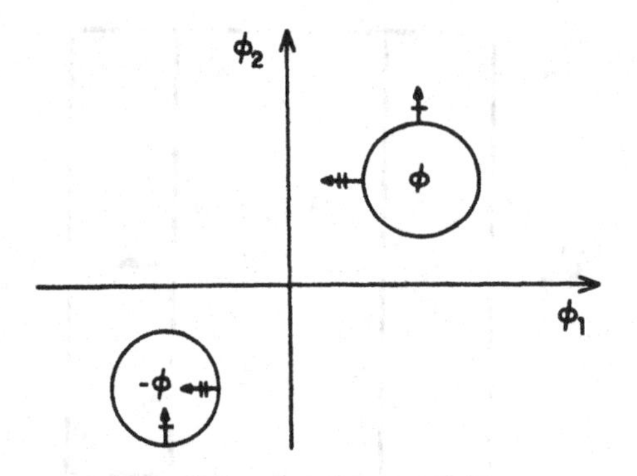

FIG. 17. The inversion on an even sphere at ϕ is a sphere at $-\phi$ but with reversed orientation.

of this fact in the next section.

Here we have considered the action of time reversal alone. Of course, time reversal is more powerful when combined with other symmetries (Herring, 1937).

X. CLASSIFICATION OF THREE-FLUX NETWORKS

The transport properties of three-flux networks with discrete $D(P_j)$'s are determined by the charges $\mathrm{ch}(P_j,S_\phi^2)$, $\phi\in D(P_j)$. This information can be arranged in tables that list the coordinates of the various charges. The corresponding matrix of average transport functions $\langle g_{\mathrm{Ad}}\rangle(P_j)$ can then be read directly from the table, as we shall proceed to explain. Toward the end of this section we introduce various notions of stability of the topological conductance and define *n*-type and *p*-type networks.

The set of Chern numbers $\{\mathrm{ch}(P,C)\}$ for all spectral projections P and all closed two surfaces C has a linear structure, so it is enough to have the periods for bases of the projections and bases, in the sense of homology, for surfaces in $\mathbb{T}^3/D(P)$. We first discuss the choice of a basis for the spectral projections. The point we want to make is that the first natural choice is actually not the best. That is, taking P_j to be the projection on the jth eigenvalue, so that j is an energy label, is bad because it contains redundant information: $D(P_j)$, $D(P_{j+1})$, and $D(P_{j-1})$ are not independent. A better choice is to take P_j to be the projection on the spectrum for all energies below the jth gap. $D(P_j)$ is then related to the set where the jth gap closes and is independent of $D(P_k)$, $k\neq j$. In other words, it is better to take j to be a gap label rather than an energy label. This choice also turns out to be the right choice in other contexts (Johnson and Moser, 1982; Thouless, 1983; Dana, Avron, and Zak, 1985; Avron and Yaffe, 1986; Bellissard, 1986b; Kunz, 1986]. By Eq. (8.3), this choice makes $\mathrm{ch}(P_j,T_{jk}(\Phi))$ the matrix of average transport in the *ground state of the j-electron system*.

In the general case, j runs over the naturals. In the special case of tight-binding models, the Hilbert space is $\mathbb{C}^{|V|}$, so j runs on $1,\ldots,|V|$. In this case $P_{|V|}=1$, and the associated curvature vanishes identically. For this reason, in tight-binding models, it is enough to consider j in $1,\ldots,|V|-1$. A similar thing occurs for free electrons, as we shall explain below.

Now we come to picking a basis for the two-chains. If we let $z_j^{(2)}$ be the basis of the second homology of the torus in flux space with $D(P_j)$ removed, then, clearly, every closed two-chain $z^{(2)}$ can be written as a linear combination with integer coefficients

$$z^{(2)}=\sum_k n_k z_k^{(2)}\ , \tag{10.1}$$

up to homology, so the basic periods $\mathrm{ch}(P_j,z_k^{(2)})$ determine all periods:

$$\mathrm{ch}(P_j,z^{(2)})=\sum_k n_k \mathrm{ch}(P_j,z_k^{(2)})\ . \tag{10.2}$$

$H_2(T^3/D(P))$ is clearly spanned by three basic sections made of two-tori, T_{lm}, $l,m=1,2,3$, and $|D(P_j)|$ spheres S_ϕ^2 that surround the points ϕ in $D(P_j)$. This seems to suggest that, given j, one needs $|D(P_j)|+3$ periods. Actually, time reversal leads to relations between the periods, so only the "charges" are needed. In fact, not even all the charges are needed, as it is enough to have those in the half-torus, as we proceed to explain.

The three periods associated with slicing the torus can be disposed of. Choose a slice that is invariant under inversion, e.g., take the slice through 0 or π. If this slice happens to intersect $D(P)$, the relevant period is that of an ϵ-translate which does not intersect $D(P)$. Then, as discussed in Sec. IX, the period is half the charge on the invariant slice. In particular, it is zero if the invariant slice does not intersect $D(P)$, so the charges determine the periods of three basic sections of the torus.

The adiabatic transport $\langle g_{\mathrm{Ad}}\rangle(P_j)$ is now determined, for all Φ, by translating the above planar slices, picking up the charges swept in this process. We conclude that the charges determine the transport.

Not all the charges are, however, independent. By Eq. (9.11), the periods of S_ϕ^2 and $S_{-\phi}^2$ are the same, so it is enough to know the periods in half of the torus, say, $0\leq\Phi_1<\pi$. Finally, even the charges in half the torus are not completely independent, for their total charge is zero. To see this, observe that the charge of half the torus is half the charge of the full torus, by Eq. (9.11). The charge in the total torus is, however, easily seen to be zero from the periodicity of $\langle g_{\mathrm{Ad}}\rangle$.

The charges provide an efficient way of displaying the transport properties. More interesting is that the charges "localize" the problem: the charge at ϕ is a local property of the bundle associated with P and can be computed from properties of the Hamiltonian near ϕ by methods of perturbation theory. In practice this means that it is only necessary to diagonalize the Hamiltonian at the point ϕ itself. This is an improvement over Eq. (6.14), which requires diagonalizing the Hamiltonian on a planar section of the torus, and so requires global information on the bundle. (Of course, one still needs to know where the eigenvalues cross, and this involves global information.)

We illustrate this with an example. Consider an isolated two-level crossing, and let $h(\Phi)$ be as in Eq. (7.2a). The map from the space of 2×2 self-adjoint matrices to $\mathbb{R}^3$ given by

$$\epsilon(\Phi)=\tfrac{1}{2}\mathrm{Tr}(h(\Phi)\sigma) \qquad (10.3)$$

has $\epsilon(\phi)=0$, but $\epsilon(\Phi)\neq 0$ for Φ on a small sphere surrounding ϕ. This is a consequence of the assumption that ϕ is an isolated degeneracy [see Eq. (7.3a)]. Equation (10.3) defines a continuous map from the two-sphere S^2_ϕ to $\mathbb{R}^3/0\sim S^2$. (Here $\sim$ denotes equivalence in homotopy.) Such maps are characterized by their degree (Dubrovin *et al.*, 1984), and, as we shall now show, the degree is $-$(*charge*).

From Eq. (7.4) the projection P_- associated with the gap is

$$P_-(\Phi)=\tfrac{1}{2}[(1-\hat{\epsilon}(\Phi)\cdot\sigma)] , \qquad (10.4)$$

and by Eq. (6.7) the curvature is

$$\omega(P_-)=\mathrm{Tr}[P_-(dP_-)(dP_-)P_-]$$

$$=-\tfrac{1}{8}\mathrm{Tr}[\sigma\cdot\hat{\epsilon}(\sigma\cdot d\hat{\epsilon})(\sigma\cdot d\hat{\epsilon})]$$

$$=-\frac{i}{4}\hat{\epsilon}\times d\hat{\epsilon}\cdot d\hat{\epsilon} . \qquad (10.5)$$

In the second step we used $\mathrm{Tr}(dP\,dP)=d\,\mathrm{Tr}(P\,dP)$ $=d\,\mathrm{Tr}(P\,dP\,P)=0$. Therefore the charge is

$$\mathrm{ch}(P_-,S^2_\phi)=\frac{1}{2\pi i}\int_{S^2_\phi}\omega(P_-)$$

$$=-\frac{1}{8\pi}\int_{S^2_\phi}\hat{\epsilon}\times d\hat{\epsilon}\cdot d\hat{\epsilon} . \qquad (10.6)$$

The right-hand side is $-$(*degree*) of the map from the two-sphere S^2_ϕ, to the two-sphere $\hat{\epsilon}(\Phi)$.

In this way, the averaged conductances can be computed by diagonalizing the Hamiltonian for a discrete set of points $D(P)$. Equation (10.6), however, still involves integration. In numerical calculations integrals may be tedious, so it is nice that, in the generic situation, calculating the degree actually reduces to computing the determinant of a single 3×3 matrix. This observation, made in a related context, is due to Simon (1983). Consider the linearized map of Eq. (7.6). Its determinant is the Jacobian of the transformation from Φ space to ϵ. The degree is the sign of the Jacobian. It follows that

$$\mathrm{ch}(P_j,S^2_\phi)=-\mathrm{sgn}\det(\nabla\otimes\epsilon) . \qquad (10.7)$$

This formula holds only if $\det(\nabla\otimes\epsilon)\neq 0$. If the determinant vanishes, the degree is not determined by the linearized map and could be *any* integer: 0, ± 1, ± 2, etc. It would be useful to have formulas that would cover some of the cases in which the degree is not determined by the linearized map.

The bundles that arise in the study of the reduced free-electron problem admit a similar analysis, but turn out to have more structure. We recall that the relevant line bundle $\psi_j(\Phi)$ satisfies Eq. (4.29) for an appropriate $\cos[k_j(\Phi)]$.

By analogy with the previous case, let $\nabla\otimes\epsilon$ denote the linearized map from (k,Φ) to (ϵ_0,ϵ) in Eq. (7.2b). If this map is of full rank, the analog of Eq. (10.7) is

$$\mathrm{charge}(k_0,\phi)=-\mathrm{sgn}\det(\nabla\otimes\epsilon) . \qquad (10.8)$$

The bands of the tight-binding model with *zero on-site potential* have an analog of the "π-shift" invariance discussed in Eq. (4.35). It says that if (E,Φ) is in the spectrum, so is $(-E,\Phi+\pi F)$, and both have the same eigenvectors. As a consequence,

$$\mathrm{charge}(E,\phi)=-\mathrm{charge}(-E,\phi+\pi F) . \qquad (10.9a)$$

The reason for the minus sign is that the "π shift" flips the sign of the energy, and this flips the "projection below" to a "projection above." The graph-theoretic tight-binding models do not have this invariance, except in the case of regular graphs, for which similar considerations give

$$\mathrm{charge}(E,\phi)=-\mathrm{charge}(-E+2|v|,\phi+\pi F) . \qquad (10.9b)$$

The band spectrum of the free-electron model also has the "π-shift" invariance, as discussed in Sec. IV, for both regular and irregular graphs. It says that $h_0(k,\phi)$ and $h_0(k+\pi,\phi+\pi|f|)$ have the same kernel. As a consequence,

$$\mathrm{charge}(k,\phi)=\mathrm{charge}(k+\pi,\phi+\pi F) . \qquad (10.9c)$$

There is no sign change, in this case, for the ordering of energies is preserved by the map. Similarly, $h_0(k,\phi)$ and $h_0(2\pi-k,\phi)$ have the same kernel. This gives yet another relation for the charges:

$$\mathrm{charge}(k,\phi)=-\mathrm{charge}(2\pi-k,\phi) . \qquad (10.9d)$$

As a consequence of Eqs. (10.9c) and (10.9d) it is clearly sufficient to list the charges with $k\in[0,\pi]$, rather than $[0,2\pi]$, and in this interval one has

$$\mathrm{charge}(k,\phi)=-\mathrm{charge}(\pi-k,\phi+\pi F) . \qquad (10.9e)$$

The topological conductances of free electrons on a network with $|V|$ vertices, and edges of equal lengths, are periodic functions of the number of electrons with period $2|V|$ (and are antiperiodic with period $|V|$). This is a consequence of the reduction to a $|V|\times|V|$ matrix problem, which makes the eigenvectors independent of the quantum number n in Eq. (4.32). [Since the metric in Eq. (4.39) is not periodic in k, there is a gap in this argument that will be patched in the next section.] The $|V|$-electron system, like the no-electron system, has trivial transport. For the classification problem this means that the charge tables have to cover band indices $j\in 1,\ldots,|V|-1$.

This periodicity is interesting from another point of view. In networks with few loops and few electrons, one may expect the topological conductances to be relatively simple functions, with few jumps from one integer to another, and of order unity. This should be the case, for there is no large number in the problem. However, in mesoscopic systems with few loops, the number of electrons can be quite large, and this could make the topological conductances complicated. In particular, large values and wild oscillations cannot be excluded *a priori.* For noninteracting electrons, we see that this is not the case, as the number of electrons is counted modulo $|V|$ and so is kept small. This suggests that the conductances should be "relatively simple."

Finally, we note that regular graphs have equivalent transport properties for the three models we have considered: the graph-theoretic tight-binding, the tight-binding with zero on-site potential, and the de Gennes–Alexander model. This follows from the fact that regular graphs, by definition, are such that $|v|$ is the same for all vertices. The two tight-binding models then differ by a constant, and the implicit and explicit eigenvalue problems are related by Eq. (4.37). Since the eigenvectors in the three dynamics are the same, the only distinction comes from the scalar product, which is different in the tight-binding and the free-electron cases. However, modulo a technical point that will be discussed in the next section, it is a basic feature of the Chern character that it is independent of the differential structure (i.e., the scalar product). This leads to the identity of the charges.[26]

An interesting aspect of the topological conductance concerns the issue of electrons and holes. In the Hall effect, electrons and holes are distinguished by comparing the direction of the actual current with the naive expectation that comes from analyzing the motion in crossed electric and magnetic fields. In networks there is also the possibility for both signs (the Chern number is not constrained to be of definite sign), and it is natural to try to classify networks by types as well, as a kind of generalization of the electron-hole concept (Avron, Seiler, and Shapiro, 1986). To do so, one has first to decide on a "naive expectation" and then to call the conductance electronlike if it agrees with that and holelike if it does not. In contrast with the Hall effect situation, where the "naive expectation" does not require any nontrivial quantum-mechanical calculation, in networks there appears to be no "naive expectation" in this sense.[27] It is natural to take the free-electron model for the network as a benchmark and call the conductances electronlike or holelike if they agree/disagree with the free-electron prediction. This definition, however, is conditioned on the stability and simplicity of the topological conductance, as we proceed to explain.

The free-electron model could, in principle, have complicated conductance functions as in Fig. 18(a), and the tight-binding model for the same graph could be as in Fig. 18(b). In such a case the tight-binding model is neither n-type nor p-type. The type is well defined if the graph of the topological conductance does not change signs too often. Simplicity of the graph is related to notions of stability that we proceed to formulate for three-flux networks.

We say that the i-j conductance is *flux stable* if it is of fixed sign for $\Phi_k \in (0,\pi)$. A network is *flux stable* if all pairs of the conductances are. If the free-electron network is nontrivial and flux stable, and if some other dynamics associated with the same graph is also flux stable and nontrivial, its type is well defined. If the dynamics leads to trivial transport, we say that it is insulating. (This definition fails in the case where the free-electron dynamics give zero topological conductance, and some other dynamics does not.)

Another stronger notion of stability is related to the stability of types when the number of particles is not fixed. We say that the i-j conductance is μ *stable* if it is flux stable and the sign is independent of the number of electrons.

A natural set of questions is the following.

(a) When are networks flux stable? General networks are expected not to be flux stable. However, as we shall see, all the nontrivial networks made of three equilateral triangles turn out to be flux stable.

(b) Are there simple rules for determining when a given dynamics leads to electronlike or holelike behavior?

(c) When are networks μ stable? Again, *a priori,* one expects that the sign of the conductance could depend on the number of electrons. As we shall see, eight out of nine three-flux networks turn out to be μ stable in the tight-binding dynamics. For such networks, the type is not affected by coupling to a bath with fixed chemical potential.

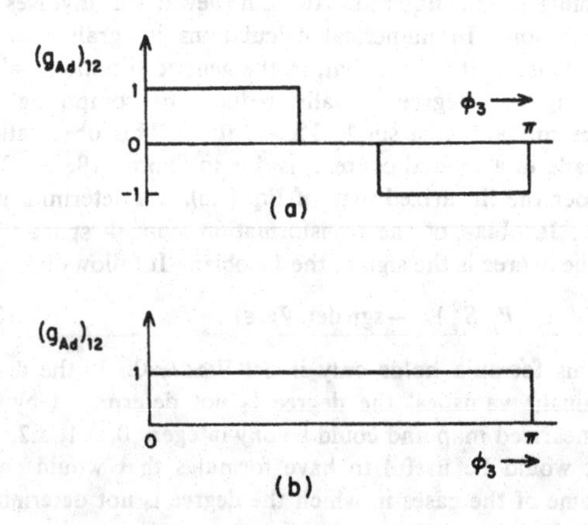

FIG. 18 The putative graphs of the topological conductances: (a) for the free-electron model; (b) for the tight-binding model. There is no natural way to decide whether the tight-binding model is n-type or p-type.

---

[26]We thank B. Simon for this observation.

[27]The "naive expectation" for a network made of ordinary resistors cannot be used to define types, for this gives a symmetric, instead of antisymmetric, matrix.

XI. CHERN NUMBERS: REDUCTION TO A MATRIX PROBLEM

One of the interesting aspects of the average conductances is that an object that arises in the study of certain partial differential operators can be computed reliably by studying finite matrices. In Sec. IV the partial differential Schrödinger equation was reduced to an ordinary differential equation in the limit of thin wires. For most questions the ordinary differential equation is only an approximation. The average conductances, being topological invariants, are insensitive to deformations, so the ordinary differential equation result is exact if the wires are sufficiently thin and one is not too close to level crossings.[28] A further reduction, from the ordinary differential equation to a matrix problem on the vertices, involves no further approximation. Since a wave function on the vertices determines the wave function on the edges, it is clear that computing the conductances of the network is a matrix problem.

Our purpose here is twofold. First, we want to describe explicit formulas for the *unaveraged* conductances that apply directly to the matrix problem. This requires the right Riemann metric (i.e., scalar product) in the matrix space, which is induced from the Hilbert space metric. Second, we want to close a gap in an argument made in the last section, that is, to complete the proof that the averaged conductances, i.e., the Chern numbers, can be computed directly from the matrix problem without regard to the "right" metric, using, for example, the usual "flat" scalar product.

Let $e=(u,v)$ be an edge with vertices u and v. Let ψ be the wave function on the vertices with components $\psi(v)$. Fix the edge e. Define a linear map $T(k)$ from $\mathbb{C}^{|V|}$ to $\oplus L^2(e)$ by

$$(T(k)\psi)(x,e)\equiv\frac{\exp[i\mathbf{a}(e,x)]}{\sin(k\,|\,e\,|\,)}\{\psi(u)\sin[k(\,|\,e\,|\,-x)]+\psi(v)\exp[-i\mathbf{a}(e)]\sin(kx)\}\;. \tag{11.1}$$

Using this map one can write the curvature two-form in Eq. (6.12) in terms of the de Gennes–Alexander $|V|$ vector. This yields direct, although somewhat longish, formulas for the curvature in terms of these $|V|$ vectors.

$T^\dagger(k)$, the adjoint of $T(k)$, is a linear operator from the range of $T(k)$ to $\mathbb{C}^{|V|}$ such that

$$T^\dagger(k)T(k)=1\;. \tag{11.2}$$

$TT^\dagger$ is not the identity, but because of Eq. (11.2) it is a projection. We apply Eq. (8.6) with

$$P'\equiv T(k)PT^\dagger(k),\quad V\equiv -iT^\dagger(k)dT(k)\;,$$
$$\mathrm{Tr}(\omega(P'))=\mathrm{Tr}(\omega(P))+d\,\mathrm{Tr}(PV)\;. \tag{11.3}$$

Equation (11.3) says that the curvature associated with the $|V|\times|V|$ projection matrix and the curvature associated with the projection in the $\oplus_E L^2(e)$ Hilbert space differ by an exact form in flux space $\mathbb{R}^h/D(P)$. In particular, to compute the *unaveraged conductance directly*, one may use Eq. (11.3) together with the $|V|\times|V|$ projection matrix $P(\Phi)$ of the Alexander–de Gennes problem. Because of the metric $M(k)$ of Eq. (4.40), P is related to the normalized eigenvector with eigenvalue 0 by

$$[P(\Phi)]_{uv}=\psi(u)\sum_{v'}\psi^*(v')[M(k)]_{v'u}\;. \tag{11.4}$$

Because of the metric $M(k)$, the actual computations are, of course, more involved. Similarly, one can write formulas in terms of eigenfunctions rather than projections.

For the computation of Chern numbers, however, one may use the flat metric. The issue at stake is whether (PV) in Eq. (11.3) is periodic in flux space. If so, the periods over the torus for the matrix, and the Hilbert space problem, are the same. The period for the matrix problem is a Chern number and by deformation argument it is independent of the metric, so it remains to show that PV in Eq. (11.3) is periodic.

From Eq. (11.1) it follows that

$$dT=i(d\mathbf{a}(x,e))(T\sigma_z)-|\,e\,|\,dk\cot(k\,|\,e\,|\,)T+dk\,T\begin{bmatrix}(\,|\,e\,|-x)\cot[k(\,|\,e\,|-x)] & 0\\ 0 & x\cot(kx)\end{bmatrix}, \tag{11.5}$$

from which we get

$$\int_e dx[T\psi]^*(x)[(dT)\varphi](x)=(\psi,[T^\dagger dT]\varphi)\;, \tag{11.6}$$

where $[T^\dagger dT]$ is a 2×2 Hermitian matrix with entries

[28]The same argument can be made about the tight-binding limit.

$$\sin^2(k\,|e|)(T^\dagger dT)_{11}=\int_e dx\,\sin^2[k(|e|-x)][i\,d\mathbf{a}(x,e)-|e|\,dk\cot(k\,|e|)$$
$$+(|e|\,x)dk\cot(k\,|e|-x)]\,,$$
$$\sin^2(k\,|e|)(T^\dagger dT)_{12}=\int_e dx\,\sin^2[k(|e|-x)]\sin(kx)\exp[-i\mathbf{a}(e)]$$
$$\times[-i\,d\mathbf{a}(x,e)-|e|\,dk\cot(k\,|e|)+x\,dk\cot(kx)]\,, \tag{11.7}$$
$$\sin^2(k\,|e|)(T^\dagger dT)_{22}=\int_e dx\,\sin^2(kx)[-i\,d\mathbf{a}(x,e)-|e|\,dk\cot(k\,|e|)-x\,dk\cot(kx)]\,.$$

The Φ dependence comes from the Φ dependence of k, dk, $d\mathbf{a}$, and $\exp(-i\mathbf{a})$. The first two are periodic in Φ by the noncrossing. The third is Φ independent, and the fourth is periodic, at least for the applications in Secs. XII and XIII. It follows that $[T^\dagger dT]$ is periodic. It is also smooth in Φ provided $\sin(k\,|e|)\neq 0$.

XII. NONTRIVIALITY: THE HOLES EFFECT

The general structure in the preceding sections is consistent with $\langle g_{Ad}\rangle=0$ identically. A basic question, therefore, is whether there are nontrivial networks. Our original interest in networks came from our interest in the Hall effect and so did much of our early intuition about what one should expect. The first networks we looked at were two-flux networks, because they are the simplest, and because the Hall effect also has two loops. Not surprisingly they were all found to be trivial for the same reason that there is no Hall conductance in zero magnetic field, namely, time reversal. One needs a third loop and a third flux to play the roles of the crystal and magnetic field. Next, we looked at three-flux networks. The first was that of Fig. 19, which looks like the Hall effect expect that a hole replaces the Hall sample and Φ_3 replaces the magnetic field B. This graph is nonplanar and the associated tight-binding model was indeed found to be nontrivial. However, as we subsequently realized, nonplanarity is not essential for nontriviality, and neither is the close correspondence with the Hall effect. In fact, nontrivial networks appear to be "generic." Here we focus on one particular planar network that is a close relative of Fig. 19 and that is shown in Fig. 20. We dub it the holes graph. We present the analysis of the tight-binding and the free-electron models for this graph.

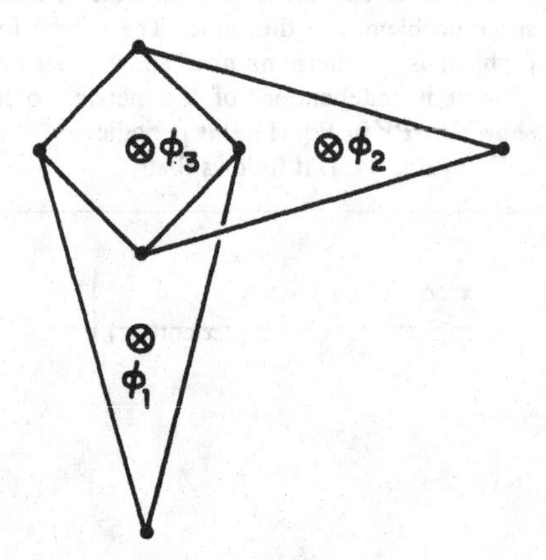

FIG. 19. A three-loop graph that mimics the Hall effect and whose corresponding tight-binding model is not trivial.

One reason for singling out the holes graph is historical: this was the first graph we analyzed in detail. From a textbook, didactic point of view this is an unfortunate choice for, as it turns out, this is a complicated model. The tetrahedron and the gasket of the next section are symmetric graphs, and this simplifies the analysis. So the holes graph is actually a typical representative of the harder, low-symmetry, models. A reader who would rather first study a model that can be analyzed in a few lines of calculation is referred to the subsection on the tetrahedron in the next section. With the graph we associate two dynamics: the (graph-theoretic) tight-binding and the free-electron dynamics. Since the graph is not regular, there is no reason why the dynamics should coincide, and a meaningful comparison of the transport properties can be made. In particular, as we shall see, both lead to nontrivial (topological) transport.

We shall first describe the numerical and analytical methods involved and the way the results will be presented. The same methods and conventions hold in the next section, where results for a whole batch of graphs are given.

As we have explained in Sec. X, the hard part of the problem is the global piece of isolating level crossings.

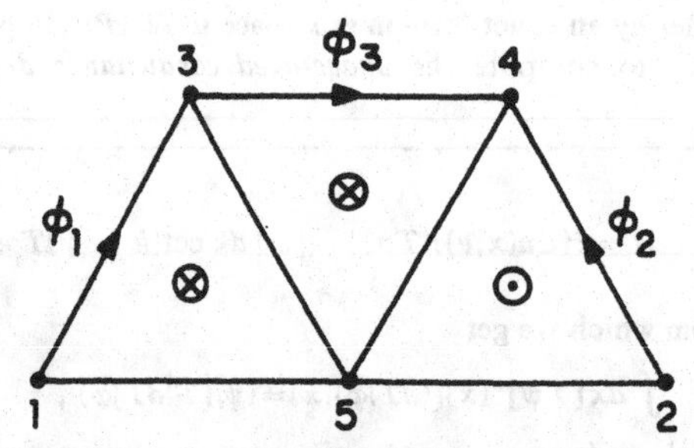

FIG. 20. The graph of the holes effect. ⊗ and ⊙ denote fluxes going into and out of the plane. The gauge field is chosen so that a phase Φ_1 is associated with the edge (1,3), Φ_2 with the edge (2,4), and Φ_3 with the edge (3,4). The graph has five vertices, seven edges, and three loops.

The two main tools are numerical method and the application of symmetry principles. The numerical method is, of course, quite powerful as it applies to general graphs, and is easily adaptable from one network to another. We have used the two methods in tandem, in the sense that the numerical results often suggested analytic derivations. Let us first describe the main features of the numerical analysis of networks.

Isolating the points of degeneracies is not trivial from a numerical point of view, because the finite numerical accuracy may blur a true degeneracy with near avoided crossing. One way to test for degeneracies is to compute the Chern number associated with small spheres or cubes. This is the complex version of an idea proposed many years ago by Herzberg and Longuet-Higgins (1963) and Longuet-Higgins (1975) in the real case. (In the real case the sphere is replaced by a circle, and the two-form by a one-form associated with adiabatic transport.) If the charge is nonzero, there is at least one degeneracy there; if it vanishes, no firm conclusion can be made. So, in principle, it is possible to divide the unit cell in flux space into small cubes and compute the charge of each cube by an appropriate surface integral. In practice this method is very expensive in computer time, for the unit cell has to be divided into many cubes if the degeneracy is to be identified with some precision. Further, each of the small cubes has to be wrapped by a fine mesh for the surface integrals. The mesh has to be fine enough so that the surface integral is close to an unambiguous integer. And finally, the integrand requires diagonalization of the Hamiltonian at every point of the mesh. A typical CPU time for such an integral, for a five-vertex network on an IBM 3081D, is on the order of 1000 sec. Instead of this systematic but time-consuming procedure we have used a simpler method that suggests where points of level crossings may lie. Once these were isolated, their charges were computed by surface integrals. We then checked that we did not miss charges by computing Chern numbers for many planar slices of the unit cell.

We looked for degeneracies by directly examining the smoothness of the projection. If we divide the cube with a mesh of size ϵ, then normalized eigenvectors away from a degeneracy satisfy

$$|\langle\psi_j(\Phi)|\psi_j(\Phi+\epsilon)\rangle|=1-O(\epsilon^2)\ . \qquad (12.1a)$$

Near a degeneracy the projection need not be close to 1. For example, if a diabolic crossing is midway between Φ and $\Phi+\epsilon$, then [see Eq. (7.4)] $\langle\psi_j(\Phi)|\psi_j(\Phi+\epsilon)\rangle=O(\epsilon^2)$. In general one has

$$|\langle\psi_j(\Phi)|\psi_j(\Phi')\rangle|=\tfrac{1}{2}[1+\hat{e}(\Phi)\cdot\hat{e}(\Phi')]\ , \qquad (12.1b)$$

with a "typical value" of order $\frac{1}{2}$. We chose a sequence of $\epsilon_j=2\pi/(3\times 2^j)$, $j=1,\ldots,10$ and computed the overlap in Eq. (12.1) for successive ϵ_j if the overlap was not close enough to 1. Once a small cube with a possible degeneracy was identified, its charge was used to decide on crossing.

The charges have been computed numerically as a surface integral using a formula that is equivalent to Eq. (8.7). [Since standard programs for diagonalization of matrices yield eigenvectors rather than projections, it proved convenient to write the analog of Eq. (8.7) in terms of eigenfunctions.]

The symmetry principles can be applied only on a case-by-case basis. For the holes graph the permutation of the vertices

$$U=(12)(34)(5)\ , \qquad (12.2a)$$

implementable as a unitary transformation on the Hamiltonian, is equivalent to

$$(\Phi_1,\Phi_2,\Phi_3)\to(\Phi_2,\Phi_1,-\Phi_3)\ . \qquad (12.2b)$$

The set of points in flux space that are left invariant under the combined action of U and time reversal defines "symmetry" points. They lie on two planes, which turn out to play a special role. The planes are

$$(\Phi_1,-\Phi_1,\Phi_3),\ \ (\pi-\Phi_1,\pi+\Phi_1,\Phi_3)\ . \qquad (12.3)$$

Another set of symmetry points are those that are left invariant under the action of U alone. These lie on two lines, and these lines also play a special role. The lines are given by

$$(\Phi,\Phi,0),\ \ (\Phi,\Phi,\pi)\ . \qquad (12.4)$$

As we shall see, the symmetry points turn out to be the loci of level crossings in both the tight-binding and the free-electron models. Moreover, the symmetry can be used to reduce the Hamiltonian to invariant subspaces,

TABLE II. The charges for the tight-binding model corresponding to Fig. 20, the holes effect. Only the charges in the half-cube of flux space with ϕ_3 in $[0,\pi]$ are listed. They have identical inversion images. $\cos(\pi\alpha)\equiv\frac{1}{2}-\sqrt{2}$ or $\alpha\sim 0.867$.

No.	ϕ/π	E	Gap	Charge	Multiplicity
1	$(\frac{2}{3},-\frac{2}{3},\frac{1}{3})$	1	1	-1	2
2	$(\frac{2}{3},\frac{2}{3},1)$	1	1	1	2
3	$(\alpha,-\alpha,\alpha/2)$	$3-\sqrt{2}$	2	1	2
4	$(\alpha,\alpha,0)$	$3-\sqrt{2}$	2	-1	2
5	$(-\frac{2}{3},\frac{2}{3},\frac{2}{3})$	3	3	-1	2
6	$(\frac{2}{3},\frac{2}{3},1)$	3	3	1	2

where the eigenvalue equations can be solved analytically, as we shall see below.

Following the classification scheme described in Sec. X, we collected the results in a single table. Table II describes the results for the tight-binding dynamics and Table III those for the free-electron dynamics. The tables are organized as follows. The coordinates of the crossings are listed under ϕ. E is the corresponding degenerate energy. The gap that closes at this degeneracy is listed under "gap." Gap 2 means that the second and third levels cross at that point. Gap index j describes the transport properties of the j-electron system in the ground state. By general principles (see Sec. X), it is sufficient to consider $j=1,\ldots,|V|-1$.

With each point listed in the table one can associate a cluster of points by symmetry operations. For the holes graph there is a quadruplet associated with U and time reversal. All the points in such a cluster have identical charges. In the tables, we denote by *multiplicity* the number of *distinct* points in the torus obtained by applying these symmetry operations. The numbering in the first column is arbitrary, but we have grouped crossings according to their gap index.

The data in the tables are a mixture of numerical results and analytic calculations in the following sense. The coordinates of degeneracies (and sometimes the charges, too) are analytic statements. However, the claim that there are no other degeneracies is based on numerical evidence.

The (graph-theoretic) tight-binding Hamiltonian corresponding to the graph of Fig. 20 is

$$H(\Phi)=\begin{bmatrix} 2 & 0 & -e^{i\Phi_1} & 0 & -1 \\ 0 & 2 & 0 & -e^{i\Phi_2} & -1 \\ -e^{-i\Phi_1} & 0 & 3 & -e^{i\Phi_3} & -1 \\ 0 & -e^{-i\Phi_2} & -e^{-i\Phi_3} & 3 & -1 \\ -1 & -1 & -1 & -1 & 4 \end{bmatrix} . \tag{12.5}$$

Because the Hamiltonian is 5×5, the eigenvalue equation, $\det[H(\Phi)-E]=0$, can be written out analytically with a finite amount of human effort. Although this computation can be done "by hand" it is actually easier, and possibly safer, to compute it mechanically using one of several available formal manipulation programs. (We have used REDUCE.) The result is

$$\det[H(\Phi)-E]=-E^5+14E^4-70E^3+152E^2-137E+38-2((2-E)(3-E)-1)[\cos(\Phi_1)+\cos(\Phi_2)]$$
$$-2(2-E)^2\cos(\Phi_3)-2(2-E)(\cos(\Phi_1+\Phi_3)+\cos(\Phi_2-\Phi_3))-2\cos(\Phi_1-\Phi_2+\Phi_3) . \tag{12.6a}$$

Points of (double) degeneracy are the simultaneous zeros of this and

$$\frac{d}{dE}\det[H(\Phi)-E]=0 ,$$
$$\frac{d}{dE}\det[H(\Phi)-E]=-5E^4+56E^3-210E^2+304E-137+2(5-2E)(\cos(\Phi_1)+\cos(\Phi_2))$$
$$+4(2-E)\cos(\Phi_3)+2(\cos(\Phi_1+\Phi_3)+\cos(\Phi_2-\Phi_3)) . \tag{12.6b}$$

TABLE III. The charges for the free-electron model associated with Fig. 20, the holes effect. Only the charges in the sections ϕ_3 and k in $[0,\pi]$ are listed. There are identical charges at flux-inverted points, and opposite ones at $2\pi-k$. $\phi_1/\pi\sim0.6824$, $\phi_2/\pi\sim0.4299$, $\phi_3/\pi\sim1.6618$, $\phi_4/\pi\sim0.051$.

No.	ϕ	$\cos(k)$	Gap	Charge	Multiplicity
1	$(\pi-\phi_1,\pi-\phi_1,0)$	$\frac{-1-\sqrt{7}}{6}$	1	1	2
2	$(\pi-\phi_1,-\pi+\phi_1,\pi-\phi_2)$	$\frac{-1-\sqrt{7}}{6}$	1	-1	2
3	(ϕ_3,ϕ_3,π)	$\frac{1-\sqrt{7}}{6}$	2	-1	2
4	$(\phi_3,-\phi_3,\phi_4)$	$\frac{1-\sqrt{7}}{6}$	2	1	2
5	$(\pi-\phi_3,-\pi+\phi_3,\pi-\phi_4)$	$\frac{-1+\sqrt{7}}{6}$	3	-1	2
6	$(\pi-\phi_3,\pi-\phi_3,0)$	$\frac{-1+\sqrt{7}}{6}$	3	1	2
7	$(\phi_1,-\phi_1,\phi_2)$	$\frac{1+\sqrt{7}}{6}$	4	1	2
8	(ϕ_1,ϕ_1,π)	$\frac{1+\sqrt{7}}{6}$	4	-1	2

Equations (12.6a) and (12.6b) for the degeneracies have the following obvious symmetries. If (ϕ_j,E_j) is a solution, so is $(-\phi_j,E_j)$, and if $(\phi_1,\phi_2,\phi_3,E_j)$ is a solution, so is $(\phi_2,\phi_1,-\phi_3,E_j)$.

The advantage of having explicit formulas like Eqs. (12.6a) and (12.6b) is not so much for finding the solutions, since, in general, this can be done only numerically. Originally, the entries in the table were derived by combining the numerical results with guesswork, which was then verified by substitution in Eq. (12.6). With reasonable numerical accuracy, the coordinates and energy of the crossings at points 1, 2, 5, and 6 suggest a natural guess. For points 3 and 4 the key turned out to be the guess for the energy. Equation (12.6b) then gives a linear equation for one unknown, $\cos\pi\alpha$.

As we subsequently realized, much of the guesswork can be replaced by group theory. This approach also revealed a remarkable property of the band functions, namely, that there are "flat bands" on the symmetry lines, Eq. (12.4), whose energy is independent of Φ. The energy values for these flat bands are given by the solution of quadratic equations. The crossing points can then be computed from Eq. (12.6b). In this way one gets the entries in the table, as we now proceed to explain.

Consider the symmetry lines in Eq. (12.4). The invariant subspaces under U are

$$\begin{aligned} &\text{I: } (\alpha_1,\alpha_1,\alpha_2,\alpha_2,\alpha_3)\ , \\ &\text{II: } (\alpha_1,-\alpha_1,\alpha_2,-\alpha_2,0)\ . \end{aligned} \tag{12.7}$$

$\alpha_{1,2,3}\in\mathbb{C}$. I is a three (complex) dimensional space and II is two dimensional. Reducing the Hamiltonian of Eq. (12.5) to II gives a 2×2 eigenvalue problem, and the band functions on the symmetry lines of Eq. (12.4) are given as the solution of quadratic equations. The result is four flat bands:

$$E_{\text{II}}(\Phi,\Phi,0)=3\pm\sqrt{2},\quad E_{\text{II}}(\Phi,\Phi,\pi)=2\pm1\ . \tag{12.8}$$

The other three bands along these lines are given by a solution of a cubic equation. We do not know what the reason is for flat bands in this problem, but because of this the energies at crossings are known for three of the six points in the table. From Eq. (12.6a) the coordinates of points 2, 4, and 6 are given by a linear equation in $\cos(\Phi)$.

Some of the properties of Table II worth noting are as follows.

(1) In agreement with the von Neumann–Wigner ansatz one finds that the simultaneous solutions of Eq. (12.6a) and Eq. (12.6b) are a discrete set $\{(\phi_j,E_j)\}$. $D(P)$ is then the union over j of $\{\phi_j\}$.

(2) For each gap index, the total charge, counting multiplicity, is zero. This is the neutrality of the cube in flux space proven in Sec. X. It gives $|V|-1$ sum rules that check for the completeness of the set of crossings.

(3) For each gap index, planes with constant Φ_1 or constant Φ_2 are neutral. Such planes either avoid all charges or contain four charges that neutralize in pairs. We discuss a consequence of this for the conductance below. This property is presumably unstable under perturbations.

(4) Generically, one expects only $\pm$ charges. This is the case here. However, because of the symmetry of the network it is not possible to argue for this on the basis of genericity alone.

As we have explained in the previous sections, once the degeneracies are known, the charges can actually be computed analytically. To make analytic calculations for the table would require further guesswork for the eigenvectors for one point in the table for each gap index. That is, three pairs of eigenvectors would need to be guessed. We have contented ourselves by making this calculation for the first gap, and verified that the analytic result agrees with the numerical result. We have chosen point 2 and describe below the computation of its charge.

Two independent eigenvectors with eigenvalue 1 of point 2, of Table II, are

$$\begin{aligned} &|\psi\rangle=(z_3,-z_3,1,-1,0)\ , \\ &|\varphi\rangle=(2,1+z_3,z_3^*,1,1)\ , \end{aligned} \tag{12.9}$$

where $z_3\equiv\exp(2\pi i/3)$ is the cubic root of unity. $|\psi\rangle$ and $|\varphi\rangle$ are neither normalized nor orthogonal. For our purpose all that matters is that they are independent. Restricting the Hamiltonian to the degenerate subspace, using Eqs. (7.8), gives

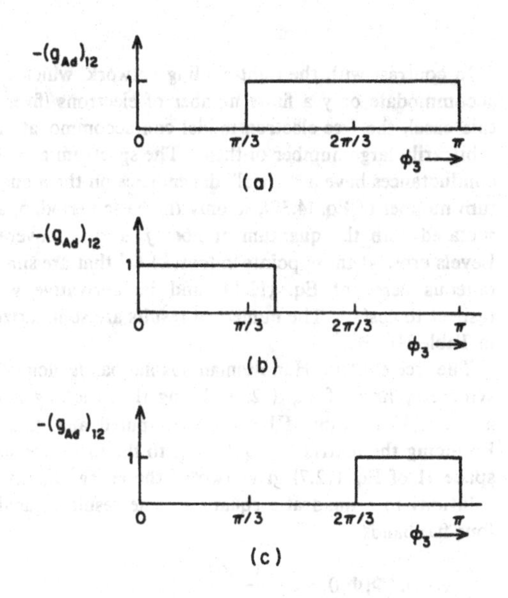

FIG. 21. The 1-2 conductances of the holes effect in the ground states for the tight-binding model of noninteracting (spinless) electrons: (a) the one-electron system; (b) the two-electron system, (c) the three-electron system. Since the graphs are periodic and antisymmetric, only half the period is shown. The four- and five-electron systems have vanishing conductances and are therefore not shown. The network is p-type, flux and μ stable.

$$d\epsilon_1=\frac{\sqrt{3}}{2}(d\Phi_1-d\Phi_2+d\Phi_3)\,,$$
$$d\epsilon_2=\tfrac{1}{2}(-d\Phi_1+d\Phi_2+d\Phi_3)\,, \tag{12.10}$$
$$d\epsilon_3=\frac{\sqrt{3}}{2}(d\Phi_2-d\Phi_3)\,.$$

The determinant of this map is $-\sin^2(2\pi/3)$, so the charge of point 2 is 1. From the neutrality of the cube we then learn that the change of point 1 is -1. This gives all the charges in the first gap.

From the table it follows that the average conductances [i.e., the periods of Eq. (8.10)] are as in Fig. 21. The three graphs correspond to the 1-2 conductances in (a) the one-electron ground state, (b) the two-electron ground state, and (c) the three-electron ground state. The four- and five-electron ground states have vanishing conductances. At most five electrons can be accommodated in the tight-binding Hilbert space for their network. That there are no charges associated with the five-electron ground state follows from general principles. However, that the four-electron ground state also has no charges is a special property of the holes graph and the tight-binding dynamics. We do not know the reason for that.

From the neutrality of the $\Phi_{1,2}$ sections of the torus, it follows that $[\langle g_{\rm Ad}\rangle]_{13}$ and $[\langle g_{\rm Ad}\rangle]_{23}$ vanish identically. This property is presumably unstable under perturbations.

Finally, we note that the conductances are flux stable, and μ stable in the sense of Sec. X.

With the holes graph we now associate the free-electron dynamics. The Alexander–de Gennes matrix $h_0(k,\Phi)$ corresponding to this dynamics, assuming a network with unit length links, is

$$\begin{bmatrix} 2\cos(k) & 0 & -e^{i\Phi_1} & 0 & -1\\ 0 & 2\cos(k) & 0 & -e^{i\Phi_2} & -1\\ -e^{-i\Phi_1} & 0 & 3\cos(k) & -e^{i\Phi_3} & -1\\ 0 & -e^{-i\Phi_2} & -e^{-i\Phi_3} & 3\cos(k) & -1\\ -1 & -1 & -1 & -1 & 4\cos(k)\end{bmatrix}. \tag{12.11}$$

The eigenvalue equation is $\det[h_0(k,\Phi)]=0$, where

$$\begin{aligned}\det[h_0(k,\Phi)]=\;&144\cos^5(k)-124\cos^3(k)+4\cos^2(k)(-2\cos(\Phi_3)-3\cos(\Phi_2)-3\cos(\Phi_1))\\&-2\cos(k)(2\cos(\Phi_3-\Phi_2)+2\cos(\Phi_1+\Phi_3)-9)\\&+2(-\cos(\Phi_1-\Phi_2+\Phi_3)+\cos(\Phi_2)+\cos(\Phi_1))\,.\end{aligned} \tag{12.12}$$

In contrast with the tight-binding network, which can accommodate only a finite number of electrons (five, in this case), the free-electron model can accommodate an arbitrarily large number of them. The spectrum and the conductances have a "trivial" dependence on the n quantum number of Eq. (4.32), so only the basic period, 5, associated with the quantum number j need be covered. Levels cross at those points in $[\cos(k),\Phi]$ that are simultaneous zeros of Eq. (12.14) and its derivative with respect to $\cos(k)$. The numerical results are summarized in Table III.

The free-electron Hamiltonian has flat bands along the symmetry lines of Eq. (12.4). Using this, the angles ϕ_j and $\cos(k)$ in Table III can be computed analytically. Reducing the matrix of Eq. (12.11) to the invariant subspace II of Eq. (12.7) gives two of the energy bands as solutions to a quadratic equation. The result is, again, four flat bands

$$\cos(k(\Phi,\Phi,0))=\frac{-1\pm\sqrt{7}}{6}\,,$$
$$\cos(k(\Phi,\Phi,\pi))=\frac{1\pm\sqrt{7}}{6}\,. \tag{12.13a}$$

From this, and the derivative of Eq. (12.12) with respect to $\cos(k)$, one derives analytic values for $\cos(\phi_{1,3})$ of Table III:

$$\cos(\phi_1)\equiv-\frac{7\sqrt{7}+1}{36}\,,$$
$$\cos(\phi_3)\equiv\frac{-7\sqrt{7}+1}{36}\,. \tag{12.13b}$$

Some further work gives

$$\cos(\phi_2)\equiv\frac{-31\sqrt{7}+164}{24(\sqrt{7}+13)}\,,$$
$$\cos(\phi_4)\equiv\frac{31\sqrt{7}+164}{24(\sqrt{7}-13)}\,. \tag{12.13c}$$

One can also compute the charges "by hand" when one does not have analytic expressions for crossing points and degenerate subspaces, provided one has approximate (i.e., numerical) vectors that span the degenerate subspaces. For example, for point 1 in Table III, a pair of degenerate vectors is

$$\begin{aligned}&|\psi\rangle=(\alpha_1,\alpha_1,\alpha_2,\alpha_2,\alpha_3)\,,\\&|\varphi\rangle=(-\alpha_4,\alpha_4,\alpha_5,-\alpha_5,0)\,,\\&\alpha_1\sim0.3301,\quad \alpha_2\sim0.046+i0.138\,,\\&\alpha_4\sim0.151-i0.234,\quad \alpha_5\sim0.338\,.\end{aligned} \tag{12.14}$$

Using Eq. (7.8) one finds for the linearized map

$$\epsilon_2+i\epsilon_1=(e^{-i\phi_1}\alpha_1\alpha_5+e^{i\phi_1}\alpha_2^*\alpha_1)(-d\Phi_1+d\Phi_2)-2\alpha_2^*\alpha_5 d\Phi_3\,,$$
$$\epsilon_3=\mathrm{Im}(e^{i\phi_1}\alpha_1\alpha_2+\alpha_4^*\alpha_5)(d\Phi_1+d\Phi_2)\,. \tag{12.15}$$

The determinant of this map is negative, so the charge is +1. (Of course, with numerical values one finds a nonzero value for the determinant "with probability one." In numerical calculations the sign is not enough, and the actual numerical value has to be sufficiently far away from zero.) Charge neutrality then determines the charge of point 2, and the "π shift" of Eq. (10.9) that of points 8 and 9. These give half the entries in the table.

From the table one concludes that the corresponding graphs for the conductances are as in Fig. 22. The four graphs describe the 1-2 conductances in the ground state of the one-, two-, three-, and four-electron systems. In the five-electron system, the conductance vanishes in the ground state. For the next five electrons, the picture repeats itself up to reversal of signs and order, as discussed in Sec. X. With more electrons present, the picture repeats itself, with the number of electrons counted mod 10.

Comparing the results for the tight-binding and the de Gennes–Alexander realizations for this graph one finds similarities and differences. Many of the basic features are common: the conductances are flux stable; points of level crossing lie on the distinguished symmetry planes and lines; all charges are ± 1; and finally, planes with fixed Φ_1 or fixed Φ_2 are neutral, leading to the identical vanishing of the 1-3 and 2-3 conductances.

There are also differences, some expected, some not. One expected qualitative difference is that the free-electron model has the "π-shift" symmetry that is not shared by the (graph-theoretic) tight-binding model. Another (unexpected) qualitative difference is that there are no charges in the four-electron ground state in the tight-binding dynamics, but there are charges in the free-electron dynamics. There are also expected differences in details: the actual positions of crossings are distinct.

The tight-binding and the free-electron Hamiltonians are not close to each other in any sense, so the fact that their conductance graphs are distinct is not in conflict with anything. This suggests that an attempt to predict what would happen in actual settings may require a reasonably accurate modeling. Both models establish nontriviality. In fact, they do more, as they show that given a graph, nontriviality is not special to one dynamics. This suggests that more realistic and more complicated models corresponding to this graph also stand a good chance of being nontrivial.

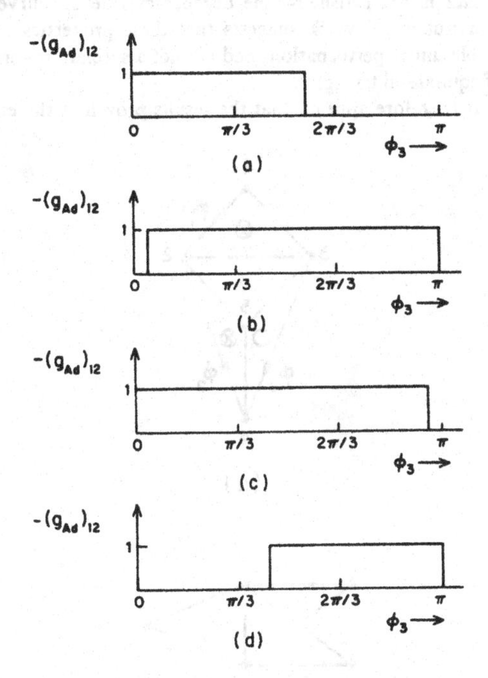

FIG. 22. The 1-2 conductances for free electrons on a network of thin wires associated with the holes effect graph: (a) The 1(mod 5)-electron systems; (b) the 2(mod 5)-electron systems; (c) the 3(mod 5)-electron systems; (d) the 4(mod 5)-electron systems. All graphs are for the ground-state conductances. The network is n-type (by definition) and is flux and μ stable.

XIII. THREE-FLUX NETWORKS OF EQUILATERAL TRIANGLES

In this section we collect results for a number of three-flux networks that correspond to the tight-binding Hamiltonians for various simple graphs. There are two reasons for this study. One is related to the question of stability of nontriviality, namely, that the graph of the holes effect considered in Sec. XII is not special. The second reason is a part of a larger program to try to gain insight into the Chern numbers that characterize various graphs and dynamics.

The simple graphs we consider are those that can be made with three equilateral triangles, 1, 2, and 3, carrying independent fluxes Φ_1, Φ_2, and Φ_3. Since, by general principles, disconnected networks are trivial, we consider only connected ones. There are six distinct planar graphs and three nonplanar ones. They are listed in Table IV. The planar graphs are shown in Fig. 23 and the nonplanar graphs in Fig. 24. The enumeration given in Table IV goes as follows. Since the network is connected, there is at least one triangle connected to the other two. Pick such a triangle and call it (tentatively) 2. Let $\#(i,j)$ be the number of common vertices of the ith and jth triangles. Clearly $\#(ij)<3$ if $i\neq j$, so

$$\#(1,2),\#(2.3)\in\{1,2\}\,.$$

TABLE IV. The nine distinct graphs that can be made with three equilateral triangles threaded by flux tubes. Six of the graphs are planar and three are not. Five have nontrivial topological conductances.

Name	Planar	$\{\#(1,2),\#(2,3),\#\}$	$\lvert V \rvert$	Trivial
windmill	Yes	1,1,1	7	Yes
Giza	Yes	1,1,2	7	Yes
gasket	Yes	1,1,3	6	No
wide kite	Yes	2,1,2	6	Yes
long kite	Yes	2,1,3	6	Yes
basket	No	2,1,4	5	No
tripod	No	2,2,2	5	Yes
holes effect	Yes	2,2,3	5	No
tetrahedron	No	2,2,4	4	No

Let 1 be the triangle such that $\#(1,2) \geq \#(3,2)$. This procedure may involve a choice for 2 if there are several triangles that connect the other two. In this case, we choose 2 so that $\#(1,2)$ is maximal.

The triplet $\{\#(1,3),\#(2,3),\#(1,2)\}$ determines a unique graph up to a possible two-fold ambiguity, which is removed by specifying $\lvert V \rvert$. A triplet that specifies a unique graph is

$$\{\#(2,3),\#(1,2),\#\},$$

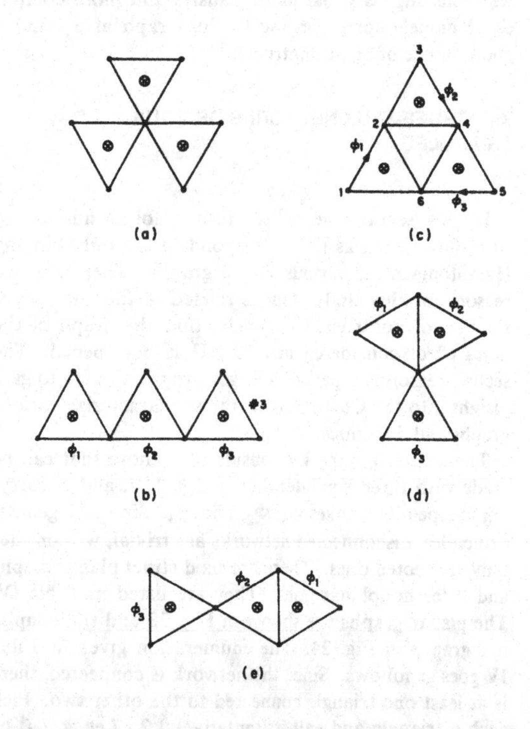

FIG. 23. The six planar graphs that can be made with three equilateral triangles: (a) the windmill; (b) the three pyramids of Giza; (c) the gasket; (d) the wide kite; (e) the long kite. The graph for the holes effect is shown in Fig. 20. We have numbered the fluxes and vertices only in those cases where such a numbering is important in the text.

where $\#$ is the number of vertices in the graph that belong to at least two triangles. The table is organized lexicographically in $\{\#(1,2)\#(2,3),\#\}$. The names are figments of our imagination.

Of the six planar networks, four are trivial in the sense that they do not have *isolated* crossing points, with nontrivial charges. They all have the property that the graphs are separable, in the graph-theoretic terminology, and one-particle reducible, in the Feynman diagram terminology. That is, the graphs are disconnected by cutting one vertex. In addition, the Wigner–von Neumann ansatz is not satisfied—the degeneracies lie on curves. Our numerical work suggests that these properties are stable under perturbations and are not a special property of equilateral triangles.

It therefore appears that the results proven at the end

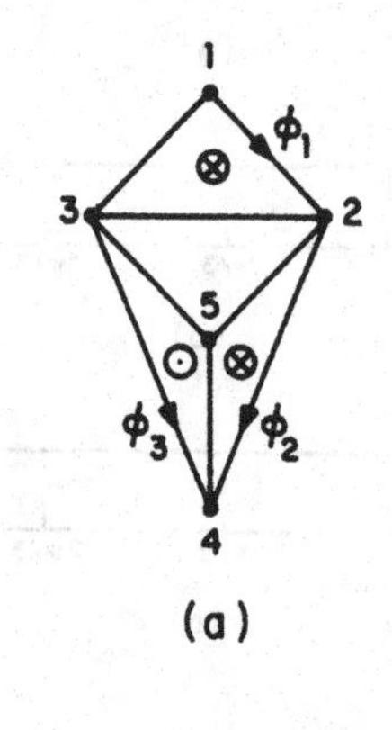

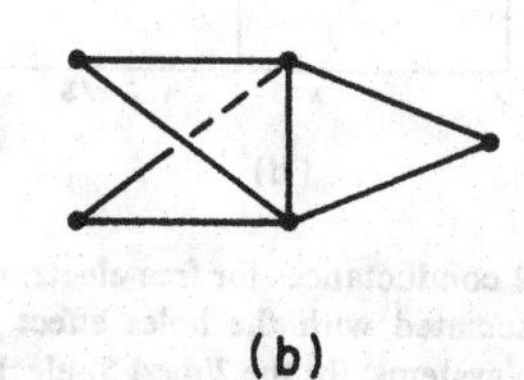

FIG. 24. Two of the three nonplanar graphs that can be made with three equilateral triangles; (a) the basket; (b) the tripod. The third, the tetrahedron, is shown in Fig. 14.

TABLE V. The charges for the tight-binding Hamiltonian for the gasket, Fig. 23(c). Only the charges in the half-cube Φ in [0,π] are listed. There are identical charges at points with the coordinates reflected about zero.

$(\phi/\pi)(1,1,1)$	E	Gap	Charge	Multiplicity
$\frac{2}{3}$	1	1	1	2
1	$\frac{7-\sqrt{21}}{2}$	1	−2	1
0	$\frac{7-\sqrt{13}}{2}$	2	2	1
0.824	1.454	2	−1	2
1	2	3	0	1
0	$\frac{7+\sqrt{13}}{2}$	5	2	1
1	$\frac{7+\sqrt{21}}{2}$	5	−2	1

of Sec. VIII about disconnected graphs has a generalization to separable graphs. However, we do not know of a proof in general. Separability together with symmetry can be used to show that *some* periods vanish. For example, for the three pyramids of Giza, $\langle g_{AD}\rangle_{12} = \langle g_{AD}\rangle_{23}=0$ is a consequence of the following argument: $\langle g_{Ad}\rangle_{12}(\Phi_3)$ is antisymmetric in Φ_3 by the Onsager relation of Sec. IX. However, $\Phi_3 \to -\Phi_3$ corresponds to permuting the two "free" vertices in triangle #3. This is implemented by a unitary operator. Unitaries leave the Chern number invariant, so $\langle g_{Ad}\rangle_{12}$ is both symmetric and antisymmetric in Φ_3 and hence zero identically. By a similar argument, $\langle g_{Ad}\rangle_{12}=0$ in the long and wide kites. For the windmill, the same argument gives $\langle g_{Ad}\rangle_{ij}=0$ for all pairs of loops i,j.

Of the nontrivial networks, the gasket and the tetrahedron have obvious symmetries; as a consequence a fair amount of analysis can be done "by hand."

The numerical procedure for analyzing these networks is the same as the one outlined in the previous section. Tables V, VI, and VII list the results for all nontrivial networks except for the holes graph, which was discussed in detail in the previous section. We have attempted to give analytic expressions in most cases.

Recall that tight-binding Hamiltonians can accommodate at most $|V|$ (spinless) electrons. The tables give the coordinates and the charges of level crossings corresponding to projections P_j with gap index j. The projection on the ground state of the one-electron system corresponds to gap index 1, the projection of the ground state of the two-electron system to gap index 2, etc. Since there are no charges in the $|V|$-electron ground state, all the tables list gap indices from 1 to $|V|-1$.

A. The gasket

The gasket looks like the first iteration in the Sierpinsky gasket (Mandelbrot, 1983), hence the name. The graph has C_3 symmetry. Among the nontrivial graphs, the gasket is unique in that the three flux-carrying triangles touch at vertices and have no common edge. All the other nontrivial graphs have at least one edge shared between the flux-carrying triangles.

As for all the other examples, there are no "accidental" level crossings, in the sense that all crossings lie on symmetry points in flux space. For the gasket, this is the line of equal fluxes (Φ,Φ,Φ). On this line, the tight-binding Hamiltonian corresponding to Fig. 23(c) can be chosen to be C_3 symmetric, for example, by putting the three fluxes on the edges (2,1), (4,3), and (6,5) in the figure. There are three two-dimensional, invariant vector subspaces of C^6:

TABLE VI. The charges for the tight-binding Hamiltonian for the basket, Fig. 24(a). Only points in the half-cube with ϕ_3 in $[0,\pi]$ are listed. There are identical charges in the other half.

No.	ϕ/π	E	Gap	Charge	Multiplicity
1	(1.222,−0.517,0.517)	1.198	1	1	2
2	(1,1,1)	$\frac{7-\sqrt{17}}{2}$	1	−2	1
3	$(\frac{2}{3})(1,1,1)$	2	2	1	2
4	$(\frac{2}{3},1,1)$	2	2	−1	2
5	(0,0,0)	5	4	−2	1
6	(1,1,1)	$\frac{7+\sqrt{17}}{2}$	4	2	1

TABLE VII. The charges for the tetrahedron. This table applies both to the tight-binding and the free-electron model. Only the charges in the half-cube are listed.

No.	$(\Phi/\pi)(1,1,1)$	E	Gap	Charge	Multiplicity
1	1	2	1	-2	1
2	$\frac{1}{2}$	$3-\sqrt{3}$	1	1	2
3	$\frac{1}{2}$	$3+\sqrt{3}$	3	-1	2
4	0	4	3	2	1

$$\begin{aligned} &\text{I: } \mathbb{C}^2\times(1,1,1)\,, \\ &\text{II: } \mathbb{C}^2\times(1,z_3,z_3^*)\,, \\ &\text{III: } \mathbb{C}^2\times(1,z_3^*,z_3)\,. \end{aligned} \tag{13.1}$$

The six energy bands for this network, which for general position are given by the zeros of a polynomial of degree 6, split on the symmetry line to three quadratic equations, associated with the invariant subspaces. One finds

$$\begin{aligned} &E_{\rm I}(\Phi)=2(1\pm\cos(\Phi/2))\,, \\ &E_{\rm II,III}(\Phi)=\tfrac{7}{2}\pm\sqrt{\tfrac{17}{4}+2\cos(\Phi\pm2\pi/3)}\,. \end{aligned} \tag{13.2}$$

The corresponding graphs of the energy bands are shown in Fig. 25. The crossing of the two bands in I occurs at $\Phi=\pi$ and $E=2$. The crossing of the bands in II and III occurs at $\Phi=\pi$ and $E=(7\pm\sqrt{21})/2$ and at $\Phi=0$ and $E=(7\pm\sqrt{13})/2$. Finally, there is a crossing between the lower band in I and the lower band in III at $\Phi=2\pi/3$ and $E=2$. There is one more crossing between the lower band in I and the lower band in II, the one associated with the second gap in the table, that gives a quartic equation in $\cos(\Phi/2)$. For this, only numerical values are given.

The charges for all nine crossing points are given in Table V. Because the degenerate subspaces are essentially identified by Eq. (13.1) one could, presumably, compute the charges analytically. We have not attempted to do so except for the zero charge at the third gap, which is computable almost by inspection: this gap closes when the two bands in I cross. The basis vectors for the degenerate subspace, $(0,1)\times(1,1,1)$ and $(1,0)\times(1,1,1)$, give a bipartition of the graph. The gauge fields do not link any two vertices in any one partition. This makes $d\epsilon_3(\Phi)$ in Eq. (7.8) vanish identically and implies zero degree, of course.

Some noteworthy features of Table V are as follows.

(1) The three-, four-, and six-electron ground states

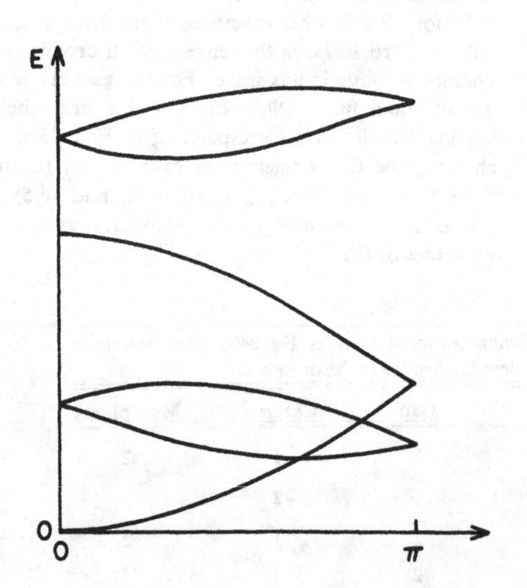

FIG. 25. The energy bands along the body diagonal in the flux spacecube, for the tight-binding model for the gasket. Only half the period is shown, since the bands are symmetric.

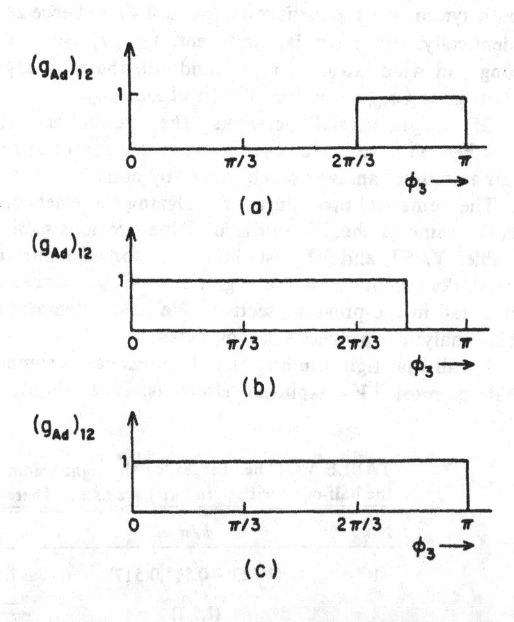

FIG. 26. The conductances of the gasket in the ground states of the tight-binding model: (a) the one-electron system; (b) the two-electron system; (c) the five-electron system. Only half the period is shown, as the graphs are antisymmetric. The three-, four-, and six-electron systems have trivial topological conductances at the ground state. The conductances are flux and μ stable.

have no charges. We do not know of a simple reason for that.

(2) In contrast with the holes effect, there are also multiple charges: 0, ±1, ±2. The appearance of 0 and ±2 is an indicator of nongenericity. Multiple charges are expected to disintegrate to elementary charges under perturbations.

(3) The total charge for fixed gap label vanishes, as it should.

(4) Note that $(1,1,1)(2\pi/3)$ has an identical image charge under inversion, i.e., at $(1,1,1)(4\pi/3)$, while $\pi(1,1,1)$ is its own image under inversion.

The topological conductances of the network are shown in Fig. 26 and, like the holes effect, are flux and μ stable.

B. The basket

This is a nonplanar graph whose symmetry is not unlike that of the holes effect. The corresponding tight-binding model can accommodate at most five electrons. There are nine points of two-level crossings. The coordinates and charges are listed in Table VI. Note that projection on the three-electron ground state is free of charges.

Unlike the holes effect, which had nontrivial conductance associated with a single pair of loops, in the basket any pair leads to nontrivial conductances. The identical vanishing of the 2-3 conductance in the holes effect is not

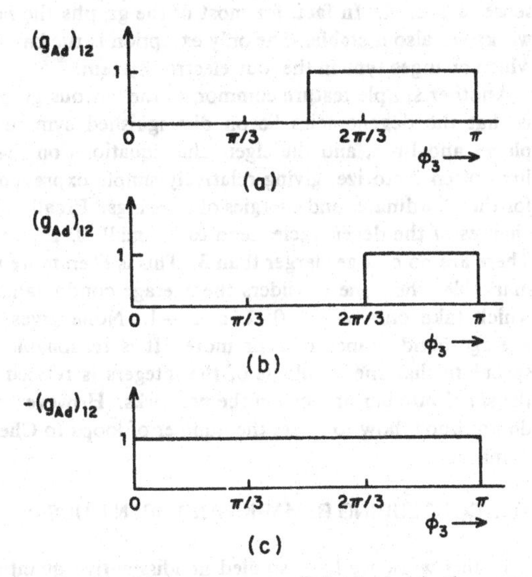

FIG. 27. The 1-2 conductances of the basket in the ground states of the tight-binding model: (a) the one-electron system; (b) the two-electron system; (c) the four-electron system. Only half the period is shown as the graphs are antisymmetric. The three- and five-electron systems have trivial topological conductances. This network is of mixed type: the four-electron system has opposite conductances to the one- and two-electron systems. The network is thus μ unstable. It is, however, flux stable.

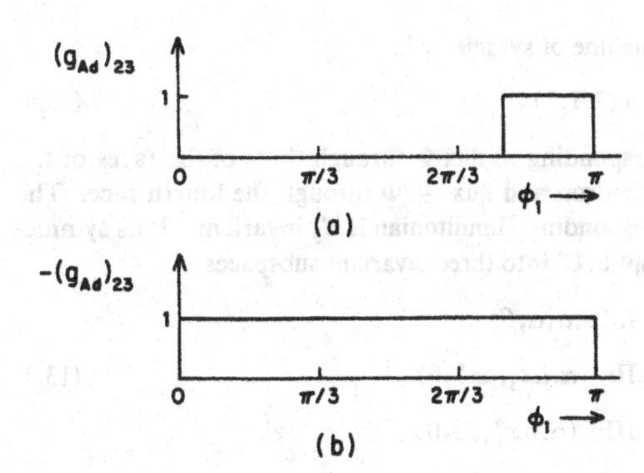

FIG. 28. The 2-3 conductances of the basket in the ground states of the tight-binding model: (a) the one-electron system; (b) the four-electron system. Only half the period is shown as the graphs are antisymmetric. The two-, three-, and five-electron systems have vanishing topological conductances. The conductance is flux stable but μ unstable.

shared by the basket. Only a remnant of this is seen in the vanishing of the 2-3 conductance in the two-electron ground state by a miraculous cancellation, analogous to the one that occurred in the holes effect for all gap indices. The topological conductances in the ground states of the 1-5 electron systems are listed in Figs. 27 and 28. This network is flux stable but not μ stable. It is the only network among those we have considered that is μ unstable.

C. The tetrahedron

Among the graphs in Table IV, the tetrahedron, Fig. 14, is noteworthy: it is the only regular graph; it is the graph with the least number of vertices; and it has a large (non-Abelian) symmetry group. A consequence of this is that the table for the charges can be computed analytically with relative ease. This is actually something we have realized with hindsight. Originally we studied the tetrahedron, as we did all other models, numerically. The analytic derivation presented below came later. There is, however, one bit of input from the numerical work that we shall need, namely, that there are no accidental degeneracies and charges. All the charges lie on the symmetry line of three equal fluxes.

Since the tetrahedron is a regular graph, its transport properties are the same in the tight-binding and the free-electron models. The energies are related by Eq. (4.37). For the sake of concreteness we use the tight-binding language.

The tight-binding Hamiltonian for the tetrahedral graph is

$$\begin{bmatrix} 3 & -e^{i\Phi_3} & -e^{-i\Phi_2} & -1 \\ -e^{-i\Phi_3} & 3 & -e^{i\Phi_1} & -1 \\ -e^{i\Phi_2} & -e^{-i\Phi_1} & 3 & -1 \\ -1 & -1 & -1 & 3 \end{bmatrix} . \tag{13.3}$$

The line of symmetry is

$$\Phi(1,1,1)\,, \tag{13.4}$$

corresponding to flux Φ through three of the faces of the tetrahedron and flux -3Φ through the fourth face. The corresponding Hamiltonian is C_3 invariant. This symmetry splits C^4 into three invariant subspaces:

$$\begin{aligned} &\text{I: } (\alpha,\alpha,\alpha,\beta)\,, \\ &\text{II: } (\alpha,\alpha z_3,\alpha z_3^*,0)\,, \\ &\text{III: } (\alpha,\alpha z_3^*,\alpha z_3 0)\,. \end{aligned} \tag{13.5}$$

I is two (complex) dimensional, $\alpha,\beta\in C$, and II and III are each one dimensional.

The energy bands on the line of symmetry can be computed analytically by reduction to invariant subspaces. This gives

$$\begin{aligned} &E_{\mathrm{I}}(\Phi)=3-\cos(\Phi)\pm\sqrt{3+\cos^2(\Phi)}\,, \\ &E_{\mathrm{II}}(\Phi)=3-2\cos(\Phi+2\pi/3)\,, \\ &E_{\mathrm{III}}(\Phi)=3-2\cos(\Phi-2\pi/3)\,. \end{aligned} \tag{13.6}$$

The bands are shown in Fig. 29.

The two bands in I never cross. Double crossings occur when one of the bands in I crosses one in II or III. This occurs at $\phi=\pm\pi/2$. Consider $\pi/2$. The lower branch of E_{I} crosses E_{II} at energy $3-\sqrt{3}$, and the upper branch of E_{I} crosses E_{III} at energy $3+\sqrt{3}$. There are also triple crossings at $\phi=0$ and $\phi=\pi$, by the general group-theoretic analysis of Sec. VII. This takes place at $E=4$ and $E=2$. This identifies the crossings and the degenerate subspaces.

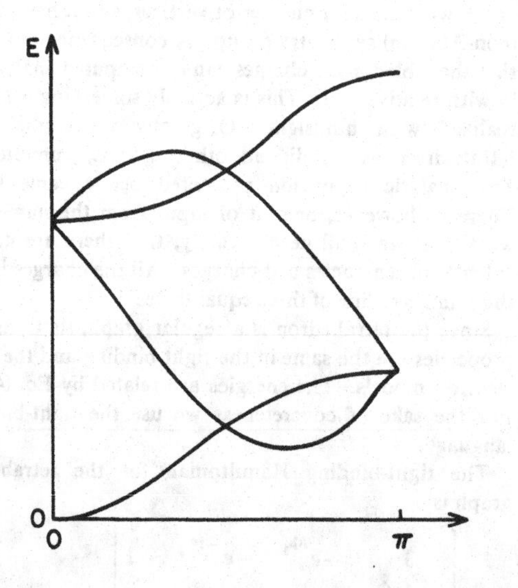

FIG. 29. The energy bands along the diagonal in flux space, for the tight-binding model for the tetrahedron. Since the bands are symmetric under reflection of Φ, only half the period is shown.

A remarkable thing about the tetrahedron is that it is enough to compute just one charge to get the full Table VII. For example, the charge of point 2 in the table determines that of point 1 by the overall neutrality. The "π shift" then determines the charges of points 3 and 4, so it remains to compute the charge of 2. Since this is a two-level crossing, the charge is a degree.

The degenerate subspace of point 2 is spanned by

$$(1,1,1,\sqrt{3})\times(1,z_3,z_3^*,0)\,. \tag{13.7}$$

One finds for the linearized map at $\phi=\pi/2$, from Eq. (7.8),

$$\begin{aligned} &d\epsilon_1=-d\Phi_1+\tfrac{1}{2}d\Phi_2+\tfrac{1}{2}d\Phi_3\,, \\ &d\epsilon_2=-\frac{\sqrt{3}}{2}d\Phi_2+\frac{\sqrt{3}}{2}d\Phi_3\,, \\ &d\epsilon_3=-\tfrac{3}{2}(d\Phi_1+d\Phi_2+d\Phi_3)\,. \end{aligned} \tag{13.8}$$

The associated determinant is negative, so the charge is $+1$.

In conclusion, we have seen several model Hamiltonians for several graphs made of three equilateral triangles, both planar and nonplanar, that have nontrivial Chern numbers. A common feature of the nontrivial bundles is simplicity: there are few charges in the j-electron ground state. There must be at least two in a nontrivial case (since the total charge is zero), and in all the networks the actual number is three. Another way of stating this is the following. All the conductances are flux stable in the sense of Sec. X. In fact, for most of the graphs the networks are also μ stable. The only exception is the basket, which changes type in the four-electron system.

Another simple feature common to the various graphs is that the degeneracies lie on distinguished symmetry planes and lines, and the eigenvalue equations on these lines often factorize, giving relatively simple expressions for the coordinates and energies of crossings. Finally, the charges of the degeneracies tend to be small, 0, 1, and 2. There are no charges larger than 3. This is even more remarkable when one considers the average conductances, which take only values 0, 1, and -1. None gives an average conductance of 2 or more. It is reasonable to speculate that the smallness of the integers is related to the small number of loops in the networks. However, we do not know how to relate the number of loops to Chern numbers.

XIV. CONCLUDING REMARKS AND OPEN QUESTIONS

In this work we have studied nondissipative quantum transport in multiply connected systems. Such transport is associated with an antisymmetric matrix of transport coefficients indexed by pairs of loops. The matrix is a function of, among other things, the fluxes threading the loops, and this dependence plays a key role in the analysis. We have focused on the topological aspects of the transport properties, and specifically on averages of the transport matrix that are quantized to be integers.

The resulting "topological conductance matrix" is an integer-valued function of some of the fluxes in the network. We have described some of its basic properties and have computed it for several networks. The situation is reminiscent of the integer quantum Hall effect, in which the Hall conductance is an integer-valued function of, say, the magnetic field acting on the Hall probe. There are two important differences, however. One is that, in the Hall effect, the integers characterize the crystal, while in networks they characterize the multiconnectivity of the graph. The second is that, in the Hall effect, the integer is directly measurable, while in networks, in general, the quantized transport arises only after averaging "by hand."

In this section we raise some of the questions that have not been adequately treated in the previous sections, in the hope that this will stimulate further work.

The most serious gap in the theoretical framework is that the self-inductance of the system has not been taken into account. The inductance matrix L_{ij}, with loop indices i and j, relates the flux change in loop i to the current change in other loops: $\delta\Phi_i=\sum L_{ij}\delta I_j$. By changing the flux in one loop to create an emf, the fluxes in the other loops change too. We have assumed that these fluxes are fixed. This is equivalent to looking at networks with no, or very small, inductances. We want to consider when this assumption is reasonable.

By dimensional arguments, the inductance L, in cgs, is of order l/c where l is a typical length scale and c the velocity of light. Typical currents are presumably on the order of the persistent currents, i.e., of order $c\partial E/\partial\Phi$. The energy scale is that of atomic units, $(e^4m/\hbar^2)$, and its dependence on Φ is like a dependence on boundary conditions and so scales like l^{-2}. Thus typical persistent currents are of the order $(c/\Phi_0)(e^4m/\hbar^2)(a_0/l)^2$, where $a_0\equiv\hbar^2/me^2$ is the unit length in atomic units (Bohr radius). $\Phi_0\equiv\hbar c/e$ is the unit flux. It follows that

$$\delta\Phi/\Phi_0\sim(e^2/\hbar c)^2(a_0/l) . \qquad (14.1)$$

In mesoscopic systems a_0/l is, of course, much smaller than unity and so is $e^2/\hbar c$, the fine-structure constant. This says that $\delta\Phi$ changes little due to typical currents. The fact that the change is proportional to the fine-structure constant is not surprising, for a change in the magnetic flux at one loop due to a change in the flux in another loop can be thought of as photon-photon scattering, mediated by the network. It is interesting, and less obvious, that the variations in the fluxes actually scale down with the size.

It is interesting and important to get an accurate handle on the conductances of networks, that goes beyond the order-of-magnitude calculation presented here. One reason is that one natural way to measure the currents is to measure the change of flux in the appropriate loop. One therefore does not want the inductance to be too small either, and so the smallness of the change in the flux is actually a mixed blessing.

There are many issues that the study of conductance in

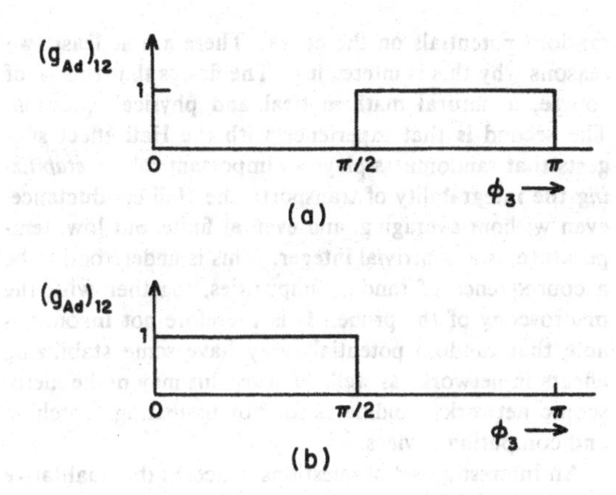

FIG. 30. The conductances of the tetrahedron in the tight-binding and free-electron models; (a) the one-electron ground state; (b) the three-electron ground state. Since the graphs are periodic and antisymmetric, only half a period is shown. The two- and four-electron ground states have vanishing topological conductances. The conductance is n-type, flux, and μ stable.

networks brings up that deserve further study, for example, the *unaveraged* conductance matrix in the adiabatic limit. With no averaging, the graph of the conductance is smooth, so that for the tetrahedron, a figure like Fig. 31 may replace Fig. 30. A detailed study of this for the graph in Secs. XII and XIII is of interest. Also of interest are corrections to adiabatic transport. Another set of problems is concerned with the study of networks with

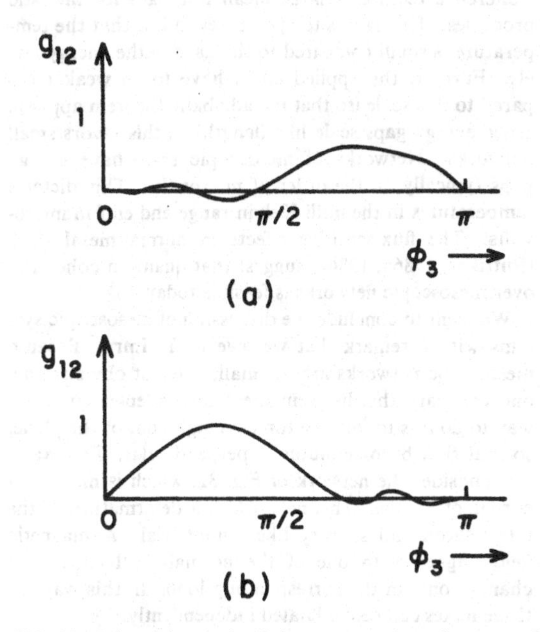

FIG. 31. Schematic graph of the unaveraged conductance of the tetrahedron network: (a) the one-electron system; (b) the three-electron system. Since the graphs are periodic and antisymmetric, only half a period is shown.

random potentials on the edges. There are at least two reasons why this is interesting. The first is that this is, of course, a natural mathematical and physical question. The second is that experience with the Hall effect suggests that randomness plays an important role *in stabilizing* the integrability of transport: the Hall conductance, even without averaging, and even at finite, but low, temperature, is a nontrivial integer. This is understood to be a consequence of random impurities, together with the macroscopy of the probe. It is therefore not inconceivable that random potentials may have some stabilizing effects in networks as well. If true, this may make mesoscopic networks candidates for nondissipating switching and computing devices.

An interesting set of questions concerns the qualitative and statistical theory of Chern numbers—understanding orders of magnitude of the charges, their distribution, etc. The study of graphs with many loops, with periodic, quasiperiodic, or hierarchical structures, etc. (Hofstadter, 1976; Rammal and Toulouse, 1982; Domany *et al.*, 1983) is completely open.

If the electron-hole concept is any guide, then one may expect that further and deeper understanding of the question of types, i.e., p-type versus n-type networks, is a useful pursuit. The related questions of flux stability and μ stability of the conductances (see Sec. X) also deserve further study.

Finally, an important subject we have not treated in any depth is the study of physical conditions. We have required coherence of the wave function over the entire network, and this may be difficult to achieve in practice. Coherence requires a large mean free path for inelastic processes. This favors temperatures so low that the temperature is small compared to the gaps in the energy levels. Further, the applied emf's have to be weak compared to this scale (so that the adiabatic theorem applies). Since energy gaps scale like $(\text{length})^{-2}$, this favors small networks. Networks of mesoscopic scale have energy gaps typically on the order of microvolts. This dictates temperatures in the milli-Kelvin range and emf in microvolts. The flux-sensitive effects in normal-metal rings (Büttiker, 1986a, 1986b) suggest that quantum coherence over mesoscopic networks is feasible today.

We want to conclude the discussion of mesoscopic systems with a remark that we owe to Y. Imry. Because mesoscopic networks are so small, it is not obvious how one can vary the fluxes in the loops independently. A way to do it is to fold the three triangles out of the plane, so that they become mutually perpendicular. For example, consider the network of Fig. 32, which is made of a corner of a cube. This network is a deformation of the tetrahedron and so very likely nontrivial. A magnetic field aligned with one of the normals will cause flux changes only in the corresponding loop. In this way the three fluxes can be modulated independently.

A different setting is that of superconducting networks. Since the superconducting wave function is coherent over macroscopic scales at temperatures of a few Kelvin, this may turn out to be a more favorable circumstance. (But

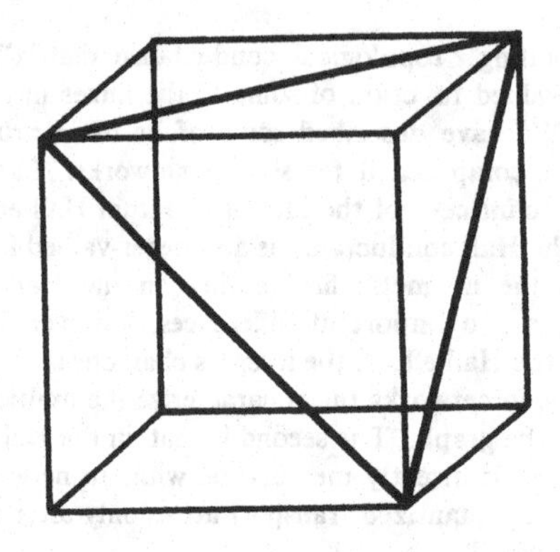

FIG. 32. A three-loop network made of the three corners of a cube. In this case, the fluxes through the mutually orthogonal triangles can be modulated independently with magnetic fields that are homogeneous on the scale of the network.

it may be that, for some other reason, more stringent conditions on temperatures and scales need to be applied.) Here the wave function is associated with the superconducting order parameter in a Landau-type theory. This theory has all the relevant formal structure needed for the geometric interpretation of transport. The two main differences are that (a) the charge is that of Cooper pairs and (b) the theory is nonlinear and has ψ^4 interaction. These features do not affect the basic structure, which relies on minimal coupling and the definition of currents. Of course, the network has to have thin enough wires so that the fluxes inside the loops are not quantized and can be varied. In other words, the network could be realized as an array of Josephson junctions. Such arrays have been fabricated and have been studied before; see, for example, Webb *et al.* (1983) and Behrooz *et al.* (1986). However, the conductance matrix of such networks, with fluxes that can be independently modulated, has not been studied as yet. It appears that the physical conditions relevant to the quantum transport properties should be similar to those relevant to, say, tunneling phenomena in macroscopic superconductors. Substantial work has been done in this field of research, most notably by J. Clarke and collaborators [see Clarke (1987) and references therein]. We suggest that the study of nondissipative quantum transport be considered a part of these general programs.

ACKNOWLEDGMENTS

This work was supported by the Israel Academy of Sciences, Minerva, and the US-Israel BSF Grant No. 84-00376 and NSF Grant No. DMS84-16049. J.E.A. acknowledges the support of the fund for the promotion of research at the Technion, and thanks M. Cross, D. Politzer, B. Simon, E. Stone, R. Vogt, and D. Wales for the hospitality extended to him at Caltech. We are indebted to M. Cross, S. Fishman, F. Gesztesy, Y. Imry, F. Kuchar, H. Kunz, B. Shapiro, S. Shtrikman, B. Simon, and J. Zak for useful discussions and comments. We thank Professor J. Goldberg for help with the transmission of files between Caltech and the Technion, and P. Weichman for a critical reading of the manuscript. This work grew out of a research program one of us (J.E.A.) has with R. Seiler, to whom is due a special acknowledgment. Parts of Secs. II, V, VI, VIII, and IX borrow from this common project. Finally we want to express our indebtedness to A. Libchaber, who, during a visit to the Technion first suggested to us that networks are like the Hall effect with no Hall probes.

REFERENCES

Aharonov, Y., and D. Bohm, 1959, "Significance of electromagnetic potentials in the quantum theory," Phys. Rev. **115**, 485.

Aharonov, Y., and D. Bohm, 1961, "Further considerations on electromagnetic potentials in the quantum theory," Phys. Rev. **123**, 1511.

Albeverio, S., F. Gesztesy, H. Holden, and R. Hoegh-Krohn, 1988, *Solvable Models in Quantum Mechanics* (Springer, New York).

Alden Mead, C., 1987, "Molecular Kramers degeneracy and non-Abelian adiabatic phase factor," Phys. Rev. Lett. **59**, 161.

Alden Mead, C., and D. G. Truhlar, 1979, "On the determination of Born-Oppenheimer nuclear motion wave function including complications due to conical intersections are identical nuclei," J. Chem. Phys. **70**, 2284.

Alexander, S., 1983, "Superconductivity of networks: A percolation approach to the effect of disorder," Phys. Rev. B **27**, 1541.

Alt'shuler, B. L., 1985, "Fluctuations in the extrinsic conductivity of disordered conductors," Pis'ma Zh. Eksp. Teor. Fiz. **41**, 530 [JETP Lett. **41**, 648 (1985)].

Alt'shuler, B. L., A. G. Aronov, and B. Z. Spivak, 1981, "The Aharonov-Bohm effect in disordered conductors," Pis'ma Zh. Eksp. Teor. Fiz. **33**, 101 [JETP Lett. **33**, 94 (1981)].

Anderson, P. W., D. J. Thouless, E. Abrahams, and D. S. Fisher, 1980, "New method for a scaling theory of localization," Phys. Rev. B **22**, 3519.

Arnol'd, V. I., 1978, *Mathematical Methods of Classical Mechanics*, Graduate Texts in Mathematics No. 60 (Springer, New York).

Avron, J. E., and R. Seiler, 1985, "Quantization of the Hall conductance for general multiparticle Schrödinger Hamiltonians," Phys. Rev. Lett. **54**, 259.

Avron, J. E., R. Seiler, and B. Simon, 1983, "Homotopy and quantization in condensed matter physics," Phys. Rev. Lett. **51**, 51.

Avron, J. E., R. Seiler, and L. Yaffe, 1987, "Adiabatic theorems and applications to the quantum Hall effect," Commun. Math. Phys. **110**, 33.

Avron, J. E., and B. Simon, 1978, "Analytic properties of band functions," Ann. Phys. (N.Y.) **110**, 85.

Avron, J. E., and L. G. Yaffe, 1986, "A diophantine equation for the Hall conductance of interacting electrons on a torus," Phys. Rev. Lett. **56**, 2085.

Avron, Y., R. Seiler, and B. Shapiro, 1986, "Generic properties of quantum Hall Hamiltonians for finite systems," Nucl. Phys. B **265** [FS15], 364.

Behrooz, A., M. J. Burns, H. Deckman, D. Lavine, B. Whitehead, and P. M. Chaikin, 1986, "Flux quantization on quasicrystalline networks," Phys. Rev. Lett. **57**, 368.

Bellissard, J., 1986a, "Ordinary quantum Hall effect and noncommutative cohomology," Lecture delivered at Bad Schandau (unpublished).

Bellissard, J., 1986b, "*K*-Theory of C^*-algebras in solid state physics, in *Statistical Mechanics and Field Theory: Mathematical Aspects*, edited by T. C. Dorlas, N. M. Hugenholtz, and M. Winnik, Lecture Notes in Physics No. 257 (Springer, Berlin/New York), p. 99.

Benoit, A. D., S. Washburn, C. P. Umbach, R. B. Laibowitz, and R. A. Webb, 1986, "Asymmetry in the magnetoconductance of metal wires and loops," Phys. Rev. Lett. **57**, 1765.

Berry, M. V., 1983, "Semiclassical mechanics of regular and irregular motion" in *Comportement Chaotique des Systemes Deterministes*, edited by G. Iooss, R. H. Helleman, and R. Stora (North-Holland, Amsterdam).

Berry, M. V., 1984, "Quantal phase factors accompanying adiabatic changes," Proc. R. Soc. London, Ser. A **392**, 45.

Berry, M. V., 1987, "Interpreting the anholonomy of coiled light," Nature (London) **326**, 277.

Berry, M. V., and M. Wilkinson, 1984, "Diabolical points in the spectra of triangles," Proc. R. Soc. London, Ser. A **392**, 15.

Biggs, N., 1974, *Algebraic Graph Theory*, Cambridge Tracts in Mathematics No. 67 (Cambridge University, London/New York).

Born, M., and R. Oppenheimer, 1927, "Zur Quantentheorie der Moleküle," Ann. Phys. (Leipzig) **84**, 457.

Büttiker, M., 1986a, "Flux sensitive effects in normal metal loops," Ann. N.Y. Acad. Sci. **480**, 114.

Büttiker, M., 1986b, "Four terminal phase coherent conductance," Phys. Rev. Lett. **57**, 1761.

Büttiker, M., Y. Imry, and R. Landauer, 1983, "Josephson behavior in small normal one-dimensional rings," Phys. Lett. A **96**, 365.

Büttiker, M., Y. Imry, R. Landauer, and S. Pinhas, 1985, "Generalized many-channel conductance formula with application to small rings," Phys. Rev. B **31**, 6207.

Byers, N. B., and C. N. Yang, 1961, "Theoretical consideration concerning quantized magnetic flux in superconducting cylinders," Phys. Rev. Lett. **7**, 46.

Chandrasekhar, V., M. J. Rooks, S. Wind, and D. E. Prober, 1985, "Observation of Aharonov-Bohm electron interference effects with periods h/e and $h/2e$ in individual micron size normal-metal rings," Phys. Rev. Lett. **55**, 1610.

Chern, S.-s., 1979, *Complex Manifolds without Potential Theory, Second Edition* (Springer, New York).

Choquet-Bruhat, Y., C. DeWitt-Morette, and M. Dillard-Bleick, 1982, *Analysis, Manifolds, and Physics, Revised Edition* (North-Holland, Amsterdam/New York/Oxford).

Clarke, J., 1987, "Quantum phenomena in superconductors, in *Proceedings of the 18th International Conference on Low Temperature Physics* (Jpn. J. Appl. Phys. **26**, 1771).

Combes, J. M., P. Duclos, and R. Seiler, 1981, "The Born-Oppenheimer Approximation," in *Rigorous Atomic and Molecular Physics*, edited by G. Velo and A. Wightman (Plenum, New York/London), p. 185.

Connes, A., 1969, "Noncommutative differential geometry," Pub. IHES **62**, 43.

Dana, I., Y. Avron, and J. Zak, 1985, "Quantised Hall conductance in a perfect crystal," J. Phys. C **18**, L679.

de Gennes, P. G., 1981, "Champ critique d'une boucle supraconductrice ramifiee," C. R. Acad. Sci., Ser. B **292**, 9; 279.

Domany, E., S. Alexander, D. Bensimon, and L. Kadanoff, 1983, "Solutions to the Schrödinger equation on some fractal lattices," Phys. Rev. B **28**, 3110.

Dubrovin, B. A., A. T. Fomenko, and S. P. Novikov, 1984, *Modern Geometry—Methods and Applications*, Graduate Texts in Mathematics Nos. 93 and 104 (Springer, New York).

Dubrovin, B. A., and S. P. Novikov, 1980, "Ground states of a two-dimensional electron in a periodic magnetic field," Zh. Eksp. Teor. Fiz. **79**, 1006 [Sov. Phys. JETP **52**, 511 (1980)].

Dyson, F., 1964, "Statistical theory of energy levels of complex systems," J. Math. Phys. **2**, 140.

Exner, P., and P. Šeba, 1987, "Free quantum motion on a branching graph," JINR Reports Nos. E2-87-213 and E2-87-214.

Flanders, H., 1963, *Differential Forms, with Applications to the Physical Sciences* (Academic, New York).

Friedland, S., J. W. Robbin, and J. H. Sylvester, 1984, "On the crossing rule," Commun. Pure Appl. Math. **37**, 19.

Gefen, Y., Y. Imry, and M. Azbel, 1984, "Quantum oscillations and the Bohm-Aharonov effect for parallel resistors," Phys. Rev. Lett. **52**, 129.

Gefen, Y., and D. J. Thouless, 1987, "Zener transition and energy dissipation in small driven systems," Phys. Rev. Lett. **59**, 1752.

Haldane, F. D. M., 1986, "Path dependence of the geometric rotation of polarization in optical fibers," Opt. Lett. **11**, 730.

Hall, E. H., 1879, "On a new action of the magnet on electric currents," Am. J. Math. **2**, 287.

Herring, C., 1937, "Effects of time reversal symmetry on energy bands of crystals," Phys. Rev. **52**, 361.

Herzberg, G., and H. C. Longuet-Higgins, 1963, "Intersection of potential energy surfaces in polyatomic molecules," Discuss. Faraday Soc. **35**, 77.

Hofstadter, D. R., 1976, "Energy levels and wave functions of Bloch electrons in rational and irrational magnetic fields," Phys. Rev. B **14**, 2239.

Imry, Y., 1986, "Physics of mesoscopic systems," in *Directions in Condensed Matter Physics*, edited by G. Grinstein and G. Mazenko (World Scientific, Singapore/Philadelphia).

Johnson, R., and J. Moser, 1982, Commun. Math. Phys. **84**, 403.

Kato, T., 1950, "On the adiabatic theorem of quantum mechanics," J. Phys. Soc. Jpn. **5**, 435.

Kato, T., 1966, *Perturbation Theory of Linear Operators* (Springer, Berlin).

Klein, M., and R. Seiler, 1988, private communication.

Kohmoto, M., 1985, "Topological invariant and the quantization of the Hall conductance," Ann. Phys. (N.Y.) **160**, 343.

Kuchar, F., R. Meisels, K. Y. Lim, P. Pichler, G. Weimann, and W. Schlapp, 1987, "Hall conductivity at microwave and submillimeter frequencies in the quantum Hall effect regime," EPS Condensed Matter Conference, Pisa Physica Scripta **T19**, 79.

Kugler, M., and M. Shtrikman, "Berry's phase, local frames and classical analogues," Weizmann Institute preprint.

Kunz, H., 1986, "Quantized currents and topological invariants for electrons in incommensurate potentials," Phys. Rev. Lett. **57**, 1095.

Kunz, H., 1987, "The quantum Hall effect for electrons in a random potential," Commun. Math. Phys. **112**, 121.

Landau, L. D., and E. M. Lifshitz, 1977, *Quantum Mechanics, 3rd Revised Edition* (Pergamon, Oxford/New York).

Landau, L. D., E. M. Lifshitz, and L. P. Pitaevskii, 1984, *Electrodynamics of Continuous Media, Second Edition* (Pergamon, Oxford/New York).

Laughlin, R. B., 1981, "Quantized Hall conductivity in two dimensions," Phys. Rev. B **23**, 5632.

Longuet-Higgins, H. C., 1975, "The intersection of potential energy surfaces of polyatomic molecules," Proc. R. Soc. London, Ser. A **344**, 147.

Lyskova, A. S., 1985, "Topological characteristics of the spectrum of the Schrödinger operator in a magnetic field and in a weak potential," Teor. Mat. Fiz. **65**, 368 [Theor. Math. Phys. (USSR) **65**, 1218 (1986)].

Mandelbrot, B., 1983, *The Fractal Geometry of Nature* (Freeman, San Francisco).

Mittra, R., and S. W. Lee, 1977, *Analytical Techniques in the Theory of Guided Waves* (Macmillan, New York).

Niu, Q., and D. J. Thouless, 1984, "Quantum adiabatic charge transport in the presence of substrate disorder and many body interaction," J. Phys. A **17**, 2453.

Niu, Q., and D. J. Thouless, 1987, "Quantum Hall effect with realistic boundary conditions," Phys. Rev. B **35**, 2188.

Niu, Q., D. J. Thouless, and Y. S. Wu, 1985, "Quantized Hall conductance as a topological invariant," Phys. Rev. B **31**, 3372.

Novikov, S. P. 1981, Dok. Akad. Nauk SSSR **257**, 538.

Onsager, L., 1931, "Reciprocal relations in irreversible processes. II," Phys. Rev. **38**, 2265.

Patterson, E. M., 1969, *Topology* (Oliver and Boyd, Edinburgh).

Platt, J. R., K. Ruedenberg, C. W. Scherr, N. S. Ham, H. Labhart, and W. Lichten, 1964, *Free-electron Theory of Conjugated Molecules: A Source Book* (Wiley, New York).

Porter, C. E., 1965, Ed., *Statistical Theories of Spectra: Fluctuations* (Academic, New York/London).

Prange, R. E., and S. T. Girvin, 1987, Eds., *The Quantum Hall Effect* (Springer, New York).

Rammal, R., and G. Toulouse, 1982, "Spectrum of the Schrödinger equation on a self-similar structure," Phys. Rev. Lett. **49**, 1194.

Ramo, S., J. R. Whinnery, and T. van Duzer, 1984, *Fields and Waves in Communication Electronics*, 2nd ed. (Wiley, New York).

Reed, M., and B. Simon, 1972–1978, *Methods of Modern Mathematical Physics*, Vols. I–IV (Academic, New York/San Francisco/London).

Rellich, F., 1969, *Perturbation Theory of Eigenvalue Problems* (Gordon and Breach, New York).

Ruedenberg, K., and C. W. Scherr, 1953, "Free-electron network model for conjugated systems. I. Theory," J. Chem. Phys. **21**, 1565.

Shapiro, B., 1983, "Quantum conduction on a Caley tree," Phys. Rev. Lett. **50**, 747.

Sharvin, D. Yu., and Yu. V. Sharvin, 1981, "Magnetic flux quantization in cylindrical films of normal metals," Pis'ma Zh. Eksp. Teor. Fiz. **34**, 285 [JETP Lett. **34**, 272 (1981)].

Simon, B., 1983, "Holonomy, the quantum adiabatic theorem and Berry's phase," Phys. Rev. Lett. **51**, 2167.

Tao, R., and F. D. M. Haldane, 1986, "Impurity effect, degeneracy and topological invariance in the quantum Hall effect," Phys. Rev. B **33**, 3844.

Thouless, D. J., 1983, "Quantization of particle transport," Phys. Rev. B **27**, 6083.

Thouless, D. J., M. Kohmoto, M. P. Nightingale, and M. den Nijs, 1982, "Quantized Hall conductance in a two-dimensional periodic potential," Phys. Rev. Lett. **49**, 405.

Thouless, D. J., and Q. Niu, 1984, "Nonlinear corrections to the quantization of Hall conductance," Phys. Rev. B **30**, 3561.

Umbach, C. P., S. Washburn, R. B. Laibowitz, and R. A. Webb, 1984, "Magnetoresistance of small quasi one-dimensional normal-metal rings and lines," Phys. Rev. B 30, 4048.
von Klitzing, K., G. Dorda, and M. Pepper, 1980, "New method for high-accuracy determination of the fine-structure constant based on quantized Hall resistance," Phys. Rev. Lett. 45, 494.
von Neumann, J., and E. Wigner, 1929, "Über das Verhalten von Eigenwerten bei Adiabatischen Prozessen," Phys. Z. 30, 467.
Webb, R., A. R. F. Voss, G. Grinstein, and P. M. Horn, 1983, "Magnetic field behavior of Josephson junction array: Two dimensional flux transport on a periodic substrate," Phys. Rev. Lett. 51, 690.
Webb, R. A., S. Washburn, C. P. Umbach, and R. B. Laibowitz, 1985 "Observation of the h/e Aharonov-Bohm oscillations in normal-metal rings," Phys. Rev. Lett. 54, 2696.
Wilczeck, F., and A. Zee, 1984, "Appearance of gauge structure in simple dynamical systems," Phys. Rev. Lett. 52, 2111.
Wilson, K., 1974, "Confinement of quarks," Phys. Rev. D 10, 2445.
Wilson, R. J., 1972, *Introduction to Graph Theory* (Oliver and Boyd, Edinburgh).
Witten, E., 1986, "Noncommutative geometry and string field theory," Nucl. Phys. 268, 253.
Yurke, B., and J. S. Denker, 1984, "Quantum network theory," Phys. Rev. A 29, 1419.

Localization, Wave-Function Topology, and the Integer Quantized Hall Effect

Daniel P. Arovas
The James Franck Institute, University of Chicago, Chicago, Illinois 60637
and
R. N. Bhatt, F. D. M. Haldane, P. B. Littlewood, and R. Rammal
Bell Laboratories, Murray Hill, New Jersey 07974
(Received 16 June 1987)

In a magnetic field, a wave function in a two-dimensional system is uniquely specified by the position of its nodes. We show that for high fields and a weak random potential, motion of the zeros of the wave function under smooth changes of the boundary conditions can be used to characterize the behavior of the one-electron states and distinguish between localized and extended states.

PACS numbers: 72.15.Gd, 71.25.Pi, 71.50.+t

Following the discovery of the quantized Hall effect[1] an argument for the quantization based on gauge invariance was given by Laughlin,[2] which also shows that changes in σ_{xy} can occur only if the Fermi energy lies in a region of extended states. Further work by Thouless and co-workers[3] and others[4-6] has related the quantized value of σ_{xy} to topological properties of wave functions; from this viewpoint, the connection with delocalization of the wave function is not clear. In this paper, we show that a simple connection between delocalization of wave functions and their topological characteristic which leads to a nonzero σ_{xy} can be made. The existence of truly delocalized states in the large field limit is to be contrasted with the behavior of 2D systems at $H=0$ where no extended states are believed to exist.

In the presence of a magnetic field, wave functions exhibit nodal points rather than nodal lines, the latter being the generic situation in systems with time reversal symmetry. In two dimensions, and in a high enough magnetic field that the electrons can be considered confined to the lowest Landau level (LLL), wave functions are *completely* determined by the location of their zeros. Such a description provides a convenient and physical approach to understand localization produced by a disordered single-body potential. When generalized periodic boundary conditions are enforced over a finite system, the zeros, now finite in number, move smoothly under continuous changes in the boundary conditions. The sensitivity of a wave function to boundary-condition changes allows a distinction between localized and extended states[7]; we shall show that there exist states whose wave function can be forced to vanish at *any* specified point in real space by a suitable choice of boundary conditions. Such a state naturally appears to be "extended"; moreover, the covering of real space by the zeros of the state is characterized by an integer, which is identical to the quantized value of the boundary-condition averaged σ_{xy}.

We consider the Hamiltonian $H=H_0+V(\mathbf{r})$ in two dimensions, with kinetic energy $H_0=(2m)^{-1}(\mathbf{p}+e\mathbf{A}/c)^2$. We work in an arbitrary gauge $\mathbf{A}=\frac{1}{2}\mathbf{r}\times\mathbf{B}+\hbar c/e\Delta\chi$, with $\mathbf{B}=-B\hat{\mathbf{z}}$ the magnetic field. For a free particle, H_0 is a harmonic oscillator in the cyclotron coordinates, with a spectrum $E_n=(n+\frac{1}{2})\hbar\omega_c$ and a natural cyclotron frequency $\omega_c=eB/mc$. Each Landau level is extensively degenerate, with a density of states per unit area $N_s/\Omega=1/2\pi l^2$, where $l=(\hbar c/eB)^{1/2}$ is the magnetic length. When an external potential $V(\mathbf{r})$ is imposed, the Landau levels will both broaden and mix; however, when the cyclotron frequency is sufficiently large, the essential physics is well described by a LLL-projected Hamiltonian. (We assume that the LLL is only partially occupied, i.e., $\nu=2\pi l^2 n<2$, where n is the areal density). The most general form of a LLL wave function is

$$\psi(\mathbf{r})=f(z)e^{-z\bar{z}/4l^2}e^{-i\chi}, \qquad (1)$$

where $z=x+iy$, and $f(z)$ is an analytic function of its

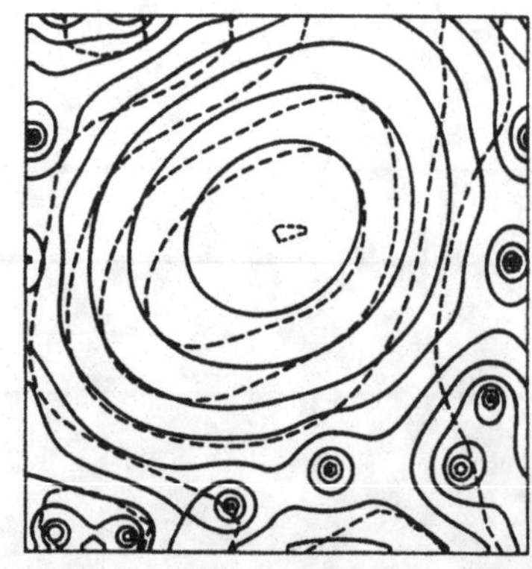

FIG. 1. Contour maps (in the primitive unit cell) of a numerically generated smooth random potential (dashed lines, linear scale), and of the probability density $|\psi_8|^2$ of the highest-energy state (solid lines, logarithmic scale).

argument. In order to treat finite systems, we impose generalized periodic boundary conditions on a square of side L containing an integral number of flux quanta $N_s = L^2/2\pi l^2$, by requiring $t(\mathbf{L}_j)|\psi\rangle = e^{i\theta_j}|\psi\rangle$ $(j=1,2)$, where $t(\mathbf{r})$ is the magnetic translation operator.[8] This leads to a general form for f (in terms of the rescaled length variable $\xi = z/L$)

$$f(\xi) = e^{\pi N_s \xi^2/2} e^{2\pi i \lambda \xi} \prod_{k=1}^{N_s} \Theta_1(\pi(\xi - \xi_k)\,|\, i), \tag{2}$$

where $\Theta_1(\omega\,|\,\tau)$ is the Jacobi Θ function, which possesses simple zeros at $\omega/\pi = j_1 + \tau j_2$ for all integers j_1, j_2. The freedom to choose arbitrary boundary-condition phases θ_1, θ_2 follows from the breaking of time-reversal symmetry by the magnetic field; fixing the θ_j leads (for N_s even) to the condition $\lambda = n_1 + \theta_1/2\pi$ as well as a constraint on the "center-of-mass" coordinate

$$\xi_{\mathrm{c.m.}} = N_s^{-1} \sum_{k=1}^{N_s} \xi_k = N_s^{-1} \left[(n_2 + \theta_2/2\pi) - i\lambda \right].$$

Here n_1, n_2 are integers, each of which may take on N_s distinct values, $0 \le n_j < N_s$. (For N_s odd, one replaces n_j by $n_j + \frac{1}{2}$ in the above formulas.) Equation (2) shows that there will be precisely N_s zeros inside the principal region Ω $[0 \le \mathrm{Re}(\xi) < 1;\ 0 \le \mathrm{Im}(\xi) < 1]$ which are

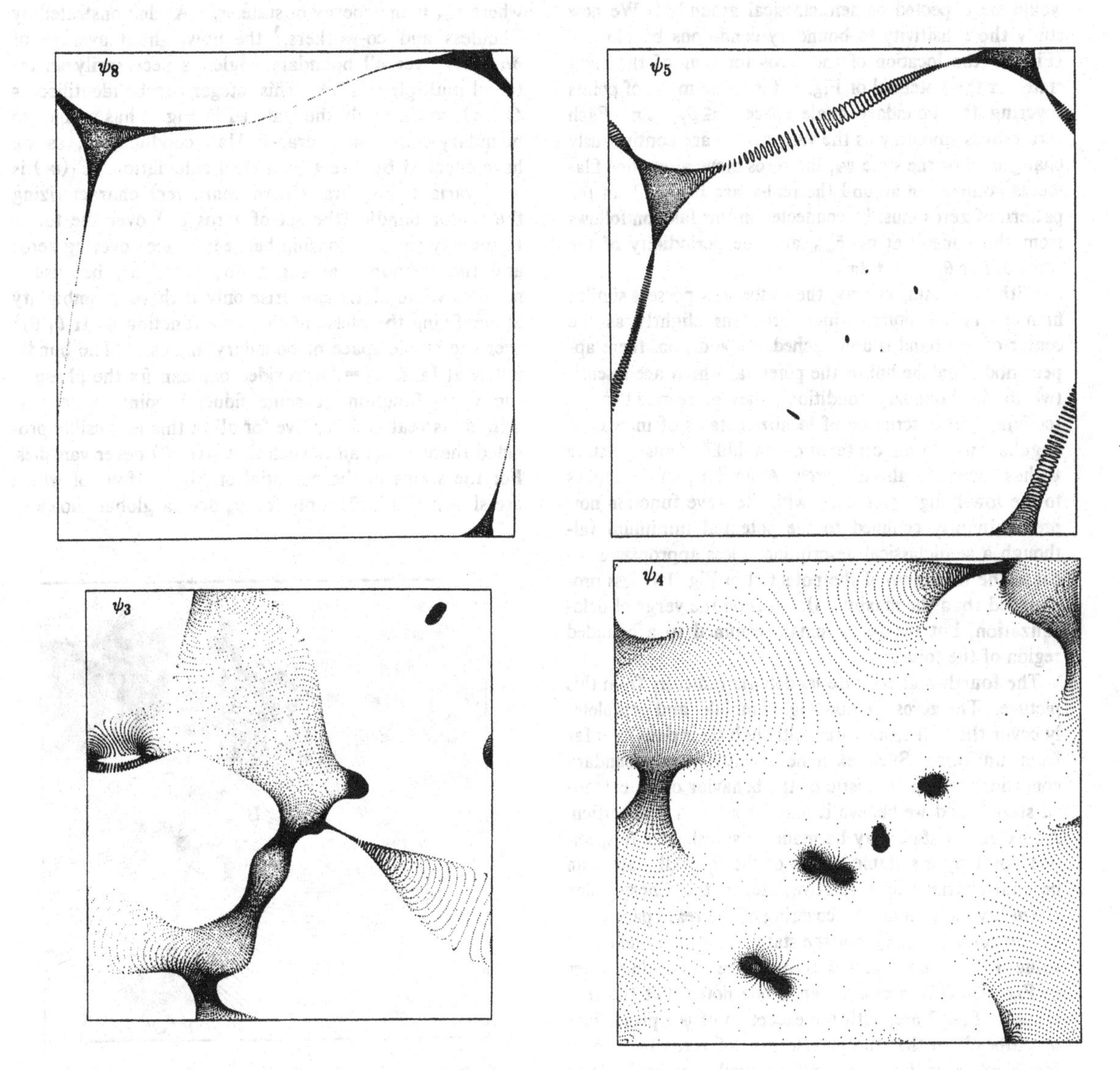

FIG. 2. Map of the nodal points of four of the $N_s(=8)$ wave functions in the potential of Fig. 1 for a fine grid of boundary conditions.

periodically repeated in every unit cell. The position of the zeros ξ_k *completely* determines the wave function. There are precisely N_s states, which are nondegenerate for a general potential $V(\mathbf{r})$, and are quasiperiodic, satisfying

$$\psi(\mathbf{r}+\mathbf{L}_j)=e^{i\theta_j}\exp(i\mathbf{r}\times\mathbf{L}_j\cdot\hat{\mathbf{z}}/2l^2)\psi(\mathbf{r}).$$

In Fig. 1, we show a contour plot of a typical smooth potential $v(\mathbf{r})$ (with a correlation length chosen equal to the magnetic length) together with a contour plot of $\log|\psi|$ for the highest energy state ψ_8 for $N_s=8$. The wave function is peaked on the hill of the potential, and the eight zeros cluster along the potential minima as would be expected on semiclassical grounds.[9] We now study the sensitivity to boundary conditions by plotting (Fig. 2) the location of the zeros for four of the eight states in the potential of Fig. 1, for a fine mesh of points covering the boundary angle space $0\le\theta_j<2\pi$. Each zero moves smoothly as the values of θ_j are continuously changed. For the state ψ_8, the zeros move along fine filaments connecting around the real-space torus. That the pattern of zeros must be connected in this fashion follows from the constraint on $\xi_{\text{c.m.}}$ and the periodicity of the zeros under $\theta_j\to\theta_j+2\pi$.

With decreasing energy, the states ψ_{7-5} possess similar filamentary structure, which broadens slightly as the center of the band is approached. In addition, there appear nodes on the hill of the potential which are insensitive to the boundary conditions; they correspond to the semiclassical description of localized states of increasing angular momentum centered on the hill.[9] Thus the state ψ_5 has three "localized" zeros. A similar picture applies to the low-lying states ψ_{1-2}, with the wave function now predominantly confined to the potential minimum (although a semiclassical description is less appropriate because the minimum of the potential in Fig. 1 is less pronounced than the maximum). ψ_3 is on the verge of delocalization, but the zeros remain confined to a bounded region of the torus.

The fourth-highest state ψ_4 departs radically from this picture. The zeros are now highly mobile, and completely cover the real-space torus, although their density is far from uniform. Such extreme sensitivity to boundary conditions is characteristic of the behavior of an extended state[7], and we believe it may be used as a definition: An extended state may be made to vanish at any specified point by a suitable choice of the θ_j. All states can be characterized by a relative integer, the Chern index $C_1(m)$, which counts the covering of the real-space torus by the zeros $\xi_k(\theta_1,\theta_2)$ for the state ψ_m. $C_1(m)$ takes the value $+1$ for $m=4$ and is zero for the remaining states in Fig. 2. This is easily seen if one notes that the zero maps of Fig. 2 are, with the exception of ψ_4, projections of "tubes," so that the inverse map of zeros in θ space for fixed ξ is typically null or double valued. It is straightforward to show that $\sum_{m=1}^{N_s}C_1(m)=1$, so that there must exist at least one nontrivial state.

We have also calculated the Hall conductivity σ_{xy} as a function of both energy ϵ and boundary-condition angles. The Kubo formula may be written as[3]

$$\sigma_{xy}(E;\theta_1,\theta_2)=\sum_{m=1}^{N_s}\delta\sigma_{xy}(m;\theta_1,\theta_2)\Theta(E-E_m),$$

$$\delta\sigma_{xy}(m;\theta_1,\theta_2)=\frac{1}{2\pi i}\frac{e^2}{h}\epsilon_{ij}\frac{\partial}{\partial\theta_i}<\tilde{\psi}_m|\frac{\partial}{\partial\theta_j}|\tilde{\psi}_m>, \tag{3}$$

$$|\tilde{\psi}_m>=\exp[-i(x\theta_1+y\theta_2]|\psi_m>,$$

where E_m is the energy of state ψ_m. As demonstrated by Thouless and co-workers,[3] the unweighted average of $\delta\sigma_{xy}(m)$ over all boundary angles is necessarily an integral multiple of e^2/h. This integer can be identified as $C_1(m)$, so that only the state ψ_4 in Fig. 2 has a nonzero boundary-condition averaged Hall conductance, as we have checked by direct numerical calculation. $C_1(m)$ is an invariant (the first Chern character) characterizing the vector bundle (the set of zeros ξ_k) over the torus. Intuitively the relationship between space-covering zeros and the boundary averaged $\delta\sigma_{xy}$ is clear, because a nonzero value of C_1 can arise only if there is ambiguity in our fixing the phase of the wave function $\psi_m(\mathbf{r},\theta_1,\theta_2)$ over the whole space of boundary angles.[10] The bundle is trivial (and $C_1=0$) provided one can fix the phase of the wave function at some fiducial point $\mathbf{r}_0$ so that $\psi(\mathbf{r}_0,\theta)$ is real and positive for all θ; this is possible provided there exists an $\mathbf{r}_0$ such that $\psi(\mathbf{r}_0,\theta)$ never vanishes. For the states in the potential of Fig. 1 (four of which are shown in Fig 2), only for ψ_4 does a global choice of

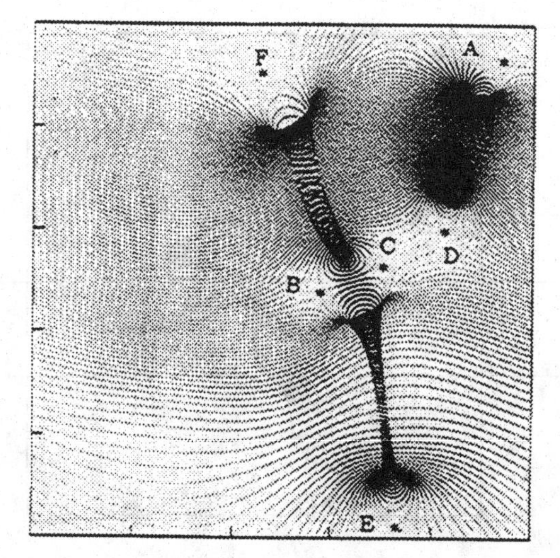

FIG. 3. Zero map of a state of Chern character +1 for $N_s=4$. The positions of the six double zeros are shown.

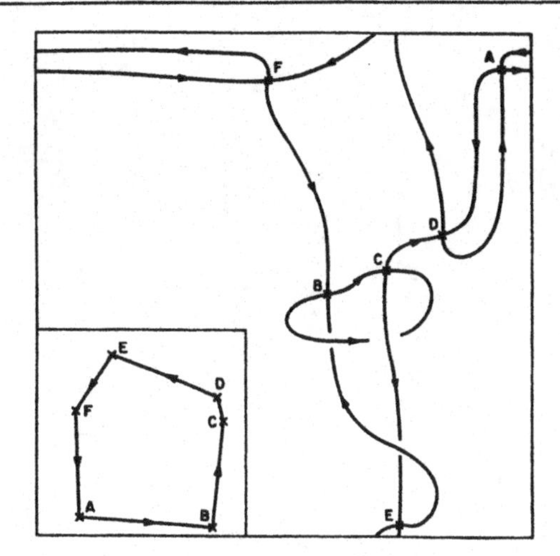

FIG. 4. Topology of the braiding of zeros associated with the closed path ABCDEFA in boundary-angle space (inset).

$\mathbf{r}_0$ not exist.

More understanding of the character of the wave functions can be obtained by the study of the braiding of paths of the zeros $\xi_k(\boldsymbol{\theta})$ under continuous paths in $\boldsymbol{\theta}$. We have found the existence of double zero ξ^* at isolated points $\boldsymbol{\theta}^*$ to be a generic occurrence for all wave functions in the band.[11] For $\boldsymbol{\theta}$ close to $\boldsymbol{\theta}^*$ the two nodes (labeled here ξ_j,ξ_k) will be distinct; under a path which circuits $\boldsymbol{\theta}^*$ (anticlockwise, say) and returns to the initial point, the two zeros will orbit each other (either anticlockwise or clockwise, i.e., with sign ± 1) and interchange. The point ξ^* is a branch point at the intersection of two sheets of the N_s-valued function $\xi(\boldsymbol{\theta})$; close to $\boldsymbol{\theta}^*$, two of the zeros are given by solutions of

$$(\xi-\xi^*)^2=K_1(\theta_1-\theta_1^*)+K_2(\theta_2-\theta_2^*),$$

with $K_{1,2}$ complex constants. Because the base space is a torus, the number of double zeros is even.

We have analyzed a state of Chern character +1 for the case $N_s=4$. The zero map for this state is shown in Fig. 3, together with the location of the six double zeros $\xi^*_{\text{A-F}}$. The value of C_1 can be determined directly from the topology of the knot produced by the motion of the zeros under the closed noncrossing path $\boldsymbol{\theta}^*_{\text{A}} \to \boldsymbol{\theta}^*_{\text{B}} \to \cdots \boldsymbol{\theta}^*_{\text{F}} \to \boldsymbol{\theta}^*_{\text{A}}$ (Fig. 4). The knot consists of lines joining the points ξ^*, with two trajectories entering, and two leaving each vertex. Those points (in real space) interior to any closed loop can be "covered" by a single zero, by contraction of the corresponding loop in boundary-angle space to a point. If there are no exterior points (which is the case of Fig. 4, because the knot is connected around the torus in both directions) the Chern character will be nonzero. If the knot is homotopic to a point, then $C_1=0$. A more detailed discussion of the braiding of zeros and the associated monodromy structure will be discussed in a forthcoming paper.[12]

To conclude, we have shown that studying the sensitivity of nodes of the wave function to changes in boundary conditions can be used to differentiate the behavior of localized and extended states. An extended state may be forced to vanish at any specified position in real space by the appropriate choice of boundary conditions. This nontrivial topological structure leads in particular to a nonzero Hall conductance.

[1]K. von Klitzing, G. Dorda, and M. Pepper, Phys. Rev. Lett 45, 494 (1980).

[2]R. B. Laughlin, Phys. Rev. B 23, 5632 (1981).

[3]D. J. Thouless, M. Kohmoto, M. P. Nightingale, and M. den Nijs, Phys. Rev. Lett 49, 405 (1982); Q. Niu, D. J. Thouless, and Y. S. Wu, Phys. Rev. B 31, 3372 (1985).

[4]J. E. Avron, R. Seiler, and B. Simon, Phys. Rev. Lett. 51, 51 (1983).

[5]M. Kohmoto, Ann. Phys. (NY) 160, 343 (1985).

[6]H. Aoki and T. Ando, Phys. Rev. Lett. 57, 3093 (1986).

[7]D. J. Thouless, Phys. Rep. 13, 93 (1974).

[8]For details of the approach, see F. D. M. Haldane and E. H. Rezayi, Phys. Rev. B 31, 2529 (1985).

[9]M. Tsukada, J. Phys. Soc. Jpn. 41, 1466 (1976).

[10]This argument is closely related to that of Kohmoto (Ref. 5), who considered zeros of the projection $\langle\phi_j|\psi\rangle$ onto a basis state $\langle\phi_j|$ *in boundary-angle space,* although his argument has been criticized recently (and we believe without justification) by Aoki and Ando (Ref. 6).

[11]In contrast to the analytic dependence of f on $\xi=\xi_1+i\xi_2$, there is no such analyticity as a function of any combination $\theta=\theta_1+\tau\theta_2$. Therefore, we denote the pair (θ_1,θ_2) by $\boldsymbol{\theta}$.

[12]D. P. Arovas, R. N. Bhatt, F. D. M. Haldane, P. B. Littlewood, and R. Rammal, to be published.

Chapter 2

THE FRACTIONAL EFFECT

2.1 Introduction

The discovery of the integer Hall effect led to considerable interest in two-dimensional electron systems, and, by 1982, theorists were beginning to feel confident that they understood why the Hall conductance *had* to be quantized in integer multiples of e^2/h. The discovery in that year of the *Fractional* Quantum Hall effect by Daniel Tsui, Horst Stormer and Arthur Gossard came rather as a surprise, and many early papers refer to the FQHE as the "anomalous" quantum Hall effect.

At the root of the quantization in the IQHE is a set of special electron densities, $\rho = neB/2\pi$, where n Landau levels are exactly filled. When one of these preferred charge densities moves at the Hall drift velocity, E/B, we find the integer QHE. What is needed for a similar understanding of the fractional effect is some mechanism that selects rational filling fractions, ν, of the Landau-level electron density. It was immediately obvious that any such mechanism requires interactions between the electrons. A beautiful and convincing explanation for a fundamental family of fractional densities was given by Laughlin [rep.13]. His "variational" wavefunction combines Fermi statistics and the analyticity [rep.12] of the lowest Landau level (LLL) wavefunction to provide an angular momentum barrier which keeps the electrons apart. This separation lowers the Coulomb energy and gives the wavefunction a large overlap with the true ground state.

The Laughlin state hosts a set of topologically interesting vortex-like quasiparticle excitations. These quasi-holes and quasi-particles have non-fluctuating fractional charges — an interesting phenomenon in its own right — but more intriguing is the observation that they have *fractional statistics* [1] [rep.14].

There is no dispute that Laughlin's wavefunctions give a good account of the fractional states at $\nu = \frac{1}{2n+1}$. More complicated rational fractions are observed, and there is not the same degree of unanimity about the best wavefunctions for describing these states. The original Hierarchy papers [reps.15,16] are here followed by a sampler of proposed variant Hierarchy constructions [reps.17,18,19,20] and an attempt at unification [rep.21].

2.2 Holomorphic Wavefunctions

In Chapter 1, we used the Landau gauge for the vector potential. We could equally well have used the *symmetric* gauge

$$\begin{aligned} A_x &= -\frac{1}{2}By \\ A_y &= \frac{1}{2}Bx. \end{aligned} \tag{2.2.1}$$

This gauge has the useful property that it is well adapted to the use of complex variables. We write $z = x + iy$ and introduce two complex derivatives

$$\begin{aligned} \partial_z &= \frac{1}{2}(\partial_x - i\partial_y) \\ \partial_{\bar{z}} &= \frac{1}{2}(\partial_x + i\partial_y). \end{aligned} \tag{2.2.2}$$

The factors of $\frac{1}{2}$ appear so that we can write

$$df(z,\bar{z}) = \partial_z f(z,\bar{z})dz + \partial_{\bar{z}} f(z,\bar{z})d\bar{z}. \tag{2.2.3}$$

We also introduce a metric tensor,

$$g_{\bar{z}z} = g_{z\bar{z}} = \frac{1}{2} \qquad g^{\bar{z}z} = g^{z\bar{z}} = 2, \tag{2.2.4}$$

in terms of which

$$ds^2 = g_{z\bar{z}}dzd\bar{z} + g_{\bar{z}z}d\bar{z}dz = d\bar{z}dz. \tag{2.2.5}$$

This tensor is useful for constructing the Laplace operator and its relatives.

The gauge-covariant derivatives appearing in the Schrödinger equation

$$\nabla_z = \partial_z - \frac{1}{4}eB\bar{z}, \qquad \nabla_{\bar{z}} = \partial_{\bar{z}} + \frac{1}{4}eBz \tag{2.2.6}$$

have the commutator

$$[\nabla_z, \nabla_{\bar{z}}] = \frac{eB}{2}, \tag{2.2.7}$$

and we may use (2.2.7) to rewrite the Schrödinger operator

$$\hat{H} = -\frac{1}{2m}(g^{z\bar{z}}\nabla_z\nabla_{\bar{z}} + g^{\bar{z}z}\nabla_{\bar{z}}\nabla_z) \tag{2.2.8}$$

as

$$\hat{H} = -\frac{2}{m}\nabla_z\nabla_{\bar{z}} + \frac{1}{2}\omega_c. \tag{2.2.9}$$

From the form (2.2.9), we deduce that any state satisfying a covariant version of the Cauchy-Riemann equation is automatically a lowest Landau level state:

$$\nabla_{\bar{z}}\psi(z,\bar{z}) = 0 \Longrightarrow \hat{H}\psi = \frac{1}{2}\omega_c\psi. \tag{2.2.10}$$

These states are of the form

$$\psi(z,\bar{z}) = f(z)e^{-\frac{eB}{4}|z|^2}, \tag{2.2.11}$$

with $f(z)$ a function of z only. The space of lowest Landau level (LLL) wavefunctions is thus equivalent to a space of *holomorphic* functions [rep.12] with inner product

$$\langle f_1|f_2\rangle = \int d^2z \bar{f}_1(z,\bar{z}) f_2(z,\bar{z}) e^{-\frac{eB}{2}|z|^2}. \tag{2.2.12}$$

This is the Bargman-Fock inner product on the Hilbert space of entire functions [2]. Perhaps confusingly, given the similarity of the names, there is a related theory of Hilbert spaces of analytic functions due to Bergman [3].

2.3 The Laughlin Wavefunction

Laughlin resolved much of the mystery of the FQHE at a single stroke by writing down his wavefunction

$$\langle z_1,\dots,z_N|0,L\rangle = \mathcal{N}\prod_{i<j}(z_i-z_j)^{2n+1} e^{-\frac{1}{4}\sum_{i=1}^N |z_i|^2} \tag{2.3.1}$$

and showing that it was a good candidate for the ground state of a system of LLL electrons with short-range repulsive interactions. It has three selling points:

- The state $|0,L\rangle$ is holomorphic in each of the electron coordinates. It is therefore a lowest Landau level many-body state.
- For n integral, $|0,L\rangle$ is antisymmetric under the interchange $z_i \rightleftharpoons z_j$, and so satisfies the Pauli principle.
- The multiple zeros, $(z_i-z_j)^{2n+1}$, ensure that the amplitude for two electrons to be close together is very small.

Despite its apparent simplicity $\langle z_1,\dots,z_N|0,L\rangle$ is not easy to work with. The electronic probability distribution

$$|\langle z_1,\dots,z_N|0,L\rangle|^2 = e^{(2n+1)\sum_{i<j}\ln|z_i-z_j|^2 - \frac{1}{2}\sum|z_i|^2} \tag{2.3.2}$$

coincides with the Boltzman factor for a two-dimensional one-component plasma with a background screening charge. Finding matrix elements for a large number of electrons requires calculating thermodynamic properties of this plasma, and this is by no means an easy task.

The one quantity that is relatively easy to determine is the mean electron density. The electrons in the plasma arrange themselves to neutralize the background charge represented by the $\frac{1}{2}\sum|z_i|^2$ term, and any deviations from local charge neutrality are strongly suppressed. By writing the exponent as

$$\frac{1}{2n+1}\left(\sum_{i<j}(2n+1)^2\ln|z_i-z_j|^2 - (2n+1)\frac{1}{2}\sum|z_i|^2\right), \tag{2.3.3}$$

we see that the "effective charge" of the plasma particles is proportional to $2n+1$, while the background charge density remains fixed at 1. Screening will occur when the plasma has a density of $1/2n+1$ of a full Landau level, so the state has filling fraction $\nu = \frac{1}{2n+1}$. The Laughlin state ties together the Pauli principle, analyticity, and the Coulomb repulsion to select these odd denominator filling fractions.

Apart from being good trial wavefunctions for the Coulomb problem, there are a class of short range pseudo-potentials for which the Laughlin states are exact ground states [4]. To see this, notice that any rotationally invariant short range interaction may be expanded as

$$V(|\mathbf{r}_1 - \mathbf{r}_2|) = \sum_n C_n (\nabla^2)^n \delta^2(\mathbf{r}_1 - \mathbf{r}_2), \tag{2.3.4}$$

where the C_n are proportional to the moments of the interaction

$$C_n \propto \int d^2\mathbf{r} (\mathbf{r}^2)^n V(r). \tag{2.3.5}$$

If we keep only the first term in the derivative expansion there are not enough derivatives to remove the multiple zeros in the $\nu = 1/3$ Laughlin state, so it is an exact ground state with zero energy. Similarly, the $\nu = \frac{1}{5}$ state is exact if we truncate the series after $(\nabla^2)^2 \delta^2(\mathbf{r}_1 - \mathbf{r}_2)$. (The expansion (2.3.4) should be treated with care, however. The individual terms are very singular, and not well defined in general scattering theory. They do make sense when used to evaluate matrix elements in the lowest Landau level.)

2.4 Statistics of Quasiholes and Holomorphic Line Bundles

The Laughlin wavefunction

$$\langle z_i | 0, L \rangle = \mathcal{N} \prod_{i<j} (z_i - z_j)^{2n+1} e^{-\frac{1}{4} \sum_i |z_i|^2} \tag{2.4.1}$$

suggests the existence of an interesting class of *quasihole* excitations. We simply insert a new polynomial factor into the wavefunction

$$|\zeta_1, \ldots \zeta_N\rangle = \mathcal{N}(\zeta_i) \prod_{i,j} (z_i - \zeta_j) \prod_{i<j} (z_i - z_j)^{2n+1} e^{-\frac{1}{4} \sum_i |z_i|^2}. \tag{2.4.2}$$

This state has all the virtues of the parent Laughlin state, but with a non-uniform charge distribution, the electrons are inhibited from approaching the quasihole locations $z = \zeta_j$. From the plasma analogy we calculate that there is a charge deficit of $e/2n+1$ in the vicinity of each of the quasiholes. The quasiholes are *fractionally charged.*

To demonstrate the fractional *statistics* of the quasiholes we can exploit the analyticity of the wavefunctions. The states $|\zeta_i\rangle$ vary holomorphically (except for possible $\bar{\zeta}_i$ terms in the normalization factor) with the ζ_i variables. As a consequence, these states form sections of a one-dimensional bundle of holomorphic states with a symmetric product of N copies of $\mathbf{C}$ as the base space. Now, holomorphic bundles have the property that the Berry connexion can be found entirely from the ζ_i dependent normalization factor. We will pause to explore this fact.

Suppose that

$$|\psi\rangle = \mathcal{N}(z, \bar{z}) |z\rangle \tag{2.4.3}$$

is a normalized state while $|z\rangle$ is unnormalized, but does not depend on $\bar{z}$. Then $\langle z|z\rangle = \mathcal{N}^{-2}$, and $\partial_{\bar{z}} |z\rangle = 0$, imply the identity

$$\begin{aligned} &(\partial_{\bar{z}_i} \langle z|) |z\rangle d\bar{z}_i + \langle z| (\partial_{z_i} |z\rangle) dz_i \\ &= -2\mathcal{N}^{-2} (\partial_{\bar{z}_i} \ln \mathcal{N} d\bar{z}_i + \partial_{z_i} \ln \mathcal{N} dz_i). \end{aligned} \tag{2.4.4}$$

The Berry connexion is then found to be

$$\mathcal{A} = \langle\psi|d|\psi\rangle = \partial_{z_i} \ln \mathcal{N} dz_i - \partial_{\bar{z}_i} \ln \mathcal{N} d\bar{z}_i, \tag{2.4.5}$$

and its curvature is

$$\mathcal{B} = d\mathcal{A} = 2\partial^2_{\bar{z}_i z_j} \ln \mathcal{N} d\bar{z}_i \wedge dz_j.$$

Analyticity also leads to additional geometric properties. When there is a unique ray for each point in the base space, it is possible to induce a natural metric on the base space from the normalization factors. To do this we define the distance between two neighbouring points as being the extent to which their associated rays differ in direction,

$$\begin{aligned} ds^2 &= \|\left(|d\psi\rangle - |\psi\rangle\langle\psi|d|\psi\rangle\right)\|^2 \\ &= -2\partial^2_{\bar{z}_i z_j} \ln \mathcal{N} d\bar{z}_i dz_j. \end{aligned} \tag{2.4.6}$$

(Unlike (2.4.5), in (2.4.6) there is no antisymmetry implicit in the $d\bar{z}_i dz_j$ product.)

An example of such a construction comes from the rotation group $SU(2)$. Consider the holomorphic states made by applying the step-down ladder operator J_- to the greatest spin state $|J, m = J\rangle$ of some given J^2:

$$|z\rangle = e^{zJ_-}|J, J\rangle. \tag{2.4.7}$$

These are coherent states with the spin pointing in the direction on the unit sphere corresponding to stereographic coordinate z. There is a one-to-one correspondence between the directions of the rays and points on S^2. As a simple example take $J = \frac{1}{2}$. Then

$$|\psi\rangle = \frac{1}{\sqrt{1+|z|^2}}\begin{pmatrix} 1 \\ z \end{pmatrix}. \tag{2.4.8}$$

The metric (2.4.8) is

$$\begin{aligned} ds^2 &= \partial^2_{\bar{z}z} \ln(1+|z|^2) dz d\bar{z} \\ &= \frac{1}{1+|z|^2} dz d\bar{z}, \end{aligned} \tag{2.4.9}$$

which is the usual metric on S^2 expressed in terms of stereographic coordinates. Such spaces, where the metric and a closed 2-form are both derived from some $\partial^2 \ln \mathcal{N}$ in this manner, are called *Kähler manifolds* and $\ln \mathcal{N}$ is called the *Kähler potential.* These spaces have many other special properties.

We now apply this machinery to the computation of the Berry connexion for the quasihole states. The normalization factor is given by

$$\mathcal{N}^{-2} = \int \left(\prod_i d^2 z_i\right) \exp\left\{\sum_i (2n+1)\ln|z_i - z_j|^2 + \sum \ln|z_i - \zeta_j|^2 - \frac{1}{2}\sum |z_i|^2\right\}. \tag{2.4.10}$$

The expression (2.4.10) is the partition function of a two-dimensional one-component plasma with some extra charges inserted at ζ_i. If the inserted charges are further apart than the Debye screening length in the gas, we can evaluate this with good accuracy as

$$\mathcal{N}^{-2} = \exp\left\{-\frac{1}{2n+1}\sum_{i<j} \ln|\zeta_i - \zeta_j|^2 + \frac{1}{2n+1}\frac{1}{2}\sum_i |\zeta_i|^2\right\}. \tag{2.4.11}$$

From (2.4.11) we read off the Berry connexion as

$$\mathcal{A} = \frac{1}{2}\frac{1}{2n+1}\sum_{i\neq j}\left(\frac{d\zeta_i}{\zeta_i - \zeta_j} - \frac{d\bar{\zeta}_i}{\bar{\zeta}_i - \bar{\zeta}_j}\right) - \frac{1}{2n+1}\sum_i \frac{1}{4}(\bar{\zeta}_i d\zeta_i - \zeta_i d\bar{\zeta}_i). \tag{2.4.12}$$

This result was first clearly stated in [rep.14]. (Although the authors of this paper seem to have ignored the ζ, $\bar{\zeta}$ dependence of the normalization $\mathcal{N}$ in describing their derivation, they end up with the correct expression). We will see in a later section that (2.4.12) is an Abelian version of the *Knizhnik-Zamolodchikov* connexion from conformal current algebra [5]. Although both (2.4.12) and the KZ connexion were discovered in the same year, the relationship between the two connexions was not appreciated until much later.

The physical significance of the last term in (2.4.12) is that, when we drag a quasihole round a closed path, the wavefunction picks up a phase factor equal to $1/2n+1$ times the area enclosed. Were the quasihole somehow attached to a heavy particle†, it would be this Berry phase that would appear in the Born-Oppenheimer hamiltonian for the particle, and so transfer the knowledge of the quasihole's fractional, $e/2n+1$, charge to the adiabatic dynamics of the slowly moving mass. The first terms give a phase that depends on how many other quasiholes have been encircled. In particular, we obtain a factor of $e^{i\theta}$ from each interchange of two holes. Here θ is the Abelian *statistics parameter*

$$\theta = \frac{2\pi}{2n+1}, \tag{2.4.13}$$

and this part of the Berry-phase turns the attached masses into *anyons*.

The thought-experiment of driving the dynamics of the quasiholes by attaching external heavy particles is necessary for the conventional interpretation of the Berry phase as the response to an adibatically changing hamiltonian or, as in the original Mead and Truhlar [6] application, the gauge potential occuring in the Born-Oppenheimer hamiltonian. It is however commonly supposed, and surely correct, that we do not need the heavy particles in order to extract physical consequences from these Berry phases. If the family of states parametrized by the location of the quasiholes exhausts the space of low lying states accessible to the electrons, we can seek solutions to the purely electron problem in the sector with a few quasiholes by taking linear combinations of these states

$$|\phi\rangle = \int \prod d^2\zeta_j \overline{\phi(\zeta_1,\ldots,\zeta_{N_h})}|\zeta_1,\ldots\zeta_N\rangle. \tag{2.4.14}$$

Since the state $|\phi\rangle$ is a conventional electron wavefunction, the Berry phases can be transfered to anyonic boundary conditions on the quasihole wavefunction $\phi(\zeta)$. This approach is discussed in the article by Laughlin in [1] on page 262.

Reprints for Chapter 2

[rep.12] ***Formalism for quantum Hall effect: Hilbert space of analytic functions***, S. M. Girvin, T. Jach, Phys. Rev. B29 (1984) 5617-5625.

[rep.13] ***Anomalous quantum Hall effect: an incompressible quantum fluid with fractionally charged excitations***, R. B. Laughlin, Phys. Rev. Lett. 50 (1983) 1395-1398.

†A repulsive interaction betwen the heavy particle and the electrons would keep the quasi-hole nearby.

[rep.14] ***Fractional statistics and the quantum Hall effect***, D. Arovas, J. R. Schrieffer, F. Wilczek, Phys. Rev. Lett. 53 (1984) 722-723.

[rep.15] ***Fractional quantization of the Hall effect: a hierarchy of incompressible quantum fluid states***, F. D. M. Haldane, Phys. Rev. Lett. 51 (1983) 605-608.

[rep.16] ***Statistics of quasiparticles and the heirarchy of fractional quantized Hall states***, B. I. Halperin, Phys. Rev. Lett. 52 (1984) 1583-1586 (2390E).

[rep.17] ***Particle-hole symmetry in the anomalous quantum Hall effect***, S. M. Girvin, Phys. Rev. B29 (1984) 6012-6014.

[rep.18] ***Hierarchical classification of fractional quantum Hall states***, N. d'Ambrumenil, R. Morph, Phys. Rev. B40 (1989) 6108-6119.

[rep.19] ***Structure of microscopic theory of the quantum Hall effect***, R. Blok, X.-G. Wen, Phys. Rev. B43 (1991) 8337-8349.

[rep.20] ***Composite fermion approach for the quantum Hall effect***, J. K. Jain, Phys. Rev. Lett. 63 (1989) 199-202.

[rep.21] ***Excitation structure of the hierarchy scheme for the fractional quantum Hall effect***, N. Read, Phys. Rev. Lett. 65 (1990) 1502-1505.

Other References for Chapter 2

[1] ***Fractional Statistics and Anyon Superconductivity***, F. Wilczek ed. World scientific 1990.

[2] V. Bargman, Rev. Mod. Phys 34,(1962) 829.

[3] S. Bergman, *The Kernal Function*, Am. Math. Soc. NY, 1950.

[4] S. A. Trugman, S. Kivelson, Phys. Rev B31 (1985) 5280.

[5] V. G. Knizhnik, A. .B. Zamolodchikov, Nucl. Phys. B247 (1984) 83.

[6] C. A. Mead, D. G. Truhlar, J. Chem. Phys. 70. (1979) 2284.

PHYSICAL REVIEW B VOLUME 29, NUMBER 10 15 MAY 1984

Formalism for the quantum Hall effect: Hilbert space of analytic functions

S. M. Girvin and Terrence Jach
Surface Science Division, National Bureau of Standards, Washington, D.C. 20234
(Received 19 December 1983)

We develop a general formulation of quantum mechanics within the lowest Landau level in two dimensions. Making use of Bargmann's Hilbert space of analytic functions we obtain a simple algorithm for the projection of any quantum operator onto the subspace of the lowest Landau level. With this scheme we obtain the Schrödinger equation in both real-space and coherent-state representations. A Gaussian interaction among the particles leads to a particularly simple form in which the eigenvalue condition reduces to a purely algebraic property of the polynomial wave function. Finally, we formulate path integration within the lowest Landau level using the coherent-state representation. The techniques developed here should prove to be convenient for the study of the anomalous quantum Hall effect and other phenomena involving electron-electron interactions.

I. INTRODUCTION

Recent experiments which have studied the quantum Hall effect[1–3] and the anomalous quantum Hall effect[4–6] have pointed up the need for a concise theoretical formulation of the two-dimensional Landau-level problem including electron-electron interactions. The remarkable phenomena associated with the anomalous quantum Hall effect are possible because of two strong constraints on the system. The very high magnetic field dominates the physics, and, as has been discussed previously,[7] severely restricts the dynamics of the particle motion. Low temperatures limit the electronic inversion layer to occupancy of the lowest spin state of the lowest Landau level. Under these conditions a commensuration energy exists which lowers the ground-state energy and introduces an excitation gap whenever the Landau-level filling is a simple rational fraction of the form p/q with q odd.[6] The origin of this commensuration effect is currently the object of intense experimental and theoretical interest.[7–15]

We wish to take advantage of the constraints on the dynamics mentioned above to present a general formalism for doing quantum mechanics within the lowest Landau level in two dimensions. The formalism we have developed is quite simple, easy to implement, and provides a useful language with which to discuss the physics of Landau levels in two dimensions. This formalism should find application in both analytic and numerical studies of the interacting electron problem.

The paper is organized as follows. Section II introduces a Hilbert space of analytic functions which spans the states of interest and with which the Schrödinger equation is very conveniently projected onto the lowest Landau level. Section III discusses the formal evaluation of the partition function. Section IV introduces coherent states and a formulation of path integrals for the lowest Landau level. A summary is presented in Sec. V. The formalism developed here is illustrated with simple analytical examples. Applications of these techniques will be presented in a sequel to the present paper.

II. HILBERT SPACE OF ANALYTIC FUNCTIONS

Consider a two-dimensional electron gas lying in the x-y plane and subject to a perpendicular magnetic field $\vec{B}=B\hat{z}$. The Hamiltonian is taken to be

$$H=H_0+V\,, \tag{1}$$

$$H_0=\sum_i \frac{1}{2m}\left[\vec{p}_i+\frac{e}{c}\vec{A}_i\right]^2 , \tag{2}$$

$$V=\sum_{i<j} v(\vec{r}_i-\vec{r}_j)\,, \tag{3}$$

where V is the Coulomb interaction or some other model interaction among the particles. The eigenvalues of the kinetic energy lie in discrete, highly degenerate Landau levels uniformly spaced in energy by $\hbar\omega_c$ where ω_c is the classical cyclotron frequency. We will assume B is sufficiently large that the magnetic energy greatly exceeds characteristic thermal and potential energies, so that mixing of Landau levels can be neglected. This (by now standard) assumption is not necessarily strictly valid for real systems, but will presumably lead only to quantitative errors not qualitative changes in the physics. Restriction of the electrons to the lowest spin state of the lowest Landau level yields a considerable advantage since the wave functions for these states have a simple analytic form. Our central purpose in this section is to develop a systematic formalism for doing quantum mechanics in the lowest Landau level.

We begin with the kinetic energy. In the symmetric gauge with vector potential $\vec{A}_i=-\frac{1}{2}\vec{r}_i\times\vec{B}$ the lowest Landau level eigenfunctions of H_0 have the form[8] (in units where $l^{-2}=eB/\hbar c=1$)

$$\psi[z]=f[z]\exp\left[-\tfrac{1}{4}\sum_i |z_i|^2\right], \tag{4}$$

where $[z]$ means $(z_1,z_2,\ldots,z_N)$ and f is a polynomial in the N variables $z_k\equiv x_k-iy_k$. The exponential factor in

(4) will be common to all wave functions, and the manipulations we wish to carry out are simplified considerably if this common factor can be removed. We do this now by formally defining a Hilbert space of analytic functions following Bargmann.[16]

Consider the set of entire functions of N complex variables,

$$\Theta \equiv \{f\} . \quad (5)$$

These functions are analytic in each of their arguments everywhere in the complex plane. Thus for $N=1$, for example, the function defined by

$$f(z)=z^3$$

is an element of Θ, but the function defined by

$$f(z)=z^*$$

is *not* analytic (since z^* cannot be expressed as a power series in z) and is thus excluded from Θ. This is a crucial point to which we shall return later.

Define an inner product on Θ via

$$(f,g)=\int d\mu[z]f^*[z]g[z] , \quad (6)$$

where the measure is

$$d\mu[z]=\prod_{i=1}^{N}\frac{1}{2\pi}e^{-|z_i|^2/2}dx_i dy_i . \quad (7)$$

We restrict Θ to include only those functions with finite norm $(f,f)<\infty$.

The Hilbert space thus defined is realized by the wave functions of the lowest Landau level since these may always be written as in Eq. (4) with f being a member of Θ. Furthermore, the inner product on Θ has been defined in such a way that wave function overlaps are given by

$$\langle \psi' | \psi \rangle = (f',f) . \quad (8)$$

The primary advantage of defining the Hilbert space so that we can work with f instead of ψ in (4) is that f is analytic while ψ is not.

We now investigate what linear operators can be defined on Θ, focusing on the case $N=1$ to obtain some useful results which are easily generalized to arbitrary N. Consider

$$g=\hat{O}f , \quad (9)$$

where $\hat{O}$ is a linear operator and f and g are in Θ. The requirement of analyticity severely restricts the form of $\hat{O}$. There are only three fundamental operations allowed: (1) multiplication by a complex constant, (2) multiplication by a power of z (but not z^*), and (3) differentiation with respect to z. Any linear operator on Θ can be expressed in terms of these fundamental operations.

In order to study these fundamental operators it is useful to define orthonormal basis functions by

$$f_n(z)=\frac{z^n}{(2^n n!)^{1/2}} . \quad (10)$$

These have the property

$$zf_m=\sqrt{2}\sqrt{m+1}f_{m+1} , \quad (11)$$

$$\frac{d}{dz}f_m=\sqrt{m/2}f_{m-1} . \quad (12)$$

Hence,

$$a^\dagger \equiv z/\sqrt{2} , \quad (13)$$

$$a\equiv\sqrt{2}\frac{d}{dz} \quad (14)$$

are boson ladder operators[7] and are easily seen to be mutually adjoint with respect to the inner product defined on Θ. Our present discussion of these operators in terms of the Bargmann space Θ makes formal Laughlin's[14] procedure of having d/dz not apply to the exponential part of the wave function in Eq. (4).

Having obtained these fundamental operators we now consider how to project the Hamiltonian onto the lowest Landau level by expressing it in terms of these operators. Because the lowest-Landau-level eigenfunctions are all degenerate, the kinetic energy commutes with any operator that has been projected on to that level. The kinetic energy is thus a constant which can be ignored. The Schrödinger equation becomes simply

$$\hat{V}\psi[z]=E\psi[z] , \quad (15)$$

where $\hat{V}$ is the projection of the potential operator onto the lowest Landau level. Returning to the case of N particles let us assume a central two-body potential which may be expanded in the form

$$V=\sum_{i<j}^{N}\sum_{n=0}^{\infty}\gamma_n(\vec{r}_{ij}\cdot\vec{r}_{ij})^n ,$$

which may be rewritten

$$V[z^*,z]=\sum_{i<j}^{N}\sum_{n=0}^{\infty}\gamma_n(z_i^*-z_j^*)^n(z_i-z_j)^n . \quad (16)$$

We are now faced with the problem that z^* takes states out of the Hilbert space (it mixes Landau levels). We need to project z^* onto our fundamental operators. Consider the matrix element

$$\alpha_{nmk}=(f_n,z_k^*f_m) . \quad (17)$$

Although $z_k^*f_m$ is outside the Hilbert space, α_{nmk} is perfectly well defined. Indeed from the definition of the inner product it is clear that

$$(f_n,z_k^*f_m)=(z_kf_n,f_m) \quad (18)$$

since z_k^* is the Hermitian conjugate of z_k. However, using (13) and (14) we have that the adjoint of z_k is

$$z_k^\dagger=2\frac{\partial}{\partial z_k} , \quad (19)$$

so that (18) becomes

$$(f_n,z_k^*f_m)=\left[f_n,2\frac{\partial}{\partial z_k}f_m\right] . \quad (20)$$

Thus z_k^* and $2\partial/\partial z_k$ have the same matrix elements within the space Θ. Despite this, the two operators are not completely equivalent since z_k^* commutes with z_k and $\partial/\partial z_k$ does not. For example, we have

$$(f, z_k z_k^* g) = (f, z_k^* z_k g) , \tag{21}$$

but

$$\left[f, z_k 2\frac{\partial}{\partial z_k} g\right] \neq \left[f, 2\frac{\partial}{\partial z_k} z_k g\right] . \tag{22}$$

Only the right-hand side of (22) is in agreement with (21), showing that occasionally it is important even for physicists to distinguish between the Hermitian conjugate and the adjoint.

The message in (21) and (22) is that z^* makes sense only when operating to the left. Hence, in order to project the potential energy onto the lowest Landau level one simply uses

$$V[z^*, z] \rightarrow \hat{N} V\left[2\frac{\partial}{\partial z}, z\right] , \tag{23}$$

where $\hat{N}$ is a normal ordering operator that keeps all the derivatives on the left. Note that if one has the product of two operators each of which is separately projected onto the lowest Landau level one has

$$A[z^*, z]B[z^*, z] \rightarrow \left[\hat{N} A\left[2\frac{\partial}{\partial z}, z\right]\right]\left[\hat{N} B\left[2\frac{\partial}{\partial z}, z\right]\right] , \tag{24}$$

whereas if only the product is to be projected one has

$$A[z^*, z]B[z^*, z] \rightarrow \hat{N}\left[A\left[2\frac{\partial}{\partial z}, z\right] B\left[2\frac{\partial}{\partial z}, z\right]\right] , \tag{25}$$

Using these rules we obtain our central result: the projection of the Schrödinger equation onto the lowest Landau level,

$$\left[\hat{N} V\left[2\frac{\partial}{\partial z}, z\right]\right]\psi[z] = E\psi[z] , \tag{26}$$

where $V[z^*, z]$ is the classical potential.

We now turn to some illustrative examples using (26). Consider the case of harmonic interaction among the particles,

$$V[z^*, z] = \tfrac{1}{2}\lambda^2 \sum_{i<j}^{N} (z_i^* - z_j^*)(z_i - z_j) . \tag{27}$$

The Schrödinger equation becomes

$$\lambda^2 \sum_{i<j}^{N} \left[\frac{\partial}{\partial z_i} - \frac{\partial}{\partial z_j}\right](z_i - z_j)\psi[z] = E\psi[z] . \tag{28}$$

Let us choose for ψ (the polynomial part of) Laughlin's wave function[9]

$$\psi_m[z] = \prod_{k<l}^{N} (z_k - z_l)^m , \tag{29}$$

where m is an odd integer. Separating terms involving only z_i or z_j individually yields

$$V\psi_m = \lambda^2 (m+1) N(N-1)\psi_m$$

$$+\lambda^2 \sum_{i<j}^{N} \sum_{k \neq i,j}^{N} m \left[\frac{z_i - z_j}{z_i - z_k} - \frac{z_i - z_j}{z_j - z_k}\right]\psi_m . \tag{30}$$

The summation in (30) may be evaluated by choosing any three electrons (e.g., 1,2,3), and noting

$$\sum_{i<j}^{3} \sum_{k \neq i,j}^{3} \left[\frac{z_i - z_j}{z_i - z_k} - \frac{z_i - z_j}{z_j - z_k}\right] = 3 . \tag{31}$$

There are

$$\begin{bmatrix} N \\ 3 \end{bmatrix}$$

distinct ways of choosing these three electrons so that (30) becomes

$$V\psi_m = N(N-1)(1 + mN/2)\psi_m , \tag{32}$$

and as previously found by other means,[7] Laughlin's wave function is an exact eigenfunction of the harmonic interaction.

As a second example, we consider how one can represent the $1/r^2$ potential,

$$V[z^*, z] = \sum_{i<j}^{N} \frac{1}{(z_i^* - z_j^*)(z_i - z_j)} . \tag{33}$$

We can rewrite this by means of the following integral representation:

$$V[z^*, z] = \sum_{i<j}^{N} \int_0^\infty d\lambda\, e^{-\lambda(z_i^* - z_j^*)(z_i - z_j)} . \tag{34}$$

Projection onto the lowest Landau level yields

$$V\left[2\frac{\partial}{\partial z}, z\right] = \sum_{i<j}^{N} \hat{N} \int_0^\infty d\lambda \exp\left[-2\lambda\left[\frac{\partial}{\partial z_i} - \frac{\partial}{\partial z_j}\right](z_i - z_j)\right] . \tag{35}$$

Similarly, the Coulomb potential may be expressed as a Gaussian integral,

$$V\left[2\frac{\partial}{\partial z}, z\right] = \sum_{i<j}^{N} \frac{1}{\sqrt{\pi}} \hat{N} \int_{-\infty}^{\infty} d\lambda \exp\left[-2\lambda^2\left[\frac{\partial}{\partial z_i} - \frac{\partial}{\partial z_j}\right](z_i - z_j)\right] . \tag{36}$$

These results are not very useful as they stand because of the presence of the normal ordering operator. Fortunately, this difficulty can be bypassed as we now demonstrate. Consider the operator (which contains no normal ordering)

$$V=\sum_{i<j}^{N} v_{ij}\ , \tag{37}$$

$$v_{ij}=\int_0^\infty d\lambda\, g(\lambda)\exp\left[-\tfrac{1}{2}\lambda(z_i-z_j)\left[\frac{\partial}{\partial z_i}-\frac{\partial}{\partial z_j}\right]\right], \tag{38}$$

where g is an arbitrary, real-valued function. To see the effect of V we note that any wave function may always be written (for a given i and j)

$$\phi[z]=\sum_{n,m=0}^{\infty} a_{nm}(z_i+z_j)^n(z_i-z_j)^m\ , \tag{39}$$

where a_{nm} depends only on coordinates other than z_i and z_j and where m is odd. Applying (38) yields

$$v_{ij}\phi[z]=\sum_{n,m=0}^{\infty}\int_0^\infty d\lambda\, g(\lambda)e^{-\lambda m}a_{nm}(z_i+z_j)^n(z_i-z_j)^m\ . \tag{40}$$

Suppose the actual potential obeys

$$v_{ij}(z_i+z_j)^n(z_i-z_j)^m=\epsilon(m)(z_i+z_j)^n(z_i-z_j)^m\ . \tag{41}$$

If we regard ϵ as a function of a continuous variable m then we can always achieve (41) by choosing g in (40) to be the inverse Laplace transform of ϵ. Thus it is possible to avoid the normal ordering problem. For example, the choice $g(\lambda)=\frac{1}{4}$ reproduces the matrix elements of $1/r_{ij}^2$ exactly. The choice $g(\lambda)=\delta(\lambda-\lambda_0)$ corresponds to a Gaussian interaction. With the use of (37) and (38) the Schrödinger equation becomes

$$\int_0^\infty d\lambda\, g(\lambda)\sum_{i<j}^{N}\exp\left[-\tfrac{1}{2}\lambda(z_i-z_j)\left[\frac{\partial}{\partial z_i}-\frac{\partial}{\partial z_j}\right]\right]\psi[z]=E\psi[z]\ . \tag{42}$$

For the remainder of the discussion we shall specialize to the case of a Gaussian interaction which is obtained by simply dropping the coupling constant integration in (42).

We see from (42) that the Schrödinger equation contains an infinite number of derivatives, but fortunately we can take advantage of the fact that $\exp(d/dz)$ is a displacement operator and $\exp(zd/dz)$ is a dilation operator. Defining

$$z_i'=z_i+Q(z_i-z_j)\ , \tag{43}$$

$$z_j'=z_j-Q(z_i-z_j)\ , \tag{44}$$

where

$$Q=(e^{-\lambda}-1)/2\ , \tag{45}$$

we see that Eq. (40) may be rewritten (dropping the coupling constant integration)

$$v_{ij}\phi[z]=\sum_{n,m=0}^{\infty} a_{nm}(z_i'+z_j')^n(z_i'-z_j')^m\ . \tag{46}$$

Hence, the Schrödinger equation becomes

$$\sum_{i<j}^{N}\phi(z_1,z_2,\ldots,z_i',\ldots,z_j',\ldots,z_N)=E\phi[z]\ . \tag{47}$$

We have thus reduced the Schrödinger equation to a purely algebraic statement about the polynomial ϕ.

We can gain further insight into the meaning of (47) if we think of it as a statement about the dilation symmetry of ϕ. Note that the effect of the potential v_{ij} on the left-hand side of (47) is to compress the distance between particles i and j by a factor $1+2Q=\exp(-\lambda)$. The multiplication by E on the right-hand side of (47) can also be related to a dilation in the following way. The wave function ϕ may always be taken to be an eigenfunction of the total angular momentum. However, ϕ has angular momentum L if and only if ϕ is homogeneous of degree L. Hence,

$$E\phi[z]=\phi[Kz]\ , \tag{48}$$

where $K=E^{1/L}$. Thus, the right-hand side of (47) corresponds to a global dilation (or contraction). The Schrödinger eigenvalue condition is simply that the sum of local contractions of individual bonds generated by the potential is equivalent to a single global contraction produced by the eigenvalue.

It is useful to consider the Gaussian interaction for various limiting values of the coupling constant λ. For infinitesimal λ we obtain the harmonic interaction discussed earlier. In the opposite limit $\lambda\to\infty$ we have $Q=-\frac{1}{2}$ so that [as can be seen from (43) and (44)] the effect of the potential is to bring the particles very close together. This is consistent with the short-range nature of the potential in this limit. If we allow for the possibility of complex coupling constants we see that for $\lambda=i\pi$, $Q=-1$ and $z_i'=z_j$, and $z_j'=z_i$. Hence,

$$\Xi_{ij} = \exp\left[(i\pi/2)(z_i - z_j)\left[\frac{\partial}{\partial z_i} - \frac{\partial}{\partial z_j}\right]\right] \tag{49}$$

is an explicit representation of the particle exchange operator. Since ϕ must be totally antisymmetric we may deduce that the energy eigenvalue obeys

$$E(\lambda \pm i\pi) = -E(\lambda) . \tag{50}$$

Hence, E must have the form

$$E(\lambda) = \sum_{m \text{ odd}}^{\infty} \alpha_m e^{-\lambda m} , \tag{51}$$

a fact which could also be deduced from (40).

In summary we have, by making use of Bargmann's Hilbert space of analytic functions, derived a simple formalism for projecting the Schrödinger equation onto the lowest Landau level. We have shown how to do this for arbitrary forms of central interactions among the particles. We found that the case of a Gaussian interaction leads to a particularly simple form in which the Schrödinger eigenvalue condition reduces to a purely algebraic property of the eigenfunction polynomial. Finally, we note that this last result suggests that it may be possible to attack the problem of solving the many-body wave equation by abstract group-theoretic methods. Of particular interest in this regard is the connection made by Bargmann[16] between polynomials in two variables, SU(2) and the rotation group. It may be possible to extend these ideas to polynomials in many variables.

III. PARTITION FUNCTION

Having established how to project the Hamiltonian onto the lowest Landau level we are in a position to discuss the formal evaluation of the partition function

$$Z \equiv \operatorname{Tr} e^{-\beta(H_0+V)} . \tag{52}$$

We begin by considering the recent work of Tosatti and Parrinello[15] (TP) which we believe treats the projection onto the lowest Landau level incorrectly. TP approximate (52) by

$$Z = \operatorname{Tr} e^{-\beta H_0} e^{-\beta V} , \tag{53}$$

and then restrict the trace to the lowest Landau level. The use of (53) ignores the fact that H_0 and V do not commute. We have been able to show that the use of this approximation destroys any possibility of observing a commensuration energy. We will demonstrate this by reviewing the derivation of TP and then extending it to obtain some new results. Following this we will present what we believe is the correct formulation of the partition function using the projection technique developed in the preceding section.

Assuming (53) to be valid we may neglect the kinetic energy since it is a constant for the lowest Landau level. Taking the trace in (53) in a coordinate representation yields

$$Z = \prod_{j=1}^{N} \int d^2 z_j e^{-\beta V[z^*, z]} P[z] , \tag{54}$$

where $d^2 z_j \equiv dx_j dy_j$ and

$$P[z] \equiv \sum_S |\psi_S[z]|^2 , \tag{55}$$

where the sum is over all Slater determinants ψ_S of N electrons in the lowest Landau level [note the ψ_S used here is the actual wave function including the exponential factor as in (4)]. Equation (54) resembles the classical partition function modified by a quantum correction P. TP argue that P contains the commensuration energy. We will now extend the analysis of TP by explicitly performing the summation in (55) to show that P does not produce a significant commensuration energy.

The Sth wave function is

$$\psi_S[z] = \frac{1}{\sqrt{N!}} \operatorname{Det} M_S , \tag{56}$$

where M_S is an $N \times N$ matrix with

$$(M_S)_{ij} = \phi_{Si}(z_j) , \tag{57}$$

and ϕ_{Si} is the ith orbital in the Sth configuration. Thus (55) becomes

$$P[z] = \frac{1}{N!} \sum_S (\operatorname{Det} M_S^\dagger)(\operatorname{Det} M_S) . \tag{58}$$

However, from the Cauchy-Binet theorem[17] we have

$$P[z] = \frac{1}{N!} \operatorname{Det}(L^\dagger L) , \tag{59}$$

where L is a rectangular matrix given by

$$L_{kj} = \phi_k(z_j) , \tag{60}$$

with j running from 1 to N, and k running over *all* orbitals (occupied or empty). Setting (60) into (59) gives

$$P[z] = \frac{1}{N!} \operatorname{Det} G[z] , \tag{61}$$

where G is the matrix defined by

$$G_{jj'} = \sum_k \phi_k^*(z_j) \phi_k(z_{j'}) . \tag{62}$$

We see that $G_{jj'}$ is simply the one-body Green's function for propagation from z_j to $z_{j'}$. In the symmetric gauge,

$$G_{jj'} = \frac{1}{2\pi} e^{-|z_j - z_{j'}|^2/4} e^{-(i/2)(x_j y_{j'} - y_j x_{j'})} , \tag{63}$$

so that G is a Gaussian with a magnetic flux phase factor. The partition function may now be written

$$Z = \frac{1}{N!} \prod_{j=1}^{N} \int d^2 z_j e^{-\beta V[z^*, z]} \operatorname{Det} G[z] . \tag{64}$$

Because of the strong Gaussian falloff of $G_{jj'}$, it is useful to separate G into a diagonal and an off-diagonal part,

$$G_{jj'} = \frac{1}{2\pi} (\delta_{jj'} + F_{jj'}) , \tag{65}$$

where $F_{ii} = 0$. We now use $\operatorname{Det}(G) = \exp(\operatorname{Tr} \ln G)$ and expand

$$Z=\frac{1}{N!}\prod_{j=1}^{N}\int\frac{d^2z_j}{2\pi}e^{-\beta V[z^*,z]}e^{\mathrm{Tr}[-(1/2)F^2+(1/3)F^3-\cdots]}. \tag{66}$$

The first term in the expansion may be viewed as a correction to the classical potential which introduces an effective interaction

$$\beta V_2=\sum_{i<j}^{N}e^{-|z_i-z_j|^2/2}. \tag{67}$$

This short-range repulsion represents the lowest-order exchange interaction. The next terms in the expansion represent cyclic three-particle exchange, etc.

We seek in this expansion a source of the commensuration energy which is observed experimentally to lower the free energy whenever the Landau-level filling is a simple rational fraction.[6] The first such term in (66) is the cyclic three-particle exchange. The three particles i, j, and k form a triangle of area A_{ijk} containing a certain magnetic flux Φ which controls the phase of the three-particle exchange term

$$\frac{1}{3}\sum_{i,j,k}e^{-(|z_i-z_k|^2+|z_k-z_j|^2+|z_j-z_i|^2)/4}\tfrac{1}{2}\cos(A_{ijk}). \tag{68}$$

The oscillating area term does produce a commensuration energy, but unfortunately the Gaussian falloff is quite severe. For a filling factor ν the quantity in (68) is of order $\exp(-\sqrt{3}\pi/\nu)$ which is approximately 10^{-7} for $\nu=\frac{1}{3}$. The higher-order contributions are even smaller. Thus, this is much too small to explain the experimentally observed commensuration energy.[4-6]

The origin of this difficulty is the approximation made by TP in using Eq. (53). We will now remedy this by going back to (52) and first projecting the Hamiltonian onto the lowest Landau level and likewise restricting the trace. One then has

$$Z=\mathrm{Tr}\,e^{-\beta(\tilde{H}_0+\tilde{V})}. \tag{69}$$

The kinetic energy $\tilde{H}_0$ is a constant for the lowest Landau level and, because the projection has already been performed, now commutes with the potential energy $\tilde{V}$.

The corrected version of (54) becomes [using (23)]

$$Z=\frac{1}{N!}\int d\mu[z]\sum_{S}\Phi_S^*[z]\exp\left[-\beta V\left[2\frac{\partial}{\partial z},z\right]\right]\Phi_S[z], \tag{70}$$

where Φ_S is the polynomial part of ψ_S defined in (56). It is assumed that V has been expressed in the manner described previously so that the normal ordering problem has been eliminated.

We would like to perform the summation in (70) to obtain the corrected version of Eq. (64). This may be done as follows. We rewrite (70) as

$$Z=\frac{1}{N!}\prod_{j=1}^{N}\int d^2z_j\sum_{S}\left[\Phi_S^*\exp\left[-\tfrac{1}{4}\sum_i|z_i|^2\right]\right]\exp\left[-\tfrac{1}{4}\sum_i|z_i|^2\right]\exp\left[-\beta V\left[2\frac{\partial}{\partial z},z\right]\right]\Phi_S. \tag{71}$$

We may move the second Gaussian weight factor to the right provided we make the transformation

$$2\frac{\partial}{\partial z}\to 2\frac{\partial}{\partial z}+\tfrac{1}{2}z^*, \tag{72}$$

and make the rule

$$\frac{\partial}{\partial z}z^*=0. \tag{73}$$

This leaves

$$Z=\frac{1}{N!}\sum_{j=1}^{N}\int d^2z_j\sum_{S}\psi_S^*[z]\exp\left[-\beta V\left[2\frac{\partial}{\partial z}+\tfrac{1}{2}z^*,z\right]\right]\psi_S[z], \tag{74}$$

$$Z=\lim_{[z']\to[z]}\frac{1}{N!}\prod_{j=1}^{N}\int d^2z_j\exp\left[-\beta V\left[2\frac{\partial}{\partial z}+\tfrac{1}{2}z^*,z\right]\right]\mathrm{Det}G[z',z], \tag{75}$$

where $G[z',z]$ is an $N\times N$ matrix elements given by the one-body propagator,

$$G_{kj}[z',z]=\frac{1}{2\pi}e^{-|z_k'-z_l|^2/4}e^{(z_k'^*z_l-z_k'z_l^*)/4}. \tag{76}$$

This formal expression for the partition function should prove to be a useful starting point for evaluation by a variety of numerical and analytic techniques, both perturbative and nonperturbative. The expression we have obtained is necessarily more complicated than that used by TP, but we have demonstrated that the latter does not yield a commensuration energy. In the next section we will discuss path-integral techniques for the evaluation of this more complicated expression for the partition function.

IV. COHERENT STATE REPRESENTATION AND PATH INTEGRATION

Consider the analytic function defined by

$$\phi_\xi[z]=\exp\left[\tfrac{1}{2}\sum_i \xi_i^* z_i\right]. \qquad (77)$$

Recalling Eq. (13), which shows that z_i is effectively a harmonic-oscillator raising operator, we see that ϕ_ξ is nothing more than a coherent state of the oscillator.[18] This is an eigenfunction of the lowering operator,

$$a_i\phi_\xi=\sqrt{2}\frac{\partial}{\partial z_i}\phi_\xi=\frac{1}{\sqrt{2}}\xi_i^*\phi_\xi . \qquad (78)$$

One can verify, in a straightforward manner, that ϕ_ξ has the norm

$$(\phi_\xi,\phi_\xi)=\exp\left[\tfrac{1}{2}\sum_i |\xi_i|^2\right], \qquad (79)$$

and corresponds to a Gaussian wave packet centered at point $[\xi]$ (when the exponential factor is restored to the wave function). These coherent states have several possible uses which we now explore.

There is currently a very active search underway for analytic wave functions to describe the ground state of interacting Landau-level electrons and to explain the energy excitation gap required by the existence of the anomalous quantum Hall effect.[9,11,12,14] One potentially useful approach to this problem is to take advantage of an integral representation of the wave function through the following identity:[16]

$$\psi[z]=\int d\mu[\xi]\psi[\xi]\phi_\xi[z] , \qquad (80)$$

where the measure is defined in Eq. (7). As Bargmann points out, the meaning of this identity is that ϕ_ξ is analogous to the Dirac δ function. Consistent with this is the fact that ϕ_ξ is the most hightly localized wave packet that can be constructed within the lowest Landau level.

As an example of the use of Eq. (80), Laughlin's wave function given by Eq. (29) may be written

$$\psi_m[z]=\int d\mu[\xi]\prod_{i<j}(\xi_i-\xi_j)^m\exp\left[\tfrac{1}{2}\sum_k \xi_k^* z_k\right]. \qquad (81)$$

With this representation it is quite easy to see how to include additional correlations in Laughlin's wave function by simply writing

$$\Psi[z]=\int d\mu[\xi]\prod_{i<j}(\xi_i-\xi_j)^m f(|\xi_i-\xi_j|)\exp\left[\tfrac{1}{2}\sum_k \xi_k^* z_k\right]. \qquad (82)$$

The form of f determines the additional correlations. For example,

$$f=|\xi_i-\xi_j|^p$$

or

$$f=1-e^{-p|\xi_i-\xi_j|^2}$$

both discourage close encounters of the particles. By using the first form of f, the integral in (82) may be explicitly carried out to yield

$$\Psi[z]=\prod_{i<j}^{N}\left[\frac{\partial}{\partial z_i}-\frac{\partial}{\partial z_j}\right]^p\prod_{k<l}^{N}(z_k-z_l)^{m+p} .$$

This seems likely to have a more favorable ground-state energy than Laughlin's wave function.

One of the attractive features of this representation is that since f is real, the total angular momentum, and hence the particle density, is independent of the form of f. This is thus a very convenient way in which to include variational freedom in the wave function while keeping the density automatically constrained. We note that Eq. (82) could also be easily generalized to include additional three-body and higher correlations as variational degrees of freedom.

A further advantage of the representation given in (80) is the complete factorization of the z dependence [as can be seen in the specific examples (81) and (82)]. We can take advantage of this factorization to cast the Schrödinger equation for the Gaussian interaction into a new form. Setting (80) into (47) yields

$$\int d\mu[\xi]\psi[\xi]\exp\left[\tfrac{1}{2}\sum_i \xi_i^* z_i\right]\sum_{k<l} e^{(Q/2)(\xi_k^*-\xi_l^*)(z_k-z_l)} =E\psi[z] , \qquad (83)$$

or, equivalently,

$$\int d\mu[\xi]\psi[\xi]\exp\left[\tfrac{1}{2}\sum_i \xi_i^* z_i\right]\left[E-\sum_{k<l} e^{(Q/2)(\xi_k^*-\xi_l^*)(z_k-z_l)}\right] =0 . \qquad (84)$$

This is simply the integral form of the Schrödinger equation conjugate to the differential form obtained previously.

Another important use of the coherent state representation is in the development of a path-integration scheme projected onto the lowest Landau level. To verify that the resolution of the identity within the coherent representation is[18]

$$I=\int d\mu[\xi]\,|\phi_\xi\rangle\langle\phi_\xi| ,$$

is a straightforward procedure. Specializing to the case $N=1$, the single-particle propagator may be rewritten using

$$G(z_f,t;z_i)=A^{-1}(\phi_{z_f},e^{-iHt}\phi_{z_i}) \quad (85a)$$

$$=\prod_{j=1}^{n+1}\int d\mu(\xi_j)(\phi_{\xi_j},e^{-iH\epsilon}\phi_{\xi_{j-1}}) , \quad (85b)$$

where

$$A\equiv(\phi_{z_i},\phi_{z_i})^{1/2}(\phi_{z_f},\phi_{z_f})^{1/2} ,$$

and $\epsilon\equiv t/(n+1)$, $\xi_0\equiv z_i$, and $\xi_{n+1}\equiv z_f$. Following Schulman[18] a series of standard manipulations leads to the path integral

$$G=\int D\xi\, e^{-iS(t)} , \quad (86)$$

where $\int D\xi$ means

$$\lim_{n\to\infty}\prod_{i=1}^{n}\int\frac{dx_i dy_i}{2\pi} ,$$

and the action S is given by

$$S(t)\equiv\int_0^t d\tau\left[\frac{1}{4i}\left[\xi^*\frac{d\xi}{d\tau}-\xi\frac{d\xi^*}{d\tau}\right]-H(\xi^*,\xi)\right] , \quad (87)$$

with

$$H(\alpha^*,\beta)\equiv\frac{(\phi_\alpha,H\phi_\beta)}{(\phi_\alpha,\phi_\beta)} . \quad (88)$$

Since the coherent states are all degenerate eigenstates of the kinetic energy (which is therefore being neglected), Eq. (88) is readily evaluated

$$H(\xi^*,\xi)=\int\frac{d^2z}{2\pi}V(z^*,z)e^{-|z-\xi|^2/2} , \quad (89)$$

where V is the classical potential. This is simply the expectation value of the potential energy for a Gaussian wave packet centered at $z=\xi$.

From the action in (87) we may find the variational equations for the extremal path

$$\frac{d\xi}{d\tau}=2i\frac{\partial H(\xi^*,\xi)}{\partial\xi^*} , \quad (90)$$

$$\frac{d\xi^*}{d\tau}=-2i\frac{\partial H(\xi^*,\xi)}{\partial\xi} . \quad (91)$$

Since H is real these are consistent. Returning to the original coordinates x and y via $\xi=x-iy$ yields

$$\dot{x}=\frac{\partial H}{\partial y} , \quad (92)$$

$$\dot{y}=-\frac{\partial H}{\partial x} . \quad (93)$$

These are simply the classical $\vec{E}\times\vec{B}$ drift equations.

We notice that the action in (87) is peculiar in that it is linear in the time derivatives. This violation of time reversal symmetry is, of course, due to the presence of the magnetic field. The form of the action suggests that we do not have an ordinary path integral, but rather something similar to a phase-space path integral.[18] Indeed if we treat ξ as a canonical coordinate and identify

$$\Pi\equiv\frac{-i}{2}\xi^* \quad (94)$$

as the canonically conjugate momentum, the path integral (86) assumes the phase-space form

$$G=\int D\xi D\Pi\, e^{iS} , \quad (95)$$

where the action is now

$$S=\int_0^t d\tau\,\tfrac{1}{2}(\Pi\dot{\xi}-\dot{\xi}\Pi)-H(\Pi,\xi) . \quad (96)$$

The variational equations for the extremal path now become

$$\dot{\xi}=\frac{\partial H}{\partial\Pi} , \quad (97)$$

$$\dot{\Pi}=-\frac{\partial H}{\partial\xi} , \quad (98)$$

which are nothing more than the usual Hamilton's equations.

There is a very nice connection between this and our previous results for the projection of the Schrödinger equation onto the lowest Landau level. To see this we consider the canonical quantization procedure for the classical theory represented by Eqs. (97) and (98). In a coordinate representation we quantize the classical theory by replacing the momentum by the operator

$$\Pi_{op}=-i\frac{\partial}{\partial\xi} . \quad (99)$$

From Eq. (94) we see that this requires the substitution

$$\xi^*\to2\frac{\partial}{\partial\xi} . \quad (100)$$

This is, however, precisely the rule given previously in Eq. (23) for projecting the quantum Hamiltonian onto the lowest Landau level. The fact that the path-integration scheme we have found is analogous to a phase-space path integral suggests that caution is required in making use of this scheme. Schulman[18] points out that phase-space path integrals are notoriously ill behaved. The problem here is that when we project the free propagator onto the lowest Landau level it becomes time independent. Thus, even for infinitesimal times the particle can propagate a finite length (on the order of the magnetic length). Formal manipulations of the path-integral expression must therefore be treated with care. Klauder has recently considered these questions in connection with the quantum Hall effect.[19]

One standard manipulation is to expand the path integral in fluctuations about the extremal path obtained in Eqs. (90) and (91). This procedure fails because of the pathological nature of the paths.[18] It is possible, however, to avoid this difficulty, and we plan to present a rigorous derivation of the propagator in the semiclassical limit elsewhere.[20]

V. SUMMARY

We have presented a general formulation of quantum mechanics within the lowest Landau level in two dimensions. This scheme involves study of the Hilbert space of functions analytic in the complex coordinate $z=x-iy$. The quantity z^* is related to the conjugate momentum and the simple replacement $z^*\rightarrow 2\,d/dz$ converts any classical quantity $f(z^*,z)$ into the associated quantum operator f_{op} properly projected onto the lowest Landau level. Within this formalism we have obtained expressions for the Schrödinger equation in both the real-space and coherent state representations. We have shown that a Gaussian interaction between the particles leads to a particularly simple form for the Schrödinger equation with the eigenvalue condition being reduced to a purely algebraic property of the polynomial wave function. We have also formulated path integration within the lowest Landau level by making use of the coherent state representation. Numerical application of the formalism presented here is currently in progress.

[1]K. v. Klitzing, G. Dorda, and M. Pepper, Phys. Rev. Lett. 45, 494 (1980).

[2]M. A. Paalanen, D. C. Tsui, and A. C. Gossard, Phys. Rev. B 25, 5566 (1982).

[3]For a recent review, see M. E. Cage and S. M. Girvin, Comments Solid State Phys. 11, 1 (1983); S. M. Girvin and M. E. Cage, *ibid.* 11, 47 (1983).

[4]D. C. Tsui, H. L. Stormer, and A. C. Gossard, Phys. Rev. Lett. 48, 1559 (1982).

[5]H. L. Stormer, D. C. Tsui, and A. C. Gossard, and J. C. M. Hwang, Physica 117-118B&C, 688 (1983).

[6]H. L. Stormer, A. Chang, D. C. Tsui, J. C. M. Hwang, A. C. Gossard, and W. Wiegmann, Phys. Rev. Lett. 50, 1953 (1983).

[7]S. M. Girvin and Terrence Jach, Phys. Rev. B 28, 4506 (1983).

[8]R. B. Laughlin, Phys. Rev. B 27, 3383 (1983).

[9]R. B. Laughlin, Phys. Rev. Lett. 50, 1395 (1983).

[10]D. Yoshioka, B. I. Halperin, and P. A. Lee, Phys. Rev. Lett. 50, 1219 (1983).

[11]B. I. Halperin, Helv. Phys. Acta 56, 75 (1983), and references therein.

[12]F. D. M. Haldane, Phys. Rev. Lett. 51, 605 (1983).

[13]R. Tao and D. J. Thouless, Phys. Rev. B 28, 1142 (1983).

[14]R. B. Laughlin, Surf. Sci. 141, 11 (1984).

[15]E. Tosatti and M. Parrinello, Lett. Nuovo Cimento 36, 289 (1983).

[16]V. Bargmann, Rev. Mod. Phys. 34, 829 (1962).

[17]A. C. Aitken, *Determinants and Matrices* (Oliver and Boyd, London, 1954), pp. 74 and 86.

[18]L. S. Schulman, *Techniques and Applications of Path Integration* (Wiley, New York, 1981), Chap. 27.

[19]J. R. Klauder (unpublished).

[20]S. M. Girvin, T. Jach, and M. Jonson (unpublished).

Anomalous Quantum Hall Effect: An Incompressible Quantum Fluid with Fractionally Charged Excitations

R. B. Laughlin
Lawrence Livermore National Laboratory, University of California, Livermore, California 94550

(Received 22 February 1983)

This Letter presents variational ground-state and excited-state wave functions which describe the condensation of a two-dimensional electron gas into a new state of matter.

PACS numbers: 71.45.Nt, 72.20.My, 73.40.Lq

The "$\frac{1}{3}$" effect, recently discovered by Tsui, Störmer, and Gossard,[1] results from the condensation of the two-dimensional electron gas in a GaAs-Ga$_x$Al$_{1-x}$As heterostructure into a new type of collective ground state. Important experimental facts are the following: (1) The electrons condense at a particular density, $\frac{1}{3}$ of a full Landau level. (2) They are capable of carrying electric current with little or no resistive loss and have a Hall conductance of $\frac{1}{3}e^2/h$. (3) Small deviations of the electron density do not affect either conductivity, but large ones do. (4) Condensation occurs at a temperature of ~ 1.0 K in a magnetic field of 150 kG. (5) The effect occurs in some samples but not in others. The purpose of this Letter is to report variational ground-state and excited-state wave functions that I feel are consistent with all the experimental facts and explain the effect. The ground state is a new state of matter, a quantum fluid the elementary excitations of which, the quasielectrons and quasiholes, are fractionally charged. I have verified the correctness of these wave functions for the case of small numbers of electrons, where direct numerical diagonalization of the many-body Hamiltonian is possible. I predict the existence of a sequence of these ground states, decreasing in density and terminating in a Wigner crystal.

Let us consider a two-dimensional electron gas in the x-y plane subjected to a magnetic field H_0 in the z direction. I adopt a symmetric gauge vector potential $\vec{A} = \frac{1}{2}H_0[x\hat{y} - y\hat{x}]$ and write the eigenstates of the ideal single-body Hamiltonian $H_{sp} = |(\hbar/i)\nabla - (e/c)\vec{A}|^2$ in the manner

$$|m,n\rangle = (2^{m+n+1}\pi m!\,n!)^{-1/2}\exp[\tfrac{1}{4}(x^2+y^2)]\left(\frac{\partial}{\partial x} + i\frac{\partial}{\partial y}\right)^m\left(\frac{\partial}{\partial x} - i\frac{\partial}{\partial y}\right)^n \exp[-\tfrac{1}{2}(x^2+y^2)], \tag{1}$$

with the cyclotron energy $\hbar\omega_c = \hbar(eH_0/mc)$ and the magnetic length $a_0 = (\hbar/m\omega_c)^{1/2} = (\hbar c/eH_0)^{1/2}$ set to 1. We have

$$H_{sp}|m,n\rangle = (n+\tfrac{1}{2})|m,n\rangle. \tag{2}$$

The manifold of states with energy $n+\frac{1}{2}$ constitutes the nth Landau level. I abbreviate the states of the lowest Landau level as

$$|m\rangle = (2^{m+1}\pi m!)^{-1/2} z^m \exp(-\tfrac{1}{4}|z|^2), \tag{3}$$

where $z = x+iy$. $|m\rangle$ is an eigenstate of angular momentum with eigenvalue m. The many-body Hamiltonian is

$$H = \sum_j \{|(\hbar/i)\nabla_j - (e/c)\vec{A}_j|^2 + V(z_j)\} + \sum_{j>k} e^2/|z_j - z_k|, \tag{4}$$

where j and k run over the N particles and V is a potential generated by a uniform neutralizing background.

I showed in a previous paper[2] that the $\frac{1}{3}$ effect could be understood in terms of the states in the lowest Landau level solely. With $e^2/a_0 \lesssim \hbar\omega_c$, the situation in the experiment, quantization of interelectronic spacing follows from quantization of angular momentum: The only wave functions composed of states in the lowest Landau level which describe orbiting with angular momentum m about the center of mass are of the form

$$\psi = (z_1 - z_2)^m (z_1 + z_2)^n \exp[-\tfrac{1}{4}(|z_1|^2 + |z_2|^2)]. \tag{5}$$

My present theory generalizes this observation to N particles.

I write the ground state as a product of Jastrow functions in the manner

$$\psi = \{\prod_{j<k} f(z_j - z_k)\}\exp(-\tfrac{1}{4}\textstyle\sum_l |z_l|^2), \tag{6}$$

and minimize the energy with respect to f. We

observe that the condition that the electrons lie in the lowest Landau level is that $f(z)$ be polynomial in z. The antisymmetry of ψ requires that f be odd. Conservation of angular momentum requires that $\prod_{j<k} f(z_j - z_k)$ be a homogeneous polynomial of degree M, where M is the total angular momentum. We have, therefore, $f(z) = z^m$, with m odd. To determine which m minimizes the energy, I write

$$|\psi_m|^2 = |\{\prod_{j<k}(z_j - z_k)^m\}\exp(-\tfrac{1}{4}\sum_l |z_l|^2)|^2 = e^{-\beta\Phi}, \tag{7}$$

where $\beta = 1/m$ and Φ is a classical potential energy given by

$$\Phi = -\sum_{j<k} 2m^2 \ln|z_j - z_k| + \tfrac{1}{2}m\sum_l |z_l|^2. \tag{8}$$

Φ describes a system of N identical particles of charge $Q = m$, interacting via logarithmic potentials and embedded in a uniform neutralizing background of charge density $\sigma = (2\pi a_0^2)^{-1}$. This is the classical one-component plasma (OCP), a system which has been studied in great detail. Monte Carlo calculations[3] have indicated that the OCP is a hexagonal crystal when the dimensionless plasma parameter $\Gamma = 2\beta Q^2 = 2m$ is greater than 140 and a fluid otherwise. $|\psi_m|^2$ describes a system uniformly expanded to a density of $\sigma_m = m^{-1}(2\pi a_0^2)^{-1}$. It minimizes the energy when σ_m equals the charge density generating V.

In Table I, I list the projection of ψ_m for three particles onto the lowest-energy eigenstate of angular momentum $3m$ calculated numerically. These are all nearly 1. This supports my assertion that a wave function of the form of Eq. (6) has adequate variational freedom. I have done a similar calculation for four particles with Coulombic repulsions and find projections of 0.979 and 0.947 for the $m = 3$ and $m = 5$ states.

ψ_m has a total energy per particle which for small m is more negative than that of a charge-density wave (CDW).[4] It is given in terms of the radial distribution function $g(r)$ of the OCP by

$$U_{\text{tot}} = \pi \int_0^\infty \frac{e^2}{r}[g(r) - 1]r\,dr. \tag{9}$$

In the limit of large Γ, U_{tot} is approximated within a few percent by the ion disk energy:

$$U_{\text{tot}} \simeq -\sigma_m \int \frac{e^2}{|r|} d^2r + \frac{\sigma_m^2}{2} \iint \frac{e^2}{|r_{12}|} dr_1^2\, dr_2^2 = (4/3\pi - 1)2e^2/R, \tag{10}$$

where the integration domain is a disk of radius $R = (\pi\sigma_m)^{-1/2}$. At $\Gamma = 2$ we have the exact result[5] that $g(r) = 1 - \exp[-(r/R)^2]$, giving $U_{\text{tot}} = -\tfrac{1}{2}\pi^{1/2}e^2/R$. At $m = 3$ and $m = 5$ I have reproduced the Monte Carlo $g(r)$ of Caillol *et al.*[3] using the modified hypernetted chain technique described by them. I obtain $U_{\text{tot}} = (-0.4156 \pm 0.0012)e^2/a_0$ and $U_{\text{tot}}(5) = (-0.3340 \pm 0.0028)e^2/a_0$. The corresponding values for the charge-density wave[4] are $-0.389e^2/a_0$ and $-0.322e^2/a_0$. U_{tot} is a smooth function of Γ. I interpolate it crudely in the manner

$$U_{\text{tot}}(m) \simeq \frac{0.814}{\sqrt{m}}\left(\frac{0.230}{m^{0.64}} - 1\right)\frac{e^2}{a_0}. \tag{11}$$

This interpolation converges to the CDW energy near $m = 10$. The actual crystallization point cannot be determined from that of the OCP since the CDW has a lower energy than the crystal described by ψ_m for $m > 71$.

I generate the elementary excitations of ψ_m by piercing the fluid at z_0 with an infinitely thin solenoid and passing through it a flux quantum $\Delta\varphi = hc/e$ adiabatically. The effect of this operation on the single-body wave functions is

$$(z - z_0)^m \exp(-\tfrac{1}{4}|z|^2) \to (z - z_0)^{m+1}\exp(-\tfrac{1}{4}|z|^2). \tag{12}$$

Let us take as approximate representations of these excited states

TABLE I. Projection of variational three-body wave functions ψ_m in the manner $\langle\psi_m|\Phi_m\rangle/(\langle\psi_m|\psi_m\rangle\langle\Phi_m|\Phi_m\rangle)^{1/2}$. Φ_m is the lowest-energy eigenstate of angular momentum $3m$ calculated with $V = 0$ and an interelectronic potential of either $1/r$, $-\ln(r)$, or $\exp(-r^2/2)$.

m	$1/r$	$-\ln(r)$	$\exp(-r^2/2)$
1	1	1	1
3	0.999 46	0.996 73	0.999 66
5	0.994 68	0.991 95	0.999 39
7	0.994 76	0.992 95	0.999 81
9	0.995 73	0.994 37	0.999 99
11	0.996 52	0.995 42	0.999 96
13	0.997 08	0.996 15	0.999 85

$$\psi_m^{+z_0} = A_{z_0}\psi_m = \exp(-\tfrac{1}{4}\sum_l |z_l|^2)\{\prod_i (z_i - z_0)\}\{\prod_{j<k}(z_j - z_k)^m\}, \tag{13}$$

and

$$\psi_m^{-z_0} = A_{z_0}^{+}\psi_m = \exp(-\tfrac{1}{4}\sum_l |z_l|^2)\left\{\prod_i \left(\frac{\partial}{\partial z_i} - \frac{z_0}{a_0^2}\right)\right\}\{\prod_{j<k}(z_j - z_k)^m\}, \tag{14}$$

for the quasihole and quasielectron, respectively. For four particles, I have projected these wave functions onto the analogous ones computed numerically. I obtain 0.998 for ψ_3^{-0} and 0.994 for ψ_5^{-0}. I obtain 0.982 for $\bar{\psi}_3^{+0} = \{\prod_i (z_i - \bar{z})\}\psi_3$, which is ψ_3^{+0} with the center-of-mass motion removed.

These excitations are particles of charge $1/m$. To see this let us write $|\psi^{+z_0}|^2$ as $e^{-\beta\Phi'}$, with $\beta = 1/m$ and

$$\Phi' = \Phi - 2\sum_l \ln|z_l - z_0|. \qquad (15)$$

Φ' describes an OCP interacting with a phantom point charge at z_0. The plasma will completely screen this phantom by accumulating an equal and opposite charge near z_0. However, since the plasma in reality consists of particles of charge 1 rather than charge m, the real accumulated charge is $1/m$. Similar reasoning applies to ψ^{-z_0} if we approximate it as $\prod_j (z_j - z_0)^{-1} P_{z_0}\psi_3$, where P_{z_0} is a projection operator removing all configurations in which any electron is in the single-body state $(z - z_0)^0 \exp(-\frac{1}{4}|z|^2)$. The projection of this approximate wave function onto $\psi_3^{-z_0}$ for four particles is 0.922. More generally, one observes that far away from the solenoid, adiabatic addition of $\Delta\varphi$ moves the fluid rigidly by exactly one state, per Eq. (12). The charge of the particles is thus $1/m$ by the Schrieffer counting argument.[6]

The size of these particles is the distance over which the OCP screens. Were the plasma weakly coupled ($\Gamma \leq 2$) this would be the Debye length $\lambda_D = a_0/\sqrt{2}$. For the strongly coupled plasma, a better estimate is the ion-disk radius associated with a charge of $1/m$: $R = \sqrt{2}\, a_0$. From the size we can estimate the energy required to make a particle. The charge accumulated around the phantom in the Debye-Hückel approximation is

$$\delta\rho = \frac{e/m}{2\pi\lambda_D^2} K_0(r/\lambda_D),$$

where K_0 is a modified Bessel function of the second kind. The energy required to accumulate it is

$$\Delta_{\text{Debye}} = \frac{1}{2}\iint \frac{\delta\rho\,\delta\rho}{|r_{12}|} = \frac{\pi}{4\sqrt{2}} \frac{1}{m^2} \frac{e^2}{a_0}. \qquad (16)$$

This estimate is an upper bound, since the plasma is strongly coupled. To make a better estimate let $\delta\rho = \sigma_m$ inside the ion disk and zero outside, to obtain

$$\Delta_{\text{disk}} = \frac{3}{2\sqrt{2}\pi} \frac{1}{m^2} \frac{e^2}{a_0}. \qquad (17)$$

For $m = 3$, these estimates are $0.062e^2/a_0$ and $0.038e^2/a_0$. This compares well with the value $0.033e^2/a_0$ estimated from the numerical four-particle solution in the manner

$$\Delta \simeq \frac{1}{2}\{E(\psi_3^{-0}) + E(\bar{\psi}_3^{+0}) - 2E(\psi_3)\}, \qquad (18)$$

where $E(\psi_3)$ denotes the eigenvalue of the numerical analog of ψ_3. This expression averages the electron and hole creation energies while subtracting off the error due to the absence of V. I have performed two-component hypernetted chain calculations for the energies of $\psi_3^{+z_0}$ and $\psi_3^{-z_0}$. I obtain $(0.022 \pm 0.002)e^2/a_0$ and $(0.025 \pm 0.005)e^2/a_0$. If we assume a value $\epsilon = 13$ for the dielectric constant of GaAs, we obtain $0.02e^2/\epsilon a_0 \simeq 4$ K when $H_0 = 150$ kG.

The energy to make a particle does not depend on z_0, so long as its distance from the boundary is greater than its size. Thus, as in the single-particle problem, the states are degenerate and there is no kinetic energy. We can expand the creation operator as a power series in z_0:

$$A_{z_0} = \sum_{j=0}^{N} A_j(z_1 \cdots, z_N) z_0^{N-j}. \qquad (19)$$

These A_j are the elementary symmetric polynomials,[7] the algebra of which is known to span the set of symmetric functions. Since every antisymmetric function can be written as a symmetric function times ψ_1, these operators and their adjoints generate the entire state space. It is thus appropriate to consider them N linearly independent particle creation operators.

The state described by ψ_m is incompressible because compressing or expanding it is tantamount to injecting particles. If the area of the system is reduced or increased by δA the energy rises by $\delta U = \sigma_m \Delta|\delta A|$. Were this an elastic solid characterized by a bulk modulus B, we would have $\delta U = \frac{1}{2}B(\delta A)^2/A$. Incompressibility causes the longitudinal collective excitation roughly equivalent to a compressional sound wave to be absent, or more precisely, to have an energy $\sim\Delta$ in the long-wavelength limit. This facilitates current conduction with no resistive loss at zero temperature. Our prototype for this behavior is full Landau level ($m = 1$) for which this collective excitation occurs at $\hbar\omega_c$. The response of this system to compressive stresses is analogous to the response of a type-II superconductor to the application of a magnetic field. The system first generates Hall currents without compressing, and then at a critical stress collapses by an area quantum $m2\pi a_0^2$

and nucleates a particle. This, like a flux line, is surrounded by a vortex of Hall current rotating in a sense opposite to that induced by the stress.

The role of sample impurities and inhomogeneities in this theory is the same as that in my theory of the ordinary quantum Hall effect.[8] The electron and hole bands, separated in the impurity-free case by a gap 2Δ, are broadened into a continuum consisting of two bands of extended states separated by a band of localized ones. Small variations of the electron density move the Fermi level within this localized state band as the extra quasiparticles become trapped at impurity sites. The Hall conductance is $(1/m) \times (e^2/h)$ because it is related by gauge invariance to the charge of the quasiparticles e^* by $\sigma_{\text{Hall}} = e^*e/h$, whenever the Fermi level lies in a localized state band. As in the ordinary quantum Hall effect, disorder sufficient to localize all the states destroys the effect. This occurs when the collision time τ in the sample in the absence of a magnetic field becomes smaller than $\tau < \hbar/\Delta$.

I wish to thank H. DeWitt for calling my attention to the Monte Carlo work and D. Boercker for helpful discussions. I also wish to thank P. A. Lee, D. Yoshioka, and B. I. Halperin for helpful criticism. This work was performed under the auspices of the U. S. Department of Energy by Lawrence Livermore National Laboratory under Contract No. W-7405-Eng-48.

[1]D. C. Tsui, H. L. Störmer, and A. C. Gossard, Phys. Rev. Lett. 48, 1559 (1982).

[2]R. B. Laughlin, Phys. Rev. B 27, 3383 (1983).

[3]J. M. Caillol, D. Levesque, J. J. Weis, and J. P. Hansen, J. Stat. Phys. 28, 325 (1982).

[4]D. Yoshioka and H. Fukuyama, J. Phys. Soc. Jpn. 47, 394 (1979); D. Yoshioka and P. A. Lee, Phys. Rev. B 27, 4986 (1983), and private communication.

[5]B. Jancovici, Phys. Rev. Lett. 46, 386 (1981). $\Gamma = 2$ corresponds to a full Landau level, for which the total energy equals the Hartree-Fock energy $-\sqrt{\pi/8}\, e^2/a_0$. This correspondence may be viewed as the underlying reason an exact solution at $\Gamma = 2$ exists.

[6]W. P. Su and J. R. Schrieffer, Phys. Rev. Lett. 46, 738 (1981).

[7]S. Lang, *Algebra* (Addison-Wesley, Reading, Mass., 1965), p. 132.

[8]R. B. Laughlin, Phys. Rev. B 23, 5632 (1981).

VOLUME 53, NUMBER 7 PHYSICAL REVIEW LETTERS 13 AUGUST 1984

Fractional Statistics and the Quantum Hall Effect

Daniel Arovas
Department of Physics, University of California, Santa Barbara, California 93106

and

J. R. Schrieffer and Frank Wilczek
Department of Physics and Institute for Theoretical Physics, University of California, Santa Barbara, California 93106
(Received 18 May 1984)

The statistics of quasiparticles entering the quantum Hall effect are deduced from the adiabatic theorem. These excitations are found to obey fractional statistics, a result closely related to their fractional charge.

PACS numbers: 73.40.Lq, 05.30.-d, 72.20.My

Extensive experimental studies have been carried out[1] on semiconducting heterostructures in the quantum limit $\omega_0\tau >> 1$, where $\omega_0 = eB_0/m$ is the cyclotron frequency and τ is the electronic scattering time. It is found that as the chemical potential μ is varied, the Hall conductance $\sigma_{xy} = I_x/E_y = \nu e^2/h$ shows plateaus at $\nu = n/m$, where n and m are integers with m being odd. The ground state and excitations of a two-dimensional electron gas in a strong magnetic field B_0 have been studied[2-4] in relation to these experiments and it has been found that the free energy shows cusps at filling factors $\nu = n/m$ of the Landau levels. These cusps correspond to the existence of an "incompressible quantum fluid" for given n/m and an energy gap for adding quasiparticles which form an interpenetrating fluid. This quasiparticle fluid in turn condenses to make a new incompressible fluid at the next larger value of n/m, etc.

The charge of the quasiparticles was discussed by Laughlin[2] by using an argument analogous to that used in deducing the fractional charge of solitons in one-dimensional conductors.[5] He concluded for $\nu = 1/m$ that quasiholes and quasiparticles have charges $\pm e^* = \pm e/m$. For example, a quasihole is formed in the incompressible fluid by a two-dimensional bubble of a size such that $1/m$ of an electron is removed. Less clear, however, is the statistics which the quasiparticles satisfy; Fermi, Bose, and fractional statistics having all been proposed. In this Letter, we give a direct method for determining the charge and statistics of the quasiparticles.

In the symmetric gauge $\vec{A}(\vec{r}) = \frac{1}{2}\vec{B}_0 \times \vec{r}$ we consider the Laughlin ground state with filling factor $\nu = 1/m$,

$$\psi_m = \prod_{j<k}(z_j - z_k)^m \exp(-\tfrac{1}{4}\textstyle\sum_l |z_l|^2), \quad (1)$$

where $z_j = x_j + iy_j$. A state having a quasihole localized at z_0 is given by

$$\psi_m^{+z_0} = N_+ \prod_i (z_i - z_0)\psi_m, \quad (2)$$

while a quasiparticle at z_0 is described by

$$\psi_m^{-z_0} = N_- \prod_i (\partial/\partial z_i - z_0/a_0^2)\psi_m, \quad (3)$$

where $2\pi a_0^2 B_0 = \phi_0 = hc/e$ is the flux quantum and $N_\pm$ are normalizing factors.

To determine the quasiparticle charge e^*, we calculate the change of phase γ of $\psi_m^{+z_0}$ as z_0 adiabatically moves around a circle of radius R enclosing flux ϕ. To determine e^*, γ is set equal to the change of phase,

$$(e^*/\hbar c)\oint \vec{A}\cdot d\vec{l} = 2\pi(e^*/e)\phi/\phi_0, \quad (4)$$

that a quasiparticle of charge e^* would gain in moving around this loop. As emphasized recently by Berry[6] and by Simon[7] (see also Wilczek and Zee[8] and Schiff[9]), given a Hamiltonian $H(z_0)$ which depends on a parameter z_0, if z_0 slowly transverses a loop, then in addition to the usual phase $\int^t E(t')\,dt'$, where $E(t')$ is the adiabatic energy, an extra phase γ occurs in $\psi(t)$ which is independent of how slowly the path is traversed. $\gamma(t)$ satisfies

$$d\gamma(t)/dt = i\langle\psi(t)|d\psi(t)/dt\rangle \;. \quad (5)$$

From Eq. (2),

$$\frac{d\psi_m^{+z_0}}{dt} = N_+ \sum_i \frac{d}{dt}\ln[z_i - z_0(t)]\psi_m^{+z_0}, \quad (6)$$

so that

$$\frac{d\gamma}{dt} = iN_+^2 \left\langle \psi_m^{+z_0} \left| \frac{d}{dt}\sum_i \ln(z_i - z_0) \right| \psi_m^{+z_0} \right\rangle. \quad (7)$$

Since the one-electron density in the presence of

the quasihole is given by

$$\rho^{+z_0}(z) = \langle \psi_m^{+z_0} | \sum_i \delta(z_i - z) | \psi_m^{+z_0} \rangle, \tag{8}$$

we have

$$\frac{d\gamma}{dt} = i \int dx\, dy\, \rho^{+z_0}(z) \frac{d}{dt} \ln[z - z_0(t)], \tag{9}$$

where $z = x + iy$. We write $\rho^{+z_0}(z) = \rho_0 + \delta\rho^{+z_0}(z)$, with $\rho_0 = \nu B/\phi_0$. Concerning the ρ_0 term, if z_0 is integrated in a clockwise sense around a circle of radius R, values of $|z| < R$ contribute $2\pi i$ to the integral while $|z| > R$ contributes zero. Therefore, this contribution to γ is given by

$$\gamma_0 = i \int_{|r| < R} dx\, dy\, \rho_0 2\pi i = -2\pi \langle n \rangle_R = -2\pi\nu\phi/\phi_0, \tag{10}$$

where $\langle n \rangle_R$ is the mean number of electrons in a circle of radius R. Corrections from $\delta\rho$ vanish as $(a_0/R)^2$, where $a_0 = (\hbar c/eB)^{1/2}$ is the magnetic length. This term corresponds to the finite size of the hole.

Comparing with Eq. (4), we find $e^* = \nu e$, in agreement with Laughlin's result. A similar analysis shows that the charge of the quasiparticle $\psi_m^{-z_0}$ is $-e^*$.

To determine the statistics of the quasiparticles, we consider the state with quasiholes at z_a and z_b,

$$\psi_m^{z_a, z_b} = N_{ab} \prod_i (z_i - z_a)(z_i - z_b) \psi_m. \tag{11}$$

As above, we adiabatically carry z_a aroound a closed loop of radius R. If z_b is outside the circle $|z_b| = R$ by a distance $d >> a_0$, the above analysis for γ is unchanged, i.e., $\gamma = -2\pi\nu\phi/\phi_0$. If z_b is inside the loop with $|z_b| - R << -a_0$, the change of $\langle n \rangle_R$ is $-\nu$ and one finds the extra phase $\Delta\gamma = 2\pi\nu$. Therefore, when a quasiparticle adiabatically encircles another quasiparticle an extra "statistical phase"

$$\Delta\gamma = 2\pi\nu \tag{12}$$

is accumulated.[10] For the case $\nu = 1$, $\Delta\gamma = 2\pi$, and the phase for interchanging quasiparticles is $\Delta\gamma/2 = \pi$ corresponding to Fermi statistics. For ν noninteger, $\Delta\gamma$ corresponds to fractional statistics, in agreement with the conclusion of Halperin.[11] Clearly, when ν is noninteger the change of phase $\Delta\gamma$ when a third quasiparticle is in the vicinity will depend on the adiabatic path taken by the quasiparticles as they are interchanged and the pair permutation definition used for Fermi and Bose statistics no longer suffices.

A convenient method for including the statistical phase $\Delta\gamma$ is by adding to the actual vector potential $\vec{A}_0$ a "statistical" vector potential $\vec{A}_\phi$ which has no independent dynamics. $\vec{A}_\phi$ is chosen such that

$$(e^*/\hbar c) \oint \vec{A}_\phi \cdot d\vec{l} = \Delta\gamma = 2\pi\nu, \tag{13}$$

when z_a encirlces z_b. One finds this fictious $\vec{A}_\phi$ to be

$$\vec{A}_\phi(\vec{r} - \vec{r}_b) = \frac{\phi_0 \hat{z} \times (\vec{r} - \vec{r}_b)}{2\pi |\vec{r} - \vec{r}_b|^2} \tag{14}$$

if the quasiparticles are treated as bosons and $\phi_0 \to \phi_0(1 - 1/\nu)$ if they are treated as fermions. Thus, the peculiar statistics can be replaced by a more complicated effective Lagrangian describing particles with conventional statistics.[12]

Finally, we note that if one pierces the plane with a physical flux tube of magnitude ϕ, the above arguments suggest that a charge $\nu e \phi/\phi_0$ is accumulated around the tube, regardless of whether ϕ/ϕ_0 is equal to the ratio of integers.

This work was supported in part by the National Science Foundation through Grant No. DMR82-16285 and No. PHY77-27084, supplemented by funds from the National Aeronautics and Space Administration. One of us (D.A.) is grateful for the support of an AT&T Bell Laboratories Scholarship.

[1]K. von Klitzing, G. Dorda, and M. Pepper, Phys. Rev. Lett. 45, 494 (1980).

[2]R. B. Laughlin, Phys. Rev. Lett. 50, 1395 (1983).

[3]F. D. M. Haldane, Phys. Rev. Lett. 51, 605 (1983).

[4]B. I. Halperin, Institute of Theoretical Physics, University of California, Santa Barbara, Report No. NSF-ITP-83-34, 1983 (to be published).

[5]W. P. Su and J. R. Schrieffer, Phys. Rev. Lett. 46, 738 (1981).

[6]M. V. Berry, Proc. Roy. Soc. London, Ser. A 392, 45–57 (1984).

[7]B. Simon, Phys. Rev. Lett. 51, 2167 (1983).

[8]F. Wilczek and A. Zee, Phys. Rev. Lett. 52, 2111 (1984).

[9]L. Schiff, *Quantum Mechanics* (McGraw-Hill, New York, 1955), p. 290.

[10]Although ψ is a variational wave function, rather than the actual adiabatic wave function, the statistical properties of the quasiparticles are not expected to be sensitive to this inconsistency. We could regard ψ to be an exact excited-state wave function for a model Hamiltonian.

[11]B. I. Halperin, Phys. Rev. Lett. 52, 1583, 2390(E) (1984).

[12]F. Wilczek and A. Zee, Institute of Theoretical Physics, University of California, Santa Barbara, Report No. NSF-ITP-84-25, 1984 (to be published).

Fractional Quantization of the Hall Effect: A Hierarchy of Incompressible Quantum Fluid States

F. D. M. Haldane
Department of Physics, University of Southern California, Los Angeles, California 90089
(Received 28 June 1983)

With use of spherical geometry, a translationally invariant version of Laughlin's proposed "incompressible quantum fluid" state of the two-dimensional electron gas is formulated, and extended to a hierarchy of continued-fraction Landau-level filling factors ν. Observed anomalies at $\nu = \frac{2}{5}, \frac{2}{7}$ are explained by fluids deriving from a $\nu = \frac{1}{3}$ parent.

PACS numbers: 71.45.Nt, 72.20.Nt, 73.40.Lq

The quantum Hall effect (quantization of the Hall resistance $\rho_{xy} = h/\nu e^2$ at simple rational values of ν at low temperatures, together with a dramatic fall in the sheet resistance ρ_{xx}) observed in GaAs-$Ga_xAl_{1-x}As$ heterostructures[1,2] may be explained (naively) if the ground state of the two-dimensional (2D) electron gas in high perpendicular magnetic fields has *no gapless excitations* (and hence no dissipation at low temperatures) when the Landau-level occupation factor takes one of the quantized values ν. This is trivially the case for free electrons when ν is integer, as seen in the earlier experiments,[1] but the effect (or its precursor anomalies) has recently been observed[2] with *fractional* quantization, to date at $\nu = \frac{2}{7}, \frac{1}{3}, \frac{2}{5}, \frac{3}{5}, \frac{2}{3}, \frac{4}{5}, \frac{4}{3}$, and $\frac{5}{3}$, all with *odd* denominators (when $\nu > 1$, the electrons are not fully spin-polarized; $\nu = \frac{4}{3}, \frac{5}{3}$ values may be understood as the $\nu = 1$ effect for majority spins, plus the $\nu = \frac{1}{3}, \frac{2}{3}$ effects for minority spins). A "Wigner solid" charge-density-wave ground state is expected[3] at low occupations, but such a state has a gapless Goldstone mode because translational and rotational symmetry (described by the Euclidean group) is broken. A state without gapless excitations may instead be characterized[4,5] as an "incompressible quantum fluid," and variational wave functions of Jastrow form that describe such states have recently been proposed by Laughlin[4] at occupations $\nu = 1/m$, m an odd integer.

The Laughlin wave functions are not translationally invariant, but describe a circular droplet of fluid, which must be confined in an external potential. Laughlin circumvented this problem by formally relating the properties of the fluid to those of the classical 2D one-component plasma, which has a thermodynamic limit, and calculating plasma properties. In this Letter, I describe a variant of Laughlin's scheme with fully translationally invariant wave functions, and extend it to describe a hierarchy of fluid states with occupation factors given by the continued fractions

$$\cfrac{1}{m + \cfrac{\alpha_1}{p_1 + \cfrac{\alpha_2}{\ddots + \cfrac{\alpha_n}{p_n}}}}$$

where $m = 1, 3, 5, \ldots$, $\alpha_i = \pm 1$, and $p_i = 2, 4, 6 \ldots$; this number will be denoted by $[m, \alpha_1 p_1, \alpha_2 p_2, \ldots, \alpha_n p_n]$, and is a rational with an *odd denominator*. The fluid state at $\nu = [m, p_1, \ldots, p_n]$ cannot occur unless its "parent" state at $\nu = [m, p_1, \ldots, p_{n-1}]$ also occurs; whether or not a given fluid state occurs will depend on the details of the interactions. The experimentally observed anomalies with $\nu < 1$ correspond to $[3,2]$, $[3]$, $[3,-2]$, $[1,2,-2]$, $[1,2]$, and $[1,4]$; they all derive from the $m = 1$ and $m = 3$ hierarchies.

The technical innovation that I make is to place a 2D electron gas of N particles on a *spherical* surface of radius R, in a radial (monopole) magnetic field $B = \hbar S/eR^2$ (>0) where $2S$, the total magnetic flux through the surface in units of the flux quantum $\Phi_0 = h/e$, is integral as required by Dirac's monopole quantization condition.[6] This device allows the construction of homogeneous states with finite N; in the limit R, N, and $S \to \infty$, the Euclidean group of the plane is recovered from the rotation group $O^+(3)$ of the sphere.

Single-particle states.—The single-particle Hamiltonian is

$$H = |\vec{\Lambda}|^2/2M = \tfrac{1}{2}\omega_c |\vec{\Lambda}|^2/\hbar S,$$

where M is the effective mass, and $\omega_c = eB/M$ is the cyclotron frequency. $\vec{\Lambda} = \vec{r} \times [-i\hbar\nabla + e\vec{A}(\vec{r})]$ is the dynamical angular momentum; $\nabla \times \vec{A} = B\hat{\Omega}$, $\hat{\Omega} = \vec{r}/R$. $\vec{\Lambda}$ has no component normal to the surface: $\vec{\Lambda} \cdot \vec{\Omega} = \vec{\Omega} \cdot \vec{\Lambda} = 0$; its commutation relations are $[\Lambda^\alpha, \Lambda^\beta] = i\hbar\epsilon^{\alpha\beta\gamma}(\Lambda^\gamma - \hbar S\Omega^\gamma)$. The generator of rotations is instead given by $\vec{L} = \vec{\Lambda} + \hbar S\hat{\Omega}$:

$[L^\alpha, X^\beta] = i\hbar\epsilon^{\alpha\beta\gamma}X^\gamma$, $\vec{X} = \vec{L}, \hat{\Omega}$, or $\vec{\Lambda}$; this *has* a normal component* $\vec{L}\cdot\hat{\Omega} = \hat{\Omega}\cdot\vec{L} = \hbar S$. This algebra implies the spectrum $|\vec{L}|^2 = \hbar^2 l(l+1)$, $l = S + n$, $n = 0,1,2,\ldots$, and that $2S$ is integral (the Dirac condition[6]); $|\vec{\Lambda}|^2 = |\vec{L}|^2 - \hbar^2S^2 = \hbar^2\{n(n+1) + (2n+1)S\}$. $\hat{\Omega}$ can be specified by spinor coordinates $u = \cos(\frac{1}{2}\theta)\exp(\frac{1}{2}i\varphi)$, $v = \sin(\frac{1}{2}\theta)\exp(-\frac{1}{2}i\varphi)$: $\hat{\Omega}(u,v) = (\sin\theta\cos\varphi, \sin\theta\sin\varphi, \cos\theta)$. To describe the wave functions, I choose the gauge $\vec{A} = (\hbar S/eR)\times\hat{\varphi}\cot\theta$; the singularities at the two poles (each admitting flux $S\Phi_0$) have no physical consequence. The Hilbert space of the lowest Landau level ($l = S$, with energy $\frac{1}{2}\hbar\omega_c$) is spanned by the coherent states $\psi_{(\alpha,\beta)}^{(S)}$ defined by $\{\hat{\Omega}(\alpha,\beta)\cdot\vec{L}\}\psi_{(\alpha,\beta)}^{(S)} = \hbar S\psi_{(\alpha,\beta)}^{(S)}$; these are polynomials in u and v of total degree $2S$:

$$\psi_{(\alpha,\beta)}^{(S)}(u,v) = (\alpha^*u + \beta^*v)^{2S}, \quad |\alpha|^2 + |\beta|^2 = 1.$$

Within this subspace, the electron may be represented as a spin S, the orientation of which indicates the point on the sphere about which the state is localized.[7] The operator $\vec{L}$ can be written as $L^+ = \hbar u\,\partial/\partial v$, $L^- = \hbar v\,\partial/\partial u$, $L^z = \frac{1}{2}\hbar(u\,\partial/\partial u - v\,\partial/\partial v)$, and $S = \frac{1}{2}(u\,\partial/\partial u + v\,\partial/\partial v)$; u and v may also be represented as independent boson creation operators, and $\partial/\partial u$ and $\partial/\partial v$ as their conjugate destruction operators.

Two-particle states.—The operator $|\vec{L}_1 + \vec{L}_2|^2$ has eigenvalues $\hbar^2 J_{12}(J_{12}+1)$, $J_{12} = 0, 1, \ldots, 2S$; the coherent states with $J_{12} = J$, $\{\hat{\Omega}(\alpha,\beta)\cdot(\vec{L}_1 + \vec{L}_2)\}\psi_{(\alpha,\beta)}^{(S,J)} = \hbar J\psi_{(\alpha,\beta)}^{(S,J)}$, have wave functions

$$\psi_{(\alpha,\beta)}^{(S,J)} = (u_1v_2 - u_2v_1)^{2S-J}\prod_{i=1,2}(\alpha^*u_i + \beta^*v_i)^J.$$

Fermi statistics requires that $2S - J_{12}$ be odd, and Bose statistics, that it be even. Note that the factor $u_1v_2 - u_2v_1$ commutes with $\vec{L}_1 + \vec{L}_2$. If Π_S is the projection operator on states of the lowest Landau level, the projection on rotationally invariant operators $V(\hat{\Omega}_1\cdot\hat{\Omega}_2)$ (such as the interparticle interaction) can be expanded as

$$\Pi_S V(\hat{\Omega}_1\cdot\hat{\Omega}_2)\Pi_S = \sum_{J=0}^{2S} V_J^{(S)} P_J(\vec{L}_1 + \vec{L}_2),$$

where $P_J(L)$ is the projection operator on states with $|\vec{L}|^2 = \hbar^2 J(J+1)$. In particular, $\Pi_S(\hat{\Omega}_1\cdot\hat{\Omega}_2)\Pi_S = \vec{L}_1\cdot\vec{L}_2/\{\hbar(S+1)\}^2$; the smaller the value of $2S - J_{12}$, the smaller the mean separation between the particles, which are precessing about their common center of mass at $\hat{\Omega}(\alpha,\beta)$.

N-particle states.—In the spirit of Laughlin,[4] I discuss the N-particle wave function,

$$\Psi_N^{(m)} = \prod_{i<j}(u_iv_j - u_jv_i)^m, \quad S = \tfrac{1}{2}m(N-1).$$

The case $m = 1$ can be alternatively expressed as the antisymmetric Slater determinant describing complete filling of the lowest Landau level, with $N = 2S + 1$. Because $\vec{L}_{\text{tot}} = \sum_i \vec{L}_i$ commutes with $u_iv_j - u_jv_i$, $\Psi_N^{(m)}$ is explicitly translationally and rotationally invariant on the surface of the sphere: $\vec{L}_{\text{tot}}\Psi_N^{(m)} = 0$. It is totally antisymmetric (Fermi statistics) for odd m, and symmetric (Bose statistics) for even m. The Laughlin droplet wave functions,[4] centered at $\hat{\Omega}(\alpha,\beta)$, can be recovered by multiplying $\Psi_N^{(m)}$ by a factor $\prod_i(\alpha^*u_i + \beta^*v_i)^n$, and taking the limit $n\to\infty$, $R\to\infty$, $R^2/2n = a_0^2$, where $a_0 = (\hbar/eB)^2$ is the Larmor radius of the lowest Landau level.

Remarks.—(1) $\Psi_{n=3}^{(m)}$ is an exact eigenstate of *any* pair interaction $\sum_{i<j}\{\Pi_S V(\hat{\Omega}_i\cdot\hat{\Omega}_j)\Pi_S\}$, because $J_{12} = J_{23} = J_{31} = S = m$; in the planar geometry, Laughlin's $N = 3$ droplet states are reportedly not exact: Overlaps with numerically calculated exact eigenstates[4] (e.g., 0.994 68 for the Coulomb interaction, $m = 5$) are close to, but *not* exactly, unity. (2) for $N \geq 4$, $m > 1$, $\Psi_N^{(m)}$ is *not* an exact eigenstate of a general interaction potential: This would require that it is an exact eigenstate with $J_{ij} = J$ of the angular momentum of any pair of particles. The spectrum of values of J_{ij} contained in $\Psi_N^{(m)}$ is easily determined by writing it as the product of three factors (i) involving coordinates i,j only, (ii) involving coordinates $k \neq i,j$ only, and (iii) the cross term $\prod_k(v_ku_i - u_kv_i)^m(v_ku_j - u_kv_j)^m$ which determines J_{ij}: $J_{ij} \leq m(N-2) = 2S - m$. The special character of the states $\Psi_N^{(m)}$ is thus that they have no components with $J_{ij} = 2S - m + 2, 2S - m + 4, \ldots \leq 2S$ that would be present in a more general wave function of the appropriate symmetry: The states of closest approach of the pair of particles are suppressed. In particular, when $S = S(N; m) \equiv \frac{1}{2}m(N-1)$, $\Psi_N^{(m)}$ may be characterized as the *exact nondegenerate ground state* of the projection-operator interaction potential

$$\Pi_S H_{m,S}^{\text{int}}\Pi_S = \sum_{i<j}\{\sum_{J>2S-m} P_J(\vec{L}_i + \vec{L}_j)\}.$$

This is essentially a kind of hard-core interaction; $\Psi_N^{(m)}$ will thus be a particularly good variational approximation for the ground state of systems with strong repulsion at close separations.

Excited states.—In this geometry, the natural excitation operators, analogous to those suggested by Laughlin,[4] are

$$A_N^\dagger(\alpha,\beta) = \prod_{i=1}^{N} (\beta u_i - \alpha v_i) \quad (\text{"holes"}),$$

$$A_N(\alpha,\beta) = \prod_{i=1}^{N} \left(\beta^* \frac{\partial}{\partial u_i} - \alpha^* \frac{\partial}{\partial v_i}\right) \quad (\text{"particles"}),$$

which, respectively, increase or decrease the flux quantum number S by $\frac{1}{2}$, and decrease or increase $\hat{\Omega}(\alpha,\beta)\cdot\vec{L}_{\text{tot}}$ by $\frac{1}{2}N\hbar$. The single-excitation states $A_N^\dagger(\alpha,\beta)\Psi_N^{(m)}$ and $A_N(\alpha,\beta)\Psi_N^{(m)}$ have $J_{\text{tot}} = \frac{1}{2}N$, and describe defects in the fluid localized around[7] $\hat{\Omega}(\alpha,\beta)$. Since

$$[A_N^\dagger(\alpha,\beta), A_N^\dagger(\alpha',\beta')] = [A_N(\alpha,\beta), A_N(\alpha',\beta')] = 0,$$

the two-hole and two-particle states are symmetric in the excitation coordinates, and the excitations thus obey Bose statistics {note, however, that $[A_N(\alpha,\beta), A_N^\dagger(\alpha',\beta')] \neq 0$}. A state with N_p^{ex} particle and N_h^{ex} hole excitations has $S = S(N;m) + \frac{1}{2}(N_h^{\text{ex}} - N_p^{\text{ex}})$; on the other hand, if the system is excited by addition or removal of an electron at fixed magnetic field, the final state has $S = S(N+1;m) \mp \frac{1}{2}m$. The comparison indicates that the hole excitations carry a fractional charge $e^* = +e/m$, and the particles a fractional charge $-e^*$, as proposed by Laughlin.[4] The degeneracy $N+1$ of the single-excitation states supports the same conclusion: In the thermodynamic limit there is one state for each unit $\Phi_m \equiv m\Phi_0 = h/e^*$ of magnetic flux through the surface.

Hierarchy of fluid states.—I will assume, following Laughlin,[4] that for some m, the ground state of the 2D electron gas with $S = S(N;m)$ is well represented by the approximate wave function $\Psi_N^{(m)}$, and that there is a gap in the excitation spectrum, the lowest-energy excitations being (bound) particle-hole pairs. Consider now a slightly different field strength so that $S = S(N;m) + \frac{1}{2}N^{\text{ex}}$; the low-energy states at this field strength can be considered as deriving from the fluid state $\Psi_N^{(m)}$ with an imbalance of particle and hole excitations, $N^{\text{ex}} = N_h^{\text{ex}} - N_p^{\text{ex}}$. Since there is, by assumption, a gap for making particle-plus-hole excitations, the lowest-energy states will belong to a manifold of purely hole states ($N^{\text{ex}} > 0$) or purely particle states ($N^{\text{ex}} < 0$), separated by a gap from higher-energy states. If the interaction energy of the *excitations* is small compared to this energy gap, the problem of constructing the collective ground state of the *excitation* fluid is precisely analogous to the original problem of constructing the ground state of the *electron* fluid, but with S replaced by $\frac{1}{2}N$, N replaced by $|N^{\text{ex}}|$, and Fermi statistics replaced by Bose statistics. A Laughlin fluid state of the excitations[8] can be constructed if

$$\tfrac{1}{2}N = S(|N^{\text{ex}}|; p),$$

where p is now *even* (Bose statistics): $p = 2, 4, 6, \ldots$. This leads to $|N^{\text{ex}}| = (N/p) + 1$; this second family of fluid states thus can occur at field strengths $S = S(N; m, \pm p) \equiv \frac{1}{2}m(N-1) \pm \frac{1}{2}[(N/p)+1]$, and requires that N be divisible by p. If this fluid state exists, with a sufficiently strong gap, the argument can be iterated by constructing a type-$[|p_1|, p_2]$ fluid state of the excitations of the primary type-$[m]$ electron fluid, and so on; the hierarchical set of equations is

$$S(N; m, p_1, \ldots, p_n) = S(N;m) + \tfrac{1}{2}|N^{\text{ex}}|\,\text{sgn}(p_1);$$

$$\tfrac{1}{2}N = S(|N^{\text{ex}}|; |p_1|, p_2, \ldots, p_n).$$

The filling factor ν is given by $N/2S$ in the thermodynamic limit; the hierarchical equations become

$$\{\nu(m, p_1, \ldots, p_n)\}^{-1} = m + \text{sgn}(p_1)\,\nu(|p_1|, p_2, \ldots p_n),$$

with the solution

$$\nu(m, p_1, \ldots, p_n) = [m, p_1, \ldots, p_n].$$

The charge of the excitations is easily found by determining how many are produced by adding an electron at fixed magnetic field: The result is that if ν is expressed as the rational P/Q, Q is odd, $e^* = e/Q$, and the Hall resistance can be written $\rho_{xy} = \Phi_0/Pe^*$, consistent with Laughlin's "gauge invariance" argument.[9]

The above analysis indicates how "incompressible fluid" states may derive from parent "incompressible fluid" states at simpler rational filling factors ν; the most stable fluid states will correspond to the simplest rationals with small values of m and p_i, where the fluid densities are highest, and hence short-range repulsion effects strongest. What is so far missing is a calculational scheme for the direct determination of *whether* a given fluid state exists for a given interaction potential, e.g., the Coulomb interaction. The new formalism based on a spherical geometry may simplify this task. From a variational viewpoint,[4] the correlation energy of $\Psi_N^{(m)}$ and its excitations $A_N^\dagger(\alpha,\beta)\Psi_N^{(m)}$ must be determined; this reduces to (i) the determination of the expansion coefficients $V_J^{(S)}$ of the interaction potential, and (ii) analysis of the wave functions to determine the relative weights of the components with a given pair angular momentum J_{ij}. Progress may be possible in this formalism. Beyond the variational approach, the problem has been reduced to a generalized version of an infinite-coordination Heisenberg problem involving N spin-S objects, and direct numerical calculation of the low-lying energy levels at an increasing sequence of values of N with $S = S(N;m,p_1,\ldots,p_n)$, coupled with a *finite-size scaling* analysis of how the gap behaves as $N\to\infty$, may prove possible at simple rationals. It may be remarked that, in this geometry, the gapless Wigner lattice would also derive from an isotropic state $L_{tot}=0$ as $N\to\infty$: The sphere cannot be tiled with a triangular lattice without introducing *disclination defects*; these will be mobile, and will restore translational and rotational invariance.

This work was brought to conclusion during a stay at the Laboratoire de Physique des Solides, Université Paris-Sud, Orsay, and the hospitality and support of R. Jullien and Professor J. Friedel is gratefully acknowledged. The assistance of M. Kolb in numerical study of the case $N=4$ was invaluable in the development of the above ideas.

[1]K. v. Klitzing, G. Dorda, and M. Pepper, Phys. Rev. Lett. 45, 494 (1980); D. C. Tsui and A. C. Gossard, Appl. Phys. Lett. 37, 550 (1981).

[2]D. C. Tsui, H. L. Stormer, and A. C. Gossard, Phys. Rev. Lett. 48, 1559 (1982); H. L. Stormer *et al.*, Phys. Rev. Lett. 50, 1953 (1983).

[3]L. Bonsall and A. A. Maradudin, Phys. Rev. B 15, 1959 (1977); D. Yoshioka and H. Fukuyama, J. Phys. Soc. Jpn. 47, 394 (1979); D. Yoshioka and P. A. Lee, Phys. Rev. B 27, 4986 (1983); K. Maki and X. Zotos, to be published.

[4]R. B. Laughlin, Phys. Rev. Lett. 50, 1395 (1983).

[5]D. Yoshioka, B. I. Halperin, and P. A. Lee, Phys. Rev. Lett. 50, 1219 (1983).

[6]P. A. M. Dirac, Proc. Roy. Soc. London, Ser. A 133, 60 (1931).

[7]The convention used here is that negative-charge coherent single-particle states (electrons, "particles") transform *contragrediently* under rotations, while positive-charge states ("holes") transform *cogrediently* ($\vec{L}$ points *away* from their position).

[8]The possibility of a hole fluid is raised, along with other suggestions such as a fluid of electron pairs, in a discussion of how Laughlin's wave functions might be generalized to other rational values of ν by B. I. Halperin, to be published.

[9]R. B. Laughlin, Phys. Rev. B 23, 5632 (1981).

Statistics of Quasiparticles and the Hierarchy of Fractional Quantized Hall States

B. I. Halperin
Physics Department, Harvard University, Cambridge, Massachusetts 02138
(Received 9 November 1983)

Quasiparticles at the fractional quantized Hall states obey quantization rules appropriate to particles of fractional statistics. Stable states at various rational filling factors may be constructed iteratively by adding quasiparticles or holes to lower-order states, and the corresponding energies have been estimated.

PACS numbers: 05.30.-d, 03.65.Ca, 71.45.Nt, 73.40.Lq

Observations of the fractional quantized Hall effect[1] show that there exist special stable states of a two-dimensional electron gas, in strong perpendicular magnetic field B, occurring at a set of rational values of ν, the filling factor of the Landau level. Laughlin[2] has constructed an explicit trial wave function (product wave function) to explain the states at $\nu = 1/m$, with m an odd integer, and has argued that the elementary excitations from the stable states are quasiparticles with fractional electric charge. Among the proposals to explain the other observed fractional Hall steps are hierarchical schemes, in which higher-order stable states ν_{s+1} are built up by adding quasiparticles to a stable state ν_s of smaller numerator and denominator.[3-5]

In the present note, we observe that the quantization rules which determine the allowed quasiparticle spacings are just those that would be expected for a set of identical charged particles that obey *fractional statistics*—i.e., such that the wave function changes by a complex phase factor when two particles are interchanged. Moreover, by assuming that the dominant interaction between quasiparticles is just the Coulomb interaction between the quasiparticle charges, we are led to a natural set of approximations for the ground-state energies and energy gaps at all levels of the hierarchy.

The appearance of fractional statistics in the present context is strongly reminiscent of the fractional statistics introduced by Wilczek to describe charged particles tied to "magnetic flux tubes" in two dimensions.[6] As in Ref. 6, the quasiparticles can *also* be described by wave functions obeying Bose or Fermi statistics, the various representations being related by a "singular gauge transformation." The boson description was, in fact, used in Refs. 3 and 4 and the fermion description in Ref. 5. However, the boson or fermion descriptions require, in effect, a long-range interaction between quasiparticles which alters the usual quantization rules. The transformation between representations is analogous to the well-known transformation between impenetrable bosons and fermions in one dimension.

As in previous discussions of the fractional quantized Hall effect, we consider a two-dimensional system of electrons in the lowest Landau level, with a uniform positive background. The filling factor ν is defined by $\nu = n/2\pi l_0^2$, where n is the density of electrons, and $l_0 = |Be/\hbar c|^{1/2}$ is the magnetic length; hence ν is the number of electrons per quantum of flux.

Let ν_s be a stable rational filling factor obtained at level s of the hierarchy. I assert that the low-lying energy states for filling factors near to ν_s can be described by the addition of a small density of quasiparticle excitations to the ground state at ν_s. The elementary quasiparticle excitations are of two types—particlelike "p excitations" and holelike "h excitations"—having charges $q_s e$ and $-q_s e$, respectively, according to a sign convention described below. For the present purposes we need only consider states with one type of excitation present. We shall describe these states by a pseudo wave function Ψ, which is a function of the coordinates $\vec{R}_k$ of the N_s quasiparticles present. I assert that the allowed pseudo wave functions can be written in the form

$$\Psi[\vec{R}_k] = P[Z_k] Q_s[Z_k] \prod_{k=1}^{N_s} \exp(-|q_s||Z_k|^2/4l_0^2), \tag{1}$$

where $Z_k = X_k \mp iY_k$ is the position in complex notation, with the sign depending on the sign of the charge of the quasiparticle, $P[Z_k]$ is a *symmetric polynomial* in the variables Z_k, and

$$Q_s = \prod_{k<l} |Z_k - Z_l|^{-\alpha/m_s}. \tag{2}$$

In Eq. (2), $\alpha = \pm 1$, according to whether we are dealing with particle- or hole-type excitations, and m_s is a rational $\geqslant 1$, to be specified by an iterative equation below. We may interpret $|\Psi[\vec{R}_k]|^2$ as the probability density for finding a quasiparticle at each of the positions $\vec{R}_1, \ldots, \vec{R}_{N_s}$, at least in the case that the $\vec{R}_k$ are not too close to each other. Since the quasiparticles have a finite size (of order l_0), however, there is no direct significance to the behavior of $|\Psi|^2$ when two positions R_k and R_l come very close together. The wave function is normalized if $\int|\Psi|^2=1$, and two wave functions Ψ and Ψ' are orthogonal if $\int\Psi^*\Psi'=0$.

The pseudo wave function (1)–(2) can be derived in different ways, starting from various microscopic descriptions that have been proposed[2–5] for the electronic state with quasiparticle or quasihole excitations. I shall give below a derivation for p excitations using the *pair model* proposed in Ref. 3.

Because there is no direct physical significance to the phase of the pseudo wave function, it is permissible to *redefine* the factor Q in Eq. (1) by *removing the absolute value sign* in Eq. (2). (This operation may be described as a singular gauge transformation.)[6] If $m_s \neq 1$, the new wave function is a multivalued function of the positions $\{\vec{R}_k\}$, and one should consider it as a function defined on the appropriate Riemann surface for $\{Z_k\}$. [Alternatively one could use a single-valued definition and specify discontinuities along cuts in the variable $(Z_k - Z_l)$.] Now if we continuously interchange the positions of two quasiparticles, the wave function will change by a complex phase factor $(-1)^{\pm 1/m_s}$, with the sign depending on the sense of rotation as the quasiparticles pass by each other. Although the extra phase factor is perhaps a complication, the pseudo wave function now has the esthetically pleasing property that it is an eigenstate of the differential operator $[\nabla_k \mp iq_s e\vec{A}(\vec{R}_k)/\hbar c]^2$ with special boundary conditions at the points $Z_k=Z_l$, where $\vec{A}$ is the vector potential in the symmetric gauge. Then Eq. (1) may be described as a general wave function appropriate to a collection of particles of charge $\pm q_s e$ obeying fractional statistics, all in the lowest Landau level. Of course, in the special case $m_s=1$, the quasiparticles are ordinary fermions.

In order to find the ground-state configuration for a given density n_s of quasiparticles, we must find the symmetric polynomial $P[Z]$ which leads to the minimum expectation value of the repulsive interaction between the quasiparticles. Using the same reasoning as Laughlin in Ref. 2, we expect that certain choices of P can lead to specially low energies, namely,

$$P[Z_k]=\prod_{k<l}(Z_k-Z_l)^{2p_{s+1}}, \qquad (3)$$

where p_{s+1} is a positive integer. The probability distribution $|\Psi|^2$ is then that of a classical one-component plasma[2] with dimensionless inverse temperature $\Gamma=2m_{2+1}$, where

$$m_{s+1}=2p_{s+1}-\alpha_{s+1}/m_s, \qquad (4)$$

and $\alpha_{s+1}=1$ or -1 as particlelike or holelike quasiparticles are involved. The density of the plasma is fixed by a charge neutrality condition,[2] so that the number of quasiparticles in an area $2\pi l_0^2$ is just $n_s=|q_s|/m_{s+1}$. Since each quasiparticle has charge $\alpha_{s+1}q_s$, we may readily calculate the electron density in the new stable state, and we find the filling factor

$$\nu_{s+1}=\nu_s+\alpha_{s+1}q_s|q_s|/m_{s+1}. \qquad (5)$$

If we multiply the pseudo wave function described above by the factor $\prod_k Z_k$, for $k=1,\ldots,N_s$, we find a deficiency near the origin of $1/m_{s+1}$ quasiparticles of level s. We identify this state as a hole excitation at level $s+1$. Similarly, we may construct a p excitation having an *excess* of $1/m_{s+1}$ quasiparticles at the origin. The iterative equation for q_s is thus

$$q_{s+1}=\alpha_{s+1}q_s/m_{s+1}. \qquad (6)$$

Together with the starting conditions $\nu_0=0$, $q_0=m_0=\alpha_1=1$, the iterative equations (4)–(6) give a sequence of rational filling factors ν_s for any choice of the sequence $\{\alpha_s,p_s\}$. At the level $s=1$, we recover Laughlin's states with $\nu_1=1/m_1=1,\frac{1}{3},\frac{1}{5},\ldots$ for various choices of p_1. If we add holes to the state $\nu_1=1$, we find at level $s=2$, the complements to the Laughlin states, $\nu_2=\frac{2}{3},\frac{4}{5},\frac{6}{7},\ldots$ (In order to stay in the lowest Landau level, we impose the restriction $\alpha_2=-1$, if $\nu_1=1$.) From the state $\nu_1=\frac{1}{3}$, we achieve such states as $\nu_2=\frac{2}{5}$ or $\frac{4}{11}$, with p excitations, and $\nu_2=\frac{2}{7}$ or $\frac{4}{13}$, with h excitations.

It can be shown, after some algebra, that the allowed values of ν_s may be expressed as continued fractions in terms of the finite sequences $\{\alpha_s,p_s\}$ and that they are identical to those of Haldane.[4] (I have used the opposite sign for α, however, and here p is one-half of Haldane's.) As noted by Haldane, every rational value of ν with odd denominator, with $0<\nu\leq 1$, is obtained once in this way. There will *not* be a quantized Hall step at *every* such rational ν, however. We know that there exists a maximum allowed value m_c for the parameter m_s, such that if at any stage of the hierarchy the calculated m_s is greater than m_c, then the quasiparticles at the density n_s will form a Wigner crystal rather than a quantum-liquid state.[2] There is then no stabilization of the electron density at the corresponding ν_s, and there will be no meaning to any further states in the hierarchy constructed from this ν_s.

The pseudo wave function (1)–(3) leads to a natural estimate of the potential energy of the system, if we assume that the dependence on the positions of the quasiparticles can be approximated by the pairwise Coulomb interaction between point particles of charge $q_s e$, in the background dielectric constant ϵ. If $E(\nu)$ is the energy per quantum of magnetic flux, we have

$$E(\nu_{s+1}) \cong E(\nu_s) + n_s \epsilon_s^{\pm} + n_s |q_s|^{5/2} u_{\rm pl}(m_{s+1}) \tag{7}$$

where $\epsilon_s^{\pm}$ is the energy to add one particlelike excitation or one holelike excitation, together with neutralizing uniform background, to the state ν_s, and $u_{\rm pl}$ is a smooth function of m_s, given (approximately) by Laughlin's interpolation formula[5]

$$u_{\rm pl}(m) = \frac{-0.814}{m^{1/2}}\left(1 - \frac{0.230}{m^{0.64}}\right)\left(\frac{e^2}{\epsilon l_0}\right). \tag{8}$$

We recall that $u_{\rm pl}(m)$ is the potential energy per particle that one would find for a system of electrons at filling factor $\nu = 1/m$ if one approximates the pair correlation function $g(r)$ for the electrons by the pair correlation function $g_{\rm pl}(r)$ for a one-component plasma at inverse temperature $\Gamma = 2m$; the factor $|q_s|^{5/2}$ in the last term of (7) reflects the smaller charge and larger magnetic length for our quasiparticles.

In order to use Eq. (7), we need an iterative formula for the quasiparticle energies $\epsilon_s^{\pm}$. It is convenient to write

$$\epsilon_s^{\pm} = \tilde{\epsilon}_s^{\pm} \pm m_s^{-1}[\epsilon_{s-1} + \tfrac{3}{2}|q_{s-1}|^{5/2} u_{\rm pl}(m_s)]. \tag{9}$$

The quantity in square brackets is the energy it would take to add one quasiparticle or quasihole of level $s-1$, if one could keep the Laughlin product form (3) for the polynomial P, and simply increase the density n_{s-1} by means of a reduction, of order $1/N$, in the magnetic length l_0 which controls the distance scale in Eq. (1).[7] The term $\tilde{\epsilon}_s^{\pm}$ in (9) may be called the *proper* excitation energy; it is relatively small, but is presumably positive for both quasiparticles and holes. For the proper hole energy, we use the approximate formula

$$\tilde{\epsilon}_s^{-} = 0.313|q_{s-1}|^{5/2} m_s^{-9/4}(e^2/\epsilon l_0). \tag{10}$$

This form has the correct dependence on the charge q_{s-1}; it passes through the exact value $0.313(e^2/\epsilon l_0)$, for $q_0 = 1$, $m_1 = 1$, and it yields $\tilde{\epsilon}_1^- = 0.264$, $\tilde{\epsilon}_1^- = 0.0837$, for $m_1 = 3$ and $m_1 = 5$, in close agreement with the values obtained by Laughlin.[5,7]

Unfortunately, there does not exist at the present time any reliable calculation of the quasiparticle excitation energy. Therefore, *for purposes of illustration,* I have made the arbitrary approximation $\tilde{\epsilon}_s^+ = \lambda \tilde{\epsilon}_s^-$, where λ is a constant independent of m_s. The resulting curve for $E(\nu)$ is plotted in Fig. 1, for the choice $\lambda = 3$, after subtraction of the "plasma approximation" $E_{\rm pl} = \nu u_{\rm pl}(\nu^{-1})$, which is a smooth function of ν. We can see that there are downward pointing cusps in the energy visible at the low-order rational ν with odd denominators. The approximation also gives *upward*-pointing cusps at all rational ν with even denominators; in fact, I find small discontinuities in E, not visible on the scale of the figure, at all these even points except for $\nu = \frac{1}{2}$, where continuity is guaranteed by the particle-hole symmetry of the cohesive energy, which is respected exactly by the present approximation.[7] Clearly the upward-pointing cusps are unphysical; the system could always lower its energy by breaking up into small regions of larger and smaller density; alternatively there may be a different type of ground state with still lower energy at these values of ν. The behavior of the approximate energy curve near the low-order rationals of odd denominator should be qualitatively and semiquan-

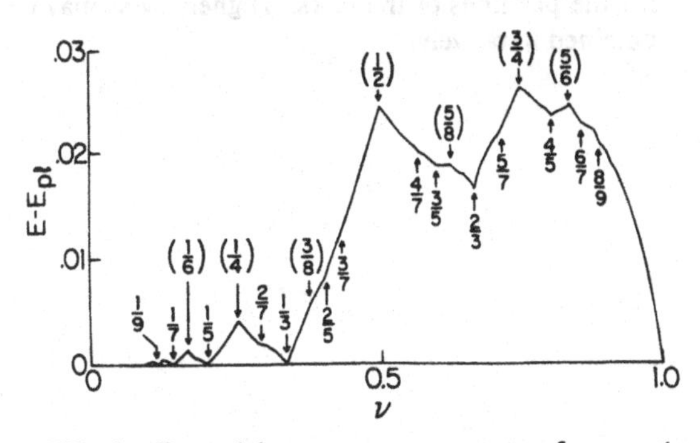

FIG. 1. Potential energy per quantum of magnetic flux, in units of $e^2/\epsilon l_0$, as a function of filling factor ν of the first Landau level, from approximate formulas (7)–(10). Smooth function $E_{\rm pl}(\nu) = \nu u_{\rm pl}(\nu^{-1})$ has been subtracted off.

titatively correct, however. More reliable estimates will be possible when p-excitation energies have been properly calculated, and when corrections are included such as the finite quasiparticle size and effects of virtual excitations of particle-hole pairs.

With the approximation described above, the energy gap $\tilde{\epsilon}_s^+ + \tilde{\epsilon}_s^-$ is equal to

$$0.313(1+\lambda)|q_s|^{5/2}m_s^{1/4}(e^2/\epsilon l_0)$$

[cf. (6) and (10)]. Except for the rather weak factor $m_s^{1/4}$, the gap is determined by the value of $|q_s|^{-1}$, which is the *denominator* of the fraction ν_s. This is in qualitative agreement with reported experimental observations on GaAs samples.[1]

Finally, we derive by induction the starting equation (1). For $s=0$, the Z_k are positions of bare electrons, and Eqs. (1) and (2) are correct, with $q_0=m_0=\alpha_1=1$. We assume that the p excitations of level $s=1$ can be formed out of *pairs* of electrons, by a generalization of Eq. (23) of Ref. 3. A system containing N_1 pairs of electrons, together with N_0-2N_1 unpaired electrons, is then described by choosing the polynomial in (1) to have the (schematic) form

$$P[Z_k]=\mathscr{S}\bar{P}[z_i]\prod_{i<j}(z_i-z_j)^{8p_1-4}\prod_{i,\gamma}(z_i-\tilde{Z}_\gamma)^{4p_1-1}\prod_{\gamma<\delta}(\tilde{Z}_\gamma-\tilde{Z}_\delta)^{2p_1}, \tag{11}$$

where z_i are the positions of the centers of gravity of the bound pairs, $\tilde{Z}_\gamma$ are the positions of the *unpaired* electrons, $\bar{P}$ is a symmetric polynomial, and $\mathscr{S}$ is an operator which symmetrizes with respect to the positions of all N_0 electrons. I have assumed that the separation between two members of a pair is small, and have dropped the variables describing these separations. To calculate the probability distribution of the pairs, we ignore the symmetrizer $\mathscr{S}$, and take the trace of $|\Psi[Z_k]|^2$ over the unpaired electron positions $\tilde{Z}_\gamma$. The result can be expressed in the form $|\bar{\Psi}[z_i]|^2\Phi[z_i]$, where $\bar{\Psi}$ has again the form of (1) and (2), with P replaced by $\bar{P}$, and with $m_1=2p_1-1/m_0$, $\alpha=1$, and $q_1=q_0/m_1$, while the remaining factor Φ is the partition function of a classical one-component plasma with sources of strength $2-m_1^{-1}$, located at the positions z_i. Now Φ will be independent of the positions z_i, provided that the sources are sufficiently separated so that their screening clouds do not overlap. Thus it is consistent to interpret $\bar{\Psi}$ as a pseudo wave function for the positions of the pairs. Higher levels may be obtained iteratively.

Derivation of the pseudo wave function for hole excitations is more complicated because of the necessity to use an integral representation, such as Eq. (25) of Ref. 3.[7]

The author is grateful for helpful discussions with R. B. Laughlin, R. Morf, H. Stormer, P. A. Lee, D. Yoshioka, S. Girvin, and P. Ginsparg. This work was supported in part by the National Science Foundation under Grant No. DMR-82-07431.

[1]See H. L. Stormer *et al.*, Phys. Rev. Lett. 50, 1953 (1983).

[2]R. B. Laughlin, Phys. Rev. Lett. 50, 1395 (1983).

[3]B. I. Halperin, Helv. Phys. Acta 56, 75 (1983).

[4]F. D. M. Haldane, Phys. Rev. Lett. 51, 605 (1983).

[5]R. B. Laughlin, in Proceedings of the Conference on Electronic Properties of Two-Dimensional Systems, Oxford, 1983 (to be published).

[6]F. Wilczek, Phys. Rev. Lett. 49, 957 (1982).

[7]Details will be given elsewhere.

PHYSICAL REVIEW B VOLUME 29, NUMBER 10 15 MAY 1984

Particle-hole symmetry in the anomalous quantum Hall effect

S. M. Girvin
Surface Science Division, National Bureau of Standards, Washington, D.C. 20234
(Received 27 February 1984)

This paper explores the uses of particle-hole symmetry in the study of the anomalous quantum Hall effect. A rigorous algorithm is presented for generating the particle-hole dual of any state. This is used to derive Laughlin's quasihole state from first principles and to show that this state is exact in the limit $\nu \to 1$, where ν is the Landau-level filling factor. It is also rigorously demonstrated that the creation of m quasiholes in Laughlin's state with $\nu = 1/m$ is precisely equivalent to creation of one true hole. The charge-conjugation procedure is also generalized to obtain an algorithm for the generation of a hierarchy of states of arbitrary rational filling factors.

I. INTRODUCTION

The anomalous quantum Hall effect[1,2] is one of the most striking many-body phenomena discovered in recent years. The Hall resistivity of a two-dimensional electron gas (inversion layer) in a high magnetic field at low temperatures exhibits quantized plateau values of the form $\rho_{xy} = h/e^2 i$, where i is a rational number $i = p/q$ with q odd. Associated with this quantization of the Hall resistivity is a marked decrease in the dissipation ($\rho_{xx} \to 0$). The latter suggests the existence of an excitation gap in the system, presumably due to many-body correlations arising from the Coulomb interaction. Laughlin[3] has proposed a set of variational wave functions which describe such collective states within the lowest Landau level with filling factor $\nu = i = 1/m$, where m is an odd integer. Several authors[4-8] have recently considered extensions which allow one to describe states with other rational fillings. The purpose of the present work is to discuss these ideas with an emphasis on the role of particle-hole symmetry and on the nature of the quasiparticle excitations of the system.

II. WAVE FUNCTIONS FOR PARTICLES AND HOLES

It is convenient to focus attention solely on the lowest Landau level for which the wave functions have a particularly simple analytic form.[9-11] A further advantage of neglecting Landau level mixing is that there then exists particle-hole symmetry between the states with filling factor ν and $1-\nu$.

A useful set of single-particle basis states is[9-11]

$$\chi_m(z) = \frac{1}{\sqrt{2\pi}} \phi_m(z) e^{-|z|^2/4} , \tag{1}$$

with

$$\phi_m(z) = (2^m m!)^{-1/2} z^m , \tag{2}$$

where z is a complex coordinate related to the position vector (x,y) via $z = (x - iy)/l$, l is the magnetic length, m ($m \geq 0$) is the angular momentum quantum number, and the symmetric gauge has been assumed. One can eliminate the exponential factor common to all wave functions and focus on the polynomial part by defining[11] a Hilbert space of functions analytic in z with inner product

$$(\psi,\phi) = \int d\mu(z) \psi^*(z) \phi(z) , \tag{3}$$

where the exponential factors have been lumped into the measure

$$d\mu(z) = \frac{dx\, dy}{2\pi l^2} e^{-|z|^2/2} . \tag{4}$$

Within (the N-particle version of) this space the variational wave functions proposed by Laughlin[3] may be written

$$\psi_m(z_1, \ldots, z_N) = \prod_{i<j} (z_i - z_j)^m , \tag{5}$$

where m is an odd integer. Laughlin has shown that ψ_m describes a state of uniform density with filling factor $\nu = 1/m$. Assuming ψ_m is a good approximation to the ground state and invoking particle-hole symmetry, the state with $\nu = 1 - 1/m$ must have *holes* described by ψ_m. It is very instructive to write down the wave function for the *electrons* in such a state. This task requires several steps which are carried out below.

The vacuum state for holes is the filled Landau level. Laughlin's ψ_1 is the exact wave function for the electrons in this state. This can be shown by noting that ψ_1 is a Vandermonde determinant[12] which has (in the limit $N \to \infty$) all single-particle states occupied. The properly normalized $(N+1)$-particle state for a filled Landau level is

$$\Phi_{N+1}(z_1, \ldots, z_{N+1}) = Q_{N+1}^{-1} \prod_{i<j}^{N+1} (z_i - z_j) , \tag{6}$$

where

$$Q_{N+1}^2 = (N+1)! \prod_{j=0}^{N} (2^j j!) .$$

In order to put a hole in the single-particle state ϕ_M one writes

$$\theta_M(z_1, \ldots, z_N) = \sqrt{N+1} \int d\mu(z_{N+1}) \phi_M^*(z_{N+1}) \times \Phi_{N+1}(z_1, \ldots, z_{N+1}) . \tag{7}$$

To prove that this is a correct description of the hole state, note that Φ_{N+1} is a single Slater determinant

$$\Phi_{N+1} = \frac{1}{Q_{N+1}} \sum_{\{P\}} (-1)^P \prod_{i=1}^{N+1} z_i^{P_i} , \tag{8}$$

where P_i is the image of i under a permutation P of $[0,1,2,\ldots,i,\ldots,N]$. Substitution of (8) in (7) yields after integration

$$\theta_M(z_1,\ldots,z_N)=\sqrt{N+1}\frac{(2^M M!)^{1/2}}{Q_{N+1}}\sum_{\{P\}}(-1)^P\prod_{l=1}^{N} z_l^{P_l}\delta_{M,P_{N+1}} . \qquad (9)$$

One sees that θ_M describes an N-particle Slater determinant with every state ϕ_j occupied for $0\leq j\leq N$ except state ϕ_M which is empty. It is essential to prove that the weighting factors are correct, that is, that the norm of θ_M is independent of M, the state of the hole. Using (9) it is straightforward to show that $(\theta_M,\theta_M)=1$ as required. Hence Eq. (7) gives an exact procedure for injecting a hole into any particular state. Furthermore, it is straightforward to show that the algorithm is still valid (ignoring normalization) when the vacuum state Φ_{N+1} is replaced by any other state. We now turn to applications of these ideas.

III. QUASIHOLE STATES

In addition to proposing the variational ground-state wave function defined in Eq. (5), Laughlin[3] has postulated that the low-lying excited states are formed by creation of quasiparticles in the system. The state having a quasihole at the point ξ is postulated to be[3]

$$\psi_m'=\prod_{j=1}^{N}(z_j-\xi)\psi_m(z_1,\ldots,z_N) . \qquad (10)$$

This form was arrived at by means of an argument involving the effect of adiabatically piercing the system at the point ξ with an extra quantum of magnetic flux. While this argument may have a certain intuitive appeal, a rigorous justification for the form of the wave function given in (10) has not been presented. Using the technique developed in Sec. II this deficiency can now be remedied.

Consider the coherent state defined by[11]

$$\phi_\xi(z)=e^{\xi^* z/2} , \qquad (11)$$

where ξ is a complex parameter. Evaluation of the probability density

$$|\phi_\xi(z)|^2 d\mu(z)=e^{-|z-\xi|^2/2}\frac{dxdy}{2\pi l^2} \qquad (12)$$

shows that (11) represents a Gaussian wave packet centered at the point ξ. This is the most localized wave packet that can be constructed within the Hilbert space and, in fact, represents the projection of a Dirac δ function onto the lowest Landau level.[11] Using this fact and particle-hole symmetry, the form of the electron wave function in the presence of a single localized hole can be readily deduced. One begins with the filled Landau level (6) and, in analogy with (7), injects a localized hole at the point ξ by writing

$$\theta_\xi(z_1,\ldots,z_N)=\sqrt{N+1}\int d\mu(z_{N+1})\phi_\xi^*(z_{N+1}) \times\Phi_{N+1}(z_1,\ldots,z_{N+1}) . \qquad (13)$$

Making use of the Bargmann identity[11,13]

$$\psi(\xi)=\int d\mu(z)e^{\xi z^*/2}\psi(z) \qquad (14)$$

allows exact evaluation of (13):

$$\theta_\xi(z_1,\ldots,z_N)=\sqrt{N+1}\Phi_{N+1}(z_1,\ldots,z_N,\xi) . \qquad (15)$$

Using (6) and ignoring the normalization constants gives

$$\theta_\xi(z_1,\ldots,z_N)=\prod_{j=1}^{N}(z_j-\xi)\Phi_N(z_1,\ldots,z_N) , \qquad (16)$$

which agrees exactly with (10) for the case $m=1$.

This result provides a rigorous basis for Laughlin's ansatz for the quasihole state and shows that it is exact in the limit $\nu\to 1$ (in disagreement with a recent suggestion to the contrary[14]). The ground state of the system in the opposite limit $\nu\to 0$ is a Wigner crystal describable by a single Slater determinant of coherent states[15] lying on a triangular lattice. Hence by particle-hole symmetry the *exact* ground state in the limit $\nu\to 1$ can be shown to be

$$\chi(z_1,\ldots,z_N)=\prod_{j=1}^{N}\prod_{K=1}^{N(1-\nu)}(z_j-\xi_K)\Phi_N(z_1,\ldots,z_N) , \qquad (17)$$

where the ξ_K lie on a triangular lattice with a unit cell of area $2\pi l^2/(1-\nu)$.

Let us now consider quasihole excitations in the state ψ_m of Eq. (10) for $m>1$. Laughlin has argued that these excitations carry fractional charge $q^*=1/m$. Anderson has stated[7] that creation of m quasiholes at a single point gives a wave function which is *locally* close to the state with an electron missing from that point, i.e., m quasiholes are locally equivalent to one true hole. We can inject one true hole at the point ξ with the analog of (13):

$$\theta_{\xi m}(z_1,\ldots,z_N)=\sqrt{N+1}\int d\mu(z_{N+1})\phi_\xi^*(z_{N+1}) \times\psi_m(z_1,\ldots,z_{N+1}) . \qquad (18)$$

Again, using the Bargmann identity yields (ignoring normalization constants)

$$\theta_{\xi m}(z_1,\ldots,z_N)=\prod_{j=1}^{N}(z_j-\xi)^m\psi_m(z_1,\ldots,z_N) . \qquad (19)$$

Comparison of (19) with (10) shows rigorously that injection of m quasiholes is *exactly* equivalent (locally and globally) to injection of one true hole.

In summary, Laughlin's proposed quasihole state given by (10) has been derived from first principles and shown to be exact for the case $m=1$. In general, injection of m quasiholes in the state ψ_m has been shown to be precisely equivalent to injection of one true hole. This does not prove that Laughlin's representation of a single quasiparticle for $m>1$ is correct but it does lend support to that conjecture.

IV. ARBITRARY RATIONAL FILLING FACTORS

There is considerable current interest[4-8] in generalizing Laughlin's wave functions to describe states of arbitrary rational filling factor ($\nu=p/q$, q odd). A natural place to begin is with states of filling factor $\nu=1-1/m$ since these are particle-hole equivalent to the known states with $\nu=1/m$. Laughlin has recently considered this approach[6] and obtained results similar to some of those described below. I have demonstrated above that the procedure in Eq. (7)

gives the correct weighting factor for injecting a hole into any single-particle state. It is therefore straightforward to extend (7) to the case of an arbitrary number of holes. For instance, the electron wave function

$$\phi(z_1,\ldots,z_N)=\prod_{k=1}^{M}\int d\mu(z_{N+k})\prod_{i<j}^{M}(z_{N+i}^*-z_{N+j}^*)^m\Phi_{N+M}(z_1,\ldots,z_{N+M}) \tag{20}$$

corresponds to the *hole* wave function being the correlated state ψ_m defined in (5). Hence this state has the filling factor $\nu=1-1/m$ and (for $N\to\infty$) is the *exact* particle-hole dual of the state with $\nu=1/m$. For large but finite M it is appropriate to take $N+M=mM$ so that the holes cover the same area as the electrons. Halperin[4] has proposed a wave function similar to that in (20) for the $\nu=\frac{2}{3}$ state but it does not obey particle-hole symmetry, and as a result seems likely to have a less favorable energy.

The question now arises as how to obtain filling factors not reachable by particle-hole symmetry arguments. One approach is to multiply the result obtained in (20) by the polynomial

$$W=\prod_{i<j}^{N}(z_i-z_j)^p\ , \tag{21}$$

where p is even to preserve the exchange symmetry. The density for this state can be obtained from the relation

$$\nu=\frac{N(N-1)}{2L}\ , \tag{22}$$

where L is the total angular momentum. Using (20) and (21) we have

$$L=\frac{(N+M)(N+M-1)}{2}+\frac{N(N-1)p}{2}-\frac{M(M-1)}{2}m \tag{23}$$

Noting $M=N/(m-1)$ as discussed above and taking $N\to\infty$ yields

$$\nu=\frac{1}{m/(m-1)+p}\ . \tag{24}$$

Thus, for instance, $m=3$, $p=2$ gives $\nu=\frac{2}{7}$. Laughlin[6] obtains the same $\frac{2}{7}$ state by a slightly different but related argument.

Haldane[5] has recently proposed a hierarchical scheme for the generation of states with arbitrary rational filling. An explicit realization of this scheme can be obtained using the procedure outlined above. Let ψ_a and ψ_b represent states with filling factors ν_a and ν_b, respectively. If these are previous members of the hierarchy then a new member ψ_c can be generated by

$$\psi_c(z_1,\ldots,z_N)=\prod_{i<j}^{N}(z_i-z_j)^p\int d\mu(z_{N+1})\cdots\int d\mu(z_{N+M})\,\psi_a^*(z_{N+1},\ldots,z_{N+M})\psi_b(z_1,\ldots,z_{N+M})\ . \tag{25}$$

Again, using (22) gives

$$\nu_c=\left[\frac{1}{\nu_b-\nu_a}+p\right]^{-1}\ . \tag{26}$$

This hierarchy can be used to generate states of any rational filling.[5] One expects states generated from Laughlin's original state to be good approximations for short-range interactions.[5,14] As Haldane has pointed out[5] the number of different hierarchical states that are actually observed experimentally will depend on the details of the true long-range interaction potential.

V. SUMMARY

This paper has explored the uses of particle-hole symmetry in the study of the anomalous quantum Hall effect. It was rigorously shown how to generate the particle-hole dual of any given state. This algorithm was then used to derive Laughlin's quasihole state from first principles and to show that this state is exact in the limit $\nu\to 1$. It was further proven that creation of m quasiholes at one location in Laughlin's state ψ_m is exactly equivalent to the creation of one true hole.

A generalization of the charge-conjugation procedure was used to produce states of arbitrary rational filling, through an explicit realization of Haldane's[5] hierarchical scheme. It would be interesting to develop numerical techniques for the evaluation of the energy for this class of wave functions.

[1] D. C. Tsui, H. L. Stormer, and A. C. Gossard, Phys. Rev. Lett. 48, 1559 (1982).
[2] H. L. Stormer, A. Chang, D. C. Tsui, J. C. M. Hwang, A. C. Gossard, and W. Wiegmann, Phys. Rev. Lett. 50, 1953 (1983).
[3] R. B. Laughlin, Phys. Rev. Lett. 50, 1395 (1983).
[4] B. I. Halperin, Helv. Phys. Acta 56, 75 (1983).
[5] F. D. M. Haldane, Phys. Rev. Lett. 51, 605 (1983).
[6] R. B. Laughlin, in *Proceedings of the Fifth International Conference on Electronic Properties of Two-Dimensional Systems, Oxford, England, 1983* [Surf. Sci. 141, 11 (1984)].
[7] P. W. Anderson, Phys. Rev. B 28, 2264 (1983).
[8] C. R. Hu (unpublished).
[9] Yu. A. Bychkov, S. I. Iordanskii, and G. M. Eliashberg, Pis'ma Zh. Eksp. Teor. Fiz. 33, 152 (1981) [JETP Lett. 33, 143 (1981)].
[10] R. B. Laughlin, Phys. Rev. B 27, 3383 (1983).
[11] S. M. Girvin and Terrence Jach, Phys. Rev. B 29, 5617 (1984) (this issue).
[12] B. Jancovici, Phys. Rev. Lett. 46, 386 (1981).
[13] V. Bargmann, Rev. Mod. Phys. 34, 829 (1962).
[14] S. A. Trugman and S. Kivelson (unpublished).
[15] Kazumi Maki and Xenophon Zotos, Phys. Rev. B 28, 4349 (1983).

PHYSICAL REVIEW B VOLUME 40, NUMBER 9 15 SEPTEMBER 1989-II

Hierarchical classification of fractional quantum Hall states

N. d'Ambrumenil
Department of Physics, University of Warwick, Coventry CV4 7AL, United Kingdom

R. Morf
Paul Scherrer Institut, c/o Laboratories RCA, Ltd., Badenerstrasse 569, CH-8048 Zürich, Switzerland
(Received 18 July 1988; revised manuscript received 2 June 1989)

We study the hierarchy of states proposed as a generalization of Laughlin's theory of the fractional quantum Hall effect at filling fractions of the lowest Landau level, $\nu=\frac{1}{3},\frac{2}{7},\frac{2}{5},\frac{3}{7}$. Our studies confirm the general validity of the hierarchy. We estimate the ground- and excited-state energies at the various filling fractions, and the "size" and "shape" of the fractionally charged excitations. The gap energies are substantially larger than would be expected if one considered the fractionally charged excitations to be point charges.

I. INTRODUCTION

Plateaus have been observed in the Hall conductivity, σ_{xy}, of some GaAs-GaAl$_x$As$_{1-x}$ heterostructures.[1,2] This phenomenon and related effects are known as the fractional quantum Hall effect.

Laughlin has given a microscopic theory of the fractional quantum Hall effect[3] for filling fractions of the lowest Landau level, ν, with $\nu=1/m$. Here, m is an odd integer. His theory has found almost universal acceptance and has already been extensively reviewed elsewhere.[4] Laughlin proposed trial ground-state and excited-state wave functions for these filling fractions. These have been studied numerically both using Monte Carlo techniques[5-8] and by direct diagonalization of the Hamiltonian for small systems.[9-12]

Laughlin's ideas were generalized by others[13,14] to account for the observation of quantized Hall states at other filling fractions: $\nu=\frac{2}{5},\frac{2}{7},\frac{3}{7}$. Haldane[13] and Halperin[14] both suggested a hierarchy of Laughlin-like ground states, with the particles at each level of the hierarchy being the (fractionally charged) excitations of the state at the level below.

However, these generalizations were perhaps more in the nature of a possible classification of the fractional quantized Hall states than full microscopic theories. Morf *et al.*[11] suggested microscopic analogs of Laughlin's trial wave functions for some of these other filling fractions. These seemed to capture the nature of the ground state at filling fractions $\nu=\frac{2}{5}$ and $\frac{2}{7}$. The wave function describing a system at $\nu=\frac{2}{5}$ could be interpreted in terms of the hierarchy proposed by Haldane[13] and Halperin.[14]

It has also been found[15] that the picture of a hierarchy might need adapting in higher Landau levels. Studies have shown that fractional quantized Hall states could exist in the $n=1$ Landau level, which, according to the hierarchy in its simplest form, should not exist. In particular, it was found that a state at $\nu_1=\frac{2}{7}$ existed even though its parent state in the hierarchy, $\nu_1=\frac{1}{3}$, was not (or only marginally) stable.

In this paper we try to quantify some of the "intuitive" arguments that have grown up around the hierarchical picture and to establish the validity of the hierarchical model in the lowest Landau level. We study both ground states and excited states of systems with up to 13 electrons confined to the surface of a sphere at filling fractions $\nu=\frac{1}{3}$, $\frac{2}{5}$, $\frac{2}{7}$, and $\frac{3}{7}$. We would like to see our study as a theoretical or computational analog of the extensive experimental study recently reported by Clark *et al.*[16] They were able to interpret their results in terms of the hierarchical classification presented by Haldane[13] and the scaling theory of Laughlin *et al.*[17]

The hierarchical classification predicts a Laughlin-like homogeneous ground state for a system of electrons in a magnetic field of strength (in units of h/e)

$$2S(N_e,\nu)=\nu^{-1}N_e+X(\nu)\ . \tag{1}$$

Here, ν is the filling fraction and N_e is the number of electrons. $X(\nu)$ is a quantity which depends on ν, and on the geometry of the system. For the systems of electrons on the surface of a sphere that we study, it is given by Eqs. (A14) and (9). Equation (1) defines what we call a natural family of states at each filling fraction ν. For these natural families we find smooth behavior of the ground-state energy per particle as a function of the number of particles, which allows reliable extrapolation to the thermodynamic limit.

According to the hierarchy, systems with magnetic field strengths different from those given by Eq. (1) should be thought of as containing defects (quasiholes and quasiparticles). We find that families of states with values for the constant $X(\nu)$ different from those predicted for the hierarchy show nontrivial finite-size corrections, and can have ground states in which the charge distribution is not homogeneous. We interpret this and the smooth behavior seen for systems with the field strengths given by Eq. (1) as strong evidence supporting the hierarchical picture.

We also study the suggested correspondence between ground states at filling fractions ν with those at ν'

$[=1/(\nu^{-1}+2)]$, which is exact for the Laughlin states and which is preserved within the hierarchical scheme for general ν. By considering a generalized hard-core limit, we find that there is only an approximate mapping between the ground state at $\nu=\frac{2}{5}$ and that at $\nu=\frac{2}{7}$.

Our studies of excitations are predictably less conclusive, showing quite sizable finite-size dependencies. However, the results strongly support the hierarchical classification of the excitations. In particular, we find that the quantum numbers describing the lowest-lying single-particle excitations are *always* correctly predicted. Treating the excitations as point particles suggests that the energy gaps should scale with the charge of the excitation.[14,18] Instead, we find that our best estimates of the gaps at $\nu=\frac{2}{5}$ and $\frac{3}{7}$ are approximately equal and around half that at $\nu=\frac{1}{3}$ (see Table VIII).

We find that the discrepancy between experimental estimates and our estimates of the energy gaps based on our calculations is larger the higher the level in the hierarchy of the system being excited. This is consistent with a picture in which the discrepancies are attributed in large part to the effects of disorder. One would expect that in higher-level fluids the excitations would be more sensitive to the presence of impurities.

We also discuss the procedures for extrapolating to the thermodynamic limit estimates for the excitation energies calculated for small systems. We believe we can clarify some of the apparent ambiguities in the literature.

II. MAPPINGS BETWEEN GROUND STATES AT DIFFERENT FILLING FRACTIONS

The hierarchy incorporates the idea of "mappings" between the trial ground and lowest-lying excited states at related filling fractions. A state describing a system at some filling fraction ν_m, with $1/(m+2)<\nu_m<1/m$, with m odd, should be mapped into a state at filling fraction ν_n with $\nu_n^{-1}=\nu_m^{-1}+n-m$ and n odd. The mapping for the case of the Laughlin states is straightforward:

$$\psi_m \rightarrow \psi_n = T_{n-m}\psi_m \tag{2}$$

with, in the symmetric gauge,

$$T_{n-m}=\prod_{1\le j<k\le N}(z_j-z_k)^{n-m} . \tag{3}$$

The quantities z_j are related to the x and y coordinates of the jth particle via $z_j=x_j-iy_j$. The obvious generalization to filling fractions ν, with $1/(m+2)<\nu<1/m$, assumes that the same operators, T, can be used. So, for example, the ground-state wave function at $\nu=\frac{2}{5}$ should be mapped into the ground-state wave function at $\nu=\frac{2}{7}$ by the operator T_2.

To investigate this mapping numerically, we need to define the Laughlin-like ground state by the relevant intermediate filling fractions. The Laughlin ground states at $\nu=1/m$ have been shown to be the exact nondegenerate ground state of a special kind of "hard-core" interaction.[13] This is easiest to show for rotationally invariant systems for which the interparticle interaction can be written

$$v(i,j)=\sum_{m'} V_{m'}P_{m'}(i,j) . \tag{4}$$

$P_{m'}(i,j)$ projects out of the many-body wave function of a system of N particles, $\psi(\mathbf{r}_1,\ldots,\mathbf{r}_N)$, those components with relative angular momentum $L_z=m'$ for particles i and j.

An important feature of Laughlin's wave function[19] ψ_m is that it contains no components with relative angular momentum less than m. At $\nu=1/m$ it is the only wave function with this property and, as such, it is the exact nondegenerate ground state for an interaction,

$$V_{m'}=\begin{cases}1, & m'<m\\ 0, & m'\ge m .\end{cases} \tag{5}$$

For the case of filling fraction ν, with $1/(m+2)<\nu<1/m$ and m odd, the natural generalization of Laughlin's wave function is the exact ground-state wave function for the potential,

$$V_{m'}=1,\quad m'\le m-2 \tag{6a}$$

$$V_m=\varepsilon , \tag{6b}$$

$$V_{m'}=0,\quad m'>m+2 \tag{6c}$$

for ε vanishingly small.

III. CALCULATIONS ON THE SPHERE

We study systems of particles confined to the surface of a sphere of radius R. This is the geometry first introduced by Haldane.[13] We consider only a Coulomb interaction, $v(r)$, between electrons on the surface of the sphere. So the interaction is given by

$$v(r)=\frac{e^2}{\epsilon r} . \tag{7}$$

Here, ϵ is assumed to be the effective dielectric constant. The distance, r, is taken to be the chord distance between particles.

The electrons move in a monopolar magnetic field $B=\hbar S/eR^2$, which gives rise to $2S+1$ cyclotron orbits in the lowest Landau level. For systems at filling fractions $\nu=1/m$,

$$2S=m(N-1) . \tag{8}$$

For systems in states at the second level of the hierarchy, the relationship is

$$2S(N_e;\nu)=\nu^{-1}N_e+\frac{\alpha_1 p_2}{p_1p_2+\alpha_2}(p_1-\alpha_2)-m$$
$$=\nu^{-1}N_e+X(\nu) . \tag{9}$$

This is derived in the Appendix. The α_i and p_i are the integers which characterize the filling fraction ν (see the Appendix).

We assume that the electrons are confined to the lowest Landau level and are spin polarized. In this case one can also treat the system as one of holes added to the filled Landau level, so that the system has a particle-hole symmetry. A system of N_e electrons at filling fraction ν, with

$2S=2S(N_e;\nu)$, has a particle-hole symmetric partner state with N_h $(=2S+1-N_e)$ holes. In certain cases this system of N_h holes belongs to a family which is predicted to have a fluidlike ground state at a lower level of the hierarchy. We call this effect aliasing.[20] An example is the case of a nine-particle system at $\nu=\frac{3}{7}$. This has $2S(9;\frac{3}{7})=16$ and $N_h=8$. As $2S(8;\frac{2}{5})=16$, the state with $N_h=8$ is actually in the family of states at $\nu=\frac{2}{5}$. Although in such cases the ground-state energies fit well into both families, this is not true for the excitation energies. We have found that these are characteristic only of the system at the lower level of the hierarchy (see Sec. IV B 2 and Table VII.

A. Finite-size effects and extrapolation to the thermodynamic limit

The determination of the bulk limit of physical quantities such as the energy per electron, E/N, in the ground state or the energy of excitations requires reliable extrapolation methods. For this one needs to know the analytic form of finite-size effects. Finite-size effects result from the curvature of the sphere as well as from the fact that the electron density ρ in a given state at filling fraction ν varies with system size.

In flat space, the electron density ρ is conventionally written in terms of the magnetic length unit l_0 and the filling fraction ν as

$$\rho=\nu/2\pi l_0^2\ , \tag{10}$$

where $l_0^2=\hbar/eB$. The Dirac monopole condition, that the total flux through the surface of the sphere must be an integer multiple, $2S$, of the flux quantum, $\Phi_0=h/e$, gives

$$4\pi R^2 B=2S\Phi_0\ . \tag{11}$$

The radius of the sphere, R, therefore satisfies

$$R=S^{1/2}l_0\ . \tag{12}$$

For a finite system of N electrons on the surface of a sphere, one also has that

$$\rho=N/4\pi R^2=(N/2S)/2\pi l_0^2\ . \tag{13}$$

As a consequence of the way the number of flux quanta, $2S$, varies with electron number N [cf. Eqs. (9) and (A14)], the electron density differs in any finite system from its flat-space value [Eq. (10)] by a factor $N/2S\nu$. This leads to large finite-size corrections to the Coulomb energy if quoted in units $e^2/\epsilon l_0$. This (trivial) size dependence can be removed by using a size and filling-factor-dependent "magnetic length unit" l_0', with

$$l_0'=(2S\nu/N)^{1/2}l_0\ , \tag{14}$$

to measure density and energy. Another useful length unit, often used in the analysis of plasmas, is the "ion disk radius" a:

$$a=(\pi\rho)^{-1/2}\ , \tag{15}$$

which is related to our modified magnetic length unit via

$$a=\left[\frac{2}{\nu}\right]^{1/2}l_0'\ . \tag{16}$$

We assume that the finite-size effects in the ground-state energy per particle scale as N^{-1} and we extrapolate to the thermodynamic limit using Padé schemes. We base this assumption on the study of Morf and Halperin.[8] They looked at the finite-size corrections to the ground-state energy per particle of a system described by Laughlin's wave function at $\nu=\frac{1}{3}$. They compared results obtained for systems with up to 144 particles in the disk geometry, for which the energy in the bulk limit was calculated directly from the pair-correlation function (i.e., without the need for any extrapolation), with those obtained for systems of up to 64 electrons in the spherical geometry. For the systems in the spherical geometry, they obtained results for systems in which the interparticle separation was taken as the chord distance and those in which the geodesic distance was taken. In all cases they obtained the same result for the ground-state energy in the thermodynamic limit, and in all cases the finite-size corrections were found to scale as N^{-1}, when the energy is measured in units of $e^2/\epsilon l_0'$. Attempts to fit the results to polynomials in the variable $x=N^{-1/2}$ lead to small or ill-defined coefficients in the odd powers of x.

We have found that if we fit the results for systems with $5<N\leq 9$ particles with a quadratic polynomial in N^{-1}, we obtain a result which agrees with the results of Ref. 8 to better than 1 part in 10^4.

We have also looked at the special case of a filled Landau level ($\nu=1$). The energy is known for this case.[21,22] Again, we find that the leading correction scales as N^{-1}. If the energy is measured in units $e^2/\epsilon l_0'$, the results for a system with $10\leq N\leq 19$ particles can be fitted accurately by a quadratic polynomial in N^{-1} and lead to a value for E/N for the bulk limit that agrees with the analytical result to better than 1 part in 10^5.

The charged excitations in a state at $\nu=r/s$ carry charge $\pm e/s$ and correspond to removing or adding $1/r$ flux units from the system. We define the proper energy, $\varepsilon^{p(h)}$, to be the thermodynamic limit of the change in Coulomb energy associated with the removal or addition of $1/r$ flux units, keeping the number of electrons and the sphere radius constant (i.e., keeping the *mean* electron density ρ constant).

The extrapolation to the bulk limit of the energy of excitations is trickier than for ground-state energies.[23] For the purpose of illustration, consider the case of a quasihole in the $\nu=1/m$ Laughlin state. This state is generated by threading an additional flux quantum through the surface of the sphere at a point $R\mathbf{\Omega}$. ($\mathbf{\Omega}$ is a unit vector pointing away from the origin.[13]) As a result, in the vicinity of the point $R\mathbf{\Omega}$, $1/m$ units of charge are expelled and uniformly distributed elsewhere in the system. Far from the defect position a density results which differs by a factor $1+O(1/mN)$ from the uniform ground-state density or, equivalently, from the density of the compensating positive background. This is observed in Monte Carlo studies of the quasihole and quasiparticle excitations at $\nu=\frac{1}{3}$; see Ref. 8. This uncompensated charge density over a surface area of $O(N)$ gives rise to an extra electrostatic energy $-(e/m)^2/\epsilon R$ for the quasihole-background interaction and $+(e/m)^2/2\epsilon R$ for the background-background interaction, and produces a net negative contribution A,

$$A=-(e/m)^2/2\epsilon R=-(1/m)^2(\nu/2N)^{1/2}(e^2/\epsilon l_0'), \quad (17)$$

which vanishes in the thermodynamic limit but only as $N^{-1/2}$. The energy per electron, E/N, thus contains an additional term scaling as $N^{-3/2}$, and this does not allow a reliable polynomial fit in terms of N^{-1}.

More generally, if a defect of charge q is added to a system at filling fraction ν, the corresponding contribution A becomes

$$A=-q^2\left[\frac{\nu}{2N}\right]^{1/2}(e^2/\epsilon l_0'). \quad (18)$$

Extrapolation of the energy of charged excitations were performed for this work by subtracting the electrostatic contribution A from the numerical values of the N-dependent excitation energies and using polynomial or Padé fits as a function of N^{-1} to these corrected values.

IV. RESULTS

A. Ground-state properties

In Table I we show the extrapolated energy per particle, E/N, at filling fractions $\nu=\frac{1}{3}, \frac{2}{5}, \frac{2}{7}$, and $\frac{3}{7}$, based on systems with up to 10, 10, 8, and 12 particles, respectively. The extrapolations to infinite system size have been made by using a Padé approximation using the variable $1/N$. The result for a system at $\nu=\frac{1}{3}$ is obtained from the data already published by Fano *et al.*,[24] but rescaled into units in which the electron density is independent of system size (see Sec. III A).

The numbers in parentheses in Table I denote the maximum deviation from the listed values in units of the last digit quoted as obtained from different Padé schemes. We restricted outselves to the [1,1], [1,0] and [0,1] Padé approximants. Where we have more data points than there are coefficients in the Padé approximants, we minimize the function

$$F(m,n)=\sum_{i=N_0}^{N_{max}}[p_m(x_i)-E(x_i)p_n(x_i)]^2 \quad (19)$$

with respect to the coefficients in the polynomials $p_{m,n}(x)$. The $E(x_i)$ are the energies of the systems with x_i^{-1} particles. N_{max} is the largest system for which we obtained results. For each family the deviation is that obtained after varying N_0 in Eq. (19) for all the Padé approximants.

TABLE I. The ground-state energy per particle at filling fractions predicted by Haldane Ref. (13), obtained by extrapolation on the results for finite systems. The numbers in parentheses represent the scatter obtained from different Padé schemes and are in units of the last quoted digit.

Units	ν = $\frac{2}{7}$	$\frac{1}{3}$	$\frac{2}{5}$	$\frac{3}{7}$
$e^2/\epsilon l_0$	−0.3810(4)	−0.4102(2)	−0.4335(9)	−0.4431(5)
$e^2/\epsilon a$	−1.008(10)	−1.0048(5)	−0.9693(20)	−0.9572(11)

The results shown in Table I are those extrapolated from families of systems for which $2S$ is given by Eqs. (8), (A17), and (A18). The dependence on system size was found to be weak and smooth and, hence, allowed for good extrapolation to infinite system size (see Sec. III A). We interpret this as strong evidence in favor of hierarchical classification of the ground states.

As a further check, we have examined a second family of states which might describe systems at filling fraction $\nu=\frac{2}{7}$. However, this is not the family predicted by the Haldane classification, Eq. (A18). Instead, we choose

$$2S=\frac{7}{2}N_e-4. \quad (20)$$

According to Haldane's classification, the ground state of this family should be thought of as a system of N_e electrons at $\nu=\frac{2}{7}$ together with either two quasiparticles in the parent fluid at $\nu=\frac{1}{3}$, each carrying effective charge $2e/7$, or with a system with four quasiparticles in the daughter fluid, each carrying effective charge $e/7$.

The energies for this second family shown in Table II are larger than those in the family predicted by the Haldane classification and show a large irregular dependence on system size. Again, we take this as evidence in favor of the classification of the ground states given by Haldane.[13] We have not tried to analyze the energies of the state in this second family in terms of excitations from the first family.

This second family, with magnetic field strength given by Eq. (20), arose during earlier studies of microscopic wave functions. Morf *et al.*[11] suggested trial wave functions, ψ_T, with (in the disk geometry)

$$\psi_T=\mathcal{A}\left[\psi_m\prod_{n=1}^{N/2}(z_{2n}-z_{2n-1})^{-t}\prod_{n,n'=1}^{N/2}(z_{2n}z_{2n-1}+z_{2n'}z_{2n'-1}-2Z_nZ_{n'})^s\right]. \quad (21)$$

Here, ψ_m is the Laughlin wave function with exponent m. $Z_n[=(z_{2n}+z_{2n-1})/2]$ denotes the center-of-mass coordinate of the pair formed by particles $2n$ and $2n-1$. The particles have coordinates x_j and y_j, and $z_j=x_j-iy_j$. To ensure that the wave function is antisymmetric with respect to interchange of particles $2n$ and $2n-1$ in the "pair n," $m-t$ must be odd. Also, $m-t\geq 1$ and $s\geq 0$. The operator $\mathcal{A}$ ensures that the wave function is an-

TABLE II. The energy per particle in the lowest-lying homogeneous (L=0) state for two families of states. The family $\frac{7}{2}N_e-2$ belongs to the hierarchical classification proposed. It shows only a small dependence on system size. The other family has higher energy and it shows large dependence on system size.

2S	4	6	8
$7N_e/2-2$	−0.386012	−0.384626	−0.383811
$7N_e/2-4$	−0.333477	−0.380028	−0.377542

TABLE III. Overlaps between various states describing a system of eight electrons (upper half) and six electrons (lower half), confined to the surface of a sphere at a filling fraction $\nu=\frac{2}{7}$.

	$\lvert\psi_C\rangle$	$\lvert\psi_{HC}\rangle$	$\lvert\psi_T\rangle$	$\lvert\psi_{2\times3/2}\rangle$
$\langle\psi_C\rvert$		0.9908	0.9963	0.9907
$\langle\psi_{HC}\rvert$	0.9942		0.9972	0.9752
$\langle\psi_T\rvert$	0.9972	0.9904		0.989
$\langle\psi_{2\times3/2}\rvert$	0.9856	0.9688	0.9938	

tisymmetric with respect to interchange of any two particles not in the same "pair."

ψ_T is a possible trial wave function for systems at $\nu=\frac{2}{7}$ when $m=3$, $s=1$, and with $t=0$ or -2. The case $t=-2$ would be for a system for which the Haldane classification predicts a hierarchical ground state [Eq. (A18)], whereas the case $t=0$ corresponds to a system at the magnetic field strength given by Eq. (20). Arguing as follows might lead one to suspect that either wave function ($t=0$ and -2) could be a good candidate ground-state wave function. Both wave functions contain ψ_3, the Laughlin wave function for systems at $\nu=\frac{1}{3}$, as a factor. They might therefore be said to have favorable distributions of zeros. This would be favorable in the sense that the zeros have been concentrated on the positions of the other electrons, thereby reducing the probability that any two electrons approach each other.[19]

The wave function with $t=0$ turns out not to have a large overlap with the exact ground-state wave function for systems with 2S given by Eq. (20). We believe that this, together with the fact that the exact ground-state energy per particle shows irregular finite-size effects as mentioned above, again supports the hierarchical classification of the ground states.

We have also studied the possible mapping between ground states at $\nu=\frac{2}{3}$ and $\frac{2}{7}$. We have taken the Laughlin state at $\nu=\frac{2}{3}$, $\psi_{3/2}$, which is equivalent to the $\nu=\frac{1}{3}$ state in holes, and applied the operator T_2 given in Eq. (3). So,

$$\psi_{2\times3/2}\equiv T_2\psi_{3/2}=\prod_{1\le i<j\le N}(z_i-z_j)^2\psi_{3/2}\ . \qquad (22)$$

We might expect the state $\psi_{2\times3/2}$ to be the exact nondegenerate ground state, ψ_{HC}, of the "generalized hard-core" interaction given by Eq. (6). The overlaps between these two states and the exact ground state of the Coulomb interaction, ψ_C, are shown for six and eight particles in Table III. In Table III we also include the overlap with the trial wave function of Eq. (21). With $m=3$, $s=1$, and $t=-2$, ψ_T is a trial wave function for a system at $\nu=\frac{2}{7}$.

Table III shows that the overlap between all four wave functions is close to 1. All three wave functions, $\psi_{2\times3/2}$, ψ_T, and ψ_{HC}, capture the essential features of the ground state at $\nu=\frac{2}{7}$. However, the overlap $\langle\psi_{2\times3/2}|\psi_{HC}\rangle$ is not identically equal to 1, implying that the mapping, implicitly assumed by many authors between the Laughlin states at filling fractions $1/(2m+3)<\nu<1/(2m+1)$ and those at filling fractions $1/(2n+3)<\nu<1/(2n+1)$, is, at best, only approximately described by the operator T_{n-m}.

B. Excitations

1. Quasiholes and quasiparticles

It is not possible to nucleate a single quasihole or quasiparticle at constant particle number, except at the lowest level of the hierarchy. This makes it necessary to say slightly more precisely how we find the wave functions describing an excited or perturbed system, and how we define the excitation energy itself.

We take the wave function describing a single quasiparticle or quasihole in a system of N_e electrons on the sphere as the ground-state wave function of the system with the corresponding magnetic charge, $2S(N_e)$, at the sphere's center. We obtain the ground state of this perturbed system by direct diagonalization of the Hamiltonian. As an example, consider the quasiparticle in a system of 13 electrons at $\nu=\frac{3}{7}$. According to Eq. (A17), a system of 13 particles on a sphere at $\nu=\frac{3}{7}$ should have $2S=25\frac{1}{3}$. A system in its ground state when $2S=25$ is the one we characterize as containing a single-quasiparticle excitation. The "missing" one-third of a flux quantum is identified with the quasiparticle carrying fractional charge—in this case, $|e|/7$.

The hierarchical model predicts the ground-state degeneracy of systems classified as containing a single quasiparticle or quasihole. For a system of N_e electrons at the lowest level of the hierarchy at filling fraction $\nu=1/m$, there are N_e+1 linearly independent ways to introduce a quasihole or quasiparticle.[13] These correspond to the $2l+1$ states with angular-momentum number $l=N_e/2$.[13] If one identifies a quasiparticle's (quasihole's) position on the sphere of radius R with $R\Omega^{p(h)}$, where $\mathbf{L}\cdot\Omega^{p(h)}=l\ (-l)$, one may think of the $2l+1$ states as $2l+1$ "quasicyclotron orbits." ($\Omega^{p(h)}$ is a unit vector pointing away from the origin.[13])

The degeneracy of excitations in systems with ground states at higher levels of the hierarchy should be determined by the number of "particles" in the hierarchical fluid that is being excited. In an nth level hierarchical fluid with N_n particles there would be N_n+1 linearly independent ways to introduce a quasihole or quasiparticle.

These should correspond to the $2l+1$ states of a multiplet characterized by angular-momentum quantum number $l=N_n/2$, and one would identify the quasiparticle position coordinate on the sphere with the direction of L. The formulas giving the relevant N_i in terms of N_e are included in large parentheses at the top of Tables V–VII.

The actual diagonalization of the Hamiltonian is not easy in the large systems we have studied. For systems at $\nu=1/m$ there are the explicit trial wave functions describing the quasihole and quasiparticle proposed by Laughlin.[3] These are relatively easy to expand in terms of Slater determinants, once one has an expansion for the ground state at $\nu=1/m$. These trial wave functions are good starting wave functions for an iterative search for the true ground state based on the Lanczos procedure. For systems at higher levels in the hierarchy, there is no simple operator generating excited-state wave functions from the ground-state wave function. This is essentially because nucleation of an excitation requires the change in particle number mentioned above. Instead, when we operate the Lanczos iterative procedure we have to start from a Slater determinant. We check that the "ground state" we obtain has well-defined total angular momentum[13] L. We would normally repeat this procedure starting from some other (combination of) Slater determinant(s) to ensure that we have not obtained an eigenstate which is not the true minimum of the energy.

We have evaluated the angular momentum of the ground state of systems classified as containing a single quasiparticle or quasihole. We find that for all systems studied the predictions of the hierarchical classification are correct. These include the systems at $\nu=\frac{1}{3}$—the "parent" fluids—systems at $\nu=\frac{2}{5}$ and $\frac{2}{7}$, which are thought of as "daughters" of the one-third states, and systems at $\nu=\frac{3}{7}$, which are the daughters of the two-fifth states. It would seem that our results confirm that the hierarchy predicts correctly not only the systems that have Laughlin-like ground states, but also the nature and degeneracy of their elementary excitations.

Identifying the position coordinates of these excitations with $R\Omega^{p(h)}$, where $\mathbf{L}\cdot\Omega^{p(h)}=l\,(-l)$, allows us to calculate the charge distribution of the excitations. In Figs. 1–3 we show the charge distribution of quasiparticle and quasihole excitations. The distance r is the chord distance of a point measured from the "center" of the excitation, $R\Omega^{p(h)}$. We take the density, ρ_0, to be the total charge of the electrons plus the charge of the excitation, q, distributed evenly over the sphere. Adding the charge of the excitation, q, is in line with previous authors.[13,8] (See also Sec. III A)

The curves shown suggest that the excitations should be considered part of a sequence which converges in the thermodynamic limit to the density profile of a quasiparticle or quasihole. The curves are nearly coincident moving out from the centers of the excitations and only deviate from one another far from the centers, where finite-size effects become apparent.

The density profiles for the quasiparticle at $\nu=\frac{2}{5}$ and $\frac{3}{7}$ show that the excess charge, q, is concentrated on a ring about the origin. We have integrated the density outwards from the origin in order to be able to estimate the radius at which the accumulated excess charge equals q ($|e|/5$ and $|e|/7$ for the cases $\nu=\frac{2}{5}$ and $\frac{3}{7}$). This leads us to a (rough) estimate of the "size" of the excitations, r_ν^p. The integrated charge actually exceeds the expected value q for certain ranges of r, so that there are, in fact, two values we could quote for r_ν^p. We indicate both possible values with arrows in the figures. Taking the larger

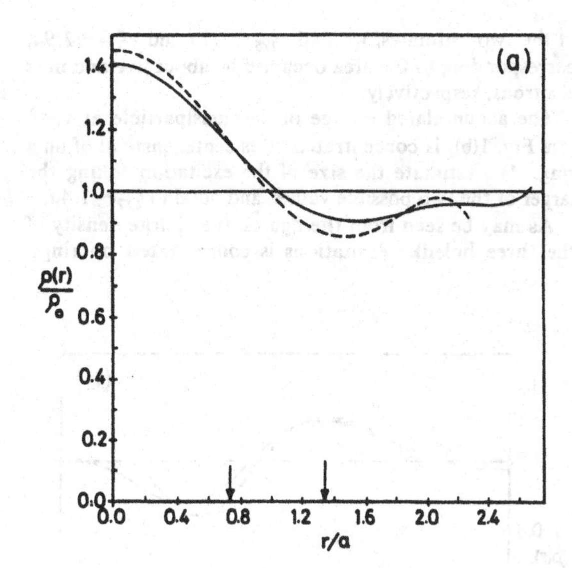

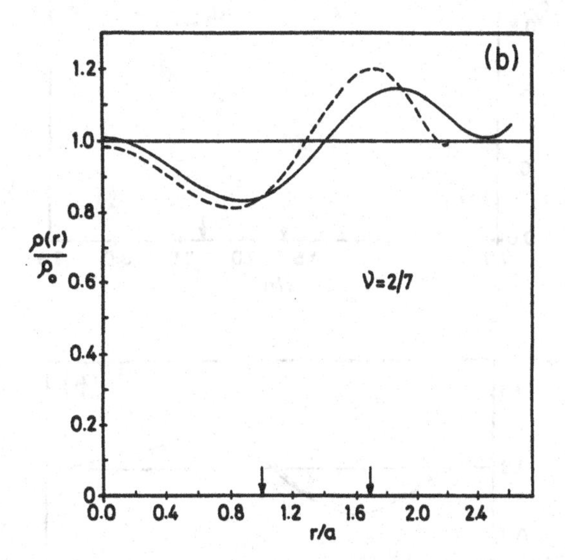

FIG. 1. The density distribution normalized to 1, $\rho(r)$, about a $\nu=\frac{2}{7}$ quasiparticle, (a), and quasihole, (b), centered at the origin. The distances are measured in units of the ion disk radius a [see Eq. (15)]. The solid lines relate to systems with seven particles and the dotted ones to systems with five particles. The two arrows indicate radii at which the accumulated charge equals $e/7$.

of the two estimates, we find $r^p_{2/5} \sim 2.3a$ and $r^p_{3/7} \sim 2.9a$, corresponding to the area occupied by about five and nine electrons, respectively.

The accumulated charge of the quasiparticle at $\nu=\frac{2}{7}$ [see Fig. 1(b)] is concentrated at its center instead of on a ring. We estimate the size of the excitation taking the larger of the two possible values, and obtain $r^p_{2/7} \sim 1.4a$.

As may be seen from the figures, the charge density of the three holelike excitations is concentrated on rings. The maxima apparent in the charge density far from the centers of the excitation are, we believe, finite-size effects, related to the fact that the systems we are able to study are not large enough to allow a uniform redistribution of the charge "expelled" when the quasihole is nucleated. We define the effective size of the quasihole, r^h_ν, as the larger of the two radii within which the accumulated charge equals q, the charge expected for the excitation. We find $r^h_{2/7} \sim 1.7a$, $r^h_{2/5} \sim 1.9a$, $r^h_{3/7} \sim 2.5a$. Although the numbers we quote for $r^{p(h)}_\nu$ are, at best, only rough

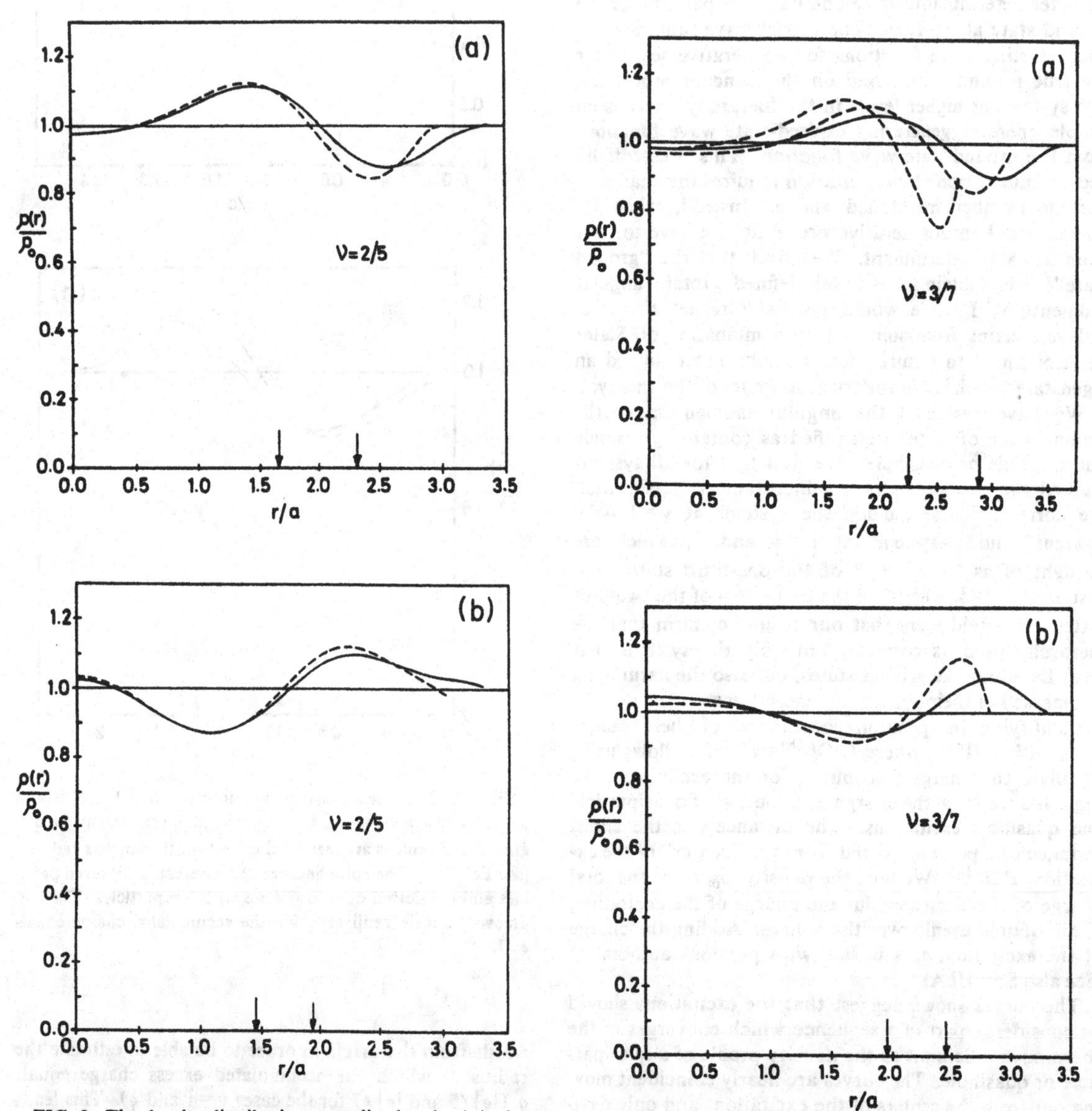

FIG. 2. The density distribution normalized to 1, $\rho(r)$, about a $\nu=\frac{2}{5}$ quasiparticle, (a), and quasihole, (b), centered at the origin. The distances are measured in units of the ion disk radius a [see Eq. (15)]. The solid lines relate to systems with 11 particles and the dotted ones to systems with 9 particles. The arrows indicate radii at which the accumulated charge equals $e/5$.

FIG. 3. The density distribution normalized to 1 for quasiparticles, (a), and quasiholes, (b), at $\nu=\frac{3}{7}$ centered at the origin. The quasiparticles are for systems with 7, 10, and 13 particles. The quasiholes are for systems with 8 and 11 particles. The arrows indicate radii at which the accumulated charge equals $e/7$.

TABLE IV. The extrapolated quasihole (quasiparticle) energies $\varepsilon^{h(p)}_{l/3}$, the gap energy $\varepsilon^{g}_{l/3}$, and the "exciton" energies $\varepsilon^{\rm exc}_{1/3}$, for N-particle systems at $\nu=\frac{1}{3}$. The numbers in parentheses are the scatter obtained from various Padé schemes in units of the last quoted digit.

Units	$\varepsilon^{h}_{1/3}$	$\varepsilon^{p}_{1/3}$	$\varepsilon^{g}_{1/3}$	$\varepsilon^{\rm exc}_{1/3}$
$e^2/\epsilon l_0$	0.027(3)	0.075(1)	0.102(4)	0.102(3)
$e^2/\epsilon a$	0.066(6)	0.183(3)	0.250(8)	0.252(7)

guides to the "size" of the excitations, they may be useful in gauging the likely success of models which treat these particles as point particles.

To estimate the energy required to nucleate an excitation, we need to know the ground-state energy of the system in its "unperturbed" state. Except at the lowest level of the hierarchy, this cannot be calculated directly, but must instead be interpolated from the values known for systems with different numbers of particles. We used a [1,2]-Padé approximation to estimate energies at $\nu=\frac{2}{5}$ and a [1,1]-Padé approximation for $\nu=\frac{2}{7}$ and $\frac{3}{7}$.

The "bare" excitation energy, $\varepsilon^{p(h)}_{\nu}$, is taken as the ground-state energy of the perturbed system minus the ground-state energy the system would have were it in its homogeneous ground state. Both energies have to be in units of $e^2/\epsilon l'_0$. The excitation energies are then "adapted" to yield the $\tilde{\varepsilon}^{p(h)}_{\nu}$ by subtracting the (negative) electrostatic contribution to the energy, A [see Eq. (18)].

In Tables IV–VII we show the results we obtain for the energies of quasiholes and quasiparticles at filling fractions $\nu=\frac{1}{3}$, $\frac{2}{7}$, $\frac{2}{5}$, and $\frac{3}{7}$. For a system at $\nu=\frac{1}{3}$ we report only the extrapolated result based on the data of Fano *et al.*[24] The extrapolations to the thermodynamic limit were made using various Padé approximations. Again, we checked that the extrapolation procedures correctly predicted the excitation energies for systems described by Laughlin's trial wave function, which are known.[8,23]

We have considered the excitations of the $\nu=\frac{2}{5}$ and $\frac{3}{7}$ states which, according to the hierarchical picture, should be considered excitations of the parent $\nu=\frac{1}{3}$ state. These are nucleated when a single flux quantum is added or removed, and carry effective charge $q=\pm 2e/5$ and $\pm 3e/7$, respectively. If the energy required to nucleate these excitations were less than 2 or 3 times the quasiparticle or quasihole energies at $\nu=\frac{2}{5}$ ($\nu=\frac{3}{7}$), then presumably the original quasiparticles and quasiholes would "coalesce" to form double (triple) excitations. This would be reminiscent of negative-U centers in semiconductors.[25] Recently, Clark *et al.*[16] claimed to have shown experimentally that this is not what happens. Unfortunately, the results we obtain appear to show large finite-size effects, and we feel they do not justify any extrapolations. Consequently, we do not include any data here.

2. *Collective excitations*

We have studied the excitation spectrum at constant magnetic field and particle number as a function of the total angular momentum L. As suggested by Haldane and Rezayi,[9] one may define an effective wave vector $k_{\rm eff}=L/R$,

$$k_{\rm eff}=L/R\ , \tag{23}$$

where R is the radius of the sphere (see Sec. III), and set $\varepsilon(k_{\rm eff})=\varepsilon_L$. This enables one to compare the spectra obtained for systems with different particle numbers.

According to Haldane and Rezayi[9] and Girvin *et al.*,[26] the excitations at small and intermediate $k_{\rm eff}$ are best thought of as analogs of the rotons in superfluid ^{4}He. A feature of the excitation spectrum of the Laughlin-like fluids at $\nu=1/m$, $\varepsilon(k_{\rm eff})$, is the rotonlike minimum expected at some value of $k_{\rm eff}$ close to the first reciprocal-lattice vector, G_ν, of a Wigner crystal of the same density.[26] Haldane[27] showed that for these intermediate wave vectors (wave vectors close to G_ν) the spectral weight is almost completely taken up by one collective excitation, and is therefore well described by the Feynman-Bijl *Ansatz* wave function proposed by Girvin *et al.*[26]

We have calculated the excitation spectrum $\varepsilon(k_{\rm eff})$ for the filling fractions $\nu=\frac{2}{7}$, $\frac{2}{5}$, and $\frac{3}{7}$. We show the results

TABLE V. The quasihole (quasiparticle, gap) energies, $\tilde{\varepsilon}^{h(p,g)}_{2/7}$ for systems at $\nu=\frac{2}{7}$. The overtildes denote energies corrected for the systematic variation, with system size described in the text. The numbers in parentheses are the scatter obtained from various Padé schemes in units of the last quoted digit. Energies in the final row are in units of $e^2/\epsilon a$. All other energies are in units of $e^2/\epsilon l_0$.

N_e	$\tilde{\varepsilon}^{h}_{2/7}$ $\left[N_1=\frac{N_e+3}{2}\right]$	$\tilde{\varepsilon}^{p}_{2/7}$ $\left[N_1=\frac{N_e+1}{2}\right]$	$\tilde{\varepsilon}^{g}_{2/7}$	$\tilde{\varepsilon}^{\rm exc}_{2/7}$ $\left[N_1=\frac{N_e+2}{2}\right]$
3	0.017 47	0.021 28	0.038 75	
4				0.029 12
5	0.015 20	0.020 30	0.035 50	
6				0.024 01
7	0.011 23	0.017 62	0.028 84	
8				0.024 99
∞	0.008(2)	0.016(2)	0.024(4)	0.022(3)
∞	0.021(5)	0.042(5)	0.064(11)	0.058(8)

TABLE VI. Energies (defined as in Table V) for a system at $\nu=\frac{2}{5}$.

N_e	$\varepsilon^h_{2/5}$ $\left(N_1=\frac{N_e+1}{2}\right)$	$\varepsilon^p_{2/5}$ $\left(N_1=\frac{N_e+3}{2}\right)$	$\varepsilon^p_{2/5}$	$\varepsilon^{exc}_{2/5}$ $\left(N_1=\frac{N_e+2}{2}\right)$
5	0.022 73	0.060 66	0.083 40	
6				0.071 57
7	0.018 28	0.056 21	0.074 48	
8				0.067 22
9	0.016 62	0.052 33	0.068 95	
10				0.067 12
11	0.016 20	0.050 97	0.067 17	
∞	0.015(5)	0.046(10)	0.061(15)	0.063(4)
∞	0.034(11)	0.103(24)	0.137(36)	0.140(9)

in Figs. 4–6. We tentatively identify a minimum in the excitation spectrum just as in the case of $\nu=\frac{1}{3}$; see Haldane and Rezayi[9] and Fano *et al.*[24] In contrast to the case at $\nu=\frac{1}{3}$, there is not always a well-defined single branch in the excitation spectra. Again, this may be evidence for the hierarchical classification of the excitations. From the hierarchical model, one might expect the existence of separate branches for each level of the hierarchy. If the gaps between adjacent levels, i and $i+1$, of the hierarchy did scale as the ratio of the effective charges times the ratio of the ion disk radii for the excitations, $|q_i/q_{i+1}|^2 a_{i+1}/a_i$, then these branches would be well separated. If, as we believe, gaps at filling fractions classified as belonging to higher levels of the hierarchy do not scale this way, so that there is no clear separation of energy scales relating to different levels of the hierarchy, then one would expect that these branches would not be distinct. This is what we observe at $\nu=\frac{2}{7}$, Fig. 4.

We remark briefly on the apparent zig-zag nature of the spectrum at $\nu=\frac{3}{7}$ for which we have no satisfactory explanation at present. Recently, though, Reynolds and d'Ambrumenil[28] have reported studies of the collective excitations in the hierarchy using the adapted Feynman-Bijl variational *Ansatz*. They find that the collective excitation spectrum predicted variationally showed no such zig-zag effect. This led them to suggest that the zig-zag effect was only a finite-size effect.

In the limit of large $k_{\rm eff}$, $\varepsilon(k_{\rm eff})$ should tend to the gap energy or to the energy to create an unbound quasihole and quasiparticle pair. One should therefore be able to make a separate estimate of the energy gap from the value of $\varepsilon(k_{\rm eff})$ in the limit of large $k_{\rm eff}$. If the ground states at the hierarchical filling fractions are indeed Laughlin-like fluids, then one might also expect to find a rotonlike minimum in the excitation spectrum $\varepsilon(k_{\rm eff})$, close to G_ν.

We have estimated the gap energy from the dispersion relations shown in Figs. 4–6. We take as our estimate the excitation energy, ε^{exc}_ν, which is the limiting value of $\varepsilon(k_{\rm eff})$ for large $k_{\rm eff}$. As Haldane and Rezayi[9] pointed

TABLE VII. Energies (defined as in Table V) for a system at $\nu=\frac{3}{7}$. The result quoted for the infinite system in $\varepsilon^{exc}_{3/7}$ is the sum of the quasihole and quasiparticle energies. The asterisks denote values that are not true excitation energies of a system at $\nu=\frac{3}{7}$ because of the aliasing affect (see text and Sec. III).

N_e	$\varepsilon^h_{3/7}$ $\left(N_2=\frac{N_e+4}{3}\right)$	$\varepsilon^p_{3/7}$ $\left(N_2=\frac{N_e+8}{3}\right)$	$\varepsilon^{exc}_{3/7}$ $\left(N_2=\frac{N_e+6}{3}\right)$
5	0.029 02		
6			0.096 96*
7		0.051 59	
8	0.019 69		
0			0.062 58*
10		0.043 46	
11	0.017 22		
12			0.050 34
13		0.041 46	
∞	0.013(8)	0.036(10)	0.049(18)
∞	0.028(15)	0.077(23)	0.105(38)

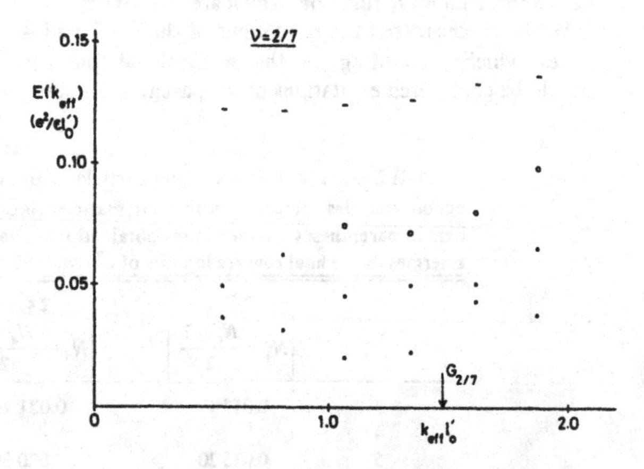

FIG. 4. The collective excitations for a system of eight particles at $\nu=\frac{2}{7}$ plotted against effective wavevector $k_{\rm eff}l_0'$ [see Eqs. (23) and (14)]. The open circles denote points which may not be fully converged. The lines denote excitations which roughly delineate a continuum of states.

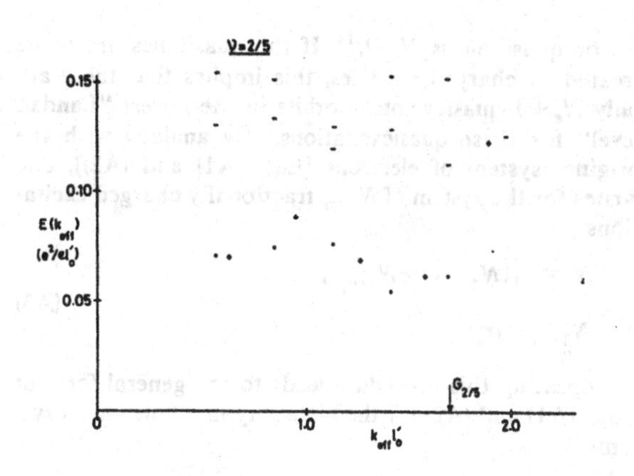

FIG. 5. The collective excitations for a system of 10 particles at $\nu=\frac{2}{5}$ plotted against effective wave vector $k_{\text{eff}}l'_0$ [see Eqs. (23) and (14)]. The circles denote excitations which we interpret as belonging to the low-lying collective excitation branch. The +'s denote the excitation energies of a system of eight particles.

out, we may identify the separation of a quasiparticle and quasihole as $2RL/N$ in a system of N particles. Clearly, this relation is only valid for $L \le N$, as two excitations cannot be further apart than $2R$ on the surface of the sphere. We therefore take $\varepsilon_\nu^{\text{exc}}$ as the energy with $L=N$.

Taking the hierarchical construction at face value, the number of particles, N, for a hierarchy terminating at the ith level is not the number of electrons, but N_i. For $\nu=\frac{2}{7}$ and $\frac{2}{5}$, N_1 is given by Eq. (A3); the systems with 8 (10) particles have $N_1=5$ (6). For $\nu=\frac{3}{7}$, N_2 is given by Eq. (A4) with $i=2$; the system with 9 (12) electrons has $N_2=5$ (6).

The estimates of the gap energies taken from the collective excitation energies are shown in Tables IV–VI. As for the single-particle excitation, we show the energy corrected for a systematic variation with system size $\bar{\varepsilon}_\nu^{\text{exc}}$. The difference between the two is just the electrostatic energy of two quasiparticles of charge $\pm q$ situated at opposite poles of the sphere. Where possible, we have then

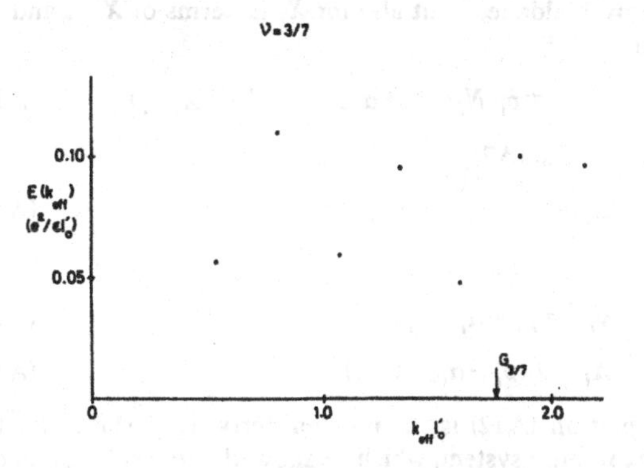

FIG. 6. The collective excitations for a system of 12 particles at $\nu=\frac{3}{7}$ plotted against effective wave vector $k_{\text{eff}}l'_0$ [see Eqs. (23) and (14)]. Only the lowest energies for each effective wave vector are shown, as the higher energies were not fully converged.

extrapolated the "corrected" energies to the thermodynamic limit.

The results in Table VII illustrate the effect of aliasing (see Sec. III). The systems with 6 and 9 particles and $2S=9$ and 16 correspond to systems of holes at $\nu=\frac{1}{3}$ and $\frac{2}{5}$, respectively. [However, the values of $\bar{\varepsilon}_\nu^{\text{exc}}$ for these two systems, 0.097 and 0.063, cannot be compared directly with those at $\nu=\frac{1}{3}$ (Table IV) and $\nu=\frac{2}{5}$ (Table VI), because they have been calculated in different units and assuming a different electrostatic correction.]

The results for the gaps obtained by extrapolation on the sum of the quasihole and quasiparticle energies and those obtained from extrapolations from the neutral excitation energies agree. This gives us added confidence in the values obtained. The results from the two independent sets of calculations clearly show that the gaps are substantially larger than would be expected if one were to treat the excitations as point particles.

C. Comparison with experiment

Taking the results obtained from the calculations for the single-particle excitations and the collective excitations, we have estimated the gaps Δ_ν^t, for a system with density $\rho=0.95\times10^{15}$ m^{-2}. We take the relative permittivity $\epsilon_r=13$ for GaAs. In Table VIII we compare the results with the experimental estimates of Clark *et al.*[16] for their sample G139, Δ_ν^e. We note that the theoretical estimates are all much larger than the experimental ones.

As has been pointed out by Girvin,[29] a difference of a factor of around 2 between the experimental and theoretical estimates of the gap energy is generally attributable to the finite width in the z direction and to inter-Landau-level mixing (although we have not checked this ourselves), but the rest may be presumed due to effects of disorder. This was borne out recently by the experiments reported in Ref. 30. There it was found that for a very-high-mobility sample there was essentially no discrepancy between theoretical and experimental estimates of the gap at $\nu=\frac{1}{3}$ after accounting for the finite width in the z direction and inter-Landau-level mixing.

The discrepancies between Δ_ν^t and Δ_ν^e are larger the higher the level of the hierarchical fluid being excited. This is what one might expect because these excitations are clearly more "fragile." They require coherence of the many-body wave function over a larger region and have lower excitation energies. We expect, though, that, as

TABLE VIII. Theoretical and experimental estimates in K of the energy gap Δ_ν in a system with density $\rho=0.95\times10^{15}$ m^{-2} and taking the relative permittivity $\epsilon_r=13$. The experimental estimates are those quoted in Ref. 16 for sample G139. The value quoted for $\Delta_{1/3}$ is a guess we have made based on the values in Ref. 16 for $\Delta_{2/3}$, $\Delta_{4/3}$, and $\Delta_{5/3}$.

	ν			
	$\frac{2}{7}$	$\frac{1}{3}$	$\frac{2}{5}$	$\frac{3}{7}$
Δ_ν^t	4.3	18	9.5	7.3
Δ_ν^e		~3.5	1.37	0.5

samples are produced with yet higher mobilities, the discrepancies between our theoretical estimates of the gap energies at higher levels of the hierarchy and the experimental ones will get smaller and eventually by attributable to finite-width effects and inter-Landau-level mixing.

V. CONCLUSIONS

All our calculations are consistent with the hierarchical picture, both in regard to ground- and excited-state properties. The hierarchy predicts families of systems whose properties should converge in the thermodynamic limit to those of a system in a fractionally quantized Hall state. We find a smooth dependence on system size of the ground-state energy for these families, which we believe allows for trustworthy extrapolation to the thermodynamic limit. Other families show a large and irregular dependence of ground-state energy on the particle number.

The hierarchical model allows for the evaluation of the "density profile" of the various excitations and the estimation of their energies. Although the extrapolation of these excitation energies requires care, we were able to estimate the energy gaps at $\nu=\frac{2}{5},\frac{2}{7},\frac{3}{7}$. We suggested that these energy gaps could be measured by experiment given the right samples and provided finite-width effects and inter-Landau-level mixing are taken into account.

The view of many has been for some time that the hierarchy correctly characterizes the fractional quantized Hall states as they are observed in the lowest Landau level. However, we hope that our results will help convince any remaining doubters of the validity of this triumphantly simple scheme. We also hope that our numerical estimates of excitation energies will be of use in future comparisons with experiment.

ACKNOWLEDGMENTS

We would like to thank P. Béran, R. G. Clark, G. Fano, B. Halperin, and A. Reynolds for many helpful conversations. This research was partly funded by the United Kingdom Science and Engineering Research Council (SERC) under Grant No. GR/D89790.

APPENDIX: THE HIERARCHY ON THE SPHERE

To study the hierarchical states on the sphere, we need to know the expression for $2S$ corresponding to Eq. (8) for the fractional filling fractions ν at which fluid ground states are predicted to occur. Following Haldane,[13] we write

$$2S(N_e)=m(N_e-1)+N_{1/m} . \tag{A1}$$

Here, N_e is the number of electrons and $N_{1/m}$ is given by

$$N_{1/m}=\alpha_1 N_1 , \tag{A2}$$

where N_1 is the number of vortices (additional or missing zeros) nucleated in the Laughlin state at $\nu=1/m$. Again, $\alpha_1=1$ for quasiholes and $\alpha_1=-1$ for quasiparticles.

The angular momentum of a state with one quasiparticle or quasihole is $N_e/2$.[13] If the quasiholes are to be treated as charged particles, this implies that there are only N_e+1 quasicyclotron orbits in the lowest "Landau level" for these quasiexcitations. By analogy with the original system of electrons [Eqs. (A1) and (A2)], one writes for the system of $N_{1/m}$ fractionally charged excitations

$$N_e=p_1(N_1-1)+N_{1/p_1} , \quad N_{1/p_1}=\alpha_2 N_2 . \tag{A3}$$

Repeating this procedure leads to the general form of Eqs. (A1) and (A2) for the hierarchy in a finite-sized system:

$$N_{i-1}=p_i(N_i-1)+N_{1/p_i} . \tag{A4}$$

The hierarchy is terminated at level n if

$$N_{1/p_n}=0 . \tag{A5}$$

Introducing "filling fractions" ν_i and constants X_i with the shorthand notation

$$\nu_i=\nu_i(p_i,\alpha_{i+1},p_{i+1},\ldots,\alpha_n,p_n) , \quad X_i=X_i(p_i,\alpha_{i+1},p_{i+1},\ldots,\alpha_n,p_n) , \tag{A6}$$

we now write

$$N_i=\nu_i N_{i-1}+\alpha_i X_i . \tag{A7}$$

The "filling fraction" ν_n and constant X_n at the level at which the hierarchy terminates are given by

$$\nu_n=1/p_n \tag{A8}$$

and

$$X_n=1 , \tag{A9}$$

while the filling fraction ν_0 is just the filling fraction of the electrons, ν. Similarly, X_0 is the constant in Eq. (1), $X(\nu)$.

Inserting Eq. (A7) into Eq. (A4) leads to recursion relations not only for ν_i in terms of ν_{i+1} as obtained originally by Haldane,[13] but also for X_i in terms of X_{i+1} and ν_i. So,

$$N_{i-1}=p_i(N_i-1)+\alpha_{i+1}(\nu_{i+1}N_i+X_{i+1}) , \tag{A10}$$

or [see Eq. (A7)]

$$N_i=\nu_i N_{i-1}+\alpha_i X_i , \tag{A11}$$

with

$$\nu_i^{-1}=p_i+\alpha_{i+1}\nu_{i+1} , \tag{A12}$$

$$X_i=\nu_i(p_i-\alpha_{i+1}X_{i+1}) . \tag{A13}$$

Equation (A12) is the relation derived by Haldane[13] for the infinite system, which he showed had a solution in the form of a continued fraction. Equation (A13) is the extra relation that must be retained when dealing with finite-size systems.

The relation for the magnetic field strength at which an

nth-level hierarchical fluid ground state might exist is then given by

$$2S(N_e;\nu)=\nu^{-1}N_e-m+\alpha_1X_1\equiv\nu^{-1}N_e+X(\nu) . \quad \text{(A14)}$$

As an illustration, we solve for X_1 for a hierarchy terminating at the second level. So [see Eq. (A9)],

$$X_2=1 . \quad \text{(A15)}$$

Then, from Eq. (A13),

$$X_1=\frac{1}{p_1+\alpha_2/p_2}(p_1-\alpha_2) . \quad \text{(A16)}$$

For a finite system with N_e particles, a hierarchical ground state is expected at $\nu^{-1}=\frac{7}{3}$ when $m=3$, $\alpha_1=\alpha_2=-1$, and $p_1=p_2=2$. The corresponding magnetic field strength is given by

$$2S(N_e;\tfrac{3}{7})=\tfrac{7}{3}N_e-5 . \quad \text{(A17)}$$

Other cases of interest would be

$$2S(N_e;\tfrac{2}{7})=\tfrac{7}{2}N_e-2 \;\; (\nu=\tfrac{2}{7}) ,$$
$$2S(N_e;\tfrac{2}{5})=\tfrac{5}{2}N_e-4 \;\; (\nu=\tfrac{2}{5}) . \quad \text{(A18)}$$

---

[1] D. C. Tsui, H. L. Störmer, and A. C. Gossard, Phys. Rev. Lett. 48, 1559 (1982).

[2] A. M. Chang, P. Berglund, D. C. Tsui, H. L. Störmer, and J. C. M. Hwang, Phys. Rev. Lett. 53, 997 (1984).

[3] R. B. Laughlin, Phys. Rev. Lett. 50, 1395 (1983).

[4] *The Quantum Hall Effect*, edited by S. M. Girvin (Springer, New York, 1987).

[5] D. Levesque, J. J. Weis, and A. H. MacDonald, Phys. Rev. B 30, 1056 (1984).

[6] R. B. Laughlin in Ref. 4.

[7] R. Morf and B. I. Halperin, Phys. Rev. B 33, 2221 (1986).

[8] R. Morf and B. I. Halperin, Z. Phys. B 68, 391 (1987).

[9] F. D. M. Haldane and E. H. Rezayi, Phys. Rev. Lett. 54, 237 (1985).

[10] F. D. M. Haldane, Phys. Rev. Lett. 55, 2095 (1985).

[11] R. Morf, N. d'Ambrumenil, and B. I. Halperin, Phys. Rev. B 34, 3037 (1986).

[12] P. A. Maksym, J. Phys. C 20, L25 (1987).

[13] F. D. M. Haldane, Phys. Rev. Lett. 51, 605 (1983).

[14] B. I. Halperin, Phys. Rev. Lett. 52, 1583 (1984).

[15] N. d'Ambrumenil and A. M. Reynolds, J. Phys. C 21, 119 (1988).

[16] R. G. Clark, J. R. Mallett, S. R. Haynes, J. J. Harris, and C. T. Foxon, Phys. Rev. Lett. 60, 1747 (1988).

[17] R. B. Laughlin, M. L. Cohen, J. M. Kosterlitz, H. Levine, S. B. Libby, and A. M. M. Pruisken, Phys. Rev. B 32, 1311 (1985).

[18] F. C. Zhang, Phys. Rev. B 34, 5598 (1986).

[19] B. I. Halperin, Helv. Phys. Acta 56, 75 (1983).

[20] N. d'Ambrumenil and R. Morf (unpublished).

[21] G. Fano and F. Ortolani, Phys. Rev. B 37, 8179 (1988).

[22] B. Jancovici, Phys. Rev. Lett. 46, 386 (1981).

[23] The extrapolation of the energy of a quasihole and a quasiparticle at $\nu=\frac{1}{3}$, as performed in Ref. 8, was incorrect, since it was based on the assumption that the energy E/N of a system containing one quasiparticle or quasihole could reliably be fitted by a polynomial in N^{-1}. Subsequently, we discovered that the presence of a term scaling as $N^{-3/2}$ had prevented reliable extrapolation of the quasiparticle and quasihole energy. We give the correct estimates here. For the Laughlin quasihole we find $\varepsilon=0.029(3)$, for the Laughlin quasiparticle we find $\varepsilon=0.080(4)$, and for the "pair" approximation to the quasiparticle (see Ref. 8) we find $\varepsilon=0.076(3)$.

[24] G. Fano, F. Ortolani, and E. Colombo, Phys. Rev. B 34, 2670 (1986).

[25] G. D. Watkins and J. R. Troxell, Phys. Rev. Lett. 44, 593 (1980).

[26] S. M. Girvin, A. H. MacDonald, and P. M. Platzman, Phys. Rev. B 33, 2481 (1985).

[27] F. D. M. Haldane, in Ref. 4, p. 303.

[28] A. M. Reynolds and N. d'Ambrumenil, J. Phys. C 21, 5643 (1988).

[29] S. M. Girvin in Ref. 4, p. 377.

[30] R. L. Willett, H. S. Störmer, D. C. Tsui, A. C. Gossard, and J. H. English, Phys. Rev. B 37, 8476 (1988).

PHYSICAL REVIEW B VOLUME 43, NUMBER 10 1 APRIL 1991

Structure of the microscopic theory of the hierarchical fractional quantum Hall effect

B. Blok and X. G. Wen
School of Natural Sciences, Institute of Advanced Study, Princeton, New Jersey 08540
(Received 27 September 1990)

The quantum numbers of the quasiparticles in the hierarchical fractional quantum Hall (FQH) states are derived with the use of microscopic electron wave functions. We discuss the relation between the effective theories of the hierarchical FQH effect and microscopic electronic wave functions first suggested by Haldane and Halperin. The rich internal structures (the topological orders) in the FQH states, which were first revealed in effective theory, are further confirmed by the microscopic theory.

I. INTRODUCTION

The characterization and the classification of all possible fractional quantum Hall effect (FQHE) states is one of the main goals of the complete theory of the fractional quantum Hall effect. As one can see from Refs. 1–3, the FQH states contain very rich topological structures that have not been noticed before. The investigation of this topological structure (the topological orders in Refs. 4 and 5) in the FQH states, in particular, the statistics and quantum numbers of the quasiparticles, is one of the most effective approaches toward achieving this goal. It is possible that the classification of all topological orders is equivalent to the classification of the possible quantum numbers of quasiparticles.

Two approaches are known toward the investigation of the FQHE states. The first one is the microscopic approach based on the Laughlin variational electron wave function (see, e.g., Ref. 6 and references therein) and its various generalizations toward the hierarchical FQH states.[7-14] The second approach is through the effective theory that describes the FQHE state in terms of the macroscopic order parameter. Such a theory was developed in Refs. 15–18 for the case of Laughlin state with the filling fraction $\nu=1/m$. The generalization to the case of the hierarchical FQH effect was made in Refs. 1, 19, 20, and 2. The effective theories reveal rich and highly nontrivial topological orders in the FQH states.

An obvious problem is the equivalence between these two approaches. The effective action that describes the Laughlin state was first derived from the microscopic theory in Ref. 17. The next obvious problem is to establish the correspondence between the microscopic and macroscopic approaches for the hierarchical FQH states.

The purpose of this paper is to derive the effective theory of the hierarchical FQHE from an analysis of the many-electron wave functions. We shall consider in this paper the standard hierarchical FQH states described by the Halperin-Haldane wave functions. The explicit form of these wave functions will be derived using fractional-statistics transformation.[6] In particular, we shall derive the quantum numbers of the quasiparticles directly from the electronic wave functions. We shall also establish the correspondence between the gauge fields of the effective theory and the operators that act on the electron wave functions.

The paper is organized in the following way. In Sec. II we review the fractional statistics transformation and explicitly construct the electronic ground-state wave function. We derive the form of the quasiparticle creation operators in the hierarchical states based on some physical considerations. We discuss two convenient choices of the quasiparticle creation operators. Using the first basis of the operators, we can easily find the quantum numbers of the quasiparticles. The operators of the second basis are in one-to-one correspondence with the fields of the effective theory proposed in Ref. 2. In Sec. III we argue that the hierarchical FQH state is related to a multicomponent plasma. The ground-state densities of the condensed quasiparticles at various levels are calculated using the plasma analog. In Sec. IV we explicitly establish the form of the quasiparticle creation operators in the second basis. We determine the charges and statistics of the quasiparticles in the FQH states described by the Haldane-Halperin wave functions using this basis of operators. We find that these quantum numbers coincide with the quantum numbers of the quasiparticles derived using the effective theory in Ref. 2. We also discuss the correspondence between the Berry phase and the electric charges of the quasiparticles. In Sec. V we identify the gauge fields of the macroscopic theory with vacuum averages of certain operators of the microscopic theory. The equations of motion for these gauge fields are obtained. In this way we derive the effective theory (in the dual form[18]) of the hierarchical FQH effect discussed in Ref. 2 from the microscopic theory. In Sec. VI we summarize our results.

II. FRACTIONAL-STATISTICS TRANSFORMATION AND THE ELECTRONIC WAVE FUNCTIONS

In this section we will review the fractional-statistics transformation[7,6] (FST) and construct the ground-state and excited-state wave functions for the hierarchical FQH state. Then we will calculate the quantum numbers of the quasiparticles.

The ground-state wave function of the hierarchical FQH states is obtained using the following physical picture.[21,7] We start from the Laughlin state with the filling fraction $1/p_1$ and assume that the quasiparticles added to this state condense after being combined with the flux tubes each carrying p_2 units of flux. As a result, we get a second-level hierarchical state described by the two components of incompressible fluids. The first component is the original electron incompressible fluid. The second component is the condensate of vortices in the Laughlin state. We can continue this procedure for an arbitrary level. We assume that quasiparticles in the nth incompressible-fluid component of the nth-level hierarchy are combined with a flux tube each carrying p_{n+1} units of flux. Then, we let the combined quasiparticles condense. This new state has $n+1$ incompressible-fluid components. The last component is the condensate of the quasiparticles in the nth component of the nth hierarchical state. The explicit form of the electron wave functions of the hierarchical states can be obtained using the FST. First, as an example, let us concentrate on the level-two hierarchical FQH states.

The first-level hierarchical state, the Laughlin state, is described by the following wave function:

$$|\Phi_1\rangle=\frac{1}{K_0}\prod_{i<j;ij=1}^{N_e}(\xi_i-\xi_j)^{p_1}\prod_{i=1}^{N_e}\exp(-\tfrac{1}{4}|\xi_i|^2)\ . \tag{1}$$

The state with N_1 quasiholes is given by

$$|z_1,\ldots,z_{N_1};\Phi_1\rangle=\frac{1}{K_{N_1}}\prod_{i=1}^{N_1}W(z_i)|\Phi_1\rangle\ ,$$
$$W(z)=\prod_{i=1}^{N_e}(z-\xi_i)\ , \tag{2}$$

where K_{N_1} is the normalization factor. With the above choice of $W(z)$, one finds that

$$K_1^2=\langle\Phi_1|W(z)^\dagger W(z)|\Phi_1\rangle$$

is independent of z. However, due to some special correlations in the FQH states K_{N_1} for $N_1>1$ is not constant. It was shown that Ref. 6 that

$$K_{N_1}\simeq\text{const}\prod_{i<j;i,j=1}^{N_1}|z_i-z_j|^{-1/p_1} \tag{3}$$

when the z_i's are well separated. We would like to emphasize that this result reflects very special correlations in the FQH states. Despite the fact that the single-particle state

$$\exp\left[-\frac{1}{4p_1}|z|^2\right]W(z)|\Phi_1\rangle$$

is properly normalized, the two-particle state

$$\exp\left[-\frac{1}{4p_1}(|z|^2+|z'|^2)\right]W(z)W(z')|\Phi_1\rangle$$

is not normalized to one. The norm is proportional to $|z-z'|^{-1/p_1}$. This is remarkable because all the correlations between the electron operators have a finite correlation length.

When the quasihole moves along a loop C, it is influenced by the Berry's phase ϕ_C:[22]

$$\phi_C\equiv i\oint_C d\mathbf{x}\cdot\langle\Phi_1;z|\partial_z z;\Phi_1\rangle=\frac{1}{p_1}S/l_B^2\ , \tag{4}$$

where l_B is the magnetic length and S is the area enclosed by the loop C. In this paper we choose our units such that $l_B=1$. The quasiholes also carry fractional statistics $\theta_1=-\pi/p_1$, i.e., in the presence of a second quasihole the Berry's phase becomes

$$\phi\equiv i\oint_C d\mathbf{x}\cdot\langle\Phi_1;z,z'|\partial_z|z,z';\Phi_1\rangle$$
$$=\frac{1}{p_1}\frac{S}{l_B^2}+\eta(z',C)2\theta_1\ , \tag{5}$$

where $\eta(z',C)$ is the number of times the loop C winds around z'. Equations (4) and (5) indicate that the dynamics of the quasiholes is governed by the following effective Hamiltonian:

$$H_{\text{eff}}=\sum_j\frac{1}{m^*}\left[-i\partial_{r_j}-\frac{1}{p_1}e\,\mathbf{A}(\mathbf{r}_j)-\mathbf{a}_j\right]^2+\sum_{i,j}V(\mathbf{r}_i-\mathbf{r}_j)\ , \tag{6}$$

where

$$a_{j,x}+ia_{j,y}=\frac{1}{p_1}\sum_{i=1;i\neq j}^{N_1}\frac{1}{z_j-z_i}\ . \tag{7}$$

$\mathbf{a}_j$ describe the fractional statistics of the quasihole. Note that the wave function of the quasiholes is single valued in this description. At the proper filling fractions, the ground state of the effective Hamiltonian is given by

$$|\Phi_2\rangle=\int\prod_i d^2z_i\tilde{\psi}_p(z_i)|z_1,\ldots,z_{N_1};\Phi_1\rangle\ ,$$

$$\tilde{\psi}_p(z_i)=\int\prod_i d^2z_i\prod_{i<j}\left[\frac{z_i-z_j}{|z_i-z_j|}\right]^{\theta_1/\pi}\psi_p(z_i)\ , \tag{8}$$

$$\psi_p=\prod_{i<j}(z_i^*-z_j^*)^{\theta_1/\pi+p_2}\prod_i\exp\left[-\frac{q_1}{4}|z_i|^2\right]\ ,$$

where p_2 is an even integer and $q_1=1/p_1$ is the charge of the quasihole. The wave function ψ_p in Eq. (8) is the so-called pseudo-wave-function of the quasiholes. Equation (8) describes the second-level hierarchical FQH state. Note $\tilde{\psi}$ is the ground-state wave function of H_{eff} if $V(z_i-z_j)$ is short ranged:

$$\prod_{i<j}\left[\frac{z_i-z_j}{|z_i-z_j|}\right]^{-\theta_1/\pi}$$

is a singular gauge transformation that removes $\mathbf{a}_j$ from the effective Hamiltonian. This allows ψ_p to have the simple standard form. From Eqs. (2), (3), and (8), we can construct the explicit electron wave function for such a hierarchical state:

$$|\Phi_2\rangle=\int\prod_{i=1}^{N_1}d^2z_i\tilde{\psi}_p(z_i)\frac{1}{K_{N_1}}\prod_{i=1}^{N_1}W(z_i)\prod_{i<j;i,j=1}^{N_e}(\xi_i-\xi_j)^{p_1}\prod_i^{N_e}\exp(-1/4|\xi_i|^2)$$

$$=\prod_{i<j;i,j=1}^{N_e}(\xi_i-\xi_j)^{p_1}\prod_i^{N_e}\exp(-1/4|\xi_i|^2)\int\prod_{i=1}^{N_1}d^2z_i\prod_{i,j=1}^{i=N_e,j=N_1}(\xi_i-z_j)$$

$$\times\prod_{i<j;i,j=1}^{N_1}|z_i-z_j|^{2/p_1}(z_i^*-z_j^*)^{p_2}\prod_i\exp\left[-\frac{q_1}{2}|z_i|^2\right].\quad(9)$$

The filling fraction of this state is $1/(p_1+1/p_2)$. The transformation from the pseudo-wave-function ψ_p to the above electron wave function is called the fractional-statistics transformation.

Now let us discuss the low-lying excitations in the hierarchy state (9). From the effective Hamiltonian (6) we see that the pseudo-wave-function for the low-lying excitations should satisfy the following two conditions: (a) the total wave function $\tilde{\psi}_p$ in (8) must be single valued and (b) the algebraic factor in the pseudo-wave-function must be antiholomorphic. The simplest excitation that satisfies conditions (a) and (b) has the pseudo-wave-function

$$\psi_p^{(2)}(z_0;z_i)=\prod_i(z_0^*-z_i^*)\prod_{i<j}(z_i^*-z_j^*)^{1/p_1+p_2}\times\prod_i\exp\left[-\frac{q_1}{4}|z_i|^2\right].\quad(10)$$

This excitation is called the second-level excitation. Applying the plasma analog to the anyon pseudo-wave-function, we find that this excitation carries the electric charge $q_2=(1/p_1p_2+1)$ and the statistics

$$\theta_2=\pi\frac{p_1}{p_1p_2+1}$$

as expected.[21,7]

We can also construct another type of excitations called the first-level excitations. We start with the state that contains N_1+1 quasiholes, $|z_0,z_1,\ldots,z_{N_1};\Phi_1\rangle$. One of the quasiholes is held at point z_0 while other N_1 quasiholes condense into a Laughlin state. The pseudo-wave-function for the excitation is given by

$$\psi_p^{(1)}(z_0;z_i)=\prod_i(z_0^*-z_i^*)^{1/p_1+m}\times\prod_{i<j}(z_i^*-z_j^*)^{1/p_1+p_2}\times\prod_i\exp\left[-\frac{q_1}{4}|z_i|^2\right].\quad(11)$$

Note that this wave function acquires a phase $-2\pi/p_1$ when other quasiholes move around the fixed quasihole. The prefactor

$$\prod_i(z_0^*-z_i^*)^{1/p_1+m}$$

is needed in order for the wave function to satisfy conditions (a) and (b). We may take $m=p_2$ for the convenience. The fixed quasihole at point z_0 is dressed by a cloud of other condensed quasiholes. The quantum numbers of the excitation come from both the bear quasihole and the cloud. Using the plasma analog we can show that the total charge and the total statistical angle of the excitation are zero. The contributions from the bear quasihole and from the cloud cancel each other. Since the density of the condensed quasiholes is

$$n_1=l_B^{-2}/(1+p_1p_2)\,,$$

the first level excitation induces no Berry's phase as it is moving around a loop. The reason is that

$$\phi_C=\left[\frac{1}{p_1}-n_1\left[\frac{1}{p_1}+p_2\right]\right]S=0\,.$$

Furthermore, one can also show that the Berry's phase induced by moving a first-level excitation around a second-level excitation is 2π and, hence, trivial. Therefore, the first-level excitations are neutral and carry no known quantum numbers.

The electron wave functions corresponding to the two types of excitations are given by

$$|z_0;\Phi_2\rangle=\exp\left[-\frac{q_2}{4}|z_0|^2\right]\int\prod_{i=1}^{N_1}d^2z_i\prod_{i,j}^{N_e}(\xi_i-\xi_j)^{p_1}\prod_i^{N_e}\exp(-\tfrac{1}{4}|\xi_i|^2)\prod_{i=1}^{N_1}(z_0^*-z_i^*)$$

$$\times\prod_{i,j=1}^{i=N_e,j=N_1}(\xi_i-z_j)\prod_{i<j;i,j=1}^{N_1}|z_i-z_j|^{2/p_1}(z_i^*-z_j^*)^{p_2}\prod_i\exp\left[-\frac{q_1}{2}|z_i|^2\right]\quad(12)$$

and

$$|z_0';\Phi_2\rangle=\int\prod_{i=1}^{N_1}d^2z_i\prod_{i<j;i,j=1}^{N_e}(\xi_i-\xi_j)^{p_1}\prod_{i=1}^{N_e}\exp(-\tfrac{1}{4}|\xi_i|^2)$$

$$\times\prod_{i=1}^{N_1}|z_0^+-z_i^*|^{2/p_1}(z_0^*-z_i^*)^{p_2}\prod_{i=1}^{N_e}(z_0'-\xi_i)\exp\left[-\frac{q_1}{2}|z_0'|^2\right]$$

$$\times\prod_{i,j=1}^{i=N_e,j=N_1}(\xi_i-z_j)\prod_{i<j;i,j=1}^{N_1}|z_i-z_j|^{2/p_1}(z_i^*-z_j^*)^{p_2}\prod_i\exp\left[-\frac{q_1}{2}|z_i|^2\right], \quad (13)$$

respectively. Thus, the operators that create the two types of the excitations are

$$W_2(z_0)=\prod_{i=1}^{N_1}(z_0^*-z_i^*),$$
$$W_1(z_0)=\prod_{i=1}^{N_1}|z_0^*-z_i^*|^{2/p_1}(z_0^*-z_i^*)^{p_2}\prod_{i=1}^{N_e}(z_0-\xi_i). \quad (14)$$

We see that the creation operator for the first-level excitations is quite complicated due to the complicated form of the fractional-statistics transformation.

We would like to make the following remark. Although we treat W operators as quasiparticle creation operators, strictly speaking, we cannot exclude the possibility that the state created by the operator W_1 is just the ground state (plus some particle-hole excitations). This is because W_1 carries trivial quantum numbers. If W_1 was proportional to identity, the first-level excitations simply would not exist. At the moment it is not clear whether the states created by W_1 are always orthogonal to the ground state or not. However, we would like to point out that in the integral quantum Hall (IQH) states, the holes created in different Landau levels are different. Those holes can be transformed into each other by neutral operators. In this case, such neutral operators do correspond to quasiparticle excitations. From this point of view and the relation[2,23] between the hierarchy FQH states and the IQH states, we expect W_1 to be a quasiparticle operator.

The above results for the quantum numbers of the excitations is consistent with the results obtained from the effective theory.[2] However, we obviously have chosen the different basis for the quasiparticles. The operators that create the quasiparticles in the effective theory are actually given by the U operators:

$$U_2(z_0)=W_2(z_0)=\prod_{i=1}^{N_1}(z_0^*-z_i^*),$$
$$U_1(z_0)=\prod_{i=1}^{N_1}|z_0^*-z_i^*|^{2/p_1}\prod_{i=1}^{N_e}(z_0-\xi_i). \quad (15)$$

U_1 is obtained by choosing $m=0$ in definition (11). Later we will show that the quantum numbers of the U operators coincide with the quantum numbers of the quasiparticles in the effective theory. Notice that U and W operators are related through

$$W_2=U_2,$$
$$W_1=U_1U_2^{p_2}. \quad (16)$$

The above discussions can be easily generalized to the higher-level hierarchical states. Let us discuss the general case inductively. Let $|\Phi_n\rangle$ be the ground state of the nth-level hierarchical state. $W_i^{(n)}$, $i=1,\ldots,n$ are creation operators of quasiparticle excitations. Let us assume that the quasiparticles created by the operators $W_i^{(n)}$, $i=1,\ldots,n-1$ carry trivial quantum numbers. Let us also assume that $W_n^{(n)}$ carries charge q_n, statistics θ_n, and has Berry's phase $\phi_C=q_nS/l_B^2$ [see Eq. (4)]. The $(n+1)$th-level hierarchical state is obtained by allowing the excitations created by $W_n^{(n)}$ to condense into a Laughlin state. We would like to show that, in the $(n+1)$th-level state, there are $n+1$ kinds of excitations, created by the operators $W_i^{(n+1)}$, $i=1,\ldots,n+1$. Among these $n+1$ kinds of excitations, only the excitations created by the operators $W_{n+1}^{(n+1)}$ carry nontrivial quantum numbers q_{n+1}, θ_{n+1}, and $\phi_C=q_{n+1}S/l_B^2$.

The effective Hamiltonian for the $W_n^{(n)}$ excitations in the nth-level hierarchical state has the form

$$H_{\text{eff}}=\sum_j\frac{1}{m^*}[-i\partial_{\mathbf{r}_j}-q_ne\,\mathbf{A}(\mathbf{r}_j)-\mathbf{a}_j]^2+\sum_{i,j}V(\mathbf{r}_i-\mathbf{r}_j), \quad (17)$$

where

$$a_{j,x}+ia_{j,y}=\frac{\theta_n}{\pi}\sum_{i\neq j}\frac{1}{z_j-z_i}. \quad (18)$$

The $(n+1)$th-level hierarchical state is given by the ground state of the Hamiltonian (17):

$$|\Phi_{n+1}\rangle=\int\prod_i d^2z_i\bar{\psi}_p(z_i)|z_i;\Phi_n\rangle$$
$$=\int\prod_i d^2z_i\prod_{i<j}\left[\frac{z_i-z_j}{|z_i-z_j|}\right]^{-\theta_n/\pi}\psi_p(z_i)|z_i;\Phi_n\rangle,$$

where

$$\psi_p=\prod_{i<j}(z_i^*-z_j^*)^{-(\theta_n/\pi)+p_{n+1}}$$
$$\times\prod_i\exp\left[-\frac{q_n}{4}|z_i|^2\right]\quad\text{for } q_n>0, \quad (19)$$

$$\psi_p=\prod_{i<j}(z_i-z_j)^{\theta_n/\pi+p_{n+1}}$$
$$\times\prod_i \exp\left[-\frac{q_n}{4}|z_i|^2\right] \quad \text{for } q_n<0 .$$

$|z_i;\Phi_n\rangle$ in (19) is the *normalized* state generated by $W_n^{(n)}(z_i)$:

$$|z_1,\ldots,z_{N_1};\Phi_n\rangle=\frac{1}{K_{N_n}}\prod_{i=1}^{N_n}\exp\left[-\frac{q_{n+1}}{4}|z_i|^2\right]\times W_n^{(n)}(z_i)|\Phi_n\rangle . \quad (20)$$

Note that, in general, K_{N_n} is a function of $|z_i-z_j|$. Since $W_n^{(n)}$ is the Laughlin quasiparticle in the last condensate, from the plasma analog one can show that

$$K_{N_n}=\prod_{i<j;i,j=1}^{N_n}|z_i-z_j|^{-1/[p_n-\text{sgn}(q_{n-1})\theta_{n-1}]} . \quad (21)$$

Starting from the Laughlin wave function (1) and assuming that the quasiholes condense at each level, we can obtain the explicit electron wave function of the nth-level state from the inductive relations (19)–(21). We find the wave function to be

$$\psi(\xi_1,\ldots,\xi_n)\sim\prod_{i,j}^{N_e}(\xi_i-\xi_j)^{p_1}$$
$$\times\int d^2\bar z_i^2\cdots\int d^2\bar z_i^n\prod_{s=2}^{s=n}\left[\prod_{i,j=1}^{N_s}|\bar z_i^s-\bar z_j^s|^{2\,\text{sgn}(q_{s-1})(\theta_{s-1}/\pi)}(\bar z_i-\bar z_j)^{p_s}\right.$$
$$\left.\times\prod_{i,j}^{N_s,N_{s+1}}(\bar z_i^s-\bar z_j^{*s+1})\exp(-\tfrac{1}{2}|q_s||\bar z_i^s|^2)\right]\exp(-1/4|\xi|^2) , \quad (22)$$

where $\bar z^s=z^s$ if s is odd and $\bar z^s=z^{*s}$ if s is even. Since we always have a quasihole condensate, we have $\text{sgn}(q_s)=-(-)^s$. Note that $q_0=-1$ is just the electron charge. The filling fraction of the above state is equal to

$$\nu=\cfrac{1}{p_1+\cfrac{1}{p_2+\cfrac{1}{\cdots+\cfrac{1}{p_n}}}} . \quad (23)$$

Now let us discuss the excitations in the general hierarchy states. The $(n+1)$th-level excitation in the $(n+1)$th-level state is just the Laughlin-type excitation in the pseudo-wave-function. It is obtained by multiplying a factor $\prod_i(z_0^*-z_i^*)$ [or $\prod_i(z_0-z_i)$] on the ground-state wave function assuming $q_n>0$ (or $q_n<0$):

$$W_{n+1}^{(n+1)}(z_0)|\Phi_{n+1}\rangle$$
$$=\int\prod_i d^2z_i\prod_i(z_0^*-z_i^*)\tilde\psi_p(z_i)|z_i;\Phi_n\rangle . \quad (24)$$

This excitation carries charge and statistics

$$q_{n+1}=-\frac{q_n}{p_{n+1}-\text{sgn}(q_n)(\theta_m/\pi)} ,$$
$$\theta_{n+1}=\frac{\text{sgn}(q_n)}{p_{n+1}-\text{sgn}(q_n)(\theta_n/\pi)} . \quad (25)$$

The density of the condensed $W_n^{(n)}$ particles is $n_n=|q_{n+1}|/2\pi l_B^2$. As the $W_{n+1}^{(n+1)}$ excitation moves around a loop, it induces a Berry's phase

$$\phi_C=2\pi n_n S=q_{n+1}S/l_B^2 .$$

Another type of excitation (called the level-n excitation) is described by the wave function

$$W_n^{(n+1)}(z_0)|\Phi_{n+1}\rangle$$
$$=\int\prod_{i=1}^{N_n}d^2z_i\tilde\psi_p(z_0,z_1,\ldots,z_{N_n})|z_0,z_1,\ldots,z_{N_n};\Phi_n\rangle . \quad (26)$$

Note, z_0 is not integrated. Following the previous calculation one can again show that the above excitation carries the trivial quantum numbers. In particular, the Berry's phase induced by moving the $W_{n+1}^{(n+1)}$ particle around the $W_n^{(n+1)}$ is 2π and hence trivial. Notice that both wave functions in the integrands of Eqs. (24) and (26) are the eigenstates of the effective Hamiltonian (17) if $V(z_i-z_j)$ is really short ranged. It is this requirement that fixes the form of the quasiparticle creation operator.

The particles created by the operators $W_i^{(n)}$, $i=1,\ldots,n-1$ behave like flux tubes with integer numbers k_i^n of flux. At the $(n+1)$th level these operators become

$$W_i^{(n+1)}(z_0)|\Phi_{n+1}\rangle=\int \prod_{i=1}^{N_n} d^2 z_i \prod_{i=1}^{N_n}\left[\frac{Z_i-z_0}{|z_i-z_0|}\right]^{k_i^n} \bar{\psi}_p(z_1,\ldots,z_{N_n})|z_0;z_1,\ldots,z_{N_n};\Phi_n\rangle,\quad i<n\ , \tag{27}$$

where

$$|z_0;z_1,\ldots,z_{N_n};\Phi_n\rangle=\frac{1}{K}W_i^{(n)}(z_0)|z_1,\ldots,z_{N_n};\Phi_n\rangle\ . \tag{28}$$

K is a normalization factor that can depend on $|z_0-z_i|$. Again, one can check that the state described by Eq. (26) is the eigenstate of the effective Hamiltonian but with a flux tube inserted at z_0. It is easy to see that the above excitations do not carry any electric charge or statistics.

The U operators that correspond to the quasiparticles in the effective theory of Ref. 2 are defined by the following inductive relations: First, $U_{n+1}^{(n+1)}=W_{n+1}^{(n+1)}$. The operator $U_n^{(n+1)}$ is given by

$$U_n^{(n+1)}(z_0)|\Phi_{n+1}\rangle=\int \prod_{i=1}^{N_n} d^2 z_i \prod_i |z_0-z_i|^{-\mathrm{sgn}(q_n)(\theta_n/\pi)}\bar{\psi}_p(z_1,\ldots,z_{N_n})|z_0,z_1,\ldots,z_{N_n};\Phi_n\rangle\ . \tag{29}$$

$U_n^{(n+1)}$ and $W_n^{(n+1)}$ only differ by a holomorphic factor $\prod(z_0-z_i)^{p_n}$ or an antiholomorphic factor $\prod(z_0^*-z_i^*)^{p_n}$ depending on the sign of q_n. The operator $U_i^{(n+1)}$, $i<n$, is given by factor $\prod(z_0^*-z_i^*)^{p_n}$ depending on the sign of q_n. The operator $U_i^{(n+1)}$, $i<n$, is given by

$$U_i^{(n+1)}(z_0)|\Phi_{n+1}\rangle=\int \prod_{i=1}^{N_n} d^2 z_i \prod_i |z_0-z_i|^{-\mathrm{sgn}(q_n)(\theta_i^n/2\pi)}\bar{\psi}_p(z_1,\ldots,z_{N_n})|z_0;z_1,\ldots,z_{N_n};\Phi_n\rangle,\quad i<n\ , \tag{30}$$

where

$$|z_0;z_1,\ldots,z_{N_n};\Phi_n\rangle=\frac{1}{K_i^{(n)}}W_i^{(n)}(z_0)|z_1,\ldots,z_{N_n};\Phi_n\rangle\ . \tag{31}$$

$K_i(n)$ is a normalization factor that can depend on z_i and z_0. In Eq. (29), θ_i^n is the Berry's phase induced by moving the $U_n^{(n)}$ particle around a $U_i^{(n)}$ particle. The wave function in the integral is an eigenstate of the effective Hamiltonian with a flux tube of the strength θ_i^n inserted at z_0. The factor

$$\prod_i |z_0-z_i|^{[-\mathrm{sgn}(q_n)\theta_i^n]/2\pi}$$

in the wave function is due to this flux tube. After the singular gauge transformation, this factor becomes holomorphic or antiholomorphic depending on the sign of q_n. Writing Eqs. (26) and (27) explicitly, we have

$$U_i^{(n+1)}(z_0)=\prod_i |z_0-z_i|^{-\mathrm{sgn}(q_n)(\theta_i^n/2\pi)} \times\frac{1}{K_i^{(n)}(z_0,z_i)}U_i^{(n)}(z_0)\ ,\quad i=1,\ldots,n\ . \tag{32}$$

Here $\theta_n^n=2\theta_n$. Later we will use Eqs. (32) to obtain the explicit form of the U operators.

Summarizing the above discussion, we have shown that the quasiparticle excitations in the nth-level hierarchical FQH state are created by n operators, $W_i^{(n)}|_{i=1,\ldots,n}$ or $U_i^{(n)}|_{i=1,\ldots,n}$, and their combinations. In the W basis, only the last operator $W_n^{(n)}$ carries nonzero electric charge and statistics. All other operators are neutral. The charge and the statistics of $W_n^{(n)}$ can be calculated inductively through Eqs. (25). This is just Halperin's results. The Berry phase of the $W_n^{(n)}$ particle is given by $\phi_C=q_n S/l_B^2$. A new result we obtained from the above discussions is that the quasiparticle $W_n^{(n)}$ of the highest level is the only quasiparticle that carries the nontrivial charge and statistics. The vortices in the previous condensates do not give rise to the quasiparticles with the new charges and statistics. This is a special nontrivial property of the hierarchical FQH states. Other FQH states, like some of Jain's states,[13,1] do not have this property.

Using (24) and (32), we can inductively calculate the explicit form of the quasiparticle operators $U_i^{(n)}$ in the electron wave function. The explicit forms of $U_i^{(n)}$ contain various normalization factors and are quite complicated. We will postpone the calculation to Sec. IV. Using the explicit form of the operators, we can calculate their quantum numbers directly. We will show that these quantum numbers coincide with those obtained in the effective theory. The quasiparticles discussed in the effective theory are created by the U operators in the microscopic theory.

III. MULTICOMPONENT PLASMA AND GROUND-STATE PROPERTIES IN THE HIERARCHICAL STATE

In this section, we would like to generalize the plasma analog given by Laughlin for the FQH states with the filling fraction $\nu=1/p_1$ to the case of the hierarchical states. Let us consider the norm of the electronic wave function (22):

$$Z=\int d^2\xi_1\cdots d^2\xi_{N_e}|\psi(\xi_1,\ldots,\xi_n)|^2$$
$$=\int d^2\xi_i\int d^2\bar{z}_i^2\cdots\int d^2\bar{z}_i^n\int d^2\bar{z}_i'^2\cdots\int d^2\bar{z}_i'^n\prod_{i,j}^{N_e}|\xi_i-\xi_j|^{2p_1}$$
$$\times\prod_{s=2}^{s=n}\left[\prod_{i,j=1}^{N_s}|\bar{z}_i^s-\bar{z}_j^s|^{2\,\mathrm{sgn}(q_s-2)(\theta_{s-1}/\pi)}(\bar{z}_i^s-\bar{z}_j^s)^{p_s}\prod_{i,j}^{N_s,N_s+1}(\bar{z}_i^s-\bar{z}_j^{*s+1})\right.$$
$$\left.\times\prod_{i,j=1}^{N_s}|\bar{z}_i'^s-\bar{z}_j'^s|^{2\,\mathrm{sgn}(q_s-2)(\theta_{s-1}/\pi)}(\bar{z}_i'^s-\bar{z}_j'^s)^{p_s}\prod_{i,j}^{N_s,N_s+1}(\bar{z}_i'^s-\bar{z}_j'^{*(s+1)})\right]$$
$$\times\prod_i^{N_s}\exp[-1/2|q_s|(|z_s^i|^2+|\bar{z}_i'^s|^2)]\prod_i\exp(-1/2|\xi_i|^2)\,. \tag{33}$$

It is easy to see that Eq. (33) can be rewritten in the form

$$Z=\int\cdots\int\prod d^2\xi_i d^2\bar{z}_i'^s d^2\bar{z}_j^s\exp[-\beta F(\xi_i,z_i^s,z_i'^s)]\,, \tag{34}$$

where β is the inverse temperature. Below we shall set β equal to one. F is given by

$$F=\sum_s\sum_{i,j}\left[\mathrm{sgn}(q_{s-2})\frac{\theta_{s-1}}{\pi}+p_s\right](\ln|\bar{z}_i^s-\bar{z}_j^s|+\ln|\bar{z}_i'^s-\bar{z}_j'^s|)$$
$$+\sum_{i,j,s}(\ln|\bar{z}_i^s-\bar{z}_i^s-\bar{z}_j^{s+1}|+\ln|\bar{z}_i'^s-\bar{z}_j'^{(s+1)}|)+\sum_i -1/2|\xi_i|^2+\sum_{i,s}-1/4|q_s|(|\bar{z}_i^s|^2+|\bar{z}_i'^s|^2)$$
$$+\sum_{i,j}\theta_{ij}(\bar{z}_i^s-\bar{z}_j^s)-\theta_{ij}(\bar{z}_i'^s-\bar{z}_j'^s)+\theta(\bar{z}_i^s-\bar{z}_j^{s+1})-\theta(\bar{z}_i'^s-\bar{z}_j'^{(s+1)})\,. \tag{35}$$

Here $\theta_{ij}(z_i-z_j)$ is the angle between the x axis and z_i-z_j:

$$\theta_{ij}=\arg(z_i-z_j)\,\mathrm{mod}2\pi\,. \tag{36}$$

Expression (35) is the partition function of a multicomponent plasma with a long-range Coulombic interaction $\ln(r_{ij})$ between particles i,j and an additional angular-dependent interaction θ_{ij}. The latter interaction is similar to the monopole-electric charge interaction in the generalized two-dimensional Coulombic gas (see, e.g., Ref. 24 for the review of the properties of the two-dimensional Coulombic gas). This multicomponent plasma has $2n-1$ components. The coordinates z_i^s belong to the ith particle in the sth component of the plasma and the coordinates $z_i'^{s'}$ belong to the ith particle in the s'th component of the plasma. Here s,s' varies from 2 to n and the first component of this plasma corresponds to the electrons. Let us note that this plasma differs from the usual Coulombic plasma by its nonuniversality. While the quasiparticles of each component interact with each other as having the electric charge

$$e_s^2=2\,sgn(q_{s-2})\frac{\theta_{s-1}}{\pi}+p_s\,, \tag{37}$$

the quasiparticles of the sth and $s+1$ types interact with each other with a Coulombic charge one (not $e_s e_{s+1}$). Moreover, the quasiparticles of the sth type (i.e., forming the sth component of the incompressible fluid) interact only with the quasiparticles of the $(s-1)$th and $(s+1)$th types.

Note also that there are two types of quasiparticles (or two plasma components) that naturally correspond to each component of the incompressible fluid (except the electronic component). The quasiparticles of the first type have coordinates z_i, the quasiparticles of the second type have coordinates z_j'. These quasiparticles look the same as far as we consider the Coulombic interaction. However, they carry opposite fluxes.

We shall now show that the main contribution to the integral (35) comes from the region $z_i'^s\sim z_i^s$ of integration. Let us begin from the case of the Laughlin state with two vortices. The correlation function of the two vortices is given by

$$G(z_a,z_b)=\int\prod_i d^2\xi_i\prod_i(z_a^*-z_i^*)\prod_i(z_b-\xi_i)\prod_{i<j}|\xi_i-\xi_j|^{2m}\prod_i\exp(-1/2|\xi_i|^2)\exp\left[-\frac{1}{4m}|z_\alpha|^2-\frac{1}{4m}|z_b|^2\right]. \tag{38}$$

We can now integrate over the electrons in Eq. (38). This was done by Laughlin in Ref. 6. He found

$$G(z_a,z_b)\sim\exp\left[-\frac{1}{4m}(|z_a|^2-|z_b|^2+2z_a^*z_b)\right]$$
$$=\exp\left[\frac{1}{4m}[-|z_a-z_b|^2+i\,\mathrm{Im}(z_bz_a^*-z_az_b^*)]\right]. \tag{39}$$

The Gaussian factor $\exp-(|z_a-z_b|^2)/4m$ indicates that the main contribution into the integral

$$Z=\int d^2z_ad^2z_bG(z_a,z_b) \tag{40}$$

comes from the region $z_a\sim z_b$ of integration. If we view Z as a partition function of a system of two particles, then Eq. (39) indicates that there is a confining interaction between these two particles. The confining potential is proportional to R^2, where R is the distance between two quasiparticles. The analogous arguments can be given for the general hierarchical state. The main contribution into the integral (35) comes from the region of integration $z_i^s\sim z_i^{\prime s}$. In other words, the particles of type s and s' are confined in pairs due to the R^2 interaction between them ($R=|z_i^s-z_i^{\prime s}|$). Hence, we can regard the $(2n-1)$-component plasma described above as the n-component plasma plasma that consists of the strongly bound molecules of the particles of the types s and s'. This new plasma has n components, each component corresponds to the incompressible fluid component of the hierarchical FQH state. Note that the phase terms $\theta_{ij}(z_i^s-z_j^{\prime s})$ cancel each other if we put $z_i^s=z_i^{\prime s}$. This plasma contains the particles that interact with each other only through the logarithmic Coulombic interaction. Below we shall always consider this n-component plasma when we speak about the plasma analog. As an approximation, we will assume the s-type and the s'-type particles are tightly bound and set $z_i^s=z_i^{\prime s}$ in Eq. (34). The partition function of the new plasma that we still denote by Z is given by

$$Z=\int\cdots\int\prod d^2\xi_id^2\bar{z}_i^s\exp\left[\sum_s\sum_{i,j}\left[2\,sgn(q_{s-2})\frac{\theta_{s-1}}{\pi}+p_s\right](\ln|z_i^s-z_j^s|+\ln|\bar{z}_i^s-\bar{z}_j^{*(s+1)}|)\right]$$
$$\times\exp\left[-\tfrac{1}{2}\sum_i|\xi_i|^2-\sum_{i,s}|q_s||z_i^s|^2\right]. \tag{41}$$

Now we shall use the plasma analog to calculate the densities of the incompressible fluid components in the hierarchical state. These densities are the densities of the different condensates of the quasiparticles that take part in the formation of the hierarchical state.

In order to calculate these densities, we note that the plasma is stable only if the total force acting on every particle in the plasma is zero. This condition gives the following system of equations:

$$p_1n_1+n_2=\rho_0\,,$$
$$\vdots$$
$$n_{s-1}+\left[2\,sgn(q_{n-2})\frac{\theta_{s-1}}{\pi}+p_s\right]n_s+n_{s+1}=2|q_{s-1}|\rho_0\,,$$
$$\vdots$$
$$n_{n-1}+\left[2\,\mathrm{sgn}(q_{n-2})\frac{\theta_{n-1}}{\pi}+p_n\right]n_n=2|q_{n-1}|\rho_0\,, \tag{42}$$

where $\rho_0=1/2\pi$ is the density of a filled first Landau level. The three terms on the left-hand side of each of Eqs. (42) correspond to the interaction of a given particle of the sth type with the other particles of the same type and with the particles of the types $s-1$ and $s+1$. The rhs of Eqs. (42) corresponds to the interaction of the particle with the neutralizing background. In particular, for the case of the second-level hierarchical state with the filling fraction $\nu=1/[p_1+(1/p_2)]$, we obtain

$$(2/p_1+p_2)n_2+n_1=2/p_1\rho_0\,,$$
$$p_1n_1+n_2=\rho_0\,. \tag{43}$$

After some simple algebra it is easy to prove that Eqs. (43) can be rewritten in the form:

$$p_1n_1+n_2=\rho_0\,,$$
$$n_1-p_2n_2=0\,. \tag{44}$$

This is exactly the equation for the ground-state condensate densities that was obtained using the effective theory for the second-level hierarchy state in Refs. 19 and 20. Solving Eqs. (44), it is easy to find that

$$n_1=\frac{p_2\rho_0}{p_1p_2+1}\,,$$
$$n_2=\frac{\rho_0}{p_1p_2+1}\,. \tag{45}$$

This result is in full agreement with the prediction of the effective theory given in Ref. 2. The general equations (42) can be written after some algebra in the form

$$\Lambda_{ij}n_j=\delta_{i1}\rho_0,\quad i=1,\ldots,n\,. \tag{46}$$

Here the matrix Λ is given by

$$\Lambda_{ij}=\begin{bmatrix} p_1' & 1 & \cdots & 0 & 0 & 0 \\ 1 & p_2' & -1 & 0 & 0 & \\ 0 & -1 & p_3' & 1 & 0 & \cdots \\ \vdots & \vdots & \vdots & \cdots & \cdots & (-1)^{n+1} \\ 0 & 0 & 0 & \cdots & (-1)^{n+1} & p_n' \end{bmatrix} . \tag{47}$$

The numbers p_i' are given by $p_i'=-(-1)^i p_i$. In order to derive Eq. (46), we need to use the explicit recurrence relations for θ_i and q_i given in Ref. 7 (see also Sec. II of this paper). These equations were already solved when we analyzed our effective theory in Ref. 2. We obtain

$$n_k=\frac{D_{n-k-1}'(p_{k+2},\ldots,p_n)}{D}p_0 , \tag{48}$$

where D_i' is the denominator of the continuous fraction (23). This result is consistent with the effective theory of Ref. 2.

IV. QUANTUM NUMBERS OF THE QUASIPARTICLES IN THE HIERARCHY STATES

In this section we will calculate the quantum numbers of the quasiparticles using the basis of operators U_i^n defined in Sec. II. First, we shall establish the explicit form of the operators U defined by Eq. (32). In order to find the explicit form of these operators, we need to find the normalization factor K in definition (32). Recall that this factor is defined from the asymptotic behavior of the correlation function

$$K^2(z_a,z_b)=\exp(-\tfrac{1}{2}q_a|z_a|^2-\tfrac{1}{2}q_b|z_b|^2)\langle\Phi_n|U_i^{\dagger n}(z_a)U_n^{\dagger n}(z_b)|U_i^n(z_a)U_n^n(z_b)|\Phi_n\rangle$$
$$\sim\frac{1}{|z_a-z_b|^{2\alpha_{in}}} . \tag{49}$$

q_a and q_b are determined from the condition

$$\exp(-\tfrac{1}{2}q_a|z_a|^2)\langle U_n^{\dagger n}(z_a)U_n^n(z_a)\rangle=\text{const}$$

and

$$\exp(-\tfrac{1}{2}q_b|z_b|^2)\langle U_i^{\dagger n}(z_b)U_i^n(z_b)\rangle=\text{const} .$$

The value of α_{in} can be found using the plasma analog discussed in the previous section. This is done by using definition (32) of the operator U and the explicit form of the operator $U_n^n=W_n^n$ that creates a quasiparticle in the last condensate. The latter operator acts on the electron wave function by the insertion of the operator $\prod(z_b-z_i^n)$ into the integrand in the definition of the electron wave function (22). Note that the quasiparticle with the coordinate z_a can be regarded as a test charge which only interacts with the particles in the last level (the nth level) and with the neutralizing background. If we write the right-hand side of (49) multiplied on $K^2(z_a,z_b)$ explicitly, we see that the free energy of the corresponding plasma gas contain terms of form

$$\sum_j \ln|z_a-z_j^n|+\alpha_{ni}\ln|z_a-z_b|-\tfrac{1}{4}q_a|z_a|^2$$

which depend on z_a. From the demand that the force that acts on the z_0 quasiparticle is zero, it is straightforward to obtain that

$$\alpha_{in}=l_n\Delta q_n^i(n) . \tag{50}$$

Here

$$\Delta q_p^i(n)=\int d^2x\,\Delta n_p^i(n) \tag{51}$$

and $\Delta n_p^i(n)$ change of the density of the pth condensate for the level-n hierarchical state when we create a quasiparticle in the ith condensate (i.e., using operator U_i^n).

From Eqs. (50) and (32) we find that the operator U has the form

$$U_s^n(z_0)=\prod_i(\bar z_i^s-\bar z_0)\prod_{p>s}\prod_i|z_i^p-z_0|^{2\Delta q_p^s(p)} . \tag{52}$$

Here we have used the fact that

$$\theta_{sp}/\pi=-\text{sgn}(q_s)\Delta q_p^s .$$

Consequently, the electronic wave function in the presence of the quasiparticle in the sth condensate has the form

$$\psi(\xi_1,\ldots,\xi_{n;0})\sim\prod_{i,j}^{N_e}(\xi_i-\xi_j)^{p_1}$$
$$\times\int d^2\bar z_i^2\cdots\int d^2\bar z_i^n\prod_{s=2}^{s=n}\left[\prod_{i,j=1}^{N_s}|\bar z_i^s-\bar z_j^s|^{2\,\text{sgn}(q_{s-2})(\theta_{s-1}/\pi)}(\bar z_i^s-\bar z_j^s)^{p_s}\prod_{i,j}^{N_s,N_{s+1}}(\bar z_2^s-\bar z_j^{*s+1})\right.$$
$$\left.\times\prod_i(\bar z_i^s-\bar z_0)\prod_{p>s}|z_i^p-z_0|^{2\Delta q_p^*(p)}\exp(-\tfrac{1}{2}|q_s||z_i^s|^2)\right]\prod_i\exp(-\tfrac{1}{4}|\xi_i|^2) . \tag{53}$$

We shall use now the quasiparticle creation operator (52) to calculate the electric charges and statistics of the quasiparticle excitations. First, we calculate the matrix Δq_p^s once again using the plasma analog. Let us insert a quasiparticle created by the operator

$$U(z_0)=\prod_i [U_i^{(n)}(z_0)]^{l_i} . \tag{54}$$

The norm of the resulting electronic wave function is

$$Z=\int \prod d^2\xi_i d^2\bar{\xi}_i \psi(\xi_i,\ldots;z_0)\psi^*(\xi_1,\ldots,\xi_n;z_0)$$

$$\sim \int \cdots \int \prod d^2\xi_i d^2 z_i^s \exp\sum_s \sum_{i,j}\left[\operatorname{sgn}q_{s-2}\frac{\theta_{s-1}}{\pi}+p_s\right](\ln|z_i^s-z_j^s|+\ln|\bar{z}_i^s-\bar{z}_j^{*(s+1)}|)$$

$$\times\exp\left[4\sum_{p>s}\sum_i \Delta q_p^s(p) l_p \ln|z_0-z_i^p|\right]$$

$$\times\prod_i \exp(-1/2|\xi_i|^2)\prod_{i,s}\exp(-|q_s|(|z_i^s|^2)) . \tag{55}$$

Z is the partition function of the plasma considered in the last section but now with a phantom particle inserted at the point z_0. This particle interacts with the other particles in the plasma through the Coulombic potential

$$V(z_0,z_i)=2\sum_{p>s}\Delta q_p^s(p) l_p \ln|z_0-z_i^p| . \tag{56}$$

We now demand the plasma to be in the equilibrium after the addition of the phantom particle. This means that a local force acting on each particle in plasma is zero. The additional force acting on the particles due to the change of density of the incompressible fluid components must be counterbalanced by the force due to the interaction with the phantom particle. This condition gives us the following system of linear equations for the parameters $\Delta q_p^s(n)$ for the nth-level hierarchy state:

$$p_1\Delta q_1+\Delta q_2=l_1 ,$$

$$\vdots$$

$$\Delta q_{s-1}+e_s^2\Delta q_s+\Delta q_{s+1}=l_s+2\sum_{p\le s} l_p\Delta q_{s-1}^p(s-1) , \tag{57}$$

$$\vdots$$

$$\Delta q_{n-1}+e_n^2\Delta q_n=l_n+2\sum_{p\le n-1} l_p\Delta q_{n-1}^p(n-1) .$$

Here $e_i^2=2\,\operatorname{sgn}(q_{s-2})(\theta_{s-1}/\pi)+p_s$. The quantity Δq_s is the change of the density of the sth condensate for the nth-level hierarchical state due to the insertion of quasiparticle (54):

$$\Delta q^p=\sum_s l_s\Delta q_p^s(n) . \tag{58}$$

The sth equation in the system of Eqs. (57) comes from the requirement that the force acting on a quasiparticle in the sth-level condensate should be zero. The rhs of Eqs. (57) corresponds to the interaction between the quasiparticle at point z_0 and the quasiparticle in the sth condensate. The terms on the lhs of Eqs. (57) correspond to the interaction between the quasiparticle in the sth condensate with other quasiparticles in the sth condensate and with the quasiparticles in the $(s-1)$th and $(s+1)$th condensates. Equations (57) are the recurrence set of linear equations. We can find the equations for Δq_p^s for the nth level of hierarchy once we know the solution of the analogous set of equations for the $(n-1)$th level. In particular, for the second level we have the following set of equations:

$$p_1\Delta q_1+\Delta q_2=l_1 ,$$
$$\Delta q_1+\left[\frac{2}{p_1}+p_2\right]=l_2+\frac{2}{p_1}l_1 . \tag{59}$$

Solving these equations we obtain

$$\Delta q_1=\frac{p_2 l_1-l_2}{1+p_1p_2} ,$$
$$\Delta q_2=\frac{p_1 l_2+l_1}{p_1p_2+1} . \tag{60}$$

We can use these results to write the equations that determine the density changes for the third level of the hierarchy:

$$p_1\Delta q_1+\Delta q_2=l_1 ,$$
$$\Delta q_1+\left[\frac{2}{p_1}+p_2\right]\Delta q_2+\Delta q_3=l_2+\frac{2}{p_1}l_1 , \tag{61}$$
$$\Delta q_2+\left[\frac{2p_1}{p_1p_2+1}+p_3\right]\Delta q_3=l_3+\frac{2p_1}{p_1p_2+1}l_2+\frac{2}{p_1p_2+1}l_1 .$$

Solving these equations we can write the equations for the fourth-level hierarchical state, etc.

The solution for the general case can be simplified

greatly after we see (after some algebra) that the system (61) can be rewritten in the form

$$\Lambda_{ij}(-1)^{j+1}\Delta q_j = l_i \ . \tag{62}$$

The matrix Λ is given by Eq. (47). The numbers p_i' are given by $p_i'=(-1)^{i+1}p_i$. This system of equations was already solved when we discussed the effective theory in Ref. 2. The matrix Δq_p^s is given by the equation

$$\Delta q_s^k(n) = (-1)^{s+1+a_k+a_s} \times \frac{D_{n-k}(p'_{k+1},\ldots,p'_n)D_{s-1}(p'_1,\ldots,p'_{s-1})}{D}, \quad s<k ,$$
$$\Delta q_s^k(n) = (-1)^{s+1+a_k+a_s} \times \frac{D_{n-s}(p'_{s+1},\ldots,p'_n)D_{k-1}(p'_1,\ldots,p'_{k-1})}{D}, \quad s\geq k . \tag{63}$$

Here $D_k(p'_{s_1},\ldots,p'_{s_k})$ is the denominator of the filling fraction

$$\nu_i = \cfrac{1}{p'_{s_1} - \cfrac{1}{p'_{s_2} - \cfrac{1}{\cdots - \cfrac{1}{p'_{s_k}}}}} \ . \tag{64}$$

The numbers D_i satisfy the recursion relation

$$D_k(p'_{i_1},\ldots,p'_{i_k}) = p'_{i_1}D_{k-1}(p'_{i_2},\ldots,p'_{i_k}) - D_{k-2}(p'_{i_3},\ldots,p'_{i_k}) ,$$
$$D_0=1, \quad D_{-1}=0 . \tag{65}$$

Note that $D\equiv D_n(p'_1,\ldots,p'_n)$. The numbers a_i are equal to 0 if $i=1,2$, mod 4 and equal to 1 if $i=3,4$ mod 4.

The electric charges of the quasiparticles are given by the change of density of the electronic (first) component of the FQH incompressible fluid:

$$Q_{\mathrm{el}} = \sum_{j=1}^{j=n} \Delta q_1^j l_j \ . \tag{66}$$

Let us note that when $l_s=1$, $l_i=0$, $i\neq s$, the electric charges given by Eq. (66) are proportional to the densities of the sth components of incompressible fluids in the FQH states [see Eq. (48)]. This relation means that the electric charges of the quasiparticles defined through the Berry phase (Ref. 22) and through the plasma calculation coincide. The Berry phase is defined as the phase that the wave function obtains when the quasiparticle is moved along the closed contour. The Berry phase can be easily calculated using adiabatic arguments similar to those used in Ref. 22 for the Laughlin state with the filling fraction $1/p_1$. Using these arguments, it is easy to find that the Berry phase one obtains moving the quasiparticle around a loop is equal to

$$\gamma = 2\pi(-1)^{s+1}n_s S \ , \tag{67}$$

where n_s is the density of the sth incompressible fluid component defined by Eq. (48) and S is the area enclosed by the loop. Comparing Eqs. (46) and (62), we see that $2\pi n_s = (-)^{s+1}\Delta q_s^1$. Thus, the electric charges given by Eq. (16) and the charges determined from the Berry phase calculation $Q_{\mathrm{el}}=\gamma/(2\pi S)$ coincide. We conclude that the Berry phase in the hierarchical states comes from the interaction of the electric charge with the external magnetic field. This is again a special property of the hierarchical FQH state which is not satisfied by more general FQH states.

Let us now determine the statistics of the quasiparticles. In order to determine the statistics we consider a state with two quasiholes at points z_a and z_b created by operator (54). Using the same adiabatic arguments, we obtain that the statistics of the quasiparticle is equal to

$$\theta = \left[\sum l_i(-1)^{i+1}\Delta q_i\right]\pi \ . \tag{68}$$

From Eqs. (62) and (58) we obtain that the statistics θ is given by

$$\theta = \pi\sum_{i,s} l_i \Lambda_{is}^{-1} l_s = \pi \sum_{i,s} l_i(-)^{i+1}\Delta q_i^s(n) l_s \ . \tag{69}$$

Let us note now that the formula for the statistics Eq. (69) where the matrix Δq_i^s is given by Eq. (63) agrees with the formula for the statistics of the quasiparticles derived using effective theory of Ref. 2. We find that quantum numbers of the quasiparticles defined using the microscopic wave functions in Eq. (22) coincide with the quantum numbers of the quasiparticles obtained using the effective theory of Ref. 2. This result suggests that this effective theory and the Halperin wave functions describe the same FQH state. More explicit correspondence between the effective theory and the microscopic wave functions will be considered in the next section.

V. EFFECTIVE THEORY AND MICROSCOPIC WAVE FUNCTIONS

In the previous section we showed that the equilibrium condition for the particles in the plasma for the ground state leads to Eq. (46) for the condensate densities. In order to derive the effective theory we shall now allow the densities to vary slowly in space and time. Equation (46) becomes

$$\Lambda_{ij} n_j(t) = \delta_{1i}\rho_0 + \frac{d}{dt} h(n_i, \mathbf{j}_i) + \partial_k h^k(n_i, \mathbf{j}_i) \ , \tag{70}$$

where $\mathbf{j}_i^s$ are the currents of the condensed quasiparticles and we have ignored the higher-order time derivatives. The matrix Λ is given by Eq. (47). Now we shall use the current conservation law

$$\frac{\partial n_s}{\partial t} + \partial_i j_s^i = 0 \ . \tag{71}$$

Equation (71) implies that h is a spatial derivative of some function and thus can be combined into the $\partial_k h^k$ term. Equation (71) and the time derivative of Eq. (70)

imply that the currents satisfy the equations

$$\Lambda_{ij}j_j^k=\epsilon^{kj}\partial_j g^i(n_i,\mathbf{j}_i)-\frac{d}{dt}h^k(n_i,\mathbf{j}_i)$$

$$(i=1,\ldots,n)\ . \quad (72)$$

The rhs of Eqs. (72) cannot be determined because the divergences of these terms are zero; hence, the continuity equation (71) remains unchanged if we add such terms to the equation. The functions h^k and g^k can depend only on the space and the time coordinates z and t through the condensate densities and currents because of the translation invariance (in both the space and the time) of the problem. Equations (70) and (72) govern the dynamics of the density fluctuations in the FQH states. Let us introduce the gauge field a_ν^s such that the strength tensor $f^{\mu\nu}$ of this field is connected to the densities and the currents of the condensates:

$$\frac{1}{4\pi}\epsilon_{\mu\nu\lambda}f^{s,\mu\nu}=j_\lambda^s\ . \quad (73)$$

It is easy to see that, if we also set $h^k=g^k=0$, Eqs. (70) and (72) acquire the form

$$\Lambda_{ss'}\partial_\mu a'_\lambda\epsilon^{\mu\nu\lambda}=0 \quad (74)$$

in terms of the gauge field. Here the matrix $\Lambda_{ss'}$ is given by Eq. (47). Equation (74) is just the equation derived from the Chern-Simons terms. The quasiparticles act like the source terms in Eq. (74). From Eq. (62) with quasiparticles, we immediately find that the Lagrangian that reproduces Eqs. (74) and (62) has the form

$$\mathcal{L}=-\sum_{s,s'}\frac{1}{4\pi}\Lambda_{s,s'}a_\nu^s\partial_\mu a_\lambda^{s'}\epsilon^{\mu\nu\lambda}+\frac{e}{2\pi}A_\mu\partial_\nu a_\lambda^1\epsilon^{\mu\nu\lambda}$$
$$+\sum_s\Phi^{s\dagger}i(\partial_0-ia_0^s)\Phi^s+\frac{1}{2M}\Phi^{s\dagger}(\partial_i-ia_i^s)^2\Phi^s\ . \quad (75)$$

The effective theory (75) and the effective theory written in Ref. 2 are identical after the trivial field redefinition $a_i\to a_i$ for $i=1,2$ mod 4 and $a_i\to -a_i$ for $i=0,3$ mod 4. Below, when we state that the effective theory (75) and that in Ref. 2 coincide, we shall always implicitly assume this trivial field redefinition. The field $\Phi^s(z)$ is a quasiparticle creation operator in the macroscopic theory. From (62) we see that this field corresponds to the quasiparticle creation operator in the microscopic theory $U_s^{(n)}(z_0)$. This correspondence can be proved by calculating the propagator of the field $\Phi(z)$ in the effective theory and the correlation function $\langle U^{s+}(z_a)U^s(z_b)\rangle$ in the microscopic approach, using the plasma analog and the incompressibility of the quasiparticle fluid. The propagators coincide (up to the irrelevant gauge transformation) confirming the identification.

The terms on the rhs of Eqs. (70) and (72) are higher-dimensional terms. They can come only from the Maxwell terms (or even higher-dimensional terms) in the effective Lagrangian and cannot influence the low-energy behavior of the theory. Hence, as far as the low-energy properties of the excitations are concerned, those terms can be ignored. They cannot modify the quantum numbers of the quasiparticles.

The Lagrangian (75) is exactly the Lagrangian of the effective theory.[2] Thus, the effective Lagrangian in Ref. 2 is derived from the microscopic theory. The strength of the gauge field $f_{\mu\nu}^s$ can be calculated from the electron wave function. $1/4\pi f_{ij}^s\epsilon_{ij}$ is the average of the density operator of the sth-level condensate

$$\rho^s(z)=\sum_i\delta^2(z-z_i^s)\ , \quad (76)$$

f_{0i}^s average of the current of the condensed quasiparticles:

$$j_j^s(z)=O\frac{1}{2m}\sum_i\{\delta^{(2)}(z-z_i^s)[\partial_j+ieq_sA(r_j)]+\text{H.c.}\}O^{-1}\ ,$$

where

$$O=\prod_{i<j}(\bar z_i^s-\bar z_j^s)^{\mathrm{sgn}(q_s-2)(\theta_{s-1}/\pi)}\prod_i\exp(-\tfrac{1}{4}|q_s||z_i^s|^2)\ . \quad (78)$$

The operator O is included such that j^s is an operator which acts on the pseudo-wave-function of quasiparticles of the sth-level condensate. The action of ρ^s and j^s on the ground state is defined by inserting (76) and (77) into the integrand in Eq. (22). This completes the correspondence of the effective theory and the microscopic theory.

The gauge fields a_ν^s are also related to the Berry phase that one obtains while moving a quasiparticle around a loop. The result for the Berry phase further confirms the coupling between the quasiparticle Φ^s and the gauge field a_μ^s. Let us consider the Berry-phase calculation in more detail. Berry's phase is defined as the contour integral of the gauge field

$$A_z^s=i\left\langle z_0^s\left|\frac{\partial}{\partial z_0^s}\right|z_0^s\right\rangle\ ,$$
$$A_{\bar z}^s=i\left\langle \bar z_0^s\left|\frac{\partial}{\partial \bar z_0^s}\right|\bar z_0^s\right\rangle\ . \quad (79)$$

Here $|z_0^s\rangle$ is the normalized electronic wave function

$$|z_0^s\rangle=\frac{1}{\sqrt{N_z}}U^s(z_0)|\Phi_n\rangle\ . \quad (80)$$

$|\Phi_n\rangle$ is the ground-state wave function (22) and U_s^n is the quasiparticle creation operator (52). The normalization factor N_z is given by

$$N_z=\langle\Phi_n|U_s^{n+}(z)U_s^n(z)|\Phi_n\rangle\ . \quad (81)$$

Using the explicit form of the quasiparticle creation operator $U_s^n(z_0)$ and the electronic wave function (22), it is easy to prove that[17]

$$A_z^s=-i/2\int\frac{d^2z'}{z-z'}\frac{1}{N_z}\langle U^{s\dagger}(z)\rho^s(z')U^s(z)\rangle\ ,$$
$$A_{\bar z}^s=-i/2\int\frac{d^2z'}{\bar z-\bar z'}\frac{1}{N_z}\langle U^{s\dagger}(z)\rho^s(z')U^s(z)\rangle\ . \quad (82)$$

Here the operator $\rho^s(z)$ is the quasiparticle density operator for the sth condensate [see Eq. (76)]. Let us write

$$\langle U^{s+}(z)\rho^s(z')U^s(z)\rangle/N_z = n_s + \Delta\rho(z-z') .$$

Here $\Delta\rho$ is the density profile of the quasiparticles. It depends only on $|z-z'|$ due to the rotation symmetry. The density n_s is the density of the s-th-level condensate in absence of the quasiparticles. It is easy to see that

$$\int d^2z' \frac{\Delta\rho^s(|z'-z|)}{z-z'} = 0 . \tag{83}$$

Hence, we obtain that the "magnetic" field strength of the field (79) is equal to

$$B^s = 2\partial_{\bar{z}} A_z^s = -i2\pi i \int \delta^{(2)}(z'-z)\langle\rho(z')\rangle$$
$$= 2\pi n_s . \tag{84}$$

Equation (84) is also valid for the case of the excited state. In this case, the function Φ_n in the definition (80) is the wave function of the excited state and n_s in Eq. (84) may depend on the space coordinates [see Eq. (51)]. Comparing Eqs. (84) and (73), we find that (up to a irrelevant gauge transformation) the gauge field (22) can be identified with the gauge field (71). Berry's phase of moving a quasiparticle along the loop is directly determined by the density of the condensates. This further confirms the coupling between Φ^s and the gauge field a^s_μ.

VI. CONCLUSIONS

In this paper we calculated the quantum numbers of the quasiparticles in the hierarchical FQH states directly from the Haldane-Halperin electron wave functions. Laughlin's plasma analog is generalized to the case of the hierarchical FQH state. We derived the explicit form of the quasiparticle creation operators (52) based on physical considerations. There are two convenient choices of the basis of the quasiparticle creation operators. The first one is the set of the W operators defined by Eqs. (24), (26), and (27). These operators permit one to easily obtain the quantum numbers of the quasiparticles in the theory. In particular, with the help of the W operators, it is straightforward to see that all the allowed quantum numbers of the quasiparticles can be expressed as

$$Q_{\text{el}} = (-1)^{n+1}\frac{l}{D}e, \quad \theta = (-1)^n \frac{D(p_1, \ldots, p_{n-1})}{D} l^2 , \tag{85}$$

where l is an integer. Another basis is given by the U operators defined by Eq. (21). We explicitly checked that in this basis not only the charges and statistics of the quasiparticles, but also the induced density changes of the condensates Δn_i, coincide with the corresponding quantities in the effective theory of Ref. 2. Hence, this choice of the basis of the quasiparticle creation operators is naturally connected with the effective theory of Ref. 2.

In Sec. V we further derived the effective theory from the microscopic calculations based on the plasma analog. We identified the gauge fields of the effective theory (in its dual form Ref. 18) as vacuum averages (79) of some density and current operators over the electron wave functions. The quasiparticle field Φ^s of the effective theory was shown to correspond to the quasiparticle creation operator U given by Eq. (52). Through the discussions presented in this paper, the relations between the effective theory and the microscopic theory is clearly demonstrated. In particular, the rich topological orders in the FQH state revealed in the effective theory are further confirmed in the microsoic theory. After this paper was completed, we became aware of a paper by Read[25] where some of the results were obtained using a different form of the hierarchical wave function.

[1]B. Blok and X. G. Wen, Phys. Rev. B **42**, 8133 (1990).
[2]B. Blok and X.-G. Wen, Phys. Rev. B **42**, 8145 (1990).
[3]X. G. Wen (unpublished).
[4]X.-G. Wen and Q. Niu, Phys. Rev. B **41**, 9377 (1990).
[5]X. G. Wen, Phys. Rev. B **40**, 7387 (1989); Int. J. Mod. Phys. B **2**, 239 (1990).
[6]R. Laughlin, in *The Fractional Quantum Hall Effect*, edited by R. E. Prange and S. M. Girvin (Springer-Verlag, Berlin, 1987).
[7]B. I. Halperin, Phys. Rev. Lett. **52**, 1583 (1984).
[8]B. I. Halperin, Helv. Phys. Acta **56**, 75 (1983).
[9]S. Girvin, Phys. Rev. B **29**, 6012 (1984).
[10]A. H. Mcdonald and D. B. Murray, Phys. Rev. B **32**, 2707 (1985).
[11]A. H. Mcdonald, G. C. Aers, and M. W. C. Dharma-Wardana, Phys. Rev. B **31**, 5529 (1985).
[12]J. K. Jain, Phys. Rev. Lett. **63**, 199 (1989).
[13]J. K. Jain, Phys. Rev. B **40**, 8079 (1989).
[14]J. K. Jain, Phys. Rev. B **41**, 7653 (1990).
[15]S. M. Girvin and A. H. Mcdonald, Phys. Rev. Lett. **48**, 1252 (1987); S. M. Girvin, in *The Fractional Quantum Hall Effect*, edited by R. E. Prange and S. M. Girvin (Springer-Verlag, Berlin, 1987).
[16]S. C. Zhang, T. H. Hansson, and S. Kivelson, Phys. Rev. Lett. **62**, 82 (1989).
[17]N. Read, Phys. Rev. Lett. **62**, 82 (1989).
[18]X. G. Wen and A. Zee, Phys. Rev. B **41**, 240 (1990).
[19]Z. P. Ezawa and A. Iwazaki, Phys. Rev. B **43**, 2637 (1991).
[20]M. P. H. Fisher and D. H. Lee, Phys. Rev. B **39**, 2756 (1989); Phys. Rev. Lett. **63**, 903 (1989).
[21]F. D. M. Haldane, Phys. Rev. Lett. **51**, 605 (1983).
[22]D. Arovas, J. R. Schrieffer, and F. Wilczek, Phys. Rev. Lett. **53**, 722 (1984).
[23]J. K. Jain, S. Kivelson, and N. Trivedi, Phys. Rev. Lett. **64**, 1297 (1990); **64**, 1993 (1990).
[24]J. Niehnius, *Critical Phenomena and Phase Transitions* (Springer-Verlag, Berlin, 1986), Vol. 8.
[25]N. Read, Phys. Rev. Lett. **65**, 1502 (1990).

Composite-Fermion Approach for the Fractional Quantum Hall Effect

J. K. Jain
Section of Applied Physics, Yale University, P.O. Box 2157 Yale Station, New Haven, Connecticut 06520
(Received 24 January 1989)

In the standard hierarchical scheme the daughter state at each step results from the fractional quantum Hall effect of the quasiparticles of the parent state. In this paper a new possible approach for understanding the fractional quantum Hall effect is presented. It is proposed that the fractional quantum Hall effect of electrons can be physically understood as a manifestation of the integer quantum Hall effect of composite fermionic objects consisting of electrons bound to an even number of flux quanta.

PACS numbers: 73.50.Jt, 73.20.Dx

Even though the experimental observations of the integer[1] and the fractional[2] quantum Hall effect[3] (QHE) are essentially identical, except for the value of the quantized Hall resistance, there are, roughly speaking, three different theoretical schemes for their explanation. While the integer QHE (IQHE) is thought of essentially as a noninteracting electron phenomenon,[4] the fractional QHE (FQHE) is believed to arise from a condensation of the two-dimensional (2D) electrons into a "new collective state of matter"[5] as a result of interelectron interactions. Even within the FQHE the "fundamental" fractions $\frac{1}{3}, \frac{1}{5}, \ldots$ play a special role and the other fractions are obtained in a hierarchical scheme[6] in which a daughter state is obtained at each step from a condensation of the quasiparticles of the parent state into a correlated low-energy state.

The purpose of this Letter is to present a theoretical framework which enables an understanding of both the IQHE and the FQHE in a unified scheme as two different manifestations of the same underlying physics. It is argued that the possibility of QHE at fractional filling factors $p/(2mp \pm 1)$, where m and p are integers, arises because the correlations in the phase factors at these filling factors are very similar to the correlations present at integer filling factors p. This approach not only gives all the observed fractions (except[7] $\frac{5}{2}$, which therefore requires some additional physics[8]), and explains in doing so why only fractions with odd denominators are observed, but also provides the order of their stability, in agreement with experiments. Furthermore, it suggests a generalization of the Laughlin wave functions to other fractions.

I start by proposing a remarkably simple picture for understanding the origin of the FQHE. The important parameter is the ratio of the total number of flux quanta ($\phi_0 = hc/e$) to the total number of electrons, which is the inverse of the filling factor ν (in the thermodynamic limit) and specifies the average number of flux quanta available to each electron. Consider a 2D electron gas in the presence of a transverse magnetic field at an integer filling factor $\nu = p$, so that there is an average flux ϕ_0/p per electron. The electronic wave function $\Psi_{\pm p}$ ($\pm$ corresponds to magnetic field in the $\mp z$ direction) in this situation is rather insensitive to the details of the interelectron interactions and is determined mainly by virtue of the fermionicity of the electrons. Thus, the long-range correlations due to the Fermi statistics provide rigidity to the electron system at integer filling factors which results in the phenomenon of IQHE. It is useful to think in the path-integral language:[9] The partition function gets contributions from the closed paths in the configuration space (for example, a path in which one electron moves in a loop while the others are held fixed, or a cooperative ring exchange path[9]). The phase associated with each closed path has two contributions: the Aharonov-Bohm phase which depends on the flux enclosed in the loop, and the statistical phase which depends on how many electrons participate in the path. An incompressible state is obtained at integer filling factors presumably because of some special correlations (which may not be easily identified) built in the phase factors corresponding to the various paths. Now attach to each electron an infinitely thin magnetic solenoid carrying a flux $\alpha\phi_0$ (pointed in the $-z$ direction). For lack of a better name, we term an electron bound to a flux tube a "composite particle." As is well known,[10] the statistics of the composite particles is in general fractional, and is such that an exchange of two composite particles produces a phase factor $(-1)^{1+\alpha}$ (Ref. 11). The relevant case here is when α is equal to an even integer ($\alpha = 2m$), and the composite particles are fermions. It is easy to see that in this case the phase factor acquired along a given closed path is identical to the phase factor acquired in the absence of the flux tubes, implying that the correlations in the phase factors for $\alpha = 2m$ are the same as those for $\alpha = 0$. Since these correlations are responsible for rigidity and QHE at integer filling factors, one can expect the composite fermion state $\Psi^{2m}_{\pm p}$, which is obtained by adding to each electron in $\Psi_{\pm p}$ a flux $2m\phi_0$, to also be rigid and show QHE.

To determine the filling factor of $\Psi^{2m}_{\pm p}$ we exploit an ingenious observation due to Arovas *et al.*[10] and Laughlin:[11] A (uniform) liquid of electrons, each carrying with it a flux $\alpha\psi_0$, is equivalent, *in a mean field sense* to a (uniform) liquid of electrons in a magnetic field of strength such that there is an average flux of $\cdot\alpha\phi_0$ per electron. A uniform electron density is required to produce a uniform flux density. Since in the state $\Psi^{2m}_{\pm p}$ there are a total of $2m\pm p^{-1}$ flux quanta per electron, we identify it with the mean-field state of *electrons* at fractional filling $\nu=p/(2mp\pm 1)$. It must be borne in mind that the true electron state is not as rigid as the composite fermion state, because in the true state the flux tubes are not strictly bound to the electrons, and the phase factors simulate IQHE only on average. However, *provided that the true electron state is also incompressible*, the composite fermion state should provide the correct description of the essential physics at the mean-field level, as it contains the correlations giving rise to the incompressibility. On the other hand, when the true electron state is not incompressible, identification of the composite fermion state with the mean-field state of the electrons is no longer valid, or meaningful.

Thus in this approach there are two types of correlations essential for FQHE. The first type of correlations, which have been widely appreciated in the field,[12,13] involve binding of electrons and zeros of the wave function, or, equivalently, of electrons and flux tubes, which is a very useful way of incorporating the effect of repulsive interactions. Thus the role of repulsive interactions in the present framework is assumed to be to generate composite fermions. The second type of correlations, that impart rigidity to the composite fermion system and thus lead to FQHE, are the correlations due to their Fermi statistics. These are included in the present scheme by mimicking the statistical correlations present in the noninteracting electron system at integer filling factors. This is the central idea of this work; it is best summarized by saying that *the FQHE of electrons is a manifestation of the IQHE of composite fermions.*

The Hall plateaus at fractional filling factors appear in this model precisely as at integer filling factors except for the trivial modification that now each electron carries with it $2m$ flux quanta. Following the argument of Laughlin and Halperin[4] consider a corbino disk geometry. The Hall resistance is related to the charge transported from one edge to the other as one flux quantum is adiabatically pierced through the center. At integer filling factor p, p electrons are transported across the sample in this process. For $\Psi^{2m}_{\pm p}$, as each electron carries $2m$ flux quanta, one must supply $2mp$ additional flux quanta (in all $2mp\pm 1$ flux quanta) to transport p electrons across the sample. This gives $R_H=h/\nu e^2$ with $\nu=p/(2mp\pm 1)$. Just as in the IQHE, sample impurities and inhomogeneities create localized states, which produce a quantized Hall plateau so long as the Fermi level lies in a mobility gap.[4]

The stable fractional filling factors obtained in this manner are $p/(2mp\pm 1)$, and due to electron-hole symmetry, $1-p/(2mp\pm 1)$. [As indicated by Haldane[8,14] this implies possible stability at fractions $n+p/(2mp\pm 1)$ and $n+1-p/(2mp\pm 1)$ in the nth Landau level (LL).] Notice that only fractions with odd denominators appear. In fact, in the present framework QHE at fractional values of ν with odd denominators is as natural as the QHE at integer values of ν. Besides explaining the "odd denominator rule," we are also able to predict the order of stability of the fractions, or the order in which new fractions should appear as the sample quality is improved. Since a collapse of the gap due to an "unbinding transition" is more likely for larger values of m, if a fraction $p/(2mp\pm 1)$ is observed for a given p then all fractions $p/(2m'p\pm 1)$ with $m'<m$ must also be observed. One also expects weaker correlations for higher values of p. Thus in Fig. 1 a given fraction in the right (left) half is more stable than the one directly above it and the one on its right (left). This is quite generally borne out in experiments.[7,15-17] This also identifies the fractions to be observed next, if any, as the sample quality is further improved. Read[18] has pointed out that the fractions obtained here are only the first level of a *new* hierarchy, and all other fractions with odd denominators can be obtained within the present formalism. However, it is interesting to note that all the observed fractions, except $\frac{4}{11}$ and $\frac{4}{13}$ (Ref. 19), are obtained in this scheme at the very first level, which is to be contrasted with the standard hierarchy in which one needs to go down many

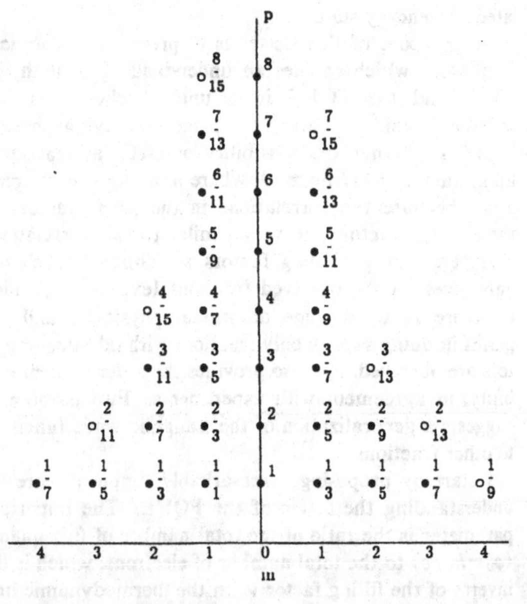

FIG. 1. The fractions $p/(2mp+1)$ and $p/(2mp-1)$ are shown in the right half and the left half, respectively. The filled circles show the fractions that have been observed in the lowest LL. The predicted values of the next most stable fractions at this level are shown near empty circles.

levels in order to obtain some of the observed fractions. It is also worth mentioning that the present scheme naturally produces the experimentally observed sequences[7,15] of fractions converging to $\frac{1}{2}$ (for $m=1$), to $\frac{1}{4}$ (for $m=2$), to $\frac{3}{4}$ (hole analog of $\frac{1}{4}$), etc.

In the following I will construct explicit trial wave functions, analogous to the Laughlin wave functions, which have the correlations discussed above. The Hamiltonian for N noninteracting electrons ($N\to\infty$) at filling factor p is given by $H_0=\sum_{j=1}^{N}(2m_e)^{-1}(\mathbf{p}_j+e\mathbf{A}_j/c)^2$, where $\mathbf{A}_j$ is chosen so as to produce a uniform magnetic field in the $\mp z$ direction of strength such that there is an average flux of $p^{-1}\phi_0$ per electron. The corresponding ground-state wave functions are $\Psi_{\pm p}$ with $\Psi_{-p}=\Psi_{+p}^*$. We first consider the fractions $p/(2mp+1)$ which are obtained by starting from Ψ_{+p}. Gauge flux tubes $2m\phi_0$ are attached to each electron by adding to the vector potential $\mathbf{A}_j$ a singular gauge potential[10,11]

$$\mathcal{A}_j=-2m\frac{\phi_0}{2\pi}\sum_{k(k\neq j)}\nabla_j\theta_{jk},$$

where θ_{jk} is defined by $(z_j-z_k)=|z_j-z_k|\exp(i\theta_{jk})$, and $z_j=x_j+iy_j$ denotes the position (x_j,y_j) of the jth particle as a complex number. The new ground-state wave function is[11]

$$\Phi_{+p}^{2m}=\prod_{j<k}\frac{(z_j-z_k)^{2m}}{|z_j-z_k|^{2m}}\Psi_{+p}.$$

Clearly this is not an appropriate wave function for describing the FQHE of electrons. Following the analogy of the Laughlin wave functions, and for the reasons mentioned below, we write instead the following closely related (unnormalized) trial wave function

$$\Psi_{+p}^{2m}=Z^{2m}\Psi_{+p},$$

where $Z^a=\prod_{j<k}(z_j-z_k)^a$. This wave function has the same topological structure as Φ_{+p}^{2m}, and also describes electrons carrying gauge flux tubes of strength $2m\phi_0$. In this state addition of flux tubes is accompanied by a change in the size of the system in such a way as to keep the total flux per unit area (i.e., the magnetic field) constant.

This state has the following properties: (i) For $p=1$, Ψ_{+1}^{2m} is identical with the corresponding Laughlin state.[5] (ii) Since Ψ_{+p} is determined almost completely by the Pauli principle, and has little dependence on interelectron interactions, Ψ_{+p}^{2m} is also largely insensitive to the interactions. This is explicitly the case for the Laughlin states[5] which have been found to be very accurate for a variety of interelectron interactions. (iii) It describes an electron gas of uniform density. This follows straightforwardly from the fact that Ψ_{+p} describes an electron gas of uniform density. It is also an eigenstate of the angular momentum. (iv) One can read off the filling factor from the wave function. Take Ψ_{+p} with p LL's completely occupied in a disk-shaped region; the number of occupied single-particle states in each LL is N/p. Since the largest power of a z_j in Z^{2m} is $2m(N-1)$, Ψ_{+p}^{2m} has $2m(N-1)+Np^{-1}$ single-particle states occupied in each LL, which immediately yields a filling factor $p/(2mp+1)$ in the thermodynamic limit. Thus the state Ψ_{+p}^{2m} (unlike Φ_{+p}^{2m}) satisfies the fundamental requirement that the filling factor obtained by counting the total number of states agrees with that obtained from the flux-counting argument (i.e., the number of flux quanta piercing the sample is equal to the number of single-particle states in each LL). (v) The factor Z^{2m} in Ψ_{+p}^{2m} partially projects the single-particle states of the higher LL's into the lowest LL. Write

$$\Psi_{+p}^{2m}=AZ^{2m}\prod_{j=1}^{N/p}\prod_{l=0}^{p-1}\zeta_{l,j-1}(z_{l+j}),$$

$$\zeta_{l,s}=(2\pi 2^{l+s}l!s!)^{-1/2}e^{|z|^2/4}\left(2\frac{\partial}{\partial z}\right)^l z^s e^{-|z|^2/2},$$

where A is the anitsymmetrization operator, $\zeta_{l,s}$ are the single-particle states, $l=0,\ldots,p-1$ is the LL index, and $s=0,\ldots,N/p-1$ is the angular momentum quantum number. Z^{2m} is a sum of terms of type $\prod_{j=1}^{N}z_j^{t_j}$ with $\sum_j t_j=mN(N-1)$, where t_j is *typically* a large power (in the thermodynamic limit) of order mN. Thus, in each term of Ψ_{+p}^{2m} the coordinate z_j of a particle appears as the product $z_j^{t_j}\zeta_{l,s}(z_j)$. For t_j of order mN, this product lies *almost entirely* in the lowest LL. Expanding it as a sum of single-particle states,

$$z_j^{t_j}\zeta_{l,s}=\sum_{k=0}^{l}a_k\zeta_{k,s+t_j+k-l},$$

one can show that the ratio a_{k+1}/a_k is of order $1/\sqrt{N}$; i.e., the amplitude of $z_j^{t_j}\zeta_{l,s}(z_j)$ is smaller by a factor of order $1/\sqrt{N}$ in each successively higher LL. Thus in Ψ_{+p}^{2m} the amplitude is expected to be in general much larger for the terms which have a greater number of the lowest LL states occupied. Furthermore, there are manifestly terms, with extremely large amplitudes, which have only the lowest LL occupied. This implies that, unless there are some very strange cancellations, the state Ψ_{+p}^{2m} lies predominantly in the lowest LL in the thermodynamic limit. (vi) Last, Ψ_{+p}^{2m} is expected to be a good variational state in the presence of repulsive interactions, because both the factors Z^{2m} and Ψ_{+p} are very efficient in keeping the electrons apart. This is a direct sense in which the correlations of the higher LL are utilized to obtain a low-energy state. Thus we believe that the states Ψ_{+p}^{2m} possess all the necessary properties of a reasonable trial state. At present we are working towards a quantitative test, which is complicated due to the complex structure and the inherent thermodynamic nature of these states. The form of the incompressible state at $\nu=p/(2mp-1)$, which is obtained starting from Ψ_{-p}, is not as obvious.

Normally the ground state is expected to be completely spin polarized, and is obtained by choosing in $\Psi_{\pm p}$ the p lowest LL's with the *same spin orientation.* However, when the spin splitting is insignificant, it may be useful to consider situations in which $\psi_{\pm p}$ has LL's with both spin orientations occupied, so that in $\Psi^{2m}_{\pm p}$ the two lowest spin-split Landau bands are occupied. Thus, for small spin splitting, there are in general many candidates for the incompressible state[12,14,20] for a given fraction: the completely spin polarized states, spin unpolarized states,[12,14] and partially spin polarized states.

There seems to be a close analogy between the composite fermion states proposed in this paper and the standard hierarchical states. To illustrate this, we consider the example of the $\frac{1}{3}$ state which is obtained by multiplying Ψ_{+1} by $\mathcal{Z}^2$. It can be shown[21] that the state Ψ_{+1} with one hole similarly produces the $\frac{1}{3}$ state with a Laughlin quasihole. By analogy, the $\frac{1}{3}$ state with a quasielectron would be obtained by multiplying by $\mathcal{Z}^2$ the state with the lowest ($l=0$) LL fully occupied and one electron in the $l=1$ LL. The state with fully occupied lowest LL and δ electrons in the $l=1$ LL then corresponds to the $\frac{1}{3}$ state with δ quasielectrons. Thus the $\frac{2}{5}$ state can be viewed in the present approach as the $\frac{1}{3}$ state with $N/2$ quasielectrons, and similarly the $\frac{3}{7}$ state can be viewed as the $\frac{2}{5}$ state with $N/3$ quasielectrons. This assignment is in exact agreement with that of the standard hierarchy theory.[6] One can also show that the quasiparticles described above have the same charge as those in the standard scheme.[21] The analogy is, however, not complete. In the standard picture stability is obtained when the quasielectrons form a Laughlin-type state, whereas in the composite-fermion scheme they derive their arrangement from the higher LL. Also, taking the above example, in the standard picture one can obtain both the $\frac{3}{7}$ and the $\frac{5}{13}$ states from the $\frac{2}{5}$ state, whereas the composite-fermion approach does not yield the experimentally unobserved $\frac{5}{13}$ state at this level.

In conclusion, this paper proposes that the FQHE can be accessed from the IQHE by adding an even number of flux quanta to each electron. This analogy between FQHE and IQHE suggests a natural generalization of the Laughlin states.

I thank N. Read, D. Stone, S. Kivelson, A. MacDonald, A. Chang, V. Goldman, and H. Störmer for helpful conversations.

[1] K. von Klitzing, G. Dorda, and M. Pepper, Phys. Rev. Lett. **45**, 494 (1980).

[2] D. C. Tsui, H. L. Störmer, and A. C. Gossard, Phys. Rev. Lett. **48**, 1559 (1982).

[3] For a review, see *The Quantum Hall Effect,* edited by R. E. Prange and S. M. Girvin (Springer-Verlag, New York, 1987).

[4] R. B. Laughlin, Phys. Rev. B **23**, 5632 (1981); B. I. Halperin, Phys. Rev. B **25**, 2185 (1982).

[5] R. B. Laughlin, Phys. Rev. Lett. **50**, 1395 (1983).

[6] F. D. M. Haldane, Phys. Rev. Lett. **51**, 605 (1983); B. I. Halperin, Phys. Rev. Lett. **52**, 1583 (1984); R. B. Laughlin, Surf. Sci. **141**, 11 (1984), and in Ref. 3. There have also been nonhierarchical states proposed for certain higher-level fractions; see, A. H. MacDonald, G. C. Aers, and M. W. C. Dharma-wardana, Phys. Rev. B **31**, 5529 (1985); Y. Yoshioka, A. H. MacDonald, and S. M. Girvin, Phys. Rev. B **38**, 3636 (1988); Phys. Rev. B **39**, 1932 (1989); R. Morf, N. D'Ambrumenil, and B. I. Halperin, Phys. Rev. B **34**, 3037 (1986).

[7] R. Willett *et al.*, Phys. Rev. Lett. **59**, 1776 (1987).

[8] F. D. M. Haldane and E. H. Rezayi, Phys. Rev. Lett. **60**, 956 (1988).

[9] S. Kivelson, C. Kallin, D. P. Arovas, and J. R. Schrieffer, Phys. Rev. Lett. **56**, 873 (1986).

[10] F. Wilczek, Phys. Rev. Lett. **49**, 957 (1982); F. Wilczek, and A. Zee, Phys. Rev. Lett. **51**, 2250 (1983); D. P. Arovas, J. R. Schrieffer, F. Wilczek, and A. Zee, Nucl. Phys. **B251**, 117 (1985).

[11] R. B. Laughlin, Phys. Rev. Lett. **60**, 2677 (1988).

[12] B. I. Halperin, Helv. Phys. Acta. **56**, 75 (1983).

[13] S. M. Girvin and A. H. MacDonald, Phys. Rev. Lett. **58**, 1252 (1987); E. H. Rezayi and F. D. M. Haldane, Phys. Rev. Lett. **61**, 1985 (1988); S. C. Zhang, T. H. Hansson, and S. Kivelson, Phys. Rev. Lett. **62**, 82 (1989); N. Read, Phys. Rev. Lett. **62**, 86 (1989).

[14] F. D. M. Haldane, in Ref. 3.

[15] R. Willett *et al.*, Surf. Sci. **196**, 257 (1988); A. M. Chang *et al.*, Phys. Rev. Lett. **53**, 997 (1984).

[16] R. G. Clark *et al.*, Surf. Sci. **196**, 219 (1988); **196**, 257 (1988).

[17] V. J. Goldman, M. Shayegan, and D. C. Tsui, Phys. Rev. Lett. **61**, 881 (1988).

[18] N. Read (private communication).

[19] V. J. Goldman, D. C. Tsui, and M. Shayegan, in Proceedings of the Nineteenth International Conference on the Physics of Semiconductors Warsaw, Poland, 1988 (World Scientific, Singapore, to be published).

[20] R. G. Clark *et al.*, Phys. Rev. Lett. **62**, 1536 (1989); J. P. Eisenstein *et al.*, *ibid.* **62**, 1540 (1989).

[21] J. K. Jain (to be published).

Excitation Structure of the Hierarchy Scheme in the Fractional Quantum Hall Effect

N. Read
Department of Applied Physics, P.O. Box 2157, and Center for Theoretical Physics, P.O. Box 6666, Yale University, New Haven, Connecticut 06520
(Received 29 June 1990)

The hierarchy schemes for the fractional quantum Hall effect are reexamined and it is shown that different schemes all give the same lattice of excitations whose statistics is determined by the norm of the corresponding vector, and hence have equivalent Ginzburg-Landau theories. Similar ideas apply to the anyon liquid. The schemes can be generalized by using different lattices; many inequivalent states can be obtained at any filling factor (or value of the statistics parameter).

PACS numbers: 73.20.Dx, 05.30.−d

Laughlin's wave functions[1] have won widespread acceptance as good model wave functions for the fractional quantum Hall effect[2] (FQHE) at filling factor $\nu=1/q$, but the situation at most other filling factors (with the exception of those related to $1/q$ by particle-hole conjugation or by filling of lower Landau levels) is somewhat less clear. Numerous schemes for extending Laughlin's ideas have been proposed, in particular what I will call the "standard" hierarchy,[3,4] a "variant" hierarchy,[5] and recently a "new" hierarchy.[6] These hierarchies are supported by different physical arguments but are alike in producing a ground state for every fraction $\nu=p/q<1$ such that q is odd, and in having fractionally charged excitations $\pm e^*=\pm e/q$.

Questions about the detailed structure of the excitation spectrum for filling factor ν have recently arisen because of its relevance to gapless excitations of an incompressible bulk FQHE state.[7] One may ask whether the hierarchies make equivalent predictions, whether there are other physically distinct incompressible states at the same ν, and how the Ginzburg-Landau (GL) theory, for $\nu=1/q$,[8-10] can be properly extended to other fillings. Similar questions may be raised about the ground states of a liquid of anyons.[11,12]

The main results of this paper are as follows. Incompressible FQHE systems will be regarded as equivalent when their filling factors are equal and they possess excitations whose quantum numbers and statistics correspond. (i) The quantum numbers of the possible "charged" excitations lie on a lattice of points in r-dimensional space for r levels of the standard hierarchy. Excitations of the same physical charge all have the same (fractional) statistics. All the hierarchy schemes are equivalent in this sense; different constructions involve different bases for the same lattice. This characterizes these systems nonhierarchically. (ii) The order parameter has r components, and the GL theory also involves r gauge potentials[12] and its structure is determined by the same lattice as the excitations. (iii) The constructions can be generalized further, in a basis-independent way, by using an arbitrary lattice, subject to certain rules. This produces other inequivalent states at *any* rational ν. (iv) Similar observations apply to spin singlet and partially polarized states, and to states of an anyon liquid.

I begin by writing the standard hierarchy electron wave function[3,4] in the form

$$\Psi(\{z_{0i}\})=\int\prod_{\alpha=1}^{r-1}\prod_{i=1}^{N_\alpha}d^2z_{\alpha i}\exp\left[-\frac{1}{4}\sum_i|z_{0i}|^2\right]\prod_{\alpha=0}^{r-1}\left[\prod_{i<j}(z_{\alpha i}-z_{\alpha j})^{a_\alpha}\prod_{ij}(z_{\alpha+1,i}-z_{\alpha j})^{b_{\alpha,\alpha+1}}\right]. \tag{1}$$

Equation (1) describes $N=N_0$ electrons at positions $z_i=z_{0i}$ and the integrals are over coordinates of quasiparticles at level $\alpha=1,\ldots,r-1$ in the hierarchy; the system contains N_α quasiparticles of level α at positions $z_{\alpha i}$. In the exponents, $a_0>0$ is odd, a_α ($\alpha>0$) is even, $b_{\alpha,\alpha+1}=\pm1$, and $b_{r-1,r}=0$. Negative exponents are unconventional; quasiholes in the electron system couple with $b_{01}=1$ as usual, but quasielectrons couple with $b_{01}=-1$. This is an acceptable alternative to the usual Laughlin quasielectron or other proposals as long as the singularity at the center is removed by projecting onto holomorphic (lowest-Landau-level) functions. Such projection only introduces a short-range interaction into the effective many-component Coulomb plasma, described below. Alternatively, the factors with negative exponents may be replaced by positive powers of the complex-conjugate factor, times additional exponential factors. This freedom of choice in the hierarchy wave functions has no influence on the following; the above form makes the structure especially clear.

In order to work with states like (1), one needs to make an orthogonality postulate. To take overlaps of two many-quasiparticle states, one must integrate over the electron coordinates. One hopes that this makes the overlap vanish unless the positions of the $\alpha=1$ quasiparticles in one state nearly coincides with those in the other. If so, then the integrations in (1) for each state can be reduced to a single set of integrals for $\alpha=1$, and the process can be iterated. For a few well-separated $\alpha=1$ quasiparticles, this can be demonstrated explicitly,[13] and so should hold for $|a_\alpha|$ large. We will assume, as is standard, that it also holds for $|a_\alpha|$ as small as 2. Then expectations in (1) behave like those of a multicomponent generalization of Laughlin's Coulomb plasma.

A homogeneous ground state in the shape of a disk is obtained if

$$a_\alpha(N_\alpha-1)+b_{\alpha,\alpha+1}N_{\alpha+1}+b_{\alpha-1,\alpha}N_{\alpha-1}=0, \qquad (2)$$

for $\alpha=0,\ldots,r-1$, where $N_r=0$, $b_{-1,0}N_{-1}=-N_\phi$, and N_ϕ is the total physical flux in the area covered by the disk. Equations (2) state that charge neutrality is satisfied (including the background $-N_\phi$) in the multicomponent Coulomb gas (1). The filling factor is given by

$$\nu=\frac{N}{N_\phi}=\cfrac{1}{a_0-\cfrac{b_{01}^2}{a_1-\cfrac{b_{12}^2}{\ddots-\cfrac{b_{r-2,r-1}^2}{a_{r-1}}}}}\equiv\frac{p}{q}. \qquad (3)$$

Since a_0 is odd and positive, a_α, $\alpha>0$, are even and of either sign, and $b_{\alpha,\alpha+1}=\pm1$, these give all the standard fractions; i.e., q is odd and p,q have no common factor.

Strengths of the logarithmic interactions in the Coulomb plasma resulting from (1) are given by the elements of

$$(G_{\alpha\beta})=\begin{pmatrix} a_0 & b_{01} & 0 & \cdots & \\ b_{01} & a_1 & b_{12} & & \\ 0 & b_{12} & a_2 & & \\ \vdots & & & \ddots & \\ & & & & a_{r-1}\end{pmatrix}. \qquad (4)$$

Then (2) becomes (neglecting 1 with respect to N_α)

$$(G_{\alpha\beta}N_\beta)=\begin{pmatrix} N_\phi \\ 0 \\ \vdots \\ 0\end{pmatrix} \qquad (5)$$

and by inversion of (5)

$$\nu=(G^{-1})_{00}=\det G'/\det G \qquad (6)$$

by Cramer's rule, where G' is the $(r-1)\times(r-1)$ matrix with elements $G_{\alpha\beta}'=G_{\alpha\beta}$ for $\alpha,\beta>0$.

A quasiparticle at z may be obtained by inserting $\prod_{\alpha i}(z_{\alpha i}-z)^{f_\alpha}$ with f_α integers into (1). The "fluxes" (or strictly, vorticities) f_α are screened by the generalized Coulomb plasma, producing screening "charges" δN_α *locally* around z,

$$G_{\alpha\beta}\delta N_\beta=-f_\alpha. \qquad (7)$$

Note that the physical electron number $\delta N=\delta N_0$ but f_0 is not the total effective physical flux because the quasiparticles δN_α constitute a backflow.

The statistics of the excitations can be found by generalizing the method of Arovas, Schrieffer, and Wilczek.[14] The phase $e^{i\theta}$ obtained by interchanging two identical excitations is

$$\theta/\pi=-f_\alpha\delta N_\alpha$$
$$=f_\alpha(G^{-1})_{\alpha\beta}f_\beta=\delta N_\alpha G_{\alpha\beta}\delta N_\beta. \qquad (8)$$

Here the direction of interchange is fixed for all ν by demanding that $\theta/\pi=1/q$ for a quasihole in the Laughlin state.[14] For a charge $\delta N=\pm1/q$ excitation in the standard hierarchy

$$\frac{\theta}{\pi}=\cfrac{1}{a_{r-1}-\cfrac{b_{r-2,r-1}^2}{\ddots-\cfrac{b_{01}^2}{a_0}}}, \qquad (9)$$

which can also be obtained from Halperin's equations.[4] Some properties of this expression are given elsewhere.[15] That (9) is independent of the type of excitation will be confirmed below.

The same calculation also gives Berry's phase per unit area due to the effective background magnetic field seen by the excitation as $-f_\alpha\bar{\rho}_\alpha=\delta N/2\pi$, where from (5) $\bar{\rho}_\alpha=(G^{-1})_{\alpha0}/2\pi$ are the average densities and so only excitations with nonzero physical charge see a field, which is the physical field, as one might have expected. These excitations therefore have Landau-level-type spectra, while the neutral excitations are propagating waves.

The set of possible excitations $\{(f_\alpha)|f_\alpha\in\mathbb{Z}\}$ may be regarded as lying on an "excitation lattice" Λ^* in a space $\mathbf{R}^r$. The coordinates f_α are the components of each lattice point in a basis $\mathbf{e}_\alpha^*$, $\alpha=0,\ldots,r-1$, of Λ^* whose *Gram matrix*[16] of scalar products is $(G^{-1})_{\alpha\beta}=\mathbf{e}_\alpha^*\cdot\mathbf{e}_\beta^*$. Thus θ/π is just the "squared length" (norm) of a vector in the lattice (not necessarily positive since G^{-1} is not necessarily positive definite). A transformation $\mathbf{e}_\alpha^*\to\mathbf{e}_\alpha^{*\prime}=S_{\alpha\beta}\mathbf{e}_\beta^*$ with S having integer matrix elements and determinant 1 changes the basis from $\mathbf{e}_\alpha^*$ to $\mathbf{e}_\alpha^{*\prime}$ but leaves the structure invariant.

For excitations (f_α) such that (δN_α) are all integers, one sees that the wave function is that obtained by adding or subtracting electrons or quasiparticles at z. Thus, as for Laughlin's states,[9] such combinations of fluxes are equivalent to adding or removing particles. Therefore a composite operator which adds such fluxes and compensating (quasi)particles has no net charge δN_α and exhibits long-range order; it is an *order parameter.* Pure states, with nonvanishing order-parameter expectations, are constructed[9] by taking linear combinations of states of different N_α with definite phases θ_α. Fluctuations in N_α change the quasiparticle distribution at the edge but leave the filling factor in the bulk unchanged.

The order parameters are in one-to-one correspondence with the integer-charged excitations that they contain, which form an r-dimensional sublattice Λ of Λ^*, which I call the "condensate lattice." By (8), the Gram matrix of Λ is G, so all scalar products of vectors in Λ are integers; i.e., Λ is an *integral lattice*. Λ^* is the dual lattice of Λ since it has the inverse Gram matrix,[16] and becomes an integral lattice if rescaled by $\sqrt{q}=\sqrt{\det G}$. The sublattice $\Lambda^\perp$ consisting of vectors of Λ having zero physical charge ($\delta N=\delta N_0=0$ in the original basis) has Gram matrix G'. $\Lambda^\perp$ is an *even* lattice (norms of these vectors are even because a_α are even for $\alpha>0$), and so these excitations have Bose statistics. It is easy to show from the form of (4) that they exhaust the neutral excitations, i.e., $(\Lambda^*)^\perp=\Lambda^\perp$, and hence that in the standard hierarchy the statistics of an excitation depends only on its charge δN.

These results imply that the form of the GL action[8-10] must be $S=\int d^2x\,dt\,L$, with

$$L=\tfrac{1}{2}(\partial_\mu\theta_\alpha-A_\mu\delta_{\alpha,0}-\mathcal{A}_{\mu\alpha})C^{\mu\nu}_{\alpha\beta}(\partial_\nu\theta_\beta-A_\nu\delta_{\beta,0}-\mathcal{A}_{\nu\beta})$$
$$+\bar{\rho}_\alpha(\partial_0\theta_\alpha-A_0\delta_{\alpha,0}-\mathcal{A}_{0\alpha})$$
$$+\frac{1}{4\pi}\varepsilon^{\mu\nu\lambda}\mathcal{A}_{\mu\alpha}(G^{-1})_{\alpha\beta}\partial_\nu\mathcal{A}_{\lambda\beta}\,, \qquad (10)$$

where $\mu,\nu,\lambda=0,1,2$ are space-time indices, the A_μ are the physical gauge potentials, $\mathcal{A}_{\mu\alpha}$ are internal gauge potentials, $C^{\mu\nu}_{\alpha\beta}=\eta^{\mu\nu}C^{(\mu)}_{\alpha\beta}$, $\eta^{\mu\nu}=\mathrm{diag}(1,-1,-1)$, and $C^{(\mu)}$ are arbitrary positive-definite matrices. The θ_α can be regarded as coordinates on a torus $\mathbf{R}^r/2\pi\Lambda^*$ and hence vortices are labeled by their flux $\int d^2x\nabla\times(\mathcal{A}_\alpha+A\delta_{\alpha,0})=2\pi f_\alpha$ and (7) and (8) follow from the final Chern-Simons term in (10). The second term gives the effective fields.

Jain's first construction[6] used wave functions χ_r for r filled Landau levels (LLs):

$$\chi_{(n_1+1/r_1)^{-1}}=(\chi_1)^{n_1}\chi_{r_1}\,, \qquad (11)$$

where $\nu=(n_1+1/r_1)^{-1}$ and n_1 is even. Even for $r_1>1$, (11) has nonzero projection to the lowest LL when $n_1>0$, the $\bar{z}$'s becoming $\partial/\partial z$'s. The resulting state may be described in terms of "fictitious LL's" or by saying that the electrons have been divided into r_1 species, each species having a different number of $\bar{z}$ factors for each electron. Thus the wave function is very close in form to a multicomponent Coulomb plasma (for χ_r itself we have r decoupled Coulomb plasmas and so the GL theory for $\nu=r$ is r copies of that for $\nu=1$). Excitations can be made by introducing holes into a single fictitious LL (or inverse powers to obtain quasielectrons). The different LL quasiholes are orthogonal in the thermodynamic limit, by a Coulomb-gas calculation, because the large number of factors of the form (z_i-z) act on distinct sets of particles. In fact, such arguments show[17] that the system exhibits a *spontaneous breakdown of permutation symmetry* and one can ignore the antisymmetrization of electrons among the species. Consequently, the $\bar{z}$ factors can be omitted and the system behaves just as an r_1 component Coulomb plasma, in which the Gram matrix G clearly has diagonal elements n_1+1 and off-diagonal n_1. These entries refer to a basis $\mathbf{e}_\alpha$ for Λ of equally charged excitations $\delta N=-1$ so the basis order parameters consist of one added electron and one of the flux combinations $\mathbf{e}_\alpha$.[18]

To make contact with the standard hierarchy, I now change basis. As the first basis vector take $\mathbf{e}_0$ which has norm n_1+1. For the remainder take $\mathbf{e}'_\alpha=\mathbf{e}_\alpha-\mathbf{e}_{\alpha-1}$, $\alpha=1,\dots,r-1$, which have $\delta N=0$ and norm 2. The off-diagonal scalar products give -1 for adjacent members of the sequence and zero otherwise. The new Gram matrix is therefore tridiagonal like (4), proving that quantum numbers and statistics of excitations are the same as those of the standard hierarchy at the same filling factor (as can also be shown by direct calculation of Λ^*). $\Lambda^\perp$ is here the root lattice A_{r-1} of SU(r),[16] and the r species behave as the fundamental representation of this group, though there is no reason why the Hamiltonian should respect all of this symmetry.

Another set of filling factors $\nu=(n_1-r_1^{-1})^{-1}$, $r_1>1$, is obtained using the conjugate of χ_{r_1} in (11), or powers n_1-1, n_1 in the Coulomb plasma, and leads in the hierarchy basis to -2 in place of 2 in G; SU(r) "symmetry" is still present. Jain has emphasized[6] that these two families include most of the experimentally observed filling factors.

Given a state χ_ν, a new filling factor is obtained[6] by adding electrons in new fictitious LLs and then attaching flux to all the particles:

$$\chi_\nu\to\chi_{\nu'}=(\chi_1)^n\chi_{r+\nu}\,, \qquad (12)$$

where n is even and $\nu'=[n+1/(r+\nu)]^{-1}$, giving a "new" hierarchy of states labeled by sequences $n_1,r_1,n_2,r_2,\dots,n_k,r_k$ for k steps. Once again there is a basis for Λ of $\delta N=-1$ excitations, one for each of the $r=\sum_{i=1}^k r_i$ species. Now take $\mathbf{e}_0$ to be one of the last set of r_k fluxes, and the $\mathbf{e}'_\alpha$ to be differences of the $\delta N=-1$ basis vectors, working back down the hierarchy. The resulting tridiagonal Gram matrix has diagonal n_k+1, 2 (r_k-1 times), $n_{k-1}+2$, 2 ($r_{k-1}-1$ times),$\dots$,2, and off-diagonal elements -1, which is the standard hierarchy form (4). Including negative entries in $n_1,\dots,r_k$ gives all the standard hierarchy states.

The variant hierarchy[5] is sufficiently similar to the standard one not to require separate discussion here; it again produces the same lattices Λ^* of excitations.

The hierarchy construction can be generalized by taking an arbitrary Gram matrix G, whose matrix elements specify a ground state as in (1). In this basis, G_{00} must be odd because of Fermi statistics, and the other diagonal elements even, and so $\Lambda^\perp$ is even. Inequivalent lattices give inequivalent FQH states. Then $\nu=p/q$ where $q=\det G$ and $p=\det G'$ may have common factors. Note that ν need not have odd denominator. Equations (5), (7), (8), and (10) continue to hold. This very large set of possible states is just those having a basis of order parameters containing a single electron since a basis for Λ of $\delta N=-1$ (or a Jain-type construction) can always be obtained. An elegant example is obtained by replacing

G' by the Gram matrix of D_{r-1}, the root lattice of $SO(2(r-1))$,[16] $r>4$. Taking $G_{00}=m$, odd and depending on how G' is extended to G, one can obtain $\nu=1/(m-1)$or $\nu=1/(m-2)$, and so reproduce $\nu=1/q$ but with a lattice of dimension r.

States with some or all of the spins of the electrons reversed can be handled similarly; one of the δN_a is identified as δS^z. As examples, Halperin's $\nu=2/(2n+1)$ spin-singlet states[19] have the same lattice structure[15] as Jain's construction (11) for $r_1=2$, while a singlet state proposed by Jain[6] for $\nu=\frac{1}{2}$ is equivalent to that in Ref. 20.

The present results should shed light on the fractionally charged edge excitations.[7,21] Also, on surfaces of nontrivial topology, like the torus, general principles[15] imply a ground-state degeneracy[21] in the thermodynamic limit, the degeneracy being given by a factor $|\Lambda^*/\Lambda| = \det G = q$ for each "handle." For the hierarchy states, p,q have no common factors, so this is just the minimal degeneracy q for the torus found by Haldane.[22]

The hierarchy for the anyon liquid[12] parallels that for the FQHE for bosons (for which Λ is even) with the statistics parameter $\alpha_s=\theta/\pi$ playing the role of ν; I find that the space of order parameters is $r+1$ dimensional for an r-level fraction.

In conclusion, I have shown the existence of previously unnoticed structure in the hierarchy schemes which characterizes these states completely at the GL level. This classifies all states having only single-electron condensates.

I am grateful to J. K. Jain, G. Moore, X.-G. Wen, and S. Sachdev for useful discussions, and the Alfred P. Sloan Foundation for a fellowship. While completing this paper I received a preprint from B. Blok and X.-G. Wen which contains some of these results.

[1]R. B. Laughlin, Phys. Rev. Lett. **50**, 1395 (1983).

[2]See, e.g., *The Quantum Hall Effect*, edited by S. Girvin and R. Prange (Springer-Verlag, New York, 1990), 2nd ed., for a review.

[3]F. D. M. Haldane, Phys. Rev. Lett. **51**, 605 (1983).

[4]B. Halperin, Phys. Rev. Lett. **52**, 1583 (1984).

[5]R. B. Laughlin, Surf. Sci. **141**, 11 (1984); S. M. Girvin, Phys. Rev. B **29**, 6012 (1984); A. H. MacDonald and D. B. Murray, *ibid.* **32**, 2707 (1985); A. H. MacDonald *et al.*, *ibid.* **31**, 5529 (1985).

[6]J. K. Jain, Phys. Rev. Lett. **63**, 199 (1989); Phys. Rev. B **40**, 8079 (1989); **41**, 7653 (1990).

[7]C. W. J. Beenakker, Phys. Rev. Lett. **64**, 216 (1990); A. H. MacDonald, Phys. Rev. Lett. **64**, 220 (1990).

[8]S. M. Girvin, in *The Quantum Hall Effect* (Ref. 2); S. M. Girvin and A. H. MacDonald, Phys. Rev. Lett. **58**, 1252 (1987).

[9]N. Read, Phys. Rev. Lett. **62**, 86 (1989).

[10]S. C. Zhang, T. H. Hansson, and S. Kivelson, Phys. Rev. Lett. **62**, 82 (1989).

[11]R. B. Laughlin, Phys. Rev. Lett. **60**, 2677 (1988).

[12]D.-H. Lee and M. P. A. Fisher, Phys. Rev. Lett. **63**, 903 (1989).

[13]R. B. Laughlin, in *The Quantum Hall Effect* (Ref. 2).

[14]D. P. Arovas, J. R. Schrieffer, and F. Wilzcek, Phys. Rev. Lett. **53**, 722 (1984).

[15]G. Moore and N. Read, Yale University report (to be published).

[16]J. H. Conway and N. J. A. Sloane, *Sphere Packings, Lattices, and Groups* (Springer-Verlag, New York, 1988).

[17]N. Read (to be published).

[18]B. Blok and X.-G. Wen, Institute for Advanced Study report (to be published).

[19]B. Halperin, Helv. Phys. Acta **56**, 75 (1983); see also F. D. M. Haldane, in *The Quantum Hall Effect* (Ref. 2).

[20]D. Yoshioka, A. H. MacDonald, and S. M. Girvin, Phys. Rev. B **38**, 3636 (1989); D.-H. Lee and C. L. Kane, Phys. Rev. Lett. **64**, 1313 (1990); see also E. H. Rezayi, Phys. Rev. B **39**, 13541 (1989).

[21]X.-G. Wen, Phys. Rev. Lett. **64**, 2206 (1990); Institute for Advanced Study reports (to be published); X.-G. Wen and Q. Niu, Phys. Rev. B **41**, 9377 (1990).

[22]F. D. M. Haldane, Phys. Rev. Lett. **55**, 2095 (1985).

Chapter 3

EFFECTS OF GLOBAL TOPOLOGY

3.1 Introduction

In Chapter 2 we discussed the physical origin of the stability of rational fraction filling-factor states. We made no attempt there to reconcile the resulting fractional Hall conductivities with the arguments of Chapter 1 which seem to legislate integral quantization of the Hall conductance. There must be a weak link in the chain of reasoning in Chapter 1, and it can be revealed by questioning the tacit assumption of a ground state with the same periodicity as the boundary conditions. If the filling fraction is $\nu = p/q$, we can obtain the correct Hall conductivity by abandoning this assumption and having at least q degenerate states that roll over into one another as we twist the boundary condition angles. The necessity of such a set of ground states was pointed out by Tao and Wu in [rep.22] and in the seminal Niu-Thouless-Wu paper [rep.6].

These ground-state multiplets were briefly controversial. A q-fold ground state had actually appeared in the numerical work of Yoshioka *et al.* [1] and Su [2], but comparable calculations by Haldane and Rezayi [3] showed no such degeneracy. There was some difference in formulation between the various calculations — Yoshioka *et al.* and Su worked with toroidal boundary conditions while Haldane and Rezayi preferred spherical geometry — but conventional wisdom asserts that boundary conditions should not affect the degeneracy in the thermodynamic limit. Haldane argued [4] that the observed degeneracy was of no physical significance. He pointed out that a q-fold multiplet is present for every state on a torus quite independently of the presence of an incompressible quantum Hall phase: the degenerate states are merely copies of each other with a shifted center of mass and, if equivalence under "large" gauge transformations is taken into acount, they should not be counted as distinct. These statements are true, but one must be careful in interpreting them: the "large" gauge transformations are precisely those disconnected parts of the gauge group that cause spectral flow in the topological calculations. When the system is incompressible, we have the additional feature that the adiabatic theorem can be applied, and then the degeneracy is exactly what is needed to make the topological arguments give the correct Hall conductivity.

It is significant that Haldane and Rezayi found no degeneracy in their spherical formulation. The "conventional wisdom" on insensitivity to boundary conditions is based on experience with spontaneous symmetry breaking. The incompressible quantum Hall states do not seem to have this (at least not in the conventional sense). They possess instead what Wen calls *topological order*: the number of degenerate ground states is sensitive to the global connectivity of the space in which the electrons move and, for any given quantum Hall phase, this number is a topological invariant. The reason for the topological invariance can be traced to the observation that, when restricted to the incompressible ground state, the bulk flow of the of the QHE fluid is irrotational and incompressible, and so may be described by a Chern-Simons action. The resulting system is the simplest example of a *topological field theory* [5,6]. In the following sections we will explore some of these topological features.

3.2 Wavefunctions on Tori

Let us explore the consequences of giving the electron gas periodic boundary conditions. To do this it is convenient to use a slightly different form for the Landau gauge in this section. We will set

$$\hat{H} = -\frac{1}{2}(\partial_x + iBy)^2 - \frac{1}{2}\partial_y^2 \tag{3.2.1}$$

on an $L_x \times L_y$ rectangle. With $p = 2\pi m/L_x$ the sum

$$\psi_m(x,y) = \sum_{n=-\infty}^{\infty} e^{-\frac{B}{2}(y+nL_y+p/B)^2 + i(p+nL_yB)x} \tag{3.2.2}$$

is a lowest Landau-level eigenstate satisfying periodic boundary conditions in the x direction, and (as expected after the discussion in section 1.6) with twisted periodicity in the y direction. If we define the integer k to be the number of flux units through the rectangle, $2\pi k = BL_xL_y$, there are k independent states of the form (3.2.2), and they span the lowest Landau level.

We can begin to make contact with the extensive literature on the theory of functions on Riemann surfaces by defining the modular parameter $\tau = iL_y/L_x$ and, as usual, setting $z = x + iy$. Then (3.2.2) can be written

$$\psi = e^{-\frac{B}{2}y^2}\theta_m^{\{k\}}(\frac{z}{L_x}|\tau). \tag{3.2.3}$$

Here $\theta_m^{\{k\}}(z|\tau)$ is a *theta function of level k*. These functions are defined by

$$\theta_m^{\{k\}}(z|\tau) = \sum_{\tilde{n}\in\mathbf{Z}+m/k} e^{i\pi\tau k\tilde{n}^2 + 2\pi i\tilde{n}kz}, \tag{3.2.4}$$

and are often seen in papers on string theory. In terms of the more familiar theta functions with characteristics [7]

$$\theta\begin{bmatrix} a \\ b \end{bmatrix}(z|\tau) = \sum_{n\in\mathbf{Z}} e^{i\pi\tau(n+a)^2 + 2\pi i(n+a)(z+b)}, \tag{3.2.5}$$

the level k functions are

$$\theta_m^{\{k\}}(z|\tau) = \theta\begin{bmatrix} m/k \\ 0 \end{bmatrix}(kz|k\tau). \tag{3.2.6}$$

Theta functions have no poles and are doubly periodic up to some extra factors

$$\begin{aligned} \theta\begin{bmatrix} a \\ b \end{bmatrix}(z+m|\tau) &= e^{2\pi i a m}\theta\begin{bmatrix} a \\ b \end{bmatrix}(z|\tau) \\ \theta\begin{bmatrix} a \\ b \end{bmatrix}(z+m\tau|\tau) &= e^{-2\pi i b m}e^{-\pi i m^2\tau - 2\pi i m z}\theta\begin{bmatrix} a \\ b \end{bmatrix}(z|\tau). \end{aligned} \tag{3.2.7}$$

Inserting these properties into the definition of the level k theta functions we find

$$\begin{aligned}\theta_m^{\{k\}}(z+1|\tau) &= \theta_m^{\{k\}}(z|\tau)\\ \theta_m^{\{k\}}(z+\tau|\tau) &= \{e^{-i\pi\tau-2\pi iz}\}^k\,\theta_m^{\{k\}}(z|\tau).\end{aligned} \tag{3.2.8}$$

Because of the extra factors, theta functions are only quasi-periodic and must be regarded as sections of a holomorphic line bundle over the 1 by τ torus, not as functions on the torus. Genuine functions on an ω_1 by ω_2 torus are truly periodic

$$f(z+\omega_1) = f(z+\omega_2) = f(z). \tag{3.2.9}$$

Liouville's theorem shows that truly periodic functions must have poles if they are not to be mere constants. They must also satisfy two obvious constraints which follow from periodicity and Cauchy's theorem: firstly, in any ω_1 by ω_2 period parallelogram, the sum of the residues at the poles must be zero; secondly, there must be exactly as many poles as zeros in this region.

There is another rather less obvious condition that doubly periodic functions must satisfy. This follows from evaluating the integral

$$I = \frac{1}{2\pi i}\int z\frac{f'(z)}{f(z)}dz \tag{3.2.10}$$

over a contour forming the boundary of a period parallelogram. If the poles of $f(z)$ in the period parallelogram are at $z = b_i$ and the zeros at $z = a_i$, then there exist integers n, m such that

$$I = \sum_i a_i - \sum_i b_i = n\omega_1 + m\omega_2. \tag{3.2.11}$$

In words, the sum of the poles minus the sum of the zeros vanishes modulo periods. These three constraints on the location of the poles and zeros provide a necessary and sufficient condition for the existance of a meromorphic function on a torus with these poles and zeros.

There is no analog of (3.2.11) for functions on the Riemann sphere. There we may construct a rational function,

$$f(z) = A\frac{\prod_{i=1}^n (z-a_i)}{\prod_{j=1}^n (z-b_j)}, \tag{3.2.12}$$

with poles and zeros in any prescribed positions. On general Riemann surfaces of genus g ($g = 0$ for a sphere and $g = 1$ for a torus), eq. (3.2.11) generalizes to a set of g conditions which are again necessary and sufficient for the existence of a function with the prescribed singularities. A meromorphic function is thus specified by: i) the position of n poles, ii) the positions of n zeros, iii) one overall multiplicative constant, and iv) g complex valued constraints. We say therefore that the complex dimension of the *moduli space* of meromorphic functions with n poles is $(2n+1-g)$. This is the content of *Abel's theorem* [7], and it is a primary example of the interplay between analyticity and topology.

The constraint (3.2.11) has consequences for our wavefunctions on a torus — even though they are sections, not functions. Take any two wavefunctions, expressed as linear combinations of $\theta_m^{\{k\}}(z|\tau)$. They will both transform under translations as in (3.2.8), and when we divide one by the other the "extra factors" will cancel. The ratio of the wavefunctions is thus a genuine meromorphic function with zeros at the zeros of the numerator and poles at the zeros of the denominator. Now each wavefunction has exactly k zeros (as may be seen by applying the principal of the argument), and it follows that the sum of these zeros is a constant, modulo periods, for all functions in the lowest Landau level. The actual value of the constant will depend on the choice of the twisted boundary condition angles.

This constraint on the "center of mass" of the zeros is important when we set out to construct many-body wavefunctions that are the toroidal analogs of Laughlin's wavefunction. We need the analytic part of the many-body wavefunction to transform as (3.2.8) in each of its arguments, and to have the same energetic advantages as the original Laughlin wavefunction we must place multiple zeros of the wavefunction at the location of the other electrons. Haldane and Rezayi found a very pretty way to do this [rep.23]. They use Jacobi's notation $\theta_1(\pi z)$ for the odd theta function

$$\theta_1(\pi z|\tau) = \theta \begin{bmatrix} 1/2 \\ 1/2 \end{bmatrix} (z|\tau) \tag{3.2.13}$$

which has its solitary zero at the origin. They try a wavefunction with Laughlin-like factors $\prod_{i<j}\{\theta_1(\pi(z_i - z_j))\}^m$ where $m = N_s/N_e$ is the inverse of the filling fraction, *i.e.*, ratio of the number of available states per electron to the number of electrons. As a function of z_1, this function has m zeros at the position of each of the $(N_e - 1)$ other electrons. The total wavefunction, however, needs N_s zeros, and their sum must be independent of the positions z_i of the other electrons. Since $N_s - m(N_e - 1) = m$, we must supply another m zeros located at points $z_1 = a_n - \sum_{i\neq 1} z_i$. This can be achieved by inserting an extra factor $F(Z)$ which has m zeros as function of $Z = \sum_i z_i$:

$$\psi(z_1, \dots, z_{N_e}) = F(Z) \prod_{i<j} \{\theta_1(\pi(z_i - z_j))\}^m. \tag{3.2.14}$$

The translation properties required of ψ shows $F(Z)$ to be a level $k = m$ theta function. There are m of these, so ψ has m degenerate center-of-mass states for each internal state.

We can continue, and so find the quasihole wavefunctions. If we create N_h quasiholes at ζ_i, we discover that we must take $N_s = mN_e + N_h$ and the wavefunction is

$$\psi(z_1, \dots, z_{N_e}; \zeta_1 \dots, \zeta_{N_h}) = F(Z + \frac{\sum_j \zeta_j}{m}) \prod \theta(\pi(z_i - \zeta_j)) \prod_{i<j} \{\theta_1(\pi(z_i - z_j))\}^m. \tag{3.2.15}$$

Now, one of the properties of the level k theta functions is that they turn into each other under translations through $1/k$ of a period

$$\theta_m^{\{k\}}(z + \frac{l\tau}{k}|\tau) = e^{-i\pi\tau l^2/k} e^{-2\pi i z l} \theta_{m+l \bmod k}^{\{k\}}(z|\tau). \tag{3.2.16}$$

The different θ_m are characterized by their response to translation in the other direction

$$\theta_m^{\{k\}}(z + \frac{1}{k}|\tau) = e^{2\pi i m/k} \theta_m^{\{k\}}(z|\tau). \tag{3.2.17}$$

It follows that, as a function of the location of the hole, the quasihole state has a period m times larger in the τ direction than the electron states that compose it. As we take a hole round one generator of the torus we find that the wavefunction acquires a constant phase factor $e^{2\pi i m/k}$, but taken around the other generator the wavefunction returns with its m quantum number increased, or decreased, by one.

As a family of states, the $\psi_l(z)$ form a bundle over the extended, 1 by $m\tau$, tori and are twisted by holomorphic transition functions

$$\begin{aligned}\psi_l(\zeta+m\tau) &= \{e^{-i\pi\tau-2\pi i\zeta}\}^{N_s}\,\psi_l(\zeta)\\ \psi_l(\zeta+1) &= \left\{e^{2\pi i l/m}\right\}\psi_l(\zeta),\end{aligned} \tag{3.2.18}$$

which are identical to the twisted periodicity of the holomorphic parts of the wavefunctions themselves. From this similarity, it follows that the Berry phase that appears as we adiabatically move the hole has total "flux", or Chern character, of N_s. The Berry-phase flux through *one* copy of the torus is N_s/m, and this is the appropriate flux for transferring knowledge of the $\frac{e}{m}$ charge to a linked adibatically moving mass.

We see that, on a torus, following the statistical phases as they accumulate from the braiding of quasihole worldlines requires us to keep track of another integer quantum number: the center-of-mass label of the background electron state. The necessity for such an additional "vacuum"-state quantum number for a consistent description of toroidal anyons was derived from the properties of braid group representations by Einarsson [8]. Einarsson also pointed out that the Haldane-Rezayi states provide a concrete example. These results were also independently obtained by Wen, Dagotton and Fradkin [9].

The appearence of extra "vacuum" quantum numbers on a torus is suggestively similar to the extra labels needed to specify the holomorphic parts correlation functions (the *conformal blocks*) of chiral field theories on tori and higher genus surfaces. Morover, the manner in which transport of quasiparticles round the generators takes us from one internal state to the next exactly parallels the action of some operators introduced into conformal field theory by Verlinde [10]. The Verlinde operations consist of carrying a conformal field round one of the two generators of the torus. This operation alters the states appearing in the partition function, and transforms the Virasoro characters into linear combinations of one another. Verlinde conjectured that the action of his operators on the characters was related to the fusion rules for the conformal fields. This was later proved by Moore and Seiberg [11], and the algebra of these global braids and fusions, the *Verlinde algebra*, has played in important role in the classification of Rational Conformal Field Theories. In our context, Verlinde's conjecture relates the effect on the degenerate ground states of taking a quasihole round the torus to the rules for adding the quasihole quantum numbers.

3.3 Non-Abelian Statistics

In the plane, adiabatic transport of the Laughlin state quasiholes gives rise to a Berry phase which forms an abelian representation of the braid group. On a torus, as we have seen, the action is more complicated, stirring up the state of the background electrons. It may be possible to find quantum Hall phases where the creation of quasiholes, having made the remainder of the electron fluid multiply connected, also mimics the phase twisting of the toroidal boundary conditions. If this happens it gives rise to the possibility of several states, $|l;\zeta_1,\dots,\zeta_{N_h}\rangle$ labeled by the quantum number l in addition to the locations of the quasiholes. The action of braiding one quasihole round the others will then include an effect of the l label

$$|l;\zeta_1,\dots,\zeta_{N_h}\rangle \to \sum_m M_{lm}|m;\zeta_1,\dots,\zeta_{N_h}\rangle. \tag{3.3.1}$$

The matrix M_{lm} will give a non-abelian representation of the braid group, and the anyon statistics of the Laughlin quasiholes has been generalized to ***non-abelian statistics***.

When there are several kinds of quasiholes, each associated with a different representation of some internal symmetry group, the braiding matrix will depend on which representation, Λ_i containing the holes. (It will not, however, depend on the particular state within the representation). Then (3.3.1) becomes

$$|l;\zeta_1,\ldots,\zeta_{N_h}\rangle \to \sum_m M_{lm}[\Lambda_1\ldots\Lambda_n]|m;\zeta_1,\ldots,\zeta_{N_h}\rangle, \tag{3.3.1a}$$

which is the starting point of the discusion of Moore and Read [15] and [rep.28].

This mixing of states as one variable circles another is reminiscent of a similar phenomenon that occurs with solutions of ordinary linear differential equations near a singularity. At a regular singularity an analytic solution will typically have a branch cut. Continuing the solution around the singularity will take us to a new sheet, but the function on this sheet still solves the equation and must be a linear combination of solution functions on the original sheet. In this way a *monodromy matrix* is associated with each singular point.

If the Berry connexion for transport of the quasiholes is flat, we can obtain our statistical monodromy from the set of linear differential equations, $\nabla_{z_i}|z_1,\ldots,z_n\rangle = 0$, defining the transport. A candidate flat connexion that generalizes the abelian adiabatic transport of holes was provided by Kniznik and Zamolodchikov in the relation obeyed by the holomorphic part of the correlators of conformal fields. The Kniznik-Zamolodchikov (KZ) equations[12], one for each z_i, are

$$\left(\partial_{z_i} + \lambda\sum_{j\neq i}\frac{\mathbf{t}_a^{(i)}\mathbf{t}_a^{(j)}}{z_i - z_j}\right)\langle\varphi_1(z_1)\ldots\varphi_N(z_N)\rangle = 0. \tag{3.3.2}$$

The independent solutions of this equation are the *conformal blocks* referred to in the previous section. The matrices $\mathbf{t}_a^{(i)}$ act on the internal indices of the i-th field. They commute for $i\neq j$, since they are acting on different spaces, and for $i = j$ they are the generators, in the representation Λ_i, of a Lie algebra, $[\mathbf{t}^a,\mathbf{t}^b] = if_{abc}\mathbf{t}^c$. (We take a basis such that $\mathrm{tr}\,(\mathbf{t}^a\mathbf{t}^b)\propto\delta_{ab}$ so f_{abc} is antisymmetric.)

Eq (3.3.2) can be written as

$$\nabla_{z_i}\langle\varphi_1(z_1)\ldots\varphi_N(z_N)\rangle = 0, \tag{3.3.3}$$

with the covariant derivatives incoporating a connexion†

$$\omega = \lambda\sum_{j\neq i}\frac{\mathbf{t}_a^{(t)}\mathbf{t}_a^{(j)}}{z_i - z_j}dz^i. \tag{3.3.4}$$

For the system (3.3.4) to be integrable we need $[\nabla_{z_i},\nabla_{z_j}] = 0$ for all pairs i and j, *i.e.*, we need the connection to be flat:

$$d\omega + \omega\wedge\omega = 0. \tag{3.3.5}$$

†The objection may be raised that the Berry connexion (2.4.12) contained both dz and $d\bar{z}$ parts while the present connexion lies only in the space of holomorphic differentials. This difference is only cosmetic: after removing the non-holomorphic normalization factors, (2.4.12), with the exception of the background curvature due to the charge of the quasiholes, can be written as an Abelian form of (3.3.4).

It is easily verified that ω does have this property at all non-coincident z_i, and for all values of λ since both $d\omega$ and $\omega \wedge \omega$ are separately zero. The latter condition follows from the antisymmetry of f_{abc} and the identity

$$0 = \frac{1}{(z_i - z_j)(z_j - z_k)} + \frac{1}{(z_j - z_k)(z_k - z_i)} + \frac{1}{(z_k - z_i)(z_i - z_j)}. \tag{3.3.6}$$

In conformal current algebra the most interesting cases occur when the parameter λ takes the special values $\lambda = -\frac{1}{2}(k + c_V)$, where k is an integer called the *level*, and c_V is defined by $\delta^{ab} c_V = f^{acd} f^{bcd}$. The level k is a parameter inherited from an underlying Kac-Moody algebra. Its origin and significance will be discussed in Chapter 5.

In field theory, the symmetry requires the correlators to be a global singlet, so the group indices carried by the field operators must combine to give invariant tensors, $T_{i_1 \ldots i_n}$. These are defined by the property

$$U_{i_1 j_1}(g) U_{i_2 j_2}(g) \ldots U_{i_n j_n}(g) T_{j_1 \ldots j_n} = T_{i_1 \ldots i_n} \tag{3.3.7}$$

for the relevant representation matrices $U(g)$. Examples of such tensors for $SU(N)$ are $f_{abc} = \mathrm{tr}\,(\mathbf{t}^a[\mathbf{t}^c, \mathbf{t}^c])$, $d_{abc} = \mathrm{tr}\,(\mathbf{t}^a\{\mathbf{t}^c, \mathbf{t}^c\})$ or the $t^a_{\alpha\beta}$ themselves. The n-point correlators thus live in the space $Inv\,(W_1 \otimes W_2 \otimes \ldots W_n)$ of invariant tensors with z_i dependent coefficients. It is a simple exercise to show that parallel transport via ω preserves this space, so the monodromy matrices act within it, and the indices l, m in (3.3.1) may be taken to label a basis of independent invariant tensors. In this way, the dimension of the space of solutions is related to the number of possible invariant tensors. If we take, for example, the four-point correlator in an $SU(2)$ theory, with all four fields carrying spin-$\frac{1}{2}$ indices, we can make singlets by coupling the spins in pairs to make either an $S = 0$ or an $S = 1$ representations, and then coupling these to make singlets. There are therefore two invariant tensors and, barring accidents, there will be two solutions. Fixing three of the operator insertions and varying the location of the last gives us a 2×2 matrix-valued first-order ODE in one variable, with three regular singular points at the location of the other z_i. This equation can always be written as a scalar second-order equation with three regular singular points. The solutions therefore are hypergeometric functions. They are discussed in [12].

Because of their connection with knots, braids, quantum groups, and Yang-Baxter equations, the KZ equations have been extensively studied by both physicists and mathematicians. The "quantum group" concept [13] and the related notion of a quasi-Hopf algebra, is related to the rules for combining the internal representations being slightly different from those of the underlying Lie algebra. The representation addition laws depend on the level k. For $SU(2)$ at level 1, for example, the spin-1 representation does not exist, and on combining two spin-$\frac{1}{2}$ representations only the singlet representation results. The space of solutions to the KZ equation is then only one dimensional. The objects with these Clebsh-Gordon series are the q-analogue of the Lie algebra where the q parameter takes the value $q = e^{-i\pi\lambda}$. A discussion of the connection between quantum groups, the fusion rules, and the KZ connexion may be found in [14].

There are several different suggestions about how to construct quantum Hall states with quasi-holes obeying these non-abelian statistics [15] [rep.27] [rep.28]. In each case, the conformal field operators are replaced by the quasiholes, and we will see in Chapter 5 how these holes acquire group representation labels as internal quantum numbers. On a closed surface, the representations again combine to form singlets, so much of the conformal field theory discussion is unchanged.

Reprints for Chapter 3

[rep.22] *Gauge invariance and the fractional quantum Hall effect*, R. Tao, Y-S. Wu, Phys. Rev. B30 (1984) 1097-1098.

[rep.23] *Periodic Laughlin-Jastrow wave functions for the fractional quantum Hall effect*, F. D. M. Haldane, E. H. Rezayi, Phys. Rev. B31 (1985) 2529-2531.

[rep.24] *Ground-state degeneracy of the fractional quantum Hall states in the presence of a random potential and on high-genus Riemann surfaces*, X-G. Wen, Q. Niu, Phys. Rev. B41 (1990) 9377-9396.

[rep.25] *Gauge invariance in Chern-Simons theory on a torus*, Y. Hosotani, Phys. Rev. Lett. 62 (1989) 2785-2788.

[rep.26] *Coulomb gas description of the collective states for the fractional quantum Hall effect*, G. Cristofano, G. Maiella, R. Musto, F. Nicodemi, Mod. Phys. Lett. A6 (1991) 1779-1786.

[rep.27] *Non-abelian statistics in the fractional quantum Hall states*, X-G. Wen, Phys. Rev. Lett. 66 (1991) 802-805.

[rep.28] *Fractional quantum Hall effect and non-abelian statistics*, N. Read, G. Moore, Prog. Theor. Phys. Suppl. 107(1992) 157.

[rep.29] *A note on braid statistics and the non-abelian Aharonov-Bohm effect*, E. Verlinde, in the proceedings of the international colloquium on Modern Quantum Field Theory (World Scientific, 1991), pp. 450–461.

[rep.30] *Singlet quantum Hall effect and Chern-Simons theories*, A. Balatsky, E. Fradkin, Phys. Rev. B43 (1991) 10622-10634.

Other References for Chapter 3

[1] D. Yoshioka, B. I. Halperin, P. A Lee, Phys. Rev. Lett 50 (1983) 1219.

[2] W. P. Su, Phys. Rev. B30 (1984) 1069.

[3] F. D. M. Haldane, E. H. Rezayi, Phys. Rev. Lett. 54 (1985) 237.

[4] F. D. M. Haldane, Phys. Rev. Lett. 55 (1985) 2095.

[5] E. Witten, Comm. Math. Phys. 121 (1989) 351.

[6] X-G. Wen, Int. Jou. Mod. Phys. B4 (1990)239-271; *Topological Orders and Chern-Simons Theory in Strongly Correlated Quantum Liquid*, X-G Wen, preprint Iassns-Hep-91/20; X. G. Wen Phys. Rev. B 40 (1989) 7387.

[7] D. Mumford, *Tata Lectures on Theta I*, Birkhauser, Boston 1983.

[8] T. Einarsson, Phys. Rev. Lett. 64 (1990) 1995.

[9] X-G Wen, E. Dagotto, E Fradkin, Phys. Rev. B42 (1990) 6110.

[10] E. Verlinde Nucl. Phys. B 300 (1988) 360.

[11] G. Moore, N. Seiberg, Phys. Lett. 212B (1988) 451.

[12] V. G. Knizhnik and A. B. Zamolodchikov, *Current Algebra and Wess-Zumina Model in Two Dimensions*, Nucl. Phys. B247 (1984) 88-103.

[13] M. Jimbo, Lett. Math. Phys. 10 (1985) 63.

[14] V. G. Drinfeld, "Quasi-Hopf Algebras and Knizhnik-Zamolodchikov Equations" in *Problems in Modern Field Theory*, A. A. Belavin, A. U. Klimyk, A. B. Zamolodchikov, eds. Springer-Verlag 1989.

[15] *Nonabelions in the Fractional Quantum Hall Effect*, G. Moore, N. Read, Nucl. Phys. B360 (1991) 362-396.

PHYSICAL REVIEW B VOLUME 30, NUMBER 2 15 JULY 1984

Gauge invariance and fractional quantum Hall effect

R. Tao and Yong-Shi Wu
Department of Physics FM-15, University of Washington,
Seattle, Washington 98195
(Received 27 April 1984)

It is shown that gauge-invariance arguments imply the possibility of the fractional quantum Hall effect; the Hall conductance is accurately quantized to a rational value. The ground state of a system showing the fractional quantum Hall effect must be degenerate; the nondegenerate ground state can only produce the integral quantum Hall effect.

The discovery of the fractional quantum Hall effect in a two-dimensional electron gas in a strong magnetic field by Tsui, Störmer, and Gossard[1,2] has prompted a series of interesting theoretical investigations.[3-7] One important and controversial problem is whether the ground state of the fractional quantum Hall system is degenerate. In our many-body theory[4,5] for this effect, the degeneracy is explicit. Anderson[8] also suggests that Laughlin's wave function may have a broken symmetry. On the other hand, Laughlin[4] and Haldane[7] claim that the ground state is nondegenerate.

In a noteworthy paper of 1981, Laughlin[9] showed that integral quantization of the Hall conductance is a consequence of gauge invariance. Can gauge-invariance arguments also imply fractional quantization of Hall conductance? And why are the Hall plateaus at the filling factor $\nu=\frac{1}{3},\frac{2}{3},\frac{2}{5},\ldots$ accurately quantized to rational values,[2] e.g., at $\nu=\frac{1}{3}$ to better than 10^{-4}?

This Rapid Communication presents some answers to the above questions. Gauge-invariance arguments also imply the possibility of fractional quantization of Hall conductance; the Hall conductance is accurately quantized to a rational value. The ground state of a system showing the fractional quantum Hall effect must be degenerate; the nondegenerate ground state can only produce the integral quantum Hall effect. The presence of an energy gap is a necessary but, perhaps, not a sufficient condition for this effect. Our gauge-invariance arguments in this Rapid Communication do not tell which value of the filling factor is more stable and have not explained the odd denominator rule observed in the experiments,[2] but they suggest another possible explanation for this rule: the ground state of the Hall system at a filling factor with an even denominator has a special topological property in Hilbert space.

We consider the geometry proposed by Laughlin:[9] a ribbon of two-dimensional system bent into a loop of circumference L, and pierced everywhere by a strong magnetic field $\vec{B}$ normal to its surface (Fig. 1). We also put a small solenoid at the center of the loop. Initially, the solenoid is not turned on. The radius of the loop is big enough so that the surface of the loop can be considered as a plane. In order to make our system capable of producing Hall current, we assume that electrons can be fed in at one edge and taken away from the other, but we only consider the electrons on the surface. The ground state, Ψ_0, of the two-dimensional electron gas on the surface satisfies

$$H(\vec{p}_1-e\vec{A}_1,\ldots,\vec{p}_N-e\vec{A}_N)\Psi_0=E_0\Psi_0 \ , \tag{1}$$

where $\vec{A}$ is the vector potential of the strong magnetic field $\vec{B}$ in a particular gauge; E_0 is the ground state energy; we set $c=1$. The dependence of the Hamiltonian, H, on $\vec{r}_1,\ldots,\vec{r}_N$ is not explicitly written in the formula. We also assume that an energy gap separates the ground state from the excited states. In the geometry considered here Ψ_0 is periodic in the y direction, with period L,

$$\Psi_0(\vec{r}_1,\ldots,\vec{r}_j+L\hat{y},\ldots)=\Psi_0(\vec{r}_1,\ldots,\vec{r}_j,\ldots) \ , \quad j=1,2,\ldots,N \ . \tag{2}$$

Now we switch the solenoid on and adiabatically increase the magnetic flux of the solenoid from zero to an arbitrary value ϕ. During this process some electrons can be transferred from one edge to the other. Because of the energy gap, the system remains in a ground state which may be different from the original one. The new wave function Ψ is given by

$$H(\vec{p}_1-e(\vec{A}_1+\vec{a}),\ldots,\vec{p}_N-e(\vec{A}_N+\vec{a}))\Psi=E'\Psi \ , \tag{3}$$

where $\vec{a}=a\hat{y}$ is the vector potential of the solenoid at the surface of the loop; E' may depend on a. Let

$$\Psi(\vec{r}_1,\vec{r}_2,\ldots,\vec{r}_N)=\exp\left[i(ea/\hbar)\sum_{j=1}^{N}y_j\right]\times u(\vec{r}_1,\vec{r}_2,\ldots,\vec{r}_N) \ , \tag{4}$$

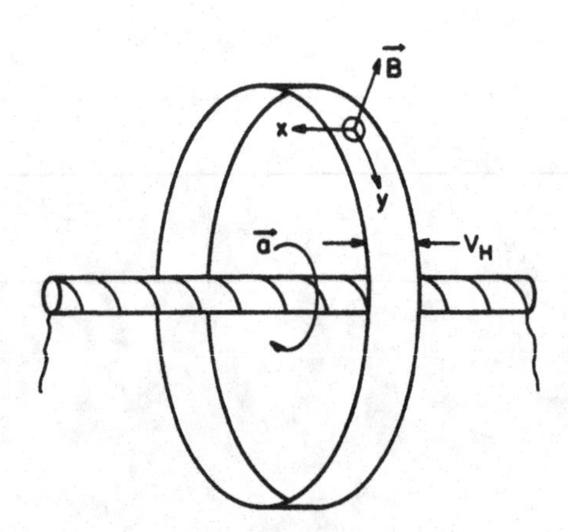

FIG. 1. Diagram of the loop and solenoid.

so that

$$H(\vec{p}_1 - e\vec{A}_1, \ldots, \vec{p}_N - e\vec{A}_N)u = E'u \ . \tag{5}$$

Since the wave function Ψ is periodic in the y direction, we have, from Eq. (4),

$$u(\vec{r}_1, \ldots, \vec{r}_j + L\hat{y}, \ldots) = \exp(-ie\phi/\hbar)u(\vec{r}_1, \ldots, \vec{r}_j, \ldots) \ , \tag{6}$$

$$j = 1, \ldots, N \ ,$$

where $\phi = aL$ is the magnetic flux of the solenoid. If $\phi \neq n\phi_0$ ($\phi_0 = h/e$, n is an integer), u cannot be periodic in the y direction, and u is different from Ψ_0.

Now let us consider the case $\phi = \phi_0$. From Eq. (6), u is also periodic in the y direction. By gauge invariance[10] we must also have $E' = E_0$. This can be proved directly. Because u is also an eigenstate of Hamiltonian $H(\vec{p}_1 - e\vec{A}_1, \ldots, \vec{p}_N - e\vec{A}_N)$ with the lowest eigenvalue and satisfies the same boundary conditions as Ψ_0, u and Ψ_0 must have the same eigenvalue, i.e., $E_0 = E'$. But this does not mean that u and Ψ_0 must be the same. We have two possibilities to consider: either the ground state of the system is nondegenerate or degenerate.

If the ground state is nondegenerate, u is the same as Ψ_0; by gauge invariance, the system simply maps back into the initial state. The net physical result is that N_0 electrons are transferred from one edge to the other. Then as Laughlin[9] showed, the energy increase due to this transfer is

$$\Delta U = N_0 e V_H \ , \tag{7}$$

where V_H is the potential drop from one edge to another. The Hall current is

$$I_H = \partial U/\partial\phi = \Delta U/\phi_0 = V_H N_0 e^2/h \tag{8}$$

and the Hall conductance is

$$\sigma_H = I_H/V_H = N_0 e^2/h \ . \tag{9}$$

Clearly only integral Hall conductance is produced.

If the ground state is degenerate, at $\phi = \phi_0$, u can be different from Ψ_0 though, by gauge invariance, they both are ground states of $H(\vec{p}_1 - e\vec{A}_1, \ldots, \vec{p}_N - e\vec{A}_N)$. Therefore, after the magnetic flux of the solenoid changes from zero to ϕ_0, the system may still not map back into the initial state. We then increase the magnetic flux of the solenoid ϕ to $2\phi_0, 3\phi_0, \ldots$. If the system maps back to Ψ_0 at a finite value of ϕ, say, $p\phi_0$, then we have fractional quantization of Hall conductance. Suppose q electrons in total are transferred from one edge to the other. The same argument as above yields

$$\sigma_H = (q/p)e^2/h \ . \tag{10}$$

The fractional Hall conductance is now accurately quantized to a rational value as in the integral case. The above discussion clearly demonstrates that if a system shows the fractional quantum Hall effect, the ground state of that system must be degenerate. On the other hand, if the system can never come back to its initial state as the magnetic flux ϕ increases, there is no quantization of Hall conductance even though there is an energy gap. For example, if the ground state of a Hall system has an infinite degeneracy, the system may never map back to its initial state.

Gauge-invariance arguments can also relate σ_H to the filling factor. Their relationship can be obtained by considering the angular momentum.[11] The charge carriers here are electrons (or holes). Our argument does not tell which value of filling factor is more stable. This should be determined by calculation of energy gaps. The above discussion does not explain the odd denominator rule observed in the experiments. Usually the guess is that this odd denominator rule is due to absence of an energy gap at fillings with an even denominator. Our gauge-invariance arguments suggest another possibility: the ground state of the Hall system at these fillings has a special topological property in Hilbert space such that it can never come back to its initial state as the magnetic flux of the solenoid increases. We consider this possibility to be more interesting since some recent numerical calculations show the possible presence of energy gaps at filling factors with even denominator.[12] We speculate there is a more profound reason for this rule.

One of us (R.T.) wishes to thank Professor D. J. Thouless for helpful discussions. This work was supported by the National Science Foundation under Grant No. DMR-83-19301 and by the U.S. Department of Energy under Contract No. DE-AC06-81ER-40048.

[1]D. C. Tsui, H. L. Störmer, and A. C. Gossard, Phys. Rev. Lett. 48, 1559 (1982).
[2]H. L. Störmer, A. Chang, D. C. Tsui, J. C. M. Hwang, and W. Wiegmann, Phys. Rev. Lett. 50, 1953 (1983).
[3]R. B. Laughlin, Phys. Rev. Lett. 50, 1395 (1983).
[4]R. Tao and D. J. Thouless, Phys. Rev. B 28, 1142 (1983).
[5]R. Tao, Phys. Rev. B 29, 636 (1984).
[6]B. I. Halperin, Phys. Rev. Lett. 52, 1583 (1984).
[7]F. D. M. Haldane, Phys. Rev. Lett. 51, 605 (1983).
[8]P. W. Anderson, Phys. Rev. B 28, 2264 (1983).
[9]R. B. Laughlin, Phys. Rev. B 23, 5632 (1981).
[10]T. T. Wu and C. N. Yang, Phys. Rev. D 12, 3845 (1975).
[11]R. Tao (unpublished).
[12]See, for example, D. Yoshioka, Phys. Rev. B (to be published).

PHYSICAL REVIEW B VOLUME 31, NUMBER 4 15 FEBRUARY 1985

Rapid Communications

The Rapid Communications section is intended for the accelerated publication of important new results. Manuscripts submitted to this section are given priority in handling in the editorial office and in production. A Rapid Communication may be no longer than 3½ printed pages and must be accompanied by an abstract. Page proofs are sent to authors, but, because of the rapid publication schedule, publication is not delayed for receipt of corrections unless requested by the author.

Periodic Laughlin-Jastrow wave functions for the fractional quantized Hall effect

F. D. M. Haldane

Department of Physics, University of Southern California, Los Angeles, California 90089-0484

E. H. Rezayi

Department of Physics and Astronomy, California State University, 5151 State University Drive, Los Angeles, California 90032

(Received 12 October 1984)

We present Laughlin-Jastrow wave functions for incompressible fluid states of two-dimensional electrons at Landau-level filling factor $1/m$ that satisfy periodic boundary conditions. This rederivation of Laughlin-type states emphasizes that it is correct short-distance behavior of the wave functions rather than angular momentum considerations that lie behind the explanation of the fractional quantized effect.

Laughlin[1] has provided the key to understanding the fractional quantized Hall effect[2] with his construction of Jastrow-type variational wave functions that describe incompressible fluid states of two-dimensional electrons in a magnetic field. As originally formulated, Laughlin's states describe circular fluid droplets containing N_e electrons that expand to provide a uniform cover of the "Hall surface" as $N_e \to \infty$. A variant formulation on a spherical surface has been described by Haldane;[3] this allows homogeneous states with *finite* N_e to be constructed. Recent finite-system studies[4] in this geometry have, we believe, conclusively confirmed that at $\frac{1}{3}$ Landau-level filling the Laughlin-Jastrow wave function describes the essential character of the ground state of systems where the interactions are sufficiently repulsive at short range, and that the Coulomb interaction belongs to this class.

Laughlin-Jastrow (LJ) states have not been previously constructed in the other popular finite-system geometry, namely, the periodic boundary conditions on the plane. In this Rapid Communication we construct such states. The philosophy of our construction is identical to that of Ref. 1; the states described here have the same thermodynamic limit as those of Ref. 1. While no new physics is being described, our construction now makes direct comparison of the LJ state with finite-size results in the periodic geometry possible. The discrete center-of-mass degeneracy of the ground state in this geometry is also made explicit.

In the Landau gauge $\mathbf{A}=-By\hat{x}$, the wave function describing a particle confined to the lowest Landau level has the analytic form

$$\psi(x,y)=\exp(-\tfrac{1}{2}y^2)f(z),\quad z=x+iy\ , \tag{1}$$

where $f(z)$ is an *entire* (holomorphic) function, and length units $\sqrt{(\hbar/eB)}=1$ are used. An essentially similar form occurs in the symmetric gauge,[1] and we emphasize that the following discussion can be carried out in any gauge. The particle translation operator that acts on the wave functions is given (using two-dimensional vector notation with a pseudoscalar cross product) by

$$t(\mathbf{L})=\exp[\mathbf{L}\cdot(\nabla-ie\mathbf{A}/\hbar)-i\mathbf{L}\times\mathbf{r}]\ . \tag{2}$$

We will impose periodic boundary conditions

$$t(\mathbf{L}_\alpha)\Psi=\exp(i\phi_\alpha)\Psi,\quad \alpha=1,2\ , \tag{3}$$

where $\mathbf{L}_1=(L_1,0)$ and $\mathbf{L}_2=(L_2\cos\theta,L_2\sin\theta)$ are two nonparallel displacements. For these boundary conditions to be simultaneously applicable, $t(\mathbf{L}_1)$ and $t(\mathbf{L}_2)$ must commute, i.e.,

$$|\mathbf{L}_1\times\mathbf{L}_2|=2\pi N_s\ , \tag{4}$$

where N_s is an integer. This means that the total magnetic flux though the parallelogram defined by $\mathbf{L}_1$ and $\mathbf{L}_2$ is exactly N_s flux quanta and integral. This region bounded by the four points $z=\frac{1}{2}L_1(\pm1\pm\tau)$, $\tau=L_2e^{i\theta}/L_1$ will be referred to as the principal region.

The boundary conditions used in the study by Yoshioka, Halperin, and Lee[5] are a special case of (3) with $\phi_\alpha=0$. However, because of the noncommutativity of translation operators when a magnetic field is present, the choice of the ϕ_α is not invariant under continuous translations of the center of mass, and the more general form (3) is more appropriate. If the periodic boundary conditions are interpreted as imposing a toroidal topology, the phases ϕ_i can be related to "solenoid fluxes" $\Phi_i=\hbar\phi_i/e$ passing through the two periodic orbits. If the phases are allowed to vary with time, this is equivalent to a uniform electric field (E^x,E^y) where the complex drift velocity $v=(E^y-iE^x)/B$ is given by $v=(d/dt)(L_1\phi_2-L_2e^{i\theta}\phi_1)/2\pi N_s$.

The periodic boundary condition on the wave function (1)

is the condition

$$\frac{f(z+L_1)}{f(z)}=e^{i\phi_1}\ ,$$
$$\frac{f(z+L_2e^{i\theta})}{f(z)}\exp\{i\pi N_s[(2z/L_1)+\tau]\}=e^{i\phi_2}\ . \quad (5)$$

Since $f(z)$ is entire, the integral of $d/dz\{\ln[f(z)]\}$ around the boundaries of the principal region counts the number of zeros of $f(z)$ inside it. The condition (5) fixes this number to be precisely N_s. The possible analytic form of $f(z)$ is thus strongly constrained, and the most general form is expressible as

$$f(z)=\exp(ikz)\prod_{\nu=1}^{N_s}\vartheta_1(\pi(z-z_\nu)/L_1|\tau)\ , \quad (6)$$

where the zeros z_ν are in the principal region, and k is real and in the range $0\leq|k|\leq\pi N_s\,\mathrm{Im}(\tau)/L_1$. $\vartheta_1(u|\tau)$ are the odd elliptic theta functions.[6] Fixing the solenoid fluxes constrains k and the sum $z_0=\sum z_\nu$ to take one of the N_s^2 sets of values satisfying

$$\exp(ikL_1)=(-1)^{N_s}\exp(i\phi_1)\ ,$$
$$\exp(2\pi iz_0/L_1)=(-1)^{N_s}\exp(i\phi_2-ikL_1\tau)\ . \quad (7)$$

If (k,z_0) is a solution of (7), the other solutions have the form

$$(k-2\pi n_1/L_1, z_0+n_1L_2e^{i\theta}+n_2L_1)\ ,$$

where n_1 and n_2 are suitable integers that keep k and z_0 in the specified ranges.

We remark that the number of linearly independent solutions of (5) is equal to the number of zeros within the principal region. The basis set of eigenstates of the translation operator $t(L_1/N_s)$ (which has the action $z\to z+L_1/N_s$) is constructed by placing the N_s zeros in a string satisfying $z_{\nu+1}=z_\nu+L_1/N_s$; there are then N_s distinct orthogonal solutions of (5). An alternative way to specify states is to construct "coherent states" by placing all the zeros at the *same* point. The wave function is then maximum at the "diametrically opposed" point $z+(1+\tau)L_1/2$; there are N_s^2 nonorthogonal solutions of (5) with this form.

We now consider the many-particle wave functions for N_e particles. Translational invariance allows these to be expressed as the product of a center-of-mass term and a factor involving only relative coordinates. We follow the arguments of Ref. 1 and seek a ground-state wave function where the relative motion is described by a *Jastrow function*, i.e., a product of pair factors

$$F(\{z_i\})=F^{\mathrm{c.m.}}(Z)\prod_{i<j}f(z_i-z_j),\quad Z=\sum_k z_k\ . \quad (8)$$

Application of the boundary condition for each particle gives

$$f(z+L_1)/f(z)=\eta_1\ ,$$
$$f(z+L_2e^{i\theta})/f(z)=\eta_2\exp[2\pi i(N_s/N_e)z/L_1]\ , \quad (9)$$

re η_1 and η_2 are constants. Integration of $d/dz\times\{\ln[f(z)]\}$ around the boundaries of the principal region shows that the number of zeros of $f(z)$ is $N_s/N_e=m$, which must be integral. We again follow Ref. 1 and seek the solution of (9) that (a) is odd under $z\to -z$, because of antisymmetry under particle exchange, and (b) has all its zeros at the point $z=0$ where the particles coincide (this eliminates "wasted" zeros). The only solutions are

$$f(z)=[\vartheta_1(\pi z/L_1|\tau)]^m,\quad m\text{ odd}\ . \quad (10)$$

As $z\to 0$, $f(z)\sim z^m$. The center-of-mass factor must then satisfy

$$\frac{F^{\mathrm{c.m.}}(Z+L_1)}{F^{\mathrm{c.m.}}(Z)}=(-1)^{(N_s-m)}e^{i\phi_1}\ ,$$
$$\frac{F^{\mathrm{c.m.}}(Z+L_2e^{i\theta})}{F^{\mathrm{c.m.}}(Z)}\exp[i\pi m(2Z/L_1)+\tau]=(-1)^{(N_s-m)}e^{i\phi_2}\ . \quad (11)$$

The general solution of this is characterized by a real wave vector K and m zeros $\{Z_\nu\}$

$$F^{\mathrm{c.m.}}(Z)=\exp(iKZ)\prod_{\nu=1}^{m}\vartheta_1(\pi(Z-Z_\nu)/L_1|\tau)\ ; \quad (12)$$

$$\exp(iKL_1)=(-1)^{N_s}\exp(i\phi_1)\ ,$$
$$\exp\left[2\pi i\sum_\nu Z_\nu/L_1\right]=(-1)^{N_s}\exp(i\phi_2-iKL_1\tau)\ . \quad (13)$$

Thus, there is an m-fold degeneracy associated with the center-of-mass coordinate in the presence of fixed solenoid fluxes. This degeneracy of the ground state with $N_s=mN_e$ was also found in the numerical study of Yoshioka *et al.*[5]

If "coherent state" center-of-mass wave functions are constructed by placing all the zeros at the same point in the principal region, there are m^2 distinct solutions of (13) compatible with the specified solenoid fluxes. For any such solution, the amplitude of the state vanishes when the center-of-mass coordinate Z/N_e lies on one of a lattice of points $z_0+n_1L_1/N_e+n_2L_2e^{i\theta}/N_e$. The charge density of the state will be essentially constant, but with a small superimposed periodic component that is minimized at these positions, and vanishes as $N_e\to\infty$.

It may be appropriate to replace the "solenoid flux" boundary condition (3) with the less restrictive condition

$$t_i(\mathbf{L}_\alpha)=t_j(\mathbf{L}_\alpha),\quad \text{all } i,j\ , \quad (14)$$

where t_i is the translation operator of the ith particle. This is a selection rule that requires all particles to satisfy the *same* boundary condition, but leaves the ϕ_α unspecified. In this case, the restrictions (13) are lifted, and z_0 can be chosen arbitrarily. Since the eigenvalue spectrum of a translationally invariant Hamiltonian is independent of the ϕ_α, there is a continuous degeneracy associated with the center-of-mass coordinate if (14) is used.

Following Laughlin's treatment on the open plane, we exhibit wave functions describing fractionally charged "hole" defects. The hole state is given by

$$\cdot(\{z_i\};\bar z)=F^{\mathrm{c.m.}}(Z)\prod_k\vartheta_1(\pi(z_k-\bar z)L_1|\tau)\prod_{i<j}[\vartheta_1(\pi(z_i-z_j)/L_1|\tau)]^m,\quad Z=\sum_k z_k+m^{-1}\bar z\ . \quad (15)$$

N_s is given by mN_e+1. $F^{\rm c.m.}(Z)$ are again solutions of (11). The defect is centered at the point $z=\bar{z}$. Since the amplitude of the wave function vanishes if any electron coordinate is at $\bar{z}$, this state has vanishing charge density at that point, which can be chosen without restriction.

As a function of the *hole* coordinate $\bar{z}$, the wave function (15) satisfies the boundary condition

$$\frac{F(\{z_i\};\bar{z}+mL_1)}{F(\{z_i\};\bar{z})}=(-1)^{m-1}e^{i\phi_1}\,,$$
$$\frac{F(\{z_i\};\bar{z}+mL_2e^{i\theta})}{F(\{z_i\};\bar{z})}\exp\{i\pi N_s[(2\bar{z}/L_1)+m\tau]\}=(-1)^{m-1}e^{i\phi_2}\,. \tag{16}$$

If $m>1$, the function is not periodic in $\bar{z}$ with the fundamental periods, but only with longer periods so the repeat distances are m times those of the electronic wave functions. The boundary conditions show that the number of zeros in the enlarged "principal region" bounded by $\bar{z}/L_1=\frac{1}{2}m\times(\pm1\pm\tau)$ which contains m^2N_s flux quanta is mN_s. Since the flux quantum for a charge q particle is $1/q$ times the flux quantum for an electron, the wave function for such a particle would have $m^2(qN_s)$ zeros in this region. The boundary conditions (16) thus indicate that the "hole" carries fractional charge[1] $|q|=1/m$. This seems to be essentially the same argument as the "adiabatic transport" argument of Arovas, Schrieffer, and Wilczek.[7]

The model "particle" defect state seems less easy to construct in the periodic geometry. The defect creation operator would have to remove one zero from the wave function as a function of each particle coordinate in each repetition of the fundamental region. An ansatz involving $\vartheta_1(\pi(d/dz_i)/L_1|\tau)$ seems the likely solution, but the choice of ordering is nontrivial, and we leave the construction of the periodic analog of Laughlin's "particle" defect as an open problem.

Numerical studies by Su[8] with square boundary conditions ($\tau=i$) identified defect states in the form of a line defect, which were eigenstates of the many-particle translation operator $\prod t_n(L_1/N_s)$. These are the analogs of the momentum-basis states usually used in Landau-gauge calculations. For direct comparison with the model states (15) which are equivalent to Laughlin's states in the thermodynamic limit, the coherent-state linear combination of the line-defect states would have to be formed. The relation between these types of states is analogous to that described above for the electron wave functions.

Finally, we note that while the above discussion is mainly formal, it does allow one physical point to be made. The original formulation[1] made use of arguments based on angular momentum conservation. The above formalism shows that Laughlin's construction can just as easily be carried out in a geometry that *does not conserve angular momentum*, and it is instead *correct behavior of the wave functions as particles approach* that is the key principle of its success. There is an analogy with the original BCS formulation of the superconducting ground state, which made use of momentum conservation, while the basic principle of pairing of time-reversed states can of course be implemented under the more general conditions of "dirty superconductivity."

We would like to acknowledge very useful conversations with R. Tao on theta functions and quasiperiodic wave functions. The work of one of us (F.D.M.H.) is supported in part by an Alfred P. Sloan Foundation Fellowship Grant, and by National Science Foundation Grant No. DMR-8405347.

[1]R. B. Laughlin, Phys. Rev. Lett. 50, 1395 (1983).
[2]D. C. Tsui, H. L. Stormer, and A. C. Gossard, Phys. Rev. Lett. 48, 1559 (1982).
[3]F. D. M. Haldane, Phys. Rev. Lett. 51, 605 (1983).
[4]F. D. M. Haldane and E. H. Rezayi, Phys. Rev. Lett. 54, 237 (1985).
[5]D. Yoshioka, B. I. Halperin, and P. A. Lee, Phys. Rev. Lett. 50, 1219 (1983).
[6]I. S. Gradshteyn and I. M. Ryzhik, *Table of Integrals, Series and Products* (Academic, New York, 1980), p. 921.
[7]D. Arovas, J. R. Schrieffer, and F. Wilczek, Phys. Rev. Lett. 53, 722 (1984).
[8]W. P. Su, Phys. Rev. B 30, 1069 (1984).

PHYSICAL REVIEW B VOLUME 41, NUMBER 13 1 MAY 1990

Ground-state degeneracy of the fractional quantum Hall states in the presence of a random potential and on high-genus Riemann surfaces

X. G. Wen*
Institute for Theoretical Physics, University of California-Santa Barbara, Santa Barbara, California 93106

Q. Niu
Department of Physics, University of California-Santa Barbara, Santa Barbara, California 93106

(Received 17 October 1989)

The fractional quantum Hall (FQH) states are shown to have $\bar{q}^g$-fold ground-state degeneracy on a Riemann surface of genus g, where $\bar{q}$ is the ground-state degeneracy in a torus topology. The ground-state degeneracies are directly related to the statistics of the quasiparticles given by $\theta=\bar{p}\pi/\bar{q}$. The ground-state degeneracy is shown to be invariant against weak but otherwise arbitrary perturbations. Therefore the ground-state degeneracy provides a new quantum number, in addition to the Hall conductance, characterizing different phases of the FQH systems. The phases with different ground-state degeneracies are considered to have different topological orders. For a finite system of size L, the ground-state degeneracy is lifted. The energy splitting is shown to be at most of order $e^{-L/\xi}$. We also show that the Ginzburg-Landau theory of the FQH states (in the low-energy limit) is a dual theory of the U(1) Chern-Simons topological theory.

I. INTRODUCTION

There are two quantum-fluid states which are known to exist at zero temperature, i.e., they may appear as the ground state of a system. One is the superfluid and the other is the incompressible fluid. The superfluid state was first disovered in He^4 (1932) (Ref. 1) and later in He^3 (1972).[2] The first example of the incompressible-fluid state is probably the superconducting state[3] discovered in 1911. The superconducting state is incompressible if we fix the positive background charge density which comes from the lattice ions (by assuming the lattice to be rigid). All excitations in the superconducting state have finite energy gaps (except the phonons which have been excluded). The incompressibility of the superconducting state comes from the long-range Coulomb interaction. In the early 1980s a new class of incompressible quantum fluids was discovered in the integer quantum Hall (IQH) effects and in the fractional quantum Hall (FQH) effects.[4] Recently, in studying high-T_c superconductors, a class of "incompressible" quantum spin-liquid states—chiral spin states—was proposed,[5,6] which does not support any gapless excitations. The time-reversal symmetry (T) and the parity (P) are broken in these spin-liquid states. Chiral spin states are closely related to the FQH states.[5,6]

The FQH states and chiral spin states are very special in the sense that their ground-state properties are not characterized by the symmetries in their ground states. The transition from one FQH state (or chiral spin state) to another is not associated with a charge in the symmetries of the states. In this paper we will demonstrate that the FQH states and chiral spin states contain nontrivial topological structures. The different FQH states and chiral spin states may be classified by topological orders.

It has been shown that the topological orders in chiral spin states can be partially characterized by the ground degeneracy of chiral spin states in compactified space.[7,8] The ground-state degeneracy depends on the topology of the space and is equal to k^g (ignoring the twofold degeneracy arising from the spontaneous T and P breaking), where g is the genus of the compactified space and k is an integer characterizing the topological order in chiral spin states. Similarly, it was known long ago that the ground-state degeneracy of the FQH states also depended on the topology of compactified space. For the simplest FQH states given by Laughlin wave function

$$\psi(z_i)=\left[\prod_{i<j}(z_i-z_j)^q\right]\exp\left[-\tfrac{1}{4}\sum_i|z_i|^2\right], \tag{1.1}$$

the ground state is found to be nondegenerate on a sphere[9] and q-fold degenerate on a torus[10] (with $g=1$). The dependence of the ground-state degeneracy on the topology of space suggests that the FQH states also contain nontrivial topological orders.

The ground-state degeneracy of the FQH states has been a puzzling problem for a long time. Especially, it is not clear whether the degeneracy arises from broken symmetries or not. There are arguments both favoring and disfavoring the symmetry-breaking picture.

According to Anderson,[11] some basic ingredients of symmetry breaking are already contained in Laughlin's description of the FQH system on a circular disc. In the $\frac{1}{3}$ filling case, for instance, the Laughlin states with zero, one, and two quasiholes are macroscopically distinct, but energetically they are the same for the bulk of the system. The Laughlin state with three quasiholes differs from that of no quasihole by only one single-particle state and they are therefore macroscopically indistinct.

Tao and Wu[12] considered the case of a cylinder

geometry and concluded, using general gauge symmetry arguments similar to those of Laughlin for the FQH case, that there must be a symmetry breaking of the system in order to exhibit the FQH effect. The same conclusion was reached by Niu *et al.*,[13] who studied the problem from a point of view of the topological invariant of the quantum Hall conductance on a torus. The latter authors also demonstrated the degeneracy of the ground states by the explicit construction of distinct Laughlin states on the torus. The existence of degeneracy was further supported by the numerical result of Su.[14] The generality of the arguments of gauge symmetry and topological invariance make the degeneracy a very robust property of the FQH system, independent of perturbations which do not close the energy gap.

There are also arguments disfavoring the idea of symmetry breaking. These are backed by the evidence of no ground-state degeneracy in the sphere geometry.[9] The kind of degeneracy found in the torus geometry was interpreted as the degeneracy of the center-of-mass motion,[10] and therefore does not qualify as degeneracy among macroscopically distinct ground states, which is essential for symmetry breaking.

In this paper we wish to resolve the above puzzle. We argue that the ground-state degeneracy of the FQH states is really a reflection of the topological order of the system. The degeneracy depends on the topology of the system geometry, and is preserved (in the thermodynamic limit) even when the translational and rotational symmetries of the system are absent. Therefore, the degeneracy should not be interpreted as a symmetry breaking of the usual type, nor should it be regarded as the center-of-mass degeneracy.

If one insists on the symmetry-breaking picture, one may attribute the ground degeneracy to broken "topological" symmetries (see Sec. IX). However, the topological symmetry can be defined only after the topology of space is specified. The very existence of the topological symmetries depends on the topology of the space. The number of topological symmetries is different for the spaces with different topologies.

The characterization of the FQH states is another unresolved problem in the FQH theory. The Hall conductance is certainly not enough to characterize the different FQH states. Two different FQH states may give rise to the same Hall conductance and yet be macroscopically distinct. Because the ground-state degeneracy of the FQH states is robust against arbitrary perturbations, the ground-state degeneracy can be used to characterize different phases in phase space. Therefore, the different FQH states with the same Hall conductance can be (at least partially) characterized (or distinguished) by their different ground-state degeneracies (on torus and high-genus Riemann surfaces). A more complete characterization of the topological orders in the FQH states can be obtained by studying the non-Abelian Berry's phases[15] associated with twisting the mass matrix of the electrons.[8]

The paper is arranged as follows. In Sec. II we study the ground-state degeneracy on a torus and its lifting by impurity potentials using the first-order perturbation theory. The selection rule of the magnetic translation group implies an energy splitting exponentially small in the shortest linear size of the system. In Secs. III and IV we study the ground-state degeneracy using the effective theory of the FQH states. The effective-theory approach not only applies to a case with a spatial dependent magnetic field, random potentials, etc., it also applies to high-genus Riemann surfaces where the magnetic translations cannot be defined. In Sec. V we study the splitting of the ground-state energies of the finite system based on the effective-theory approach. In comparison with the results obtained in Sec. II, the results obtained here are nonperturbative (but qualitative). The energy split is found to be of order $e^{-L(m^*\Delta)^{1/2}}$ for generic random potentials. Here m^* and Δ are the effective mass and the energy gap of the fractionally charged quasiparticles and quasiholes, and L is the size of the system. We also demonstrate explicitly that the ground-state degeneracy of the FQH state (or any other system) is determined directly by the fractional statistics of the quasiparticles, instead of the filling fraction $\nu=p/q$. In Sec. VI the ground-state degeneracy of the hierarchy FQH states is discussed. In Sec. VII a duality picture of the Ginzburg-Landau (GL) theory of the FQH states is developed. The results obtained in Sec. III–VI are rederived in the dual picture. The dual picture allows us to directly apply our results on the FQH states to the chiral spin states. In Sec. VIII we show that the ground degeneracy on a genus g Riemann surface is given by $\bar{q}^{\,g}$ if the quasiparticle excitations have statistics $\theta=\pi\bar{p}/\bar{q}$. In Sec. IX we discuss the concept of topological symmetry and conclude the paper.

II. GROUND-STATE DEGENERACY AND ITS LIFTING BY IMPURITY POTENTIALS

In this section we discuss the ground-state degeneracy of a FQH system on a torus geometry using elementary methods. We show how and to what extent the degeneracy is lifted by weak impurity potentials. This is done by projecting the impurity potentials onto the subspace of the ground states and by applying the degenerate perturbation theory. This approach was first taken by Tao and Haldane.[16,17] Here we give a more detailed analysis. A very simple effective form of the impurity potentials is derived, from which the dependences of the impurity effects on the system size and the phases of the boundary conditions are clearly seen. In the end of the section, we remark on the practical significance of the degeneracy lifting.

We first give a brief review of the magnetic translation group. Consider an electron of charge $-e$ on a rectangular plane of size $L_1\times L_2$, with a magnetic field B in the perpendicular direction ($\hat{z}$). In the absence of impurities, the Hamiltonian is

$$H=\frac{1}{2m}\left[\left|-i\hbar\frac{\partial}{\partial x}+eA_x\right|^2+\left|-i\hbar\frac{\partial}{\partial y}+eA_y\right|^2\right], \tag{2.1}$$

where (A_x, A_y) is the vector potential such that

$$\frac{\partial}{\partial x}A_y-\frac{\partial}{\partial y}A_x=B\ . \tag{2.2}$$

The Hamiltonian has a symmetry of magnetic translations

$$t(\mathbf{a})=e^{i\mathbf{a}\cdot\mathbf{k}/\hbar}\ , \tag{2.3}$$

where $\mathbf{a}$ is a vector in the plane, and $\mathbf{k}$ is an operator (pseudomomentum) defined by

$$k_x=-i\hbar\frac{\partial}{\partial x}+eA_x+eBy,$$
$$k_y=-i\hbar\frac{\partial}{\partial y}+eA_y-eBx\ . \tag{2.4}$$

It can be easily shown that $\mathbf{k}$ [and therefore $t(\mathbf{a})$] commutes with the dynamical momenta:

$$\Pi_x=-i\hbar\frac{\partial}{\partial x}+eA_x,$$
$$\Pi_y=-i\hbar\frac{\partial}{\partial y}+eA_y\ , \tag{2.5}$$

and therefore with the Hamiltonian (2.1). From the commutator

$$[k_x,k_h]=i\hbar eB\ , \tag{2.6}$$

we have

$$t(\mathbf{a})t(\mathbf{b})=t(\mathbf{b})t(\mathbf{a})\cdot e^{-i(\mathbf{a}\times\mathbf{b})/l^2}\ , \tag{2.7}$$

where $l\equiv(\hbar/eB)^{1/2}$ is the magnetic length, and $\mathbf{a}\times\mathbf{b}$ means $\hat{\mathbf{z}}\cdot(\mathbf{a}\times\mathbf{b})$.

when there are N_e electrons, each with a kinetic energy of (2.1), interacting mutually via a potential

$$V(\mathbf{r}-\mathbf{r}')\ , \tag{2.8}$$

the many-body magnetic translation

$$T(\mathbf{a})\equiv\prod_{j=1}^{N_e}t_j(\mathbf{a}) \tag{2.9}$$

leaves the Hamiltonian of the system invariant, where t_j acts on the jth electron. In order to utilize this symmetry, we impose on the many-body wave function, the periodic boundary conditions:

$$t_j(\mathbf{L}_1)\psi=\psi,$$
$$t_j(\mathbf{L}_2)\psi=\psi\ , \tag{2.10}$$

where $\mathbf{L}_1=L_1\mathbf{x}$, $\mathbf{L}_2=L_2\mathbf{y}$. This means that the wave function is the same when an electron is magnetically translated $\mathbf{L}_1$ or $\mathbf{L}_2$ across the plane.

We assume there are N_s (integer) magnetic flux quanta through the surface

$$N_s=\frac{L_1L_2}{2\pi l^2}\ , \tag{2.11}$$

which is also the total number of single-particle states in a Landau level. Corresponding to a fractional filling of the lowest Landau level, we have

$$N_e=\frac{p}{q}N_s\ , \tag{2.12}$$

where p and q are mutually prime integers. The translations which also leave the boundary conditions (2.11) invariant are

$$T_1\equiv T(\mathbf{L}_1/N_s)\ ,$$
$$T_2\equiv T(\mathbf{L}_2/N_s)\ , \tag{2.13}$$

and their integral powers. We can thus choose a ground state ψ_0 to be an eigenstate of T_2, i.e.,

$$T_2\psi_0=e^{i\lambda}\psi_0\ , \tag{2.14}$$

where λ is a real number because of the unitarity of T_2. Moreover, since

$$T_1T_2=T_2T_1e^{-iN_e(\mathbf{L}_1\times\mathbf{L}_2)/(N_s^2l^2)}$$
$$=T_2T_1e^{-i2\pi p/q}\ , \tag{2.15}$$

there will be $q-1$ more states degenerate (in energy) with ψ_0. These are

$$\psi_n\equiv T_1^n\psi_0,\quad n=1,2,\ldots,q-1\ , \tag{2.16}$$

and they are eigenstates of T_2,

$$T_2\psi_n=e^{i\lambda}e^{i2\pi p_n/q}\psi_n \tag{2.17}$$

with different eigenvalues, and therefore they are orthogonal to ψ_0 and to one another. This implies that the ground states are at least q-fold degenerate.

In the remaining part of this section, we assume there are exactly q ground states. Then T_1^q and T_2^q are constants within the ground-state subspace. We now consider a weak impurity potential

$$U=\sum_j U(\mathbf{r}_j)$$
$$=\sum_{\mathbf{k}}\bar{U}(\mathbf{k})\sum_j e^{i\mathbf{k}\cdot\mathbf{r}_j}\ , \tag{2.18}$$

where $\mathbf{k}=[(2\pi n_1/L_1),(2\pi n_2/L_2)]$ is a Fourier wave vector, with n_1,n_2 being integers. Since the many-body states are antisymmetric in the electron labels, we can effectively write

$$U=N_e\sum_{\mathbf{k}}\bar{U}(\mathbf{k})e^{i\mathbf{k}\cdot\mathbf{r}}\ , \tag{2.19}$$

where $\mathbf{r}$ now stands for the coordinate of any one electron. We assume that the potential is weaker compared with the energy gap above the ground-state energy, so that we can use first-order degenerate perturbation within the ground-state subspace.

The fact that the ground states form an irreducible representation of T_1 and T_2 implies a number of selection rules. Consider the commutation relations

$$T_1^qe^{i\mathbf{k}\cdot\mathbf{r}}=e^{i\mathbf{k}\cdot\mathbf{r}}T_1^qe^{i2\pi n_1q/N_s}\ ,$$
$$T_2^qe^{i\mathbf{k}\cdot\mathbf{r}}=e^{i\mathbf{k}\cdot\mathbf{r}}T_2^qe^{i2\pi n_2q/N_s}\ . \tag{2.20}$$

Taking the matrix element of both sides of these equations, we have

$$\langle \psi_{n'}|e^{i\mathbf{k}\cdot\mathbf{r}}|\psi_n\rangle = \langle \psi_{n'}|e^{i\mathbf{k}\cdot\mathbf{r}}|\psi_n\rangle e^{i2\pi n_1 q/N_s}\,,$$
$$\langle \psi_{n'}|e^{\mathbf{k}\cdot\mathbf{r}}|\psi_n\rangle = \langle \psi_{n'}|e^{i\mathbf{k}\cdot\mathbf{r}}|\psi_n\rangle e^{i2\pi n_2 q/N_s}\,, \tag{2.21}$$

where we have used the fact that T_1^q and T_2^q are effectively constants. The matrix element is zero unless

$$n_1 = l_1 N_s/q\,,$$
$$n_2 = l_2 N_s/q\,, \tag{2.22}$$

where l_1, l_2 are integers. The impurity potential can therefore be written effectively as

$$U = N_e \sum_{l_1 l_2} \bar{U}(l_1K_1, l_2K_2)e^{i(l_1K_1x - l_2K_2y)}\,, \tag{2.23}$$

where $K_1 = L_2/(ql^2)$, $K_2 = L_1/(ql^2)$, and they are proportional to the system size. Thus, if the potential is smooth within a linear scale of much larger than ql^2/L_1 and ql^2/L_2, then $\bar{U}(l_1K_1, l_2K_2)$ will be exponentially small [except for $(l_1, l_2) = (0,0)$]. Furthermore, if the ground states are made of primarily the single-particle states in the lowest Landau levels, then, as has been shown in Ref. 16, we can write (2.23) effectively as

$$U = N_e \sum_{l_1 l_2} \bar{U}(l_1K_1, l_2K_2)\exp\left\{-\frac{1}{2}\left[\left(\frac{l_1L_2}{ql}\right)^2 + \left(\frac{l_2L_1}{ql}\right)^2\right]\right\} t\left(\frac{l_2}{q}\mathbf{L}_1 - \frac{l_1}{q}\mathbf{L}_2\right)\,, \tag{2.24}$$

where t is a magnetic translation acting on an electron.

The extra Gaussian factor makes the potential extremely small, even if the potential is not smooth. To lowest order of $e^{-(1/2)(L_2/ql)^2}$ or $e^{-(1/2)(L_1/ql)^2}$, we can write

$$U = u_1 t\left(\frac{\mathbf{L}_1}{q}\right) + U_1^* t\left(\frac{-\mathbf{L}_1}{q}\right)$$
$$+ u_2 t\left(\frac{\mathbf{L}_2}{q}\right) + U_2^* t\left(\frac{-\mathbf{L}_2}{q}\right)\,, \tag{2.25}$$

where

$$u_1 = N_e e^{-(1/2)(L_1/ql)^2}\bar{u}(0, K_2)\,,$$
$$u_2 = N_e e^{-(1/2)(L^2/ql)^2}\bar{u}(-K_1, 0)\,,$$

and we have also ignored the constant part $N_e\bar{u}(0,0)$.

We now proceed to derive the effective forms of $t(L_1/q)$ and $t(L_2/q)$. We define an integer r satisfying

$$+pr + qm = 1,\quad |r| < \frac{q}{2},\quad m = \text{integer}\,, \tag{2.26}$$

which has a unique solution, if q is odd. Then it can be shown that

$$T_1^{-r} t\left(\frac{\mathbf{L}_1}{q}\right)$$

and

$$T_2^{-r} t\left(\frac{\mathbf{L}_2}{q}\right) \tag{2.27}$$

commute with both T_1 and T_2. They must be constants within the subspace of the ground states which form an irreducible representation of the group generated by T_1 and T_2. We can thus write

$$t\left(\frac{\mathbf{L}_1}{q}\right) = e^{i\phi_1}T_1^r\,,$$
$$t\left(\frac{\mathbf{L}_2}{q}\right) = e^{i\phi_2}T_2^r\,. \tag{2.28}$$

The potential (2.25) can then be written effectively as

$$U = \bar{U}_1 T_1^r + \bar{U}_1^* T_1^{-r} + \bar{U}_2 T_2^r + \bar{U}_2^* T_2^{-r}\,, \tag{2.29}$$

wher $\bar{U}_1 = U_1 e^{i\phi_1}$ and $\bar{U}_2 = U_2 e^{i\phi_2}$.

Next, we consider the effective changing the boundary condition (2.10) to

$$t_j(\mathbf{L})\psi(\boldsymbol{\alpha}) = e^{i\boldsymbol{\alpha}\cdot\mathbf{L}}\psi(\boldsymbol{\alpha})\,. \tag{2.30}$$

The states satisfying different boundary conditions can be connected by the twister $b(\boldsymbol{\alpha})$ defined as

$$b(\boldsymbol{\alpha}) = T(\boldsymbol{\alpha}\times\hat{\mathbf{z}}l^2)\,. \tag{2.31}$$

If $\psi(0)$ is an eigenenergy state saisfying (2.10), then

$$\psi(\boldsymbol{\alpha}) = b(\boldsymbol{\alpha})\psi(0) \tag{2.32}$$

is an eigenstate of the same energy (in the absence of impurity potentials) satisfying (2.30). It must be kept in mind, that (2.32) is not a gauge transformation unless the operators are also transformed accordingly. In other words, (2.32) is a change of boundary condition, if the operators of observables remain unchanged.

We can, of course, keep the wave function unchanged, but transform the operators by $b(\boldsymbol{\alpha})$. Then we have

$$U = \bar{U}_1 e^{i\alpha_1 L_1 rp/q}T_1^r + \bar{U}_2 e^{i\alpha_2 L_2 rp/q}T_2^r + \text{H.c.} \tag{2.33}$$

The Schrödinger equation becomes

$$\varepsilon\psi_n = \bar{U}_1 e^{i\alpha_1 L_1 rp/q}\psi_{n+r} + \bar{U}_1^* e^{-i\alpha_1 L_1 rp/q}\psi_{n-r}$$
$$+ (\bar{U}_2 e^{i\alpha_2 L_2 rp/q}e^{ir\lambda}e^{i2\pi prn/q} + \text{c.c.})\psi_n\,,$$
$$\psi_{n+q} = e^{i\delta}\psi_n\,, \tag{2.34}$$

where $e^{i\delta}$ is the constant of T_1^q. This can be transformed to the standard Harper's equations

$$\varepsilon\phi_n = R_1(\phi_{n+1}+\phi_{n-1})$$
$$+2R_2\cos\left[\frac{2\pi r}{q}n+\frac{\alpha_2 L_2 rp}{q}+r\lambda+\vartheta_2\right]\phi_n , \quad (2.35)$$
$$\phi_{n+q}=e^{i[\vartheta_1 q+\alpha_1 L_1 rp+r\delta q]}\phi_n ,$$

where

$$R_1 e^{i\vartheta_1}=\bar{U}_1 ,$$
$$R_2 e^{i\vartheta_2}=\bar{U}_2 ,$$

and

$$\phi_n = e^{-i[\vartheta_1+\alpha_1 L_1 rp/q]n}\psi_{nr} .$$

The band structure of (2.35) is well known. There are q bands, $\varepsilon_j(\alpha_1,\alpha_2)$, with a periodicity of $\alpha_1=(2\pi/rpL_1)$ and $\alpha_2=(2\pi/rpL_2)$. This periodicity implies that there are $(rp)^2$ inequivalent boundary conditions giving rise to the same ground-state energy splittings. For a large aspect ratio ($L_1 \ll L_2$), we have $R_2 \ll R_1$. The band widths are of order R_1/q, and the gaps are about R_2. In the other limit ($L_1 \gg L_2$), the roles of R_1 and R_2 are exchanged.

In any case, there is a unique ground state for each (α_1,α_2). It is then tempting to conclude that the Hall conductance should be an integer using the theory of topological invariant. This conclusion is wrong for two reasons. First, the arguments of topological invariant are only applicable to a state separated from others by energy gaps which do not become zero in the thermodynamic limit.[13] Secondly, linear-response theory does not apply when the gaps are small such that Zener tunneling becomes important. However, both the linear-response theory and the topological invariant arguments are valid for a group of states which are separated from others by finite-energy gaps, even though the energy gaps among themselves are infinitesimal. At a temperature larger than the energy splittings, the q states are equally populated. The total Hall conductance can be calculated as the average of the contribution from each state[18] as if the Kubo formula is applicable to each of them. A more direct way is to invoke the topological invariance, and to calculate it in the absence of the impurity potential. Both methods should, of course, give the same result: $\sigma_H=(p/q)(e^2/h)$.

III. GROUND-STATE DEGENERACY OF THE FQH STATES: AN EFFECTIVE-THEORY APPROACH

In this section we are going to give a simple heuristic argument about the ground-state degeneracy of the FQH states. The approach is based on the effective Ginzburg-Landau theory of the FQH effects.[19,20] More rigorous proof will be given in the next section. We will first consider the case with translation symmetry and reobtain the results in Ref. 10 and in the previous section.

The GL theory for the FQH states can be written as

$$\mathcal{L}_{GL}=\left[\phi^*(i\partial_0-a_0-eA_0)\phi-\frac{1}{2m}\phi^*(i\partial_i-a_i-eA_i)^2\phi +\mu|\phi|^2-\lambda|\phi|^4\right]+\left[\frac{-1}{4\pi q}\varepsilon^{\mu\nu\lambda}a_\mu\partial_\nu a_\lambda\right]+\cdots$$
$$=\mathcal{L}_\phi+\mathcal{L}_a+\cdots , \quad (3.1)$$

where q is an odd integer and $f_{\mu\nu}$ is the field strength of a_μ. A_μ is the electromagnetic field and a_μ is a $U(1)$ gauge field introduced in Ref. 20. In this paper we will always regard A_μ as a fixed classical background field. We will not discuss the dynamics of the electromagnetic field.

The precise meaning of the GL effective theory (3.1) is the following. An interacting (spinless) electron system in presence of electromagnetic field is described by the Lagrangian

$$L_0=\int d^2x\left[\psi^*(i\partial_0-A_0)\psi-\frac{1}{2m_e}\psi^*(i\partial_i-eA_i)^2\psi\right]$$
$$-\int d^2x\,d^2x'|\psi(x)|^2V(x-x')|\psi(x')|^2 , \quad (3.2)$$

where $V(x-x')$ describes the interaction between electrons. After integrating out the electron field ψ we obtain an effective Lagrangian for the electromagnetic field A_μ:

$$\exp\left[i\int dt\, L_{\text{eff}}(A_\mu)\right]=\int D\psi^* D\psi \exp\left[i\int dt\, L_0(\psi,A_\mu)\right] . \quad (3.3)$$

We say (3.1) is an effective theory of the electron system (3.2) if the same effective Lagrangian $L_{\text{eff}}(A_\mu)$ can be obtained after we integrate out ϕ and a_μ in (3.1):

$$\exp\left[i\int dt\, L_{\text{eff}}(A_\mu)\right]=\int D\phi^* D\phi\, Da_\mu \exp\left[i\int d^3x \mathcal{L}_{GL}(\phi,a_\mu,A_\mu)\right] . \quad (3.4)$$

To satisfy (3.4) the GL effective Lagrangian may be very complicated and contain high-derivative terms. In (3.1) we only keep the lowest-derivative terms because we are only interested in low-energy and long-wavelength properties of the system.

All the physical properties of the electron system are measured by an electromagnetic field A_μ. Therefore, we may use the effective theory to study the physical properties of the effectron system. The GL effective theories are useful because some states, like FQH, have simple forms in terms of the effective theories. Some physical properties of those states are more transparent when expressed in terms of the effective theories. The effective-theory approach is more general. It may apply to high-genus Riemann surfaces where the ordinary magnetic translations cannot be defined and used to study the ground-state degeneracy. It also applies to the case with a spatial-dependent magnetic field.

Certainly the effective theories are not unique. Different ground states have simple forms only in the different effective theories. To study different FQH states we may use different effective theories to simplify the problem. The equivalence between (3.1) and (3.2) has not been proven in the sense that (3.4) is satisfied. However, it is demonstrated that (3.1) reproduces all known long-distance and low-energy properties of the FQH state. Therefore, we will assume (3.1) is the effective theory of the FQH states, at least in the low-energy, long-wavelength limit. The effective-theory approach is less rigorous because (3.4) has not been rigorously established yet.

According to Ref. 20, the FQH state is given by a mean-field vacuum of (3.1):

$$\langle\phi\rangle=\sqrt{n}=\left[\frac{e^2B}{2\pi q}\right]^{1/2},\qquad eA_\mu+a_\mu=0\,, \tag{3.5}$$

where n is the electron density. The filling factor is given by $\nu=1/q$ and the Hall conductance $\sigma_H=(1/q)(e^2/h)$.

Now let us consider the FQH state on a torus of size $L_1\times L_2$. Notice that all the local quasiparticle fluctuations around the mean-field vacuum $\langle\phi\rangle=\sqrt{n}$ have finite-energy gaps. Therefore, the vacuum degeneracy (excitations with zero energy) can only come from the global excitations. On the torus we may separate the local and the global excitations by writing

$$a_i+eA_i=\frac{\vartheta_i(t)}{L_i}+\delta a_i(x,t)\,, \tag{3.6}$$

where $\delta a_i(x,t)$ satisfies

$$\int \delta a_i(x,t)d^2x=0\,.$$

δa_i corresponds to the local excitations and ϑ_i global excitations. The effective theory of the global excitations ϑ_i is obtained by integrating out a_0, δa_i, and ϕ:

$$\exp\left[i\int dt\,L_{\rm eff}(\vartheta_i)\right]=\int Da_0\,D\delta a_i\,D\phi\exp\left[i\int d^3x\,\mathcal{L}_{\rm GL}(a_\mu,\phi)\right]. \tag{3.7}$$

Substituting (3.6) into (3.1) we find that (3.7) can be rewritten as

$$\exp\left[i\int dt\,L_{\rm eff}(\vartheta_i)\right]=\exp\left[i\int dt\frac{1}{4\pi q}(\vartheta_1\dot\vartheta_2-\vartheta_2\dot\vartheta_1)\right]\int Da_0\,D\delta a_i\,D\phi\exp\left[i\int d^3x[\mathcal{L}_\phi(a_\mu,\phi)+\mathcal{L}_a(\delta a_\mu)]\right]. \tag{3.8}$$

$L_{\rm eff}(\vartheta_i)$ can be shown to have the following form:

$$L_{\rm eff}(\vartheta_i)=\frac{1}{4\pi q}(\vartheta_1\dot\vartheta_2-\vartheta_2\dot\vartheta_1)+f_i(\vartheta_i)\dot\vartheta_i+\tfrac{1}{2}M(\dot\vartheta_1^2+\dot\vartheta_2^2)-V_1(\vartheta_1,\vartheta_2)+(\text{higher-derivative terms})\,. \tag{3.9}$$

The first term in (3.9) comes from the Chern-Simons term. The second and the third terms come from the quantum fluctuations of ϕ and δa_i. (See Fig. 1). The potential term $V_1(\vartheta_i)$ is nonzero because the ϕ field condenses. V_1 is of order n/m. Notice that ϕ carries a unit charge of the a_μ gauge field and the path integral in (3.8) is invariant under the following transformations:

$$\phi\to\phi'=\exp\left[-i2\pi\left[\frac{p_1x_1}{L_1}+\frac{p_2x_2}{L_2}\right]\right]\phi\,,\qquad (\vartheta_1,\vartheta_2)\to(\vartheta_1+2\pi p_1,\vartheta_2+2\pi p_2)\,, \tag{3.10}$$

where p_1 and p_2 are integers such that ϕ' is a single valued function on the torus. In other words, (p_1,p_2) are the winding number of ϕ on the torus coordinated by (x_1,x_2). The symmetry (3.10) implies that the potential $V_1(\vartheta_i)$ and the function $f_i(\vartheta)$ are periodic functions

$$V_1(\vartheta_1+2\pi p_1,\vartheta_2+2\pi p_2)=V_1(\vartheta_1,\vartheta_2)\,,\qquad f_i(\vartheta+2\pi p_1,\vartheta_2+2\pi p_2)=f_i(\vartheta_1,\vartheta_2)\,. \tag{3.11}$$

The explicit form of $V_1(\vartheta_i)$ may be obtained in a semiclassical approximation. In this approximation we assume $\delta a_i=0$ (in this case $f_i=0$). The integration of $a_0(x_\mu)$ imposes a constraint

$$|\phi|^2=n=\frac{e^2B}{2\pi q}\,.$$

The integration of ϕ is truncated to a summation of stationary points given by

$$\phi_{p_1p_2}(x)=\sqrt{n}\exp\left[-i2\pi\left[\frac{p_1x_1}{L_1}+\frac{p_2x_2}{L_2}\right]\right], \tag{3.12}$$

where p_1 and p_2 are integers. Now (3.8) becomes

$$\exp\left[i\int dt\,L_{\rm eff}(\vartheta_i)\right]=\exp\left[i\int dt\frac{1}{4\pi q}(\vartheta_1\dot\vartheta_2-\vartheta_2\dot\vartheta_1)\right]\sum_{p_1p_2}\exp\left[-i\int dt\frac{n}{2m}\left[\frac{L_2}{L_1}(\vartheta_1+2\pi p_1)^2+\frac{L_1}{L_2}(\vartheta_2+2\pi p_2)^2\right]\right].$$

We find that

$$V_1(\vartheta_1,\vartheta_2)=\frac{n}{2m}\left[\frac{L_2}{L_1}\vartheta_1^2+\frac{L_1}{L_2}\vartheta_2^2\right]\Bigg|_{-\pi\le\vartheta_1,\vartheta_2\le\pi}. \tag{3.13}$$

For other values of ϑ_i, $V_1(\vartheta_1)$ is determined by (3.11).

We would like to emphasize that the specific forms of the potential $V_1(\vartheta_i)$ and the function $f_i(\vartheta_i)$ are not important in our discussions. Our discussions (followed

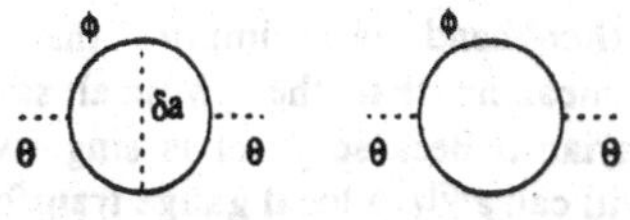

FIG. 1. Some of the Feynman diagrams which contribute to the second and the third terms in (3.9).

below) only depend on the periodic condition (3.11). The periodic condition is a consequence of the charge-*one*-boson (ϕ field) condensation and is very robust. The periodic condition can be changed only through phase transitions in which, for example, the charge-one-boson condensation changes to charge-N-boson condensation.

Now we are ready to study the dynamics of the global excitation governed by (3.9). The Lagrangian in (3.9) effectively describes a charged particle moving in a ϑ space. There is a periodic potential in the ϑ space $V_1(\vartheta_i)$ with period 2π in both the ϑ_1 and ϑ_2 directions. The first and the second terms in (3.9) imply, that there is a "magnetic" field in the θ space with $(2\pi/q)$ flux per plaquette. This system has been studied in detail.[21,18] The Hamiltonian is given by

$$H=-\frac{1}{2M}\left[\left[\frac{\partial}{\partial\vartheta_1}+i\bar{A}_1\right]^2+\left[\frac{\partial}{\partial\vartheta_2}+i\bar{A}_2\right]^2\right]+V_1(\vartheta_1,\vartheta_2) \tag{3.14}$$

with

$$\frac{\partial}{\partial\vartheta_1}\bar{A}_2-\frac{\partial}{\partial\vartheta_2}\bar{A}_1=\frac{1}{2\pi q}+\varepsilon^{ij}\frac{\partial}{\partial\vartheta_i}f_j .$$

The ground state of (3.14) is found to be q-fold degenerate. One way to prove this result is to notice that H in (3.14) commutes with the magnetic translations T_1 and T_2:

$$\begin{aligned}&T_1:\vartheta_1\to\vartheta_1+2\pi ,\\&T_2:\vartheta_2\to\vartheta_2+2\pi .\end{aligned} \tag{3.15}$$

T_1 and T_2 satisfy an algebra

$$T_2T_2=e^{i(2\pi/q)}T_2T_1 \tag{3.16}$$

whose irreducible representation has a dimension of q. The ground states of H must form a representation of (3.16) and hence have to be at least q-fold degenerate. Sometimes the ground states may be nq-fold degenerate if different irreducible representations of (3.16) *happen* to have the same lowest energy. However, this is not a generic situation.

We would like to remark that the magnetic translations T_1 and T_2 are the quantum realization of the classical transformations in (3.10) with (m,n) equal to (1,0) and (0,1), respectively. In the absence of the Chern-Simons term the transformations (3.10) should be regarded as the gauge symmetry. This means that we should identify ϑ_1 with $\vartheta_1+2\pi$ and ϑ_2 with $\vartheta_2+2\pi$. The ϑ space is actually *finite*. However, in the presence of the Chern-Simons term the quasiparticles (and the quasiholes) carry a fractional charge. The quasiparticle-quasihole tunneling process described in Sec. V produces physical operators proportional to T_1 or T_2. Noticing that T_1 and T_2 do not commute, we cannot regard T_1 and T_2 as the gauge transformations, because the gauge transformations should leave all physical operators invariant. Noticing that T_1^q and T_2^q commute with T_1 and T_2 [see (3.16)], we can still regard T_1^q and T_2^q as gauge transformations. This implies that we may identify ϑ_1 with $\vartheta_1+2\pi q$ and ϑ_2 with $\vartheta_2+2\pi q$. The ϑ space is still finite.

We would like to emphasize that the above result does not depend on the particular simple form of the approximated Hamiltonian (3.14). The ground states remain q-fold degenerate as long as there exists the magnetic translations which satisfy the algebra (3.16) and commute with the Hamiltonian (3.14). This only requires that the physics described by (3.14) is periodic and there is a $2\pi/q$ flux in an area of period square. Our result holds even for the following general Hamiltonian:

$$H=K\left[i\left[\frac{\partial}{\partial\vartheta_1}+i\bar{A}_1\right],i\left[\frac{\partial}{\partial\vartheta_2}+i\bar{A}_2\right]\right]+V(\vartheta_1,\vartheta_2) , \tag{3.17}$$

where $K(x,y)$ is an arbitrary positive function and $V(\vartheta_1,\vartheta_2)$ an arbitrary periodic potential satisfying (3.11). The "magnetic" field

$$\bar{B}(\vartheta_i)=\frac{\partial}{\partial\vartheta_1}\bar{A}_2-\frac{\partial}{\partial\vartheta_2}\bar{A}_1$$

is periodic with period 2π in both ϑ_1 and ϑ_2. $\bar{B}(\vartheta_i)$ further satisfies

$$\int_0^{2\pi}\int_0^{2\pi}d\vartheta_1\,d\vartheta_2\bar{B}(\vartheta_i)=\frac{2\pi}{q} . \tag{3.18}$$

The periodicity in ϑ_1 and ϑ_2 is a consequence of the gauge symmetry (3.10) and the $2\pi/q$ flux is determined by the coefficient of the Chern-Simons term. Therefore, we expect that the Hamiltonian for ϑ_1 and ϑ_2 has the form of (3.17) even if we include all the quantum corrections (except for a nonperturbative effect which vanishes exponentially in thermodynamic limit). (See Sec. V.)

Now let us include the impurity potential in our system. In the framework of the effective theory, the effects of the impurity potential may be include by allowing the various coefficients in the effective GL theory to have a spatial dependence, except that the coefficient in front of the Chern-Simons terms which must be a constant as required by the gauge symmetry. We also allow the magnetic field B to have a spatial dependence. In this general situation, the above discussions are still valid. The transformations (3.10) remain a symmetry of the path integral in (3.8) and the Hamiltonian for θ_1 and θ_2 still takes the form in (3.17). Thus, the ground states remain q-fold degenerate.

We would like to stress that the derivation presented in this section is not strictly correct. To obtain the effective theory for ϑ_i we have assumed that ϑ_i are slow variables. But the ϑ_i and other local fluctuations actually have a similar energy scale. The separation of the global and the local fluctuations is quite artificial in this case. However, the effective theory of ϑ_i does contain correct algebraic structure. This is the reason why we obtain the correct result. A more rigorous and abstract derivation will be presented in the next section.

IV. TRANSFORMATION-ALGEBRA ANOMALY AND THE GROUND-STATE DEGENERACY OF THE FQH STATES

In this section we are going to give a general proof of the ground-state degeneracy of the FQH states on a torus. We will construct operators similar to, but more general than, the magnetic translation operators introduced in Sec. II. These operators commute with the Hamiltonian of the system, but not with each other, implying the degeneracy of the energy eigenstates of the systems. The proof given here is general enough to apply to situations with random potential, spatial-dependent magnetic field, and many other perturbations, as long as all quasiparticle excitations have finite-energy gaps.

The essence of the approach in the last section to the ground-state degeneracy is the magnetic translations (3.15) (in ϑ space), which is nothing but the gauge transformations (3.10). Therefore, the better approach is to directly use the algebra of the gauge transformations (3.10) to calculate the ground-state degeneracy without deriving the effective Lagrangian (3.9) for the global excitations. To do so we first need to quantize the Lagrangian (3.1). In the following we will allow μ, m, λ, and the magnetic field in (3.1) to have a spatial dependence.

We may quantize the gauge field[22,23] a_μ in the gauge

$$a_0=0\ . \tag{4.1}$$

The equation of motion for a_0 serves as a constraint:

$$\begin{aligned}G[f]&=i\int d^2x\, f(x)\frac{\delta L_{\text{eff}}}{\delta a_0}\\&=i\int d^2x\, f(x)\left[-\phi^*\phi-\frac{1}{4\pi q}\varepsilon^{ij}f_{ij}\right]\\&=i\int d^2x\, f(x)G(x)=0\ ,\end{aligned}\tag{4.2}$$

where $f(x)$ is an arbitrary globally defined real function on the torus. After quantization the constraint (4.2) is met by demanding all the states in the physical Hilbert space to satisfy (the Gauss law)

$$\hat{G}[f]|\Psi_{\text{phy}}\rangle=0\ . \tag{4.3}$$

From (3.1) we see that a_1 and a_2 canonically conjugate to each other

$$[\hat{a}_1(x),\hat{a}_2(y)]=i2\pi q\delta^2(x-y)\ . \tag{4.4}$$

Similarly,

$$[\hat{\phi}^+(x),\hat{\phi}(y)]=i\delta^2(x-y)\ . \tag{4.5}$$

Using (4.4) and (4.5) one can easily check that $\hat{G}$ generates a gauge transformation

$$\begin{aligned}e^{-i\hat{G}[f]}\hat{a}_i e^{i\hat{G}[f]}&=\hat{a}_i+i\partial_i f\ ,\\ e^{-i\hat{G}[f]}\hat{\phi}e^{i\hat{G}[f]}&=e^{-if}\hat{\phi}\ .\end{aligned}\tag{4.6}$$

On the other hand, (4.3) implies that $e^{-i\hat{G}[f]}|\Psi_{\text{phy}}\rangle=|\Psi_{\text{phy}}\rangle$, meaning that the physical state should be gauge invariant. Because $f(x)$ is single valued on the torus we will call $e^{i\hat{G}[f]}$ a local gauge transformation.

Using $\hat{G}(x)$ we can also construct so-called large transformations. Consider the operators

$$\begin{aligned}T_1&=\exp\left[i\int d^2x\, f_1(x)\hat{G}(x)\right]\ ,\\ T_2&=\exp\left[i\int d^2x\, f_2(x)\hat{G}(x)\right]\ ,\end{aligned}\tag{4.7}$$

where $f_i(x)$ have a 2π jump along a loop in the x_i direction which goes all the way around the torus (Fig. 2). One can check

$$\begin{aligned}T_j^{-1}\hat{a}_i T_j&=\hat{a}_i+\partial_i f_j(x)\ ,\\ T_j^{-1}\hat{\phi}T_j&=e^{-if_j(x)}\hat{\phi}\ .\end{aligned}\tag{4.8}$$

Notice that $\partial_i f_j(x)$ and $e^{if_j(x)}$ are smooth functions. T_j generate nonsingular transformations and are well-defined operators. From (4.4) and (4.5) T_1 and T_2 can be shown to satisfy the famous algebra

$$T_1T_2=e^{i(2\pi/q)}T_2T_1\ . \tag{4.9}$$

The algebra (4.9) is very important. The noncommunitivity of T_1 and T_2 is purely a quantum effect. Classically the transformations generated by f_1 and f_2 definitely commute with each other. Due to the algebra (4.9), there is no state which is invariant under both T_1 and T_2. Despite the classical Lagrangian being invariant under the transformation (4.8), the quantum states cannot be invariant under T_1 and T_2. We will call this phenomenon transformation-algebra anomaly. T_1 and T_2 are physical operators in the sense that they are generated by the physical tunneling process discussed in Sec. V. The physical Hilbert space is defined as a representation of the physical operators. In particular, the physical Hilbert space forms a representation of the algebra (4.9). Because the gauge transformation must commute with all the physical operators, we cannot regard T_1 and T_2 as gauge transformations. However, from (4.9) we find that T_1^q and T_2^q commute with T_1 and T_2. We may regard T_1^q and T_2^q as generators of (large) gauge transformations. Since T_1^q and T_2^q commute, we may require the physical

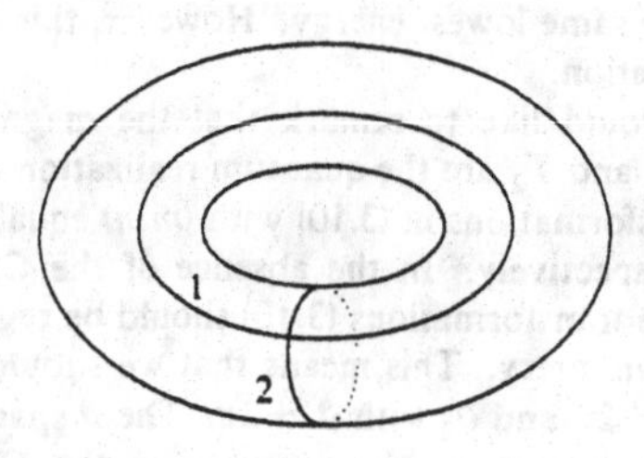

FIG. 2. $f_1(x)$ and $f_2(x)$ have a 2π jump along the two loops 1 and 2, respectively.

states to be invariant under these large gauge transformations

$$T_1^q|\Psi_{phy}\rangle=|\Psi_{phy}\rangle ,$$
$$T_2^q|\Psi_{phy}\rangle=|\Psi_{phy}\rangle . \tag{4.10}$$

Because T_1 and T_2 commute with T_1^q and T_2^q, T_1 and T_2 naturally act on the physical Hilbert space, i.e., a physical state when acted on by T_1 and T_2 still remains a physical state.

The Hamiltonian of the system (3.1) is given by

$$H=-\tfrac{1}{2}\hat{\phi}^\dagger(\partial_i+i\hat{a}_i+ieA_i)\frac{1}{m(x)}(\partial_i+ia_i+ieA_i)\hat{\phi}$$
$$-\mu(x)\hat{\phi}^\dagger\hat{\phi}+\lambda(x)(\hat{\phi}^\dagger\hat{\phi})^2 . \tag{4.11}$$

The Hamiltonian commutes with $\hat{G}(f)$, T_1^q, and T_2^q. Therefore, H acts on the physical Hilbert space. H also commutes with T_1 and T_2. Therefore, each energy level of H is q-fold degenerate and forms an irreducible representation of (4.9). In particular, the ground states of the FQH state are at least q-fold degenerate on the torus.

We would like to remark that once written in the GL form, the system has a degeneracy, despite the disorders in the coefficients (A_i,μ,λ,m) of the theory, as has been proven in Sec. III and this section. However, the reader should be warned that the effect of disorder in the original theory is a different story and is discussed in Secs. II and V. Disorder in the original theory may give rise, in addition to the disorder in the GL theory, to corrections in the GL Hamiltonian (4.11), which breaks the symmetries of T_1 and T_2.

In the following we would like to discuss the ground state wave functions of the FQH state. In the boson ϕ condensed phase, the Hilbert space is divided into sectors. The states in each sector describe the quantum fluctuations around the stationary point

$$\phi=\phi_{p_1p_2} ,$$
$$a_i=-eA_i+p_i\frac{2\pi}{L_i} , \tag{4.12}$$

where $\phi_{p_1p_2}$ is given by (3.12). Thus, different sectors are labeled by two integers (p_1,p_2). The states in the sector (p_1,p_2) are given by the wave functional

$$\Psi[a_i,\phi]=\Psi\left[-eA_i+\frac{2\pi p_i}{L_i}+\delta a_i,\phi_{p_1p_2}+\delta\phi\right] , \tag{4.13}$$

where δa_i and $\delta\phi$ are small fluctuations around the stationary point. In the thermodynamic limit, the states in the different sectors do not mix, i.e., the quantum fluctuations cannot connect a state in one sector to another state in a different sector. The Hamiltonian does not contain off-diagonal terms (in the thermodynamic limit) which mix two different sectors. Let $|p_1,p_2\rangle$ denote the lowest-energy state (the ground state) in the sector (p_1,p_2). Notice that T_1 and T_2 map a state in one sector to a state in a different sector. Since T_1 and T_2 commute with the Hamiltonian, they map the ground state of one sector to the ground state of another sector:

$$T_1|p_1,p_2\rangle=e^{i\alpha(p_1,p_2)}|p_1+1,p_2\rangle ,$$
$$T_2|p_1,p_2\rangle=e^{i\beta(p_1,p_2)}|p_1p_2+1\rangle . \tag{4.14}$$

The ground states in different sectors all have the same energy.

In the above discussion we did not consider the gauge symmetries. The ground states in different sectors, in general, are not the physical states, i.e., they do not satisfy (4.3) and (4.10). Only some particular superpositions of those ground states correspond to the physical ground states. In the absence of the Chern-Simons term, all physical operators commute with T_1 and T_2. Since T_1 and T_2 commute, we may require the physical states to be invariant under T_1 and T_2:

$$T_1|\Psi_{phy}\rangle=|\Psi_{phy}\rangle ,$$
$$T_2|\Psi_{phy}\rangle=|\Psi_{phy}\rangle . \tag{4.15}$$

We may choose the phase of $|p_1p_2\rangle$ such that $\alpha(p_1p_2)$ and $\beta(p_1p_2)$ in (4.14) are equal to zero. This is possible because T_1 and T_2 commute. It is not difficult to see that, in using $|p_1p_2\rangle$, we can only construct one physical ground state satisfying (4.15):

$$|0_{phy}\rangle=\sum_{p_1p_2}|p_1p_2\rangle . \tag{4.16}$$

Therefore, in the absence of the Chern-Simons term, the ground state is nondegenerate as we expected.

In the presence of the Chern-Simons term, T_1 and T_2 are the physical operators. The physical ground states form a representation of (4.9). In terms of $|p_1p_2\rangle$, the q physical ground states satisfying (4.10) and forming a representation of (4.9) are given by

$$|n_{phy}\rangle=\sum_{p_1p_2}e^{i(2\pi/q)np_2}|p_1p_2\rangle , \tag{4.17}$$

where $n=1,\ldots,q$. The phases of $|p_1p_2\rangle$ have been chosen such that

$$\alpha(p_1p_2)=\frac{2\pi}{q}p_2 ,$$
$$\beta(p_1p_2)=0 . \tag{4.18}$$

One can check that this choice of the phases is consistent with the algebra (4.9). The states $|n_{phy}\rangle$ satisfy

$$T_1|n_{phy}\rangle=|(n+1)_{phy}\rangle ,$$
$$T_2|n_{phy}\rangle=e^{-i(2\pi/q)n}|n_{phy}\rangle ,$$
$$|n_{phy}\rangle=|(n+q)_{phy}\rangle . \tag{4.19}$$

In the thermodynamic limit the perturbative Hamiltonian does not mix different physical ground states, and all the physical ground states have the same energy. However, for finite system the tunnel process described in Sec. V induces a term in the Hamiltonian which mixes the different ground states and lifts the ground-state degeneracy.

V. ENERGY SPLIT OF THE GROUND STATES OF FINITE SYSTEM

In Sec. IV we show that the ground states of the FQH state are q-fold degenerate even in presence of random potentials. This is because the generators of the algebra (4.9) commute with the Hamiltonian of the system. However, the above result is only valid in the thermodynamic limit. For systems with finite size there is a nonperturbative effect. After including the nonperturbative effect, the Hamiltonian obtains a small correction proportional to $\gamma e^{-L(m^*\Delta)^{1/2}}$ which does not commute with T_1 and T_2. Therefore, the energy of the ground states can be shown to have a split of order $\gamma e^{-L(m^*\Delta)^{1/2}}$.

The nonperturbative effect comes from the following tunneling process. A pair of quasiparticles and quasiholes is virtually created at a time t_0. The quasiparticle and quasihole move in opposite directions and propagate all the way around the torus. When they meet on the opposite side of the torus, they annihilate at time t_0+T. The resulting new ground state is different from the old ground state. The magnitude of the tunneling amplitude is given by

$$|A|=e^{-S},\qquad S=T\Delta+2\tfrac{1}{2}m^*\left[\frac{L}{2T}\right]^2 T, \tag{5.1}$$

where Δ is the energy gap of the quasiparticle-quasihole pair creation, m^* is the effective mass of the quasiparticle, and L is the size of the torus. (In this section we will assume $L_1=L_2=L$.) S is minimized at $T=\frac{1}{2}L(m^*/\Delta)^{1/2}$ with the minimum value $S=L(\Delta m^*)^{1/2}$. Hence,

$$|A|=e^{-L(\Delta m^*)^{1/2}}. \tag{5.2}$$

Therefore, the magnitude of the nonperturbative correction is exponentially small.

In order to obtain the explicit form of the nonperturbative corrections and to show that the corrections do not commute with T_1 and T_2, we need to study the tunneling process in more detail. The quasiparticle in the FQH state is given by a vortex in the ϕ field. The ansatz of the quasiparticle may be chosen to be

$$\frac{\phi(z)}{\sqrt{n}}=\frac{z-z_0(t)}{|z-z_0(t)|+\xi},\qquad a_0=0,\qquad a_1(z)+ia_2(z)=\frac{i(z-z_0)}{|z-z_0|^2+\xi'^2}-eA_1-ieA_2, \tag{5.3}$$

where $z=x_1+ix_2$ and z_0 is the position of the quasiparticle. ξ and ξ' in (5.3) are positive, which determines the size of the quasiparticle. A pair of the quasiparticles and quasiholes is describes by the ansatz

$$\frac{\phi(z)}{\sqrt{n}}=\frac{[z-z_0(t)][z-\bar{z}_0^*(t)]+f^2(|z_0-\bar{z}_0|/\xi)}{|z-z_0(t)||z-\bar{z}_0(t)|+\xi^2},\qquad a_1(z)=ia_2(z)=\frac{i(z-z_0)}{|z-z_0|^2+\xi'^2}-\frac{i(z-\bar{z}_0)}{|z-\bar{z}_0|^2+\xi'^2}-eA_1-ieA_2, \tag{5.4}$$

where z_0 and $\bar{z}_0$ are the positions of the quasiparticle and the quasihole, respectively. The function $f(x)$ in (5.4) satisfies $f(0)=\xi$ and $f(x)=0|_{x>1}$. When $|z_0-\bar{z}_0|$ is large, (5.4) describes a vortex and an antivortex. When $z_0-\bar{z}_0=0$, (5.4) describes a mean-field vacuum state. Thus, by separating z_0 and $\bar{z}_0$, (5.4) describes a process of creation of a quasiparticle and quasihole.

In order to construct the quasiparticle and the quasihole on the torus, ϕ and a_i must satisfy the periodic boundary conditions. We find that, on the torus, a pair of the quasiparticles and the quasihole is given by

$$\frac{\phi(z)}{\sqrt{n}}=\frac{F(z|z_0,\bar{z}_0)L+f(z_0,\bar{z}_0)\exp\left[-i\frac{2\pi}{L^2}\mathrm{Re}(z_0-\bar{z}_0)\mathrm{Im}z\right]}{|F(z|z_0,\bar{z}_1)|+\xi},$$
$$a_1(z)+ia_2(z)=\sum_{mn}\left[\frac{i(z-z_0-Z_{mn})}{|z-z_0-Z_{nm}|^2+\xi'^2}-\frac{i(z-\bar{z}_0-Z_{mn})}{|z-\bar{z}_0-Z_{mn}|^2+\xi'^2}\right]-eA_1-ieA_2, \tag{5.5}$$

where $Z_{mn}=mL+inL$. $f(z_0,\bar{z}_0)$ is a positive periodic function of z_0 and $\bar{z}_0$:

$$f(z_0+mL+inL,\bar{z}_0+\bar{m}L+i\bar{n}L)=f(z_0,\bar{z}_0). \tag{5.6}$$

$f(z_0,\bar{z}_0)$ is nonzero only when $|z_0-\bar{z}_0-Z_{mn}|<\xi$ for some m and n. $f(z_0,\bar{z}_0)=\xi$. $F(z|z_0,\bar{z}_0)$ in (5.8) is a periodic function in z:

$$F(z+L|z_0,\bar{z}_0)=F(z+iL|z_0,\bar{z}_0)=F(z|z_0,\bar{z}_0). \tag{5.7}$$

F has a zero at z_0,

$$F(z|z_0,\bar{z}_0)\sim(z-z_0)|_{z\to z_0}, \tag{5.8}$$

and an "antizero" at $\bar{z}_0$,

$$F(z|z_0,\bar{z}_0)\sim(z^*-\bar{z}_0^*)|_{z\to\bar{z}_0} . \tag{5.9}$$

An order $O(1)$ function F satisfying (5.7)–(5.9) is given by

$$F(z|z_0,\bar{z}_0)=\exp\left[-i\frac{\pi}{L^2}\mathrm{Re}(z_0-\bar{z}_0)[\mathrm{Im}2z+L]\right]\exp\left[-\frac{\pi}{L^2}[\mathrm{Im}(z-z_0)]^2-\frac{\pi}{L^2}[\mathrm{Im}(z-\bar{z}_0)]^2\right]\times\vartheta_1\left[\frac{z-z_0}{L}\Big|i\right]\vartheta_1^*\left[\frac{z-\bar{z}_0}{L}\Big|i\right] , \tag{5.10}$$

where $\vartheta_1(u|\tau)$ is the odd elliptic ϑ function[24] satisfying

$$\frac{\vartheta_1(u+1|\tau)}{\vartheta_1(u|\tau)}=-1 ,$$
$$\frac{\vartheta_1(u+\tau|\tau)}{\vartheta_1(u|\tau)}=-e^{-i\pi(2u+\tau)} , \tag{5.11}$$
$$\vartheta_1(u|\tau)\sim u|_{u\to0} ,$$

when $z_0=\bar{z}_0$ or $z_0=\bar{z}_0+L$, F satisfies

$$F(z|z_0,z_0)=|F(z|z_0,z_0)| ,$$
$$F(z|z_0,z_0+L)=e^{-i(2\pi/L)\mathrm{Im}z}|F(z|z_0,z_0)| . \tag{5.12}$$

When z_0 and $\bar{z}_0$ are well separated, f in (5.5) can be dropped and (5.5) describes a quasiparticle at z_0 and a quasihole at $\bar{z}_0$. When z_0 and $\bar{z}_0$ are close to each other (5.5) describes creation or annihilation of the quasiparticle and the quasihole.

The tunneling process described at the beginning of this section is obtained by choosing $z_0(t)$ and $\bar{z}_0(t)$ in (5.5) to be

$$z_0(t)=\bar{z}_0(t)=0,\quad t<t_0 ,$$
$$z_0(t)=-\bar{z}_0(t)=\frac{L}{2T}t,\quad t_0<t<t_0+T , \tag{5.13}$$
$$z_0(t)=-\bar{z}_0(t)=\frac{L}{2},\quad t>t_0+T .$$

Before the tunneling ($t<t_0$) the vacuum state is given by (5.5) with $z_0=\bar{z}_0$:

$$\frac{\phi}{\sqrt{n}}=1,\quad a_1+ia_2=-eA_1-ieA_2 . \tag{5.14}$$

After the tunneling ($t>t_0+T$) we have $z_0=-\bar{z}_0=L/2$ and the vacuum state is given by

$$\frac{\phi}{\sqrt{n}}=e^{-i(2\pi/L)x_2} ,$$
$$a_1=-eA_1,\quad a_2=\frac{2\pi}{L}-eA_2 . \tag{5.15}$$

The two states (5.14) and (5.15) are related by the transformation T_2. This result is easy to understand because from Fig. 3 one can see that the quasiparticle-quasihole tunneling adds a unit a_μ flux quantum to the hole of the torus.

Let us use two integers (p_1,p_2) to label different mean-field vacua

$$(p_1,p_2)\begin{cases}\dfrac{\phi}{\sqrt{n}}=e^{-i(2\pi/L)(p_1x_1+p_2x_2)} ,\\[2mm] a_i=\dfrac{2\pi}{L}p_i-eA_i .\end{cases} \tag{5.16}$$

The different vacua are connected by transformations T_1 and T_2. From (5.14) and (5.15) we see that the tunneling process for the quasiparticle moving in the x_1 direction changes the (0,0) vacuum to the (0,1) vacuum. Similarly, the tunneling in the x_2 direction changes the (0,0) vacuum to the $(-1,0)$ vacuum. We may define the amplitudes of the above tunnelings, $(0,0)\to(0,1)$ and $(0,0)\to(-1,0)$ to have zero phase (i.e., the amplitudes are real and positive). The tunneling from, say, (p_1,p_2) to (p_1,p_2+1) can be obtained by making a gauge transformation. The configuration describing the tunneling

$$(p_1,p_2)\to(p_1,p_2+1)$$

is given by

$$\phi'=e^{-i(2\pi/L)(p_1x_1+p_2x_2)}\phi ,$$
$$a_i'=a_i+\frac{2\pi}{L}p_i , \tag{5.17}$$

where ϕ and a_i are given by (5.5). The phase of the tunneling amplitude is

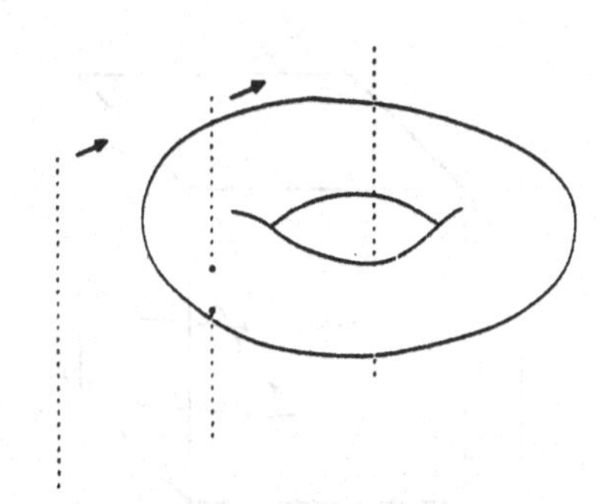

FIG. 3. A solinoid (represented by the dotted lines) creates a quasiparticle and a quasihole when pearing through the torus. The particle-hole tunnel process discussed here can be viewed as a solinoid cutting through the torus. Such a process adds unit flux to the hole, which changes the winding number of ϕ going around the hole.

$$\varphi_{(p_1,p_2)(p_1,p_2+1)}=\int d^3x[\mathcal{L}_{\rm GL}(a_i',\varphi')-\mathcal{L}_{\rm GL}(a_i,\varphi)]$$
$$=\int d^3x\frac{-1}{4\pi q}\left[\frac{2\pi p_1}{L}\partial_0 a_2-\frac{2\pi p_2}{L}\partial_0 a_1\right], \tag{5.18}$$

where a_i is given by (5.5). After some calculations we find that

$$\varphi_{(p_1,p_2)(p_1,p_2+1)}=\frac{2\pi p_1}{2q}. \tag{5.19}$$

Similarly, we find that for the tunneling $(p_1,p_2)\to(p_1-1,p_2)$:

$$\varphi_{(p_1,p_2)(p_1-1,p_2)}=\frac{2\pi p_2}{2q}. \tag{5.20}$$

$\varphi_{(p_2,p_2)(p_1',p_2')}$ as the phase of the hopping amplitude between *two* different states is not a physically observable quantity. Physically observable quantities are the phases of tunneling with the same initial and final states. Let us consider the following tunneling process:

$$(p_1,p_2)\to(p_1,p_2+1)\to(p_1-1,p_2+1)\to(p_1-1,p_2)\to(p_1,p_2),$$

(see Fig. 4). The total phase of the tunneling is given by

$$\varphi_{(p_1,p_2)(p_1,p_2+1)}+\varphi_{(p_1,p_2+1)(p_1-1,p_2+1)}-\varphi_{(p_1-1,p_2)(p_1-1,p_2+1)}-\varphi_{(p_1,p_2)(p_1-1,p_2)}=\frac{2\pi}{q}. \tag{5.21}$$

The tunneling in the x_1 direction, $(p_1,p_2)\to(p_1,p_2+1)$, changes one ground state to another and defines a unitary matrix U_1 acting on the ground states. The tunneling in the x_2 direction, $(p_1,p_2)\to(p_1-1,p_2)$, defines a unitary matrix U_2 acting on the ground states. The result (5.21) implies that U_1 and U_2 satisfy the algebra

$$U_2^{-1}U_1^{-1}U_2U_1=e^{i(2\pi/q)} \tag{5.22}$$

which is identical to the algebra satisfyed by T_1 and T_2. Noticing that U_1 (U_2) changes a state to its transformed state by T_2 (T_1), we may conclude that, in the subspace spanned by the degenerate ground states, U_2 and U_1 are proportional to T_1 and T_2, respectively,

$$U_1=\gamma_2T_2,$$
$$U_2=\gamma_1T_1. \tag{5.23}$$

After including the nonperturbative effects, the Hamiltonian (4.11) receives a correction

$$\Delta H=A(U_1+U_1^\dagger+U_2+U_2^\dagger)$$
$$=A(\gamma_1T_1+\gamma_2T_2+\text{H.c.}), \tag{5.24}$$
$$A=\gamma e^{-L(\Delta m^*)^{1/2}}.$$

ΔH does not commute with T_1 and T_2. The ground-state degeneracy is lifted by the nonperturbative effects. The energy split is of order $\gamma e^{-L(\Delta m^*)^{1/2}}$.

We would like to point out that the tunneling process described by Fig. 4 can be deformed into two linking loops (Fig. 5). Therefore, the phase in (5.21) is equal to the phase we obtained by moving one quasiparticle around another. This phase is given by 2θ where θ is the statistical angle of the quasiparticle. Thus, (5.22) can be rewritten as

$$U_2^{-1}U_1^{-1}U_2U_1=e^{i2\theta}. \tag{5.25}$$

Because the ground states form a representation of the algebra (5.25), the ground-state degeneracy is *directly* determined by the statistics of the quasiparticles.

We would like to remark that the tunnelings along two different tuneling paths given by, say, $x_2=0$ and $x_2=\Delta x_2$

FIG. 4. The four particle-hole tunnel processes are represented by the four directed paths in the space time. *ABCD* represents the torus. *AB* is identified with *CD* and *BC* is identified with *AD*.

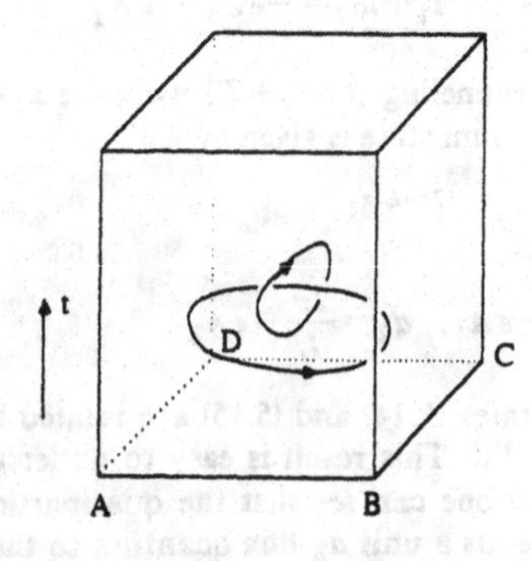

FIG. 5. The four tunnel paths in Fig. 4 can be deformed into two linked loops.

have a phase difference

$$\Delta\varphi = \frac{e^2 B}{q} \Delta x_2 L$$

because the quasiparticle carries the electrical charge e/q. Therefore, after summing up all the tunneling paths associated with different Δx_2, the factor γ in the tunneling amplitude (5.24) takes a form

$$\gamma \propto \int d\Delta x_2\, e^{i(e^2B/q)\Delta x_2 L} e^{-L(m^*\Delta)^{1/2}} . \quad (5.26)$$

If our system respects translation symmetry, m^* and Δ in (5.26) do not depend on Δx_2 and we find that $\gamma=0$. The ground-state degeneracy is exact even in system with finite size. This agrees with the result in Ref. 10. Only when the translation symmetries are broken can the ground-state degeneracy by really lifted by the nonperturbative effects.

Strictly speaking, the total tunneling amplitude is given by the sum of the amplitudes of all different tunneling paths C:

$$A \propto \int D\mathbf{x}(t) \exp\left[i\frac{e}{q} \oint_C d\mathbf{x}\cdot\mathbf{A} \right] \times \exp\left[-\oint_C dt \left[\frac{m^*}{2}\dot{\mathbf{x}}(t) + \Delta + V(\mathbf{x}(t)) \right] \right] , \quad (5.27)$$

where $\mathbf{x}(t)$ describes the tunneling path C and $V(\mathbf{x})$ is the random potential. Or equivalently, we may express the tunneling amplitude A in terms of the Green functions of the quasiparticle and the quasihole, G^p and G^h:

$$A \sim \int d^2x\, dt\, G^h\left[x, x+\frac{L}{2}; t_0, t_0+t \right] \times G^p\left[x, x-\frac{L}{2}; t_0, t_0+t \right] . \quad (5.28)$$

If we ignore the phase factor $\oint_C d\mathbf{x}\cdot\mathbf{A}$, the second exponential in (5.27) gives rise to the factor $e^{-L(\Delta m^*)^{1/2}}$ in (5.24). The summation of the phase factor $\exp[i(e/q)\oint_C d\mathbf{x}\cdot\mathbf{A}]$ corresponds to the reduction factor γ in (5.24). Because the phase factor changes extremely fast from path to path, the factor γ itself may be exponentially small.

In a potential produced by a single impurity (i.e., a potential which is nonzero only in a finite region), the quasiparticle can only do circular motion due to the strong magnetic field. The propagator of the quasiparticle is localized and takes the form

$$|G(x,x';\omega)| \sim e^{-\alpha[(x-x')^2/l^2]} ,$$

where l is the magnetic length and α an $O(1)$ constant. Therefore, we expect the total tunneling amplitude A to be a quadratic exponential in L:

$$A \sim e^{-\alpha(L^2/l^2)} . \quad (5.29)$$

When the potential V is periodic, the situation is very different because of possible resonance effects. Let us consider a periodic potential V such that there is a multiple of $2\pi q$ flux going through each plaquette. Such a potential changes the Landau levels of the quasiparticles into energy bands with a finite width. The nontrivial dispersion relation $E(k)$ (k is the crystal momentum) implies that the quasiparticles are delocalized by the periodic potential. In other words, the wave packet of the quasiparticle moves in a straight line in the presence of the periodic potential. In this case we expect the tunneling amplitude to be a linear exponential in L:

$$A \sim e^{-\alpha(L/l)} . \quad (5.30)$$

A more direct way to understand the above result is to notice that the easy tunneling paths alpha by the periodic potential have phase factors which only differ from each other by a multiple of 2π (Fig. 6.) There is no cancellation between the amplitudes of the easy paths. All the easy paths together contribute to the total tunneling amplitude A, a term of order $e^{-\alpha(L/l)}$.

For generic random potentials, the qusiparticle Green function is shown to have a form[25]

$$|G(x,x';\omega)| \sim e^{-\alpha(|x-x'|/\xi)} .$$

In this case the tunneling amplitude is expected to be given by (5.30).

The point of the above discussion is the following. The strength of the tunneling amplitude A depends on whether the quasiparticles are localized or not in the potential V. If the quasiparticles are not localized (e.g., in the periodic potential), the amplitude A is expected to be of order $e^{-\alpha(L/l)}$. If the quasiparticles are localized (e.g., in the single impurity potential), the amplitude A is expected to be smaller than $e^{-\alpha(L/l)}$, or more precisely, $\ln|A|/L \to -\infty|_{L\to\infty}$. In case of single impurity potential we further expect $A \sim e^{-\alpha(L^2/l^2)}$.

Before ending this section we would like to mention that Haldane[26] has suggested that the tunneling process discussed in this section may change one ground state of the FQH system to another ground state. In the topolog-

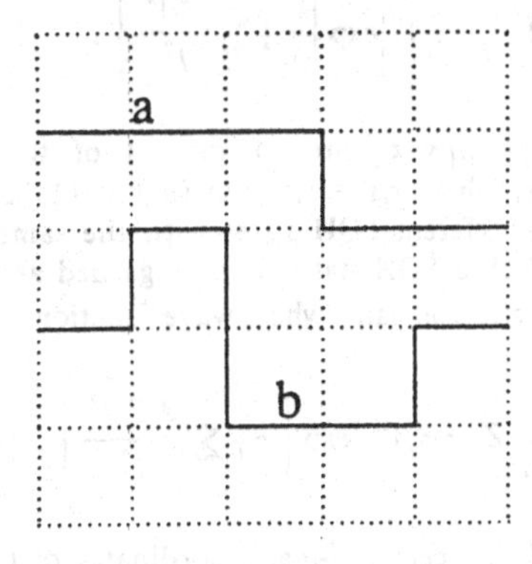

FIG. 6. The dotted lines represent the minima of the potential. The two easy paths (a) and (b) favored by the potential enclose an integer number of plaquettes. The phase of tunneling amplitudes of (a) and (b) only differ by a multiple of 2π.

ical Chern-Simons theory the algebra of the tunneling loops (or Wilson lines) (5.22) has been used to construct all the ground states.[27] Read has also pointed out that the tunneling loops satisfy the algebra (5.22) which may be used to construct the ground states.[28] These physical pictures and ideas are demonstrated explicitly in this section in the frame work of the effective GL theory of the FQH effects.

VI. THE GROUND-STATE DEGENERACY OF THE HIERARCHY FQH STATES

For the general filling fraction $\nu=p/q$, the FQH states are given by the hierarchy scheme suggested in Ref. 29. The hierarchy FQH states may be described phenomenologically by the following effective GL theory:

$$\mathcal{L}_{GL}=\Phi^*(i\partial_0-a_0-e^*A_0)\Phi+\frac{1}{2m}\Phi^*(i\partial_i-a_i-e^*A_i)^2\Phi + V(\Phi)-\frac{\bar{p}}{4\pi\bar{q}}\epsilon^{\mu\nu\lambda}a_\mu\partial_\nu a_\lambda \tag{6.1}$$

with $e^*=(r/s)e$ where (r,s) and $(\bar{p},\bar{q})$ are two pairs of incommensurable integers. The Hall conductance is given by

$$\sigma_{xy}=\frac{\bar{p}}{\bar{q}}\frac{e^{*2}}{\hbar}=\frac{\bar{p}r^2}{\bar{q}s^2}\frac{e^2}{\hbar}$$

and the filling fraction $\nu=\bar{p}r^2/\bar{q}s^2=p/q$. We always rescale a_μ such that Φ carries a unit a_μ charge. To obtain (6.1) we include the possibility that the FQH state is given by an n-boson condensed state $\langle\phi^n\rangle\neq 0$. Thus, Φ in (6.1) may correspond to ϕ^n in (3.1). In this case $\bar{q}$ in (6.1) can be an even integer.[28] However, an electron system may not be able to produce the general GL theory (6.1) with all possible integer pairs (r,s) and $(\bar{p},\bar{q})$. It is possible that only a subset of the integer pair (r,s) and $(\bar{p},\bar{q})$ is realized by electron systems.

When $(r,s)=(1,1)$ and $(\bar{p},\bar{q})=(1,3)$, (6.1) describes a Laughlin state with filling factors $\nu=\frac{1}{3}$. The wave function of this FQH state is given by

$$\left[\prod_{i<j}(z_i-z_j)^3\right]\exp\left[-\frac{1}{4}\sum\frac{|z_i|^2}{l^2}\right],$$

where $z_i=x_{i1}+ix_{i2}$ are coordinates of the electrons. However, when $(r,s)=(2,1)$ and $(\bar{p},\bar{q})=(1,12)$, (6.1) describes a different FQH states with the same fill factor $\nu=\frac{1}{3}$. Such a FQH state can be regarded as a Laughlin state for electron pairs, whose wave function is given by

$$\left[\prod_{i<j}(Z_i-Z_j)^{12}\right]\exp\left[-\frac{1}{4}\sum\frac{2|Z_i|^2}{l^2}\right],$$

where Z_i are center-of-mass coordinates of the electron pairs.

We would like to remark that a more general effective GL theory of the QH states may contain several boson fields and gauge fields, for example, it may take a form

$$\mathcal{L}_{GL}=\sum_{I=1}^{N}\left[\Phi_I^*(i\partial_0-a_{I0}-e_I^*A_0)\Phi_I+\frac{1}{2m_I}\Phi_I^*(i\partial_i-a_{Ii}-e_I^*A_i)^2\Phi+V_I(\Phi_i)-\frac{\bar{p}_I}{4\pi\bar{q}_I}\epsilon^{\mu\nu\lambda}a_{I\mu}\partial_\nu a_{I\lambda}\right]. \tag{6.1a}$$

If we choose $e_I^*=e$ and $\bar{p}_I=\bar{q}_I=1$, (6.1a) describes an IQH state with N-filled Landau levels. For simplicity we will concentrate on the effective GL theory in (6.1). Most of the results obtained for (6.1) can be easily generalized so that they also apply to (6.1a).

The discussions in Secs. II and III can be directly generalized to apply to (6.1). The transformations T_1 and T_2 defined in (3.7) now obey an algebra

$$T_1T_2=e^{i(2\pi\bar{p}/\bar{q})}T_2T_1\ . \tag{6.2}$$

The ground states of (6.1) defined on the torus have $\bar{q}$-fold degeneracy in the thermodynamic limit.

The quasiparticle in (6.1) has a fractional statistics given by $\theta=\pi\bar{p}/\bar{q}$. The denominator of the statistical angle is directly related to the ground-state degeneracy. The statistical angle θ, however, is not directly related to the Hall conductance σ_{xy} or the filling fraction $\bar{p}r^2/\bar{q}s^2=p/q$.

We would like to remark that the two hierarchy FQH states given by the same $(\bar{p},\bar{q})$ and different (r,s) have the same (low-energy) topological structure or topological order. They only differ by a rescaling of the electric charge. On the other hand, two FQH states with the same Hall conductance (and filling fraction p/q) may have a different topological order corresponding to different $(\bar{p},\bar{q})$ [and (r,s)].

We would like to make a side remark here. We know that in the mean-field approach the anyon superfluid states[30] are closely related to the QH states. The filling fraction ν of the associated QH problem is determined by the statistical angle θ of the anyons:

$$\nu=\frac{\pi}{\theta}\ .$$

We know that the QH states with the same filling fraction may have different topological orders. This fact suggests that the anyon superfluid state may have different phases.[31] Each phase has a different topological order. The quasiparticle excitations in different superfluid phases may have different statistics. In particular, the semion superfluid state obtained from the QH state of two filled Landau levels does not support quasiparticles with fractional statistics, while a different semion superfluid state obtained from the (tide binding) semion pair condensation does support semionic quasiparticle excitations.[31]

We would like to emphasize that in this paper we only show that the ground states of the FQH state are at least $\bar{q}$-fold degenerate on the torus. Our proof does not exclude the possibility that the ground-state degeneracy may be large than $\bar{q}$. However, our results do imply that the ground-state degeneracy must be a multiple of $\bar{q}$.

We would like to mention that the FQH state for example, at filling fraction $\nu=\frac{2}{5}$, may be described by (6.1) with $(\tilde{p},\tilde{q})=(1,10)$ and $(r,s)=(2,1)$. The quasiparticle carries a charge $e/5$ and has a statistic $\theta=\pi/10$. It is not clear whether the integer pair $(\tilde{p},\tilde{q})=(2,5)$ and $(r,s)=(1,1)$ can be realized by (spinless) electron systems or not.

VII. DUALITY PICTURE AND APPLICATIONS TO THE CHIRAL SPIN STATES

The GL theory (6.1) has a dual form[32,33] in which the order parameter Φ is replaced by a U(1) gauge field $\tilde{a}_\mu$. Some discussions in the previous sections become more transparent in the dual theory.

To give a simple heuristic derivation of the dual theory of the GL theory (6.1), let us first turn off a_μ and A_μ in (6.1). Now (6.1) described a superfluid state. However, the low-lying excitations have a spectrum of the form $\varepsilon_k=k^2/2m$ corresponding to the free-boson condensation. For interacting bosons the low-lying spectrum is linear $\varepsilon_k=c_s k$ which describes a phonon mode. Therefore, the low-lying excitation of the superfuid is rather described by

$$L=\int d^2x\frac{1}{2g^2}(\partial_\mu\chi)^2 \tag{7.1}$$

after including the interactions. In (7.1) we have set the phonon velocity $c_s=1$. g^2 in (7.1) is the rigidness constant. χ is the phase of Φ. The superfluid current J_i is given by

$$J_i=\frac{1}{g^2}\partial_i\chi . \tag{7.2}$$

Therefore, $g^2=m/n$ (in the limit $c_s=1$).

It is pointed out in Refs. 32 and 34 that (7.1) is equivalent to a U(1) gauge theory described by

$$L=\int d^2x\frac{g^2}{16\pi^2}\tilde{f}^2_{\mu\nu} , \tag{7.3}$$

where $\tilde{f}_{\mu\nu}=\partial_\mu\tilde{a}_\nu-\partial_\nu\tilde{a}_\mu$, if we identify the superfluid current J_i and the sperfluid density $J_0=n$ with $\varepsilon_{\mu\alpha\beta}\tilde{f}^{\alpha\beta}$:

$$J_\mu=\frac{1}{4\pi}\varepsilon_{\mu\alpha\beta}\tilde{f}^{\alpha\beta} . \tag{7.4}$$

A vortex in the superfluid can be viewed as a particle carrying $\tilde{a}_\mu$ charge. Including the vortex-antivortex excitations, the effective theory may be written as

$$L=\int d^2x\left[\frac{g^2}{16\pi^2}\tilde{f}^2_{\mu\nu}+\tfrac{1}{2}|(\partial_0+i\tilde{a}_0)\Psi|^2 -\tfrac{1}{2}c_v^2|(\partial_i+i\tilde{a}_i)\Psi|^2-\tfrac{1}{2}m_v^2|\Psi|^2\right] . \tag{7.5}$$

The vortex density is given by $\mathrm{Re}(i\Psi^*\partial_0\Psi)$. A Ψ particle creates an "electric" field $\tilde{f}_{i0}$ around it. From (7.4) we see that the "electric" field in radial direction corresponds to a superfluid current circulating around the Ψ particle. Thus, the Ψ particle indeed generates a vortex in the superfluid. We have assigned a unit $\tilde{a}_\mu$ charge to Ψ particle such that it creates a minimum quantized vortex [see (7.4)]. There is no particle carrying fractional $\tilde{a}_\mu$ charge because the circulation of a vortex is quantized. The fact that the $\tilde{a}_\mu$ charge of the excitations in the dual theory is quantized as an integer reflects that the superfluid state is a single-boson Φ condensed state, i.e., $\langle\Phi\rangle\neq 0$. Had the superfluid state come from the N-boson condensation, $\langle\Phi^N\rangle\neq 0$, the $\tilde{a}_\mu$ charge would be quantized as a multiple of $1/N$.

The GL theory (6.1) of the FQH effects is obtained by coupling the superfluid current J_μ to $a_\mu+A_\mu$ and including the Chern-Simons term of a_μ. We may do the same thing to the dual theory (7.3) of the superfluid to obtain the dual theory of (6.1). After including

$$J_\mu(a^\mu+e^*A^\mu)-\frac{\tilde{p}}{4\pi\tilde{q}}a_\mu\partial_\nu a_\lambda\varepsilon^{\mu\nu\lambda}$$

to (7.5) we obtain the dual theory of the GL theory

$$L_{d\mathrm{GL}}=\int d^2x\left[\frac{g^2}{16\pi^2}\tilde{f}^2_{\mu\nu}+\frac{1}{4\pi}(a_\mu+e^*A_\mu)\tilde{f}_{\nu\lambda}\varepsilon^{\mu\nu\lambda} -\frac{\tilde{p}}{4\pi\tilde{q}}a_\mu\partial_\nu a_\lambda\varepsilon^{\mu\nu\lambda}+\tfrac{1}{2}|(\partial_0+i\tilde{a}_0)\Psi|^2 -\tfrac{1}{2}c_v^2|(\partial_i+i\tilde{a}_i)\Psi|^2-\tfrac{1}{2}m_v^2|\Psi|^2\right] . \tag{7.6}$$

After integrating out a_μ we get

$$L_{d\mathrm{GL}}=\int d^2x\left[\frac{g^2}{16\pi^2}\tilde{f}^2_{\mu\nu}+\frac{e^*}{4\pi}A_\mu\tilde{f}_{\nu\lambda}\varepsilon^{\mu\nu\lambda} +\frac{\tilde{q}}{4\pi\tilde{p}}\tilde{a}_\mu\partial_\nu\tilde{a}_\lambda\varepsilon^{\mu\nu\lambda}+\tfrac{1}{2}|(\partial_0+i\tilde{a}_0)\Psi|^2 -\frac{c_v^2}{2}|(\partial_i+i\tilde{a}_i)\Psi|^2-\tfrac{1}{2}m_v^2|\Psi|^2\right] . \tag{7.7}$$

Ψ describes the quasiparticle (quasihole) excitations above the FQH state. Due to the Chern-Simons term in (7.7), the Ψ particle (the quasiparticle) generates $2\pi\tilde{p}/\tilde{q}$ flux of the $\tilde{a}_\mu$ gauge field. As a bound state of charge and flux, the quasiparticle has a fractional statistics[35] $\theta=\pi(\tilde{p}/\tilde{q})$. The quasiparticle carries fractional electric charge $(\tilde{p}/\tilde{q})e^*$ which can be derived from the coupling $(e^*/4\pi)A_\mu\tilde{f}_{\nu\lambda}e^{\mu\nu\lambda}$ in (7.7).

To rigorously prove that (7.7) is effective theory of the FQH state, we need to prove that, after integrating out $\tilde{a}_\mu$ and Ψ, (7.7) produces the same effective Lagrangian $L_{\mathrm{eff}}(A_\mu)$ as the electron system does. A relation similar to (3.4) should be satisfied. Although here we cannot show that (3.4) is satisfied by the dual theory (7.7), the dual effective theory does reproduce (at least qualitatively) all known low-energy properties of the FQH states. Therefore, we expect that the FQH states are correctly described by (7.7) at low energies and we may use (7.7) to study another (unknown) low-energy properties of the FQH states.

In order to use the dual theory (7.7) to study the ground-state degeneracy of the FQH states, we first need to quantize (7.7).[27] At the moment let us ignore the quasiparticle field Ψ. Following the approach in Sec. IV, we may quantize (7.7) in the gauge $\tilde{a}_0=0$. The constraint associated with the equation of motion is

$$G(x)=\frac{\delta L_{dGL}}{\delta a_0}=\frac{g^2}{4\pi^2}\partial_i \tilde{f}^{0i}+\frac{\bar{q}}{2\pi\bar{p}}\varepsilon^{ij}\partial_i\tilde{a}_j$$
$$=\partial_i\pi^i+\frac{\bar{q}}{4\pi\bar{p}}\varepsilon^{ij}\partial_i\tilde{a}_j=0 . \quad (7.8)$$

π^i in (7.8) is the canonical momentum conjugated to a_i:

$$\pi^i=\frac{\delta L_{dGL}}{\delta \dot{a}_i}=\frac{g^2}{4\pi^2}\tilde{f}^{0i}+\frac{\bar{q}}{4\pi\bar{p}}\varepsilon^{ij}\tilde{a}_j . \quad (7.9)$$

After the quantization the operators $\hat{a}_i$ and $\hat{\pi}_i$ satisfy

$$[\hat{\pi}_i(x),\hat{a}_j(y)]=i\delta^2(x-y) . \quad (7.10)$$

Under the gauge transformations $\hat{a}_i$ and $\hat{\pi}_i$ transform as

$$\hat{a}_i\to\hat{a}_i+\partial_i f ,$$
$$\hat{\pi}_i\to\hat{\pi}_i+\frac{\bar{q}}{4\pi\bar{p}}\varepsilon^{ij}\partial_j f , \quad (7.11)$$

where f is a single-valued function on the torus. Using (7.10) we see that the gauge transformation (7.11) is generated by the operator

$$G[f]=\exp\left[i\int d^2x\,\partial_i f\left[\pi^i+\frac{\bar{q}}{4\pi\bar{p}}\varepsilon^{ij}\tilde{a}_j\right]\right]$$
$$=\exp\left[-i\int d^2x\,fG(x)\right] . \quad (7.12)$$

Once again the constraint (7.8) generates the gauge transformation. Due to the gauge invariance of the theory, all physical operators commute, with $G[f]$. Noticing that $G[f]$ and $G[f']$ commute, we may require the physical states (which form a representation of physical operators) to satisfy

$$G[f]|\Psi_{\text{phy}}\rangle=|\Psi_{\text{phy}}\rangle . \quad (7.13)$$

The constraint (7.8) is satisfied by the physical states, $G(x)|\Psi_{\text{phy}}\rangle=0$. The condition (7.13) defines the physical Hilbert space.

The operator $G[f]$ given by (7.12) is well defined even when f is a multivalued function. In particular, $G[\alpha f_1]$ and $G[\alpha f_2]$ are well-defined operators, where f_1 and f_2 are defined in (4.7) and α is a constant. Using (7.10) and (7.12) one can check that $G[\alpha f_1]$ and $G[\beta f_2]$ satisfy an algebra

$$G[\alpha f_1]G[\beta f_2]=e^{i\alpha\beta 2\pi\bar{q}/\bar{p}}G[\beta f_2]G[\alpha f_1] . \quad (7.14)$$

Because the $\tilde{a}_\mu$ charge is quantized as integers, this is equivalent to say that $G[f_1]$ and $G[f_2]$ generate the (large) gauge transformations and commute with all the physical operators. However, because $G[f_1]$ and $G[f_2]$ do not commute

$$G[f_1]G[f_2]=e^{i(2\pi\bar{q}/\bar{p})}G[f_2]G[f_1] \quad (7.15)$$

if $\bar{p}\neq 1$, we cannot require the physical states to be invariant under both $G[f_1]$ and $G[f_2]$. But we can further restrict the physical Hilbert space by requiring the physical states to satisfy, for example,

$$G[f_1]|\Psi_{\text{phy}}\rangle=|\Psi_{\text{phy}}\rangle ,$$
$$G^{\bar{p}}[f_2]|\Psi_{\text{phy}}\rangle=|\Psi_{\text{phy}}\rangle , \quad (7.16)$$

because $G[f_1]$ and $G^{\bar{p}}[f_2]$ commute.

Notice that $G[\alpha f_1]$ and $G[\alpha f_2]$ commute with $G[f]$. When $\alpha=\bar{p}/\bar{q}$, $G[\alpha f_1]$ and $G[\alpha f_2]$ also commute with $G[f_1]$ and $G[f_2]$. Therefore,

$$T_1\equiv G\left[\frac{\bar{p}}{\bar{q}}f_1\right] ,$$
$$T_2\equiv G\left[\frac{\bar{p}}{\bar{q}}f_2\right] , \quad (7.17)$$

act on the physical Hilbert space defined by (7.13) and (7.16). Later we will show that T_1 and T_2 are generated by the quasiparticle tunneling described in Sec. V and they are physical operators. T_1 and T_2 satisfy the algebra

$$T_1T_2=e^{i(2\pi\bar{p}/\bar{q})}T_2T_1 \quad (7.18)$$

and the physical states form a representation of the algebra (7.18). The Hamiltonian of the dual theory (7.7) is given by (after ignoring Ψ field)

$$H=\int\left[\frac{g^2}{8\pi^2}(f^{0i})^2+\frac{g^2}{8\pi^2}(f^{12})^2\right]d^2x . \quad (7.19)$$

The Hamiltonian commutes with the gauge generators $G[f]$, $G[f_1]$, and $G[f_2]$. Therefore, H acts on the physical Hilbert space. The Hamiltonian (7.19) also commutes with the *physical* operators T_1 and T_2. Hence, the ground states of H must form a representation of the algebra (7.18) and are (at least) $\bar{q}$-fold degenerate.

We would like to remark that the above discussions demonstrate that the topological Chern-Simons theory of compact U(1) gauge field can be (mathematically) consistently quantized, even when the coefficient in front of the Cern-Simons is a rational number. This is true at least when the space-time metrics is kept fixed.

We would like to point out that if we separate the local and the global excitations by writing

$$\tilde{a}_i=\frac{2\pi\theta_i}{L_i}+\delta\tilde{a}_i \quad (7.20)$$

from (7.4), we see that ϑ_i (ϑ_2) corresponds to a constant current density in x_2 (x_1) direction. Therefore, $\varepsilon^{ij}\vartheta_i$ are proportional to the center-of-mass coordinate x_{ci}. The operator T_1 and T_2 shift ϑ_i:

$$\vartheta_i\to\vartheta_i+2\pi\frac{\bar{p}}{\bar{q}} \quad (7.21)$$

if we choose $f_i=2\pi(x_i/L_i)$. Thus, the operator T_1 (T_2) discussed in this section corresponds to the magnetic translation T_2 (T_1) discussed in Sec. II which also shifts the center-of-mass coordinates. However, the operators T_i discussed in this section have a local definition and can be easily generalized to the high-genus Riemann surface.

Now let us consider the effects of the quasiparticle fluctuations Ψ. First we notice that, for finite torus, two operators

$$W_1=e\left[-i\int_0^{L_1}dx_1\hat{a}_1\right],$$
$$W_2=e\left[+i\int_0^{L_2}dx_2\hat{a}_2\right], \tag{7.22}$$

are invariant under the gauge transformations generated by $G[f]$, $G[f_1]$, and $G[f_2]$. There is no reason to exclude the gauge invariant term

$$\Delta H=(c_1W_1+c_2W_2+\text{H.c.}) \tag{7.23}$$

from the effective Hamiltonian. Indeed, after we integrate out the Ψ field (with fixed $\bar{a}_\mu$), ΔH is induced by the quasiparticle fluctuations. It precisely comes from the quasiparticle-quasihole tunneling process discussed in Sec. V. Under T_i the operators W_i transform as

$$T_1^{-1}W_1T_1=e^{-i(2\pi p/q)}W_1,$$
$$T_2^{-1}W_1T_2=W_1,$$
$$T_1^{-1}W_2T_1=W_2,$$
$$T_2^{-1}W_2T_2=e^{i(2\pi p/q)}W_2. \tag{7.24}$$

When restricted to the subspace spanned by the ground states [which are assumed to form an irreducible representation of (7.18)], W_1 (W_2) can be shown to be proportional to T_2 (T_1). Because ΔH does not commute with T_i, the ground-state degeneracy is lifted by the quasiparticle tunneling effects.

Using the approach in Sec. V we can show explicitly that the quasiparticle tunneling generates physical operators T_i.

On the torus the quasiparticle-quasihole tunneling discussed in Sec. V is given by the following ansatz:

$$\bar{a}_1+i\bar{a}_2=\frac{\bar{p}}{\bar{q}}\sum_{mn}\left[\frac{i(z-z_0-Z_{mn})}{|z-z_0-Z_{mn}|^2+\xi'^2}-\frac{i(z-\bar{z}_0-Z_{mn})}{|z-\bar{z}_0-Z_{mn}|^2+\xi'^2}\right], \tag{7.25}$$

where z_0 and $\bar{z}_0$ satisfying (5.13) are the coordinates of the quasiparticle and the quasihole. After the tunneling in the x_1 direction), an initial configuration $(\bar{a}_1,\bar{a}_2)$ is changed to a final configuration $[\bar{a}_1,\bar{a}_2+(\bar{p}/\bar{q})(2\pi/L)]$. The tunneling in the x_2 direction changes the configuration $(\bar{a}_1,\bar{a}_2)$ to $[\bar{a}_1-(\bar{p}/\bar{q})(2\pi/L),\bar{a}_2]$. Let us use operators U_1 to U_2 to denote the above transformations:

$$U_2:(\bar{a}_1,\bar{a}_2)\to\left[\bar{a}_1,\bar{a}_2+\frac{p}{q}\frac{2\pi}{L}\right],$$
$$U_1:(\bar{a}_1,\bar{a}_2)\to\left[\bar{a}_1-\frac{p}{q}\frac{2\pi}{L},\bar{a}_2\right]. \tag{7.26}$$

Using the similar calculation performed in Sec. V [see (5.16)–(5.22)], we find that U_1 and U_2 satisfy

$$U_1^{-1}U_2^{-1}U_1U_2=e^{i(2\pi p/q)}. \tag{7.27}$$

From (7.26) and (7.27) we see that U_i are proportional to T_i.

According to Ref. 5, after setting $A_\mu=0$, (7.7) with $\bar{p}=1$ and $\bar{q}$ an even integer is precisely the effective theory of the chiral spin states. The Ψ field now describes the spinon excitations. Therefore, the discussions in this paper about the FQH states also apply to the chiral spin states. In particular, we find that the ground-state degeneracy of the chiral spin is very robust as suggested in Ref. 8. The degeneracy persists even when the translation symmetry is broken, e.g., when the spin-spin coupling J_{ij} has a spatial dependence.

VIII. GROUND-STATE DEGENERACY OF THE FQH STATES ON ARBITRARY RIEMANN SURFACE

In Ref. 7 the ground-state degeneracy of the chiral spin states [described by (7.7) with $\bar{p}=1$] is shown to be $\bar{q}^g$ (for a given chirality) on a Riemann surface with genus g. In this section we will derive a similar result for the FQH state. We will take the Lagrangian (6.1) as our starting point. However, on an arbitrary Riemann surface Σ_g with genus $g\neq1$, (6.1) needs to be generalized to

$$\mathcal{L}_{GL}=\Phi^* iD_0\Phi-\frac{1}{2m}g^{ij}D_i\Phi^*D_j\Phi-V(\Phi)-\frac{\bar{p}}{4\pi\bar{q}}\varepsilon^{\mu\nu\lambda}a_\mu\partial_\nu a_\lambda, \tag{8.1}$$

where

$$D_\mu\Phi=(\partial_\mu+ia_\mu+A_\mu)\Phi$$

and

$$D_\mu\Phi^*=(\partial_\mu-ia_\mu-ie^*A_\mu)\Phi^*.$$

g^{ij} in (8.1) is a two-dimensional metrics which, in general, has a spatial dependence. The matrices g^{ij} is necessary because we cannot choose a single coordinate patch to cover the whole Riemann surface Σ_g with $g\neq1$. On the Riemann surface Σ_g with $g<1$ the translation symmetry is bound to be broken.

We will use the method developed in Sec. IV to derive our result. On a Riemann surface Σ_g there are $2g$ canonical one-cycle denoted as α_a and β_a, $a=1,\ldots,g$ (Fig. 7). We choose $2g$ functions f_b ($b=1,\ldots,2g$) on Σ_g such that f_a has a 2π jump along α_a and f_{g-a} has a 2π jump along β_a, here $a=1,\ldots,g$. However, we require $\partial_i f_a$ to be a smooth vector field on Σ_g. Using f_a we define unitary operators T_a as the following:

$$T_a=\exp\left[i\int d^2x\,f_a(x)\hat{G}(x)\right]$$
$$=\exp\left[-i\int d^2x\,\hat{\Phi}^+\hat{\Phi}f_a\right]\times\exp\left[-i\frac{\bar{p}}{2\pi\bar{q}}\int d^2x\,a_i\partial_jf_a\varepsilon^{ij}\right],\quad a=1,\ldots,2g. \tag{8.2}$$

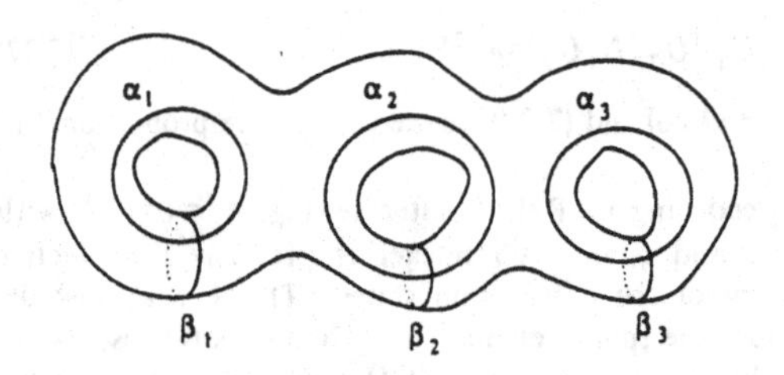

FIG. 7. A Riemann surface and its canonical one-cycles α_a and β_a (for $g=3$).

After a transformation by T_a, $\hat{\Phi}\to\hat{\Phi}'=e^{if_a}\hat{\Phi}$. $\hat{\Phi}'$ remain a smooth function on Σ_g. Using the commutation relation

$$[\hat{a}_1(x),\hat{a}_2(y)]=i2\pi\frac{\bar{q}}{\bar{p}}\delta^2(x-y)\,, \tag{8.3}$$

we find that

$$T_aT_b=\exp\left[i\frac{p}{2\pi q}\int d^2x\,\partial_i f_a\partial_j f_b\varepsilon^{ij}\right]T_bT_A\,. \tag{8.4}$$

The exponent in (8.4) can be evaluated and we find

$$\int d^2x\,\partial_i f_a\partial_j f_b\varepsilon^{ij}=(2\pi)^2\eta_{ab}\,,$$
$$(\eta_{ab})=\begin{bmatrix}0 & 1\\ -1 & 0\end{bmatrix}\otimes I_{g\times g}\,, \tag{8.5}$$

where $I_{g\times g}$ is a $g\times g$ unit matrix. Therefore, (8.4) can be rewritten as

$$T_aT_{g+a}=e^{i(2\pi\bar{p}/\bar{q})}T_{g+a}T_a,\quad a=1,\ldots,g\,,$$
$$[T_a,T_b]=0,\quad b\neq a+g,\quad a,b=1,\ldots,2g\,, \tag{8.6}$$

where we have assumed $T_{a+2g}=T_a$. The pairs of operators T_a and T_{g+a} generate g copies of the algebra (6.2), which commute with each other. Each copy of the algebra contributes a factor $\bar{q}$ to the ground-state degeneracy. The total ground-state degeneracy is $\bar{q}^{\,g}$.

If we compactify the space into g copies disconnected tori, the ground states of (6.1) are obviously $\bar{q}^{\,2}$-fold degenerate. The result in this section implies that the ground-state degeneracy is unchanged after we connect the g tori by tubes to form a genus g Riemann surface (Fig. 8).

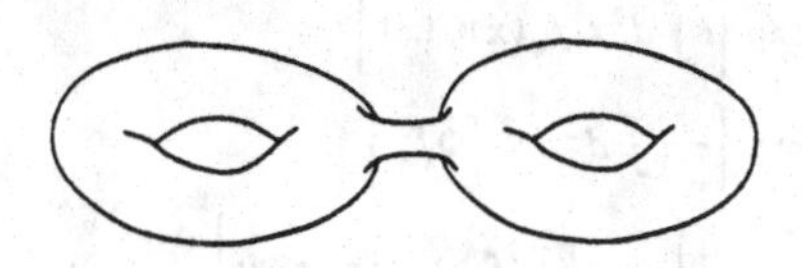

FIG. 8. A genus-two Riemann surface is formed by connecting two tori by a tube.

IX. DISCUSSIONS

In this paper we show that the FQH states on the Riemann surface Σ_g have $\bar{q}^{\,2}$-fold degenerate ground states if the quasiparticles in the FQH states have fractional statistics $\theta=\pi\bar{p}/\bar{q}$. The fact that the ground-state degeneracy depends on the topology of the space suggests that the degeneracy is not due to the broken symmetry. We also show that the ground-state degeneracy (in the thermodynamic limit) is robust against arbitrary perturbations. This means that the ground-state degeneracy remains a constant in a finite region in the phase space. Therefore, we may use the ground-state degeneracy to characterize different phases in the phase space. We may say that the phases with different ground-state degeneracy have different topological orders. As we change the coupling constants in the theory, the ground-state degeneracy may jump which signals a phase transition between two phases with different topological orders.

If one insists on a symmetry-breaking picture, one may regard the ground-state degeneracy considered in this paper as a result of broken "topological" symmetries. The topological symmetry transformation is defined as the following. Consider a FQH state on a torus. We adiabatically add a unit flux through the hole of the torus [Fig. 9(a)]. The Hamiltonian is invariant after adding a unit flux. Therefore, the adiabatic process changes one

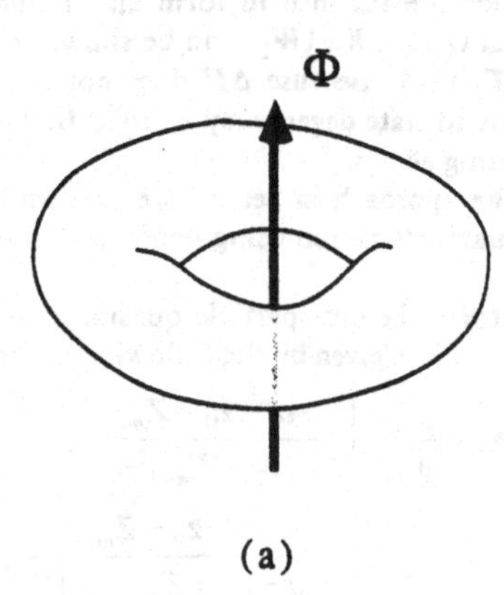

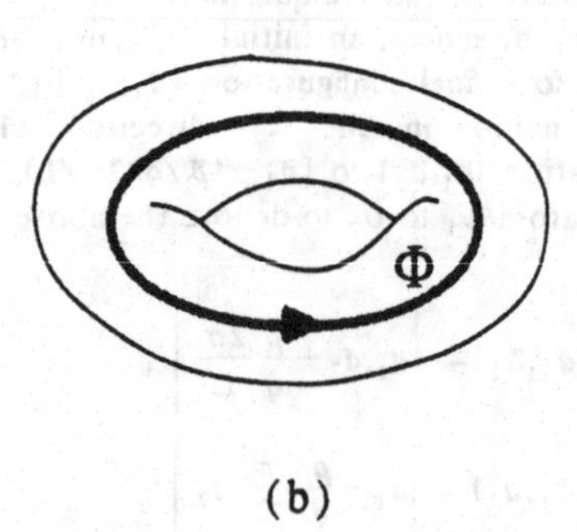

FIG. 9. A torus with flux going (a) through the hole and (b) through the tube.

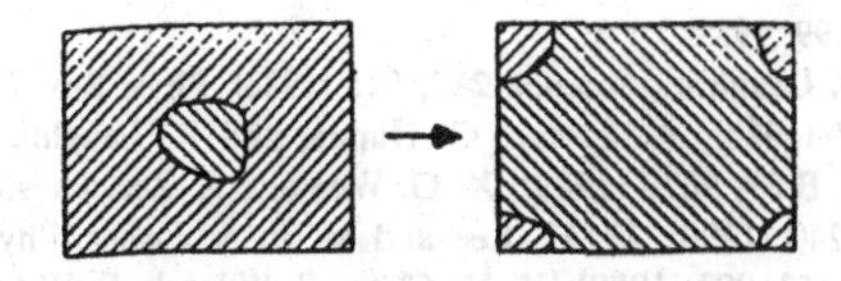

FIG. 10. Two ground states resulting from a broken symmetry can be connected by a domain-wall-tunneling process in which a domain sweeps over the whole system.

ground state of the FQH state to another. Such a transformation can be represented by a unitary operator U_1 which acts on the ground states. Similarly, the adiabatic turning on a unit flux going through the tube of the torus [Fig. 9(b)] generates an operator U_2 acting on the ground states. We call the operators U_1 and U_2 the topological symmetry transformations. Notice that the topological symmetry transformations can be defined only after we specify the topology of the space. The very existence of the topological symmetry depends on the topology of the space. On the spheres there is no topological symmetry. That is why the ground state of the FQH states is nondegenerate on the sphere. On the Riemann surface Σ_g of genus g, there are $2g$ topological symmetry transformations. From Ref. 13 we find that the operators U_1 and U_2 satisfy the algebra

$$U_1^{-1}U_2^{-1}U_1U_2=e^{i(2\pi p/q)}, \tag{9.1}$$

where p/q is the filling fraction. Therefore, U_1 and U_2 cannot be the identity in the subspace spanned by the ground states. This implies that the topological symmetry is spontaneously broken.

On a finite system, the ground-state degeneracy may be lifted by finite-size effects. For the degenerate ground states associated with ordinary symmetry breaking, the energy split is expected to be of order e^{-L^2/ξ^2}, where ξ is a microscopic length scale of the theory and L is the size of system. This is because the different ground states associated with the broken symmetry can only be connected by a tunneling process in which a domain wall sweeps over the whole system (Fig. 10). Such a domain-wall-tunneling process has an amplitude of order e^{-L^2/ξ^2}. However, the different ground states associated with the broken *topological* symmetry can be connected by the particle tunneling process (see Sec. V). In this case the energy split is given by $e^{-L/\xi}$. Such an energy split also indicates that the ground-state degeneracy of the FQH state is not due to the ordinary broken symmetry.

ACKNOWLEDGMENTS

We would like to thank D. Arovas, F. D. M. Haldane, and D. J. Thouless for many helpful discussions. XGW would like to thank D. Arovas for his invitation and the Physics Department of University of California-San Diego (UCSD) for hospitality, where part of the work was performed. X.G.W. was supported in part by the National Science Foundation under Grant No. PHY82-17853, supplemented by funds from the National Aeronautics and Space Administration, at the University of California at Santa Barbara. Q.N. was supported in part by the National Science Foundation (NSF) under Grant No. DMR-87-03434 and by the U.S. Office of Naval Research (ONR) under Grant No. N00014-84-K-0548.

*Present address: School of Natural Science, Institute for Advanced Study, Princeton, NJ 08540.

[1]P. L. Kapitza, Nature 141, 74 (1932).

[2]D. D. Osheroff, R. C. Richardson, and D. M. Lee, Phys. Rev. Lett. 28, 885 (1972).

[3]H. K. Onnes, Comments Phys. Lab. Univ. Leiden, Nos. 119, 120, 122 (1911).

[4]K. von Klitzing, G. Dorda, and M. Pepper, Phys. Rev. Let. 45, 494 (1980); D. C. Tsui, H. L. Störmer, and A. C. Gossard, *ibid.* 48, 1599 (1982).

[5]X. G. Wen, F. Wilczek, and A. Zee, Phys. Rev. B 39, 11 413 (1989).

[6]V. Kalmayer and R. Laughlin, Phys. Rev. Lett. 59, 2095 (1988); Phys. Rev. B 39, 11 879 (1989); X. G. Wen and A. Zee, Phys. Rev. Lett. 63, 461 (1989); P. Wiegmann, Physica C 153-155, 103 (1988); P. W. Anderson (unpublished); D. Khveshchenko and P. Wiegmann (unpublished); R. Laughlin, Ann. Phys. 191, 163 (1989); G. Baskaran (unpublished); R. Laughlin and Z. Zou, Phys. Rev. B 41, 664 (1990).

[7]X. G. Wen, Phys. Rev. B 40, 7387 (1989).

[8]X. G. Wen, Int. J. Mod. Phys. B 4, 239 (1990).

[9]F. D. M. Haldane, Phys. Rev. Lett. 51, 605 (1983); F. D. M. Haldane and E. H. Rezayi, *ibid.* 54, 237 (1985).

[10]F. D. M. Haldane and D. Rezayi, Phys. Rev. B 31, 2529 (1985); F. D. M. Haldane, Phys. Rev. Lett. 55, 2095 (1985).

[11]P. W. Anderson, Phys. Rev. B 28, 2264 (1983).

[12]R. Tao and Y. S. Wu, Phys. Rev. B 30, 1097 (1984).

[13]Q. Niu and D. J. Thouless, J. Phys. A 17, 2453 (1984); Q. Niu, D. J. Thouless and Y. S. Wu, Phys. Rev. B 31, 3372 (1985); J. Avron and R. Seiler, Nucl. Phys. B265, 364 (1986).

[14]D. Yoshioka, B. I. Halperin, and P. A. Lee, Phys. Rev. Lett. 50, 1219 (1983); D. Yoshioka, Phys. Rev. B 29, 6833 (1984); W. P. Su, *ibid.* 30, 1069 (1984).

[15]F. Wilczek and A. Zee, Phys. Rev. Lett. 52, 2111 (1984).

[16]R. Tao and F. D. M. Haldane, Phys. Rev. B 33, 3844 (1986); Q. Li and D. J. Thouless (unpublished).

[17]D. J. Thouless, Phys. Rev. B 40, 12 034 (1989).

[18]D. J. Thouless, M. Kohmoto, M. P. Nightingale, and M. den Nijs, Phys. Rev. Lett. 49, 405 (1982); Q. Niu, Phys. Rev. B 34, 5093 (1986).

[19]S. M. Girvin and A. H. MacDonald, Phys. Rev. Lett. 58, 1252 (1987); N. Read, *ibid.* 62, 86 (1989).

[20]S. C. Zhang, T. H. Hansson, and S. Kivelson, Phys. Rev. Lett. 62, 82 (1989).

[21]D. R. Hofstadter, Phys. Rev. B 14, 2239 (1976); G. H. Wannier, Phys. Status Solidi B 88, 757 (1978); M. Ya. Azbel, Zh. Eksp. Teor. Fiz. 46, 939 (1964) [Sov. Phys.-JETP 19, 634 (1964)]; J. Zak, Phys. Rev. 134A, 1602 (1964).

[22]P. Ramond, *Field Theory—A Modern Primer* (Benjamin/Cummings, New York, 1981), p. 280; L. Susskind, in *Weak*

and Electromagnetic Interactions at High Energies, Proceedings of the Les Houches Summer School of Theoretical Physics, 1976 (North-Holland, Amsterdam, 1977), p. 207.

[23]E. Witten, Comments Math. Phys. **121**, 351 (1989); G. V. Dunne, R. Jackiw, and C. A. Trugenberg, Phys. Rev. D **41**, 661 (1990); J. M. F. Labastida and A. V. Romallo (unpublished); Y. Hosotani, Phys. Rev. Lett. **62**, 2785 (1989); S. Elitzur, G. Moore, A. Schwimmer, and N. Seiberg (unpublished).

[24]D. Mumford, *Tata Lectures on Theta* (Birkhäuser, Boston, 1983); see also, Ref. 10.

[25]B. I. Shklovskii and A. L. Efros, Zh. Eksp. Teor. Fiz. **84**, 811 (1983) [Sov. Phys.—JETP **57**, 470 (1983)]; Q. Li and D. J. Thouless, Phys. Rev. B **40**, 9738 (1989).

[26]F. D. M. Haldane (private communication).

[27]The quantization of (7.7) with $\bar{p}=1$ on compactified spaces has been studied by many people. See Ref. 23.

[28]N. Read (private communication).

[29]F. D. M. Haldane, Phys. Rev. Lett. **51**, 605 (1983); B. I. Halperin, *ibid.* **52**, 1583 (1984); **52**, 2390 (1984); Helv. Phys. Acta **56**, 75 (1983); R. B. Laughlin, Surf. Sci. **141**, 11 (1984); S. M. Girvin, Phys. Rev. B **29**, 6012 (1984); J. K. Jain, Phys. Rev. Lett. **63**, 199 (1989).

[30]R. B. Laughlin, Science **242**, 525 (1988); Phys. Rev. Lett. **60**, 1057 (1988); A. Fetter, C. Hanna, and R. Laughlin, Phys. Rev. B **39**, 9679 (1989); X. G. Wen and A. Zee, Phys. Rev. B **41**, 240 (1990); D. H. Lee and M. P. A. Fisher, Phys. Rev. Lett. **63**, 903 (1989); Y. H. Chen, F. Wilczek, E. Witten, and B. I. Halperin; Int. J. Mod. Phys. B **3**, 1001 (1989); T. Bank and J. Lykken (unpublished); Y. Hosotani and S. Chakravarty (unpublished); Y. Kitazawa and H. Murayama (unpublished).

[31]X. G. Wen (unpublished).

[32]X. G. Wen and A. Zee, Phys. Rev. B **41**, 240 (1990).

[33]M. P. A. Fisher and D. H. Lee, Phys. Rev. B **39**, 2756 (1989); D. H. Lee and M. P. A. Fisher, Phys. Rev. Lett. **63**, 903 (1989); A. Polychronakos (unpublished); Soo-Jong Rey, Phys. Rev. D **40**, 3396 (1989).

[34]A. M. Polyakov, Nucl. Phys. B **120**, 429 (1977); I. Affleck, J. Harvey, and E. Witten, *ibid.* **206**, 413 (1982).

[35]Y. S. Wu and Z. Zee, Phys. Lett. B **207**, 39 (1988); **147B**, 325 (1984); H. C. Tze and S. Nam, *ibid.* **210B**, 76 (1988); X. G. Wen and Z. Zee, J. Phys. **50**, 1623 (1989); A. Goldhaber, R. Mackenzie, and F. Wilczek (unpublished).

Gauge Invariance in Chern-Simons Theory on a Torus

Yutaka Hosotani[a]
Institute for Advanced Study, Princeton, New Jersey 08540
(Received 3 February 1989)

In Chern-Simons gauge theory on a manifold $T^2 \times R^1$ (two-torus×time) the unitary operators, which induced large gauge transformations shifting the nonintegrable phases of the two distinct Wilson-line integrals on the torus by multiples of 2π, do not commute with each other unless the coefficient of the Chern-Simons term is quantized. In U(1) theory this condition gives the statistics phase $\theta = \pi/n$ (n is an integer). The condition coincides with the one previously derived on a manifold S^3 (three-sphere) for SU($N \geq 3$) theory but differs by a factor of 2 for SU(2) theory. The requirement of the Z_N invariance in pure SU(N) gauge theory imposes a stronger constraint.

PACS numbers: 11.15.−q, 05.30.−d, 74.65.+n

In 2+1 dimensions one can always add to the Lagrangian the Chern-Simons term

$$\mathcal{L}^1_{CS} = \tfrac{1}{2}\mu\epsilon^{\lambda\nu\rho}A_\lambda \partial_\nu A_\rho \tag{1}$$

in U(1) gauge theory, or

$$\mathcal{L}^2_{CS} = \mu\epsilon^{\lambda\nu\rho}\mathrm{Tr}A_\lambda(\partial_\nu A_\rho + \tfrac{2}{3}igA_\nu A_\rho) \tag{2}$$

in non-Abelian gauge theory, where $A_\mu = A_\mu^a T^a$ and $[T^a, T^b] = if^{abc}T^c$ with the trace in the fundamental representation $\mathrm{Tr}T^aT^b = \frac{1}{2}\delta^{ab}$. It was previously introduced to generate a topological mass of gauge bosons.[1-4] More recently, it has been argued that the addition of (1) in U(1) theory leads to fractional statistics,[5] and could be essential to construct an effective theory for high-T_c superconductivity.[6] Also it has been shown that pure non-Abelian Chern-Simons theory is a powerful tool in exploring knot theory in mathematics,[7] and provides a new way of formulating theory of gravity in 2+1 dimensions.[8]

It is known that on a manifold S^3 (a three-sphere) the coefficient μ in (2) in non-Abelian gauge theory must be quantized in the unit of $g^2/4\pi$ so that the action may change only by multiples of 2π under large gauge transformations.[2] We consider a theory on a manifold $T^2 \times R^1$ (two-torus×time) and derive a quantization condition for μ in both Abelian and non-Abelian theories. In addition to academic curiosity about properties of gauge theory on multiply connected space, putting a gauge theory on a torus has the advantage of eliminating the infrared ambiguity which quite often plagues analysis of gauge theory in Minkowski spacetime.

In gauge theory on a multiply connected space nonintegrable phases of the Wilson-line integrals along noncontractable loops become physical degrees of freedom.[9,10] Dynamics of such phases lead to rich physical consequences,[9-11] which, in general, do not disappear even in the infinite-volume limit. As an example, in QED on $S^1 \times R^1$ (circle×time) the nonintegrable phase couples through the anomaly to the zero mode of fermion-antifermion bound states, leading to the θ vacuum.[11] In other words the structure of the θ vacuum is a direct consequence of the invariance of the theory under large gauge transformations. It is our hope that the analysis of Chern-Simons gauge theory on a torus, in its infinite-volume limit, gives crucial information on fractional statistics and high-T_c superconductivity.

We start to analyze a U(1) theory with the Lagrangian

$$\mathcal{L}_{tot} = -\tfrac{1}{4}\kappa F_{\mu\nu}F^{\mu\nu} + \mathcal{L}^1_{CS} + \mathcal{L}_{mat}[A_\mu, \psi], \tag{3}$$

on a torus ($0 \leq x_j \leq L_j$, $j = 1,2$). Since the space is multiply connected, one has to specify boundary conditions for the fields A_μ and ψ. After translations along noncontractible loops the fields need to return to their original values up to gauge transformations:

$$A_\mu[h_j(x)] = A_\mu[x] + \frac{1}{e}\partial_\mu\beta_j(x),$$
$$\psi[h_j(x)] = e^{i\beta_j(x)}\psi[x], \tag{4}$$

where $h_1(x) = (t, x_1 + L_1, x_2)$ and $h_2(x) = (t, x_1, x_2 + L_2)$. The most general β_j which is t independent and linear in x is given, up to gauge transformations, by

$$\beta_j(x) = -\epsilon^{jk}\pi a x_k / L_k, \tag{5}$$

where $\epsilon^{jk} = -\epsilon^{kj}$ ($\epsilon^{12} = 1$). To guarantee $\psi[h_2(h_1(x))] = \psi[h_1(h_2(x))]$, the constant a must be an integer. It leads to the flux-quantization condition[12] $\Phi = \int dx F_{12} = -2\pi a/e$.

The integer a is related, through one of the equations of motion,

$$\kappa\partial_\nu F^{\mu\nu} - \tfrac{1}{2}\mu\epsilon^{\mu\nu\lambda}F_{\nu\lambda} = eJ^\mu, \tag{6}$$

to the total charge

$$Q = \int dx J^0 = -\frac{\mu}{e}\Phi = \frac{2\pi\mu}{e^2}a. \tag{7}$$

As we shall see below, $2\pi\mu/e^2$ must be an integer ($\equiv n$) so that $Q = q$ must be a multiple of n ($q = an$). Gauge transformations, which respect (4), are

$$A'_\mu = A_\mu + \frac{1}{e}\partial_\mu \Lambda, \quad \psi' = e^{i\Lambda}\psi, \quad \Lambda = 2\pi\left[\frac{m_1 x_1}{L_1} + \frac{m_2 x_2}{L_2}\right] + \tilde{\Lambda}(t,\mathbf{x}). \tag{8}$$

Here m_1 and m_2 are integers, and $\tilde{\Lambda}(t,\mathbf{x})$ is a periodic function of $\mathbf{x}$.

First we consider the case $\kappa=0$, in which there exists no photon degree of freedom.[4] In the div$\mathbf{A}=0$ gauge,

$$A_0 = -\frac{e}{\mu}\int d\mathbf{y}\, D(\mathbf{x}-\mathbf{y})(\partial_1 J^2 - \partial_2 J^1)(t,\mathbf{y}),$$
$$eL_j A_j = \theta_j(t) + \epsilon^{jk}\frac{e^2 q x_k}{2\mu L_k} + \frac{e^2 L_j}{\mu}\int d\mathbf{y}\, D(\mathbf{x}-\mathbf{y})\epsilon^{jk}\partial_k\left[J^0(t,\mathbf{y}) - \frac{q}{L_1 L_2}\right], \tag{9}$$

where $\nabla^2 D(\mathbf{x}) = \delta(\mathbf{x})$ and $\int d\mathbf{x}\, D(\mathbf{x}) = 0$. θ_j's, the nonintegrable phases of the Wilson-line integrals $\exp(ie \times \int_0^{L_j} dx_j A_j)$, are the only physical gauge-field degrees of freedom. The residual gauge invariance in the $Q=0$ sector, for instance, is given by

$$\theta_j(t) \to \theta_j(t) + 2\pi m_j, \quad \psi_{n_1,n_2}(t) \to \psi_{n_1-m_1,n_2-m_2}(t), \tag{10}$$

where m_1 and m_2 are integers, and $\psi_{n_1,n_2}(t)$'s are Fourier components of $\psi(t,\mathbf{x})$.

Substitution of (9) into (3) yields the Lagrangian $= \mu\theta_2\dot{\theta}_1/e^2 + \cdots$ so that $\mu\theta_2/e^2$ is canonically conjugate to θ_1: $[\theta_1,\theta_2] = ie^2/\mu$. Therefore, the unitary operators, which generate the residual gauge transformations $(m_1,m_2)=(1,0)$ and $(0,1)$, are

$$U_j = \exp\left[+\epsilon^{jk}\frac{2\pi i\mu}{e^2}\theta_k\right]U_j^{\rm mat}. \tag{11}$$

Here $U_j^{\rm mat}$'s induce the shift in the matter fields. U_1 and U_2 commute with the Hamiltonian. However, since $U_1 U_2 = \exp(-4\pi^2 i\mu/e^2)U_2 U_1$, they commute with each other and states can be gauge invariant only if

$$\mu = \frac{e^2}{2\pi}n \quad (n \text{ is an integer}). \tag{12}$$

It is known[13] that in the presence of the Chern-Simons term the interchange (π rotation) of two identical particles gives Schrödinger wave functions an extra phase factor $e^{i\theta}$, where $\theta = e^2/2\mu$. Therefore, $\theta_{\rm stat} = \pi/n$. A similar quantization condition has been previously derived[14] from the requirement of the gauge invariance in the presence of magnetic monopoles in R^3. Also it has been recently shown[15] that the modular invariance in (θ_1,θ_2) space is achieved only for an even integer n in (12).

The presence of the F^2 term in (3) does not affect the result. The relevant part of the Lagrangian is

$$\frac{\kappa}{2e^2}\left[\frac{L_2}{L_1}\dot{\theta}_1^2 + \frac{L_1}{L_2}\dot{\theta}_2^2\right] + \frac{\mu}{2e^2}(\theta_2\dot{\theta}_1 - \theta_1\dot{\theta}_2) + \cdots. \tag{13}$$

Conjugate momenta to θ_j's are

$$p_j = \frac{\kappa L_1 L_2}{e^2 L_j^2}\dot{\theta}_j + \epsilon^{jk}\frac{\mu}{2e^2}\theta_k. \tag{14}$$

They satisfy $[\theta_j, p_k] = i\delta_{jk}$. All other commutators vanish. This time

$$U_j = \exp\left[2\pi i\left[p_j + \epsilon^{jk}\frac{\mu}{2e^2}\theta_k\right]\right]U_j^{\rm mat}. \tag{15}$$

The commutativity of U_1 and U_2 leads to the same quantization condition (12). In view of (14), (15) reduces to (11) in the $\kappa=0$ limit.

In SU(N) gauge theory we focus on a particular boundary condition $A_\mu[h_j(x)] = A_\mu[x]$. More general boundary conditions have been analyzed in Ref. 10. Then our boundary condition is invariant under gauge transformations $A_\mu \to \Omega A_\mu \Omega^\dagger - (i/g)\Omega\partial_\mu\Omega^\dagger$, provided that $\Omega[h_j(x)] = \Omega[x]$, or, in pure gauge-field theory, $\Omega[h_j(x)]\Omega[x]^\dagger$ is an element of the center of SU(N).

Let us consider pure SU(N) Chern-Simons theory: $\mathcal{L}_{\rm tot} = \mathcal{L}_{\rm CS}^2$. One of the equations gives a constraint $F_{12} = 0$. Given an arbitrary single-valued A_1 in this subspace, the gauge transformation,

$$\Omega(x)^\dagger = W(x)\exp[igx_1 B(t,x_2)],$$
$$W(x) = P\exp\left[-ig\int_0^{x_1} dy_1 A_1(t,y_1,x_2)\right],$$
$$\exp[-igL_1 B(t,x_2)] = W(t,L_1,x_2),$$

which satisfies $\Omega[h_j(x)] = \Omega[x]$, brings $A_1(x)$ to $B(t,x_2)$, which in turn is diagonalized by a second x_1-independent gauge transformation. Then the constraint $F_{12}=0$ implies that A_1 is x_2 independent and A_2 also is diagonal and x_1 independent. A third gauge transformation with diagonal $\Omega = \Omega(t,x_2)$ can eliminate the x_2 dependence of A_2. Therefore, one can take without loss of generality,

$$gL_j A_j = \begin{pmatrix}\theta_{j1}(t) & & \\ & \ddots & \\ & & \theta_{jN}(t)\end{pmatrix}, \tag{16}$$

where $\sum_{a=1}^N \theta_{ja}(t) = 0$. A_0 is a dependent variable. Indeed, parts of the equations $F_{0j}=0$ with (16) imply that A_0 also is diagonal and depends only on t. A fourth gauge transformation with diagonal $\Omega = \Omega(t)$ then can gauge away A_0 entirely ($A_0 = 0$).

There are two kinds of residual gauge invariances.

One is

$$\Omega_{ab}=\delta_{ab}\exp\left[2\pi i\left(\frac{m_{1a}x_1}{L_1}+\frac{m_{2a}x_2}{L_2}\right)\right],\quad \theta_{ja}\to\theta_{ja}+2\pi m_{ja}, \tag{17}$$

where m_{ja}'s are integers satisfying $\sum_{a=1}^{N} m_{ja}=0$. The other is the Z_N transformation for which $m_{ja}=(1-N\delta_{ab})l_j/N$ $[a,b=1\text{-}N,\ l_j=1\text{-}(N-1)]$:

$$\theta_{ja}\to\theta_{ja}+2\pi l_j\left(\frac{1}{N}-\delta_{ab}\right). \tag{18}$$

This is a special symmetry in pure gauge-field theory.

Substitution of (16) and $A_0=0$ into $\mathcal{L}_{CS}^2$ yields, in terms of θ_{ja} $[a=1\text{-}(N-1)]$,

$$L=\frac{2\mu}{g^2}\left[\sum_a{}'\theta_{2a}\dot\theta_{1a}+\sum_a{}'\theta_{2a}\sum_b{}'\dot\theta_{1b}\right],$$

where $\sum_a' = \sum_{a=1}^{N-1}$. Therefore, $p_{ja}=\epsilon^{jk}(2\mu/g^2)(\theta_{ka}+\sum_b'\theta_{kb})$ satisfies

$$[\theta_{ja},p_{kb}]=i\delta_{jk}\delta_{ab},\quad [\theta_{1a},\theta_{2b}]=i\frac{g^2}{2\mu}\left(\delta_{ab}-\frac{1}{N}\right),\quad [p_{1a},p_{2b}]=i\frac{2\mu}{g^2}(\delta_{ab}+1), \tag{19}$$

with all other commutators vanishing.

The unitary operators $U_{ja}=\exp(2\pi i p_{ja})$ $[a=1\text{-}(N-1)]$, which generate (17), satisfy

$$U_{1a}U_{2b}=\exp\left[-\frac{8\pi^2 i\mu}{g^2}(\delta_{ab}+1)\right]U_{2b}U_{1a}, \tag{20}$$

so that the commutativity of U_{ja}'s leads to

$$\mu=\begin{cases}(g^2/8\pi)n, & \text{for SU(2)},\\ (g^2/4\pi)n, & \text{for SU}(N\ge 3),\end{cases} \tag{21}$$

where n is an integer. The condition (21) is the same as the one derived on a manifold S^3 in Ref. 2 for SU($N \ge 3$), but is weaker than that by a factor of 2 for SU(2). It is to be seen how the additional factor of 2 constraint arises in SU(2) theory on a torus.[15,16]

Equation (18) is generated by combinations of $\bar U_j = \exp[(2\pi i/N)\sum_a' p_{ja}]$ and U_{ja}. The requirement of the commutativity of these unitary operators leads to a stronger constraint:

$$\mu=\frac{Ng^2}{4\pi}n\quad (n \text{ is an integer}). \tag{22}$$

In other words, if μ satisfies (21) but not (22), then the Z_N symmetry is spontaneously broken.

In the presence of the F^2 term one cannot simultaneously diagonalize A_1 and A_2 in general. If one freezes all gauge-field degrees of freedom but the nonintegrable phases of the Wilson-line integrals, then one finds that conjugate momenta to θ_{ja} $(a=1,\dots,N-1)$ are

$$p_{ja}=\frac{L_1L_2}{L_j^2}\frac{2\kappa}{g^2}\left[\dot\theta_{ja}+\sum_b{}'\dot\theta_{jb}\right]+\epsilon^{jk}\frac{\mu}{g^2}\left[\theta_{ka}+\sum_b{}'\theta_{kb}\right].$$

They satisfy $[\theta_{ja},p_{kb}]=i\delta_{jk}\delta_{ab}$. All other commutators vanish. The unitary operators generating (17) are

$$U_{ja}=\exp\left\{2\pi i\left[p_{ja}+\epsilon^{jk}\frac{\mu}{g^2}\left(\theta_{ka}+\sum_b{}'\theta_{kb}\right)\right]\right\}.$$

The commutativity of these operators leads to the same results as (20) and (21).

When μ obeys the quantization condition (12) or (21), it is meaningful to consider states which are gauge invariant up to a phase. In U(1) theory,

$$U_j\Psi_{\alpha_1\alpha_2}=e^{i\alpha_j}\Psi_{\alpha_1\alpha_2}. \tag{23}$$

In the $Q=0$ sector of nonrelativistic theory $\Psi=\Psi_{\rm gauge}\otimes|0\rangle_F$. For $\kappa=0$, θ_1 and θ_2 are canonically conjugate to each other, and the wave function of the state $\Psi_{\rm gauge}$ is

$$u(\theta_1)=\langle\theta_1|\Psi_{\rm gauge}\rangle,$$

$$v(\theta_2)=\langle\theta_2|\Psi_{\rm gauge}\rangle=\frac{\sqrt{n}}{2\pi}\int_{-\infty}^{+\infty}d\theta_1\,e^{-in\theta_1\theta_2/2\pi}u(\theta_1).$$

For $\mu=e^2n/2\pi$, wave functions satisfying (23) are given by

$$u_{\alpha_1\alpha_2}(\theta_1)=e^{i\alpha_1\theta_1/2\pi}\delta_{2\pi}\left[\theta_1+\frac{1}{n}(\alpha_2+2\pi l)\right], \tag{24}$$

where $l=0,1,\dots,n-1$ and $\delta_{2\pi}(\theta)$ is a periodic δ function with a period 2π. It is easy to check that $v(\theta_2)$ takes the same form as $u(\theta_1)$.

In the $Q=1$ sector of nonrelativistic theory (with neutralizing uniform background charge $Q_{\rm bg}=-1$) states satisfying (23) are

$$|\Psi\rangle=\int_{-\infty}^{+\infty}d\theta_1\sum_{m_1,m_2}e^{i\alpha_1\theta_1/2\pi+im_2(n\theta_1+\alpha_2)}h(\theta_1-2\pi m_1)\,|\theta_1\rangle\otimes|m_1,m_2\rangle,$$

where $|m_1,m_2\rangle=\psi^\dagger_{m_1m_2}|0\rangle_F$. $h(\theta)$ is an arbitrary function, and should be determined so as to solve the Schrödinger equation.

In SU(2) theory with $\mu = g^2 n/8\pi$ the structure of the commutation relations and the unitary operators are the same as in U(1) theory so that wave functions are given by (24) with the substitution $\theta_1 \to \theta_{11}$, $\alpha_j \to \alpha_{j1}$. In SU(3) theory with $\mu = g^2 n/4\pi$, states satisfying $U_{ja} | \Psi \rangle = e^{i\alpha_{ja}} | \Psi \rangle$ $(a = 1,2)$ are

$$u(\theta_{11},\theta_{12}) = e^{i(\alpha_{11}\theta_{11}+\alpha_{12}\theta_{12})/2\pi} \delta_{2\pi}\left[\theta_{11} + \frac{1}{3n}(2\alpha_{21} - \alpha_{22} - 2\pi r) - \frac{2\pi q}{n}\right] \delta_{2\pi}\left[\theta_{12} + \frac{1}{3n}(-\alpha_{21} + 2\alpha_{22} - 2\pi r)\right],$$

where $r = 0,1,\ldots,3n-1$ and $q = 0,1,\ldots,n-1$. There are $3n^2$ states with given α_{ja}'s.

In this paper we have explored implications of the gauge invariance in Chern-Simons theory on a torus. The requirement of the gauge invariance has led to the quantization condition for the coefficient of the Chern-Simons term.

It is an interesting fact that the quantization condition follows even in U(1) theory, the case probably most important in physical applications. It is, however, a dynamical question whether or not the gauge invariance remains unbroken in the ground state. Moreover, one might wonder how the quantization condition derived on a torus has any relevance in physics in the Minkowski spacetime. It is quite likely that something very special happens in the Minkowski spacetime when the quantization condition is satisfied.[17] The experience in the analysis of QED on a circle[11] also suggests that as a consequence of the gauge invariance the wave function of the ground state with matter has the θ-vacuum structure, which should remain intact in the infinite-volume limit. If this is the case, the notion of the gauge invariance has to play an important role in discussing fractional statistics and high-T_c superconductivity.

This research was supported in part by DOE Contract No. DE-AC02-83ER-40105 and by the McKnight-Land grant at the University of Minnesota. The author would like to thank the Institute for Advanced Study for its hospitality where this work was done, and J. Hetrick, C-L. Ho, N. Seiberg, and F. Wilczek for stimulating discussions.

[a]On leave of absence from School of Physics and Astronomy, University of Minnesota, Minneapolis, MN 55455.

[1]J. Schonfeld, Nucl. Phys. **B185**, 157 (1981); R. Jackiw and S. Templeton, Phys. Rev. D **23**, 2291 (1981).

[2]S. Deser, R. Jackiw, and S. Templeton, Phys. Rev. Lett. **48**, 975 (1983); Ann. Phys. (N.Y.) **140**, 372 (1984).

[3]A. N. Redlich, Phys. Rev. Lett. **52**, 18 (1984); Phys. Rev. D **29**, 2366 (1984); R. Jackiw, Phys. Rev. D **29**, 2375 (1984).

[4]C. R. Hagen, Ann. Phys. (N.Y.) **157**, 342 (1984).

[5]F. Wilczek, Phys. Rev. Lett. **49**, 957 (1982); F. Wilczek and A. Zee, Phys. Rev. Lett. **51**, 2250 (1983); D. P. Arovas, R. Schrieffer, F. Wilczek, and A. Zee, Nucl. Phys. **B251**, 117 (1985); A. Goldhaber, R. MacKenzie, and F. Wilczek, Harvard University Report No. HUTP-88/A044 (to be published); X. G. Wen and A. Zee, Santa Barbara Report No. NSF-ITP-88-114 (to be published); J. Fröhlich and P. A. Marchetti, "Quantum field theories of vortices and anyons," Eidgenössische Technische Hochschule University (to be published).

[6]P. W. Anderson, Science **235**, 1196 (1987); S. A. Kivelson, D. S. Rokhsar, and J. P. Sethna, Phys. Rev. B **35**, 8865 (1987); V. Kalmeyer and R. B. Laughlin, Phys. Rev. Lett. **59**, 2095 (1987); "Theory of the spin liquid state of the Heisenberg antiferromagnet," Stanford University (to be published); I. Dzyaloshinskii, A. Polyakov, and P. Wiegmann, Phys. Lett. A **127**, 112 (1988); A. M. Polyakov, Mod. Phys. Lett. A **3**, 325 (1988); J. March-Russel and F. Wilczek, Phys. Rev. Lett. **61**, 2066 (1988); R. B. Laughlin, Phys. Rev. Lett. **60**, 2677 (1988); X. G. Wen, F. Wilczek, and A. Zee, Santa Barbara Report No. NSF-ITP-88-179 (to be published).

[7]E. Witten, Institute for Advanced Study Report No. IASSNS-HEP-88/33 (to be published).

[8]E. Witten, Institute for Advanced Study Reports No. IASSNS-HEP-88/32, No. -88/55, and No. -89/1 (to be published).

[9]Y. Hosotani, Phys. Lett. B **126**, 309 (1983); D. Tom, Phys. Lett. B **126**, 445 (1983).

[10]Y. Hosotani, University of Minnesota Report No. UMN-TH-662/88 (to be published), and references therein.

[11]N. S. Manton, Ann. Phys. (N.Y.) **159**, 220 (1985); J. E. Hetrick and Y. Hosotani, Phys. Rev. D **38**, 2621 (1988).

[12]F. D. M. Haldane and E. H. Rezayi, Phys. Rev. B **31**, 2529 (1985).

[13]A. S. Goldhaber *et al.*, in Ref. 5.

[14]O. Alvarez, Commun. Math. Phys. **100**, 279 (1985); M. Henneaux and C. Teitelboim, Phys. Rev. Lett. **56**, 689 (1986); R. D. Pisarski, Phys. Rev. D **34**, 3851 (1986).

[15]S. Elitzer, G. Moore, A. Schwimmer, and N. Seiberg, Institute for Advanced Study Report No. IASSNS-HEP-89/20 (to be published).

[16]G. Moore and N. Seiberg, Institute for Advanced Study Report No. IASSNS-HEP-89/6 (to be published); G. V. Dunne, R. Jackiw, and C. A. Trugenberger, Massachusetts Institute of Technology Report No. CTP-1711 (to be published).

[17]F. Wilczek (private communication).

Modern Physics Letters A, Vol. 6, No. 19 (1991) 1779–1786

A COULOMB GAS DESCRIPTION OF THE COLLECTIVE STATES FOR THE FRACTIONAL QUANTUM HALL EFFECT†

G. CRISTOFANO, G. MAIELLA, R. MUSTO, and F. NICODEMI
Dipartimento di Scienze Fisiche, Università di Napoli, and
INFN Sezione di Napoli, Mostra d'Oltremare Pad. 19 – 80125 Napoli, Italy

Received 6 April 1991

Many anyons wavefunctions relevant for the fractional Quantum Hall Effect at filling $\nu=1/m$ are obtained by using Coulomb gas conformal Vertex operators. They provide irreducible representations of a subgroup of the magnetic translation group on the torus and their degeneracy is related to the allowed set of anyonic charges.

The characteristic feature of the Quantum Hall Effect[1] (QHE) is the appearance of plateaux in the Hall conductance at integer and fractional values for the filling factor ν. Such behavior, independent of the specific condensed matter sample, reflects universal properties of the quantum collective motion for a 2-dimensional electron system in a transverse magnetic field.

The basic features of the integer QHE are already transparent in a simple description first introduced by Laughlin,[2] in which the quantized Hall conductance is seen as due to the exact matching between an integer charge passing through the sample and the adiabatic change of magnetic flux of one elementary unit, $\phi_0 = hc/e$, inducing the electromotive force. This simple picture can be formalized leading to an understanding of the integer QHE in terms of topological arguments[3,4] and of the presence of delocalized states allowed for the system.[1] However, a model based on single particle states can explain the integer QHE but not the fractional one.

Two different paths have been followed in the attempt of explaining the fractional QHE. The first consists in the introduction of *anyons*, i.e., of "particles" of fractional charge and statistics, giving an effective description of the interacting electron system.[5–7] The second, based on a discussion of the invariance properties under the magnetic translation group, emphasizes the necessity of having a degenerate ground state when doubly periodic boundary conditions are imposed.[8]

In a previous paper,[9] hereafter denoted as **I**, we have shown that these two points of view appear naturally related when an appropriate description of the electron system on a torus is achieved. This description is based on the existence

†Work partially supported by "Ministero della Università e Ricerca Scientifica".

of a set of Vertex operators, each corresponding to an electron or to an anyonic "particle," with a given value of the electric charge and of the associated integer magnetic flux. The allowed values of these "charges" are completely determined by the consistency requirement of translation invariance on the torus in the presence of the external magnetic field. In fact, for a given basic set of Vertex operators, using standard conformal field theory techniques[10,11] one can construct a set of wavefunctions for an appropriate number of electrons and/or anyons and show that they provide a basis for an irreducible representation of an appropriate magnetic translation group. The degeneracy of the wavefunctions can then be seen from two different but equivalent points of view, either as due to translation invariance on the torus[12] or to the presence of a set of basic Vertex operators, each corresponding to an anyonic particle.

Furthermore we have shown that our description of the anyonic and electronic many-body states of the system by means of Coulomb gas Vertex operators formalizes the Laughlin "one component plasma" interpretation of the wavefunctions.[13]

In this paper we show how the formal structure introduced in I leads to a natural description of the fractional QHE. The fractional values of the filling, $\nu = N_e/N_s$, relative to the observed plateaux, will correspond to the different number of electrons, N_e, for which translation invariance is restored in the presence of a given magnetic flux, N_s, while the corresponding value of the Hall conductance will be fixed by the "charges" of the anyonic "particles" associated to the vertex operators. More specifically we will limit ourselves to the case $\nu = 1/m$, while the case $\nu = p/m$ will be discussed in a forthcoming publication.[14]

To this aim we recall that in I we have introduced a square torus of length L in magnetic units, $\lambda^2 = \hbar c/eB$, and assumed that the total flux through the surface, measured in quantum flux units, hc/e, is an integer

$$L^2 = 2\pi N_s \,. \tag{1}$$

We have then shown that for each filling $\nu = 1/m$ (m odd) it is natural to introduce a set of Vertex operators,[15] typical of the Coulomb gas approach to conformal field theory

$$V_{\alpha_l}(z) =: e^{i\alpha_l \phi(z)} : , \tag{2}$$

where $\alpha_l = l/\sqrt{m}$, $l = 1, 2, \ldots, m$. The scalar field ϕ is defined on S_1, i.e., it is compactified on a circle of radius R,

$$R^2 = m \tag{3}$$

and has the standard mode expansion:

$$\phi(z) = q - ip \ln z + \sum_{n \neq 0} \frac{a_n}{n} z^{-n} \,. \tag{4}$$

The coefficients satisfy the commutation relations:

$$[a_n, a_{-n'}] = n\delta_{n,n'} \qquad [q, p] = i \tag{5}$$

and the normal ordering is defined in the usual sense for the creation and annihilation operators and by taking q to the left of p for the zero modes.

The anyonic nature of these operators can be easily seen by recalling their "braiding properties." The effect of interchanging two generic Vertex operators, as defined in Eqs. (2) and (4), is indeed given by

$$V_{\alpha_1}(z_1)V_{\alpha_2}(z_2) = e^{i\pi\alpha_1\alpha_2}V_{\alpha_2}(z_2)V_{\alpha_1}(z_1)\,. \qquad (6)$$

We see that for $\alpha_1 = \alpha_2 = \sqrt{m}$, i.e., $l = m$, the two operators anticommute, m being odd, and the relative Vertex can be associated with the electron. Since the braiding factor in this case is given by $\exp[i\pi m]$ and as the electron charge is taken to be one, we can associate[5] a magnetic flux m with the electron. We will then more simply say that there is a *magnetic charge* m associated with the electron.

In general the braiding factor of two vertices will be $\exp[il_1l_2\pi/m]$, that can be thought as due to a charge l_1/m going around a magnetic flux l_2 or vice versa. To a generic anyonic vertex it is then associated an electric charge l/m and a magnetic charge l, so that for any "particle," both electron or anyon, the ratio of the electric to the magnetic charge, is always $1/m$. As we shall see later this interpretation is consistent with the one particle and many particles wavefunctions and with their plasma description.

Note that the compactification radius given by Eq. (3), corresponds to the radius of the lowest cyclotron orbit, measured in magnetic units, relative to an anyon of charge $1/m$.

We can now use standard methods of 2-dimensional conformal field theory to evaluate correlation functions on the torus for an arbitrary set of M "particles" obtaining[a]

$$\left\langle V_{\frac{p_1}{\sqrt{m}}}(w_1)\cdots V_{\frac{p_M}{\sqrt{m}}}(w_M)\right\rangle_l^{g=1} = \prod_{i<j=1}^{M}\left[\frac{\Theta_1(w_{ij}|i)}{\Theta_1'(0|i)}\right]^{\frac{p_ip_j}{m}}\Theta\begin{bmatrix} l/m \\ 0 \end{bmatrix}\left(\frac{W_m}{L}|im\right), \qquad (7)$$

where w_i are the torus variables related to the plane variables z_i by $w_i = (L/2\pi i) \times \ln z_i$, $w_{ij} = (w_i - w_j)/L$ and $W = \sum_{i=1}^{M}\frac{p_iw_i}{m}$ is the center of "charge" coordinate.

We remind that the theta-functions with characteristics are defined as[16]:

$$\Theta\begin{bmatrix} a \\ b \end{bmatrix}(\zeta|\tau) = \sum_{k\in\mathbb{Z}}\exp\{i\tau\pi(k+a)^2 + i2\pi(k+a)(\zeta+b)\}\,, \qquad (8)$$

while Θ_1 is the odd theta-function corresponding to $a = b = 1/2$.

It is useful to discuss in detail the one electron case, for which $m = N_s$. Then, from Eq. (7), we recover the correct quantum mechanical result for the analytic part of the wavefunctions[12,17]:

[a]A precise definition of $\left\langle V_{\frac{p_1}{\sqrt{m}}}(w_1)\cdots V_{\frac{p_M}{\sqrt{m}}}(w_M)\right\rangle_l^{g=1}$ is given in I.

$$\left\langle V_{\sqrt{N_s}}(w)\right\rangle_l^{g=1} = \Theta\begin{bmatrix} l/N_s \\ 0 \end{bmatrix}\left(w\frac{N_s}{L}|iN_s\right) = f_l(w)\,, \tag{9}$$

where $l = 1, 2, \ldots, N_s$.

These functions form a basis[16] in the space of entire functions satisfying the periodicity conditions

$$f(w+L) = f(w)\,, \tag{10}$$

$$f(w+iL) = f(w)e^{\frac{1}{2}L^2}e^{-iLw} \tag{11}$$

required by the presence of the magnetic field. Then, in the gauge $\mathbf{A} = y\hat{x}$, a basis for the one-particle states in the first Landau level is given by

$$\psi_l(x,y) = e^{-y^2/2}f_l(w)\,. \tag{12}$$

In the language of conformal field theory the N_s-fold degeneracy of the one-electron wavefunctions can be explained if we note that the different states relative to the topology of the torus can be interpreted as due to the interaction of the external electron with each possible virtual anyonic state. At the same time, defining the "magnetic" translation group,[3,12] by the action of the operators $\mathcal{S}$ and $\mathcal{T}$:

$$\mathcal{S}_a f(w) = f(w+a) \qquad \mathcal{T}_b f(w) = e^{\frac{-b^2}{2}+ibw}f(w+ib) \tag{13}$$

which satisfy the commutation relations

$$\mathcal{S}_a\mathcal{T}_b = e^{iab}\mathcal{T}_b\mathcal{S}_a$$

we see that the set of functions given by Eq. (9) provides a basis for a N_s-dimensional irreducible representation of the discrete subgroup generated by $\mathcal{S}_{L/N_s}$ and $\mathcal{T}_{L/N_s}$.

In fact the action of these generators on the functions $f_l(w)$ is easily obtained by using Eqs. (8) and (9), leading to:

$$\begin{aligned} \mathcal{S}_{\frac{L}{N_s}} f_l(w) &= e^{2\pi i\frac{l}{N_s}}f_l(w)\,, \\ \mathcal{T}_{\frac{L}{N_s}} f_l(w) &= f_{l+1}(w)\,. \end{aligned} \tag{14}$$

Note that $\mathcal{T}_{\frac{L}{N_s}}$ provides the step-up operator for such a representation.

Moreover the set of functions given by Eq. (9) transforms in the appropriate way under the modular transformation corresponding to the exchange of the x and y axes, i.e., of the two non-trivial cycles of the torus. Indeed, when $\tau \to -1/\tau$, one has

$$\Theta\begin{bmatrix} 1/N \\ 0 \end{bmatrix}\left(\frac{\zeta\sqrt{N}}{\tau}\Big| -\frac{N}{\tau}\right) = e^{i\pi\zeta^2/\tau}\sqrt{\frac{\tau}{iN}}\sum_{l'=1}^{N} e^{-2i\pi\frac{ll'}{N}}\Theta\begin{bmatrix} l'/N \\ 0 \end{bmatrix}(\zeta\sqrt{N}/N\tau)\,.$$

In our case the factor $e^{i\pi\zeta^2/\tau} = e^{w^2/2}$ is exactly what is needed to build the correct Gaussian prefactor of the wavefunctions, Eq. (12), once the appropriate gauge transformation is also performed.

One can also evaluate the one-particle wavefunction for the generic anyon, relative to the vertex $V_{p/\sqrt{m}}$ obtaining the correct quantum mechanical result for a particle of charge $|e^*| = p/m$ moving in a magnetic field p/m-times smaller than the one of the electron. This is consistent with the picture we have obtained by means of the braiding relations, Eq. (6), of anyons as particles associated with an electric charge p/m and an elementary flux- or magnetic charge-p.

Turning now to the multi-particle states, let us start with the case of the N_e-electron wavefunctions. These functions can be obtained by taking in Eq. (7) all α_i equal to $\sqrt{m}$ recovering the proposal made in Refs. 12 and 17

$$f_l(w_1, w_2, \dots, w_{N_e}) = \prod_{i<j=1}^{N_e} \left[\frac{\Theta_1(w_{ij}|i)}{\Theta_1'(0|i)}\right]^m \Theta\begin{bmatrix} l/m \\ 0 \end{bmatrix} \left(W\frac{m}{L}|im\right). \tag{15}$$

As already stressed in I, the condition $mN_e = N_s$, corresponding to a filling factor $\nu = 1/m$, ensures that the N_e-electron wavefunctions verify in each electron variable w_i, the boundary conditions given by Eqs. (10) and (11). Note that the condition $N_s = mN_e$ can be seen as the conservation of the magnetic charge.

As a consequence of this condition, one can see that the above set of functions provide a m-dimensional representation of the subgroup of *total* magnetic translation group, generated by the operators relative to the step $a = b = \frac{L}{N_s}$:

$$S_a = \prod_{i=1}^{N_e} S_a^i \qquad T_b = \prod_{i=1}^{N_e} T_b^i, \tag{16}$$

where S_a^i and T_b^i are defined as in Eq. (13), relative to the variable w_i of the i-th electron.

To prove this statement it is enough to consider the translations corresponding to the elementary step $a = b = L/N_s$. As in the N_e-electron wavefunctions each term, but the center of charge theta-function, depends on the differences of electron variables, we only have to check the transformation properties of these functions. But, under the elementary displacements, Eq. (16), the center of charge variable transforms as

$$W = \sum_{i=1}^{N_e} w_i \to W + N_e\frac{L}{N_s} = W + \frac{L}{m} \tag{17}$$

and

$$W = \sum_{i=1}^{N_e} w_i \to W + iN_e\frac{L}{N_s} = W + i\frac{L}{m}. \tag{18}$$

Then, simply by using Eq. (8), we see that

$$\begin{aligned} S_{\frac{L}{N_s}} f_l(w_1, w_2, \dots, w_{N_e}) &= e^{2\pi i \frac{l}{m}} f_l(w_1, w_2, \dots, w_{N_e}), \\ T_{\frac{L}{N_s}} f_l(w_1, w_2, \dots, w_{N_e}) &= f_{l+1}(w_1, w_2, \dots, w_{N_e}). \end{aligned} \tag{19}$$

The total magnetic translation operators relative to the elementary finite displacements therefore play exactly the same role as those given by Eq. (14).

The functions given by Eq. (15) also have the correct covariance properties under $\tau \to -1/\tau$, provided that $N_s = mN_e$.

The value $\nu = 1/m$ of the filling factor corresponds then to a plateau for the QHE, as the restoration of the translational symmetry allows the existence of delocalized states in the center of charge variable.[12] This can be seen more explicitly by rewriting the center of charge dependence appearing in Eq. (15) in the form

$$\Theta\begin{bmatrix} l/m \\ 0 \end{bmatrix}\left(W\frac{m}{L}|im\right) = e^{-iKW}\prod_{\nu=1}^{m}\Theta_1\left(\frac{W-W_\nu}{L}|i\right) \tag{20}$$

corresponding to a propagating wave of momentum $K = \pi(m-2l)/L$. Here $W_\nu(l)$ are the locations of the m zeros.

A similar discussion can also be made in the case of N_a anyons[b] of electric charge $1/m$, provided the electric charge conservation, $N_a = mN_e$, or equivalently the magnetic charge conservation, $N_a = N_s$ holds. Indeed by taking in Eq. (7) all α_i equal to $1/\sqrt{m}$ we get:

$$\left\langle V_{\frac{1}{\sqrt{m}}}(w_1)\ldots V_{\frac{1}{\sqrt{m}}}(w_{N_a})\right\rangle_l^{g=1} = \sum_{i<j=1}^{N_a}\left[\frac{\Theta_1(w_{ij}|i)}{\Theta_1'(0|i)}\right]^{\frac{1}{m}}\Theta\begin{bmatrix} l/m \\ 0 \end{bmatrix}\left(\frac{Wm}{L}|im\right). \tag{21}$$

Then the displacement of the center of charge variable is exactly the same as for N_e electrons and we implement again the same total magnetic translation group, as in Eq. (19).

We can also argue that the value of the Hall conductance at filling $\nu = 1/m$ is given exactly by $1/m$ in the natural Hall units e^2/h. To this purpose we recall the simple heuristic description due to Laughlin for the case of the integer filling $\nu = 1$. In this picture the electromotive force is realized by means of an adiabatic variation of an external magnetic flux, and the correct value of the Hall conductance corresponds to the exact matching between the transport of one electron and a change in the magnetic flux of an elementary quantum $\phi_0 = hc/e$. This description emphasizes that at the integer filling $\nu = 1$ there is exactly one elementary magnetic flux associated to each electron filling the Landau level:

$$\sigma_H = \frac{\text{(electric charge)}}{\text{(magnetic flux)}} = 1\,. \tag{22}$$

This is completely consistent with our description by means of vertex operators, as for $\nu = 1$, there exists only the electron vertex, with an associated unit of magnetic flux. On the other hand, for generic filling $\nu = 1/m$, to a change of flux p in magnetic units, on the basis of the same argument, there will be associated an anyon of electric charge p/m, leading to a conductance $\sigma_H = \frac{p/m}{p} = \frac{1}{m}$.

[b] An extensive analysis of the Hilbert space of anyons on a torus has been given, in the framework of the Chern–Simons approach, in Ref. 18.

The above qualitative argument can be formalized on the torus, also for a *degenerate* ground state, expressing the conductance σ_H as a topological invariant.[19]

We emphasize that our description formalizes the plasma interpretation of the wavefunction.[13] In this framework we will have a further understanding of the role of the magnetic charge associated with our "particles." As already noted in I, a more direct and intuitive picture is reached by working on the plane, where the relative wavefunction can be easily evaluated, by usual conformal techniques.[10,11] For example in the case of N_a anyons of charge p/m we obtain the wavefunction:

$$\psi(x_1, y_1, \ldots, x_{N_a}, y_{N_a}) = e^{-\frac{1}{4}\frac{p}{m}\sum_{i=1}^{N_a}|z_i|^2} \prod_{i<j=1}^{N_a} (z_i - z_j)^{\frac{p^2}{m}}, \tag{23}$$

where we have explicitly introduced its non-analytic part.

If we then define the interaction energy of the plasma through the identification $|\psi|^2 = e^{-\beta U}$, with $\beta = \nu = 1/m$, that is equal to the inverse square of the compactification radius, we get:

$$U = -2p^2 \sum_{i<j=1}^{N_a} \ln|z_i - z_j| + \frac{p}{2}\sum_{i=1}^{N_a} |z_i|^2 . \tag{24}$$

The first term in the above equation represents the interaction energy between "charges" of value p, while the second their interaction with a uniform background.

The plasma interpretation of the wavefunctions emphasizes the role of the quantum scalar field $\phi(z)$, given by Eq. (4), because its propagator gives the long-range interaction between the "charges," as it has been discussed in I, in the case of both the plane and the torus topology. The coincidence of the value of the "charge," entering in the plasma description, with the magnetic charge of the anyons, and the structure of Eq. (23) is strongly suggestive of a gas of 2-dimensional vortices, of "stength" p.[20] This picture may lead to a better understanding of the universality properties of the QHE, both integer and fractional, that appear to be due to the "topological order" described by the field ϕ.

Our construction, based on Coulomb-gas Vertex operators, appears then to be a natural framework in which one can exploit the analogy, already noted, between the QHE and other collective phenomena showing a 2-dimensional nature.[21]

References

1. For the general aspects of QHE, see e.g., *The Quantum Hall Effect*, eds. R. E. Prange and S. M. Girvin (Springer, 1987); A. H. MacDonald, *Quantum Hall Effect: a Perspective* (Jaca Book, 1989).
2. R. B. Laughlin, *Phys. Rev.* **B23** (1981) 5632.
3. Yong-Shi Wu, *Topological aspects of the quantum Hall effect*, IASSNS-HEP-90/33 preprint.
4. For a general discussion on the topological aspects of QHE see also G. Morandi, *Quantum Hall Effect- Topological Problems in Condensed Matter Physics* (Bibliopolis, 1989).
5. F. Wilczek, *Phys. Rev. Lett.* **49** (1982) 957; F. Wilczek and A. Zee, *Phys. Rev. Lett.* **51** (1983) 2250.

6. See R. B. Laughlin, article contained in R. E. Prange and S. M. Girvin, eds., cited in Ref. 1.
7. J. Fröhlich and T. Kerler, *Universality in quantum Hall systems*, preprint ETH-TH/90-26.
8. D. Arovas, J. R. Schrieffer, and F. Wilczek, *Phys. Rev. Lett.* **53** (1984) 722.
9. G. Cristofano, G. Maiella, R. Musto, and F. Nicodemi, *Coulomb gas approach to quantum Hall effect*, Naples preprint DSF-T INFN-NA 90/14.
10. For a review on the subject see: P. Ginsparg, *Les Houches*, 1988, eds. D. Brezin and J. Zinn-Justin, Vol. XLIX.
11. G. Cristofano, G. Maiella, R. Musto, and F. Nicodemi, *Phys. Lett.* **B237** (1990) 379; G. Cristofano, G. Maiella, and F. Nicodemi, *Proc. 3rd Hellenic School on Elementary Particle Physics* (World Scientific, 1990).
12. F. D. M. Haldane and E. H. Rezayi, *Phys. Rev.* **B31** (1985) 2529.
13. R. B. Laughlin, *Phys. Rev. Lett.* **50** (1983) 1395.
14. G. Cristofano, G. Maiella, R. Musto, and F. Nicodemi, in preparation.
15. The Vertex operators were first introduced in the framework of the Dual Models by S. Fubini and G. Veneziano, *Nuovo Cimento* **A67** (1970) 29; *Ann. Phys.* **63** (1970) 12. Their relevance for the QHE has been recently advocated by S. Fubini, see e.g. S. Fubini and C. A. Lütken, *Mod. Phys. Lett.* **A6** (1991) 487. See also G. V. Dunne, A. Lerda, and C. A. Trugenberger, MIT preprint CTP# 1938, 1991.
16. D. Mumford, *Tata Lectures on Theta* I (Birkhäuser, 1983).
17. R. B. Laughlin, *Ann. Phys.* **191** (1989) 163.
18. R. Iengo and K. Lechner, SISSA preprint, 182/90/EP.
19. Q. Niu, D. Thouless, and Yong-Shi Wu, *Phys. Rev.* **B31** (1985) 3372.
20. J. M. Kosterlitz and D. J. Thouless, *J. Phys.* **C6** (1973) 1181.
21. Y. H. Chen, F. Wilczek, E. Witten, and B. I. Halperin, *Int. J. Mod. Phys.* **B3** (1989) 1001; S. M. Girvin, in *The quantum Hall effect* cited in Ref. 1.

Non-Abelian Statistics in the Fractional Quantum Hall States

X. G. Wen
School of Natural Sciences, Institute of Advanced Study, Princeton, New Jersey 08540

(Received 5 October 1990)

The fractional quantum Hall states with non-Abelian statistics are studied. Those states are shown to be characterized by non-Abelian topological orders and are identified with some of the Jain states. The gapless edge states are found to be described by non-Abelian Kac-Moody algebras. It is argued that the topological orders and the associated properties are robust against any kind of small perturbations.

PACS numbers: 73.20.Dx, 05.30.-d

It has become clearer and clearer that the ground states of strongly interacting electron systems may contain very rich structures[1-6] which cannot be characterized by broken symmetries and are called the topological orders.[2] Physical characterizations of the topological orders are discussed in Refs. 2 and 6. It has been shown that the fractional quantum Hall (FQH) states, the chiral spin states, and the anyon superfluid states contain nontrivial topological orders characterized by the Abelian Chern-Simons (CS) theories.[2,3,5,7] It would be interesting to know whether or not the non-Abelian topological orders characterized by the non-Abelian CS theories[8] can be realized in strongly interacting electron systems. In this paper we will construct some FQH states which contain non-Abelian topological orders. The effective theory of these states is shown to be non-Abelian CS theory and the quasiparticles carry non-Abelian statistics.[8] We will also discuss how to understand the non-Abelian statistics in terms of the electron wave function.

Different electron wave functions with filling fraction $1/2n$ have been constructed in Ref. 4. The quasiparticles in these states were shown to have non-Abelian statistics, provided that these states are incompressible and the quasiparticles have finite size for a local Hamiltonian.

The spirit of our discussion is very similar to that in the mean-field approach to the spin-liquid states.[9,10] A similar construction is also used to study the SU(N) spin chains.[11] Consider a two-dimensional spinless (i.e., spin-polarized) electron system in strong magnetic field with filling fraction $\nu = M/N$. For convenience we will put the electron system on a lattice; thus the electron Hamiltonian has the form

$$H=\sum_{ij}[t_{ij}e^{ieA_{ij}}c_i^\dagger c_j+V_{ij}n_in_j]\,, \tag{1}$$

where A_{ij} is the electromagnetic gauge potential on the lattice and $n_i = c_i^\dagger c_i$. To construct a FQH state with a non-Abelian topological order, we would like to break each electron into N partons[12] ψ_a each carrying electric charge e/N:

$$c=\psi_1\psi_2\cdots\psi_N=\frac{1}{N!}\sum_{ab\cdots c}\epsilon_{ab\cdots c}\psi_a\psi_b\cdots\psi_c\,, \tag{2}$$

where ψ_a are fermionic fields and N is odd. After substituting (2) into (1) and making a mean-field approximation we reach the following mean-field Hamiltonian:

$$H_{\text{mean}}=\sum_{ij}t_{ij}e^{ieA_{ij}/N}U_{ij,ab}\psi_{ia}^\dagger\psi_{jb}\,, \tag{3}$$

where

$$U_{ij,aa'}=e^{ieA_{ij}(N-1)/N}(1/N!)^2 \times\langle\epsilon_{ab\cdots c}(\psi_b\cdots\psi_c)_i^\dagger\epsilon_{a'b'\cdots c'}(\psi_{b'}\cdots\psi_{c'})_j\rangle\,. \tag{4}$$

The mean-field solution U_{ij} can be obtained by minimizing the average of the Hamiltonian (1) on the ground state of H_{mean} in (3) (i.e., $E=\langle\Phi_{\text{mean}}|H|\Phi_{\text{mean}}\rangle$). Let us assume that there exists a Hamiltonian H such that the mean-field solution takes the most symmetric form $U_{ij,ab}=\eta\delta_{ab}$. In this case the mean-field Hamiltonian (3) describes N kinds of free partons in magnetic field, each with a filling fraction $\nu = M$. Thus the mean-field ground-state wave function is given by $\Phi_{\text{mean}}\{z_i^a\} = \prod_{a=1}^N\chi_M(z_i^a)$, where z_i^a is the coordinate of the ath kind of parton and $\chi_M(z_i)$ is the fermion wave function of M filled Landau levels.

Notice that the mean-field theory (3) contains many unphysical degrees of freedom arising from the breaking of the electrons into partons. In order to use the mean-field theory to describe the original electron system we need to project onto the physical Hilbert space which satisfies the constraint

$$\psi_{1i}^\dagger\psi_{1i}=\cdots=\psi_{Ni}^\dagger\psi_{Ni}\,. \tag{5}$$

In the physical Hilbert space, different kinds of partons always move together. The bound states of the partons correspond to the original electrons. The electron ground-state wave function can be obtained by doing the projection on the mean-field wave function Φ_{mean} by setting $z_i^1=z_i^2=\cdots=z_i$, where z_i is the electron coordinate. The electron wave function obtained this way is just one of the FQH wave functions proposed by Jain.[12] In the following we will call such a state the NAF (non-Abelian FQH) state.

The wave function of the NAF state, $[\chi_M(z_i)]^N$, is the exact ground state of the following local Hamiltonian.[13] The kinetic energy in the Hamiltonian is such that the first $NM-N+1$ Landau levels have zero energy and other Landau levels have finite positive energies. One such kinetic energy is given by $\prod_{l=0}^{NM-N}[K-(l-\frac{1}{2})\omega_c]$, where $K=-(1/2m)(\partial_i - ieA_i)^2$. The two-body potential in the Hamiltonian has the form $V(r)\propto\partial^{N-1}\delta(r)$. One can easily see that the Hamiltonian is positive definite, and the NAF state has zero energy because the electrons in the ground state all lie in the first $NM-N+1$ Landau levels and the ground-state wave function has Nth-order zeros as $z_i \to z_j$. However, it is not clear whether the state has the ***highest*** filling fraction among the zero-energy states (this is related to the incompressibility). We can only show that among the Jain states[12] the NAF state is the zero-energy state with highest filling fraction. We do not know whether it is sufficient to consider only the Jain states. It would be interesting to numerically test the incompressibility of the NAF state for the above Hamiltonian. Numerical calculations have been done only for the projection onto the first two Landau levels.[14] In this case one indeed finds the Jain $\frac{2}{5}$ state to be the exact *incompressible* ground state.

The projection, or the constraint (5), can be realized by including a gauge field. Notice that under local SU(N) transformations $\psi_{ai}\to W_{i,ab}\psi_{bi}$, $W_i\in$ SU(N), the electron operator c_i in (2) is invariant. Thus the Hamiltonian contains *a local* SU(N) symmetry after we substitute (2) into (1). The local SU(N) symmetry manifests itself as a gauge symmetry in the mean-field Hamiltonian (3). Notice that (3) is invariant under the SU(N) gauge transformation W_i: $\psi_i\to W_i\psi_i$ and $U_{ij}\to W_iU_{ij}W_j^\dagger$. The gauge fluctuation in the mean-field theory can be included by replacing the mean-field value $U_{ij}=\eta$ by $U_{ij}=\eta\exp(ia_{ij})$, where a_{ij} is a $N\times N$ Hermitian matrix. a_{ij} is just the SU(N) gauge potential on the lattice. The time component of the SU(N) gauge field can be included by adding a term[10] $\psi_i^\dagger a_0(i)\psi_i$ to the mean-field Hamiltonian. The constraint (5) is equivalent to the following constraint:[10]

$$J^l_\mu(i)=0,\quad l=1,\ldots,N^2-1\,, \tag{6}$$

where J^l_μ are the SU(N) charge and the current density. The constraint (6) can be enforced in the mean-field theory by integrating out the gauge-field fluctuation a_μ.[10] After the projection, the only surviving states are those which are invariant under the local SU(N) transformations. Those states correspond to the physical electron states.

The effective theory of the NAF state described above can be obtained by first integrating out the ψ_a field:

$$\mathcal{L}_{0\text{eff}}=\frac{M}{4\pi N}A_\mu\partial_\nu A_\lambda\epsilon^{\mu\nu\lambda}+\frac{M}{8\pi}\text{Tr}a_\mu f_{\nu\lambda}\epsilon^{\mu\nu\lambda}\,, \tag{7}$$

which is just the level-M SU(N) CS theory.[8] $f_{\mu\nu}$ in (7) is the strength of the SU(N) gauge field. The quasiparticle excitations in the NAF state correspond to the holes in various Landau levels of the partons. Those excitations are created by the parton fields ψ_a. After including the gauge fields, the properties of the quasiparticles are described by the following effective Lagrangian:

$$\mathcal{L}_{q\text{eff}}=\sum\psi^\dagger\left[\left(i\partial_t+\frac{e}{N}A_0+a_0\right)-\frac{1}{2m}\left(\partial_i-i\frac{e}{N}A_i-ia_i\right)^2\right]\psi\,. \tag{8}$$

Equations (7) and (8) describe the low-energy properties of the NAF state.

The non-Abelian CS theory given by (7) and (8) has been studied in detail in Ref. 8. The quasiparticles ψ_a (which are called the Wilson lines in Ref. 8) are found to have non-Abelian statistics. In the following we will summarize some special properties associated with the non-Abelian statistics and discuss their relation to the microscopic electron wave function. Let us put the NAF state on a sphere. The ground state of (7) is found to be nondegenerate on the sphere. (On genus-g Riemann surfaces the ground states are degenerate.) Now let us create m quasiparticles and m' quasiholes using the operators ψ_{a_i} and $\psi_{a_j}^\dagger$. If we have ignored the gauge field a_μ (setting $a_\mu=0$), the Hilbert space generated by $\psi_{a_i}|_{i=1}^m$ and $\psi_{a_j}^\dagger|_{j=1}^{m'}$ would be $(\mathcal{H}_R)^m\times(\mathcal{H}_{\bar R})^{m'}$ which has $N^{m+m'}$ dimensions. Here $\mathcal{H}_R$ is the fundamental representation of SU(N) and $\mathcal{H}_{\bar R}$ is the dual of $\mathcal{H}_R$. However, after we include the gauge fluctuations and do the projection $z_i^a=z_i$, only the gauge-invariant states can survive the projection and appear as the physical states of the original electron system. In particular, all the states that transform nontrivially under global SU(N) are projected away. Thus the Hilbert space $\mathcal{H}_{mm'}$ of the physical states is contained in the SU(N)-invariant subspace of $(\mathcal{H}_R)^m\times(\mathcal{H}_{\bar R})^{m'}$:[8] Inv$[(\mathcal{H}_R)^m\times(\mathcal{H}_{\bar R})^{m'}]$. In the above we have only used the global SU(N) gauge symmetry. The local gauge symmetry may further reduce the dimension of the Hilbert space. Not every (global) SU(N) singlet state can survive the projection and become a physical state. Thus the dimension of $\mathcal{H}_{mm'}$ can be less than that of Inv$[(\mathcal{H}_R)^m\times(\mathcal{H}_{\bar R})^{m'}]$.

When $m=1$ and $m'=0$, there is no invariant state and the dimension of $\mathcal{H}_{10}$ is zero. When $m=m'=1$ there is only one invariant state. It is shown that such a state is always physical and $\mathcal{H}_{11}$ is one dimensional. In this case moving one particle around the other induces a Berry phase $\exp[i2\pi(N+1)/N(N+M)]$. When $m=m'=2$, Inv$[(\mathcal{H}_R)^2\times(\mathcal{H}_{\bar R})^2]$ is two dimensional. It turns out that $\mathcal{H}_{22}$ is two dimensional if $M>1$ and one dimensional if $M=1$.[8] As we interchange the two particles created by ψ_{a_i}, $i=1,2$, we obtain a non-Abelian Berry phase for $M>1$. The 2×2 matrix describing the non-Abelian Berry phase is found[8] to have eigenvalues $-\exp[i\pi \times(-N+1)/N(N+M)]$ and $\exp[i\pi(N+1)/N(N+M)]$. For $M=1$ the Hilbert space $\mathcal{H}_{22}$ is one dimensional and the corresponding Berry phase is $\exp(i\pi/N)$. The later result is expected because the $M=1$ NAF state is just the Laughlin state with filling fraction $1/N$. The reproduction of the well-known results of the Laughlin states is a nontrivial self-consistency check of our theory.

Some of the above results can be easily understood in terms of the microscopic electron wave function. First we notice that the mean-field state Φ_{mean} is a (global) SU(N) singlet and the NAF wave function can be expressed as $\langle 0|\prod_i c(z_i)|\Phi_{\text{mean}}\rangle=[\chi_M(z_i)]^N$, where $c(z_i)$ is given by (2). The quasiparticles discussed above are described by the following electron wave function: $\langle 0|\prod_i c(z_i)\prod_{l,l'}\psi_{a_l}\psi^\dagger_{b_{l'}}|\Phi_{\text{mean}}\rangle$. Since $\langle 0|\prod_i c(z_i)$ is an SU(N) singlet, it is clear that only the states in $\text{Inv}[(\mathcal{H}_R)^m\times(\mathcal{H}_{\bar{R}})^{m'}]$ can survive the projection and give rise to nonzero electron wave functions. The dimension of the physical Hilbert space may be smaller than that of the invariant space because the electron wave functions induced from different mean-field singlet states may not be orthogonal to each other. For $m=m'=1$ the electron wave function can be obtained by the projection of the mean-field state $\psi_1(Z_1)\psi_1^\dagger(Z_2)|\Phi_{\text{mean}}\rangle$. The electron wave function is nonzero and is given by $\chi_M(z_i;Z_1;Z_2)[\chi_M(z_i)]^{N-1}$, where $\chi_M(z_i;Z_1;Z_2)$ has one hole and one particle at Z_1 and Z_2. Thus $\mathcal{H}_{11}$ is one dimensional. For $m=m'=2$ the two electron wave functions $\Phi_{1,2}$ can be obtained by the projection of mean-field states $\psi_1(1)\psi_1(2)\psi_1^\dagger(3)\psi_1^\dagger(4)|\Phi_{\text{mean}}\rangle$ and $\psi_1(1)\psi_2(2)\times\psi_1^\dagger(3)\psi_2^\dagger(4)|\Phi_{\text{mean}}\rangle$ (which contain two singlets). Notice that locally the electron wave functions are the same near each quasiparticle no matter whether the quasiparticle is created by ψ_1 or ψ_2. More precisely the physical correlation functions, like the density correlation, are the same around each quasiparticle when the quasiparticles are well separated. This is because ψ_1 can be rotated into ψ_2 by a global SU(N) transformation, while the density correlation, being a SU(N)-invariant quantity, will not be changed. The effects of the other quasiparticles can be ignored since the correlation in the NAF states is short ranged and the other particles are far away. Thus the two electron states Φ_1 and Φ_2 should have the same local correlations and hence the same energy. Such a degeneracy is a bulk property just like the degeneracy of the FQH states on a torus.

When $M=1$ each kind of parton fills only the first Landau level. The action of $\psi_1(1)\psi_1(2)\psi_1^\dagger(3)\psi_1^\dagger(4)$ on the first-Landau-level wave function corresponds to multiplication by a factor

$$A_{22}=\sum_{i_1,i_2}\delta(Z_3-z_{i_1})\delta(Z_4-z_{i_2})\prod_{i,j\neq i_1,i_2}\frac{(z_i-Z_1)(z_j-Z_2)}{(z_i-z_{i_1})(z_j-z_{i_2})}$$

and the action of $\psi_1(1)\psi_1^\dagger(3)$ corresponds to a factor

$$A_{11}=\sum_{i_1}\delta(Z_3-z_{i_1})\prod_{i\neq i_1}\frac{z_i-Z_1}{z_i-z_{i_1}}.$$

After the projection the two resulting electron wave functions are given by $A_{22}(\chi_M)^N$ and $(A_{11})^2(\chi_M)^N$ which describe the same state since $A_{22}\propto(A_{11})^2$. Similar derivations apply to other values of m and m', and the physical Hilbert space $\mathcal{H}_{mm'}$ can be shown to be at most one dimensional for $M=1$. This is just the result of the non-Abelian CS theory. More detailed discussions of the structures of excitations in the NAF state will appear elsewhere.

We would like to remark that although the gauge field mediates no long-range interactions between quasiparticles due to the CS term, the quasiparticles ψ_a are not really equivalent to the "free" quarks in the absence of the gauge field. This is because the quasiparticles are dressed by non-Abelian flux which carries the SU(N) charge. Thus it is conceivable that when $M=1$ the quasiparticles behave like Abelian anyons with no internal degree of freedom, as has been shown in the above discussion.

Now let us discuss another fascinating property of the NAF state—the gapless edge excitations[15,6] in the NAF state. We will follow the discussions in Ref. 6. First let us ignore the constraint (6) and set $a_\mu=0$ in the mean-field theory. In this case the edge excitations are those of the integer quantum Hall states[15] described by

$$\mathcal{L}=\sum_{a\alpha}i\lambda^{a\alpha\dagger}(\partial_0-v\partial_x)\lambda^{a\alpha}, \tag{9}$$

where $\lambda^{a\alpha}$ is a fermion field describing the edge excitations of the αth Landau level of the ath kind of parton. The Hilbert space of (9) can be represented[11] as a direct product of the Hilbert spaces of a U(1) Kac-Moody (KM) algebra,[16] a level-N SU(M) KM algebra, and a level-M SU(N) KM algebra. This decomposition is a generalization of the spin-charge separation in the 1D Hubbard model. Notice that the total central charge of the above three KM algebras is

$$1+\frac{N(M^2-1)}{M+N}+\frac{M(N^2-1)}{N+M}=MN$$

which is equal to the central charge of (9). The above three KM algebras are generated by currents $J_\mu=eN^{-1}\times\lambda^{a\alpha\dagger}\partial_\mu\lambda^{a\alpha}$ (the electric current), $j^l_\mu=t^l_{\alpha\beta}\lambda^{a\alpha\dagger}\partial_\mu\lambda^{a\beta}$, and $J^l_\mu=T^l_{ab}\lambda^{a\alpha\dagger}\partial_\mu\lambda^{b\alpha}$, where t^l [T^l] are the generators of the SU(M) [SU(N)] Lie algebra. The currents in the SU(N) KM algebra are just the currents in (6) which couple to the SU(N) gauge field a_μ. To obtain the physical edge excitations in the electron wave function, we need to do the projection to enforce the constraint (6). Because of the above decomposition, the projection can be easily done by removing from the Hilbert space of (9) the states associated with the SU(N) KM algebra.[6] The remaining physical edge states are generated by the

U(1)×SU(M) KM algebra. The central charge of the U(1)×SU(M) KM algebra is given by $c=M(MN+1)/(M+N)$ are the specific heat (per unit length) of the edge excitations[17] is $C=c(\pi/6)T/v$. The electron creation operator[6] on the edge is given by $c=\lambda^{1a_1}\cdots\lambda^{Na_N}$ which has a propagator $(x-vt)^{-N}$ along the edge. We would like to point out that in general the edge excitations may have several different velocities in contrast to what was implicitly assumed above.

The above construction can be easily generalized in a number of directions: (a) We may decompose electrons into partons with different electric charge. (b) We may choose a different mean-field ground state which breaks the SU(N) gauge symmetry. Actually (a) is a special case of (b).[18] The effective CS theory for (a) and (b) will in general contain several Abelian and non-Abelian gauge fields. In particular, the FQH states studied in Refs. 3 and 6 correspond to breaking the SU(N) gauge symmetry into $[U(1)]^{N-1}$ gauge symmetry. One interesting NAF state in case (a) is the $\nu=(1+\frac{1}{2}+\frac{1}{2})^{-1}=\frac{1}{2}$ state. Its non-Abelian statistics are described by the level-2 SU(2) CS theory. The electrons in such a state lie in the first three Landau levels.

We would like to argue that the NAF states studied in this paper are generic states and their non-Abelian topological orders are robust against small perturbations. (i) The non-Abelian structures in the NAF states come from the SU(N) gauge symmetry of the mean-field ground state. To destroy the non-Abelian topological orders (and the associated non-Abelian statistics) we must break the SU(N) gauge symmetry through the Higgs mechanism. This cannot be achieved unless we add *finite* perturbations. (ii) All excitations in the NAF have finite energy gap and the interactions between them have finite range. Therefore the NAF states do not have infrared divergences and it is self-consistent to assume the interactions between the excitations do not destabilize the NAF states. Thus we expect the properties studied in this paper are universal properties of the NAF states which are robust against any small perturbations. The NAF states are a new type of the infrared fixed points and the NA topological orders should appear as a general possibility for the ordering in the ground states of strongly interacting electron systems.

It is not clear under what conditions the NAF states might be realized in nature. However, since the NAF states are generic states, they may appear in experiments under the right conditions, especially when the electron density is low and higher Landau levels are important. The low-density FQH states are largely unexplored in experiments.

This research was supported by DOE Grant No. DE-FG02-90ER40542.

[1] F. D. M. Haldane, Phys. Rev. Lett. **51**, 605 (1983); B. Halperin, Phys. Rev. Lett. **52**, 1583 (1984).

[2] X. G. Wen, Int. J. Mod. Phys. B **2**, 239 (1990); Phys. Rev. B **40**, 7387 (1989); X. G. Wen and Q. Niu, Phys. Rev. B **41**, 9377 (1990).

[3] B. Blok and X. G. Wen, Phys. Rev. B **42**, 8133 (1990); **42**, 8145 (1990); Institute for Advanced Study Report No. IASSNS-HEP-90/66 (to be published).

[4] G. Moore and N. Read (to be published).

[5] N. Read, Phys. Rev. Lett. **65**, 1502 (1990).

[6] X. G. Wen, Institute for Advanced Study Report No. IASSNS-HEP-90/42 (to be published).

[7] X. G. Wen and A. Zee, Nucl. Phys. B (Proc. Suppl.) **15**, 135 (1990).

[8] E. Witten, Commun. Math. Phys. **121**, 351 (1989).

[9] X. G. Wen, F. Wilczek, and A. Zee, Phys. Rev. B **39**, 11413 (1989); R. Laughlin and Z. Zou, Phys. Rev. B **41**, 664 (1990).

[10] I. Affleck, Z. Zou, T. Hsu, and P. W. Anderson, Phys. Rev. B **38**, 745 (1988); E. Dagotto, E. Fradkin, and A. Moreo, Phys. Rev. B **38**, 2926 (1988).

[11] I. Affleck, Nucl. Phys. **B265**, 409 (1986).

[12] J. K. Jain, Phys. Rev. B **40**, 8079 (1989).

[13] A similar observation has also been made by J. K. Jain (private communication).

[14] A. H. MacDonald (private communication).

[15] B. I. Halperin, Phys. Rev. B **25**, 2185 (1982); X. G. Wen, Institute for Theoretical Physics Report No. NSF-ITP-89-157 (to be published); A. H. MacDonald, Phys. Rev. Lett. **64**, 220 (1990); X. G. Wen, Phys. Rev. Lett. **64**, 2206 (1990); Phys. Rev. B **41**, 12838 (1990); M. Stone, University of Illinois, Urbana, Report No. IL-TH-8; University of Illinois, Urbana, Report No. IL-TH-90-32 (to be published); F. D. M. Haldane, Bull. Am. Phys. Soc. **35**, 254 (1990).

[16] A general review of the KM algebra can be found in D. Gepner and E. Witten, Nucl. Phys. **B278**, 493 (1986); P. Goddard and D. Olive, Int. J. Mod. Phys. **1**, 303 (1986); V. G. Kac, *Infinite Dimensional Lie Algebra* (Birkhauser, Boston, 1983).

[17] I. Affleck, Phys. Rev. Lett. **56**, 746 (1986).

[18] I would like to thank E. Fradkin for pointing this out to me.

Fractional quantum Hall effect
and
nonabelian statistics

N. READ

Departments of Applied Physics and Physics, P.O. Box 2157
Yale University, New Haven, CT 06520

and

G. MOORE

Department of Physics, P.O. Box 6666
Yale University, New Haven, CT 06511

It is argued that fractional quantum Hall effect wavefunctions can be interpreted as conformal blocks of two-dimensional conformal field theory. Fractional statistics can be extended to nonabelian statistics and examples can be constructed from conformal field theory. The Pfaffian state is related to the 2D Ising model and possesses fractionally charged excitations which are predicted to obey nonabelian statistics.

Talk presented by N. Read at the 4th Yukawa International Symposium, Kyoto, Japan, 28 July to 3 August, 1991.

I. Introduction: Particle Statistics and Conformal Field Theory

This paper is a brief overview of some aspects of the relationship between the theories of the fractional quantum Hall effect (FQHE) [1] and two-dimensional conformal fields (CFT) [2], which has been explored in more detail in [3]. The present section describes general aspects of particle statistics, especially nonabelian statistics; the second section uses the Laughlin wavefunctions as an example of the relation with CFT; the third section presents a case of nonabelian statistics in a FQHE system (the Pfaffian state).

The notion of particle statistics in quantum mechanics usually refers to the action of the permutation group S_n on the wavefunction for a collection of n identical, indistinguishable particles: the wavefunction is taken to transform as a definite representation of this group, which interchanges particles. The usual examples are Bose statistics and Fermi statistics, which are the trivial and alternating representations, respectively. A more modern approach prefers to exchange particles along some definite paths in space, and allows for the possibility that the wavefunction be not well-defined when particles coincide, so that intersecting exchange paths must be omitted. Topology then enters the subject, and while it turns out for spatial dimension $d > 2$ that the group of exchanges still reduces to S_n, for $d = 2$ the group of topologically distinct exchanges is Artin's braid group $\mathcal{B}_n$ [4]. The infinite group $\mathcal{B}_n$ can be generated by a set of elementary exchanges B_i, $i = 1, \ldots n-1$ which simply exchange particles i, $i+1$ along a path not enclosing any other particles. The most familiar representation of $\mathcal{B}_n$ is fractional statistics, where each B_i acts on the wavefunction as $e^{i\theta}$ (θ real) and so is a one-dimensional, abelian representation; this includes Bose and Fermi statistics as special cases.

We will use the term "nonabelian statistics" when particle wavefunctions transform as a nonabelian representation of the permutation or (especially) of the braid group, and we have introduced the term "nonabelions" for such particles [3]. In this case, not all the representatives of the B_i commute, and must be matrices acting on vector wavefunctions. Thus, even when the positions and quantum numbers of the particles have been specified, the wavefunction is not unique but is a member of a vector space (a subspace of the Hilbert space). The vector space may be finite dimensional but its dimension will grow with the number of particles. Physical observables are invariant under exchanges, so cannot distinguish which state in this space the system is in, just as the phase of the wavefunction cannot be directly measured in the usual (one dimensional) case. In the latter case, differences or changes of phase, and interference effects, can, however, be observed, and analogously, nonabelian statistics can manifest itself in physical effects.

For the permutation group, nonabelian statistics is an old idea known as "parastatistics". It appears, however, that parastatistics cannot be realized in a *local* theory in $d > 2$ in a nontrivial way (the literature containing this result is reviewed in [5]). The issue is whether the global degeneracy and nonlocal effects of exchanges are compatible with the natural physical idea of locality of interactions, as realized for example in quantum field theory governed by a local Lagrangian density.

An analogous analysis for the braid group in $d = 2$ has been carried out in [5]. It involves not only the matrices B_i but also the notion of "fusion rules". A theory will usually contain several distinct types of particles, each with its own B matrices, and taking a particle of one type around one of another type will produce a matrix effect like that of B^2. Fusion of two or more particles into a composite produces either annihilation of the particles or else a particle of some new type with its own statistics properties. The matrix elements for these processes, denoted F, describe the fusion rules and will have to satisfy consistency relations involving B's. For example, taking a third particle around a pair before or after they fuse should give the same result, since the third particle is far from the fusing pair and so cannot distinguish the close pair from their composite (locality). Frohlich *et al* [5] conclude that nonabelian statistics can be acceptable in two dimensions. Indeed, local actions are known that produce nonabelian properties: they are Chern-Simons terms for nonabelian gauge fields, and the particle worldlines are represented as Wilson lines [6]. Note that as in the abelian case, the statistics effects can be transferred between the wavefunctions and the Hamiltonian by a singular gauge transformation; the general discussion above was for the gauge choice where statistics is exhibited in the wavefunction, *i.e.* no Chern-Simons-type vector potentials in the Hamiltonian.

As a partial example of nonabelian statistics, we present the following system. In the example, we may create particles σ, but only in even numbers. The σ particles have no internal quantum numbers. For two particles, the vector space for fixed positions z_1, z_2 is one dimensional, and the wavefunction is $(z_1 - z_2)^{-1/8}$, so one might imagine that we have abelian fractional statistics $\theta/\pi = -1/8$. However, for four particles, the vector space is two-dimensional. If we place particles at 0, 1 and ∞ (which in this example can always be done through a global conformal transformation), the wavefunctions are functions only of the remaining coordinate z, and (choosing a basis) are

$$\psi_{\pm}(z) = (z(1-z))^{-1/8}\sqrt{1 \pm \sqrt{1-z}}. \tag{1.1}$$

In the z-plane, there are branch points at $z = 0, 1$. As z is analytically continued around $z = 1$, we see that $\psi_{\pm}$ are interchanged. This operation corresponds to two exchanges of the particle at z with that at 1, *i.e.* to B^2. Other double exchanges involving z are diagonal in this basis. To exhibit B itself we would have to perform a conformal transformation to obtain $z \mapsto 1$, $1 \mapsto z$. Thus we have a single multibranched function whose two sheets are linearly independent functions of z, and so the σ particles behave as nonabelions. For general number $2n$ of σ particles the wavefunctions form a 2^{n-1}-dimensional space.

The example arises from a particular conformal field theory (CFT), the two dimensional Ising model at its critical point. We next describe some general features of CFT [7]. In a $1+1$-dimensional system with short range interactions (or a local lagrangian) at a critical point, one has not only scale invariance but also conformal invariance, an infinite-parameter group under which correlation functions transform covariantly. Exactly at the critical point (so that corrections due to slowly decaying irrelevant or marginal operators are omitted) the correlation functions of a collection of fields $\{\phi_{i_r}(\mathrm{x}_r)\}$ can be split in the form (taking the "diagonal" case for simplicity):

$$\left\langle \prod_{r=1}^{n} \phi_{i_r}(\mathrm{x}_r) \right\rangle = \sum_p |\mathcal{F}_{p;i_1 \ldots i_n}(z_1, \ldots z_n)|^2 \tag{1.2}$$

where $z_r = x_r + iy_r$. The conformal block functions $\mathcal{F}_p$ are multibranched functions, analytic in their arguments, except possibly when two z's coincide. The variable p labelling different functions runs over a finite set in a *rational* CFT. As the z's are varied so as to exchange some ϕ_i's, the functions $\mathcal{F}_p$ are analytically continued to different sheets, but can be expressed as z independent linear combinations of the original functions through some braiding matrices B as we saw above. In this way, the correlation function can be single valued.

Another operation that can be performed on the correlation functions is the operator product expansion. As the arguments z_1, z_2 of two fields approach one another, the operators merge into a linear combination of single operators:

$$\phi_i(z)\phi_j(w) \sim \sum_k C_{ij}^k(z-w)\phi_k(w) \tag{1.3}$$

as $z \to w$, where ϕ_k is some new field of type k and $C_{ij}^k(z-w)$ is a singular coefficient function. This operation can be used to define some new matrices F, the fusion matrix that describes which fields k appear in the product of i and j. In [8] the consistency conditions that must be satisfied by B, F were analyzed; these same conditions emerged later in the work of Frohlich *et al.* Thus CFT provides many examples of nonabelian statistics, when conformal blocks are interpreted as wavefuctions. In the following we show that this idea applies directly to FQHE states.

II. Example: Laughlin States

In this section we show that both the fractional statistics of quasiparticles in the Laughlin states [9] and the actual wavefunctions are related to CFT conformal blocks. The construction given here is from [3], but parts of it have been obtained independently by others.

In $1+1$-dimensional Euclidean spacetime, define a free scalar field by its correlator,

$$\langle \varphi(z)\varphi(z')\rangle = -\log(z-z') \tag{2.1}$$

where the log is complex, so the field creates right-moving excitations only. All correlators can be obtained from this one using Wick's theorem. Now consider the function

$$\left\langle \prod_{i=1}^{N} e^{i\sqrt{q}\varphi(z_i)}\, e^{-i\sqrt{q}\int d^2z'\,\bar{\rho}\varphi(z')} \right\rangle \tag{2.2}$$

where $\bar{\rho} = 1/2\pi q$, q is an integer, and the integral is taken over a disk of area $2\pi qN$ centered at the origin. We expand and contract using (2.1). Each exponential is assumed normal ordered, *i.e.* φ's from the same exponential are not to be contracted together. The result is [10]

$$\prod_{i<j}(z_i - z_j)^q\, e^{-\frac{1}{2\pi}\sum_i \int d^2z' \log(z_i - z')}. \tag{2.3}$$

In the last factor, the real part of the log gives $e^{-\frac{1}{4}\sum_i |z_i|^2}$ for z_i inside the disk. Apart from the remaining phase, we now recognize the function as Laughlin's wavefunction for particles in the lowest Landau level at filling factor $\nu = 1/q$. Since the vertex operators $\exp i\sqrt{q}\varphi(z)$ are associated with Coulomb charges, it is clear that we have a holomorphic version of Laughlin's 2D plasma picture [9], which gives the wavefunction and not just its modulus squared!

The imaginary part of the log in (2.3) winds by $2\pi d^2z'$ as each z_i goes round each point z' in the integration region, so is highly singular [11]. It contributes a pure phase to the function, which can be gauged away by an equally singular gauge transformation. This phase describes a uniform magnetic field, which we identify as the physical magnetic field in the FQHE problem. A nice way to see this is to consider not the phase itself but the *change* as one particle describes a closed loop C. Clearly the exponent changes by

$$-\frac{1}{2\pi}\int d^2z'\,2\pi i \tag{2.4}$$

where the integral is over the area enclosed by C, so the phase change is $1/2\pi$ times the area of the loop, which is just the Berry phase for adiabatic transport of a particle of charge 1 in a field of strength $1/2\pi$. In all the following equations, the singular phase will be implicitly gauged away.

In a similar way, one can obtain quasihole wavefunctions as (for two quasiholes):

$$\begin{aligned}&\left\langle e^{\frac{i}{\sqrt{q}}\varphi(z)} e^{\frac{i}{\sqrt{q}}\varphi(w)} \prod_{i=1}^{N} e^{i\sqrt{q}\varphi(z_i)}\, e^{-i\sqrt{q}\int d^2z'\,\bar{\rho}\varphi(z')} \right\rangle \\ &\quad = (z-w)^{1/q}\prod_k (z-z_k)(w-z_k)\prod_{i<j}(z_i-z_j)^q\, e^{-\frac{1}{4}\sum_i |z_i|^2-\frac{1}{4}|z|^2-\frac{1}{4}|w|^2}. \end{aligned} \tag{2.5}$$

The first factor gives explicitly the fractional statistics, $\theta/\pi = 1/q$. Apparently, the constructions from CFT correlators give results in the gauge where all Berry phases—background magnetic field as well as fractional statistics—appear in the wavefunction, not as a vector potential (connection) in the Hamiltonian. This will be important to us in the next section when we move on to nonabelian statistics (adiabatic transport).

We now return to the $2+1$-dimensional world of electrons in two dimensions in a strong magnetic field and describe briefly the order parameter picture of the FQHE, following [12] (see also [13,14]). In words, a composite of 1 electron and q quasiholes (or "vortices", or "flux quanta") at filling factor $\nu = 1/q$, (i) is a boson; (ii) sees no net effective magnetic field. When both (i) and (ii) are true, the composite can "Bose condense", *i.e.* have long-range order. For the Laughlin states, (i) and (ii) follow from the calculations in [15].

More formally, let $\psi^\dagger$ create an electron in the lowest Landau level (LLL), and let

$$U(z) = \prod_{i=1}^{N}(z - z_i) \tag{2.6}$$

be Laughlin's quasihole operator (acting in the LLL). If $|0_{\rm L}; N\rangle$ is the normalized Laughlin state for N particles, then one can show [12]

$$\lim_{|z-z'|\to\infty} \lim_{N\to\infty} \langle 0_{\rm L}; N|\, U^\dagger(z)^q \psi(z)\psi^\dagger(z')U(z')^q\, |0_{\rm L}; N\rangle\, e^{-\frac{1}{4}|z|^2-\frac{1}{4}|z'|^2} = \bar{\rho}. \tag{2.7}$$

Given our choice of order parameter operator, the Laughlin state is the state with the most order. In fact, we have [12]

$$|0_{\rm L}; N\rangle = \left(\int d^2z\, \psi^\dagger(z)U(z)^q e^{-\frac{1}{4}|z|^2}\right)^N |0\rangle \tag{2.8}$$

where $|0\rangle$ is the vacuum (no electrons). This says that the Laughlin state is precisely a Bose condensation of the composite bosons.

The order parameter in a general state obeys [12] a system of Landau-Ginzburg-Chern-Simons equations [16], which also involve internal vector and scalar potentials. Similar equations are found in another, more field theoretic approach [14,17] which has been reviewed recently by Zhang [18]. The ideas have been extended [17,19,20] to the hierarchy scheme [21,22,23,24] and its generalizations [19], which give states with abelian statistics for all filling factors $p/q < 1$. These also have interpretations as CFT correlators [3,19].

III. Pfaffian State and Nonabelions

Consider the wavefunction [3]

$$\Psi_{\rm Pf}(z_1,\cdots,z_N) = \mathrm{Pfaff}\left(\frac{1}{z_i - z_j}\right)\prod_{i<j}(z_i - z_j)^q\, e^{-\frac{1}{4}\sum|z_i|^2} \tag{3.1}$$

The Pfaffian is defined by

$$\mathrm{Pfaff}\, M_{ij} = \frac{1}{2^{L/2}(L/2)!}\sum_{\sigma\in S_L}\mathrm{sgn}\,\sigma\prod_{k=1}^{L/2} M_{\sigma(2k-1),\sigma(2k)} \tag{3.2}$$

for an $L\times L$ antisymmetric matrix whose elements are M_{ij}, or as the square root of the determinant of M; S_n is the permutation group on n objects. $\Psi_{\rm Pf}$ can be regarded as a wavefunction for spinless or spin polarized electrons in the LLL if $q>0$ is even, since then it is antisymmetric; the filling factor is $\nu = 1/q$. Note that $\mathrm{Pfaff}\,((z_i - z_j)^{-r})$, r odd, $< q$ would also give a valid wavefunction.

The idea behind the construction of this wavefunction was the following. From the calculations in [15], it follows that, in an incompressible fluid state of electrons at filling factor $1/q$, the composite $\psi^\dagger(z)U(z)^q$ is a neutral boson if q is odd (as mentioned in §II) and is a neutral fermion if q is even. Hence, for q odd the operator can Bose condense, which gives the Laughlin states. For q even, on the other hand, it cannot condense singly, but pairs of fermions can condense, as in the BCS theory of superconductivity. For the spinless or spin-polarized case, the pairing function must be of odd parity to satisfy Fermi statistics. Thus a possible function is

$$\left(\int d^2z\,d^2w\,\frac{1}{z-w}\,\psi^\dagger(z)U^q(z)\psi^\dagger(w)U^q(w)e^{-\frac{1}{4}(|z|^2+|w|^2)}\right)^{N/2}|0\rangle. \tag{3.3}$$

Writing out this function in co-ordinate representation yields (3.1). The Pfaffian is precisely the sum of products of fermion pairs, antisymmetrized over all distinct ways of pairing. More generally, the pairing function $(z-w)^{-1}$ can be replaced by $(z-w)^{-r}$ with r odd, provided

$r < q$, as noted above. This construction was also inspired by noticing that the Haldane-Rezayi (HR) spin-singlet state for electrons at $\nu = 1/q$, q even [25], can be written in the form involving $\det(z_i^{\uparrow} - z_j^{\downarrow})^{-2}$. The real-space form of the spin singlet BCS wavefunction for spin-$\frac{1}{2}$ fermions is just such a determinant of an even parity pairing function, as is well known (it is perhaps less well known that the analogous result for spinless or spin-polarized fermions is a Pfaffian). Thus the HR state is a condensate of pairs of spin-$\frac{1}{2}$ composite neutral fermions, $\psi_\sigma^\dagger(z)U(z)^q$ [3]!

The pairing picture suggests two kinds of possible excitations. One is the analogue of the BCS quasiparticle, obtained by adding composite fermions, or by "breaking pairs". Thus a state with such a neutral fermion excitation localized at z is obtained by acting on the ground state with $\psi^\dagger(z)U(z)^q$. (In the HR state, one likewise obtains spin-$\frac{1}{2}$ fermions.) These excitations have no analogue in the Laughlin states since there the bosons are already condensed; the only excitations, other than the collective density mode [26], are Laughlin's quasiparticles, which are fractionally charged. In the order parameter picture, these are vortices, *i.e.* the order parameter winds in phase by a multiple of 2π around each quasiparticle, which thus resemble flux quanta in a superconductor, the flux quantum being $\Phi_0 = hc/e$ since the order parameter carries charge 1 from the single electron that it contains. The flux Φ determines the charge e^* through the quantized Hall relation, $e^* = \nu(\Phi/\Phi_0)$. Similarly, in the paired states, one expects to find excitations corresponding to flux quantized in multiples of $\frac{1}{2}\Phi_0$, since the order parameter contains two electrons, and these will have charge in multiples of $(2q)^{-1}$.

These ideas motivated the following wavefunction [3] for a pair of quasiholes in the Pfaffian state:

$$\Psi_{\mathrm{Pfaff+qholes}}(z_1, \ldots, z_N; v_1, v_2) = \tag{3.4}$$

$$\left\{ \sum_{\sigma \in S_N} \frac{\operatorname{sgn} \sigma \prod_{k=1}^{N/2} [(z_{\sigma(2k-1)} - v_1)(z_{\sigma(2k)} - v_2) + (v_1 \leftrightarrow v_2)]}{(v_1 - v_2)^{\frac{1}{8} - \frac{1}{4q}} (z_{\sigma(1)} - z_{\sigma(2)}) \cdots (z_{\sigma(N-1)} - z_{\sigma(N)})} \right\} \prod_{i<j} (z_i - z_j)^q \mathrm{e}^{-\frac{1}{4} \sum_i |z_i|^2 - \frac{1}{4}|v_1|^2 - \frac{1}{4}|v_2|^2}$$

The factor in brackets can be rewritten as

$$(v_1 - v_2)^{\frac{1}{4q} - \frac{1}{8}} \mathrm{Pfaff} \left(\frac{(z_i - v_1)(z_j - v_2) + (v_1 \leftrightarrow v_2)}{z_i - z_j} \right). \tag{3.5}$$

(The reason for the assumed form of the exponent of $(v_1 - v_2)$ will be explained below.) The extra factor in each term of the Pfaffian resembles a pair of Laughlin quasihole operators, except that in each term each factor acts on only one member of each pair of fermions. Therefore, in an average sense, each quasihole is a half flux and carries charge $1/2q$. Note that this construction cannot produce one quasihole. One way to see that this is impossible is that working on a compact geometry like the sphere, the total flux is quantized in units of Φ_0. Another feature of (3.5) is that as $v_1 \to v_2$, we recover a Laughlin quasihole of charge $1/q$.

While our construction of the Pfaffian state was motivated by order parameter considerations, we can also give an interpretation using conformal field theory, which then suggests the full structure of the system of excited quasihole states. Introduce free, massless real (Majorana) fermions χ in $1+1$ dimensions:

$$\langle \chi(z)\chi(z') \rangle = \frac{1}{z - z'} \tag{3.6}$$

Then using also the free scalar field as before, construct

$$\left\langle \prod_{i=1}^{N} \chi(z_i) \mathrm{e}^{i\sqrt{q}\varphi(z_i)} \, \mathrm{e}^{-i\sqrt{q}\int d^2z' \, \bar{\rho}\varphi(z')} \right\rangle. \tag{3.7}$$

This reproduces the FQHE wavefunction (3.1). To reproduce the two quasihole function (3.5), we should understand the CFT we are dealing with. Majorana fermions arise naturally in the two-dimensional Ising model at its critical point. The fermions, in conjunction with their leftmoving partners $\bar{\chi}(\bar{z})$, represent the energy density fluctuation $\varepsilon(z,\bar{z}) = \bar{\chi}(\bar{z})\chi(z)$ which hence has dimension $x = 1$ and so the correlation length exponent is $\nu = (2-x)^{-1} = 1$. The other fields in the Ising model are the spin fields, *i.e.* the Ising spin itself. These are operators which produce a square root branch point in the fermi field [7]. With the latter we can reproduce the two quasihole state:

$$\Psi_{\mathrm{Pfaff+qholes}} = \left\langle \sigma(v_1)\mathrm{e}^{\frac{i}{2\sqrt{q}}\varphi(v_1)}\sigma(v_2)\mathrm{e}^{\frac{i}{2\sqrt{q}}\varphi(v_2)} \prod_{i=1}^{N} \chi(z_i)\mathrm{e}^{i\sqrt{q}\varphi(z_i)}\, \mathrm{e}^{-i\sqrt{q}\int d^2z'\,\bar{\rho}\varphi(z')} \right\rangle. \tag{3.8}$$

The square roots produced by the spin fields σ are cancelled by those produced by the "half fluxes" $\exp(\frac{i}{2\sqrt{q}}\varphi(v_1))$, so the electron wavefunctions are single-valued. The factor $(v_1-v_2)^{-\frac{1}{8}}$ in (3.5) is now explained; it was inserted to make the equality with the correlator hold, and reflects the fact that the conformal weight of the spin field is $1/16$ (and hence $\eta = 1/4$ in the Ising model).

The generalization to $2n$ of our "half flux" quasiholes now seems self-evident: we should insert $2n$ of the combinations $\sigma\exp(\frac{i}{2\sqrt{q}}\varphi)$ into the correlator. However, it is known [7] that the Ising model part of such an expression is ambiguous; for fixed positions $v_1, \ldots v_{2n}$ of the spin fields there are many different possible correlators (conformal blocks), forming a vector space of dimension 2^{n-1} (over the complex numbers) independent of how many fermions χ are present. To resolve the ambiguity of notation, the spin fields should be replaced by "chiral vertex operators" [8]. Furthermore, the monodromy of these functions is nonabelian, *i.e.* analytically continuing v_i around v_j produces a linear combination of the original branches of the functions, the coefficients being some braiding matrices that don't commute. At present we do not know the explicit functions for $2n$ spin fields and N fermions in general, except in the case $N = 0$. The case $N = 0$ may seem strange from the point of view of electron wavefunctions, but it does at least provide an explicit realization of nonabelian statistics, which is just that given in §I, where the wavefunctions for the "σ

particles" are just $N = 0$ Ising conformal blocks. In fact, the braiding properties of the spin fields are independent of N. Note that, when discussing the N fermion functions as electron wavefunctions, we must show that they are 2^{n-1} linearly independent functions *of the electron co-ordinates* but this is easily done using the operator product expansion of the spin fields as the co-ordinates $v_1, \ldots v_{2n}$ approach each other in pairs $v_1 \to v_2$, $v_3 \to v_4$, etc. [27], even without full knowledge of these functions. The same operator product expansion also shows that as any v_i approaches any v_j, the leading term is equivalent to a Laughlin quasihole of charge $1/q$, which is clearly a desirable property. This completes our construction, though it remains to check that adiabatic transport of our quasiholes does give the same nonabelian statistics as the monodromy of the conformal blocks. We are confident that this is true because of the existence of 2^{n-1} functions and because analogous properties hold in the abelian examples of §II.

Before closing, some comments on recent related work. Wen [28] has argued that excitations of wavefunctions arising in one of Jain's later constructions [24] possess nonabelion excitations, using the point of view of ref [3] of wavefunctions as conformal blocks. These examples involve SU(N) symmetry. Greiter, Wen and Wilczek [29] have studied numerically the vicinity of $\nu = 1/2$ for certain Hamiltonians in spherical geometry, and find evidence for an incompressible state with the excitations (neutral fermions and charge 1/4 quasiholes) of the Pfaffian state. However, they claim that the quasiholes obey abelian $\theta/\pi = 1/8$ statistics. Their arguments are based on an "adiabatic heuristic" principle that connects the FQHE state to a paired state of fermions in zero magnetic field given by the Pfaffian alone. While this picture is extremely similar to the order parameter or condensation picture put forward in previous papers [12,13,14,3] and herein, we do not agree with the conclusion that the statistics is abelian. The full structure of the Landau-Ginzburg-Chern-Simons theory of the state must account for the fermion excitations, and not only the paired order parameter. Their conclusion is what one would obtain by neglecting the Pfaffian completely when calculating the statistics of the quasiholes by adiabatic transport *à la* [15]. In fact, while a direct search for nonabelian statistics must use wavefunctions for 4 quasiholes (which are not given by Greiter *et al*), there is already a difference in the apparent abelian statistics for two quasiholes, as shown by our eq. (3.5), which implies $\theta/\pi = 1/8 - 1/8 = 0$; this effect would be due to the Pfaffian. These authors also point out that the Pfaffian state is the unique highest density zero energy eigenstate of a certain *three-body* Hamiltonian.

Clearly, once the possibility of nonabelions is accepted, many questions remain. How would we confirm their statistics in an experimental setting? Is a gas of nonabelions a superfluid, as for anyons [30]?

Acknowledgements: The work of NR was partially supported by an A.P.Sloan Foundation Fellowship and by NSF PYI DMR-9157484. The work of GM was supported by DOE grant DE-AC02-76ER03075 and by a PYI.

Bibliography

[1] See, *e.g.*, R.E. Prange and S.M. Girvin, eds., *The Quantum Hall Effect*, 2nd Edition, Springer (New York), 1990, for a review.

[2] See *e.g.* P. Ginsparg, "Applied Conformal Field Theory", in Les Houches lecture volume, 1988 (North Holland) for a review.

[3] G. Moore and N. Read, Nucl. Phys. **B360**, 362 (1991).

[4] For a simple discussion, see, *e.g.*, Y.-S. Wu, Phys. Rev. Lett. **52**, 2013 (1984).

[5] J. Frohlich, F. Gabbiani and P.-A. Marchetti, Banff lectures, 1989.

[6] E. Witten, Commun. Math. Phys. **121**, 351 (1989).

[7] A. Belavin, A. Polyakov and A. Zamolodchikov, Nucl. Phys. **B241**, 33 (1984).

[8] G. Moore and N. Seiberg, Commun. Math. Phys. **123**, 177 (1989).

[9] R.B. Laughlin, Phys. Rev. Lett. **50**, 1395 (1983).

[10] S. Coleman, Phys. Rev. **D11**, 2088 (1975).

[11] G.W. Semenoff, Phys. Rev. Lett. **63**, 1026 (1989).

[12] N. Read, Phys. Rev. Lett. **62**, 86 (1989).

[13] S.M. Girvin and A.H. MacDonald, Phys. Rev. Lett. **58**, 1252 (1987).

[14] S.-C. Zhang, T.H. Hanssen and S. Kivelson, Phys. Rev. Lett. **62**, 82 (1989).

[15] D.J. Arovas, J.R. Schrieffer and F. Wilczek, Phys. Rev. Lett. **53**, 722 (1984).

[16] S.M. Girvin, in Ref [1].

[17] D.-H. Lee and M.P.A. Fisher, Phys. Rev. Lett. **63**, 903 (1989).

[18] S.-C. Zhang, IBM Almaden preprint.

[19] N. Read, Phys. Rev. Lett. **65**, 1502 (1990).

[20] B. Blok and X.-G. Wen, Phys. Rev. B**42**, 8133, 8145 (1990); *ibid* **43**, 8337 (1991).

[21] F.D.M. Haldane, Phys. Rev. Lett. **51**, 605 (1983).

[22] B. Halperin, Phys. Rev. Lett. **52**, 1583 (1984).

[23] R.B. Laughlin, Surf. Sci. **141**, 11 (1984); S.M. Girvin, Phys. Rev. B**29**, 6012 (1984); A.H. MacDonald and D.B. Murray, *ibid* **32**, 2707 (1985); A.H. MacDonald *et al*, *ibid* **31**, 5529 (1985).

[24] J.K. Jain, Phys. Rev. Lett. **63**, 199 (1989); Phys. Rev. B**40**, 8079 (1989); *ibid* **41**, 7653 (1990).

[25] F.D.M. Haldane and E.H. Rezayi, Phys. Rev. Lett. **60**, 956 (1988); **60**, 1886 (E) (1988).

[26] S.M. Girvin, A.H. MacDonald and P. Platzman, Phys. Rev. B**33**, 2481 (1986).

[27] N. Read, unpublished.

[28] X.-G. Wen, Phys. Rev. Lett. **66**, 802 (1991); B. Blok and X.-G. Wen, IAS preprint (1991).

[29] M. Greiter, X.-G. Wen and F. Wilczek, Phys. Rev. Lett. **66**, 3205 (1991); IAS preprint (1991).

[30] R.B. Laughlin, Phys. Rev. Lett. **60**, 2677 (1988).

A Note on Braid Statistics and the Non-Abelian Aharonov-Bohm Effect

Erik Verlinde

School of Natural Sciences,
Institute for Advanced Study,
Princeton, NJ 08540

Abstract: We give a first quantized formulation of particles obeying braid statistics and study the non-abelian analog of the Aharonov-Bohm effect. We find that the S-matrix for two-body scattering is essentially given by the braid matrix.

Exotic statistics in two spatial dimensions has been actively studied in recent years from many different points of view. Particles with exotic abelian statistics, usually called anyons, have been considered for quite some time [1], and, as has been discussed at this conference by F. Wilczek, may play an important role in explaining high-T_C superconductivity [2]. The general theoretical frame work for exotic statistics, as described in J. Fröhlich's talk, also allows for particles whose statistics is characterized by a non-abelian representation of the braid group [3,4]. A possible mechanism by which particles can acquire non-abelian braid statistics is when they are coupled to a non-abelian gauge field whose dynamics is described by a Chern-Simons term [5,6].

In this note we will present a Hamiltonian formulation of first quantized point-like particles obeying braid statistics, but without other interactions. In our formulation we make use of the relation between braid statistics and the monodromy properties of 2-dimensional conformal field theory. A crucial ingredient will be the so-called Knizhnik-Zamolodchikov connection [7], which in our model governs the monodromy behavior of the wave-functions. Our description can in principle be obtained by quantizing point-like particles coupled to Witten's Chern-Simons theory [6]. However, we will not discuss this quantization procedure here, but instead we start directly from the resulting Hamiltonian formulation. One of our aims is to study the scattering problem for particles obeying braid statistics. This scattering problem is a non-abelian analog of the Aharonov-Bohm effect and has many similarities with the scattering problem in 2 + 1-dimensional gravity [8]. It may also have applications to cosmic string/black hole physics in 3 + 1 dimensions, see e.g. [9].

In two spatial dimensions the quantum states for a system of n particles can be represented by wave-functions $\Psi(x_1, \ldots, x_n)$ defined on the configuration space $\mathcal{C}_n$, which for distinguishable particles is given by

$$\mathcal{C}_n = \Big\{ (x_1, \ldots, x_n) \in (R^2)^n \ ; \ x_i \neq x_j \text{ for } i \neq j \ \Big\}. \tag{1}$$

What makes two spatial dimensions special is the fact that the configuration space $\mathcal{C}_n$ is non-simply connected. This has important consequences for the n-particle Hilbert space: it allows the particles to obey braid statistics. The first homotopy group $\pi_1(\mathcal{C}_n)$ of the configuration space is given by what is known as the pure braid group $\mathcal{B}_n$. An (over-complete) set of generators of $\mathcal{B}_n$ is provided by the operations $\{M_{ij}, i \neq j\}$ which take the ith particle and move it over a closed path clockwise around the jth particle.

$$M_{ij} : \quad i \ \bullet \circlearrowleft \ j \ \bullet \tag{2}$$

In a first quantized formulation particles obeying braid statistics can be characterized by the fact that the monodromies M_{ij} have a non-trivial effect on the Hilbert states. This can be described by representing the Hilbert states as multi-valued wave-functions transforming in some (possibly non-abelian) representation of the braid group $\mathcal{B}_n$. In this note we will first present an alternative description involving a non-abelian connection on the configuration space $\mathcal{C}_n$.

We consider a system of n particles each of which carries a 'statistical charge' corresponding to a representation $\mathcal{R}_{l_i}$, $i = 1, \ldots, n$, of a non-abelian group G which for definiteness we take to be $G = SU(2)$. The wave-functions take their values in the tensor product of these representations

$$\Psi \in \mathcal{R}_{l_1} \otimes \mathcal{R}_{l_2} \otimes \ldots \otimes \mathcal{R}_{l_n}. \tag{3}$$

The $SU(2)$-generators in the representation R_{l_i} will be denoted by T_i^a. Now, in order to give the wave-function non-trivial monodromy properties, we couple the particles to a non-abelian connection on $\mathcal{C}_n$. For this we will take the so-called Knizhnik-Zamolodchikov connection [7], which first appeared in the context of conformal field theory. As we will see it is *not* simply a $SU(2)$-gauge field: it depends explicitly on the positions of the other particles and also acts non-trivially in the other representation spaces $\mathcal{R}_{l_j}$ with $j \neq i$. In complex coordinates $x_i = (z_i, \bar{z}_i)$ the connection $\nabla_i = (\nabla_{z_i}, \nabla_{\bar{z}_i})$ has the form

$$\begin{aligned} \nabla_{z_i} &= \frac{\partial}{\partial z_i} - \frac{1}{k+2} \sum_{j \neq i} \frac{T_i^a T_j^a}{z_i - z_j}, \\ \nabla_{\bar{z}_i} &= \frac{\partial}{\partial \bar{z}_i}, \end{aligned} \tag{4}$$

where the constant k is one of the parameters characterizing the theory.

We can now use the covariant derivatives ∇_i to build the observables of our theory. In this note we will consider non-relativistic spinless particles, which except for obeying braid statistics, have no interactions. For a system of n such particles the Hamiltonian is simply given by

$$H = -\sum_{i=1}^{n} \frac{1}{2m_i} \nabla_i^2. \tag{5}$$

At this point the reader may worry that this Hamiltonian H is not hermitian because $\nabla_i^\dagger \neq -\nabla_i$. However, we can take an inner product of the form

$$\langle \Psi_1 | \Psi_2 \rangle = \int \Psi_1^\dagger \cdot G \cdot \Psi_2, \tag{6}$$

where $G(x_1, \ldots, x_n)$ is determined by the condition $\nabla_i^\dagger G + G \nabla_i = 0$. Then with respect to (6) our Hamiltonian H is self-adjoint.

As mentioned before, it is possible to derive the Hamiltonian (5) and the form of the inner product (6) by quantizing non-relativistic particles coupled to $SU(2)$-Chern-Simons theory for example following [6,10]. The constant k in (4) is directly related to the Chern-Simons coupling constant and for consistency needs to be integer. The function G appearing in the inner product turns out to be given by a correlation function of a particular two-dimensional field theory, the $SL(2,\mathbb{C})/SU(2)$-WZW model.

The connection (4) has the important property that its curvature vanishes on $\mathcal{C}_n$

$$F_i = \left[\nabla_{z_i}, \nabla_{\bar{z}_i}\right] = 0 \quad \text{for } x_i \neq x_j, \tag{7}$$

and $[\nabla_i, \nabla_j] = 0$ for $i \neq j$. This condition implies that locally the connection A is trivial, but since $\mathcal{C}_n$ is non-simply connected it can have non-trivial holonomies around closed loops. Indeed when we make $\mathcal{C}_n$ simply connected by adding the points $x_i = x_j$ the curvature F has δ-function singularities at these points

$$(k+2)F_i = \sum_{j \neq i} T_i^a T_j^a \, \delta(x_i - x_j). \tag{8}$$

The intuitive interpretation of this equation is that each particle carries a non-abelian flux which is determined by its statistical charge, *i.e.*its representation R_l. This is precisely what causes the particles to acquire braid statistics. It is well known that when a charged particle is transported around a flux the wave-function picks up an, in this case, non-abelian Aharonov-Bohm phase [11]. To show that this is the only physical effect of the connection, we will now reformulate the theory described by (3)-(6) in such a way that the connection is eliminated from the Hamiltonian but instead shows up in the monodromy properties of the wave-functions.

Thanks to the fact that the curvature F_i vanishes on $\mathcal{C}_n$ there locally exist functions $\mathcal{F}_\alpha \in \mathcal{R}_{l_1} \otimes \ldots \otimes \mathcal{R}_{l_n}$ which are covariantly constant

$$\nabla_i \mathcal{F}_\alpha = 0. \tag{9}$$

Using the explicit form of the covariant derivatives equation (9) tells us that the $\mathcal{F}_\alpha$ are holomorphic and satisfy the Knizhnik-Zamolodchikov equation

$$(k+2)\frac{\partial}{\partial z_i}\mathcal{F}_\alpha(z_1, \ldots, z_n) = \sum_{j \neq i} \frac{T_i^a T_j^a}{z_i - z_j} \mathcal{F}_\alpha(z_1, \ldots, z_n). \tag{10}$$

We will refer to the solutions $\mathcal{F}_\alpha$ to (10) as the conformal blocks, although strictly speaking the conformal blocks in conformal field theory only form a subset of all solutions.

An important property of the conformal blocks $\mathcal{F}_\alpha$ is that, due to the fact that the Knizhnik-Zamolodchikov connection has curvature singularities at the positions of the other particles, they are not single-valued under the monodromy operations M_{ij}. Instead under M_{ij} the conformal blocks transform into a linear combination of themselves

$$M_{ij}: \mathcal{F}_\alpha \longrightarrow \sum_\beta (M_{ij}^{-1})_{\alpha\beta} \mathcal{F}_\beta. \tag{11}$$

The next step is to expand the wave-function Ψ in terms of the conformal blocks

$$\Psi = \sum_\alpha \psi_\alpha \mathcal{F}_\alpha. \tag{12}$$

In this equation Ψ and $\mathcal{F}_\alpha$ carry the group indices of the representations $R_{j_1}, \ldots, R_{j_n}$ while new wave-functions ψ_α are group singlets. From (9) we deduce that the covariant derivatives act as

$$\nabla_i \Psi = \sum_\alpha \frac{\partial \psi_\alpha}{\partial x_i} \mathcal{F}_\alpha. \tag{13}$$

In this way we find that the Hamiltonian (5), when represented on the new wave-functions ψ_α, indeed becomes just the free Hamiltonian, since all the covariant derivatives ∇_i in H are replaced by ordinary derivatives. What makes the theory non-trivial is the fact that wave-functions ψ_α are multi-valued and behave under the monodromy operations M_{ij} as

$$M_{ij}: \psi_\alpha \longrightarrow \sum_\beta (M_{ij})_{\alpha\beta} \psi_\beta. \tag{14}$$

This follows from the fact that Ψ in (13) is single-valued, while the conformal blocks $\mathcal{F}_\alpha$ transforms as in (11). So we find that the wave-functions transform in a non-abelian representation of the braid group defined by the Knizhnik-Zamolodchikov equation.

For our further discussion we need to know a bit more about the monodromy matrices M_{ij}. First let us consider its possible eigenvalues. From the Knizhnik-Zamolodchikov equation (10) it is not difficult to see that the spectrum of M_{ij} has the form

$$\text{spectrum of } M_{ij} \subseteq \text{spectrum of } e^{\frac{2\pi i}{k+2} T_i^a T_j^a}. \tag{15}$$

Since the matrices M_{ij} in general do not commute it is not possible to diagonalize them simultaniously. The relations satisfied by the monodromies M_{ij} are most conveniently expressed by rewriting them as

$$M_{ij} = (B_{ij})^2 \tag{16}$$

in terms of half-monodromies or braid operations B_{ij} which interchange the particles i and j. Graphically one can represent B_{ij} as

$$B_{ij} : \quad \overset{j \quad i}{\times} \tag{17}$$

These braid operations B_{ij} generate the full braid group $\mathcal{B}_n^{\text{full}} = \pi_1(\mathcal{C}_n/S_n)$, where S_n is the group of permutations of the n positions x_i. They satisfy the well-known 'Yang-Baxter equation'

$$B_{ij}B_{ik}B_{jk} = B_{jk}B_{ik}B_{ij}, \tag{18}$$

$$B_{ij}B_{kl} = B_{kl}B_{ij} \tag{19}$$

where i, j, k and l are all distinct. The behavior of the wave-function when two particles are interchanged is represented by matrices $(B_{ij})_{\alpha\beta}$ satisfying (18). The strings in the graphical representation (17) of B_{ij} can be interpreted as Wilson lines (Mandelstam strings) or fluxtubes attached to the particles. The intuitive explanation of why the particles obey braid statistics is that the wave-functions ψ not only depend on the positions x_i, but also on the topological configuration of these strings.

The fact that these particles obey braid statistics undoubtedly has many physical consequences. Actually, the term 'statistics' in this situation is a bit misleading, because the wave-function ψ_α can satisfy (14) even when the particles are distinguishable. The effect of (14) is more that of a long-range 'topological' interaction, which unlike conventional Bose- or Fermi-statistics leads to non-trivial scattering.

Let us now turn our attention to the scattering of particles obeying braid statistics. We will mainly focus on the problem of two-particle scattering. We like to show that the S-matrix describing the scattering of the i-th and j-th particle can be expressed in terms of the braid matrix as

$$\mathbf{S}_{ij} = B_{ij}^{\epsilon(L_{ij})}, \tag{20}$$

where $\epsilon(L_{ij})$ denotes the sign of the helicity operator associated with the two particles

$$L_{ij} = \epsilon_{\mu\nu\lambda} p_i^\mu p_j^\nu (x_i - x_j)^\lambda. \tag{21}$$

Note that $\mathbf{S}_{ij}$ is relativistic invariant. Before discussing the derivation of (20) let us first make some comments on this result. The intuitive interpretation of the expression for the S-matrix $\mathbf{S}_{ij}$ is quite simple. Classically the sign $\epsilon(L_{ij})$ equals $+1$ or -1 depending on whether the two particles pass each other on the left or on the right. The form of the S-matrix (20) then just tells us that in the first case the behavior of the wave-function is described by the braid matrix, while in the second case we need its inverse, precisely as expected. It is interesting to note that the braid matrix satisfies the relation (18), so at first one may hope that also the two-particle S-matrices satisfy the factorization condition

$$\mathbf{S}_{ij}\mathbf{S}_{ik}\mathbf{S}_{jk} \stackrel{?}{=} \mathbf{S}_{jk}\mathbf{S}_{ik}\mathbf{S}_{ij}. \tag{22}$$

If such a relation would be true, it would suggest that the n-particle S-matrix can be factorized into the product of two-body scattering matrices. On second thought however it seems unlikely that equation (22) is correct: first of all, the braid relation (18) does not hold when one of the braid matrices is replaced by its inverse, and secondly, the operators L_{ij}, L_{jk} and L_{ik} do not mutually commute. Could it be that these two objections miraculously cancel each other?

As a last related remark, we note that there is a corresponding scattering problem in one spatial dimension, which roughly speaking is obtained by restricting the particles to a circle with large radius. One can write down a S-matrix for this problem of the same form as (20) but where now $L_{ij} = \epsilon_{\mu\nu} p_i^\mu p_j^\nu$. In this case (22) is satisfied, and so this dimensionally reduced model appears to have a factorizable S-matrix.

Let us now explain how one arrives at the expression (20). The derivation of (20) is based on the simple observation that for two particles we can diagonalize the braid matrix, and thus the calculation reduces to the abelian case, which has been studied more than 30 years ago by Aharonov and Bohm [12]. In the following we will drop the indices i and j and if needed label the two particles with $i = 1$ and $j = 2$. We may consider these two particles in their centre of mass frame and study the reduced wave-function $\psi_\alpha(r, \theta)$ as a function of the relative distance r

and an angular variable θ. In this frame $\epsilon(L)$ just gives us the sign of the angular momentum $i\partial_\theta$. The wave-function satisfies the Schrödinger equation

$$\left[(r\partial_r)^2 + \partial_\theta^2 + (pr)^2\right]\psi_\alpha(r,\theta) = 0 \tag{23}$$

with boundary conditions

$$\psi_\alpha(r,\theta + 2\pi) = M_{\alpha\beta}\psi_\beta(r,\theta), \tag{24}$$

where $M = B^2$ is the monodromy matrix. Next we decompose ψ in eigenfunctions of M, say with eigenvalue $e^{2\pi i\Delta}$. From (15) we deduce that these eigenvalues are of the form

$$e^{2\pi i\Delta} = e^{2\pi i(\Delta_{l_0} - \Delta_{l_1} - \Delta_{l_2})}, \tag{25}$$

where $(k+2)\Delta_l = l(l+1)$ is the value of the Casimir $T_l^a T_l^a$ and $|l_1 - l_2| \leq l_0 \leq l_1 + l_2$. From (24) it then follows that the usual integral spectrum of the angular momentum $i\partial_\theta$ is shifted by an amount Δ. The solutions to the Schrödinger equation (23) are expressed in generalized Bessel functions $J_\lambda(pr)$, which are regular at $r = 0$ when $\lambda \geq 0$. In this way one finds that the wave-function is of the general form

$$\psi(r,\theta) = \sum_{n,\Delta} c_{n,\Delta} J_{|n+\Delta|}(pr) e^{i(n+\Delta)\theta}. \tag{26}$$

Then, from the asymptotic formula

$$i^\lambda J_\lambda(y) \underset{y\to\infty}{\sim} \frac{e^{-iy}}{\sqrt{y}} + e^{i\pi\lambda}\frac{e^{iy}}{\sqrt{y}} \tag{27}$$

one reads off that the wave-function asymptotically behaves as

$$\psi(r,\theta) \underset{r\to\infty}{\sim} \psi^{\rm in}(\theta)\frac{e^{-ipr}}{\sqrt{pr}} + (\mathbf{S}\psi^{\rm in})(\theta)\frac{e^{ipr}}{\sqrt{pr}}, \tag{28}$$

where $\psi^{in}(\theta)$ and $(\mathbf{S}\psi^{\rm in})(\theta)$ represent the in- and outcoming wave functions, and $\mathbf{S}$ is the S-matrix. In the angular momentum eigenbasis we find that $\mathbf{S}$ is simply given by a phase $e^{i\pi|n+\Delta|}$, and thus in this basis coincides with the braid matrix B or its inverse depending on the sign of the angular momentum. This last statement can be cast in a form which is independent of the basis or reference frame, and then leads to the S-matrix given in (20).

In order to compute the cross section we represent our incoming state as $\psi^{in}_{\alpha}(\theta) = \epsilon_{\alpha}\delta(\theta)$ describing a plane wave coming from the direction $\theta = 0$. The differential cross section for scattering angle θ is given by

$$\frac{d\sigma}{d\theta} = \frac{1}{2\pi p}|(S\psi^{in})(\theta)|^2 = \frac{1}{2\pi p}\left|\frac{(B - B^{-1})\epsilon}{1 - e^{i\theta}}\right|^2, \tag{29}$$

which can be written as

$$\frac{d\sigma}{d\theta} = \frac{1}{2\pi p}\frac{1}{\cos^2\theta/2}\Big(1 - \mathrm{Re}\,\langle\psi^{in}|M|\psi^{in}\rangle\Big). \tag{30}$$

Remains to compute the matrix element of the monodromy matrix M. From (26) we know that it must be of the general form

$$\langle\psi^{in}|M|\psi^{in}\rangle = \sum |c_{\Delta}|^2 e^{2\pi i\Delta}, \tag{31}$$

where the possible values for Δ are given in (25). Note that when we choose ψ^{in} to be in an eigenstate of M that (30) coincides with the well-known result of Aharonov and Bohm, as it should. In general, however, the incoming state will be a mixture of M-eigenstates. In this situation we want the wave function to describe two particles whose flux tubes or Mandelstam strings separately go off to infinity in the direction from where each particle is coming. For this particular case one can actually determine the 'fusion' coefficients $|c_{\Delta}|^2$ and compute the cross section explicitly using some identities from conformal field theory, or the general theory of superselection sectors. Since an exposition of either of these theories goes beyond the scope of this note, we can only give a heuristic sketch of this computation.

An important aspect of our theory is that there is no distinction between charge or flux. This means in particular, that the scattering problem for two charged particles can alternatively be thought of as describing a charge scattering off a flux, or even a flux scattering off a flux (see eg.[9]). A non-abelian flux is specified by the holonomy of the connection, and may upto conjugation be represented by a group element $e^{i\Phi T_3} \in SU(2)$ with $\Phi \in [0, 2\pi]$. The matrix element $\langle\psi^{in}|M|\psi^{in}\rangle$ precisely describes the effect of taking say the first charge l_1 around the flux Φ_{l_2} associated with the second particle. One may expect therefore that the result can be expressed as a $SU(2)$-character $\chi_l(\Phi) = \mathrm{tr}_{\mathcal{R}_l}(e^{i\Phi T_3})$ which equals

$$\chi_l(\Phi) = \frac{\sin(2l+1)\Phi/2}{\sin\Phi/2}. \tag{32}$$

Note however that $\langle\psi^{\rm in}|M|\psi^{\rm in}\rangle$ has to be symmetric in l_1 and l_2. Graphically it may be represented as

$$\langle\psi^{\rm in}|M|\psi^{\rm in}\rangle_{l_1,l_2} = \tag{33}$$

ℓ_1 ℓ_2

This matrix element, which coincides with the (normalized) expectation value of two linked Wilson lines in Chern-Simons theory, can be computed by various means; for example, using arguments involving surgery of three-manifolds [6], using Moore and Seiberg's polynomial equations [13], or using the general theory of superselection sectors in (2+1)d QFT [4,14]. The result is the same in all cases and reads

$$\langle\psi^{\rm in}|M|\psi^{\rm in}\rangle_{l_1,l_2} = \frac{\chi_{l_1}(\Phi_{l_2})}{\chi_{l_1}(\Phi_0)} = \frac{\chi_{l_2}(\Phi_{l_1})}{\chi_{l_2}(\Phi_0)} \tag{34}$$

where Φ_l denotes the flux associated with a particle with charge l, which turns out to be given by

$$\Phi_l = 2\pi\left(\frac{2l+1}{k+2}\right) \tag{35}$$

This relation between charge and flux can be understood from requirement that the expression (34) is symmetric in l_1 and l_2 and by looking at the δ-function contributions in the curvature (8).

In calculating the matrix element $\langle\psi^{\rm in}|M|\psi^{\rm in}\rangle_{l_1,l_2}$ using the polynomial equations one finds an intermediate expression of the expected form (31), namely

$$\langle\psi^{\rm in}|M|\psi^{\rm in}\rangle_{l_1,l_2} = \sum_{l_0} \frac{\chi_{l_0}(\Phi_0)}{\chi_{l_1}(\Phi_0)\chi_{l_2}(\Phi_0)} e^{2\pi i(\Delta_{l_0}-\Delta_{l_1}-\Delta_{l_2})} \tag{36}$$

where l_0 runs over the usual range with the extra condition $l_0 \leq k-l_1-l_2$. We leave it to the reader to perform the sum over l_0 and verify that this gives the result (34).

Finally, notice that even a charge-less particle carries a non-zero flux Φ_0, but of course must have a vanishing cross section. Indeed, as a final answer for the cross section for the scattering of two particles with statistical charges l_1 and l_2 we find

$$\left(\frac{d\sigma}{d\theta}\right)_{l_1,l_2} = \frac{1}{2\pi p}\frac{1}{\cos^2\theta/2}\left(1 - \frac{\chi_{l_1}(\Phi_{l_2})}{\chi_{l_1}(\Phi_0)}\right). \tag{37}$$

An interesting limiting case of this formula is when we take $k \to \infty$ and $l_2 \to \infty$ while letting the flux Φ_{l_2} approach some finite value Φ and keeping $l_1 = l$ fixed. In this limit the relevant factor in the cross section becomes just $(1 - \chi_j(\Phi)/d_j)$, which is what one expects for the scattering of a charge j off a flux Φ [9].

References

[1] J.M. Leinaas and J. Myrheim, Nuovo Cimento **37B** (1977) 1; G. A. Goldin, R Menikoff, and D. H. Sharp, J. Math. Phys. **22** (1981) 1664; F. Wilczek, Phys. Rev. Lett. **49** (1982) 957; Y.S. Wu and A. Zee, Nucl. Phys. **B 251** (1985) 117.

[2] For a review, see F. Wilczek, 'Lectures on Fractional Statistics and Anyon Superconductivity', IAS-preprint IASSNS-HEP-89/59.

[3] J. Fröhlich, "Statistics of Fields, the Yang-Baxter Equation and the Theory of Knots and Links", in "Non-perturbative quantum field theory", G. 't Hooft at al. (eds.), New York: Plenum 1988.

[4] K.H. Rehren and B. Schroer, Nucl. Phys. **B 312** (1989) 715; K. Fredenhagen, K.H. Rehren, and B. Schroer, Comm. Math. Phys **125** (1989) 201.

[5] J. Schonfeld, Nucl. Phys. **B185** (1981) 157; R. Jackiw and S. Templeton, Phys. Rev. **D23** (1981) 2291; S. Deser, R. Jackiw and S. Templeton, Phys. Rev. Lett. **48** (1982) 975.

[6] E. Witten, Comm. Math. Phys. **121** (1989) 351.

[7] V.G. Knizhnik and A.B. Zamolodchikov, Nucl. Phys. **B 247** (1984) 83; T. Kohno, Ann. Inst. Fourier (Grenoble) **37** (1987) 139.

[8] G. 't Hooft, Comm. Math. Phys. **117** (1988) 685; S. Deser and R. Jackiw, Comm. Math. Phys. **118** (1988) 495.

[9] M. Alford, J. March-Russel, F. Wilczek, Nucl. Phys. **B337** (1990) 695; F. Wilczek and Y.S. Wu, Phys. Rev. Lett. **65** (1990) 13.

[10] S. Elitzur, G. Moore, A. Schwimmer, N. Seiberg, Nucl. Phys. **B326** (1989), 104; S. Axelrod, S. Dellapietra, and E. Witten, 'Geometric Quantization of Chern-Simons Theory" IAS-preprint IASSNS-HEP-89/57; K. Gawedski and A. Kupiainen, '$SU(2)$ Chern-Simons theory at genus zero' preprint IHES/P/90/18.

[11] T.T. Wu and C.N. Yang, Phys. Rev. **D12** (1975) 3845.

[12] Y. Aharonov and D. Bohm, Phys. Rev. **119** (1959) 485.

[13] A. Tsuchiya and Y. Kanie. Adv. Stud. Pure Math. 16 (1988) 297; G. Moore and N. Seiberg, Phys. Lett. **B 212** (1988) 451, Comm. Math. Phys. **123** (1989) 177; L. Alvarez-Gaumé, C. Gomez and G. Sierra, Nucl. Phys. **B 319** (1989) 1153.

[14] J. Fröhlich, F. Gabbiani and P.-A. Marchetti: 'Superselection structure and statistics in three-dimensional local quantum field theory', Proc. 12th John Hopkins Workshop, G. Lusanna (ed.), 'Braid statistics in three dimensional quantum theory', Proc. Banff Summer School in Thoer. Phys. 'Physics, Geometry and Topology', August 1989.

Singlet quantum Hall effect and Chern-Simons theories

Alexander Balatsky
Department of Physics, University of Illinois at Urbana-Champaign, 1110 W. Green St., Urbana, Illinois 61801 and Landau Institute for Theoretical Physics, Moscow, U.S.S.R.

Eduardo Fradkin
Department of Physics, University of Illinois at Urbana-Champaign, 1110 W. Green St., Urbana, Illinois 61801
(Received 31 August 1990)

In this paper, we present a theory of the singlet quantum Hall effect (SQHE). We show that the Halperin-Haldane SQHE wave function can be written in the form of a product of a wave function for charged semions in a magnetic field and a wave function for the chiral spin liquid of neutral spin-$\frac{1}{2}$ semions. We introduce a field-theoretic model in which the electron operators are factorized in terms of charged spinless semions (holons) and neutral spin-$\frac{1}{2}$ semions (spinons). Holons and spinons are described in terms of a spinless charged fermion field coupled to U(1) (charge) Chern-Simons gauge field and a spin-$\frac{1}{2}$ neutral fermion coupled to SU(2) (spin) Chern-Simons gauge fields, respectively. Only the holons couple to the magnetic field. The physics that we find agrees with the results obtained using the Haldane-Halperin wave function: the spectrum of excitations has a gap and the quantum Hall conductance σ_{xy} equals $\nu/2\pi$, where ν is the filling fraction. The entire spectrum of physical states is shown to factorize into a charge and a spin contribution. Our picture makes the SU(2) spin symmetry manifest. The spin sector of the wave function is shown to behave like a conformal block of primary fields of the SU(2) Wess-Zumino-Witten model. The conformal dimensions of primary fields unambiguously dictates the semion statistics of the spinons. We find a generalization of the Fock cyclic condition for singlet semion wave functions.

I. INTRODUCTION

In this paper we consider the physical properties of the singlet quantum Hall effect (SQHE) states, given by the Halperin-Haldane wave function:[1,2]

$$\Psi_m([z_i],[\eta_\alpha]) = \prod_{i<j,\alpha<\beta} (z_i-z_j)^{m+1}(\eta_\alpha-\eta_\beta)^{m+1}(z_i-\eta_\alpha)^m \times\exp\left[-\tfrac{1}{4}\sum_i |z_i|^2-\tfrac{1}{4}\sum_\alpha |\eta_\alpha|^2\right] \quad (1.1)$$

where the set of coordinates z_i, $i=1,\ldots,N$ corresponds to the spin $\uparrow$ electrons, and η_α, $\alpha=1,\ldots,N$ corresponds to the spin $\downarrow$ electrons and m is an even integer. In this case $\Psi_m([z_i],[\eta_\alpha])$ satisfies the Fock cyclicity condition. In this state, the eigenvalue of the total spin operator is $S=0$ and the z component of the spin also has eigenvalue $S_z=0$. This kind of wave function naturally appears in consideration of the spin unpolarized states in the quantum-Hall-effect (QHE) phase. We will call this particular state the singlet QHE in order to stress its singlet nature.

In contrast to the spin polarized states, in this case we need to describe the charge sector of the SQHE phase as well as the spin sector. By inspecting the structure of this wave function one finds that it has the simple but very important property that the spin and charge degrees of freedom are factorized. The total wave function $\Psi_m([z_i],[\eta_\alpha])$ can be written as a product of the charge wave function $\Psi_m^{(1)}([z_i],[\eta_\alpha])$ and spin wave function $\Psi^{(2)}([z_i],[\eta_\alpha])$. Below we will discuss the properties of the charge and spin wave functions separately. At the end we will put them together again by imposing the constraint that the position of the charges coincides with the position of the spins. This property is strongly reminiscent of the charge and spin separation present in models of strongly correlated electron systems in the context of theories of high temperature superconductors.[3]

The wave function is factorized in the following manner:

$$\Psi_m([z_i],[\eta_\alpha])=\Psi^{(2)}([z_i],[\eta_\alpha])\Psi_m^{(1)}([z_i],[\eta_\alpha]) \quad (1.2)$$

with

$$\Psi_m^{(1)}([z_i],[\eta_\alpha])= \prod_{i<j,\alpha<\beta} (z_i-z_j)^{m+1/2}(\eta_\alpha-\eta_\beta)^{m+1/2}(z_i-\eta_\alpha)^{m+1/2}\exp\left[-\tfrac{1}{4}\sum_i |z_i|^2-\tfrac{1}{4}\sum_\alpha |\eta_\alpha|^2\right], \quad (1.3)$$

$$\Psi^{(2)}([z_i],[\eta_\alpha])= \prod_{i<j,\alpha<\beta} (z_i-z_j)^{1/2}(\eta_\alpha-\eta_\beta)^{1/2}(z_i-\eta_\alpha)^{-1/2}\,. \quad (1.4)$$

Why does this decomposition make sense? The plasma analogy, when applied to $\Psi_m^{(1)}([z_i],[\eta_\alpha])$, shows that this state is described by a one-component plasma, in which the particles at points z_i and η_α have equal charge:

$$|\Psi_m^{(1)}([z_i],[\eta_\alpha])|^2=\exp\left[(2m+1)\left[\sum_{i<j}\ln|z_i-z_j|+\sum_{\alpha<\beta}\ln|\eta_\alpha-\eta_\beta|+\sum_{i\alpha}\ln|z_i-\eta_\alpha|\right]-\tfrac{1}{2}\sum_i|z_i|^2-\tfrac{1}{2}\sum_\alpha|\eta_\alpha|^2\right]. \tag{1.5}$$

We regard $\Psi_m^{(1)}([z_i],[\eta_\alpha])$ as the wave function for the charge degrees of freedom.

If we apply the same plasma analogy to the wave function $\Psi^{(2)}([z_i],[\eta_\alpha])$ we get:[4]

$$|\Psi^{(2)}([z_i],[\eta_\alpha])|^2=\exp\left[\sum_{i<j}\ln|z_i-z_j|+\sum_{\alpha<\beta}\ln|\eta_\alpha-\eta_\beta|-\sum_{i\alpha}\ln|z_i-\eta_\alpha|\right] \tag{1.6}$$

and we can easily see that $\Psi^{(2)}([z_i],[\eta_\alpha])$ corresponds to a two-component plasma, where the effective charge of the particles q is given by the spin projection $q=2s_z=\pm1$. It is natural to consider $\Psi^{(2)}([z_i],[\eta_\alpha])$ as the wave function of the spin degrees of freedom.

We will show below that $\Psi_m^{(1)}([z_i],[\eta_\alpha])$ can be regarded as a wave function for semions in an external magnetic field. From Eq. (1.3) we conclude that, for any m, $\Psi_m^{(1)}([z_i],[\eta_\alpha])$ describes particles with semion statistics: any exchange of two of them leads to a change of phase of $\pi(m+\frac{1}{2})$ and, if m is even, this particles are semions. From the same considerations it follows that $\Psi^{(2)}([z_i],[\eta_\alpha])$ represents a two-component semion gas. The sign of the spin projection s_z determines the effective phase change in any interchange of two particles $q_1q_2\pi/2$, where q_1,q_2 are ±1 for spin $\uparrow,\downarrow$. This model with two-component semions was considered in Refs. 5 and 4. In particular, Girvin *et al.*[4] have pointed out that the state described by the wave function $\Psi^{(2)}([z_i],[\eta_\alpha])$ is a local spin singlet due to the plasma screening of any charge.

The decomposition of Eq. (1.2) can be represented in terms of the slave-*semion* operators:

$$\psi_\sigma(r)=\varphi(r)\xi_\sigma(r)\,, \tag{1.7}$$

where $\psi_\sigma(r)$ is the electron operator, $\varphi(r)$ is a charge e spinless semion operator, ξ_σ is a spin-$\frac{1}{2}$ charge-neutral semion operator, σ is a spin index, and we assume that $[\varphi(r),\xi(r)]=0$.

In principle this decomposition is neither better nor worse than any other slave boson or slave fermion factorization, like the ones that are commonly used in theories of strongly correlated systems. The choice of any particular decomposition of the initial electron operator is purely a matter of convenience. Our choice is motivated by the simplicity of the physical picture that we get in the end.

In Mott-Hubbard insulators, the strong correlations force the constraint of single particle occupancy. In the case of the SQHE, the origin of the strong correlations is the drastic reduction of phase space due to the presence of a strong magnetic field: the kinetic energy is quenched and the interactions dominate. In close analogy with the Mott-Hubbard problem we argue that in the singlet quantum Hall effect the spin and charge degrees of freedom are separated in the sense of the decomposition of Eq. (1.7). Here too, a gauge symmetry arises as a result of this factorization. This gauge symmetry means that the *relative* phase between charge and spin states is not a physically observable degree of freedom. The SQHE wave function is a singlet under this gauge symmetry. However, the decomposition equation (1.7) requires that the entire spectrum of states must be singlets under this gauge symmetry. Given the close analogy with the Mott-Hubbard problem, we will refer to this symmetry as the RVB gauge symmetry. The presence of this RVB gauge symmetry gives rise to an RVB gauge field which puts the charge and spin semions together to form the allowed physical states. Thus, although the wave functions of all the states can be factorized as a product of a *charge* and *spin* wave functions, there is no separation of spin and charge in this system. In consequence, the system has a gap to *all* excitations and it is *incompressible.* The factorized form of the SQHE wave function, Eq. (1.2), appears to suggest that there may be a gapless neutral spin excitation which would lead to *compressibility.* Because the RVB gauge charge is confined, these excitations are no part of the physical spectrum. It is important to stress that the incompressibility results entirely from the charge sector.

The plan of this paper is as follows. In Sec. II we consider the charge sector, given by $\Psi_m^{(1)}([z_i],[\eta_\alpha])$. We show that the wave function $\Psi_m^{(1)}([z_i],[\eta_\alpha])$ represents an incompressible fractional quantum Hall state of spinless charged semions. We show that this state, and its entire excitation spectrum, can be described in terms of a field theory of semions in an uniform magnetic field. The generating functional of the correlation functions of this theory is given in the form of a functional integral of fermions in an external uniform magnetic field, coupled to an Abelian Chern-Simons statistical gauge field with coupling constant $\theta=1/\pi$. We consider the spectrum of fluctuations and show that this state is incompressible since all of its excitations have a gap. We further calculate the Hall conductance and show that it has the Halperin-Haldane values. In the Sec. III we discuss the properties of the spin sector and $\Psi^{(2)}([z_i],[\eta_\alpha])$. We will show that effective low-energy spin dynamics is given by the SU(2) Chern-Simons (CS) theory coupled to the spinor field ξ_σ. The effective statistics of ξ_σ then becomes semionic, if we consider the spin-$\frac{1}{2}$ representation and use the coefficient in the CS term $k=1$. Using Witten's ideas about relation between the CS model and the Wess-Zumino-Witten (WZW) model,[6] we argue that $\Psi^{(2)}([z_i],[\eta_\alpha])$ is described in terms of the Wilson lines corresponding to the positions of the particles. This naturally leads to the description of $\Psi^{(2)}([z_i],[\eta_\alpha])$ in terms of the correlators of the SU(2) WZW model. This result was also found by one of us in the framework of the theory of chiral spin liquid.[7] Moore and Read[8] have recently considered the problem of the SQHE, following a similar line of thought, and derived the same result. In addition, we derive a generalization of the Fock cyclic condition which the many-body semion wave functions must satisfy in order to be spin singlet states (the details are given in Appendix B) and prove that $\Psi^{(2)}([z_i],[\eta_\alpha])$ satisfies this condition. We also give a brief discussion of a generalization of the theory to the case $k \geq 2$ which has the interesting feature of exhibiting non-Abelian statistics. In Sec. IV we put the charge and spin degrees of freedom together by projecting onto the space of physical states by introducing an additional gauge field. In this section we show the equivalence of the descriptions involving Chern-Simons gauge fields and the conventional problem of spin-$\frac{1}{2}$ interacting electrons in a magnetic field. Section V is devoted to the conclusions. In Appendix A we show that the charge wave function $\Psi_m^{(1)}([z_i],[\eta_\alpha])$ represents an incompressible state. In Appendix B we give the derivation of the generalized Fock condition for many-body wave functions for spin-$\frac{1}{2}$ semions.

After this work was submitted for publication we received a preprint of a paper by Moore and Read[9] in which they discuss the SQHE. These authors have also found the factorization of the Halperin-Haldane wave function which we discussed above. They stress the connection between the spin part of the wave function and the theory of conformal blocks. In particular, Moore and Read present an extensive discussion of the quasiparticle states with non-Abelian statistics.

II. THE QUANTUM HALL EFFECT FOR SEMIONS

In Sec. I we showed that the Halperin-Haldane wave functions Ψ_m represent incompressible spin-singlet states. We argued that these wave functions can be written as a product of two wave functions, $\Psi_m^{(1)}$ and $\Psi^{(2)}$, which represent the charge and spin degrees of freedom, respectively. This factorized structure of the wave function suggests that the systems which are well described by these states must exhibit separation between spin and charge degrees of freedom. In this section we discuss the properties of the *charge* sector. In Sec. III we discuss the spin sector.

The wave function $\Psi_m^{(1)}$ of Eq. (1.3), can be written in the form

$$\Psi_m^{(1)}(z_1,\ldots,z_{2N})=\prod_{i<j=1,\ldots,2N}(z_i-z_j)^{m+1/2}\exp\left[-\frac{1}{4}\sum_{i=1}^{2N}|z_i|^2\right], \tag{2.1}$$

where $z_1,\ldots,z_N$ are the coordinates of the up spins and $z_{N+1},\ldots,z_{2N}$ are the coordinates of the down spins and m is an *even* integer. We will use here the Roman notations for coordinates of both spin components since we are interested in the charge sector. There is however another possible interpretation of $\Psi_m^{(1)}$: a Laughlin-Halperin type wave function[10,1] for a fractional quantum Hall effect (FQHE) for $2N$ *semions*. This is the main claim of this section.

A simple way to see that $\Psi_m^{(1)}$ describes semions is to note that, if any pair of labels are exchanged, the wave function picks up a phase factor of $e^{\pm i\pi/2}$. Furthermore, the structure of this wave function suggests that each semion has an even number (m) of flux quanta attached to it. From this point of view, the wave function $\Psi_m^{(1)}$ can be regarded as a generalization of the standard Laughlin wave function for the case of semions. As was shown in Sec. I, a calculation which uses Laughlin's plasma analogy reveals that the state Ψ state is incompressible. As a matter of fact, only $\Psi_m^{(1)}$ is incompressible. $\Psi^{(2)}$ can be described as the correlator of *primary fields* in SU(2) WZW conformal field theory and, hence, like a representation of the group of local conformal transformations. This corresponds to the compressibility of the state given by $\Psi^{(2)}([z_i],[\eta_\alpha])$.

In order to prove our claim that the state $\Psi_m^{(1)}$ represents the Laughlin ground state of a system of $2N$ charged spinless semions in an uniform magnetic field, with filling fraction $1/\nu=\frac{1}{2}+m$ we will construct a field theory whose ground state is $\Psi_m^{(1)}$. We take the point of view that systems described by wave functions of the Laughlin type are generally described by field theories of matter coupled to Chern-Simons gauge fields. The matter fields describe the dynamics of particles. The Chern-Simons gauge fields represent fluxes which are attached to the particles and are responsible both for their fractional statistics and for the FQHE. The Chern-Simons description of the *phenomenology* of the FQHE was first introduced by Girvin and MacDonald.[11] Shortly afterward, Zhang, Hansson, and Kivelson[12] observed that the FQHE was equivalent to a set of bosons (with hard-cores) coupled to a Chern-Simons statistical gauge field. They further showed that Laughlin's mean field theory, alternatively, could be understood as the condensation of the bosons. However, the bosons are coupled to the fluctuations of the statistical gauge field and, hence, it is unclear in what sense are the bosons really condensed. In fact, Zhang *et al.* showed that the *phase* mode of the bosons is actually absorbed by the statistical gauge field. These authors finally argued that the Chern-Simons description amounts to an effective theory for the low energy degrees of freedom in the manner of Landau-Ginzburg. Recent studies of the anyon gas[13-17] have revealed that the Chern-Simons theory plays a central role in theories of particles with fractional statistics. Also recently, Jain[18] has reconsidered Laughlin's theory of the FQHE. In one of Jain's descriptions, the Laughlin state can be viewed as an integer quantum Hall effect of some effective fermions, each an electron bound to an even number of flux quanta. Lopez and Fradkin[19] (LF) have recently shown that the FQHE for fermions is *exactly* equivalent to a theory in which the fermions are coupled to the same external magnetic field and to a Chern-Simons gauge field which attaches an *even* number of flux quanta to each fermion. In this description, Laughlin's theory appears as the semiclassical (i.e., mean-field) limit of the Chern-Simons theory. It is important to stress the fact that the identification is exact, not just an identity in the low-energy limit. Blok and Wen[20] have also reached similar conclusions in terms of effective theories at low energies. In what follows we are going to follow the methods of LF.[19]

We are interested in studying the behavior of a set of $2N$ charged spinless semions moving in the presence of a uniform magnetic field B perpendicular to the plane in which they move. Studies of anyons[14,17] have shown that the simplest local description of an assembly of particles which exhibit fractional statistics consists of a theory of (say) *fermions* coupled to a dynamical Chern-Simons gauge field. The Chern-Simons coupling constant θ and the statistical phase δ are related by $\delta=1/2\theta$. The wave functions pick up a factor of $-e^{\pm i\delta}$ whenever two anyons are exchanged. Thus, a shift of the phase $\delta\to\delta+2\pi n$, with n any integer, has no observable consequences. In other words, the theory must be *periodic* in δ with period 2π. While this is true at the level of the Hilbert space description, it is far from obvious at the level of the path integral of the Chern-Simons theory. LF observe that, by going back to a description in which the number is fixed (i.e., "first quantization"), the periodicity is indeed recovered. Furthermore, a shift of δ by $2\pi n$ is equivalent to attaching $2n$ flux quanta to each particle. Hence, the periodicity can be used to construct a mean-field theory in the manner of Jain. LF make the choice $\theta=1/2\pi(m-1)$, which is appropriate for the standard FQHE. We will show below that, for the problem of *semions* in a uniform magnetic field, the convenient choice of θ now is

$$\frac{1}{\theta}=2\pi(\tfrac{1}{2}-m)\,, \tag{2.2}$$

where m is the *even* integer that appears in the wave function of Eq. (2.1).

We begin by constructing a path integral for semions in an external magnetic field. Let φ, A_ν, and $\mathcal{A}_\nu$ represent a Fermi field, the external electromagnetic field and the statistical (i.e., Chern-Simons) gauge field, respectively, with $\nu=0,1,2,\ldots$. The system is *assumed* to obey the dynamics determined by the Lagrangian density $\mathcal{L}_{\text{charge}}$

$$\mathcal{L}_{\text{charge}}=\varphi^*(x)(iD_0+\mu)\varphi(x)-\varphi^*(x)\hat{h}\varphi(x)$$
$$-\frac{\theta}{2}\epsilon_{\mu\nu\lambda}\mathcal{A}^{\mu}\partial^{\nu}\mathcal{A}^{\lambda} \tag{2.3}$$

where D_ν is the covariant derivative

$$D_\nu=\partial_\nu+ieA_\nu+i\mathcal{A}_\nu \tag{2.4}$$

and μ is the chemical potential. The last term in Eq. (2.3) is the Chern-Simons term. The (gauge invariant) one-particle Hamiltonian $\hat{h}$ represents the motion of the charged particles and it depends on the vector potentials only through the covariant derivatives D_μ. A simple choice of $\hat{h}$ is

$$\hat{h}=\frac{\mathbf{P}^2}{2m}\,, \tag{2.5}$$

where $\mathbf{P}$ is the gauge covariant momentum operator $\mathbf{P}=i\mathbf{D}$, in units in which $e=\hbar=c=1$. To simplify matters we will not include any other interactions, at least not explicitly. However, we will assume that the short-range Coulomb repulsive forces favor the local stability of the mean field state. This can be checked by direct calculation.[19]

The easiest way to see that the Lagrangian density $\mathcal{L}_{\text{charge}}$ of Eq. (2.3) represents semions in a magnetic field is to go back to a first quantized picture. The case of anyons is discussed at great length by one of us in Ref. 21. In this picture we have a set of $2N$ point particles coupled to both statistical and electromagnetic fields. The path-integral representation of the evolution operator of the system consists of a sum over all possible particle histories, i.e., trajectories in a $2+1$-dimensional Euclidean space-time. According to Feynman's prescription, each history is represented by an amplitude which is the exponential of (minus) the total Euclidean action $\mathcal{S}$. The action $\mathcal{S}$ can be split as the sum of three terms: a term $\mathcal{S}_p$ representing the motion of free particles through their kinetic energy, a term $\mathcal{S}_{\text{gauge}}$ which represents their coupling to the gauge fields and a term $\mathcal{S}_{\text{CS}}$ which is equal to the Chern-Simons action for the statistical gauge fields. $\mathcal{S}_{\text{gauge}}$ can always be written in the form of a line integral of the gauge fields along the trajectories Γ of all the particles in space-time:

$$\mathcal{S}_{\text{gauge}}=\int_\Gamma dx_\mu(eA^\mu+\mathcal{A}^\mu)\,. \tag{2.6}$$

In the imaginary-time formulation, $\mathcal{S}_p$ is real while the other two terms are pure imaginary. The results of Refs. 19 and 21 show that all processes in which any pair of particles are exchanged are represented by histories in which the *linking numbers* of the trajectories of the particles are increased or decreased by one unit per exchange and that the amplitudes of these processes differ by factors of the form $\exp[(i/2\theta)\Delta\nu_L]$, where $\Delta\nu_L$ is the change in linking number. This result follows after averaging over all possible configurations of the statistical gauge fields.[22,6,21] With the choice of θ of Eq. (2.2), this factor is equal to $\pm i$ per exchange. Thus the particles are semions. Since they are also coupled to the external electromagnetic fields, we have semions in a magnetic field.

(i) Mean-field-approximation: We already pointed out (see Appendix) that the state $\Psi_m^{(1)}([z_i],[\eta_\alpha])$ describes the incompressible one-component plasma. In order to prove that the state $\Psi_m^{(1)}([z_i],[\eta_\alpha])$ is an incompressible state of semions in a uniform magnetic field we calculate the effective action $\mathcal{S}_{\text{eff}}$ for all the gauge fields by integrating out the fermions first. It will be sufficient to calculate $\mathcal{S}_{\text{eff}}$ by means of a saddle-point expansion of the functional integral at the quadratic (Gaussian) level. This approximation is exactly equivalent to a random-phase approximation (RPA) and has proven to be very useful in theories of anyons without a field.[13,14,17] Thus, we consider the partition function

$$Z=\int \mathcal{D}\varphi^*\mathcal{D}\varphi\mathcal{D}\mathcal{A}\exp\left[i\int d^3x\,\mathcal{L}\right], \tag{2.7}$$

where $\mathcal{L}$ is the Lagrangian of Eq. (2.3) and θ is given by Eq. (2.2). Since the Lagrangian is a quadratic form in the Fermi fields, they can be integrated out explicitly to yield

$$Z=\int \mathcal{D}\mathcal{A}\,\mathrm{Det}[iD_0+\mu-\hat{h}(A_\mu,\mathcal{A}_\mu)]\times\exp\left[-i\int d^3x\frac{\theta}{2}\epsilon_{\mu\nu\lambda}\mathcal{A}^\mu\partial^\nu\mathcal{A}^\lambda\right]. \tag{2.8}$$

The mean-field approximation for this problem is simply the saddle-point approximation of this functional integral. The partition function Z is thus a functional integral over the statistical gauge fields with the effective action $S_{\rm eff}$ given by

$$S_{\rm eff}=-i\ln\mathrm{Det}[iD_0+\mu-\hat{h}]-\int d^3x\,i\frac{\theta}{2}\epsilon_{\mu\nu\lambda}\mathcal{A}^\mu\partial^\nu\mathcal{A}^\lambda . \tag{2.9}$$

The condition $\delta S_{\rm eff}=0$ that the effective action be stationary against local changes of the statistical gauge fields yields the equation ($\mu=0,1,2$)

$$\langle j_\mu\rangle=\frac{\theta}{2}\epsilon_{\mu\nu\lambda}\langle\mathcal{F}^{\nu\lambda}\rangle , \tag{2.10}$$

where $\langle j_\mu\rangle$ is the average charge and current of the fermions moving in the presence of both an external uniform magnetic field B and an average statistical field with field strength $\langle\mathcal{F}^{\mu\nu}\rangle$. The ground state will be chosen to be time-independent and as translation invariant as possible. These last conditions require that the only nonzero components of $\langle\mathcal{F}^{\mu\nu}\rangle$ to be spacelike with an average statistical flux $\langle\mathcal{B}\rangle$ of

$$\langle\mathcal{B}\rangle=\frac{\langle\rho\rangle}{\theta}, \tag{2.11}$$

where $\langle\rho\rangle$ is the average particle density. Thus, at the mean-field level, this problem happens to be equivalent to a set of $2N$ fermions (with $2N=\langle\rho\rangle L^2$) moving in the presence of an *effective* magnetic field $B_{\rm eff}$ equal to

$$B_{\rm eff}=B+\langle\mathcal{B}\rangle . \tag{2.12}$$

Thus, the total *effective* flux piercing the system is

$$2\pi N_\phi^{\rm eff}=B_{\rm eff}L^2=L^2[B+(2N)2\pi(\tfrac{1}{2}-m)] . \tag{2.13}$$

In terms of the filling fraction $\nu=2N/N_\phi$, we can write $N_\phi^{\rm eff}$ in the form

$$N_\phi^{\rm eff}=\frac{1}{2\pi}\left[\frac{2\pi(2N)}{\nu}+(2N)2\pi\left[\frac{1}{2}-m\right]\right]. \tag{2.14}$$

Since the filling fraction $1/\nu=\frac{1}{2}+m$ we find

$$N_\phi^{\rm eff}=2N . \tag{2.15}$$

In other words, the effective fermions fill up exactly their lowest Landau level created by the effective field

$$B_{\rm eff}=\frac{2\pi(2N)}{L^2}=2\pi\langle\rho\rangle .$$

This is precisely Jain's construction.[18]

(ii) Effective action, incompressibility, and Hall conductance: Now we want to derive an expression for the effective action for the Gaussian fluctuations of the statistical gauge field around the mean-field state. This is done in the usual manner, by expanding the action $S_{\rm eff}$ of Eq. (2.9) around the configuration $\langle\mathcal{A}_\mu\rangle$ of the mean-field equations (2.10). The effective action has the form

$$S_{\rm eff}^{(2)}=\int d^3x\int d^3y\tfrac{1}{2}\Pi_{\mu\nu}(x,y)\tilde{\mathcal{A}}^{\mu}(x)\tilde{\mathcal{A}}^{\nu}(y)-\frac{\theta}{4}\mathcal{S}_{\rm CS}(\tilde{\mathcal{A}}_\mu-A_\mu) , \tag{2.16}$$

where $\Pi_{\mu\nu}(x,y)$ is the polarization operator of this mean-field state (i.e., fermions in a magnetic field) and the last term is the Chern-Simons term for the fluctuations. Notice that we have shifted the integration variable $\tilde{\mathcal{A}}_\mu$ by an amount equal to the fluctuating component of the electromagnetic field A_μ which now appears only in the last term. We will use this last fact to determine the electromagnetic response of the system.

By gauge invariance $\Pi_{\mu\nu}$ has to be conserved. Since we are interested primarily in the low-energy (i.e., frequencies much smaller than the effective Landau gap $\hbar\omega_c^{\rm eff}$) and long-distance (i.e., length scales much longer than the effective cyclotron radius $l_0^{\rm eff}$) behavior of the theory it will be sufficient to expand $S_{\rm eff}^{(2)}$ in gradients and to keep the lowest terms. In this limit, gauge invariance forces $S_{\rm eff}^{(2)}$ to have the local form

$$S_{\rm eff}^{(2)}\approx\int d^3x\left[\frac{\epsilon}{2}\mathcal{E}^2(x)-\frac{\chi}{2}\mathcal{B}^2(x)+\frac{\sigma_{xy}^0}{4}\mathcal{L}_{\rm CS}(\tilde{\mathcal{A}}_\mu)-\frac{\theta}{4}\mathcal{L}_{\rm CS}(\tilde{\mathcal{A}}_\mu-A_\mu)\right]. \tag{2.17}$$

The first three terms of Eq. (2.17) result from the gradient expansion of the polarization operator. The first two terms come from the symmetric part of $\Pi_{\mu\nu}$ and are given in terms of $\mathcal{E}$ and $\mathcal{B}$, the "electric" and "magnetic" fields associated with the fluctuating component of the *statistical* gauge fields $\mathcal{A}_\mu$. The coefficients ϵ and χ can be determined from the mean-field theory but are heavily renormalized by the higher-order corrections to the mean field. Thus, we will keep them as effective parameters of the theory. The third term in Eq. (2.17) comes from the antisymmetric part of $\Pi_{\mu\nu}$ and represents the Hall response of fermions in a magnetic field. The coefficient σ_{xy}^0 is the quantum Hall conductance of the fermions. Since the ground state is such that only the lowest Landau level is filled we find that $\sigma_{xy}^0=1/2\pi$, in units in which $e^2/\hbar=1$. Unlike ϵ and χ, σ_{xy}^0 does not get renormalized away from its mean-field value by higher-order corrections.[17] Hence, mean-field theory gives the exact value for this coefficient. Also notice that σ_{xy}^0 is the Hall conductance of the effective fermions in the presence of the statistical gauge field and not the true Hall conductance of the electrons. The true quantum Hall conductance σ_{xy} has to be calculated as the response of the whole system to the presence of the electromagnetic field.

The incompressibility of the mean-field state follows from reading off the spectrum of collective modes. Since in the effective action of Eq. (2.17) we got a nonzero Chern-Simons term for the statistical gauge fields, we find that their fluctuations have a Chern-Simons "topological" mass,[23] or gap, $\Delta=\theta/\nu\chi$ and propagate with a speed $v=\sqrt{\chi/\epsilon}$. Notice that this is not equal to the, much large, Landau gap $\omega_c^{\rm eff}\sim B_{\rm eff}$. Since all states have a gap, the ground state is incompressible. This result agrees with the criterion derived by one of us in Ref. 17.

We can determine the value of σ_{xy} by integrating out the statistical gauge fields and finding the effective action for the electromagnetic fields. Using the results of Lopez and Fradkin,[19] up to higher order terms, we get

$$S_{\rm eff}(A_\mu)\approx\int d^3x\frac{\sigma_{xy}}{4}\mathcal{L}_{\rm CS}(A_\mu)+\cdots, \tag{2.18}$$

where σ_{xy} is given by[19]

$$\frac{1}{\sigma_{xy}}=\frac{1}{\sigma_{xy}^0}+\frac{1}{(-\theta)}. \tag{2.19}$$

Since $\sigma_{xy}^0=1/2\pi$ and $1/\theta=2\pi(\frac{1}{2}-m)$, we find

$$\frac{1}{\sigma_{xy}}=2\pi(m+\tfrac{1}{2})\equiv\frac{2\pi}{\nu}, \tag{2.20}$$

where ν is the filling fraction. This is of course the expected result $\sigma_{xy}=\nu/2\pi$ which, as was first argued by Laughlin,[24] is a consequence of electromagnetic gauge invariance.

III. THE SPIN SINGLET WAVE FUNCTION AND SU(2) CHERN-SIMONS THEORY

In this section we will discuss the properties of the spin wave function $\Psi^{(2)}([z_i],[\eta_\alpha])$. We will show that this wave function can be obtained from the SU(2) Chern-Simons (CS) gauge theory,[6] treating the spin of the particles as the sources for the gauge field.

First we will consider the physical reasons for why the spin wave function $\Psi^{(2)}([z_i],[\eta_\alpha])$ has to be described by the SU(2) CS theory. There are two different ways to incorporate the spin quantum number of the electron into the wave function. One is "Abelian." In this procedure the single-particle states are described in terms of the spin projection on the z axis, and for simplicity, this axis is assumed to be in the same direction everywhere. Thus, we deal only with the U(1) diagonal subgroup of the full SU(2) spin group. Also, the plasma analogy for $\Psi^{(2)}([z_i],[\eta_\alpha])$ leads to the correspondence now with the two-component plasma with effective charge $q=+q_0$ for spin $\uparrow$ and $q=-q_0$ for spin $\downarrow$ particles. This analogy suggests that we should attach different fluxes to particles with opposite spin and deal with them much in the same way as we did with the charge sector in Sec. II.

However, there is a problem with this approach. So far as we neglect the Zeeman term in the consideration of the SQHE,[2] there is no spin anisotropy in this state. The "Abelian" approach breaks the SU(2) spin symmetry from the outset. Its recovery is a highly nontrivial matter. In principle one has to be able to formulate the SQHE wave function while keeping the full SU(2) invariance and to allow for a quantization axis that is varying in space. Girvin *et al.*[4] have pointed out that $\Psi^{(2)}([z_i],[\eta_\alpha])$ leads to a partition function for a two-component plasma and that any extra charge = spin is screened. The screening in the two component anyon gas, in the context of the spin coupled to a gauge field, was found in Ref. 5. Thus, what is needed is a procedure to attach different fluxes to particles with $\uparrow$ and $\downarrow$ spins in a manner that is compatible with the SU(2) spin symmetry. Fortunately such an approach does exist: it is the non-Abelian SU(2) CS theory. A non-Abelian CS term, much like the Abelian CS theory used in the description of the spin polarized QHE,[11,12,25,19] attaches fluxes to particles. But, unlike the Abelian approach mentioned above, the non-Abelian CS theory is invariant under SU(2) rotations of the spin. Furthermore, this invariance is local and this is a gauge theory. It turns out that the CS theory represents the only possible local way to attach particles to SU(2) fluxes. Below we will follow this second way in considering the spin wave function.

Consider the set of coordinates $[z_i],[\eta_\alpha]$ of a set of some spinors with the spin up components, located at points $[z_i]$, and spin down at points $[\eta_\alpha]$. The points $[z_i],[\eta_\alpha]$ will be regarded as the positions of sources of an SU(2) field $\xi_\sigma(x)$, taken in the fundamental representation. It corresponds to the spin $\frac{1}{2}$ of the electrons, constituting the QHE state. Let us now consider the coupling of the spinor field $\xi_\sigma(x)$ to the SU(2) gauge field A_μ^a ($a=1,2,3$). The Lagrangian $\mathcal{L}_{\rm spin}$ of the spin sector is taken to be given by

$$\mathcal{L}_{\rm spin}=\xi^+ iD_0\xi-\frac{1}{2m}\xi^+D_\mu^2\xi-\mathcal{L}_{\rm CS},$$
$$\mathcal{L}_{\rm CS}=\frac{k}{4\pi}{\rm tr}(A_\alpha\partial_\beta A_\gamma+\tfrac{2}{3}A_\alpha A_\beta A_\gamma)\epsilon^{\alpha\beta\gamma}, \tag{3.1}$$

where $\mathcal{L}_{\rm CS}$ is the non-Abelian Chern-Simons term. We assume that ξ_σ obeys fermionic anticommutation rules $[\xi_\sigma(r),\xi_{\sigma'}(r')]_+=\delta_{\sigma\sigma'}\delta(r-r')$, $D_0=\partial_0-iA_0^a t^a$ is the time component of the covariant derivative. Analogously, the spatial components are defined as D_μ, $\mu=x,y$, and $A_i=A_i^a t^a$, $i=0,x,y$. The generators of the SU(2) t^a in the spin-$\frac{1}{2}$ representation are chosen to be $t^a=\sigma^a/\sqrt{2}$, and σ^a are the Pauli matrices. The integer coefficient k defines the topological structure of the model.

The points where the ξ_σ particles are located are the sources for the gauge field, as it can be seen from the variation of the Lagrangian (3.1) over A_0^a:

$$\frac{\delta\mathcal{L}_{\rm spin}}{\delta A_0^a}=\xi^+t^a\xi-\frac{k}{2\pi}F_{xy}^a=0. \tag{3.2}$$

The strength of the gauge field is given by $F_{xy}^a=\partial_x A_y^a-\partial_y A_x^a+[A_x,A_y]^a$. Let us assume that the particles have a mass m. The path-integral representation of a matrix element of the evolution operator, among states in which the particles are located at ξ_σ, is given as a sum over all possible particle trajectories and gauge field histories. The constraint of Eq. (3.2) requires that each term in this amplitude should contain a factor representing a path-ordered exponential of the SU(2) gauge field along each particle trajectory. These path-ordered exponentials are usually referred to as Wilson lines. In first quantization, the time evolution during the time interval t of the heavy sources will be given by the amplitude

$$\Psi([z_i'],[\eta_\alpha'],t)=\sum_{\text{Paths}} \exp\left[-i\int dt\left[\sum_i m/2|dz_i/dt|^2+\sum_\alpha m/2|d\eta_\alpha/dt|^2\right]\right]$$
$$\times\int D[A]\otimes_{i,\alpha}W_i(z_i',z_i)W_\alpha(\eta_\alpha',\eta_\alpha)\exp\left[i\int d^2x\, dtL_{\text{CS}}\right]\Psi([z_i],[\eta_\alpha],0)\ , \tag{3.3}$$

where z',η' are the set of final positions of the sources, and

$$W_i(z_i',z_i)=P\exp\left[i\int_{z_i}^{z_i'}A_l dx^l\right]. \tag{3.4}$$

are Wilson lines evaluated on the three-dimensional paths from z_i to z_i'. We will consider the 2D disc geometry pierced by the Wilson lines. The coordinate space is $D\times R$, where R is the time. The integral in the exponent in $W_i(z_i',z_i)$ is the quasiclassical expression for the spin-current-gauge potential coupling $\int A_\mu^a j_\mu^a d^2x\, dt$, assuming that $j_\mu^a=t^a(dx_\mu/dt)\delta(x-x_i(t))$ and $x_i(t)$ parametrize the quasiclassical path of the particle.

The CS action for the gauge field leads to the effective semion statistics of Wilson lines. Let us fix two Wilson lines, corresponding, for example, to particles at z_1 and η_1. And let us consider two processes which represent evolutions with the same final state and only differ by the presence of an extra knot in their histories $W(z_i',z_1)$, $W(\eta_1',\eta_1)$. Then the final amplitudes $\Psi([z_i'],[\eta_\alpha'])$ will gain different phases in these processes. One can find,[26] that the amplitudes are related by

$$\Psi_{\text{knotted}}([z_i'],[\eta_\alpha'])=\exp(i\gamma)\Psi_{\text{unknotted}}([z_i'],[\eta_\alpha'])\ , \tag{3.5}$$

where γ is the *conformal weight* of the primary field for the SU(2) level k group, and is given by

$$\gamma=\frac{4\pi j(j+1)}{k+2}\ . \tag{3.6}$$

In our case, $k=1$, $j=\frac{1}{2}$, the phase difference between two configurations is π what brings a phase of $\pi/2$ per particle. If we assume that the evolution between two configurations is adiabatic, the kinetic energy does not modify the value of γ because it is quadratic in time derivative. The only contribution to the phase comes from the CS action and it leads to the semion statistics of the ξ_σ particles, exhibited in the spin wave function $\Psi^{(2)}([z_i],[\eta_\alpha])$.

It is interesting to note that, for an SU(2) CS gauge theory, in addition to the Abelian statistics that we have discussed, it is also possible to have non-Abelian statistics. The spin wave function $\Psi^{(2)}([z_i],[\eta_\alpha])$ is multivalued in the complex plane due to branch cuts. If we introduce the permutation operator, defined by $x_i-x_j\to e^{i\pi}(x_i-x_j)$ (where x_i can be either z_i or η_α in our case), one can show that, generally speaking, the multivalued functions in the complex plane transform nontrivially under this operation.

Suppose that we have some multivalued function $F([x_i])$, which is a wave function like $\Psi^{(2)}([z_i],[\eta_\alpha])$. We can define its value for fixed configuration of x_i. For any new configuration, the new value of the wave function can be obtained by an analytic continuation procedure. Since $F([x_i])$ is multivalued, we conclude that the translation of "probe" particle along some path correspond to a nontrivial monodromy, i.e., different paths with the same initial and final states give different phases for the new value of the wave function. The permutation operation, which corresponds to a half-monodromy (braiding)[27,28] cannot be realized in terms of the Abelian group of permutations $\mathbb{P}$, and corresponds rather to the non-Abelian braid group on x_i: $\mathbb{B}$. The natural consequence of this fact is that the correlation function of the SU(2) WZW model for level $k\geq 2$ $F([x_i])$ exhibits non-Abelian statistics. We define statistics to be non-Abelian if the permutation operation leads to the multiplication of the wave function by some matrix and these matrices cannot be diagonalized simultaneously. This case is a more general situation than the Abelian statistics, where the wave function is multiplied by phase factors under the permutations.

We can illustrate these facts by considering the correlation function of SU(2) WZW model at level k, which for $k=1$ was identified as a spin wave function $\Psi^{(2)}([z_i],[\eta_\alpha])$. In general, the correlation functions of the primary fields of the SU(2) WZW model obey the Knizhnik-Zamolodchikov equations.[29] Let us fix the coordinates x_i, $i\neq 1$ and let us calculate the new value of $F([x_i])$ after transporting first coordinate from x_1 to x_1'. From the Knizhnik-Zamolodchikov equations we find

$$F(x_1')=\exp\left[\frac{1}{k+2}\sum_{i\neq 1}\int_{x_1}^{x_1'}\frac{t_1^a t_i^a}{x-x_i}dx\right]F(x_1)\ . \tag{3.7}$$

This equation tells us that, after circling the particle x_1 around, for example, x_2, the wave function $F([x_i])$ should be multiplied by a matrix given by the exponential factor. Only for $k=1$ and spin-$\frac{1}{2}$ particles we can rewrite the above formula as

$$F(x_1')=\exp\left[\sum_{i\neq 1}\int_{x_1}^{x_1'} dx \frac{q_1 q_i}{x-x_i}\right] F(x_1)$$

$$=\exp\left[iq_1 q_i \sum_{i\neq 1} \Delta \arg \ln(x-x_i)|_{x_1}^{x_1'}\right] F(x_1) . \quad (3.8)$$

In this case the exchange of two particles is equivalent to the multiplication of the wave function by a phase factor. Hence, the statistics of the excitations in the SQHE is Abelian, at least for the major fractions. However, we think that it is possible to find QHE-type models whose excitations have non-Abelian statistics. Work on this subject is in progress.

The wave functions $\Psi([z_i],[\eta_\alpha])$ correspond to states in the Hilbert space of the CS theory with the sources. The kinetic term in the exponent (3.3) does not change this feature. Witten[6] has pointed out that the states in the Hilbert space of the CS theory are in natural correspondence with the conformal blocks of a two-dimensional conformal field theory, the Wess-Zumino-Witten (WZW) model.[29] The integer k in the CS action dictates the level of the conformal Kac-Moody current algebra in the corresponding WZW model.[29]

We will stop here for a moment and introduce some notation from conformal field theory.[30] If we define the components of the current via the expansion $J_n^a=\int_C dz\, J^a(z) z^n$ with the contour C around the origin, then the SU(2) at level k Kac-Moody algebra takes the form

$$[J_n^a, J_m^b]=\sqrt{2}\,\epsilon^{abc} J_{n+m}^c + k\delta^{ab}\delta_{0,n+m} n . \quad (3.9)$$

At $m=n=0$ this equation reduces to the ordinary SU(2) current algebra. It is also important for further consideration to define the multiplet primary fields $\phi^a(z)$ which satisfy the conditions

$$J_n^a \phi=0, \quad n>0, \quad J_0^a\phi=t^a\phi . \quad (3.10)$$

The field ϕ transforms as the spin-$\frac{1}{2}$ representation of SU(2), where the rhs of Eq. (3.10) is the shorthand for $t_{l,k}^a \phi^k l, k=1,2$.

Now we want to return to the main line of our consideration. The proof of Witten's conjecture for the general case was considered in the large number of articles, see, for example, Refs. 27 and 26. For the case at hand, this means that the *wave function* $\Psi([z_i],[\eta_\alpha])$ can be written as a *correlation function* of the WZW model of the form, found in[7]

$$\Psi([z_i],[\eta_\alpha])=\langle \otimes_i \phi^{\uparrow}(z_i) \otimes_\alpha \phi^{\downarrow}(\eta_\alpha)\rangle , \quad (3.11)$$

where $\phi^{\uparrow(\downarrow)}$ are the SU(2)$k=1$ primary fields of weight $j=\frac{1}{2}$, and with the s_z projection up (down), points z_i corresponds to the spin-up field and η_α to the spin-down field. The average is computed in a two-dimensional Euclidean theory weighted with the WZW action:

$$S_{\rm WZW}=\frac{1}{2\pi}\int_D d^2x\, \mathrm{tr}(g^{-1}\partial_\mu g)^2 + \frac{1}{4\pi}\int_{D\times R} d^2x\, dt\, \mathrm{tr}(g^{-1}\partial g)^3 , \quad (3.12)$$

where $g\in$ SU(2). Notice that we put $k=1$ explicitly in the WZW term in Eq. (3.12). To prove Eq. (3.11) we will use the boson representation of the current algebra equation (3.9) $J_z, J^{\pm}=1/\sqrt{2}(J_1\pm J_2)$ in terms of the free-boson field $\Phi(z)$,[31] $\langle \Phi(0)\Phi(z)\rangle=-\ln z$:

$$J^{\pm}=:e^{\pm i\sqrt{2}\Phi(z)}:, \quad J^z=:\partial\Phi(z): \quad (3.13)$$

the primary fields with spin-up and spin-down projections are given by

$$\phi^{\uparrow\downarrow}=:e^{\pm(i/\sqrt{2})\Phi(z)}: . \quad (3.14)$$

It can be checked that the representation of Eqs. (3.13) and (3.14) satisfies the current algebra of Eq. (3.9). The calculation of the correlation function becomes trivial and it yields the result[7]

$$\langle \otimes_i \phi^{\uparrow}(z_i) \otimes_\alpha \phi^{\downarrow}(\eta_\alpha)\rangle = \prod_{i<j}(z_i-z_j)^{1/2} \prod_{\alpha<\beta}(\eta_\alpha-\eta_\beta)^{1/2} \prod_{i,\alpha}(z_i-\eta_\alpha)^{-1/2} . \quad (3.15)$$

By comparing Eq. (3.15) with Eq. (3.11) we get the main result of this section

$$\Psi^{(2)}([z_i],[\eta_\alpha])=\Psi([z_i],[\eta_\alpha]) , \quad (3.16)$$

where $\Psi^{(2)}([z_i],[\eta_\alpha])$ is given by the rhs of Eq. (3.15), as was discussed in the Introduction, and the wave function $\Psi([z_i],[\eta_\alpha])$ is given in terms of the evolution equation (3.3) for the SU(2) CS theory.

The wave function $\Psi^{(2)}([z_i],[\eta_\alpha])$ should be interpreted as the orbital part of the many-body wave function $|\Psi^{(2)}\rangle$ for spin-$\frac{1}{2}$ semions, which also depends on the configuration of the spins of the particle. So far we have dropped the spin indices in our discussion such as, for example, the right-hand side of Eq. (3.15). The many-body semion wave function $|\Psi^{(2)}\rangle$ is a spin singlet. There are two ways to see that. On the one hand, we can explicitly check that

$$S^+S^-|\Psi^{(2)}\rangle\equiv 0, \quad S^z|\Psi^{(2)}\rangle\equiv 0 . \quad (3.17)$$

The proof of this statement is presented in Appendix B. As a by-product of this proof we found a new cyclic condition satisfied by the orbital wave function $\Psi^{(2)}([z_i],[\eta_\alpha])$:

$$\left[1+e^{-i\pi/2}\sum_\alpha e(i,\alpha)\right]\Psi^{(2)}([z_i],[\eta_\alpha])=0 , \quad (3.18)$$

where $e(i,\alpha)$ is the braiding operator for semions defined in Appendix B. Equation (3.18) is a generalization of the Fock cyclic condition[32] to the case of semions. On the other hand, the singletness of the $|\Psi^{(2)}\rangle$ can also be checked by simply realizing that it carries the full spin content of the total *electron* SQHE wave function. The details of this argument are given in Appendix B.

In summary, in this section we have showed that the spin part of the SQHE wave function can be described in terms of the wave functions of an effective non-Abelian CS theory. In turn, the wave functions of the non-Abelian CS theory were argued to form a *conformal block*. This construction is also valid for the entire spectrum of excitations, not just the ground state. In fact, in Sec. IV we argue that the wave functions for the elementary excitations of the SQHE also have the factorized structure exhibited by the ground state. Note that the spin wave function $\Psi^{(2)}([z_i],[\eta_\alpha])$ is precisely the Kalmeyer-Laughlin wave function for the chiral-spin-liquid (CSL) phase, written in the specific representation.[33] This fact means that there is a nontrivial correspondence between the dynamics of the CSL phase and the spin dynamics of the SQHE. In particular, it was argued that in the SQHE state can support *spinon* excitations,[7] which, in principle, can be observed experimentally. This issue will be discussed elsewhere.

IV. SEWING TOGETHER THE CHARGE AND SPIN SECTORS

In the past two sections we discussed separately the charge and spin sectors. Now we have to put them back together again. We already indicated in the Introduction that charge and spin states cannot exist separately in this system, at least as states in the bulk. Some subtleties arise at the edges of the system where "spinon" (i.e., spin-$\frac{1}{2}$ charge neutral) or "holon" [i.e., spin-0 charge $1/(2m+1)$] states may occur. This issue will not be discussed further in this paper.

The Lagrangians $\mathcal{L}_{\text{charge}}$ of Eq. (2.3) and $\mathcal{L}_{\text{spin}}$ of Eq. (3.1) describe the separate dynamics of the charge and spin sectors. The factorization of Eq. (1.7) of the electron operator indicates that only the bound states of holons and spinons are present in the physical spectrum. Indeed, the *relative phase* of the spinon and holon states is not an observable degree of freedom. Thus, the system must be invariant under local changes of their relative phase

$$\begin{aligned} \varphi(x) &\to \exp(i\alpha(x))\varphi(x) , \\ \xi_\sigma(x) &\to \exp(-i\alpha(x))\xi_\sigma(x) . \end{aligned} \tag{4.1}$$

In other words, there is an additional gauge symmetry in this problem as a result of the factorization of the fermion operator. This is, of course, the same gauge symmetry which arises in models of strongly correlated systems, in particular to the resonating valence bond (RVB) picture.[3] For these reasons we call this additional gauge symmetry RVB gauge invariance. Consequently, there is an additional gauge field, the RVB gauge field.

Thus, there are three gauge fields in our model: a CS gauge field A_μ^a in the triplet representation of SU(2), another CS gauge field $\mathcal{A}_\mu$ in U(1), and an RVB U(1) gauge field $\mathcal{B}_\mu$. Each has a distinct role here. While the role of the Chern-Simons gauge fields A_μ^a and $\mathcal{A}_\mu$ is to attach SU(2) and U(1) fluxes to spin and charges separately, the role of the RVB gauge field $\mathcal{B}_\mu$ is to enforce the RVB gauge invariance of Eq. (4.1) and to project outside the Hilbert space all the states which are not invariant. Hence, unless the RVB gauge symmetry could somehow be broken, exact RVB gauge invariance will require that only the bound states of Eq. (1.7) will be present in the spectrum of physical states: spinons and holons cannot exist separately, they are *confined*. In this paper we *assume* that the RVB gauge symmetry is exact. Thus, while spinons and holons *cannot* exist separately, the wave functions of the elementary excitations still factorize into a spinon wave function times a holon wave function. The coordinates of holons and spinons must be the same so as to respect the RVB gauge symmetry.

The problem that we have to solve is to find the Lagrangian which describes the dynamics of spinons and charges moving together. One possible choice would be to simply write down the Lagrangian for interacting *electrons* in an external uniform magnetic field. But, if we chose to do that, we would have lost all the insight that we have gained so far. Thus, we must look for a Lagrangian $\mathcal{L}$ which combines $\mathcal{L}_{\text{charge}}$ and $\mathcal{L}_{\text{spin}}$ and, at the same time, is manifestly invariant under the RVB gauge symmetry. Clearly, neither $\mathcal{L}_{\text{charge}}$ nor $\mathcal{L}_{\text{spin}}$ are invariant. Further, the dynamical system described by $\mathcal{L}$ must be equivalent to the problem of interacting electrons in a magnetic field. The easiest way to find this Lagrangian is to make *both* $\mathcal{L}_{\text{charge}}$ and $\mathcal{L}_{\text{spin}}$ invariant under the RVB gauge symmetry.

Let us redefine the covariant derivatives for charge D_μ^{charge} and spin D_μ^{spin} in the following manner:

$$\begin{aligned} D_\mu^{\text{charge}} &= \partial_\mu + ieA_\mu^{\text{EM}} + i\mathcal{A}_\mu + i\mathcal{B}_\mu , \\ D_\mu^{\text{spin}} &= \partial_\mu + iA_\mu^a t^a - i\mathcal{B}_\mu , \end{aligned} \tag{4.2}$$

where we have used the same notation as in Secs. II and III and A_μ^{EM} is the external electromagnetic field. The Lagrangian $\mathcal{L}$ that we are looking for is simply the sum of $\mathcal{L}_{\text{charge}}$ and $\mathcal{L}_{\text{spin}}$ but with D_μ^{charge} as the covariant derivative for $\mathcal{L}_{\text{charge}}$ and D_μ^{spin} for $\mathcal{L}_{\text{spin}}$

$$\mathcal{L} = \mathcal{L}_{\text{charge}}(\varphi, A_\mu^{\text{EM}}, \mathcal{A}_\mu, \mathcal{B}_\mu) + \mathcal{L}_{\text{spin}}(\xi_\sigma, A_\mu^a, \mathcal{B}_\mu) . \tag{4.3}$$

Thus, φ and ξ_σ couple to the RVB gauge field $\mathcal{B}_\mu$ with opposite RVB charge and the invariance (4.1) is satisfied. Notice that the RVB gauge field $\mathcal{B}_\mu$ does not have any dynamics of its own. This Lagrangian satisfies all the requirements posed above. The states determined by $\mathcal{L}$ have wave functions which exhibit spin-charge factorization and are RVB gauge invariant.

But, we now have to establish a connection between the problem described by $\mathcal{L}$ and our original problem, namely, that of unpolarized interacting electrons in a magnetic field. Fortunately, these two problems happen to be *identical*. The simplest way to prove it is to go back to first quantization and to show that all possible quantum mechanical amplitudes of one problem are identical to another one in the other problem. Consider an arbitrary matrix element of the evolution operator of the system represented by the Lagrangian $\mathcal{L}$. Such matrix elements can be labeled by the coordinates of the particles both in the initial and in the final states. Since the only states in the spectrum have to be gauge invariant, the coordinates represent the presence of a charged particle with either spin orientation. When these matrix elements are represented in terms of path integrals we get a problem in which we have to sum over all possible histories of both particles and gauge fields. This is exactly what we did in Secs. II and III for both charge and spin sectors separately. When we put them back together, each path will be represented by a weight which is the product of the separate contributions of charge and spin separately. In a Euclidean representation we get the usual factor which involves the kinetic energy of the individual particles moving in the presence of the external electromagnetic field times a factor which represents their mutual interaction. But, in addition we get extra factors which represent the motion of the particles in the presence of the various gauge fields times a contribution from the Chern-Simons terms. If we average each contribution to the total amplitude over all the configurations of all the gauge fields we find the following results.

(a) As a result of averaging over the RVB gauge fields, the only paths that carry a nonzero weight represent trajectories in which charge and spin move together. This is so because each weight has the form of a Wilson line of RVB gauge fields. Since the charge φ and spin ξ_σ fields carry opposite RVB charge, the average is zero unless the trajectories coincide, in which case the Wilson lines are exactly canceled.

(b) The average over the U(1) charge and SU(2) Chern-Simons fields leads to the same weights discussed in Secs. II and III, namely, to semion statistics. But, since the only nonzero contributions come from paths with the charge and spin trajectories exactly on top of each other, the phase factors associated with the semion statistics of spinons and holons cancel each other out. Hence, we find that the weight carried by each configuration is identical to the weight that we would find in the original problem and, in a rigorous sense, the Chern-Simons gauge fields do not contribute to the amplitudes. Thus, the Lagrangian $\mathcal{L}$, which describes the dynamics of semion spinons and holons (in a magnetic field) yields the same quantum mechanical amplitudes as the Lagrangian for interacting fermions in a magnetic field in the spin singlet sector. This argument completes our description of the SQHE in terms of Chern-Simons theories. It is important to stress, once again, that what we have developed is not just a description of the ground state, but of the entire spectrum of excitations above the SQHE state. In particular, the factorized structure of the SQHE wave function is a property of all the states in the spectrum.

Perhaps the simplest way to see this is to consider the wave function of quasiparticle (QP) in the first quantized representation, as it has been done in Ref. 2. For example, for the QP of spin $\frac{1}{2}$ with $s_z=-\frac{1}{2}$ at point z_0:

$$\Psi_{z_0,\downarrow}([z_i],[\eta_\alpha])=\prod_i (z_i-z_0)\Psi_m([z_i],[\eta_\alpha]) \ . \quad (4.4)$$

The form of this wave function indicates that the creation of the QP is equivalent to the creation of the extra zero at point z_0 for the wave function of the particles with the spin $s_z=+\frac{1}{2}$ projection. By using the plasma analogy it is easy to conclude that this zero is equivalent to the QP of spin $s_z=-\frac{1}{2}$ with charge $e=1/2m+1$.

Now we will explicitly show that the wave function of the QP in the SQHE can be represented as a composite excitation of neutral spinon with $s=\frac{1}{2}$ and of the spinless holon with charge $e=1/(2m+1)$. We can rewrite $\Psi_{z_0,\downarrow}$ as

$$\Psi_{z_0,\downarrow}([z_i],[\eta_\alpha])=\prod_{i,\alpha}(z_i-z_0)^{1/2}(\eta_\alpha-z_0)^{1/2}\Psi_m^{(1)}([z_i],[\eta_\alpha])\prod_{i,\alpha}(z_i-z_0)^{1/2}(\eta_\alpha-z_0)^{-1/2}\Psi^{(2)}([z_i],[\eta_\alpha]) \ . \quad (4.5)$$

The first product $\Psi_{z_0}^{(1)}=\prod_{i,\alpha}(z_i-z_0)^{1/2}(\eta_\alpha-z_0)^{1/2}\Psi_m^{(1)}([z_i],[\eta_\alpha])$ is nothing more than the holon excitation in the one component plasma, corresponding to the $\Psi_m^{(1)}([z_i],[\eta_\alpha])$. From what follows that the effective charge of the holon is $e=\frac{1}{2}/(m+\frac{1}{2})=1/(2m+1)$. The second product $\Psi_{z_0\downarrow}^{(2)}=\prod_{i,\alpha}(z_i-z_0)^{1/2}(\eta_\alpha-z_0)^{-1/2}\Psi^{(2)}([z_i],[\eta_\alpha])$ is the spinon excitation, corresponding to the extra spin $s_z=-\frac{1}{2}$ excitation, created at point z_0.

There is an apparent problem with the identification of the sign of the spin projection for the excitation $\Psi_{z_0,\downarrow}^{(2)}$. The fictitious spin $\frac{1}{2}$ at point z_0 by the plasma analogy has the same projection as the spinons at points z_i, i.e., $s_z=+\frac{1}{2}$. But then, due to the plasma screening in the two-component plasma, the real spinons will screen out this fictitious spin, thus creating the $s_z=-\frac{1}{2}$ cloud of real spinons, centered at point z_0. This is precisely the reason why the spin projection of the excitation $\Psi_{z_0\downarrow}^{(2)}$ is down.

Once this confusing point has been clarified, we come to the statement that the spin-$\frac{1}{2}$ charge $e=1/(2m+1)qp$ can be represented as a product of the spinon and holon QP created at the same point z_0:

$$\Psi_{z_0\downarrow}=\Psi_{z_0}^{(1)}\Psi_{z_0\downarrow}^{(2)} \ . \quad (4.6)$$

Equation (4.6) is a decomposition (1.7) written in the first-quantized representation.

We find that the slave semion decomposition (1.7) for the SQHE is valid not only on the level of the ground state but on the level of QP excitations as well. Clearly the argument given above can be generalized trivially for the case of n QP. The fact that we need to put our spinon and holon on the same place explicitly indicates that these excitations with opposite RVB charge are confined to form an RVB neutral object, only allowed as the physical state.

Thus we showed that the slave semion decomposition equation (1.7) is a quite natural way to distinguish the physics in the charge and spin sector of SQHE. This *factorization* (but not *separation*) can be observed for any state in the Hilbert space of SQHE.

V. CONCLUSIONS

In this paper we have presented a field-theoretic description of the single quantum Hall effect. We have shown that the SQHE wave function, as well as the entire spectrum of excitations which stands on it, exhibits a spin-charge factorized structure. Spin and charge degrees of freedom are shown to be naturally described in terms of holons and spinons both with *semion* statistics. The holons interact with the external magnetic field and their ground state is a generalization of the Laughlin state for semions. This state is shown to have a gap for all elementary excitations and hence to be incompressible. The quantum Hall conductance is calculated and shown to obey Laughlin's argument, i.e., $\sigma_{xy}=\nu/2\pi$ where ν is the filling fraction. The spinons have a ground state which is identical to the chiral spin liquid state of Kalmeyer and Laughlin. This wave function is identified with the correlator of spin-$\frac{1}{2}$ primary fields of the SU(2) Wess-Zumino-Witten conformal field theory. We showed that the spinon wave function is a spin singlet and, as a by-product, we derived a generalization of the Fock cyclic conditions for many-body spin-$\frac{1}{2}$ semion wave functions. We have also briefly discussed a generalization of the theory in which the spinons have non-Abelian statistics. We also showed that spinons and holons must necessarily be bound to each other, at least for bulk states. Since the wave function for the SQHE is written in terms of *electron* coordinates, it follows that it must be a single valued analytic function. In particular, as stressed by Haldane, the QP wave function $\Psi_{z_0,\downarrow}$ should have a simple zero. Thus, the branch cuts of holons and spinons must cancel each other out. This feature is enforced by putting holons and spinons on top of each other as required by their confinement. We conclude that all the excitations of this system have a finite energy gap and do not exhibit separation of spin and charge. It is interesting to note that since the wave function $\Psi^{(2)}([z_i],[\eta_\alpha])$ is a conformal block, it is covariant under dilatations. This feature of $\Psi^{(2)}([z_i],[\eta_\alpha])$ appears to suggest that it may describe a *compressible* state. However, compressibility is a dynamical property which, therefore, depends explicitly on the choice of Hamiltonian. This feature may be useful to the theory of the chiral spin liquid, but is not relevant to the SQHE due to the lack of separation of spin and charge. Finally, the description of the problem in terms of the Chern-Simons gauge fields was found to be identical to the problem of interacting electrons in a magnetic field, in its spin singlet sector.

ACKNOWLEDGMENTS

We are grateful to M. Stone for useful discussions and to S. Girvin for making an unpublished version of his work available to us. We are also grateful to N. Read for an early discussion of his results. We wish to thank V. Kalmeyer for useful discussions on the single nature of the wave functions for spin-$\frac{1}{2}$ semions. This work was supported in part by the National Science Foundation through Grants NSF-DMR88-18713 and NSF-DMR88-17613 at the University of Illinois at Urbana-Champaign. E.F. acknowledges partial support from the Center for Advanced Study of The University of Illinois.

APPENDIX A

In this appendix we will prove that the state described by the wave function $\Psi_m^{(1)}([z_i],[\eta_\alpha])$ is incompressible. We need to use the plasma analogy, well known in the analysis of the QHE state properties:[34]

$$|\Psi_m^{(1)}([z_i],[\eta_\alpha])|^2=e^{-\beta H_{\rm pl}}\ , \tag{A1}$$

where we chose $\beta=m+\frac{1}{2}$, then we find for the effective plasma Hamiltonian

$$H_{\rm pl}=2\sum_{i<j}\ln|z_i-z_j|+2\sum_{\alpha<\beta}\ln|\eta_\alpha-\eta_\beta|+2\sum_{i\alpha}\ln|z_i-\eta_\alpha|+\frac{1}{2m+1}\sum_i|z_i|^2+\frac{1}{2m+1}\sum_\alpha|\eta_\alpha|^2\ . \tag{A2}$$

The Hamiltonian $H_{\rm pl}$ describes the one-component plasma with the effective charge $q=1$ and with the neutralizing background density $\rho=1/\pi(2m+1)$, so that the filling factor $\nu=2\pi\rho=2/(2m+1)$. These filling factors are well known for the SQHE.[2]

Now we can prove that the state given by (A2) is incompressible. In the one-component plasma the only collective mode is a plasmon, which has a gap. So we can use the singlet-mode approximation (SMA) in the sum rule.[35] Let us define the wave-vector-dependent gap:

$$\Delta(k)=f(k)/s(k)\ , \tag{A3}$$

where

$$f(k)=\frac{1}{2N}\langle\Psi^{(1)}|\rho_k^+(H-E_0)\rho_k|\Psi^{(1)}\rangle \tag{A4}$$

is the energy of the density fluctuations,

$$s(k)=\frac{1}{2N}\langle\Psi^{(1)}|\rho_k^+\rho_k|\Psi^{(1)}\rangle \tag{A5}$$

is the structure factor of the fluctuations, and $\rho_k=\sum_i e^{ikr_i}$ is the density operator, $2N$ is the total number of particles, E_0 is the ground state energy. H is the QHE Hamiltonian:

$$H=\sum_i\frac{(p_i-e/cA_i)^2}{2m}+\frac{1}{2}\sum_{i<j}V(r_i-r_j)\ , \tag{A6}$$

where A_i is the gauge potential of the real external field, V is the Coulomb interaction between particles.

By using SMA we can find that

$$f(k)=\frac{k^2}{2m} \tag{A7}$$

and the structure factor for the one-component plasma is[35]

$$s(k)\sim k^2,\ k\to 0 . \tag{A8}$$

We see from (A3), (A7), and (A8) that at $k\to 0$ the excitation spectrum has a gap, and the state described by (A1) is incompressible:

$$\Delta(k\to 0)=\Delta_0 . \tag{A9}$$

APPENDIX B

In this appendix we give a sufficient condition that the semion wave function $\Psi^{(2)}[(z_i],[\eta_\alpha])$ should satisfy the total wave function [see Eq. (B2)] to be a singlet, i.e., $S=S_z=0$. This condition is a generalization of the Fock cyclic condition obeyed by the orbital part of the electron many-body wave function for the state to be a spin singlet.[32]

For the case of semions, with statistical angle $\delta=\pi/2$, we demand that the orbital part of the semion wave function $\Psi^{(2)}([z_i],[\eta_\alpha])$ to meet the following requirements.

(a) The wave function must transform like the same one-dimensional representation of the braid group **B** for the braiding of coordinates within the same spin sector:

$$e^{i\pi/2}\Psi^{(2)}(z_1\cdots z_i\cdots z_j\cdots z_N;[\eta_\alpha]) =\Psi^{(2)}(z_1\cdots z_j\cdots z_i\cdots z_N;[\eta_\alpha]) . \tag{B1a}$$

(b) The wave function $\Psi^{(2)}([z_i],[\eta_\alpha])$ should further satisfy an extra condition for braiding of coordinates of particles with opposite spin projection:

$$\left[1+e^{-i\pi/2}\sum_\alpha e(i,\alpha)\right]\Psi^{(2)}([z_i],[\eta_\alpha])=0 , \tag{B1b}$$

where $e(i,j)$ is the braiding operation of the ith and jth objects and, in order to avoid ambiguity, we define braidings as a clockwise winding by π followed by a translation. For fermions and bosons, $e(i,j)$ becomes the familiar exchange operator.

The proof of the Fock condition holds for any fractional statistics parameter δ. It turns out however, that for spin-$\frac{1}{2}$ anyons this condition for the Jastrow form of the anyon wave function can be satisfied for $\delta=\pi/2$ only. This can be easily illustrated on the case of two anyons.

The true anyon many-body wave function is a product of $\Psi^{(2)}([z_i],[\eta_\alpha])$ with the spin wave function $\Phi_s([u_i];[d_\alpha])$ summed over all possible braidings B of $2N$ elements:

$$|\Psi^{(2)}\rangle=\sum_B\frac{(e^{i\pi/2})^B}{\sqrt{(2N)!}}\Psi^{(2)}([z_{Pi}];[\eta_{P\alpha}])\Phi_s([u_{Pi}];[d_{P\alpha}]) . \tag{B2}$$

Here we use the notations u_i,d_α to describe the spinor variables corresponding to spin up at points z_i and spin down at points η_α.

The condition (B1a) states that, upon the braiding of any pair of anyons with the same spin projection, the spin part of the anyon wave function is even, and the multivalued orbital wave function $\Psi^{(2)}([z_i],[\eta_\alpha])$ gets multiplied by an overall factor $e^{i\pi/2}$. Equation (B1b) results from demanding that $|\Psi^{(2)}\rangle$ be a spin singlet:

$$S^+S^-|\Psi^{(2)}\rangle\equiv 0 . \tag{B3}$$

It is obviously true that $S^z|\Psi^{(2)}\rangle=0$.

The proof of this condition can be obtained by repeating the derivation of the original Fock condition for ordinary wave functions.

An alternative way to see that $\Psi^{(2)}([z_i],[\eta_\alpha])$ corresponds to a singlet anyon many-body wave function is to start from the Fock condition for the electron wave function $\Psi_m([z_i],[\eta_\alpha])$:

$$\left[1-\sum_\alpha e(i,\alpha)\right]\Psi_m([z_i],[\eta_\alpha])=0 , \tag{B4}$$

which is satisfied. By using the following property of the wave function of the spinless anyons $\Psi_m^{(1)}([z_i],[\eta_\alpha])$:

$$e(i,\alpha)\Psi_m^{(1)}([z_i],[\eta_\alpha])=e^{i\pi/2}\Psi_m^{(1)}([z_i],[\eta_\alpha]) . \tag{B5}$$

We can substitute (B5) into (B4) and find that singlet condition for $\Psi^{(2)}([z_i],[\eta_\alpha])$ is given by (B1b).

[1]B. I. Halperin, Phys. Rev. Lett. **52**, 1583 (1984).
[2]F. D. M. Haldane, in *The Quantum Hall Effect*, edited by R. Prange and S. Girvin (Springer, Berlin, 1990), Chap. 7.
[3]G. Baskaran, Z. Zou, and P. W. Anderson, Solid State Commun. **63**, 973 (1987); S. Kivelson, D. Rokhsar, and J. Sethna, Phys. Rev. B **35**, 8865 (1987).
[4]S. M. Girvin, A. H. MacDonald, M. P. A. Fisher, S.-J. Rey, and J. Sethna, Phys. Rev. Lett. **65**, 1671 (1990).
[5]A. V. Balatsky and V. Kalmeyer, Phys. Rev. B **43**, 6228 (1991).
[6]E. Witten, Commun. Math. Phys. **121**, 351 (1989).
[7]A. V. Balatsky, Phys. Rev. B **43**, 1257 (1991); Phys. Rev. Lett. **66**, 814 (1991).
[8]N. Read, private communication.
[9]G. Moore and N. Read (unpublished).
[10]R. B. Laughlin, Phys. Rev. B **23**, 3383 (1983).
[11]S. Girvin and A. MacDonald, Phys. Rev. Lett. **58**, 1252 (1987).
[12]S. C. Zhang, T. Hansson, and S. Kivelson, Phys. Rev. Lett. **62**, 82 (1989).
[13]A. Fetter, C. Hanna, and R. B. Laughlin, Phys. Rev. B **39**, 9679 (1989).
[14]H. Y. Chen, F. Wilczek, E. Witten, and B. I. Halperin, Int. J. Mod. Phys. B **3**, 1001 (1989).
[15]M. P. A. Fisher and D. H. Lee, Phys. Rev. Lett. **63**, 903 (1989).
[16]E. Fradkin, Phys. Rev. Lett. **63**, 322 (1989).

[17]E. Fradkin, Int. J. Mod. Phys. B 3, 1965 (1989); Phys. Rev. B **42**, 570 (1990).
[18]J. Jain, Phys. Rev. Lett. **63**, 199 (1989); Phys. Rev. B **40**, 8079 (1989); **41**, 7653 (1990).
[19]A. Lopez and E. Fradkin (unpublished).
[20]B. Blok and X. G. Wen (unpublished).
[21]E. Fradkin, "Anyons for Beginners"; to be published in "Symposium in honor of Prof. J. J. Giambiag," edited by F. Schaposnik (World Scientific, Singapore, 1990).
[22]A. M. Polaykov, Int. J. Mod. Phys. A 3, 325 (1988).
[23]S. Deser, R. Jackiw, and S. Templeton, Phys. Rev. Lett. **48**, 372 (1982).
[24]R. B. Laughlin, Phys. Rev. B **27**, 5632 (1981); B. I. Halperin, *ibid.* **25**, 2185 (1982).
[25]N. Read, Phys. Rev. Lett. **62**, 86 (1989).
[26]J. M. Labastida and F. Ramallo (unpublished).
[27]G. Moore and N. Seiberg, Commun. Math. Phys. **123**, 177 (1989); Phys. Lett. B **220**, 422 (1989).
[28]L. Alvarez-Gaumé, G. Sierra, and C. Gomez, in *Topics in Conformal Field Theory,* Contribution to the Kniznik Memorial Volume, edited by L. Brink, D. Friedan, and A. Polyakov (World Scientific, Singapore, 1991).
[29]V. Knizhnik and A. Zamolodchikov, Nucl. Phys. **B247**, 83 (1984).
[30]P. Ginsparg, in Lecture in Les-Houches, Session XLIX, 1988, edited by E. Brezin and J. Zinn-Justin (Elsevier, New York, 1990).
[31]A. Zamolodchikov and V. Fateev, Yad. Fiz. **43**, 1031 (1986) [Sov. J. Nucl. Phys. **43**, 657 (1986)].
[32]M. Hammermesh, *Group Theory and its Application to Physical Problems* (Dover, New York, 1989).
[33]V. Kalmeyer and R. Laughlin, Phys. Rev. B **39**, 11 879 (1989).
[34]R. Laughlin, in *The Quantum Hall Effect* (Ref. 2), Chap. 6.
[35]S. Girvin, in *The Quantum Hall Effect* (Ref. 2), Chap. 8.

Chapter 4

EFFECTIVE THEORIES

4.1 Introduction

In superconductivity, which the Hall effect somewhat resembles, the Landau-Ginzburg effective action provided a phenomenological theory which preceded the discovery of the BCS wavefunction by several years. This effective action neatly encodes the most important macroscopic properties of the superconductor and forms a starting point for investigating further phenomena, such as vortex creep and critical currents. In the case of the FQHE the historical sequence was reversed. There was the Laughlin wavefunction, and consequently a good understanding of microscopic properties, but no useful macroscopic description. The absence of such an order-parameter based phenomenology was stressed by Girvin in the summary of [1], and he provided an outline of a Landau-Ginzburg theory which included a Chern-Simons gauge field. This initial sketch, however, explained neither the meaning of order parameter nor the origin of the Chern-Simons field. In an expanded version [rep.33], Girvin and MacDonald explicitly exhibited an order-parameter constructed by combining the fermion field with a singular gauge transformation. They showed that this operator had quasi-long-range order, its correlations having only a power-law decay instead of the more common exponential fall off. They also partially explained why there was a Chern-Simons term, but their coupling to the electromagnetic field was still not quite satisfactory.

In 1989, two "rival" derivations of rather similar effective actions appeared in the same issue of Physical Review Letters. The first, [rep.34], evolved from early ideas of Kivelson *et al.* [rep.32]. They had argued that when the electron density is carefully adjusted it is possible for a magnetic field to nullify the Fermi statistics of the electrons, allowing them to "Bose condense" through a proliferation of long exchange paths. The freedom to move along these paths would then melt a Wigner crystal into a "supersolid" quantum Hall phase. Unfortunately, the observed FQHE states do not seem to be supersolids. This interesting idea had also to compete with the very successful Laughlin wavefunction picture, and, as a result, it was not widely accepted. After Girvin and MacDonald [rep.33] suggested including a Chern-Simons field, Zhang, Hanssen and Kivelson (ZHK) [rep.34] realized that the statistics transformation implicit in the Chern-Simons field was just what was needed to formalize the electronic Bose condensation in a way that was independent of any Wigner solid parent state. The resulting scalar field theory plus Chern-Simons term accounts very succesfully for most QHE phenomenology.

Meanwhile, Read had been seeking an order parameter with true long-range order, a quantity whose non-zero expectation value would be characteristic of an incompressible quantum Hall phase. He suggested the use of an operator similar to that of Girvin-Macdonald, but distinct from it. Read's operator simultaneously creates an electron and $n = \frac{1}{\nu}$ quasiholes. Because it changes fermion number, its acquisition of a non-vanishing expectation value will signal the presence of off-diagonal (in a fermion-number basis) long-range order. Using a version of Read's operator, Rezayi and Haldane [rep.35] found numerical evidence for ODLRO in quantum Hall states. They also knew that by varying the relative strengths of the short and long-range parts of the interaction they could destroy the incompressibility. They did this, and saw the long-range order disappear at the same moment. The ODLRO does seem to characterize the Hall fluid phase. Encouraged by this, Read went on to use the expectation value of his operator to parametrize deviations from the energetically favoured Laughlin state. He found [rep.36] that a Chern-Simons theory emerged, one that appears almost identical to the action of Zhang, Hansson and Kivelson. We will soon see that this similarity is deceptive.

4.2 Chern-Simons Actions

4.2.1 Mathematical Origins

In the quantum Hall effect, Chern-Simons actions arise quite naturally in two different, but related, ways. Before reviewing these applications of Chern-Simons forms it is worthwhile to glance at their origin in mathematics.

We have already met the family of topologically interesting objects called the *Chern character* forms. These are gauge-invariant $2n$-forms constructed as the trace of the n-th exterior power of the curvature 2-form, $F = dA + A^2$, of some connexion

$$ch_n(A) = \mathrm{tr}\,\{F^n\}. \tag{4.2.1}$$

The Bianchi identity

$$dF + [A, F] = 0 \tag{4.2.2}$$

shows that they are closed, $d(ch_n(A)) = 0$. If we deform the connexion $A \to A + \delta A$, the Chern character forms change by a total derivative,

$$ch_n(A) \to ch_n(A) + nd\{\mathrm{tr}\,(\delta A F^{n-1})\}, \tag{4.2.3}$$

so an integral of these forms over homologically non-trivial submanifolds yields topological invariants, the *characteristic classes* that serve to classify the topology of the gauge bundle.

In a contractable region, we can use Poincaré's theorem to write $ch_n(A) = d\Xi_n(A)$ for some $(2n-1)$-form $\Xi_n(A)$. In the case of $ch_2(A) = \mathrm{tr}\,(F^2)$, for example, we find

$$\mathrm{tr}\,(F^2) = d\{\mathrm{tr}\,(dA + \frac{2}{3}A^3)\} \tag{4.2.4}$$

The Ξ_n are the *Chern-Simons secondary characteristic classes.* Under gauge transformations they change by total derivatives so, when integrated over a region with a boundary, they give an expression that is gauge invariant up to boundary terms. When we integrate $ch_n(A)$ over a non-contractable submanifold, we can always cover the submanifold with a patchwork of contractable regions. By then using the Ξ_n to write $\int ch_n$ as a boundary term, we may express the characteristic class in terms of the gauge transformations (in mathematical language the *transition functions*) between the different regions. Describing this in any more detail would lead us into the delights of Čech cohomology and *sheaf theory*. For an account of the rudiments of sheaf theory accessible to physicists, read the chapter *Cohomology and Field Theory* by Orlando Alverez, in [2].

4.2.2 Connection with Anyons

Chern-Simons terms were introduced into physics in the context of anyons, and a general reference for this section is the reprint volume [3]. With anyons we need to monitor the phases accumulated by the braiding worldlines of the particles. A way to keep track of the winding is to use Ampère's law. A closed loop, C_1, of wire carrying a current density $\mathbf{J}_1$ produces a magnetic field $\mathbf{B}_1$ where

$$\nabla \wedge \mathbf{B}_1 = \mathbf{J}_1. \tag{4.2.5}$$

When a second closed loop, C_2 is introduced, the number of times C_2 winds round C_1 is a topological invariant of the resulting link, and is called *the Gaussian linking number*. It is denoted by $\Gamma(C_1, C_2)$. By giving the second loop a current density $\mathbf{J}_2$ we find that $\Gamma(C_1, C_2)$ is given by

$$I_1 I_2 \Gamma(C_1, C_2) = \int \mathbf{B}_1 \cdot \mathbf{J}_2 d^3x = \int \mathbf{B}_2 \cdot \mathbf{J}_1 d^3x. \tag{4.2.6}$$

A convenient and automatic way of generating such expressions is by the use of an abelian Chern-Simons action

$$CS = \frac{1}{4\theta} \int \epsilon^{\mu\nu\sigma} a_\mu \partial_\nu a_\sigma d^3x. \tag{4.2.7}$$

If we couple a fluctuating a_μ field to some conserved current, $\mathbf{J}$,

$$\mathcal{Z} = \int d[a] \exp\left(i \int J^\mu a_\mu + \frac{i}{4\theta} \int \epsilon^{\mu\nu\sigma} a_\mu \partial_\nu a_\sigma \right), \tag{4.2.8}$$

we can perform the gaussian functional integral to get

$$\mathcal{Z} = \exp i\theta \int d^3x J^\mu B_\mu(\mathbf{J}), \tag{4.2.9}$$

where $B_\mu(\mathbf{J})$ is the classical field produced by the source. This classical field obeys $\nabla \wedge \mathbf{B} = \mathbf{J}$. It is usual to call θ the *statistics parameter*.

Regarding the worldlines of two point particles as wires carrying a unit current,

$$\mathbf{J}(x) = \mathbf{J}_1 + \mathbf{J}_2 = \int ds \left(\dot{\mathbf{x}}_1 \delta^3(x - x_1(s)) + \dot{\mathbf{x}}_2 \delta^3(x - x_s(s)) \right), \tag{4.2.10}$$

we substitute in (4.2.9) and, provided we ignore the *self windings*, we find the phase is 2θ times the number of times one worldline has wound round the other. To interchange the particles we go only half way round and so pick up a phase factor $e^{i\theta}$. This is why we define θ with the factor of four. The phase is (-1) if $\theta = \pi$ and provides the factor needed to change bosons to fermions and vice versa. Clearly physical consequences should not be changed if θ is incremented by an integer multiple of 2π.

Ignoring the self-winding terms $\int \mathbf{B}_1 \cdot \mathbf{J}_1 + \int \mathbf{B}_2 \cdot \mathbf{J}_2$ where the field and source are on the same curve, is not entirely innocent. The integrals in the self-linking numbers are convergent, but the resulting quantities are neither integers nor topological invariants of the curve. One might try to redefine them by splitting the wordline in two, always taking one end of the a_μ propagator on one part and one end on the other. This gives an integer, the twist of the ribbon edged by the split curve, and it is an invariant — but in the limit when the two curves coincide, this integral does not go smoothly over to the integral with both ends of the propagator on one curve. There is a δ-function contribution from the region where the propagator is very short, and this gives a residual part equal to the integral of the Serret-Frenet torsion along the curve. In a relativistic theory, where the space and time directions are linked, the torsion is proportional to the angle of Thomas precession, and hence is associated with the spin of the particles. In a non-relativistic context, where the spin-statistics conection is looser, we can "stretch" the time axis, make the torsion vanishingly small, and so "justify" the omission of these terms.

By substituting a non-abelian Chern-Simons theory for the abelian one, and replacing the current loops with Wilson lines, Witten was able to obtain many other knot invariants [4]. He also demonstrated the deep connection of the Chern-Simons theories with conformal field theories.

It should not be surprising that Chern-Simons actions provide topological information: as the integrals of differential forms, they make no reference to any metrical properties of the space in which they live. The space can be pulled and twisted, or, more technically, subjected to diffeomorphisms, without altering the outcome of any calculations (except for the regularization effects mentioned above).

4.2.3 Role in the Hall Effect

The occurence of anyons in the Hall effect suggests that an application of Chern-Simons technology might be useful. A more direct hint comes from looking at the response of the Hall fluid to an electromagnetic field. The minimum information we want to extract from an effective action description of the FQHE is

$$j_\mu = \frac{\delta S_{eff}(A)}{\delta A_\mu} = \frac{1}{2n+1}\epsilon_{\mu\nu\sigma}F_{\nu\sigma}, \tag{4.2.11}$$

or, in other words, the relations between the Hall current and the external electric field $F_{\mu 0}$, and between the electron density and the magnetic field F_{12}. This equation is precisely what we get from taking

$$S_{eff}(A) = \frac{1}{2n+1}\int d^3x \epsilon^{\mu\nu\sigma} A_\mu \partial_\nu A_\sigma. \tag{4.2.12}$$

Zhang, Hansson and Kivelson combined these two ideas by describing the electrons as boson fields and using a C-S term to recover the correct statistics. They found they could cancel the effect of the statistics field by the magnetic field at exactly $\nu = 1/2n+1$, and that, at these densities, boson field could condense. They obtained the couplings to the external field in the form of (4.2.12), and found vortex-like quasiholes and a gapped phonon mode — all the phenomenology that flowed from the Laughlin wavefunction. There is however a minor technical problem, one that was not at first noticed. To appreciate it we must dwell a little on the phonon-like modes expected in the Hall fluid.

4.3 Phonon-Like Modes in the 2DEG

The superfluidity exhibited by the liquid helium isotopes, and by the electrons in a superconductor, depends crucially on the paucity of low-lying excitations into which collective motion might decay. In some of the fermion superfluids there is an energy gap between the ground state and the excitations, while in others there is merely a very small number of low energy modes. In the quantum Hall fluids there are expected to be two distinct energy scales associated with gaps. First, there is the separation $\hbar\omega_c$ between the Landau levels, and second, there must be a Coulomb induced gap above the FQHE ground state to ensure that it is not destroyed by its flow against impurities and boundaries.

4.3.1 Magneto-Plasmons

The excitations made by promoting an electron from the lowest Landau level to the next were named ***magneto-plasmons***, or ***inter-Landau level*** modes, by Girvin, MacDonald, and Platzman [rep.31][5]. If we place the electron some distance from the hole it leaves behind, the energy of the resultant excitation is reduced somewhat from $\hbar\omega_c$ by the Coulomb attraction between the promoted electron and its relic hole, but at long wavelength the magneto-plasmons become bulk cyclotron motions of the electron center of mass and, by a theorem due to Kohn, must have energy exactly equal to $\hbar\omega_c$. [6,7]. Another consequence of this theorem will be described in more detail in the next chapter. The energy associated with these modes is in the order of a hundred times larger that the energy gap above the Laughlin state, so they are not relevant for determining whether or a not a particular FQHE state exists.

4.3.2 Magneto-Phonons

The important excitation is the ***magneto-phonon***, or ***intra-Landau level*** mode. Its minimum energy occurs away from $k = 0$ at the ***magneto-roton*** minimum k_{min}. It is the magneto-roton gap, $\omega(k_{min})$, that sets the energy and temperature scale for the observed FQHE phases. These low-energy collective modes are clearly visible in numerical studies, but their physics is more complicated than that of the magneto-plasmons. Girvin *et al.* draw an analogy with conventional plasma physics and describe the magneto-plasmon as an "upper hybrid mode" and the magneto-phonon as a "lower hybrid" mode. These are two-component plasma resonances close to the electron (upper mode) and ion (lower mode) cyclotron frequencies, but it is not clear to me what is playing the role of the ions here. At large k, the magneto-phonon recapitulates the large k magneto-plasmon behavior and evolves into a separated particle-hole pair, this time a fractionally charged quasihole and quasiparticle. The essential physics at longer wavelengths must be a distortion of the Laughlin wavefunction so that each electron is no longer located at an n-fold zero of the other electron amplitudes. Girvin, MacDonald, and Platzman [rep.31] use a variational wavefunction to create these distortions as a coherent superposition of slightly separated pairs, and this gives a good fit to the numerical data.

The magneto-roton minimum is probably a symptom of a softening of the magneto-phonon mode, a precursor to the formation of a Wigner crystal with a lattice spacing of k_{min}^{-1}.

As I have just observed, there is something of a recapitulation of the integer effect excitations in the FQHE magnetophonons. At first sight this is well described by the Chern-Simons actions of the ZHK variety: for the FQHE states the statistics parameter, θ, is adjusted so that there is a larger "statistical flux" associated with each electron than is absolutely necessary for turning the boson into an electron. In the commonly used mean-field picture, the extra 2π's of flux cancel part of the external magnetic field and make the system look like the integer effect with an exactly filled Landau level. Surely, then, magneto-plasmon of this pseudo-IQHE must be the magneto-phonon of the FQHE? This is a tempting idea, but it does not work! Although the induced statistical magnetic field looks like an ordinary magnetic field, it *exerts no forces!* The statistics field $f_{\mu\nu} = \partial_\mu a_\nu - \partial_\nu a_\mu$ obeys

$$j_\mu \epsilon_{\mu\nu\sigma} = \frac{1}{4\theta} f_{\nu\sigma}, \tag{4.3.1}$$

so the Lorentz force exerted by it on the matter vanishes:

$$f_{\nu\sigma} j_\sigma = \frac{1}{4\theta} j_\mu j_\sigma \epsilon_{\mu\nu\sigma} = 0. \tag{4.3.2}$$

This should not be surprising. The statistics field was carefully constructed to produce only Bohm-Aharonov phase effects. It cannot, and does not, alter frequencies. As a consequence, the gapped phonon excitations derived from the ZHK Chern-Simons action always lie near $\hbar\omega_c$. This point was made, with some referee-induced equivocation, in [rep.38], and stated forcefully by Zhang and Lee in [rep.37]. To find the true magneto-phonon in these actions, one needs to look at non-perturbative vortex excitations [rep.37].

4.3.3 Wigner Solids

A discussion of modes in two-dimensional electron gases would not be complete without a discussion of the solid phases. At low enough filling-fraction, the Coulomb repulsion between electrons will dominate over the delocalizing effects of the residual kinetic energy. Under these conditions, the Hall fluid is expected to freeze into a charge-density wave, or *Wigner Crystal.* There are various ways of detecting this. A crystaline phase breaks translation invariance so the electron solid can be pinned, or locked in place, by inhomogeneities and impurities. When this occurs, charge transport can only take place by tearing the crystal structure and this will be signalled by complex nonlinear phenomenon. A cleaner diagnostic is provided by the appearance of new phonon modes. In the quantum Hall fluid phases the phonon-like compressive modes are all gapped, but a crystal can sustain *shear modes*, and these transverse modes turn out to be gapless even in a magnetic field.

Suppose we describe the elastic properties of the crystal by a pair of Lamé constants, λ and μ, which relate the stress tensor, $\sigma_{\alpha\beta}$, to the strain tensor, $e_{\alpha\beta}$:

$$\sigma_{\alpha\beta} = 2\mu e_{\alpha\beta} + \lambda e_{\gamma\gamma}\delta_{\alpha\beta}. \tag{4.3.3}$$

The strain tensor is in turn determined by the displacement field η_α as

$$e_{\alpha\beta} = \frac{1}{2}(\partial_\alpha \eta_\beta + \partial_\beta \eta_\alpha). \tag{4.3.4}$$

Now the body force due to the stress is equal to the divergence of the stress tensor and, including the Lorentz force, the equation of motion becomes

$$\rho \partial_t^2 \eta_\alpha = \mu \nabla^2 \eta_\alpha + (\lambda + \mu)\partial_\alpha(\partial_\beta \eta_\beta) + \frac{e}{m}\rho \epsilon_{\alpha\beta\gamma}(\partial_t \eta_\beta B_\gamma). \tag{4.3.5}$$

In the absence of a magnetic field there would be a longitudinal mode with velocity c_L, and a transverse mode with velocity c_T:

$$\begin{aligned} \rho c_L^2 &= (\lambda + 2\mu) \\ \rho c_T^2 &= \mu. \end{aligned} \tag{4.3.6}$$

Introducing $\mathbf{B} = (0, 0, B)$ and setting

$$\eta = \begin{bmatrix} \eta_1 \\ \eta_2 \end{bmatrix} e^{ikx - i\omega t}, \tag{4.3.7}$$

we find mixed longitudinal-transverse waves with a dispersion law determined by the secular equation

$$\begin{vmatrix} c_L^2 k^2 - \omega^2 & -i\omega\omega_c \\ +i\omega\omega_c & c_T^2 k^2 - \omega^2 \end{vmatrix} = 0, \tag{4.3.8}$$

i.e.,

$$\omega^4 - \omega^2[(c_L^2 + c_T^2)k^2 + \omega_c^2] + c_L^2 c_T^2 = 0. \tag{4.3.9}$$

One branch of the $\omega(k)$ dispersion curve starts out at $k = 0$ with the cyclotron frequency ω_c, and evolves into longitudinal waves at large k. A second branch is ungapped, $\omega = 0$ at $k = 0$, and evolves into the transverse shear modes. The shear wave curve always lies below the compression mode curve. The gapless shear waves are a characteristic of the Wigner solid, and they have been reported to occur experimentally on either side of the filling fraction $\nu = \frac{1}{5}$ quantum Hall state [8-13].

Of course the simple Lamé form for the elastic constants is not strictly applicable to the Wigner crystal because of the long-range nature of the Coulomb forces. In reality, at small k, one expects the ungapped branch to disperse as $\omega = k^{\frac{3}{2}}$ rather than as the $\omega = k^2$ that follows from (4.3.9).

4.4 Field Theory in the Lowest Landau Level

The inter-Landau-level energy-gap, $\hbar\omega_c$, is usually much larger than the energy scales relevant for the FQHE, whose essential physics results from the rearrangement of the nearly degenerate states in a single Landau level. In obtaining the dispersion curve for the magneto-phonons, Girvin, MacDonald and Platzman [rep.31] found it essential to ignore the existence of other Landau levels and restrict their Hilbert space to the lowest level. If they had not, they would have been swamped by the magneto-plasmons that saturate the oscillator strength sum-rule at low k. In [rep.35], Read also works with states entirely in the lowest Landau level, so there is no possibility of contamination by magneto-plasmons, and his derivation of the effective action and its collective modes is a position-space parallel to the variational calculations of Girvin *et al.* As a result, it is clear that Read's collective mode is the true magnetophonon. All is not solved, however. While the order parameter in this action looks almost identical to the bosonized fermions of [rep.34], unlike them it is a charge neutral operator, and its minimal coupling to the magnetic field in the covariant derivatives is derived only for constant fields. This coupling cannot therefore be used to infer the effects of induction produced electric fields. The coupling to scalar potentials is sufficiently complicated that it is omitted from the final formulae in [rep.35]. Indeed, it is a challenge to the reader to derive the Hall effect from Read's action. As a helpful hint, this section derives some formalism that should be useful in this task.

Some of the machinery for the first quantized approach to the lowest-Landau-level states was worked out in [rep.12], and more properties of the momentum-space many-body density operators are constructed as they were needed in [rep.31]. Read uses a field-theory version of the same calculus in [rep.36]. Here I give an introduction to some of the unconventional features of lowest-Landau-level (LLL) projected fields. Since Girvin, Platzman, and Jach have done a good job in describing the formalism in momentum space, I will concentrate on the configuration space properties of the density and current-density operators. My understanding of this subject has benefited greatly from discussions with Juan Martinez, and any novel results in this section have been obtained jointly with him.

4.4.1 Lowest Landau Level Field Operators

The LLL formalism works most naturally when we exploit the analyticity of the wavefunctions. We will use an orthonormal basis of holomorphic LLL states and construct a projected electron field operator

$$\psi = \sum_{n=0}^{\infty} \hat{a}_n \frac{1}{\sqrt{2\pi 2^n n!}} z^n e^{-\frac{1}{4}|z|^2}. \tag{4.4.1}$$

The operators $\hat{a}_n$ and $\hat{a}_m^\dagger$ annihilate and create normalized LLL states so they have the usual Fermi canonical anticommutation relations

$$\{\hat{a}_n, \hat{a}_m^\dagger\} = \delta_{mn}. \tag{4.4.2}$$

Because the LLL functions are not complete in the whole Hilbert space, we have somewhat unconventional equal time commutators for the fields

$$\{\psi^\dagger(z_1), \psi(z_2)\} = \frac{1}{2\pi} e^{-\frac{1}{4}|z_1-z_2|^2} e^{\frac{1}{4}(\bar{z}_1 z_2 - \bar{z}_2 z_1)} = \{z_1|z_2\}. \tag{4.4.3}$$

A bilocal kernel $\{z_1|z_2\}$ has replaced the usual $\delta^2(\mathbf{r}-\mathbf{r}_2)$ function. This kernal is a LLL analogue of the delta function and retains its reproducing property

$$\int d^2 z_1 \psi(z_1)\{z_1|z_2\} = \psi(z_2), \tag{4.4.4}$$

where ψ can be any function of the form (4.4.1), irrespective of whether the a_n are operators or c-numbers. (Note that $\{z'|z\}$ has this form as a function of z, so it reproduces itself.) Being bilocal, the kernel is not gauge invariant — the second exponential factor is a phase, and would change if we selected a different gauge.

The operators $\psi^\dagger(z)$ insert electrons in coherent states centered on z. We may use them to make a many-electron state out of any antisymmetric holomorphic function

$$|f\rangle = \int \prod_1^N d^2 z_i f(z_1, \dots, z_N) e^{-\frac{1}{4}\sum_1^N |z_i|^2} \psi^\dagger(z_1)\dots\psi^\dagger(z_n)|0\rangle. \tag{4.4.5}$$

The anticommutation relations (4.4.3) and the reproducing kernel, enable us to express the inner product of two such states as

$$\langle f|g\rangle = N! \int \prod d^2 z_i \overline{f(z_i)} g(z_i) e^{-\frac{1}{2}\sum_i |z_i|^2}. \tag{4.4.6}$$

In particular, defining the state

$$|z_1, \ldots, z_N\rangle = \psi^\dagger(z_1)\ldots\psi^\dagger(z_n)|0\rangle, \tag{4.4.7}$$

we recover the wavefunction description

$$\langle z_1, \ldots, z_N|f\rangle = f(z_1, \ldots, z_N)e^{-\frac{1}{4}\sum_1^N |z_i|^2}. \tag{4.4.8}$$

We get the Laughlin wavefunction by taking $f = \prod(z_i - z_j)^{2n+1}$.

4.4.2 Charge Density and Current Operators

It is natural to define the density operator as

$$\hat{\rho}(z) = \psi^\dagger(z)\psi(z), \tag{4.4.9}$$

since the integral of $\hat{\rho}(z)$ gives the total fermion number. The action of $\hat{\rho}(z)$ on the state $|f\rangle$ is

$$\hat{\rho}(z)|f\rangle = \sum_i \int \left(\prod d^2 z_i\right)' f(z_1, \ldots, z, \ldots, z_n)|z_1, \ldots, z, \ldots, z_N\rangle, \tag{4.4.10}$$

where the z appears in the i'th slot and the prime on the product indicates that the d^2z_i factor is to be omitted. The *projected* operator $\hat{\rho}(z)$ therefore acts on the wavefunctions in a manner similar to the *unprojected* density operator $\hat{\rho}_{full}(\mathbf{r}) = \sum_i \delta^2(\mathbf{r} - \mathbf{r}_i)$.

It is now reasonable to conjecture that the LLL current operator would also be given by its usual form

$$\begin{aligned} \hat{j}_z(z) &= \frac{1}{2mi}(\psi^\dagger \nabla_z \psi - (\nabla_z \psi^\dagger)\psi) \\ \hat{j}_{\bar{z}}(z) &= \frac{1}{2mi}(\psi^\dagger \nabla_{\bar{z}} \psi - (\nabla_{\bar{z}} \psi^\dagger)\psi), \end{aligned} \tag{4.4.11}$$

with the covariant derivatives containing the symmetric gauge A_μ fields

$$\begin{aligned} \nabla_z\psi &= (\partial_z - \bar{z}/4)\psi, \qquad \nabla_z\psi^\dagger = (\partial_z + \bar{z}/4)\psi^\dagger \\ \nabla_{\bar{z}}\psi^\dagger &= (\partial_{\bar{z}} - z/4)\psi^\dagger, \qquad \nabla_{\bar{z}}\psi = (\partial_{\bar{z}} + z/4)\psi. \end{aligned} \tag{4.4.12}$$

Unfortunately, things are not so simple. The operator identities,

$$\nabla_{\bar{z}}\psi = 0 = \nabla_z\psi^\dagger, \tag{4.4.13}$$

which express the vanishing of the non-diagonal part of the kinetic energy in the LLL states, can be used to simplify (4.4.12) and we see that this form of the current depends only on the density

$$\hat{j}_z(z) = \frac{1}{2mi}\partial_z(\psi^\dagger\psi(z)), \qquad \hat{j}_{\bar{z}}(z) = -\frac{1}{2i}\partial_{\bar{z}}(\psi^\dagger\psi(z)). \tag{4.4.14}$$

The current in (4.4.14) is the diamagnetic current due to the cyclotron motion of the electrons. When the density is uniform, the currents from adjacent cyclotron orbits cancel each other; with a density gradient, such as at the edges of the electron gas, the cancellation is imperfect and a net diamagnetic current appears.

Clearly the definition (4.4.11) contains sensible physics, but not enough. A glance at (4.4.14) shows that the density induced current is *solonoidal*

$$\partial^\mu j_\mu = 2(\partial_{\bar z} j_z + \partial_z j_{\bar z}) = 0, \tag{4.4.15}$$

and if $\hat{j}_\mu$ were the entire story it would imply that $\partial_t \hat\rho = 0$. This is certainly not correct.

With the definition (4.4.11), we have managed to lose the drift currents due to potential gradients. The reason why the usual formula for $\hat{j}_\mu$ fails in the LLL is that the equation of motion for ψ is not obtained by simply replacing the operators in the unprojected equation of motion by their projected versions. This is so even when the equation of motion is linear in ψ. As an example, consider an interaction with an external potential V. The kinetic part of the hamiltonian is diagonal in the LLL basis and may be ignored, so we take

$$\hat H = \int d^2z V(z,\bar z)\psi^\dagger \psi(z). \tag{4.4.16}$$

To find $\partial_t \psi$ we commute $\hat H$ through ψ. We find non-local terms that must be tidied up before they can be interpreted.

A useful gauge invariant tool for doing such tidying is the covariant form of Taylor's theorem. For a covariantly constant field $(\partial_\mu + A_\mu)\varphi \equiv 0$ we can write

$$\varphi(x_2) = \mathcal{P}\exp\left\{-\int_{x_1}^{x_2} A_\mu dx^\mu\right\}\varphi(x_1). \tag{4.4.17}$$

The path-ordered exponential may be evaluated along any path between x_1 and x_2. Even when the covariant derivative is not identically zero, there is still a gauge covariant relation

$$\varphi(x_2) = \mathcal{P}\exp\left\{-\int_{x_1}^{x_2} A_\mu dx^\mu\right\}\sum_0^\infty \frac{1}{n!}\left((x_2-x_1)^\mu \nabla_\mu\right)^n \varphi(x_1). \tag{4.4.18}$$

Here the path-ordering must be along a straight line connecting x_1 and x_2. This identity is most easily proved by gauge-transforming A_μ so that its components along the path vanish, and (4.4.18) reduces to Taylor's theorem. Then note that the full expression is gauge covariant, so if (4.4.18) is true in any one gauge it is true in any gauge.

From this result, we see that

$$\psi(z_2) = (phase)e^{(z_2-z_1)\nabla_z + (\bar z_2 - \bar z_1)\nabla_{\bar z}}\psi(z_1). \tag{4.4.19}$$

We may simplify this by using the commutator

$$[\nabla_z, \nabla_{\bar z}] = \frac{1}{2}, \tag{4.4.20}$$

in conjunction with the Baker-Hausdorff identity, to rearrange the exponentials. We get

$$\begin{aligned}\psi(z_2) =& (phase)e^{-\frac{1}{4}|z_2-z_1|^2} e^{(z_2-z_1)\nabla_{z_1}} e^{(\bar z_2 - \bar z_1)\nabla_{\bar z_1}}\psi(z_1)\\ =& (phase)e^{-\frac{1}{4}|z_2-z_1|^2}\sum_n \frac{(z_2-z_1)^n}{n!}\nabla_{z_1}^n \psi(z_1).\end{aligned} \tag{4.4.21}$$

Comparison with (4.4.1) shows that, up to gauge factors, the covariant derivatives of ψ are the expansion coeficients, $\hat{a}_n(z_1)$, for an expansion of ψ about some point other than $z = 0$.

Now $[\psi(z_1), \psi^\dagger(z_2)\psi(z_2)] = \{z_2|z_1\}\psi(z_2)$ is gauge covariant, so the phase factor cancels in this combination and

$$\{z_2|z_1\}\psi(z_2) = e^{-\frac{1}{2}|z_2-z_1|^2}\sum_n \frac{1}{n!}(z_2 - z_1)^n(\nabla_z)^n\psi(z_1). \tag{4.4.22}$$

As an aside, note that we may use (4.4.13) to write

$$\psi^\dagger(z_1)\psi(z_2) = \{z_1|z_2\}\sum_{n=0}^{\infty} \frac{1}{n!}(z_2 - z_1)^n\partial_z^n\hat{\rho}(z), \tag{4.4.23}$$

i.e., the one-particle density matrix is determined entirely by its diagonal part, the density operator.

We now return to our problem, and use (4.4.22) to calculate the time rate of change of the density due to the potential V:

$$\partial_t\hat{\rho}(z_1) = i[\hat{H}, \psi^\dagger\psi(z_1)]. \tag{4.4.24}$$

We find

$$\begin{aligned}\partial_t\hat{\rho}(z_1) =& i\int\frac{d^2z}{2\pi}V(z,\bar{z})e^{-\frac{1}{2}|z-z_1|^2}\{\sum\frac{1}{n!}(\bar{z}-\bar{z}_1)^n\partial_{\bar{z}}^n\hat{\rho}(z_1)\\ &-\sum\frac{1}{n!}(z-z_1)^n\partial_z^n\hat{\rho}(z_1)\}.\end{aligned} \tag{4.4.25}$$

In (4.4.25) we have again used (4.4.13) and the derivation property of covariant derivatives.

By introducing an apodized potential

$$\tilde{V}(z_1,\bar{z}_1) = \int\frac{d^2z}{2\pi}e^{-\frac{1}{2}|z-z_1|^2}V(z,\bar{z}), \tag{4.4.26}$$

we can write (4.4.25) in a way that makes it appear relatively simple, and, moreover, a sum of *local* terms

$$\partial_t\hat{\rho}(z) = i\sum_{n=1}^{\infty}\frac{2^n}{n!}\left\{\partial_{\bar{z}}^n\tilde{V}(z,\bar{z})\partial_z^n\hat{\rho}(z) - \partial_z^n\tilde{V}(z,\bar{z})\partial_{\bar{z}}^n\hat{\rho}(z)\right\}. \tag{4.4.27}$$

In this form, by integrating by parts, it is easy to confirm that the total charge is conserved.

For potentials varying slowly on the scale of the magnetic length, we need keep only the lowest derivatives

$$\partial_t\hat{\rho} \approx 2i(\partial_{\bar{z}}\tilde{V}\partial_z\hat{\rho} - \partial_z\tilde{V}\partial_{\bar{z}}\hat{\rho}). \tag{4.4.28}$$

We can view the right-hand side of (4.4.28) as the divergence of a current $\rho\mathbf{v}$ with $\mathbf{v}$ being the usual drift velocity

$$\begin{aligned}v^x &= -\,\partial_y\tilde{V}\\ v^y &= +\,\partial_x\tilde{V}.\end{aligned} \tag{4.4.29}$$

It is therefore possible to find the correct drift motion even when we restrict the basis states to the LLL, but one pays a price of some complexity in the formulation, and I do not know a simple general formula for the current that gives rise to the $\hat{\rho}(z)$ evolution.

4.4.3 The Quasihole Operator

In developing his effection action, Read uses a composite operator containing the fermion creation operator and odd powers of the Laughlin's quasihole creation operator, $U(\zeta)$. In first-quantized language $U(\zeta)$ acts by multiplying the wavefunction by $\prod(z_i - \zeta)$. How should it be defined as a second-quantized LLL operator? It is tempting to try

$$U(\zeta)\psi^\dagger(z) = (z-\zeta)\psi^\dagger(z)U(\zeta) \quad (?) \tag{4.4.30}$$

and

$$U(\zeta)|0\rangle = |0\rangle, \tag{4.4.31}$$

as this will have the same effect as the first-quantized operator on any state of the form

$$|z_1,\ldots,z_n\rangle = \psi^\dagger(z_1)\ldots\psi^\dagger(z_n)|0\rangle. \tag{4.4.32}$$

Unfortunately, this set of states is *overcomplete* and our requirements overly restrictive, making the definition (4.4.30) inconsistent.

To see the inconsistency, note that (4.4.30) implies that

$$U(0)\psi^\dagger(0)|0\rangle = 0 \quad (?), \tag{4.4.33}$$

but $\psi^\dagger(0)|0\rangle = \frac{1}{2\pi}a_0^\dagger|0\rangle$ should not be annihilated by $U(0)$, it merely rolled over into $a_1^\dagger|0\rangle$, as can be seen by applying (4.4.30) to the identity

$$a_n^\dagger = \frac{1}{\sqrt{2\pi 2^n n!}}\int d^2z z^n e^{-\frac{1}{4}|z|^2}\psi^\dagger(z). \tag{4.4.34}$$

A better definition is

$$\begin{aligned} U(\zeta)\psi^\dagger(z) &= (2\partial_{\bar{z}} + \frac{1}{2}z - \zeta)\psi^\dagger(z)U(\zeta) \\ &= (2\nabla_{\bar{z}} + z - \zeta)\psi^\dagger(z)U(\zeta) \end{aligned} \tag{4.4.30a}$$

while retaining (4.4.31) unchanged. In states $|f\rangle$, of the form (4.4.5), we can integrate by parts and replace the $2\partial_{\bar{z}}$ by $\frac{1}{2}z$. So on this complete set (4.4.30a) has the same effect as the original (4.4.30). Further, we now find that that

$$U(0)a_n^\dagger = \sqrt{2n+2}\,a_{n+1}^\dagger U(0), \tag{4.4.35}$$

showing that the $U(\zeta)$ operator has the desired effect of pushing the electrons out by one orbit. The consistency of the new definition, (4.4.30a), is established by noting that we could have taken (4.4.35) as the definition of the action of $U(\zeta)$ on a complete set of states, and that (4.4.30a) may then be deduced from (4.4.35).

Read now observes that the combination $\Phi^\dagger(z) = \psi^\dagger(z)U^{2n+1}$ is a Bose-like operator, and that the N particle Laughlin state can be written as

$$|0,L\rangle = \left(\int d^2z\psi^\dagger(z)U^{2n+1}(z)e^{-\frac{1}{4}|z|^2}\right)^N|0\rangle, \tag{4.4.36}$$

very suggestive of a Bose condensate. A superposition of states with different N can satisfy the cluster decomposition property and possess a non-vanishing constant expectation value for $\Phi^\dagger(z)$. If $\Phi^\dagger(z)$ is varying in space, its covariant f $\langle\Phi^\dagger(z)\rangle$ gives the amplitudes for the electrons to be in orbitals near the zeros of the wavefunction rather than exactly at them. In this way, $\Phi^\dagger(z)$ serves to parametrize deformations of the Laughlin state.

Reprints for Chapter 4

[rep.31] *Collective excitation gap in the fractional quantum Hall effect*, S. M. Girvin, A. H. MacDonald, P. M. Platzman, Phys. Rev. Lett. 54 (1981) 581-583.

[rep.32] *Cooperative-ring-exchange theory of the fractional quantized Hall effect*, S. Kivelson, C. Kallin, D. P. Arovas, J. R. Schrieffer, Phys. Rev. Lett. 56 (1986) 873-876.

[rep.33] *Off-diagonal long range order, oblique confinement, and the fractional quantum Hall effect*, S. M. Girvin, A. H. MacDonald, Phys. Rev. Lett. 58 (1987) 1252-1255.

[rep.34] *Effective-field-theory model for the fractional quantum Hall effect*, S. C. Zhang, T. H. Hansson, S. Kivelson, Phys. Rev. Lett. 62 (1989) 82-85.

[rep.35] *Off-diagonal long range order in fractional quantum-Hall-effect states*, E. Rezayi, F. D. M. Haldane, Phys. Rev. Lett. 61 (1988) 1985-1988.

[rep.36] *Order parameter and Ginzburg–Landau theory for the fractional quantum Hall effect*, N. Read, Phys. Rev. Lett. 62 (1989) 86-89.

[rep.37] *Collective excitatations in the Ginzburg–Landau theory of the fractional quantum Hall effect*, D-H. Lee, S-C. Zhang, Phys. Rev Lett 66 (1991) 1220-1223.

[rep.38] *Superfluid dynamics of the fractional quantum Hall state*, M. Stone, Phys. Rev. B42 (1990) 212-217.

Other References for Chapter 4

[1] *The Quantum Hall Effect* edited by E. Prange and S. M. Girvin (Springer 1987).

[2] Proceedings of the *Symposium on Anomalies, Geometry, and Topology*, March 28-30 1985, W. A. Bardeen, A. R. White, editors, World Scientific, 1985.

[3] *Fractional Statistics and Anyon Superconductivity*, F Wilczek, ed. World Scientific 1990.

[4] E. Witten, Comm. Math. Phys, 121 (1989) 351.

[5] S. M. Girvin, A. H. MacDonald, P. M. Platzman Phys. Rev. B 33 (1986) 2481.

[6] W. Kohn, Phys. Rev. 123 (1961)1242.

[7] S-K. Yip, Phys. Rev. B 43 (1991) 1707.

[8] E. Y. Andrei, G. Deville, D. C. Glattli, F. I. B. Williams, Phys. Rev. Lett. 60 (1988) 2765.

[9] D. C. Glattli, E. Y. Andrei, R. G. Clark, G. Deville, C. Dorin, B. Etienne, C. T. Foxon, J. J. Harris, E. Paris, O. Probst, F. I. B. Williams, P. A. Wright, Physica B169 (1991) 328-335.

[10] H. W. Jiang, R. L. Willet, H. L. Stormer, D. C. Tsui, L. N. Pfeiffer, K. W. West, Phys. Rev. Lett. 65 (1990) 633.

[11] R. L. Willet, M. A. Paalanen, R. R. Ruel, K. W. West, L. N. Pfeiffer, D. J. Bishop, Phys. Rev. Lett. 65 (1990) 112.

[12] H. Buhmann. W. Joss, K. von Klitzing, I. V. Kukshkin, G. Martinez, A. S. Plaut, K. Ploog, V. B. Timofeev, Phys. Rev. Lett. 65 (1990) 1056.

[13] M. Goldman. J. E. Cunningham, M. Shayegan, M. Santos, Phys. Rev. Lett. 65 (1990) 2189.

Collective-Excitation Gap in the Fractional Quantum Hall Effect

S. M. Girvin
Surface Science Division, National Bureau of Standards, Gaithersburg, Maryland 20899

and

A. H. MacDonald
National Research Council of Canada, Ottawa, K1A 0R6, Canada

and

P. M. Platzman
AT&T Bell Laboratories, Murray Hill, New Jersey 07974
(Received 25 October 1984)

We present a theory of the collective excitation spectrum in the fractional quantum Hall-effect regimes, in analogy with Feynman's theory for helium. The spectrum is in excellent quantitative agreement with the numerical results of Haldane. *Within this approximation* we prove that a finite gap is generic to any liquid state in the extreme quantum limit and that in this single-mode *approximation* gapless excitations can arise only as Goldstone modes for ground states with broken translation symmetry.

PACS numbers: 72.20.My, 71.45.Gm, 73.40.Lq

The fractional quantum Hall effect[1] (FQHE) is one of the most remarkable many-body phenomena discovered in recent years. Associated with the quantization of the Hall resistance is a nearly complete freedom from dissipation. The latter suggests the existence of an excitation gap, presumably due to many-body correlations arising from the Coulomb interaction. Considerable theoretical effort has been made to understand the nature of the ground state which, at least for values of the Landau-level filling factor of the form $\nu = 1/m$, where m is an odd integer, seems to be quite well described by Laughlin's variational wave function.[2] In this Letter we present a theory of the excitation spectrum in the FQHE analogous to Feynman's theory for the excitation spectrum of superfluid ^{4}He.[3]

The Feynman argument for the excitation energy is equivalent to the assumption that the dynamic structure factor[4]

$$S(k,\omega) = \sum_n |\langle n|\rho_k|0\rangle|^2 \delta(E_n - E_0 - \omega) \tag{1}$$

is of the form

$$S(k,\omega) = Ns(k)\delta(\omega - \Delta(k)) \tag{2}$$

where $\rho_k \equiv \sum_j \exp(i\mathbf{k}\cdot\mathbf{r}_j)$. In this single-mode approximation (SMA) the excitation energy is

$$\Delta(k) = f(k)/s(k), \tag{3}$$

where the oscillator strength is

$$f(k) = N^{-1}\int d\omega\, \omega S(k,\omega), \tag{4}$$

and $s(k)$ is the static structure factor.

We may gain some insight into the validity of the SMA by noting that it works well in a variety of systems. In superfluid ^{4}He it is exact at long wavelengths and gives a good approximation to the entire phonon-roton excitation curve.[3] For the three- and two-dimensional electron gas (no magnetic field) it is again an excellent approximation to the plasmon at long wavelengths and a rough fit to the entire single-particle plasmon continuum at shorter wavelengths. For the two-dimensional electron gas in a large magnetic field it gives an accurate description at long wavelengths of the magnetoplasmon mode near $\omega_c \equiv eH/mc$. In general the SMA is accurate at long wavelengths where the oscillator strength in continuum modes is small or wherever these continuum modes do not exist.

For the FQHE high-energy cyclotron modes are not of primary interest. Of more relevance to the experiment and the nature of ground-state correlation are the low-lying excitations. Equation (3) tells us very little about such modes. However, if we insist that the excited states $|n\rangle$ in Eq. (1) lie within the lowest Landau level we get a version of Eq. (3) (the projected SMA), which describes these low-lying excitations. To do this we replace ρ_k by its projection $\bar{\rho}_k$ (bars indicate projected quantities), i.e.,

$$\Delta(k) = \bar{f}(k)/\bar{s}(k). \tag{5}$$

With use of the projected density operator[5] ($z_j = x_j + iy_j$, $k = k_x + ik_y$)

$$\bar{\rho}_k = \sum_{j=1}^{N} \exp(ik\partial/\partial z_j)\exp(ik^* z_j/2), \tag{6}$$

$\bar{s}(k)$ is easily shown to be

$$\bar{s}(k) = s(k) - (1 - e^{-|k|^2/2}) \tag{7}$$

$[\hbar = l \equiv (ec/H)^{1/2} = 1]$.

To find the projected oscillator strength we manipulate Eqs. (1) and (4) in the standard way,[4] i.e.,

$$\bar{f}(k) = N^{-1}\langle 0|\bar{\rho}_k^\dagger[\bar{H},\bar{\rho}_k]|0\rangle, \tag{8}$$

where the projected Hamiltonian is

$$\bar{H} = \tfrac{1}{2}\int [d^2q/(2\pi)^2]v(q)[\bar{\rho}_q^\dagger \bar{\rho}_1 - \rho e^{-q^2/2}]. \tag{9}$$

Under the assumption that the electrons are embedded in a solid with a static dielectric ϵ_0, $v(q) = 2\pi e^2/(\epsilon_0 q)$. The commutation in Eq. (8) may be computed to yield

$$\bar{f}(k) = (\nu/2\pi)\int [d^2q/(2\pi)^2]v(q)\int d^2r[g(r)-1][e^{-|k|^2/2}e^{iq\cdot r}(e^{(kq^*-k^*q)/2}-1) + e^{i(k+q)\cdot r}(e^{k\cdot q}-e^{k^*q})], \tag{10}$$

where $g(r)$ is the two-point correlation function related to $s(k)$ (for a homogeneous and isotropic system) by

$$s(k) = 1 + \rho\int d^2r e^{ik\cdot r}[g(r)-1] + \rho(2\pi)^2\delta^2(k). \tag{11}$$

We have thus succeeded in expressing $\Delta(k)$, the excitation energy, in terms of quantities dependent solely on the ground state. Since the kinetic energy has been quenched by the magnetic field the scale of energy is set solely by the scale of the interaction $[e^2/(\epsilon_0 l)]$.

Expansion of Eq. (10) shows that $\bar{f}(k)$ vanishes as $|k|^4$. Examination of (7) and (11) shows that $\bar{s}(k)$ also vanishes as $|k|^4$ if $M_0 = M_1 = -1$, where

$$M_n = \rho\int d^2r(r^2/2)^n[g(r)-1]. \tag{12}$$

In the symmetric gauge, at any filling factor ν, $g(r)$ may be written as

$$\rho[g(r)-1] = (2\pi\nu)^{-1}\sum_{s=0}^{\infty}(r^2/2)^s\exp(-r^2/2)/s![\langle n_s n_0\rangle - \langle n_s\rangle\langle n_0\rangle - \nu\delta_{s0}], \tag{13}$$

where n_s is the occupation number for the sth angular momentum state. Substitution of (13) into (12) yields

$$M_0 = \nu^{-1}[\langle N n_0\rangle - \langle N\rangle\langle n_0\rangle] - 1, \tag{14}$$

$$M_1 = \nu^{-1}[\langle (L+N)n_0\rangle - \langle L+N\rangle\langle n_0\rangle] - 1, \tag{15}$$

where $N = \sum_{s=0}^{\infty} n_s$ and $L = \sum_{s=0}^{\infty} s n_s$ are the total particle number and angular momentum. Since L and N are constants of the motion, their fluctuations vanish leaving $M_0 = M_1 = -1$. This general result implies that for any homogeneous and isotropic ground state $\bar{s}(k) \approx |k^4|$ for $k \to 0$. In order to relate this ground-state property to the excitation spectrum we must use the SMA which may only be approximate. However, it seems *plausible* that in this system there can be no low-lying single-particle excitations to defeat this gap. The kinetic energy necessary to produce such excitations has been quenched by the magnetic field. Hence we can conjecture that the only way that gapless excitations can occur is as Goldstone modes in systems with broken translation symmetry (e.g., the Wigner crystal[6,7]). In this approximation, it would appear that the existence of a gap for liquid ground states is the *rule* rather than the exception. Whether or not liquid ground states must have rational ν is an entirely separate question, about which nothing has been proved.

In order to evaluate Eq. (5) using (10) and (11) we need a specific model for the ground state. We have chosen to use the Laughlin ground state (LGS) for $\nu = \frac{1}{3}, \frac{1}{5}$.[2] For the LGS $\bar{s}(k)$ does vanish as $|k|^4$ with a coefficient which may be calculated exactly, i.e.,[8,9]

$$\bar{s}(k) = |k|^4(1-\nu)/8\nu. \tag{16}$$

For the LGS we chose to fit an analytic parametrization[10] for $g(r)$ to the Monte Carlo simulation data of Ref. 7. The resulting gap functions from Eq. (5) for $\nu = \frac{1}{3}, \frac{1}{5}$ are plotted in Fig. 1. The deep minimum in the gap dispersion is caused by a peak in $\bar{s}(k)$ and is, in this sense, quite analogous to the roton minimum in helium.[3] We interpret the deepening of the minimum in going from $\nu = \frac{1}{3}$ to $\nu = \frac{1}{5}$ to be a precursor of the collapse of the excitation gap which occurs at the critical density for Wigner crystallization[6] ($\nu \sim \frac{1}{6.5}$). Further evidence for this interpretation is provided by the fact that the magnitude of the primitive reciprocal lattice vector for the crystal lies close to the roton minimum, as indicated by the arrows in the figure.

Note that the $\nu = \frac{1}{3}$ results are in excellent agreement with the small system ($N = 6, 7$ particles) numerical calculations of Haldane.[11] Note also that in contrast to the case of helium the SMA works well without explicit backflow corrections even in the region of the roton minimum.[3] This can be understood from a semiclassical point of view (which can be shown to be valid for the lowest Landau level) in which the local current density has the form $\mathbf{J}(\mathbf{r}) = \rho(\mathbf{r})\nabla\phi(\mathbf{r})\times\hat{\mathbf{B}}$, where ϕ is the local potential. The current density around a particular charge then satisfies $\nabla\cdot\mathbf{J} = 0$ so that the backflow condition[3] (con-

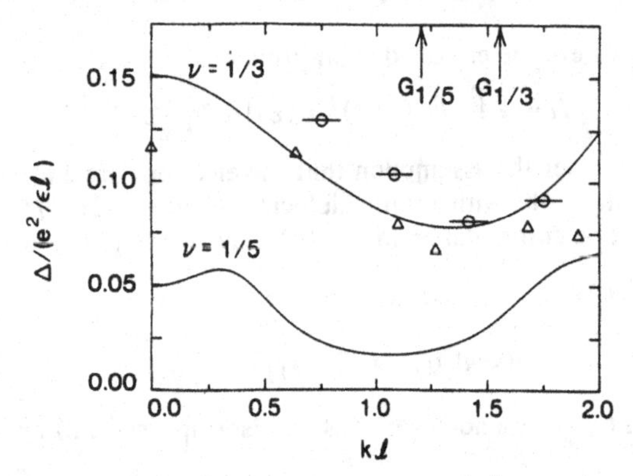

FIG. 1. Gap Δ vs wave vector for $\nu = \frac{1}{3}, \frac{1}{5}$. Circles are from $N = 7$ small spherical system calculations of Ref. 10. Horizontal error bars indicate uncertainty in conversion of angular momentum on a sphere to linear momentum. Triangles are for $N = 6$ periodic boundary condition calculations with a hexagonal unit cell of Ref. 10. The arrows indicate the magnitude of the primitive reciprocal-lattice vectors of the hexagonal Wigner crystal, for $\nu = \frac{1}{3}, \frac{1}{5}$. No small system calculations exist for $\nu = \frac{1}{5}$.

tinuity equation) is automatically satisfied.

For wave vectors beyond the roton minimum the SMA rapidly breaks down and it can be shown that the exact first moment of the excitation spectrum saturates at a large finite value $2|E_c|/(1-\nu)$, where E_c is the cohesive energy. Nevertheless, it is possible for us to estimate the excitation gap at $k = \infty$ by supposing that the lowest collective mode crosses over at the roton minimum from being a pure density oscillation to a bound quasiparticle-quasihole exciton.[9] The asymptotic exciton dispersion is[9] $\Delta_\mu(k) = \Delta_\mu - \nu^3/k$. Equating this to the SMA approximation to the gap at the minimum yields $\Delta_{1/3,1/5} = 0.106, 0.025$. These values lie considerably above the results of hypernetted chain calculations of Laughlin,[2] $\Delta_{1/3,1/5} = 0.057, 0.014$, and of Chakraborty,[12] $\Delta_{1/3,1/5} = 0.053, 0.014$. However, preliminary Monte Carlo results of Morf and Halperin[13] yield a larger value, $\Delta_{1/3} = 0.094 \pm 0.005$. Haldane's small system calculations[11] yield a value (extrapolated to $N = \infty$) of $\Delta_{1/3} = 0.105 \pm 0.005$, in excellent agreement with the present result. Meaningful comparison of these results with experimental activation energies[14] must await a deeper understanding of the role of disorder.

The authors are grateful to F. D. M. Haldane for supplying them with his numerical results prior to publication. One of us (S.M.G.) would like to acknowledge several conversations with R. B. Laughlin concerning the implications of the present results and would also like to thank D. K. Kahaner, S. Leigh, and J. Filliben for assistance with the numerical analysis and graphics. This work was begun at the Aspen Center for Physics during the 1984 workshop on the quantum Hall effect.

[1]H. L. Stormer, *Advances in Solid State Physics*, edited by P. Grosse (Vieweg, Braunschweig, 1984), Vol. 24, pp. 25–44, and references therein.

[2]R. B. Laughlin, Phys. Rev. Lett. **50**, 1395 (1983).

[3]R. P. Feynman, *Statistical Mechanics* (Benjamin, Reading, Mass., 1972), Chap. 11, and references therein.

[4]D. Pines and P. Nozieres, *Theory of Quantum Liquids* (Benjamin, New York, 1966).

[5]S. M. Girvin and Terrence Jach, Phys. Rev. B **29**, 5617 (1984).

[6]Pui K. Lam and S. M. Girvin, Phys. Rev. B **30**, 473 (1984).

[7]D. Levesque, J. J. Weis, and A. H. MacDonald, Phys. Rev. B **30**, 1056 (1984).

[8]J. M. Caillol, D. Levesque, J. J. Weis, and J. P. Hansen, J. Stat Phys. **28**, 235 (1982).

[9]R. B. Laughlin, in *Proceedings of the Seventeenth International Conference on Low-Temperature Physics*, edited by U. Eckern, A. Schmid, W. Weber, and H. Wühl (North-Holland, Amsterdam, 1984).

[10]S. M. Girvin, Phys. Rev. B **30**, 558 (1984).

[11]F. D. M. Haldane, unpublished.

[12]T. Chakraborty, unpublished.

[13]R. Morf and B. I. Halperin, unpublished.

[14]A. M. Chang, P. Berglund, D. C. Tsui, H. L. Stormer, and J. C. M. Huang, Phys. Rev. Lett. **53**, 997 (1984); S. Kawaji, J. Wakabayashi, J. Yoshino, and H. Sakaki, J. Phys. Soc. Jpn. **53**, 1915 (1984).

Cooperative-Ring-Exchange Theory of the Fractional Quantized Hall Effect

Steven Kivelson,[a] C. Kallin, Daniel P. Arovas,[b] and J. R. Schrieffer
Institute for Theoretical Physics, University of California, Santa Barbara, California 93106
(Received 3 September 1985)

A semiclassical path-integral approach is used to calculate the contribution of large-correlated-ring exchanges to the energy of a two-dimensional Wigner crystal in a strong magnetic field. This correlation energy $E_c(\nu)$ shows cusps at fractional fillings $\nu_c = n/m$ of the lowest Landau level. The uniform Wigner crystal is locally unstable for $\nu \neq \nu_c$ and the theory predicts the existence of fractionally charged quasiparticles to accommodate the extra density $\nu - \nu_c$.

PACS numbers: 71.45.Gm, 73.40.Lq

Recent experiments on the low-temperature, large-magnetic-field (B_0) conductivity of high-mobility electron layers[1] provide evidence of the existence of a family of novel condensed phases of the two-dimensional (2D) electron gas at "special" rational values of the dimensionless density ν, where ν is the mean number of electrons in the area $2\pi l_0^2 = \phi_0/B_0$ covered by one flux quantum $\phi_0 = hc/e$. Specifically, (1) the anomalously small value of σ_{xx} suggests[2] that there is a gap in the spectrum of current-carrying states, and hence a cusp in the correlation energy $E_c(\nu)$ at the special values of ν; (2) the fractional quantization of σ_{xy} may be evidence[3] of the existence of fractionally charged quasiparticles. We have studied this system in the absence of impurities, in the high-magnetic-field limit where the spacing $\hbar\omega_c = e\hbar B_0/m^*c$ between Landau levels as well as the Zeeman energy $\mu_B^* g B_0$ are much larger than the Coulomb repulsion between electrons $\sim V_0 = e^2/\epsilon l_0$, where ϵ is the dielectric constant. Thus, for $\nu < 1$, we assume that only spin-up electrons in the lowest Landau level (LLL) need be considered. Since the noninteracting system is highly degenerate, correlation effects are clearly very important.

Attempts to describe this system as a Wigner crystal (WC), which is the correct ground state for low densities, have been hampered by the fact that such descriptions give no evidence of cusps in the energy.[4,5] Maki and Zotos[5] used an *Ansatz* WC wave function, including Gaussian quantum fluctuations, and found densities $\nu = 1/m$ (m odd) favored, but only extremely weakly as a result of the small overlap between neighboring sites. Laughlin[3] proposed a Jastrow-type trial wave function which innovatively treated pairwise electronic correlations. He found fractionally charged excitations ($q = e/m$) above the $\nu = 1/m$ (m odd) ground state. At intermediate densities, Laughlin's wave function describes a liquid state with lower energy than that of the WC. Numerical studies on small numbers of electrons give support for such a state.[6] Recently, Chui, Hakim, and Ma[7] proposed a solidlike trial wave function, which they find to have a lower energy than that of Laughlin.

We have studied this problem, also beginning from a WC state, but using a systematic, semiclassical approximation.[8,9] Our results may be summarized as follows. We find exchange effects in which L electrons in a ring coherently rotate to an equivalent configuration leading to contributions to $E_c(\nu)$ which can be orders of magnitude larger than pair-exchange contributions because of the reduced tunneling barrier. In addition, these contributions exhibit nonanalytic cusplike behavior for certain rational values of ν.[10] Rings with large L play a dominant role for two reasons. Firstly, although the contribution from any single ring decreases exponentially with L, there are a very large number of rings with large L ($\sim K^L$ where K is the connectivity). Secondly, they can make a nonanalytic contribution to $E_c(\nu)$ as a result of interference between different exchange rings. The contribution from each ring contains a phase factor $\theta = 2\pi \times B_0 A(\nu)/\phi_0$ (Bohm-Aharonov effect), where $A(\nu)$ is the enclosed area. For arbitrary ν, the contributions from large rings add incoherently. However, because $A(\nu)$ is always approximately equal to an integer multiple of the area per elementary plaquette of the WC, for certain rational densities ν_c the different rings add in phase. It is this effect which makes these densities energetically favorable and which leads to cusps in $E_c(\nu)$ at $\nu = \nu_c$ when arbitrarily large rings are included.

Our model derives from a LLL path-integral representation for the partition function $Z = \mathrm{Tr}\, e^{-\beta H_N}$, with

$$H_N = \sum_{i=1}^{N} \frac{1}{2m^*}\left[\mathbf{p}_i + \frac{e}{2c} B_0 \hat{\mathbf{z}} \times \mathbf{r}_i\right]^2 + \sum_{j<k} V_2(\mathbf{r}_j - \mathbf{r}_k).$$

The single-particle Hamiltonian admits a continuous representation for LLL eigenstates, $\phi_{\mathbf{R}}(\mathbf{r}) = \langle \mathbf{r} | \mathbf{R} \rangle$:

$$\langle \mathbf{r} | \mathbf{R} \rangle = (2\pi)^{-1/2} \exp\{-\tfrac{1}{4}(\mathbf{r} - \mathbf{R})^2 + \tfrac{1}{2} i (\mathbf{r} \times \mathbf{R}) \cdot \hat{\mathbf{z}}\}$$

with $H_1 |\mathbf{R}\rangle = \frac{1}{2}\hbar\omega_c |\mathbf{R}\rangle$ and where we set $l_0 \equiv 1$. The resolution of the LLL projection operator, $P_0 = (1/2\pi)\int d^2R\, |\mathbf{R}\rangle\langle\mathbf{R}|$, may be used to develop a path-integral expression[11] for Z:

$$Z(\nu)=\frac{1}{N!}\sum_{P\in S_N}(-1)^P\int d^{2N}r\langle\{\mathbf{r}_i\}|e^{-\beta H_N}|\{\mathbf{r}_{P(i)}\}\rangle,$$

where $N=\nu B_0/\phi_0$ and $|\{\mathbf{r}_i\}\rangle\equiv|\mathbf{r}_1\ldots\mathbf{r}_N\rangle$. This yields

$$Z(\nu)=\mathcal{N}\sum_{P\in S_N}(-1)^P\int\prod_{j=1}^{N}\mathcal{D}\mathbf{R}_j(\tau)e^{-S[R(\tau)]},\tag{1}$$

where $\mathcal{N}$ is a normalization constant and the boundary conditions require $\mathbf{R}_j(0)=\mathbf{R}_{P(j)}(\beta)$. The action for continuous paths is[11,12]

$$S[R]=\frac{1}{2}\int_0^\beta d\tau\left\{-i\sum_{j=1}^{N}(\dot{\mathbf{R}}_j\times\mathbf{R}_j)\cdot\hat{\mathbf{z}}+\sum_{j\neq k}V(\mathbf{R}_j-\mathbf{R}_k)\right\},\tag{2}$$

where V is the matrix element of the Coulomb potential between coherent states, $V(\mathbf{R})=\frac{1}{2}\sqrt{\pi}(e^2/\epsilon)\times\exp(-\frac{1}{8}R^2)I_0(\frac{1}{8}R^2)$. We will refer to the integration variable τ as the (imaginary) time.

The partition function Z is evaluated within the semiclassical approximation. This entails the finding of all paths $R^c(\tau)$ which extremize the action [R^c is a vector function with $2N$ components $\mathbf{R}_j(\tau)$] and then the inclusion of quantum fluctuations by expansion of the action to second order in $R-R^c$. In this way Z can be expressed as a sum over classical paths,

$$Z=\sum_c D[R^c]e^{-S[R^c]},\tag{3}$$

where $D[R^c]$ is the fluctuation determinant. The extremal (classical) paths satisfy the equations of motion

$$i\dot{X}_j=\partial V_j/\partial Y_j,\quad i\dot{Y}_j=-\partial V_j/\partial X_j,\tag{4}$$

where $V_j\equiv\sum_{i\neq j}V(\mathbf{R}_i-\mathbf{R}_j)$. These are simply the imaginary-time $\mathbf{E}\times\mathbf{B}$ drift equations. To find solutions to Eq. (4) which satisfy the boundary conditions $\mathbf{R}_j(0)=\mathbf{R}_{P(j)}(\beta)$, we analytically continue the path integral to complex values of X_j and Y_j.[12]

The path with the smallest action is the triangular WC. This path with pairwise exchange and Gaussian (phonon) fluctuations about it makes the leading-order contribution to Z, $Z_0\equiv D_0e^{-S_0}=\exp[-\beta\times NE_0(\nu)]$, where E_0 is the energy per site of the static WC as computed by Maki and Zotos.[5] $E_0(\nu)$ is a smooth, monotonic function of ν for $\nu<\frac{1}{2}$. Since the shear modulus is negative for $\nu>\nu_+\simeq0.45$,[5] we restrict our analysis to $\nu<\nu_+$. (For $\nu>1-\nu_+$, the same considerations apply to the hole lattice.) So as not to have prohibitively large action, the important classical paths must resemble the WC at most points in space and time. Therefore, we consider only classical paths whose initial configuration is the static WC. In addition, as discussed earlier, we focus on those paths which are most likely to lead to structure in $E_c(\nu)$ because of their systematic dependence on ν; that is, we consider paths, such as those illustrated in Fig. 1, which consist of a cyclic, coherent superposition of nearest-neighbor exchanges.

For convenience, we factor out the leading-order contribution that is common to all classical paths by writing Eq. (3) as

$$Z=Z_0\sum_c\tilde{Z}_c=Z_0\sum_c\tilde{D}[R_c]e^{-\tilde{S}[R_c]},\tag{5}$$

where $\tilde{D}=D/D_0$ and $\tilde{S}=S-S_0$. Let us consider the contribution of a single large exchange ring to Z. The real part of the classical action is approximately proportional to the number of electrons in the ring L and the imaginary part is $\theta=\pm2\pi(\phi/\phi_0)+\pi(L-1)$, where ϕ is the enclosed flux, the $\pm$ refers to positive or negative sense of rotation, and the term $\pi(L-1)$ reflects the Fermi statistics. If the classical path exactly followed the straight-line segments between sites (as shown in Fig. 1) then $\theta=\pm\pi(\nu^{-1}-1)N_A+\pi(\mathrm{mod}2\pi)$, where N_A is the number of enclosed plaquettes of area $\pi l_0^2/\nu$, and we have used the fact that N_A is even (odd) when L is even (odd). Similarly, the real part of the action would be $\alpha_0(\nu)L$, where α_0 is independent of path.[13] We have argued that the net effect of deviations from such linear paths is to renormalize α_0.[8] The fluctuation determinant is $\tilde{D}[R_c]=-\tau_0^{-1}d\tau\exp[-\Delta\alpha L+O(\ln L)]$, where τ_0 is the cooperative tunneling time.[14] Therefore, we approximate the contribution of a large-ring exchange as

$$\Delta\tilde{Z}_c(\nu)=\tau_0^{-1}d\tau\exp[-\alpha(\nu)L+ihN_A+O(\ln L)],$$

where $\alpha=\alpha_0+\Delta\alpha$ and $h=\pi(\nu^{-1}-1)$.

Whether or not large-ring exchanges contribute significantly to $Z(\nu)$ is determined by the numerical value of $\alpha(\nu)$. We have estimated $\alpha(\nu)$ for the simple case in which a single line of electrons exchange,

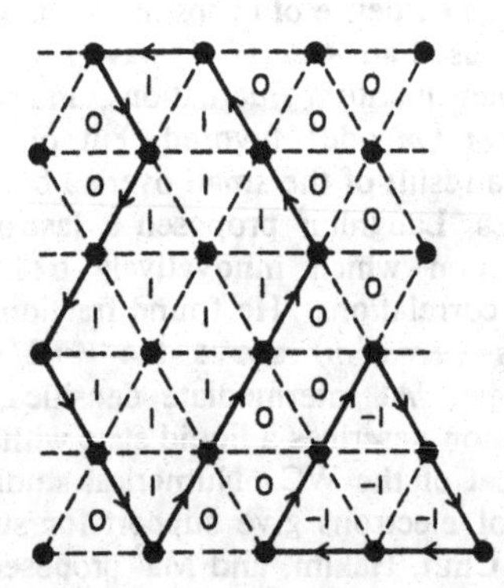

FIG. 1. Examples of exchange paths and their representation in terms of a configuration of the dual lattice.

so that $X_j(\beta)=X_j(0)+a_\nu$, $Y_j(\beta)=Y_j(0)$, where $a_\nu=(4\pi/\sqrt{3}\nu)^{1/2}$ is the lattice constant of the WC. We assume that only the electrons in this one line move in the background of the static potential of all other electrons. Thus we *overestimate* α; however, we believe that the relaxation corrections are small. We find that the extremal path corresponds to rigid motion of the line $[\mathbf{R}_j(\tau)-\mathbf{R}_j(0)=\mathbf{R}(\tau)$ for j on the line] and that $\alpha(\frac{1}{3})\sim 0.81$ (as a function of the density, α is approximately proportional to $1/\nu$). As we shall see, this indicates that large-ring exchanges are important at densities of experimental interest.

A general path in Eq. (5) contains many exchange rings and for small α, they are sufficiently dense that the exchange events do not form a dilute gas. Therefore, we include what we believe to be the most important interactions between exchange events that overlap in space and time. The time interval β is divided into slices of width τ_0 and, because the classical paths are exponentially localized in time, we ignore interactions between exchange events which occur in different time slices; i.e., $\tilde{Z}_c\simeq[Z_{\rm slice}]^{\beta/\tau_0}$, where $Z_{\rm slice}$ is the trace over all exchanges in a given time slice. The exchange paths in a given time slice are enumerated in terms of integer-valued spin variables S_λ; S_λ is defined to be the number of clockwise minus the number of counterclockwise exchange rings that encircle the plaquette λ. Hence, λ labels a site on the dual (honeycomb) lattice. We associate with each spin configuration an energy

$$H_{\rm DG}=\alpha\sum_{\langle\lambda,\gamma\rangle}(S_\lambda-S_\gamma)^2+ih\sum_\lambda S_\lambda, \tag{6}$$

where $\langle\lambda,\gamma\rangle$ denotes nearest-neighbor sites. Then we make the approximation $Z_{\rm slice}\simeq{\rm Tr}\exp(-H_{\rm DG})$ which is exact for all configurations of isolated rings and includes a repulsion between rings that share one or more nearest-neighbor bonds. Equation (6) is the Hamiltonian of the discrete Gaussian (DG) model in an imaginary field, where α^{-1} plays the role of temperature. This model is known to have a phase transition at a critical value of $\alpha=\alpha_c(h)$.[15] For $h=2\pi m$ $[\nu=1/(2m+1)]$, α_c takes on its maximum value, which we estimate to be $\alpha_c\simeq 1.1$. For $\alpha(\nu)>\alpha_c$, the system behaves like a classical WC, while for $\alpha(\nu)<\alpha_c$, the system is highly quantum mechanical and arbitrarily large exchange rings dominate the behavior of the system.

To study the $\alpha<\alpha_c$ phase it is useful to exploit the exact equivalence of the DG model and the Coulomb gas (CG)[15]:

$$H_{\rm CG}=\frac{2\pi^2}{\alpha}\sum_{\lambda\gamma}\left(q_\lambda-\frac{h}{2\pi}\right)G_{\lambda\gamma}\left(q_\gamma-\frac{h}{2\pi}\right),$$

where $G_{\lambda\gamma}\sim\ln|R_\lambda-R_\gamma|$ is the (honeycomb) lattice Green's function and q_λ is an integer charge. The small-α phase can be analyzed by a study of the ground-state properties of the CG. The $h=2\pi m$ ground state of $H_{\rm CG}$ has $q_\lambda=m$ and zero energy. The ground state for $|h-2\pi m|\equiv\delta h\ll 2\pi$ has a fraction $\delta h/2\pi$ of sites with charge $1-\delta h/2\pi$, which themselves form a Wigner lattice. The remaining sites have charge $-\delta h/2\pi$. Thus the free energy $F_{\rm CG}$ of the CG at small α is proportional to $|\delta h|\ln|2\pi/\delta h|$, which for our problem implies $E_c(\nu)\sim|\delta\nu|\ln|\delta\nu|$ for $|\delta\nu|\ll 1/(2m+1)$. Since $\partial E_c/\partial\nu$ diverges as $\delta\nu\to 0$, the uniform WC state is thermodynamically unstable in the open neighborhood of $\nu=1/(2m+1)$. This motivates the need for quasiparticles discussed below.

From studies of Josephson junction arrays in a transverse field,[16] $F_{\rm CG}$ is believed to have cusps of the form $|\delta h|\ln|2\pi/\delta h|$ at all rational $h/2\pi$ for $\alpha(\nu)<\alpha_c(h(\nu))$.[17] From our estimate of $\alpha(\nu)$ and from Monte Carlo calculations of Shih and Stroud,[16] we find that this inequality is most likely to be satisfied by $\nu=\frac{1}{3},\frac{1}{5},\frac{2}{5},\frac{2}{7},\frac{1}{4},\frac{3}{7}$, and $\frac{4}{9}$, although some of these phases may be unstable with respect to competing phases.[18] Since the $\nu=\frac{1}{3}$ uniform-density state is much more stable than any other special density, one might also consider a hierarchy of states formed by starting with a WC of quasiparticles and repeating our analysis for ring exchange of quasiparticles with fractional charge and fractional statistics.[19,20] In this case, one obtains a sequence of stable densities in agreement with previous hierarchical analyses.[20,21] We will discuss the relative stability of these states elsewhere.[8]

Our model is thus incompressible since it is rigid with respect to uniform dilations. However, there are quasiparticle (qp) excitations of charge Q^* which correspond to local dilations of the Wigner crystal by an amount δA. As a result of the deformation of the lattice, all rings that enclose the qp acquire an extra phase $\Delta\theta=2\pi B_0\delta A/\phi_0$ relative to those that do not. One can see immediately from the CG representation that this will cost a logarithmically divergent energy unless $\Delta\theta=2\pi\times$integer, and hence the elementary excitations have charge $Q^*=\pm\nu e$. We have constructed an approximate qp with a core size $\sim l_0$ and estimate its creation energy $E_{\rm qp}(\nu)\sim 0.5\nu^2e^2/\epsilon l_0$. When ν is near a favored density ν_c, the system can reduce its energy by forming a nonuniform state with density (averaged over a plaquette) equal to ν_c everywhere except in the core of the qp. The cusps in the energy are thus $E(\nu)\sim\nu_c^{-1}|\nu-\nu_c|E_{\rm qp}(\nu_c)$ for ν near ν_c. (We have not evaluated the creation energies with sufficient accuracy to distinguish between quasiparticle and quasihole energies.)

Of great interest is the magnetophonon (mp) spectrum. In the sparse-ring phase, the mp dispersion resembles the standard $\omega_k\sim k^{3/2}$ form.[5] In the dense-ring phase, however, nonlinear interactions between density fluctuations strongly alter the mp spectrum.[8] This effect and the issue of dissipation are presently under consideration.

We conclude with some brief comments on the consistency of our approach. In the small-α phase, the gas of exchange loops is quite dense. Hence, the time-slice path decomposition is not well defined. However, the major role of the time-slice approximation is only to provide an ultraviolet cutoff corresponding to a repulsion between overlapping exchange loops. One particular feature of the dense-ring phase is the proliferation of intersecting rings in the dual spin model. Some of these configurations correspond to high-action paths in real space (e.g., crossings), and should be discouraged by the inclusion of additional short-ranged spin-spin interactions.[22] We expect that such terms should lead to a renormalization of α, and that they will not change the universality class of the spin model.

An interesting open question concerns the existence of long-ranged charge-density-wave (CDW) order. At finite temperature, the magnetophonons will destroy any long-ranged order. At zero temperature, none of the terms that we have computed explicitly destroy the CDW. However, we have yet to establish fully whether or not such order is actually present in this limit, and the relation between our theory and that of Laughlin remains unclear. Nevertheless, it seems likely that our results do not depend essentially on the answer to this question, since the vanishing compressibility at rational ν derives from the coherent addition of many large exchange loops, an effect that could persist in the absence of CDW order.

We are grateful to A. J. Berlinsky, V. J. Emery, F. D. M. Haldane, T. C. Halsey, D. Stroud, and E. Tosatti for useful conversations. This work was supported in part by the National Science Foundation through Grants No. DMR 82-16285 and No. DMR 83-18051. One of us (D.P.A.) acknowledges support by an AT&T Bell Laboratories Scholarship, the work of another of us (C.K.) is partially supported by a Canadian Natural Sciences and Engineering Research Council Postdoctoral Fellowship, and one of us (S.K.) is the recipient of an Alfred P. Sloan Fellowship.

[(a)]Permanent address: Department of Physics, State University of New York at Stony Brook, Stony Brook, N.Y. 11794.

[(b)]Permanent address: Department of Physics, University of California, Santa Barbara, Cal. 93106.

[1]See, for example, D. C. Tsui, H. L. Stormer, and A. C. Gossard, Phys. Rev. Lett. **48**, 1559 (1982); H. L. Stormer, A. M. Chang, D. C. Tsui, J. C. M. Hwang, A. C. Gossard, and W. Wiegmann, Phys. Rev. Lett. **50**, 1953 (1983).

[2]R. B. Laughlin, Phys. Rev. B **23**, 5632 (1981); B. I. Halperin, Helv. Phys. Acta **56**, 75 (1983).

[3]R. B. Laughlin, Phys. Rev. Lett. **50**, 1395 (1983).

[4]D. Yoshioka and P. A. Lee, Phys. Rev. B **27**, 4986 (1983), and references therein.

[5]K. Maki and X. Zotos, Phys. Rev. B **28**, 4349 (1983).

[6]F. D. M. Haldane and E. H. Rezayi, Phys. Rev. Lett. **54**, 237 (1985); D. Yoshioka, B. I. Halperin, and P. A. Lee, Phys. Rev. Lett. **50**, 1219 (1983).

[7]S. T. Chui, T. M. Hakim, and K. B. Ma, to be published; see also S. T. Chui, Phys. Rev. B **32**, 1436 (1985).

[8]S. Kivelson, C. Kallin, D. P. Arovas and J. R. Schrieffer, to be published.

[9]This approximation is justified by the fact that the quantum parameter is small for $\nu < \frac{1}{2}$. See Ref. 8

[10]The possibility that large-ring exchanges might lead to a phase transition was discussed in a quite different context by R. P. Feynman, Phys. Rev. **91**, 1291 (1953). See also S. Chakravarty and D. B. Stein, Phys. Rev. Lett. **49**, 582 (1982).

[11]See, for example, L. Schulman, *Techniques and Applications of Path Integration* (Wiley, New York, 1981), Chap. 27.

[12]When discontinuous paths are important, one must use the discrete version of Eq. (1). See Ref. 11 and J. R. Klauder, in *Path Integrals and Their Applications in Quantum Statistical and Solid State Physics*, edited by G. Papadopoulos and J. T. Devreese (Plenum, New York, 1977), p. 5.

[13]The effective interaction between particles on the exchange path falls off sufficiently rapidly that α_0 is independent of path. See Ref. 8.

[14]The fluctuation determinant has been evaluated approximately. It contains a factor -1 per ring-exchange event. The τ_0^{-1} term arises due to the zero mode associated with time translation.

[15]S. T. Chui and J. D. Weeks, Phys. Rev. B **14**, 4978 (1976).

[16]W. Y. Shih and D. Stroud, Phys. Rev. B **32**, 158 (1985), and references therein.

[17]T. C. Halsey, unpublished.

[18]Note that the plateau at $\nu = \frac{1}{4}$ does not appear in the usual hierarchy scheme. In this context, we note that a plateau in σ_{xy} at $\nu = \frac{3}{4}$ as well as a broad minimum in σ_{xx} at $\nu = \frac{1}{2}$ have been reported by G. Ebert *et al.*, J. Phys. C **17**, L775 (1984).

[19]D. P. Arovas, J. R. Schrieffer, and F. Wilczek, Phys. Rev. Lett. **53**, 722 (1984).

[20]B. I. Halperin, Phys. Rev. Lett. **52**, 1583, 2390(E) (1984).

[21]F. D. M. Haldane, Phys. Rev. Lett. **51**, 605 (1983); R. B. Laughlin, Surf. Sci. **142**, 163 (1984).

[22]At the level of the spin model, several unphysical paths have already been discarded or suppressed. See Ref. 8.

Off-Diagonal Long-Range Order, Oblique Confinement, and the Fractional Quantum Hall Effect

S. M. Girvin
Surface Science Division, National Bureau of Standards, Gaithersburg, Maryland 20899

and

A. H. MacDonald
National Research Council, Ottawa, Ontario, Canada K1A OR6
(Received 24 November 1986)

We demonstrate the existence of a novel type of off-diagonal long-range order in the fractional-quantum-Hall-effect ground state. This is revealed for the case of fractional filling factor $v=1/m$ by application of Wilczek's "anyon" gauge transformation to attach m quantized flux tubes to each particle. The binding of the zeros of the wave function to the particles in the fractional quantum Hall effect is a (2+1)-dimensional analog of *oblique confinement* in which a condensation occurs, not of ordinary particles, but rather of composite objects consisting of particles and gauge flux tubes.

PACS numbers: 72.20.My, 71.45.Gm, 73.40.Lq

A remarkable amount of progress has recently been made in our understanding of the fractional quantum Hall effect (FQHE)[1] following upon the seminal paper by Laughlin.[2] There remains, however, a major unsolved problem which centers on whether or not there exists an order parameter associated with some type of symmetry breaking.[3-6] The apparent symmetry breaking associated with the discrete degeneracy of the ground state in the Landau gauge[5] is an artifact of the toroidal geometry[6,7] and is not an issue here. Rather, the questions that we are addressing have been motivated by the analogies which have been observed to exist[4,8] between the FQHE and superfluidity and by recent progress towards a phenomenological Ginsburg-Landau picture of the FQHE.[4] Further motivation has come from the development of the correlated-ring-exchange theory of Kivelson *et al.*[9] (see also Chui, Hakim, and Ma,[10] and Chui,[10] and Baskaran[11]). The existence of infinitely large ring exchanges is a signal of broken gauge symmetry in superfluid helium[12] and is suggestive of something similar in the FQHE. However, the concept of ring exchanges on large length scales has not as yet been fully reconciled with Laughlin's (essentially exact[7]) variational wave functions which focus on the short-distance behavior of the two-particle correlation function. Furthermore it is clear that we cannot have an ordinary broken gauge symmetry since the particle density (which is conjugate to the phase) is ever more sharply defined as the length scale increases. The purpose of this Letter is to unify all these points of view by demonstrating the existence of a novel type of off-diagonal long-range order (ODLRO) in the FQHE ground state.

In second quantization the one-body density matrix is given by

$$\rho(z,z')=\sum_{m,n}\varphi_m^*(z)\varphi_n(z')\langle 0|c_n^\dagger c_m|0\rangle, \quad (1)$$

where $\varphi_n(z)$ is the nth lowest-Landau-level single-particle orbital[1] in the symmetric gauge, and z is a complex representation of the particle position vector in units of the magnetic length.[1] It is an unusual feature of this problem that there is a unique single-particle state for each angular momentum. Hence by making only the assumption that the ground state is isotropic and homogeneous we may deduce $\langle 0|c_n^\dagger c_m|0\rangle = v\delta_{nm}$, and thereby obtain the powerful identity:

$$\rho(z,z')=vg(z,z')=(v/2\pi)\exp(-\tfrac{1}{4}|z-z'|^2)\exp[\tfrac{1}{4}(z^*z'-zz'^*)], \quad (2)$$

where $g(z,z')$ is the ordinary single-particle Green's function.[13]

We see from (2) that the density matrix is short ranged with a characteristic scale given by the magnetic length, just as occurs in superconducting films in a magnetic field.[14] The same result can be obtained within first quantization via the expression

$$\rho(z,z')=\frac{N}{Z}\int d^2z_2\cdots d^2z_N\Psi^*(z,z_2,\ldots,z_N)\Psi(z',z_2,\ldots,z_N), \quad (3)$$

where Z is the norm of Ψ.

If the lowest Landau level has filling factor $v=1/m$ and the interaction is a short-ranged repulsion, then in the low-electron mass limit,[7] the *exact* ground-state wave function is given by Laughlin's expression:

$$\Psi(z_1,\ldots,z_N)=\prod_{i<j}(z_i-z_j)^m\exp\left(-\tfrac{1}{4}\sum_k|z_k|^2\right). \quad (4)$$

Laughlin's plasma analogy[2,15] proves that the ground state is a liquid of uniform density so that Eq. (2) is valid. The rapid phase oscillations of the integrand in (3) cause ρ to be short ranged. There is, nevertheless, a peculiar type of long-range order hidden in the ground state. For reasons which will become clear below, this order is revealed by considering the singular gauge field used in the study of "anyons"[16,17]:

$$\mathcal{A}_j(z_j)=\frac{\lambda\Phi_0}{2\pi}\sum_{i\neq j}\nabla_j \mathrm{Im}\ln(z_j-z_i), \tag{5}$$

where $\Phi_0=hc/e$ is the quantum of flux and λ is a constant. The addition of this vector potential to the Hamiltonian is not a true gauge transformation since a flux tube is attached to each particle. If, however, $\lambda=m$ where m is an integer, the net effect is just a change in the phase of the wave function:

$$\Psi_{\mathrm{new}}=\exp\left(-im\sum_{i<j}\mathrm{Im}\ln(z_i-z_j)\right)\Psi_{\mathrm{old}}. \tag{6}$$

Application of (6) to the Laughlin wave function (4) yields

$$\Psi(z_1,\ldots,z_N)=\prod_{i<j}|z_i-z_j|^m\exp\left(-\tfrac{1}{4}\sum_k|z_k|^2\right), \tag{7}$$

which is purely real and is symmetric under particle exchange for both even and odd m. Hence we have the remarkable result that both fermion and boson systems map into bosons in this singular gauge.

Substituting (7) into (3) and using Laughlins's plasma analogy,[2,15] a little algebra shows that the singular-gauge density matrix $\tilde{\rho}$ can be expressed as

$$\tilde{\rho}(z,z')=(\nu/2\pi)\exp[-\beta\Delta f(z,z')]\,|z-z'|^{-m/2}, \tag{8}$$

where $\beta\equiv 2/m$, and $\Delta f(z,z')$ is the difference in free energy between two impurities of charge $m/2$ (located at z and z') and a single impurity of charge m (with arbitrary location). Because of complete screening of the impurities by the plasma, the free-energy difference $\Delta f(z,z')$ rapidly approaches a constant as $|z-z'|\rightarrow\infty$. This proves the existence of ODLRO[18] characterized by an exponent $\beta^{-1}=m/2$ equal to the plasma "temperature." For $m=1$ the asymptotic value of Δf can be found exactly: $\beta\Delta f_\infty=-0.03942$. For other values of m, $\beta\Delta f(z,z')$ can be estimated either by use of the ion-disk approximation[2,15] or the static (linear response) susceptibility of the (classical) plasma calculated from the known static structure factor[8] (see Fig. 1).

The rigorous and quantitative results we have obtained above are valid for the case of short-range repulsive interactions for which Laughlin's wave function is exact. We now wish to use these results for a qualitative examination of more general cases and to deepen our understanding of the ODLRO. We begin by noting that $\tilde{\rho}$ can be rewritten in the ordinary gauge as

$$\tilde{\rho}(z,z')=\frac{N}{Z}\int d^2z_2\cdots d^2z_N\exp\left[-i\frac{e}{\hbar c}\int_{\mathbf{z}}^{\mathbf{z}'}d\mathbf{r}\cdot\mathcal{A}_1\right]\Psi^*(z,z_2,\ldots,z_N)\Psi(z',z_2,\ldots,z_N), \tag{9}$$

where $\mathbf{z}$ and $\mathbf{z}'$ are vector representations of z and z'. The line integral in (9) is multiple valued but its exponential is single valued because the flux tubes are quantized. The additional phases introduced by the singular gauge transformation will cancel the phases in Ψ nearly everywhere, and produce ODLRO in $\tilde{\rho}$ if and only if the zeros of Ψ (which must necessarily be present because of the magnetic field[19]) are bound to the particles. Thus ODLRO in $\tilde{\rho}$ *always* signals a condensation of the zeros onto the particles (independent of whether or not the composite-particle occupation of the lowest momentum state diverges[18]). Because the gauge field $\mathcal{A}_1$ depends on the positions of *all* the particles, $\tilde{\rho}$ differs not just in the phase but in *magnitude* from ρ. This multiparticle object, which explicitly exhibits ODLRO, is very reminiscent of the topological order parameter in the *XY* model[20] and related gauge models[21,22] and is intimately connected with the frustrated *XY* model which arises in the correlated-ring-exchange theory.[9]

For short-range interactions, the zeros of Ψ are directly on the particles and the associated phase factors are exactly canceled by the gauge term in (9) [see Eq. (7)]. As the range of the interaction increases, $m-1$ of the zeros move away from the particles but remain nearby

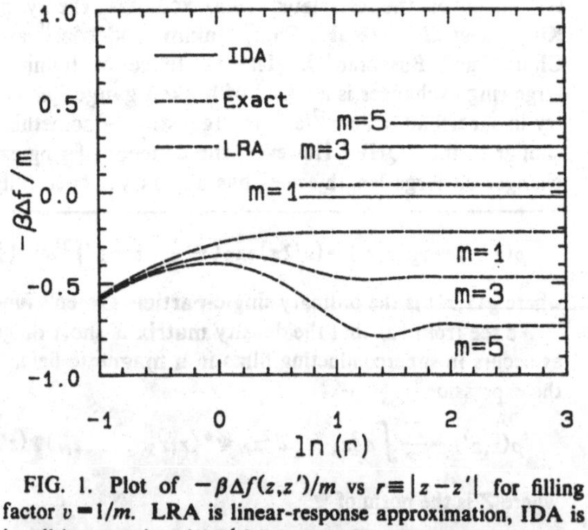

FIG. 1. Plot of $-\beta\Delta f(z,z')/m$ vs $r\equiv|z-z'|$ for filling factor $\nu=1/m$. LRA is linear-response approximation, IDA is ion-disk approximation (shown only for separations exceeding the sum of the ion-disk radii). Because the plasma is strongly coupled, the IDA is quite accurate at $m=1$ and improves further with increasing m. The LRA is less accurate at $m=1$ and worsens with increasing m.

and bound to them.[7,23] The gauge and wave function phase factors in (9) now appear in the form of the bound vortex-antivortex pairs. We expect such bound pairs *not* to destroy the ODLRO and speculate (based on our understanding of the Kosterlitz-Thouless transition[20]) that the effect is at most to renormalize the exponent of the power law in (8). As the range of the potential is increased still further, numerical computations[7] indicate that a critical point is reached at which the gap rather suddenly collapses and the overlap between Laughlin's state and the true ground state drops quickly to zero. We propose that this gap collapse corresponds to the unbinding of the vortices and hence to the loss of ODLRO and the onset of short-range behavior of $\tilde{\rho}(z,z')$. Recall that the distinguishing feature of the FQHE state is its long wavelength excitation gap. At least within the single-mode approximation,[8] this gap can only exist when the ground state is homogeneous and the two-point correlation function exhibits perfect screening:

$$M_1 \equiv (\nu/2\pi)\int d^2r(r^2/2)[g(r)-1] = -1.$$

In the analog plasma problem, the zeros of Ψ act like point charges seen by each particle and the M_1 sum rule implies that electrons see each other as charge-$(m=1/\nu)$ objects; i.e., that m zeros are bound to each electron. Thus (within the single-mode approximation) there is a one-to-one correspondence between the existence of ODLRO and the occurrence of the FQHE.

The exact nature of the gap-collapse transition, which occurs when the range of the potential is increased,[7] is not understood at present. However, it has been proven[8] that the M_1 sum rule is satisfied by every homogeneous and isotropic state in the lowest Landau level. Hence the vortex unbinding should be a first-order transition to a state which breaks rotation symmetry (like the Tao-Thouless state[24]) and/or translation symmetry (like the Wigner crystal[4,8]). We know that as a function of *temperature* (for fixed interaction potential) there can be no Kosterlitz-Thouless transition[20] since isolated vortices (quasiparticles) cost only a finite energy in this system[4,25] (see, however, Ref. 10).

Further insight into the gap collapse can be obtained by considering the exceptional case of Laughlin's wave function with $m>70$. In this case the zeros are still rigorously bound to the particles so that the analog plasma contains long-range forces (and $\tilde{\rho}$ exhibits ODLRO), but the plasma "temperature" has dropped below the freezing point.[2,15] If such a state exhibits (sufficiently[10]) long-range positional correlations, the FQHE would be destroyed by a gapless Goldstone mode associated with the broken translation symmetry. Hence in this exceptional case the normal connection between ODLRO and the FQHE would be broken by a gap collapse due to positional order at a finite wave vector.

The existence of ODLRO in $\tilde{\rho}$ is the type of infrared property which suggests that a field-theoretic approach to the FQHE would be viable. It is clear from the results presented here that the binding of the zeros of Ψ to the particles can be viewed as a condensation,[18] not of ordinary particles, but rather of composite objects consisting of a particle and m flux tubes. (We emphasize that these are *not* real flux tubes, but merely consequences of the singular gauge. The assumption that electrons can bind real flux tubes[26] is easily shown to be unphysical.[27]) The analog of this result for hierarchical daughter states of the Laughlin states[7,15] would be a condensation of composite objects consisting of n particles and m flux tubes (cf. Halperin's "pair" wave functions[19]). This seems closely analogous to the phenomenon of *oblique confinement*[22] and it ought to be possible to derive the appropriate field theory from first principles by use of this idea.

Since the singular gauge maps the problem onto interacting bosons, coherent-state path integration[28] may prove useful. A step in this direction has been taken recently in the form of a Landau-Ginsburg theory which was developed on phenomenological grounds.[4] In the static limit, the action has the "θ vacuum" form

$$S=\int d^2r\,|(-i\nabla+\mathbf{a})\psi(\mathbf{r})|^2+i\phi(\mathbf{r})(\psi^*\psi-1)-i(\theta/8\pi^2)(\phi\nabla\times\mathbf{a}+\mathbf{a}\times\nabla\phi), \quad (10)$$

where $\mathbf{a}$ is not the physical vector potential but an effective gauge field[4] representing frustration due to density deviations away from the quantized Laughlin density and ϕ is a scalar potential which couples both to the charge density and the "flux" density. From (10) the equation of motion for $\mathbf{a}$ is (in the static case):

$$\theta\nabla\times\mathbf{a}=(\psi^*\psi-1). \quad (11)$$

This equation and the parameter θ, which determines the charge carried by an isolated vortex, originally had to be chosen phenomenologically.[4] Now, however, it can be justified by examination of Eq. (5) which shows that the curl of $\mathcal{A}_j$ is proportional to the density of particles. If we identify $\mathbf{a}$ in (10) and (11) as

$$\mathbf{a}=\mathcal{A}_1+\mathbf{A}, \quad (12)$$

where $\mathbf{A}$ is the physical vector potential and we take $\psi^*\psi$ as the particle density relative to the density in the Laughlin state, then Eq. (11) follows from (5) with the θ angle being given by $\theta=2\pi/m$. This yields[4] the correct charge of an isolated vortex (Laughlin quasiparticle) of $q^*=1/m$. The connection between this result and the Berry phase argument of Arovas *et al.*[29] should be noted (see also Semenoff and Sodano[30]). To summarize, it is the strong phase fluctuations induced by the frustration associated with density variations [Eq. (11)] which pin the density at rational fractional values and account for the differences between the FQHE and ordinary superfluidity.[4]

We believe that these results shed considerable light on the FQHE, unify the different pictures of the effect, and emphasize the topological nature of the order in the zero-temperature state of the FQHE. The present picture leads to several predictions which can be tested by numerical computations by use of methods very similar to those now in use.[31] ODLRO will be found only in states exhibiting an excitation gap. The decay of the singular-gauge density matrix will be controlled by the distribution of distances of the zeros of the wave function from the particles. This distribution, which can be artificially varied by changing the model interaction, directly determines the short-range behavior of the density-density correlation function and hence the ground-state energy.[7,23]

The authors would like to express their thanks to C. Kallin, S. Kivelson, and R. Morf for useful conversations and suggestions.

[1] *The Quantum Hall Effect,* edited by R. E. Prange and S. M. Girvin (Springer-Verlag, New York, 1986).

[2] R. B. Laughlin, Phys. Rev. Lett. **50**, 1395 (1983).

[3] P. W. Anderson, Phys. Rev. B **28**, 2264 (1983).

[4] S. M. Girvin, in Chap. 10 of Ref. 1.

[5] R. Tao and Yong-Shi Wu, Phys. Rev. B **30**, 1097 (1984).

[6] D. J. Thouless, Phys. Rev. B **31**, 8305 (1985).

[7] F. D. M. Haldane, in Chap. 8 of Ref. 1.

[8] S. M. Girvin, A. H. MacDonald, and P. M. Platzman, Phys. Rev. Lett. **54**, 581 (1985), and Phys. Rev. B **33**, 2481 (1986); S. M. Girvin in Chap. 9 of Ref. 1.

[9] S. Kivelson, C. Kallin, D. P. Arovas, and J. R. Schrieffer, Phys. Rev. Lett. **56**, 873 (1986).

[10] S. T. Chui, T. M. Hakim, and K. B. Ma, Phys. Rev. B **33**, 7110 (1986); S. T. Chui, unpublished.

[11] G. Baskaran, Phys. Rev. Lett. **56**, 2716 (1986), and unpublished.

[12] R. P. Feynman, Phys. Rev. **91**, 1291 (1953).

[13] S. M. Girvin and T. Jach, Phys. Rev. B **29**, 5617 (1984).

[14] E. Brézin, D. R. Nelson, and A. Thiaville, Phys. Rev. B **31**, 7124 (1985).

[15] R. B. Laughlin, in Chap. 7 of Ref. 1.

[16] F. Wilczek, Phys. Rev. Lett. **49**, 957 (1982).

[17] D. P. Arovas, J. R. Schrieffer, F. Wilczek, and A. Zee, Nucl. Phys. B **251**, 117 (1985).

[18] We refer to this as ODLRO or condensation because of the slow power-law decay even though the largest eigenvalue $\lambda \equiv \int d^2z\, \tilde{\rho}(z,z')$ of the density matrix diverges only for $m \leq 4$ [see C. N. Yang, Rev. Mod. Phys. **34**, 694 (1962)].

[19] B. I. Halperin, Helv. Phys. Acta **56**, 75 (1983).

[20] J. V. José, L. P. Kadanoff, S. Kirkpatrick, and D. R. Nelson, Phys. Rev. B **16**, 1217 (1977).

[21] J. B. Kogut, Rev. Mod. Phys. **51**, 659 (1979).

[22] J. L. Cardy and E. Rabinovici, Nucl. Phys. B **205**, 1 (1982); J. L. Cardy, Nucl. Phys. B **205**, 17 (1982).

[23] D. J. Yoshioka, Phys. Rev. B **29**, 6833 (1984).

[24] R. Tao and D. J. Thouless, Phys. Rev. B **28**, 1142 (1983). The symmetric gauge version of this state exhibits threefold rotational symmetry.

[25] A. M. Chang, in Chap. 6 in Ref. 1.

[26] M. H. Friedman, J. B. Sokoloff, A. Widom, and Y. N. Srivastava, Phys. Rev. Lett. **52**, 1587 (1984), and **53**, 2592 (1984).

[27] F. D. M. Haldane and L. Chen, Phys. Rev. Lett. **53**, 2591 (1984).

[28] L. S. Schulman, *Techniques and Applications of Path Integration* (Wiley, New York, 1981).

[29] D. Arovas, J. R. Schrieffer, and F. Wilczek, Phys. Rev. Lett. **53**, 722 (1984).

[30] G. Semenoff and P. Sodano, Phys. Rev. Lett. **57**, 1195 (1986).

[31] F. D. M. Haldane and E. H. Rezayi, Phys. Rev. Lett. **54**, 237 (1985), and Phys. Rev. B **31**, 2529 (1985); F. C. Zhang, V. Z. Vulovic, Y. Guo, and S. Das Sarma, Phys. Rev. B **32**, 6920 (1985); G. Fano, F. Ortolani, and E. Colombo, Phys. Rev. B **34**, 2670 (1986).

Effective-Field-Theory Model for the Fractional Quantum Hall Effect

S. C. Zhang
Institute for Theoretical Physics, University of California, Santa Barbara, California 93106

T. H. Hansson and S. Kivelson
Physics Department, State University of New York at Stony Brook, Stony Brook, New York 11794
(Received 26 July 1988)

Starting directly from the microscopic Hamiltonian, we derive a field-theory model for the fractional quantum Hall effect. By considering an approximate coarse-grained version of the same model, we construct a Landau-Ginzburg theory similar to that of Girvin. The partition function of the model exhibits cusps as a function of density and the Hall conductance is quantized at filling factors $\nu=(2k-1)^{-1}$ with k an arbitrary integer. At these fractions the ground state is incompressible, and the quasiparticles and quasiholes have fractional charge and obey fractional statistics. Finally, we show that the collective density fluctuations are massive.

PACS numbers: 73.50.Jt

Despite the successes of the microscopic theories[1-3] of the fractional quantum Hall effect (FQHE),[4] it is still important to develop an effective-field-theory model analogous to the Landau-Ginzburg theory of superconductivity. An important step in this direction was taken by Girvin[5] and by Girvin and MacDonald,[6] who proposed a field-theory model, containing a complex scalar field ϕ coupled to a vector field $(a_0,\mathbf{a})$ with a Chern-Simons action (or topological mass term). This model exhibits vortex solutions with finite energy and fractional charge which can be identified with Laughlin's quasiparticles and quasiholes. The amplitude fluctuations of the ϕ field are massive and are identified with the density-fluctuation modes of the single mode approximation.[7,8] There is, however, no explanation for why the Hall conductance is quantized at certain specific fractional values, and in Ref. 6 it is also argued that the phase-fluctuation modes remain massless, contrary to the belief (based on the microscopic models) that *all* elementary excitations above the ground state have a finite gap, corresponding to an incompressible quantum liquid. Despite this, the model in Refs. 5 and 6 provides an important step towards a complete macroscopic description.

In this Letter, we derive a related model directly from the microscopic Hamiltonian, and demonstrate that it explains almost all known aspects of the FQHE. The coefficient of the Chern-Simons term in our case is determined by demanding that the elementary quanta of the ϕ field obey Fermi statistics, and the model exhibits cusps in the partition function at densities $n=n_B/(2k-1)$ (where n_B is the density of states in the lowest Landau level) corresponding to uniform solutions. For densities near an odd-integer filling fraction, the homogeneous state has an energy $\sim B|\delta n|$. In fact, the lowest-lying charged excitations involve a nonuniform charge density; they are spatially localized vortices with the same charge and statistics as the quasiparticles in Laughlin's approach.[1,9,10] We also show that the amplitude fluctuation of the ϕ field has a gap and can be identified with the collective density fluctuations[7,8] while the phase fluctuation of the ϕ field, or the Goldstone boson, is "eaten" by the vector field $(a_0,\mathbf{a})$ as a result of the Anderson-Higgs mechanism and disappears entirely from the spectrum.

We start from the following second-quantized many-body Hamiltonian

$$H=\int d^2r\,\phi^*(\mathbf{r})\left[\frac{1}{2m}[-i\nabla-e\mathbf{A}(\mathbf{r})-e\mathbf{a}(\mathbf{r})]^2+eA^0(\mathbf{r})\right]\phi(\mathbf{r})+\tfrac{1}{2}\int d^2r\,d^2r'\,\phi^*(\mathbf{r})\phi^*(\mathbf{r}')V(\mathbf{r}-\mathbf{r}')\phi(\mathbf{r})\phi(\mathbf{r}'),\quad(1)$$

where

$$a^i(\mathbf{r})=\frac{\theta}{\pi e}\epsilon^{ij}\int d^2r'\frac{r_j-r_j'}{|\mathbf{r}-\mathbf{r}'|^2}\phi^*(\mathbf{r}')\phi(\mathbf{r}'),\quad(2)$$

and where we have set $\hbar=c=1$. This Hamiltonian describes a system of identical particles with mass m which are created by the complex field operator ϕ. These particles interact via a two-dimensional gauge potential $\mathbf{a}$, and a two-body potential V. They are also coupled to an external electromagnetic field A^μ. From (2) it is clear that $\mathbf{a}$ is nothing but the "statistical" gauge potential employed in Refs. 11 and 12. Thus, if we take the ϕ field to be bosonic, the above Hamiltonian describes "anyons" obeying θ statistics (i.e., the wave function changes by a phase θ under the interchange of particles). For $\theta=(2k-1)\pi$ with k an arbitrary integer, this is simply the Hamiltonian for spin-polarized electrons in an external electromagnetic field interacting via the two-body potential $V(\mathbf{r})$.

From (2) we immediately get the following expression for the "statistical" gauge field b

$$b(\mathbf{r}) = -\epsilon^{ij}\partial_i a_j(\mathbf{r}) = \frac{2\theta}{e}|\phi(\mathbf{r})|^2 \equiv s|\phi(\mathbf{r})|^2\left(\frac{2\pi}{e}\right), \tag{3}$$

which corresponds to associating $\theta/\pi = s$ units of flux to each particle. Changing between different k's corresponds to singular gauge transformations.

We can incorporate the constraint (3) by means of a Lagrange multiplier field a_0, and we add a chemical potential μ, which leads to the following coherent-state path-integral representation for the partition function:

$$Z[A^\mu] = \int [d\phi][da_i^T][da_0]e^{iS[\phi,a_i^T,a_0]}, \tag{4}$$

where a_i^T is a transverse gauge field (i.e., satisfying $\partial^i a_i^T = 0$), and $S = \int dt\, d\mathbf{r}\, \mathcal{L}$ with

$$\mathcal{L} = i\phi^*\partial_0\phi - H(\phi) + \mu\phi^*\phi - a_0[(e^2/2\theta)\epsilon^{ij}\partial_i a_j^T + e\phi^*\phi]. \tag{5}$$

The term $\sim a^0$ in this expression is nothing but the Chern-Simons action in the radiation gauge.[13-15] The usual form of the Chern-Simons term can be obtained by reintroducing the (infinite) gauge volume (i.e., by reversing the usual Faddeev-Popov gauge fixing procedure[16]). We want to emphasize this rather interesting result. Several authors have pointed out that the excitations of two-dimensional field theories with Chern-Simon terms exhibit fractional statistics, and our derivation starting from the anyon formulation of Wilczek clearly demonstrates this. Also note that the size of the topological mass term is the one obtained both in our previous analysis of topologically massive (2+1)-dimensional QED,[17] and in the work of Semenoff.[18,19] So far we have made no approximations, other than those involving the intrinsic ambiguities of the coherent-state path integral itself.

In order to apply mean-field theory, we must first integrate out the short-distance fluctuations of the $\phi(\mathbf{r})$ field to obtain an effective action which describes the physics at distance scales larger than the magnetic length. We are currently engaged in carrying out such a calculation. For now, we make a simple *Ansatz* which we think is valid in the quantum Hall regime: The effective action is of the same form as the microscopic action, but with a renormalized stiffness constant κ replacing the bare mass $1/m$, and an effective interaction strength λ replacing the nonlocal interaction $V(r)$. Since the Chern-Simons term embodies the statistics of the bare particles, we do not expect it to renormalize. Higher-order derivative terms in a_μ, such as $f_{\mu\nu}f^{\mu\nu}$, which are probably generated upon coarse graining, should not affect the long-distance behavior of the theory since the Chern-Simons term involves only one derivative and renders the a_μ field massive.[15] The resulting partition function in this approximation is of the form

$$Z[A_\mu] = \int [d\phi\, da_\mu]e^{i(S_\phi[\phi,a_\mu,A_\mu]+S_a[a_\mu])} \tag{6}$$

with

$$\mathcal{L}_a = -(e^2/4\theta)\epsilon^{\mu\nu\sigma}a_\mu\partial_\nu a_\sigma, \tag{7}$$

$$\mathcal{L}_\phi = \phi^*[i\partial_0 - e(A_0+a_0)]\phi - \tfrac{1}{2}\kappa\phi^*[-i\nabla - e(\mathbf{A}+\mathbf{a})]^2\phi + \mu|\phi|^2 - \lambda|\phi|^4, \tag{8}$$

and $\mu = 2\lambda n$, where n is the density.

This action is similar to the one introduced by Girvin.[5] There are some differences in that we have a $\lambda\phi^4$ term for the scalar field, and we have also incorporated time dependence in the action. The essential difference, however, is that Girvin's Chern-Simons action is for the sum of both the statistical gauge field and the real external electromagnetic field, while in our case, only the statistical gauge field appears in the Chern-Simons term. In Girvin's case, the partition function corresponding to (4) is independent of the external electromagnetic field, and therefore cannot be used to derive the fractional quantum Hall conductance. Furthermore, the vortex does not carry charge with respect to the real U(1) electromagnetic gauge group. In deriving (8) we have assumed that $V(r)$ is short ranged; if it is long ranged the expression can be generalized to include a renormalized interaction $\tilde{V}(r)$ which will be equal to $V(r)$ at long distances. In the spirit of the conventional Landau-Ginzburg theory, we ignore terms with higher powers of ϕ and higher derivatives that are generated by the coarse-graining procedure, and we treat κ and λ as phenomenological parameters. This completes the derivation of our model; we now proceed to demonstrate that this effective field theory correctly describes the phenomena related to the FQHE.

First consider the case $A_0 = 0$ and $\epsilon^{ij}\partial_i A_j = -B = \text{const}$. It is immediately clear that S will be minimized by the trivial constant solution $\phi = \sqrt{n}$, $\mathbf{a} = -\mathbf{A}$, $a_0 = 0$. Since the statistical gauge field is related to the density via (3), this solution exists only for $\nu = n/n_B = \pi/\theta = 1/(2k-1)$. This does not necessarily mean that there is a solution only for the particular fraction corresponding to the θ chosen in the Lagrangian, since the different choices of k are connected via singular gauge transformations which induce an $\mathbf{r}$-dependent phase in ϕ.

To calculate the Hall conductance, we apply an external scalar potential A_0 with $\partial_i A_0 = -E_i$, in addition to the vector potential $\epsilon^{ij}\partial_i A_j = -B$. The observable (gauge invariant) current is given by

$$j^i = \frac{\delta S}{\delta A_i} = -\frac{\delta S_\phi}{\delta a_i} = +\frac{\delta S_a}{\delta a_i} = \frac{e^2}{2\theta}\epsilon^{ij}(\partial_0 a_j - \partial_j a_0), \tag{9}$$

where we used the equation of motion $\delta S/\delta a_i = 0$. Thus,

$$\langle j^i \rangle = \frac{1}{Z} \int [d\phi\, da_\mu] \frac{e^2}{2\theta} \epsilon^{ij} (\partial_0 a_j - \partial_j a_0) e^{iS(\phi, a_0, a_i, A_0, A_i)} . \tag{10}$$

By shifting the integration variable $a_0 = -A_0 + a_0'$, and expanding to linear order in the electric field, we find that the induced current is

$$\langle j^i \rangle_{\text{ind}} \equiv -i \left[\frac{\delta \ln Z}{\delta A_i} \bigg|_{A_0, A_i} - \frac{\delta \ln Z}{\delta A_i} \bigg|_{A_0 = 0, A_i} \right] = \frac{e^2}{2\theta} \epsilon^{ij} E^j = \sigma_H^{ij} E^j , \tag{11}$$

and since $\theta = \pi(2k-1)$, this demonstrates that the Hall conductance is quantized in odd fractions of e^2/h.

Let us now analyze what happens when we move away from the odd-integer filling fractions. Since the b field is locked at $eb = s2\pi n$, where now $n = n_B + \delta n$, the particles will feel a net field $e\,\delta b = e(b - B) = s2\pi \delta n$. Each particle (or hole) will acquire a cyclotron energy $= \kappa e\, \delta b/2$, which implies an energy density

$$\mathcal{E} = (2k-1)\pi e B\, |\delta n|\, \kappa . \tag{12}$$

For large B it is thus natural to assume that by moving away from the good filling fractions one creates localized density disturbances. As we shall see, our model exhibits such quasiparticle and quasihole excitations in the form of vortices similar to those found in Refs. 1, 2, and 5.

From the equation of motion derived from (6)–(8), one easily finds that at $n = \pi n_B/\theta$ there are static, nonuniform, finite-energy vortex solutions. If (r,φ) are polar coordinates with the center of the vortex at $r = 0$, the $r \to \infty$ behavior of the solution is

$$\phi(r,\varphi) = \sqrt{n}\, e^{\pm i\varphi} , \tag{13}$$

$$\mathbf{a}(r,\varphi) = \pm \hat{\varphi}/er , \tag{14}$$

and $a_0(r,\varphi) = 0$, corresponding to one unit of statistical flux per vortex. The equation of motion for a^0 implies

$$j^0 \equiv -\frac{\delta S}{\delta A_0} = -\frac{\delta S_\phi}{\delta a_0} = \frac{\delta S_a}{\delta a_0} = \frac{e^2}{2\theta} b , \tag{15}$$

and so the total charge carried by the vortex is

$$q^1 = \int d^2x\, j^0 = \frac{e^2}{2\theta} \oint \mathbf{a} \cdot d\mathbf{r} = \pm \frac{\pi}{\theta} e = \pm \frac{e}{2k-1} . \tag{16}$$

These field configurations can thus be identified as the fractionally charged quasiparticle and quasiholes above the ground state. According to the results of Refs. 10–12, 17, and 18, this implies that the quasiparticles obey fractional statistics with $\theta_1 = q_1 \Phi_1 / 2 = \pi/(2k-1)$, where $\Phi_1 = 2\pi/e$ is the flux of the vortex. Note that since $\phi(\mathbf{r})$ must vanish at the center of the vortex, there is necessarily a difference in the profile and creation energies of the quasiparticles and quasiholes. (This also illustrates the intrinsic problems with the Landau-Ginzburg approach at distances of the order of the magnetic length, since the true quasiparticle density certainly does not vanish in the core.)

We have thus shown that the vortices in our model have the same charge and statistics as the quasiparticles in Laughlin's approach.[1,9,10] The presence of these excitations naturally leads to the so-called hierarchy scheme,[9,20] which has been proposed to explain the quantization of the Hall conductance at fractions other than $1/(2k-1)$.

Finally let us turn to the collective excitations. As a result of the symmetry-breaking potential in L_ϕ, the ϕ fields acquire a nonvanishing vacuum expectation value $|\langle \phi \rangle| = \sqrt{n}$. The ϕ field can thus be parametrized by $\phi(x) = [\phi_0 + \delta\phi(x)] e^{ie\eta(x)}$, $\mathbf{a}(x) = \delta\mathbf{a}(x) + \nabla\eta(x)$, and $a_0(x) = \delta a_0(x) - \partial_0 \eta(x)$, describing the amplitude and phase fluctuations about the classical vacuum. We see that the phase fluctuation $\eta(x)$ is "gauged away" in accordance with the standard Anderson-Higgs mechanism. Since the statistical gauge field is nondynamical, there is no propagating mode (massive or not) corresponding to phase fluctuations. This reflects that there is a unique ground-state in the quantum Hall effect and, we believe, implies that there will be no Josephson-type effects. Only the amplitude fluctuation remains, and by expanding the Lagrangian, up to terms quadratic in $\delta\phi$, $\delta\mathbf{a}$, and δa_0, about the constant solution, we find the following dispersion relation:

$$\omega(q)^2 = (e\kappa B)^2 + \tfrac{1}{4} \kappa q^2 (\kappa q^2 + 8\lambda\phi_0^2) . \tag{17}$$

Note that the mass of the amplitude mode is $\sim B$, and for negative λ the dispersion curve has the same shape as that derived in Refs. 7 and 8. Note that even for negative λ, as long as $|\lambda|/\kappa$ is sufficiently small, the quasiparticle creation energy is positive, and the Hamiltonian is bounded from below.

In conclusion, we have derived a field theory for the FQHE directly from the microscopic Hamiltonian, and find that a coarse-grained version of this theory describes almost all the known phenomenology of the FQHE including incompressibility, fractional Hall conductance with odd denominators, and the fractional charge and statistics of the quasiparticles. It is to be warned, however, that this coarse-grained theory certainly makes errors on the magnetic length scale, and it treats the statistical gauge field $\mathbf{a}$ within mean-field theory; i.e., the particles feel the b field which produces the statistics, whereas in fact the exact $\mathbf{a}$ is pure gauge. Despite these shortcomings, we believe that the long-wavelength properties of the quantum-Hall system are correctly reproduced by this Landau-Ginzberg theory. We note that the same physical idea, i.e., that the long-wavelength effects of the physical magnetic field are canceled by the statistical field, is the basic result of the cooperative-ring-exchange theory[2,3] of the quantum Hall effect, and of a mean-field theory recently introduced by Laughlin.[21] There are some similarities between the present results and some independent recent work of Read.[22]

We would like to thank V. Emery, S. Girvin, D. H. Lee, N. Read, and X. G. Wen for encouraging discussions. The work of S.C.Z. is partially supported by the National Science Foundation under Grant No. PHY82-17853, supplemented by funds from the National Aeronautics and Space Administration, at the University of California at Santa Barbara. The work of T.H.H. is partially supported by the U.S. Department of Energy under Grant No. DE-FG02-88ER40388. The work of S.K. is partially supported by the National Science Foundation under Grant No. NSF-DMR-87-06250.

[1]R. B. Laughlin, Phys. Rev. Lett. **50**, 1395 (1983).

[2]S. Kivelson, C. Kallin, D. P. Arovas, and J. R. Schrieffer, Phys. Rev. Lett. **56**, 873 (1986).

[3]D. H. Lee, G. Baskaran, and S. Kivelson, Phys. Rev. Lett. **59**, 2467 (1987).

[4]*The Quantum Hall Effect*, edited by R. E. Prange and S. M. Girvin (Springer-Verlag, New York, 1986).

[5]S. M. Girvin, in Ref. 4, Chap. 10.

[6]S. M. Girvin and A. H. MacDonald, Phys. Rev. Lett. **58**, 1252 (1987).

[7]S. M. Girvin, A. H. MacDonald, and P. M. Platzman, Phys. Rev. Lett. **54**, 581 (1985).

[8]S. M. Girvin, A. H. MacDonald, and P. M. Platzman, Phys. Rev. B **33**, 2481 (1986).

[9]B. I. Halperin, Phys. Rev. Lett. **52**, 1583 (1984).

[10]D. P. Arovas, J. R. Schrieffer, and F. Wilczek, Phys. Rev. Lett. **53**, 722 (1984).

[11]F. Wilczek, Phys. Rev. Lett. **49**, 957 (1982).

[12]D. P. Arovas, J. R. Schrieffer, F. Wilczek, and A. Zee, Nucl. Phys. **B251**, 117 (1985).

[13]W. Siegel, Nucl. Phys. **B156**, 135 (1979).

[14]J. F. Schonfeld, Nucl. Phys. **B185**, 157 (1981).

[15]S. Deser, R. Jackiw, and S. Templeton, Ann. Phys. (N.Y.) **140**, 372 (1982).

[16]L. D. Faddeev and V. N. Popov, Phys. Lett. **25B**, 29 (1967).

[17]T. H. Hansson, M. Roček, I. Zahed, and S. C. Zhang, State University of New York at Stony Brook Report No. ITP-SB-88-32, 1988 [Phys. Lett. B (to be published)].

[18]G. W. Semenoff, Phys. Rev. Lett. **61**, 517 (1988).

[19]We should mention that conflicting claims have been made concerning the statistics of field excitations in the presence of a Chern-Simons term. For a discussion see Ref. 17.

[20]F. D. M. Haldane, Phys. Rev. Lett. **51**, 605 (1983).

[21]R. B. Laughlin, Phys. Rev. Lett. **60**, 2677 (1988).

[22]N. Read, following Letter [Phys. Rev. Lett. **61**, 86 (1988)].

Off-Diagonal Long-Range Order in Fractional Quantum-Hall-Effect States

E. H. Rezayi
Department of Physics and Astronomy, California State University, Los Angeles, California 90032

F. D. M. Haldane
Department of Physics, University of California, San Diego, La Jolla, California 92093
(Received 17 May 1988)

We present evidence of off-diagonal long-range order in the incompressible fractional quantum Hall states at $\nu=\frac{1}{3}$ and $\frac{2}{5}$ Landau-level fillings. We construct correlation functions that exhibit this property only for incompressible quantum-Hall-effect states. In a compressible state no off-diagonal long-range order is observed.

PACS numbers: 73.40.Kp, 73.20.Dx, 73.50.Jt

Recently, there has been much interest[1-3] in identifying an "off-diagonal long-range order" (ODLRO) associated with the fractional quantum Hall effect (QHE).[4,5] Girvin and MacDonald[2] interpret the Laughlin model wave functions[6] for the Landau-level filling factor $\nu=1/q$ QHE states as being related to a Bose condensate of composite charge plus flux-tube objects. Read[3] has also pointed out that a property special to the Laughlin state can be interpreted as a sign of ODLRO. These formulations have been mainly focused on properties specific to the Laughlin states[6]; in this Letter, we report a more general formulation of an ODLRO correlation function that is not tied to any specific model wave function or filling factor, and which is easily adapted to numerical finite-size calculations.

We also describe such numerical studies at $\nu=\frac{1}{3}$ and $\frac{2}{5}$, which demonstrate the existence of ODLRO in the incompressible QHE state, and its disappearance if incompressibility is destroyed by varying the interaction parameters. This is the first direct demonstration that an ODLRO property can be related to incompressibility of the QHE ground state, allowing the presence or absence of a QHE to be determined by study of the ground state itself, rather than of the excitation spectrum.

Our formulation of the ODLRO order parameter is based on the idea that at Landau-level filling $\nu=p/q$, local injection of q flux quanta with an infinitesimally thin solenoid combined with local injection of p charge quanta (electrons) leaves a QHE state essentially locally unchanged.

To formulate ODLRO correlation functions, we introduce composite "flux-charge creation operators" $A_q^\dagger(\mathbf{r};\{n_{m\alpha}\})$, where $\{n_{m\alpha}\}$ is the set $\{n_{0\alpha},n_{1\alpha},\ldots,n_{(q-1)\alpha}\}$. These are defined through their action on eigenstates of the operators $\hat{n}_{m\alpha}(\mathbf{r})=c_{m\alpha}^\dagger(\mathbf{r})c_{m\alpha}(\mathbf{r})$, $0\le m<\infty$, which measure occupation of orbitals in the angular momentum ("symmetric gauge") basis centered at the point $\mathbf{r}$, where α represents the Landau-level index and any internal quantum numbers, such as spin.

If $|\mathbf{r};\{n_{m\alpha}\}\rangle$ are such occupation-number eigenstates, then

$$A_q^\dagger(\mathbf{r};\{n_{m\alpha}\})\,|\mathbf{r};\{n'_{m\alpha}\}\rangle=|\mathbf{r};\{n''_{m\alpha}\}\rangle, \tag{1}$$

where $n''_{m\alpha}=n_{m\alpha}$ for $0\le m<q$, and $n''_{m\alpha}=n'_{(m-q)\alpha}$ for $q\le m$. The operator $A_q^\dagger(\mathbf{r};\{n_{m\alpha}\})$ adds q quanta of magnetic flux to the system through the action of an ideal infinitesimal solenoid piercing the 2D plane at $\mathbf{r}$. The flux injection process is carried out on a time scale τ so that it is *adiabatic* as far as Landau-level index quantum numbers α are concerned (e.g., $\hbar\omega_c\gg\hbar/\tau$, so no inter-Landau-level transitions can occur), but *instantaneous* with respect to intra-Landau-level processes (e.g., $\hbar/\tau\gg e^2/\epsilon l$). The flux-charge injection process thus increases the angular momentum m of existing orbitals by q units, but otherwise leaves their occupations unchanged; it also introduces q new orbitals with $m<q$, and adds particles to give them occupation numbers $\{n_{m\alpha}\}$. In what follows, we mainly specialize to fully spin-polarized lowest-Landau-level states and drop the index α.

In this formalism, Laughlin's quasihole operator[6] $h^\dagger(\mathbf{r})=\Pi_i(z-z_i)$, where $z=x+iy$ is the complex coordinate, can be represented as $\hat{\Lambda}(\mathbf{r})A_1^\dagger(\mathbf{r};\{0\})$, with

$$\hat{\Lambda}(\mathbf{r})=\prod_{m=1}^{\infty} m^{(1/2)\hat{n}_m(\mathbf{r})}. \tag{2}$$

The operator $A_1(\mathbf{r};\{0\})$ has been used by MacDonald and Girvin[7] as an alternative to Laughlin's model quasiparticle creation operator.

The N-particle Laughlin-Jastrow wave functions $|\Psi_N^{LJ}\rangle$ for $\nu=1/q$ QHE states have the property[8,9]

$$[h^\dagger(\mathbf{r})]^q\,|\Psi_N^{LJ}\rangle\propto c_0(\mathbf{r})\,|\Psi_{N+1}^{LJ}\rangle, \tag{3}$$

i.e., the state produced by the addition of q quasiholes to the N-particle Laughlin state at the point $\mathbf{r}$ is the same as that produced by acting on the $N+1$ particle Laughlin-Jastrow state with the destruction operator of a Gaussian lowest-Landau-level state centered at $\mathbf{r}$. Read[3] has pointed out that this implies

$$\langle\Psi_{N+1}^{LJ}|\,c_0^\dagger(\mathbf{r})[h^\dagger(\mathbf{r})]^q\,|\Psi_N^{LJ}\rangle\neq 0, \tag{4}$$

which can be interpreted as an ODLRO property analogous to $\langle\Psi_{N+1}^{GS}|\psi^{\dagger}(\mathbf{r})|\Psi_N^{GS}\rangle\neq 0$ in Bose condensed systems.

If the factor $\Lambda(\mathbf{r})$ in $h^{\dagger}(\mathbf{r})$ is dropped, so $h^{\dagger}(\mathbf{r}) \to A_1^{\dagger}(\mathbf{r};\{0\})$, and

$$c_0^{\dagger}(\mathbf{r})[h^{\dagger}(\mathbf{r})]^q \to A_q^{\dagger}(\mathbf{r};\{1,0,0,0,\ldots\}),$$

these two operators change global quantum numbers (charge, flux, and angular momentum about $\mathbf{r}$) in the same way and thus obey the same selection rules, suggesting that $\langle\Psi_{N+1}^{LL}|A_q^{\dagger}(\mathbf{r};\{1,0,0,\ldots\})|\Psi_N^{LL}\rangle\neq 0$ also.

For more general QHE states with $\nu=p/q$, we seek the condition for a flux-charge creation operator that adds q flux quanta and p particles to have a nonzero matrix element, i.e.,

$$\langle\Psi_{N+p}|A_q^{\dagger}(\mathbf{r};\{n_m\})|\Psi_N\rangle\neq 0, \tag{5}$$

where $|\Psi_N\rangle$ is an N-particle $\nu=p/q$ QHE ground state. Since QHE states are translationally and rotationally invariant, this imposes a condition on the total planar angular momentum $M=\sum_{m<q} m n_m$ of the injected particles. For technical reasons, it is convenient to derive this condition by our considering the finite-size spherical geometry,[10] and then taking the infinite-size limit.

In the spherical geometry, the hierarchy scheme[10] fixes the value of flux and charge at which QHE states can occur. These states have total angular momentum $L=0$ reflecting rotational and translational invariances on the surface of the sphere, and thus also have azimuthal angular momentum $\hat{\Omega}_r\cdot\mathbf{L}\equiv L_{\parallel}=0$. In units of the flux quantum $\Phi_0=h/e$, the total flux leaving the sphere is an integer N_{Φ}. For the p-particle, q-flux quanta injection process into the $(N_{\Phi}+1)$-fold-degenerate lowest Landau level with N particles

$$\Delta L_{\parallel}=\tfrac{1}{2}(pN_{\Phi}-qN+pq)-M.$$

A QHE state with filling factor ν has $N_{\Phi}=\nu^{-1}N+O(1)$, where N_{Φ} and N are proportional to the system size. The requirement $\Delta L_{\parallel}=0$ can thus only be satisfied if $\nu=p/q$. For the principal ($\nu=1/q$) states of the fractional QHE hierarchy for spinless fermions (odd q) or bosons (even q), $N_{\Phi}=q(N-1)$ and $M=0$. Thus only $\langle\Psi_{N+1}|A_q^{\dagger}(\mathbf{r};\{1,0,0,\ldots\})|\Psi_N\rangle$ can have nonzero values. At the second level of the hierarchy,[10] $\nu=(q_0\pm 1/p_1)^{-1}$, with p_1 even (so $p=p_1$ and $q=p_1q_0\pm 1$), and the secondary QHE condensate is built out of quasiparticles ($-$) or quasiholes ($+$). The hierarchy equations[10] give $N_{\Phi}=q_0(N-1)\pm N'$, and $N=p_1(N'-1)$, where N' is the number of quasiholes or quasiparticles making up the secondary fluid. This leads to the selection rule $M=\tfrac{1}{2}q_0p_1(p_1-1)\pm p_1$ for a nonzero matrix element (5) in second-level hierarchy states. Some examples are given in Table I.

TABLE. I. Allowed flux-charge injection patterns for ODLRO in selected QHE states. Top left: Principal (one-component) states for spinless bosons (even q) and spinless fermions (odd q). Bottom left: Some spinless fermion states from the second level of the hierarchy. Right: Some second-level spinless boson states.

ν	M	$(n_0,n_1,\ldots,n_{q-1}\vert$	ν	M	$(n_0,n_1,\ldots,n_{q-1}\vert$
$1/q$	0	$(1,0,0,\ldots\vert$	2/3	0	$(2,0,0\vert$
2/3	3	$(0,1,1\vert$	2/5	4	$(1,0,0,0,1\vert$
2/5	1	$(1,1,0,0,0\vert$			$(0,1,0,1,0\vert$
					$(0,0,2,0,0\vert$
2/7	5	$(1,0,0,0,0,1,0\vert$			
		$(0,1,0,0,1,0,0\vert$	2/7	2	$(1,0,1,0,0,0,0\vert$
		$(0,0,1,1,0,0,0\vert$			$(0,2,0,0,0,0,0\vert$

To avoid calculating matrix elements between states of different size systems, we numerically studied the two-point ground-state (GS) correlations

$$C_q(\mathbf{r}-\mathbf{r}';\{n_m\})=\langle\Psi_{GS}|A_q^{\dagger}(\mathbf{r}';\{n_m\})A_q(\mathbf{r};\{n_m\})|\Psi_{GS}\rangle. \tag{6}$$

The flux-charge operators have the property

$$A_q^{\dagger}(\mathbf{r};\{n_m\})A_q(\mathbf{r};\{n_m\})=\prod_{m<q}\delta_{\hat{n}_m(\mathbf{r}),n_m}, \tag{7}$$

therefore

$$C_q(0;\{n_m\})=\langle\Psi_{GS}|\prod_{m<q}\delta_{\hat{n}_m(\mathbf{r}),n_m}|\Psi_{GS}\rangle$$

is the amplitude of the local configuration $\{n_m\}$ in the ground state, and obeys a sum rule

$$\sum_{\{n_m\}}C_q(0;\{n_m\})=1. \tag{8}$$

Note that $C_1(r;\{1\})=1$ (i.e., perfect ODLRO) for the *filled* Landau level ($\nu=1$).

The calculations were performed in the spherical geometry[10,11] in which fully translationally and rotationally invariant ground states occur in finite systems. The operator $A_q^{\dagger}(\mathbf{r};\{n_m\})$ can be adapted to the finite-size spherical geometry[10] in an obvious way: In this geometry, the lowest Landau level becomes an angular momentum multiplet with $L=\tfrac{1}{2}N_{\Phi}$. The states $|\mathbf{r};\{n_m\}\rangle$ are now occupation-number eigenstates in the azimuthal angular momentum basis where $\mathbf{r}$ represents a direction $\hat{\Omega}_r$ and $\hat{\Omega}_r\cdot\mathbf{L}=\tfrac{1}{2}N_{\Phi}-m, m=0,\ldots,N_{\Phi}$. $A_q^{\dagger}$ increases N_{Φ} by q.

The technique used was to construct $|\Psi(\mathbf{r})\rangle=A_q(\mathbf{r};\{n_m\})|\Psi_{GS}\rangle$ with $\hat{\Omega}_r$ at the north pole of the sphere, and (by projection) separates it out into angular momentum components $|\Psi_{LM}\rangle$

$$|\Psi(\mathbf{r})\rangle=\textstyle\sum_L a_{LM}|\Psi_{LM}\rangle, \tag{9}$$

where $\langle\Psi_{LM}|\Psi_{LM}\rangle=1$, and M is fixed by the configuration $\{n_m\}$ of the flux-charge destruction operator. The translated state $|\Psi(\mathbf{r}')\rangle$ is obtained by action of

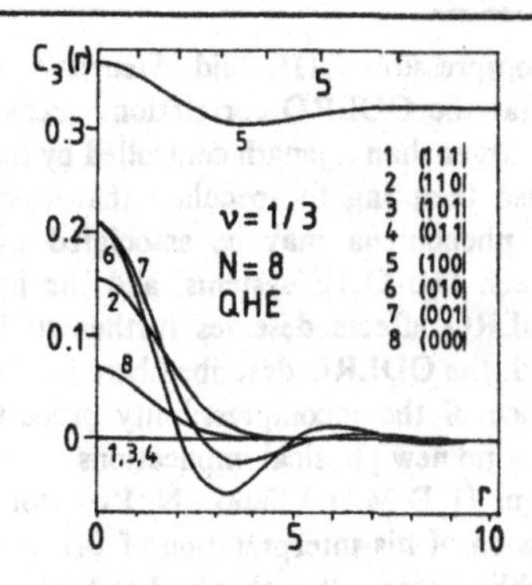

FIG. 1. The correlation functions of the charge-flux creation operators $A_3^\dagger(\mathbf{r};\{n_0,n_1,n_2\})$ in the incompressible $\nu=\frac{1}{3}$ QHE phase (lowest Landau level, Coulomb interactions) for $N=8$. The asymptotic approach of the correlations to a constant for $(n_0,n_1,n_2|=(1,0,0|$ is an indication of ODLRO. The lower scale is the arc distance (in magnetic length units) measured along a great circle while the upper scale is the chord distance.

the (great circle) rotation Θ that takes $\hat{\Omega}_r$ into $\hat{\Omega}_{r'}$.

$$|\Psi(\mathbf{r}')\rangle=\sum_{LM'}\alpha_{LM}D^L_{MM'}(\Theta_{\mathbf{r},\mathbf{r}'})|\Psi_{LM'}\rangle, \qquad (10)$$

where $D^L_{MM'}(\Theta)$ are rotation matrices. $C_q(\mathbf{r}-\mathbf{r}')$ is then given by

$$\langle\Psi(\mathbf{r}')|\Psi(\mathbf{r})\rangle=\sum_L|\alpha_{LM}|^2D^L_{MM}(\Theta_{\mathbf{r},\mathbf{r}'}).$$

Two possible definitions of distance on the sphere are the arc length $R|\Theta|$ and the chord length $2R\sin\frac{1}{2}|\Theta|$, where R is the sphere radius; both are shown in our figures.

Figure 1 shows the eight correlation functions $C_3(\mathbf{r};\{n_m\})$ for the $N=8$, $\nu=\frac{1}{3}$ fractional QHE system with Coulomb interactions. This system has a ground state very close to the Laughlin-Jastrow state.[6,11] It can be seen that all correlations decay rapidly except in the case $\{n_m\}=(1,0,0|$ for which ODLRO is possible. [The notation $(1,0,0|$ is intended to indicate that the left-most orbital is under the solenoid.] Figure 2 shows the corresponding data for a compressible system, obtained by our lowering the short-range pseudopotential V_1 from its Coulomb value of 0.487 to 0.350 while keeping the other pseudopotentials at their Coulomb values. (In Ref. 11 it is shown that the system has a transition to a compressible non-QHE state for $V_1\lesssim 0.37$.) The striking ODLRO of $C_3(\mathbf{r};\{1,0,0\})$ seen in Fig. 1 is not present in the compressible state.

The dependence of the asymptotic value of the $(1,0,0|$ correlation function on V_1 is shown in Fig. 3. (Here "asymptotic" means the limiting case where r and r' are diametrically opposed on the sphere.) A rapid decay of ODLRO is seen for $V_1\lesssim 0.37$ which coincides with the

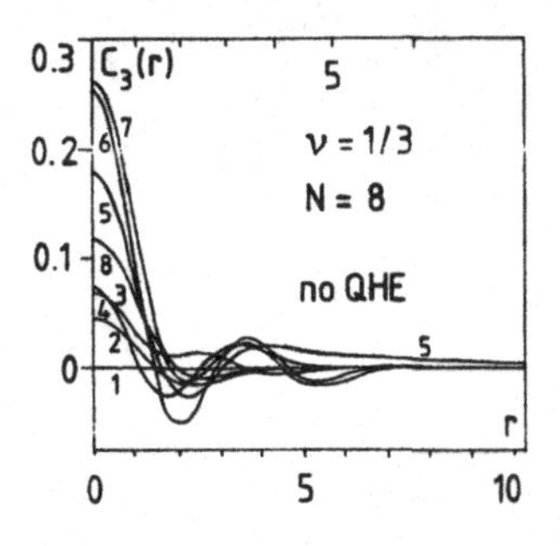

FIG. 2. As in Fig. 1, but for the non-QHE compressible state with $V_1=0.35$ (modified Coulomb interaction) showing the absence of ODLRO.

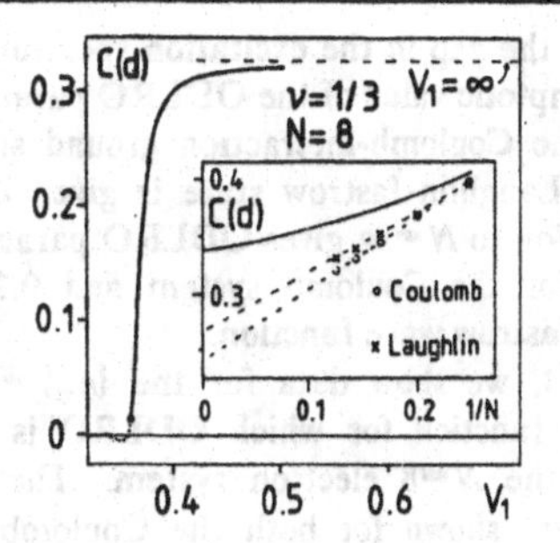

FIG. 3. Asymptotic value of $C_3(d;\{1,0,0\})$ ($d\equiv$sphere diameter) as the "hard-core" component V_1 of the Coulomb-interaction potential is varied in an $N=8$, $\nu=\frac{1}{3}$ system. A sharp transition where ODLRO is lost is seen near $V_1\approx 0.37$; this also corresponds to the collapse of the excitation gap. For continuity we plot $C_3(d)$ for the lowest isotropic ($L=0$) state in the compressible region, even when this is not the ground state (dotted line). Inset: $1/N$ dependence of the asymptotic value of $C_3(d)$ at $\nu=\frac{1}{3}$ for both the Coulomb potential ground state (circles) and the model Laughlin-Jastrow wave function (crosses). The full curve shows the value $\langle n_0\rangle=N/(N_\phi+1)$, which is an upper bound to $C_3(d)$.

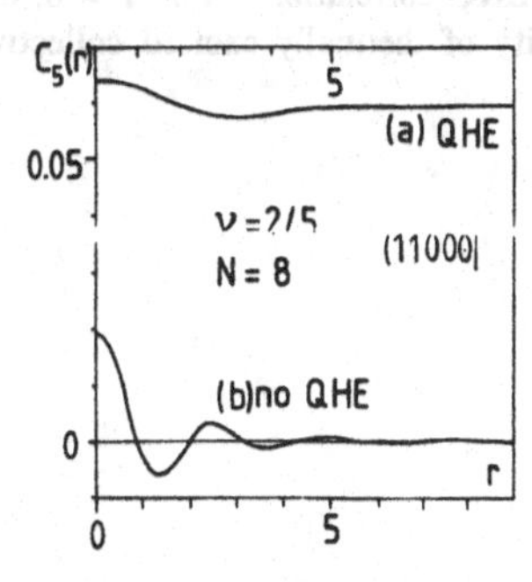

FIG. 4. As in Figs. 1 and 2, but for $C_5(\mathbf{r};\{1,1,0,0,0\})$ in the $N=8$, $\nu=\frac{2}{5}$ system: (a) QHE incompressible state; (b) a compressible (zero gap) ground state of a modified Coulomb interaction.

collapse of the gap in the excitation spectrum.[11] A plot of the asymptotic value of the ODLRO parameter versus $1/N$, for the Coulomb-interaction ground state and for the exact Laughlin-Jastrow state is given in the inset. Extrapolation to $N=\infty$ gives ODLRO parameter values of 0.281 for the Coulomb system and 0.290 for the Laughlin-Jastrow wave function.

In Fig. 4, we show data for the $\{n_m\}=(1,1,0,0,0|$ correlation function for which ODLRO is possible at $\nu=\frac{2}{5}$ for the $N=8$ electron system. The correlation functions are shown for both the Coulomb-interaction system ($V_1=0.476$), which is incompressible and manifestly exhibits ODLRO, and a compressible state without ODLRO (modified Coulomb interaction, $V_1=0.25$). We observed collapse of the excitation gap at $V_1\approx 0.29$.

The correlation functions $C_q(\mathbf{r},\mathbf{r}')$ presented here are real because translations on the sphere were made along a great circle between $\mathbf{r}$ and $\mathbf{r}'$. A more general correlation function combining a rotation about $\mathbf{r}$ with a great-circle translation would be complex.

Table I shows that ODLRO in more general QHE states will involve a (Hermitian) *matrix* of flux-charge operator correlation functions; for example, a 3×3 matrix in the case of $\nu=\frac{2}{7}$ spinless fermions. We have not yet studied such cases numerically, but speculate that only one eigenvalue of the asymptotic matrix correlation function will remain finite in the thermodynamic limit, with an eigenvector that corresponds to the linear combination of flux-charge operators that most nearly recreates the local correlations of the QHE state.

We have also considered ODLRO in the spin-*singlet* half-integral QHE state recently constructed by us.[12] While $\nu=\frac{1}{2}$, the allowed flux-charge addition is of a singlet *pair* of electrons and *four* flux quanta in the $M=0$ configuration $(\uparrow\downarrow,0,0,0|$. This implies that ν in this case is perhaps best represented as the *unreduced* fraction $\frac{2}{4}$.

It is natural to consider the effect of finite temperature on the ODLRO correlations. For $T>0$, there will be a finite density of thermally excited collective excitations of the incompressible QHE fluid. Heuristic arguments[13] suggest that the ODLRO correlations decay rapidly at distances larger than a length controlled by this density.

It is also tempting to speculate that Josephson-type tunneling phenomena may be associated with a weak link between two QHE systems, and the possibility of novel ODLRO effects deserves further study. On the other hand, the ODLRO described here may merely be a consequence of the incompressibility property of QHE states, with no new physical implications.

One of us (F.D.M.H.) thanks N. Read for a very useful discussion of his interpretation of (4) as ODLRO of the Laughlin state. We thank the Aspen Center for Physics, where a portion of this work was completed. F. D. M. H. would like to thank the Alfred P. Sloan Foundation for financial support. E. H. R. thanks the California State University, Los Angeles, Intermediate Energy Group for use of its computing facility.

---

[1] S. M. Girvin, in *The Quantum Hall Effect*, edited by R. E. Prange and S. M. Girvin (Springer-Verlag, New York, 1986), Chap. 10.

[2] S. M. Girvin and A. H. MacDonald, Phys. Rev. Lett. **58**, 1252 (1987).

[3] N. Read, unpublished.

[4] D. C. Tsui, H. L. Störmer, and A. C. Gossard, Phys. Rev. Lett. **48**, 1559 (1982).

[5] *The Quantum Hall Effect*, edited by R. E. Prange and S. M. Girvin (Springer-Verlag, New York, 1986).

[6] R. B. Laughlin, Phys. Rev. Lett. **50**, 1395 (1983).

[7] A. H. MacDonald and S. M. Girvin, Phys. Rev. B **33**, 4414 (1986), and **34**, 5639 (1986).

[8] S. M. Girvin, Phys. Rev. B **29**, 6012 (1984).

[9] E. H. Rezayi, Phys. Rev. B **35**, 3032 (1987).

[10] F. D. M. Haldane, Phys. Rev. Lett. **51**, 605 (1983), and in Ref. 5, Chap. 8.

[11] F. D. M. Haldane and E. H. Rezayi, Phys. Rev. Lett. **54**, 237 (1985).

[12] F. D. M. Haldane and E. H. Rezayi, Phys. Rev. Lett. **60**, 956, 1886(E) (1988).

[13] F. D. M. Haldane and E. H. Rezayi, unpublished.

Order Parameter and Ginzburg-Landau Theory for the Fractional Quantum Hall Effect

N. Read
Department of Physics, Massachusetts Institute of Technology, Cambridge, Massachusetts 02139, and Section of Applied Physics, Yale University, New Haven, Connecticut 06520[(a)]

(Received 15 August 1988)

A new order parameter with a novel broken symmetry is proposed for the fractional quantum Hall effect, with the Laughlin state as the mean-field ground state. The classical Ginzburg-Landau theory of Girvin is derived microscopically from this starting point and exhibits all the phenomenology of the fractional quantum Hall effect.

PACS numbers: 73.20.Dx, 03.50.Kk, 05.30.Fk

While there is now a good understanding of the properties of the states responsible for the fractional quantum Hall effect (FQHE) in the lowest Landau level,[1] a completely general characterization of these states has not yet been given. Girvin[2] has suggested that this might be done by invoking a superfluid analogy, in which the fluid is described by a complex scalar order parameter obeying a Ginzburg-Landau equation, and the vortex excitations are identified with the fractionally charged quasiparticles of Laughlin's theory.[1] In a later Letter, Girvin and MacDonald[2] (GM) showed that a certain modified density matrix exhibits algebraic off-diagonal long-range order in the Laughlin state, providing further evidence for the superfluid analogy.

In this Letter, I construct the superfluid analogy explicitly on a microscopic basis. An order parameter that shows genuine long-range order in the Laughlin state is constructed, related to, but distinct from, that of GM. The broken symmetry is identified, and the Ginzburg-Landau action is derived at the classical, linearized level. All the phenomenology of the FQHE at filling factors $\nu = 1/q$ follows, and generalization to other filling factors can be made at least in principle. Physically, the order parameter describes the special correlations of the Laughlin state (binding of zeroes to particles).

We first exhibit a correlation function which possesses off-diagonal long-range order, indicating that the usual Laughlin state[3] is not a pure phase[4] and that a symmetry is broken. We use (i) a lowest-Landau-level-projected second-quantized field operator in the symmetric gauge,[5]

$$\psi(z) = \sum_{n=0}^{\infty} a_n u_n(z), \quad u_n = \frac{z^n e^{-|z|^2/4}}{(2\pi 2^n n!)^{1/2}}, \tag{1}$$

where a_n is a destruction operator for the nth single-particle basis state u_n, and (ii) Laughlin's quasihole operator,[3] in first quantization,

$$U(z) = \prod_{i=1}^{N} (z_i - z), \tag{2}$$

in the N-particle subspace. Note that while $\psi(z)$ removes a particle bodily from the fluid, leaving a hole of charge 1, $U(z)$ moves particles outwards by increasing their angular momentum about z, leaving a deficiency of charge $1/q$ there if the state is a *fluid state* of slowly varying density ρ close to $\rho_0 = 1/2\pi q$ with no positional long-range order. $U(z)^q$ thus leaves the same charge deficiency as $\psi(z)$, and the essence of the present approach is that these two types of hole states are physically equivalent, so that they have a nonzero overlap.[6]

In the normalized Laughlin ground state $|0_L;N\rangle$ for N particles, whose (unnormalized) coordinate representation is

$$\prod_{i<j} (z_i - z_j)^q \exp\left[-\tfrac{1}{4}\sum_i |z_i|^2\right],$$

we can show that

$$\langle 0_L;N|\tilde{U}^\dagger(z)^q \psi(z)\psi^\dagger(z')\tilde{U}(z')^q|0_L;N\rangle$$
$$= \rho_0^{-1}\langle 0_L;N+1|\rho(z)\rho(z')|0_L;N+1\rangle \to \rho_0 \tag{3}$$

as $|z-z'| \to \infty$ with $|z|,|z'| > N$, which is not equal to

$$|\langle 0_L;N|\tilde{U}^\dagger(z)^q \psi(z)|0_L;N\rangle|^2$$

which vanishes identically. Here and below we denote

$$\tilde{U}(z)^q|\alpha\rangle = U(z)^q|\alpha\rangle / \langle\alpha|\,|U(z)|^2\,|\alpha\rangle^{1/2},$$

where $|\alpha\rangle$ is a normalized fluid state.

Equation (3) shows that the Laughlin state is not a pure state.[4] A pure state, in which $\psi^\dagger U^q$ has a nonzero expectation value, can be constructed as

$$|0_L;\theta\rangle = \sum_{N=1}^{\infty} |a_N| e^{-iN\theta} |0_L;N\rangle, \tag{4}$$

where $\{|a_N|^2\}$ is a binomial distribution function for N with mean $\bar{N} \gg 1$ and variance of order $\bar{N}$, and θ is arbitrary. For this state and arbitrary z,

$$\langle \Psi^\dagger(z)\rangle \equiv \langle \psi^\dagger(z)\tilde{U}(z)^q\rangle \to \rho_0^{1/2} e^{i\theta}$$

as $\bar{N} \to \infty$, and this defines our order parameter. From now on, fluid states $|\alpha\rangle$ will be taken to be pure states with nonzero order parameter. Physical properties will be more transparent when working with pure states.

Since $\psi^\dagger$ increases N, the particle number, by 1, and U^q increases $M(z)$, the angular momentum about z, by qN, $\psi^\dagger U^q$ breaks the symmetry generated by $\frac{1}{2}N+M/qN$, while $\frac{1}{2}N-M/qN$ is unbroken. Ψ characterizes the Laughlin state, since

$$|0_L;N\rangle = \left(\int d^2z\, \psi^\dagger(z)U(z)^q e^{-|z|^2/4}\right)^N |0\rangle$$

up to a normalization factor, in exact analogy with the ground state of a Bose gas or BCS superconductor. Thus the Laughlin state as in (4) is precisely the mean-field theory of the FQHE.

Note that $\langle\Psi^\dagger(z)\rangle$ is a *local* order parameter, even though $U(z)$ acts on particles far from z, because an (in principle distinct) value can be associated with each point z; this allows it to have thermodynamic significance in a Ginzburg-Landau description, as will be shown.

While the present order parameter resembles that of GM in involving a particle bound to a flux tube (here U^q), it differs in that we find true long-range order in the Laughlin state whereas GM find only algebraic order. The algebraic order of GM is apparently an artifact of their choice of flux operator. We note that any choice of flux operator in place of U^q gives a candidate order parameter for some fluid ground state of filling factor $1/q$, since by a Berry-phase calculation[7] the flux operator will be fermionic if q is odd, and the counting of charge makes the combination, like $\psi^\dagger U^q$, a locally neutral Bose-type operator, which may Bose condense, giving a liquid state.

In constructing the Ginzburg-Landau action, we will use states

$$|\alpha;z,n\rangle = (2^n n!)^{-1/2}(2\partial/\partial z + i\mathcal{A}_-)^n \bar{U}(z)^q|\alpha\rangle, \quad (5)$$

where $|\alpha\rangle$ is a (pure) fluid state. Equation (5) is the generalization to the normalized hole state $\bar{U}(z')^q|\alpha\rangle$ of the expansion of the unnormalized state $U(z')^q|\alpha\rangle$ in powers of $z'-z$ about some point z. The vector potential $\mathcal{A}_- = \mathcal{A}_x - i\mathcal{A}_y$ accounts for the normalization and generalizes[8] analyticity $\partial/\partial\bar{z}\equiv 0$,

$$(2\partial/\partial\bar{z} + i\mathcal{A}_+)\bar{U}(z)^q|\alpha\rangle \equiv 0,$$

giving us

$$i\mathcal{A}_-(z) = q\int \frac{d^2z'}{z'-z}\langle\bar{U}^\dagger(z)^q \rho(z')\bar{U}(z)^q\rangle, \quad (6)$$

where $\rho(z') = \psi^\dagger(z')\psi(z')$ is the density in the lowest Landau level. Equation (6) can equivalently be obtained from adiabatic transport of the hole.[7] $\mathcal{A}_-(z)$ can be calculated approximately by first (exactly) commuting $\rho(z')$ to the right to give

$$i\mathcal{A}_-(z) = q\int \frac{d^2z'}{z'-z}\frac{\langle\alpha|U^\dagger(z)^q U(z)^q R_q(z',z)|\alpha\rangle}{\langle\alpha|U^\dagger(z)^q U(z)^q|\alpha\rangle}, \quad (7)$$

where $R_n(z',z)$ is a one-body operator. Equation (7) is now approximated by the insertion of $|\alpha\rangle\langle\alpha|$ between U^q and R_q, in which case, remarkably,

$$\tfrac{1}{2}\epsilon_{\alpha\beta}\partial_\alpha\mathcal{A}_\beta \equiv i\partial\mathcal{A}_-/\partial\bar{z} = -\pi q\langle\rho(z)\rangle, \quad (8)$$

where $\alpha = x,y$, $\partial_x = \partial/\partial x$, etc. For a circular droplet of density $\langle\rho\rangle = \rho_0 = (2\pi q)^{-1}$, we find $\mathcal{A}_- = \frac{1}{2}i\bar{z}$, the symmetric gauge.

Because of the existence of the order parameter, we can relate hole states by the approximate expansion

$$\psi(z)|\alpha\rangle = \sum_{m=0}^{\infty}\beta_m|\alpha;z,m\rangle,$$
$$\beta_m = (2^m m!)^{-1/2}(2\partial/\partial\bar{z} + ia_+)^m\langle\Psi(z)\rangle, \quad (9)$$

where $\mathbf{a} = \mathbf{A} - \boldsymbol{\mathcal{A}}$ satisfies[2]

$$\epsilon_{\alpha\beta}\partial_\alpha a_\beta = 2\pi q(\langle\rho\rangle - \rho_0). \quad (10)$$

In Eq. (9) the $|\alpha;z,m\rangle$ for different m were treated as orthonormal; if $|\alpha\rangle$ is the ground state, this is correct; otherwise, there are overlaps between different n values because $\epsilon_{\alpha\beta}\partial_\alpha\mathcal{A}_\beta$ is not constant. These overlaps and the norms of the states may be calculated recursively; the definition of $\mathcal{A}_-$ implies that $|\alpha;z,0\rangle$ is normalized and orthogonal to $|\alpha;z,1\rangle$ in general. An orthonormal set of states can be constructed by the Gram-Schmidt method. This introduces corrections to the $|\alpha;z,n\rangle$ which, however, can be neglected in the *linearized* calculation described below, because a derivative of a slowly varying expectation value of either ρ or Ψ always appears, which is certainly already of first order in the deviation from the ground-state value. Hence, we may use (5) in (9). We see that

$$\langle\rho\rangle \simeq |\langle\Psi\rangle|^2 + \cdots.$$

We now derive the Ginzburg-Landau theory for $\langle\Psi\rangle$ by first obtaining approximate equations of motion for $(id/dt)\langle\Psi^\dagger\rangle$ and then writing down an action whose variation gives these equations.

The Hamiltonian projected into the lowest Landau level contains only potential-energy terms:

$$H = -\int d^2z\, V_{\text{ext}}(z)\rho(z) + \tfrac{1}{2}\int d^2z_1 d^2z_2 V(z_1-z_2)\psi^\dagger(z_1)\psi^\dagger(z_2)\psi(z_2)\psi(z_1), \quad (11)$$

where $V(z)$ is a function of $|z|$ only.

Working in the Heisenberg picture, we straightforwardly obtain

$$\left\langle i\frac{d\psi^{\dagger}(z)}{dt}\tilde{U}(z)^{q}\right\rangle = \sum_{n=0}^{\infty}(2\pi n!)^{-1}\left[\frac{\partial^{n}}{\partial z^{n}}\int d^{2}z' V_{\text{ext}}(z')e^{-|z'-z|^{2}/2}\right]\left[2\frac{\partial}{\partial\bar{z}}-ia_{+}\right]^{n}\langle\Psi^{\dagger}(z)\rangle$$
$$-V_{H}(0)\langle\Psi^{\dagger}(z)\rangle+\nabla^{2}V_{H}(0)\left[\frac{\partial}{\partial\bar{z}}-\tfrac{1}{2}ia_{+}\right]\left[\frac{\partial}{\partial z}-\tfrac{1}{2}ia_{-}\right]\langle\Psi^{\dagger}(z)\rangle. \quad (12)$$

In the interparticle potential terms, use has been made of (9), linearized in the deviation of $\langle\Psi^{\dagger}\rangle$ from its ground-state value, and higher derivatives have been dropped. The "Hartree-type potential,"

$$V_{H}(z_{1}-z) = \int d^{2}z_{2}V(z_{1}-z_{2})\langle\tilde{U}^{\dagger}(z)^{q}\rho(z_{2})\tilde{U}(z)^{q}\rangle, \quad (13)$$

is evaluated in the ground state and is then a function of $|z_{1}-z|$ only. I have neglected in (12) terms arising from taking $\langle\Psi^{\dagger}\rangle$ to be its ground-state value, and keeping the change in V_{H} to linear order; these "exchangelike" terms might give mass, quartic interaction, or additional gradient-squared terms in the Ginzburg-Landau long-wavelength action. The omission of these terms, which has no effect on the physics derived here, is our main dynamical approximation.

The remainder of the equation of motion is obeyed by the projection of $(i\,d/dt)\tilde{U}(z)^{q}|a\rangle$ onto the basis set (5); one finds

$$i\frac{d}{dt}\tilde{U}(z)^{q}|a\rangle = \Phi(z)|a;z,0\rangle+\sum_{n=1}^{\infty}(2^{n}n!)^{-1/2}\left[\left[2\frac{\partial}{\partial\bar{z}}\right]^{n}\Phi_{C}(z)\right]|a;z,n\rangle, \quad (14)$$

$$\Phi_{C}(z) = \frac{\langle U^{\dagger}(z)^{q}(i\,d/dt)U(z)^{q}\rangle}{\langle U^{\dagger}(z)^{q}U(z)^{q}\rangle} \simeq \int d^{2}z'\,2\frac{\partial V_{\text{ext}}}{\partial\bar{z}'}(z')\sum_{r=0}^{q-1}\frac{\langle R_{r+1}(z',z)\rangle}{z'-z}$$
$$-\int d^{2}z_{1}d^{2}z_{2}\,2\frac{\partial V}{\partial\bar{z}_{1}}(z_{1}-z_{2})\sum_{r=0}^{q-1}\frac{\langle\psi^{\dagger}(z_{2})R_{r+1}(z_{1},z)\psi(z_{2})\rangle}{z_{1}-z}, \quad (15)$$

and $\Phi = \text{Re}\Phi_{C}$; I have approximated by decoupling as in (7) and (8) and also dropped a term in the two-body part that involves both $[\rho(z_{2}),U(z)^{q}]$ and R_{r+1}, which is a higher-order correlation. Then, by manipulations similar to those used in (8) and (12), we obtain

$$2\frac{\partial\Phi_{C}}{\partial\bar{z}} = -\int\frac{d^{2}z'}{2\pi q}\,2\frac{\partial V_{\text{ext}}}{\partial\bar{z}'}(z')\sum_{r=0}^{q-1}\frac{|z'-z|^{2r}e^{-|z-z'|^{2}/2}}{2^{r}r!}-2\pi q\nabla^{2}V_{H}(0)\langle\Psi^{\dagger}(z)\rangle\left[\frac{\partial}{\partial\bar{z}}+\tfrac{1}{2}ia_{+}\right]\langle\Psi(z)\rangle, \quad (16)$$

up to higher gradients of $\langle\Psi\rangle$. From the exact expressions for a,Φ,

$$da_{+}/dt+2\,\partial\Phi/\partial\bar{z} \equiv 2\,\partial\Phi_{C}/\partial\bar{z}, \quad (17)$$

and the right-hand side of (16) can be interpreted as the drift-motion current due to the external and interparticle potentials.

Finally,

$$i\frac{d}{dt}\langle\Psi^{\dagger}(z)\rangle = \sum_{n=0}^{\infty}(2\pi n!)^{-1}\left[\frac{\partial^{n}}{\partial z^{n}}\int d^{2}z' V_{\text{ext}}(z')e^{-|z-z'|^{2}/2}\right]\left[2\frac{\partial}{\partial\bar{z}}-ia_{+}\right]^{n}\langle\Psi^{\dagger}(z)\rangle$$
$$-\sum_{n=1}^{\infty}(2\pi qn!)^{-1}\left[\frac{\partial^{n-1}}{\partial\bar{z}^{n-1}}\int d^{2}z'\frac{\partial V_{\text{ext}}}{\partial\bar{z}'}(z')\sum_{r=0}^{q-1}\frac{|z'-z|^{2r}}{2^{r}r!}e^{-|z-z'|^{2}/2}\right]\left[2\frac{\partial}{\partial z}-ia_{-}\right]^{n}\langle\Psi^{\dagger}(z)\rangle$$
$$+[\Phi(z)-V_{H}(0)]\langle\Psi^{\dagger}(z)\rangle+\nabla^{2}V_{H}(0)\left[\frac{\partial}{\partial\bar{z}}-\tfrac{1}{2}ia_{+}\right]\left[\frac{\partial}{\partial z}-\tfrac{1}{2}ia_{-}\right]\langle\Psi^{\dagger}(z)\rangle \quad (18)$$

to linear order in deviations from the ground state. Note the similar structure of the first two terms.

Since we are working at indefinite particle number and area, we must add chemical potential and pressure terms to the Hamiltonian, which can be incorporated in V_{ext}; this takes the form of a constant potential in the interior of the droplet, with slowly rising confining walls near the edge, and can be arranged to cancel in the ground-state case the terms on the right-hand side of (18) with no gradients of $\langle\Psi^{\dagger}\rangle$. Thus $(i\,d/dt)\langle\Psi^{\dagger}\rangle$ vanishes in the interior of the droplet in the ground-state case where $\langle\Psi^{\dagger}\rangle$ is uniform in space.

With the remainder of V_{ext} omitted for clarity, Eqs. (10) and (16)-(18) can be obtained by variation of the Lagrangian density

$$L=\Psi^{\dagger}i\frac{d}{dt}\Psi-\tfrac{1}{2}C\left[\left(2\frac{\partial}{\partial z}-ia_{-}\right)\Psi^{\dagger}\right]\left(2\frac{\partial}{\partial\bar{z}}+ia_{+}\right)\Psi+\Phi[|\Psi|^{2}-\rho_{0}-(2\pi q)^{-1}\epsilon_{\alpha\beta}\partial_{\alpha}a_{\beta}]-\frac{1}{4\pi q}\epsilon_{\alpha\beta}a_{\alpha}\frac{d}{dt}a_{\beta};\qquad(19)$$

where $C=\frac{1}{2}\nabla^2 V_H(0)$ (as usual, L is determined by the equations of motion only up to total time derivatives).

The same linearized equations of motion can be solved for plane-wave excitations to yield a dispersion relation

$$\omega^2=C^2+\tfrac{1}{2}Ck^2(C+\tfrac{1}{2}Ck^2);$$

thus this collective mode has a gap $\omega=C$ as $k\rightarrow 0$ which is due to the long-range gauge forces in the action (19) (the Anderson-Higgs mechanism). The roton minimum[9] at a larger wave vector $k\sim\rho_0^{1/2}$ is not obtained within the present approximation. The fractional statistics[7,10] of the vortices and the quantized Hall conductance also follow from (19).

The physical meaning of the order parameter $\langle\Psi^{\dagger}(z)\rangle$ is that it is the amplitude for finding a particle at z at the zeroes of the many-particle wave function, and by (9) its gradients represent the amplitudes for displacements from the zeroes. A nonzero displacement leads to a higher Hartree energy (13) (which just involves the two-particle correlation function) and hence to the stiffness constant $C=\frac{1}{2}\nabla^2 V_H(0)$. The long-range gauge forces are related to those in Laughlin's plasma analogy,[3] but here take on a dynamic as well as static role.

It should be possible to describe quantum fluctuations about the Laughlin state by quantizing the action (19), but this should be done with care to ensure a connection with the microscopic description. Since the Laughlin state is exact for the pseudopotential Hamiltonian[11] with $V_1, V_3, \ldots, V_{q-2}\neq 0$, $n\geq q$, quantum fluctuations will be controlled by the size of V_n, $n\geq q$.

A similar order parameter can be constructed for general filling factors $\nu=p/q$ with use of $\psi^{\dagger p}U^q$, and also extended to spin-singlet states;[12] these extensions and details of the present work will be given elsewhere.[13]

A brief report of part of this work was given previously.[14] After this work was completed, we received a preprint from Rezayi and Haldane,[15] who studied related order parameters numerically, and showed that they are nonzero in FQHE states at $\nu=\frac{1}{3},\frac{2}{5}$ but vanish in compressible states. We also learned of work by Zhang, Hansson, and Kivelson[16] on the Ginzburg-Landau action.

I thank S. Libby, F. D. M. Haldane, P. A. Lee, S. Zhang, and S. Kivelson for helpful discussions. Work at MIT was supported by NSF Grant No. DMR-85-21377.

[(a)]Present address.

[1]*The Quantum Hall Effect*, edited by R. E. Prange and S. M. Girvin (Springer-Verlag, New York, 1986).

[2]S. M. Girvin, in Ref. 1, Chap. 10; S. M. Girvin and A. H. MacDonald, Phys. Rev. Lett. **58**, 1252 (1987).

[3]R. B. Laughlin, Phys. Rev. Lett. **50**, 1395 (1983), and in Ref. 1.

[4]See, e.g., J. Glimm and A. Jaffe, *Quantum Physics* (Springer-Verlag, New York, 1981), p. 72.

[5]The conventions here are that **B** points in the negative $\hat{z}$ direction, coordinates in the plane are labeled by $z=x+iy$, $\partial/\partial z\equiv\frac{1}{2}(\partial/\partial x-i\partial/\partial y)$, the magnetic length $l=\hbar=1$, and the magnitude but not the sign of the electric charge is absorbed into the potentials. All results are gauge covariant.

[6]The equivalence of q quasiholes to one real hole in the Laughlin ground state was probably first noted by P. W. Anderson, Phys. Rev. B **28**, 2264 (1983).

[7]D. P. Arovas, J. R. Schrieffer, and F. Wilczek, Phys. Rev. Lett. **53**, 722 (1984).

[8]P. Griffiths and J. Harris, *Principles of Algebraic Geometry* (Wiley, New York, 1978), p. 73.

[9]S. M. Girvin, A. H. MacDonald, and P. M. Platzman, Phys. Rev. B **33**, 2481 (1986).

[10]B. I. Halperin, Phys. Rev. Lett. **52**, 1583 (1984).

[11]F. D. M. Haldane, Phys. Rev. Lett. **51**, 645 (1983), and in Ref. 1.

[12]F. D. M. Haldane and E. Rezayi, Phys. Rev. Lett. **60**, 956,1886(E) (1988).

[13]N. Read, to be published.

[14]N. Read, Bull. Am. Phys. Soc. **32**, 923 (1987).

[15]E. Rezayi and F. D. M. Haldane, Phys. Rev. Lett. **61**, 1985 (1988).

[16]S. C. Zhang, T. H. Hansson, and S. Kivelson, preceding Letter [Phys. Rev. Lett. **61**, 82 (1988)].

Collective Excitations in the Ginzburg-Landau Theory of the Fractional Quantum Hall Effect

Dung-Hai Lee
IBM Research Division, T. J. Watson Research Center, Yorktown Heights, New York 10598

Shou-Cheng Zhang
IBM Research Division, Almaden Research Center, San Jose, California 95120

(Received 10 December 1990)

The collective excitations of the fractional-quantum-Hall liquid are studied within the Ginzburg-Landau theory. We show that (1) Gaussian fluctuations of the phase of the order parameter correspond to the cyclotron mode with an energy gap of $\hbar\omega_c$ at $\vec{q}=0$ and a contribution to the static structure factor proportional to $|\vec{q}|^2$ as $\vec{q}\to 0$, in accordance with Kohn's theorem, and (2) vortex-antivortex fluctuations give rise to the lowest-Landau-level collective mode with an energy gap that depends only on the Coulomb energy and a static structure factor that vanishes as $|\vec{q}|^4$ as $\vec{q}\to 0$.

PACS numbers: 73.50.Jt, 05.30.-d, 74.20.-z

In their seminal works, Girvin and MacDonald[1] and Read[2] discovered a hidden order parameter in the Laughlin state.[3] They have shown that if one views an electron as a hardcore boson with an odd-integer number of Dirac flux quanta attached, the equal-time boson-boson correlation function acquires long-range behavior in the Laughlin state. Subsequently, Zhang, Hansson, and Kivelson[4] and Read[2] constructed a Ginzburg-Landau (GL) theory for the fractional quantum Hall effect (FQHE). It was shown that all the essential features of the FQHE can be derived from the GL theory by looking at the saddle-point solution and the Gaussian fluctuation around it. These successes certainly make the GL approach extremely appealing. However, the GL theory is also faced with some serious problems (in the following we shall focus our discussion on Ref. 4): (1) The GL theory predicts the wrong gap for collective excitations at long wavelength; (2) the static structure factor computed in this theory vanishes like $|\vec{q}|^2$, not like $|\vec{q}|^4$ as obtained from a variational approach;[5] and (3) it fails to explain the existence of the "roton minimum" in the dispersion of the collective excitations.

In general, there are two types of collective excitations at fractional filling factors: One is the cyclotron-resonance mode of the center of mass, which can be viewed as an inter-Landau-level particle-hole excitation, and the other is the intra-Landau-level excitations. On the one hand, by invoking Galilean invariance, Kohn's theorem[6] shows that the frequency of the first mode is exactly $\hbar\omega_c$ at $\vec{q}=0$ and has an oscillator strength in the static structure factor that vanishes as $|\vec{q}|^2$ in the long-wavelength limit. On the other hand, the frequency of the second mode is solely determined by the Coulomb interaction, and at the magic filling factors of the FQHE it gives rise to a correction to the static structure factor proportional to $|\vec{q}|^4$ in the long-wavelength limit. Following Girvin, MacDonald, and Platzman[5] (GMP), we shall call the intra-Landau-level collective mode the single-mode-approximation (SMA) mode. As shown by GMP, the dispersion of the SMA mode displays a magnetoroton minimum at a wavelength roughly equal to the interparticle distance.

In this Letter, we reexamine the problem associated with the collective excitations in the GL theory. Here we summarize our main results as follows. Following Ref. 4, we transform the original fermionic problem into a bosonic problem by introducing a statistical gauge field that has a Chern-Simons action. We then perform a duality transformation,[7] and arrive at an action containing both the Gaussian phase and the vortex degrees of freedom. We first show that the Gaussian-phase fluctuation at $\vec{q}=0$ corresponds to the center-of-mass cyclotron mode with an energy gap of $\hbar\omega_c$. At small $|\vec{q}|$, its contribution to the static structure factor vanishes as $|\vec{q}|^2$. We then integrate out the Gaussian degrees of freedom and obtain an effective action for the vortices. This effective action describes anyonlike vortices that interact via the Coulomb interaction and obey the guiding-center equations of motion. We then quantize this anyon problem and show that both the net vorticity and the total dipole moment of the vortices are conserved. These conservation laws state that only quadrupole (and higher) moment fluctuations can contribute to the long-wavelength dynamical structure factor $S(\vec{q},\omega)$, which leads to a static structure factor proportional to $|\vec{q}|^4$ as $\vec{q}\to 0$. We also show that the creation energy Δ of the vortices is determined purely by the Coulomb interaction, and predict the SMA dispersion relation for large $|\vec{q}|$. We also present a physical picture that explains the existence of the magnetoroton minimum.

We start with the GL Lagrangian (in units in which $c=\hbar=e=1$) written in Euclidean space-time:[4]

$$\mathcal{L}=i\bar{\Psi}\left(\frac{\partial_0}{i}+a_0-A_0\right)\Psi+\frac{1}{2m}\left|\left(\frac{\vec{\partial}}{i}+\vec{a}-\vec{A}\right)\Psi\right|^2+\tfrac{1}{2}(\bar{\Psi}\Psi-\bar{\rho})V(\bar{\Psi}\Psi-\bar{\rho})-\frac{i}{2}\sigma_{xy}\mathbf{a}\cdot\nabla\times\mathbf{a}\,, \quad (1)$$

where $\sigma_{xy}=1/k$ with k being an odd integer, and $\vec{A}=(H_0/2)(x\hat{y}-y\hat{x})$ with H_0 being the external magnetic field. We have also defined $\mathbf{A}$ and $\vec{A}$ as a three-vector and a two-vector, respectively, $\bar{\rho}$ as the average particle density, V as the two-body interaction, and

$$(\bar{\Psi}\Psi-\bar{\rho})V(\bar{\Psi}\Psi-\bar{\rho})\equiv\int d^2r'[\bar{\Psi}(\vec{r})\Psi(\vec{r})-\bar{\rho}]V(\vec{r}-\vec{r}')[\bar{\Psi}(\vec{r}')\Psi(\vec{r}')-\rho].$$

The last term in (1) is the so-called Chern-Simons term; its effect is to attach fluxes in $\vec{a}$ to particles. We then separate the modulus and the phase degrees of freedom by writing $\Psi=\rho^{1/2}\phi$, where $\phi\equiv e^{i\theta}$ is a unimodular field. Substituting this back into (1) and performing a Hubbard-Stratonovich decoupling of the kinetic-energy term we obtain

$$\mathcal{L}=i\mathbf{J}\cdot\left(\bar{\phi}\frac{\nabla}{i}\phi+\mathbf{a}-\mathbf{A}\right)+\frac{m}{2\rho}|\vec{J}|^2+\frac{1}{2m}|\vec{\partial}\rho^{1/2}|^2+\tfrac{1}{2}(\rho-\bar{\rho})V(\rho-\bar{\rho})-\frac{i}{2}\sigma_{xy}\mathbf{a}\cdot\nabla\times\mathbf{a}, \tag{2}$$

where $\vec{J}$ is the auxiliary field, and $\mathbf{J}\equiv(\rho,\vec{J})$. Here we note that by differentiating (2) with respect to $\mathbf{A}$ we obtain $\mathbf{J}$ as the physical three-current. To calculate the partition function, one has to perform path integrals over $\mathbf{J}$, ϕ, and $\bar{\phi}$.

We split θ into the topologically trivial and nontrivial parts via $\bar{\phi}(\nabla/i)\phi=\nabla\theta_g+\bar{\phi}_v(\nabla/i)\phi_v$, where θ_g is the topologically trivial Gaussian phase and ϕ_v contains the vortex configurations. Of course, in order to have a well-defined Ψ field, ρ should vanish inside the cores of the vortices. We then integrate over θ_g which produces a constraint on $\mathbf{J}$, namely, $\nabla\cdot\mathbf{J}=0$. We explicitly satisfy this constraint by writing $\mathbf{J}=\nabla\times\mathbf{b}$ (where $\mathbf{b}$ is an unconstrained field). Substituting this back into (2) and integrating out $\mathbf{a}$ we obtain

$$\mathcal{L}=\frac{m}{2(\nabla\times\mathbf{b})_0}|(\nabla\times\mathbf{b})_\perp|^2+\frac{1}{2m}|\vec{\partial}(\nabla\times\mathbf{b})_0^{1/2}|^2+\tfrac{1}{2}[(\nabla\times\mathbf{b})_0-\bar{\rho}]V[(\nabla\times\mathbf{b})_0-\bar{\rho}]+i\mathbf{b}\cdot(\mathbf{J}_v-\nabla\times\mathbf{A})+\frac{i}{2\sigma_{xy}}\mathbf{b}\cdot\nabla\times\mathbf{b}. \tag{3}$$

Here $\mathbf{J}_v\equiv\bar{\phi}_v(\nabla/i)\phi_v$ is the vortex three-current and $(\nabla\times\mathbf{b})_0$ and $(\nabla\times\mathbf{b})_\perp$ denote the time and space components of the three-vector $\nabla\times\mathbf{b}$, respectively. Two observations should be made here: (1) $\nabla\cdot\mathbf{J}_v=0$ and (2) since the winding of ϕ is always an integer, $\mathbf{J}_v$ describes the three-current of integer-quantized point particles. These two facts enable us to write $\rho_v=\sum_i q_i\delta(\vec{r}-\vec{r}_i)$ and $\vec{J}_v=\sum_i q_i\dot{\vec{r}}_i\delta(\vec{r}-\vec{r}_i)$, where q_i is the integer vorticity.

Because of the fact that $\vec{J}=(\nabla\times\mathbf{b})_\perp=\vec{J}_l(\equiv\hat{z}\times\dot{\vec{b}})+\vec{J}_t(\equiv-\hat{z}\times\vec{\partial}b_0)$ (the first term is the longitudinal current and the second term is the transverse current), it is particularly convenient to work in the Coulomb gauge ($\vec{\partial}\cdot\vec{b}=0$) where $|\vec{J}|^2=|J_l|^2+|J_t|^2$. By integrating out b_0 we obtain

$$\mathcal{L}=\frac{m}{2(\vec{\partial}\times\vec{b})}|\delta\dot{\vec{b}}|^2+\frac{\vec{\partial}\times\vec{b}}{2m}\left|\vec{\partial}G\left(\rho_v+\frac{1}{\sigma_{xy}}\vec{\partial}\times\delta\vec{b}\right)\right|^2+\frac{1}{2m}|\vec{\partial}(\vec{\partial}\times\vec{b})^{1/2}|^2$$
$$+\frac{1}{2}(\vec{\partial}\times\delta\vec{b})V(\vec{\partial}\times\delta\vec{b})+\frac{i}{2\sigma_{xy}}\delta\vec{b}\times\delta\dot{\vec{b}}+i\vec{J}_v\cdot\vec{b}-iA_0(\vec{\partial}\times\vec{b}), \tag{4}$$

where $G\equiv1/|\vec{\partial}|^2$, and we have defined $\vec{b}\equiv\delta\vec{b}+\langle\vec{b}\rangle$ with $\langle\vec{b}\rangle=\tfrac{1}{2}\bar{\rho}(x\hat{y}-y\hat{x})$.

We first set $\mathbf{J}_v,A_0=0$ and study the Gaussian fluctuations of $\vec{b}$. The physical meaning of $\delta\vec{b}$ becomes clear if we define the displacement field $\vec{u}\equiv\bar{\rho}\hat{z}\times\delta\vec{b}$. Substituting this expression into (4) and linearizing the result at long wavelength we obtain

$$\mathcal{L}=\frac{m}{2}\bar{\rho}|\dot{\vec{u}}|^2=\frac{m}{2}\bar{\rho}\omega_c^2(\vec{\partial}\cdot\vec{u})G(\vec{\partial}\cdot\vec{u})+\frac{\bar{\rho}^2}{2}(\vec{\partial}\cdot\vec{u})V(\vec{\partial}\cdot\vec{u})+\frac{i}{2}\frac{\bar{\rho}^2}{\sigma_{xy}}\vec{u}\times\dot{\vec{u}}. \tag{5}$$

Here $\omega_c\equiv\bar{\rho}/m\sigma_{xy}$ is the cyclotron frequency, and the Coulomb-gauge constraint becomes $\vec{\partial}\times\vec{u}=0$, i.e., $\vec{u}$ describes purely longitudinal displacements. Let us first consider the uniform displacement. In that case, $\vec{\partial}\times\vec{u}=0$ is automatically satisfied; hence there is no further constraint imposed on the uniform displacement. The effective Lagrangian for that case describes the motion of the center of mass and is given by $\mathcal{L}=(M/2)|\dot{\vec{u}}|^2+(i/2)M\omega_c\vec{u}\times\dot{\vec{u}}$, where $M\equiv m\int d^2r\bar{\rho}$ is the total inertial mass of the fluid. This is precisely the Euclidean space-time Lagrangian for a particle of mass M, coordinate $\vec{u}$, moving in an external magnetic field $H=M\omega_c$. Such a Lagrangian implies cyclotron motion of the center of mass with the cyclotron frequency equal to ω_c.

At nonzero wave vector, the term proportional to $\vec{u}\times\dot{\vec{u}}$ in (5) is ineffective due to the constraint $\vec{\partial}\times\vec{u}=0$. The remaining effective Lagrangian gives normal-mode dispersion $\omega_q^2\equiv\omega_c^2+(\bar{\rho}/m)|\vec{q}|^2V(\vec{q})$. Hence for a short-range two-body interaction V, the normal-mode frequency starts out from $\omega=\omega_c$ and disperses quadratically, whereas if $V(r)\propto1/r$ it disperses linearly, in agreement with the random-phase-approximation calculation by Kallin and Halperin.[8] Next we calculate the dynamical structure factor $S(\vec{q},\omega)$ due to this collective mode. To do that we turn A_0 back on and integrate out the displacement field $\vec{u}$. In the resulting action, $\tfrac{1}{2}S(\vec{q},\omega)$ is the coefficient of the A_0^2 term. Explicit calculation of

$S(\vec{q},\omega)$ gives (after Wick's rotation) $S(\vec{q},\omega)=\bar{\rho}|\vec{q}|^2/m(\omega^2-\omega_q^2)$, which exhibits poles at $\omega=\omega_q$ and has an oscillator strength that vanishes like $|\vec{q}|^2$ as $\vec{q}\to 0$, in accordance with Kohn's theorem. This mode has been mistaken in the literature[4,7] as the lowest-Landau-level collective mode. This misidentification is corrected by our results.

Now we restore the vortex degrees of freedom and study the effects of vortex-antivortex fluctuation on the dynamical structure factor. By assuming that the vortex motion is much slower than the cyclotron frequency, we integrate out $\delta\vec{b}$ in Eq. (4) to obtain

$$\mathcal{L}=-iA_0\bar{\rho}+\tfrac{1}{2}A_0SA_0+i\sigma_{xy}\sum_j q_jA_0(\vec{r}_j)+\sum_j\Delta(q_j)+\tfrac{1}{2}\sigma_{xy}^2\sum_{i\neq j}q_iq_jV(\vec{r}_i-\vec{r}_j)$$
$$+\frac{i}{2}\bar{\rho}\sum_j\vec{r}_j\times\dot{\vec{r}}_j-i\sigma_{xy}\sum_{i\neq j}q_iq_j\dot{\vec{r}}_i\cdot(\hat{z}\times\vec{\partial}_i)G(\vec{r}_i-\vec{r}_j)\,, \tag{6}$$

where $\Delta(q_j)$, the quasiparticle (a quasiparticle is a vortex plus a screening cloud produced by $\vec{\partial}\times\delta\vec{b}$) creation energy, is given by $\Delta(q)\equiv\frac{1}{2}\int d^2r_1d^2r_2V(\vec{r}_{12})g_q(\vec{r}_1,\vec{r}_2)$, in which $g_q(\vec{r}_1,\vec{r}_2)\equiv\langle\vec{\partial}\times\delta\vec{b}(\vec{r}_1)\vec{\partial}\times\delta\vec{b}(\vec{r}_2)\rangle$ is the pair-distribution function in the presence of a vortex with vorticity q situated at the origin. In this work knowing that $\Delta(q)$ is determined solely by the interaction energy is sufficient.

Now we briefly outline the steps leading from (4) to (6). We first consider the case of a single static vortex. Because of the nonlinearity of Eq. (4) and the constraint $\vec{\partial}\times\vec{b}\geq 0$, brute-force integration over $\delta\vec{b}$ is formidable. What we have done instead are the following: (1) recognize that Eq. (4) is the action for a problem of a long-range interacting boson in the presence of an impurity; (2) write down the corresponding Hamiltonian; (3) show[9] that the modulus of the Laughlin quasihole wave function is the ground state of $H_{1/m}$, the part of Hamiltonian that is proportional to $1/m$, and the associated eigenenergy is $N\hbar\omega_c/2$, where N is the total number of particles; and (4) compute $\Delta(+1)$ and show that the total induced charge in the vicinity of the vortex is $+\sigma_{xy}$. The situation is more complicated with an antivortex. This is because in order to diagonalize $H_{1/m}$ and maintain the eigenenergy as $N\hbar\omega_c/2$ a localized vortex-antivortex cloud is induced. However, in regions of space far away from the antivortex, ρ_v is well approximated by a δ function and the asymptotic form of the wave function can be written down easily. Fortunately, knowing the asymptotic wave function is sufficient to show that the total induced charge is $-\sigma_{xy}$. Finally, (5) by considering far separated multivortex configurations and allowing the vortices to move adiabatically, we can show that Eq. (6) yields the action for the path integral for the vortices.

The last term in Eq. (6) describes the Berry phase that vortices experience when they move around each other; hence it describes quasiparticles carrying a fraction (σ_{xy}) of charge with σ_{xy} statistics moving in external magnetic field $\bar{\rho}$, i.e., Laughlin quasiparticles. The goal now is to integrate out the vortex degrees of freedom to obtain the correction to the dynamical structure factor due to the quasiparticles. To do that, it is most convenient to write down the corresponding quantum-mechanical description of the vortices.[10] Equation (6) can also be viewed as the coherent-state path-integral representation of the following quantum Hamiltonian:

$$\mathcal{H}=\tfrac{1}{2}\sigma_{xy}^2\sum_{i\neq j}q_iq_jV'(x_i,p_i)\,, \tag{7}$$

with the commutation relation $[x_i,p_j]=i\delta_{ij}$. Here V' is a normal-ordered operator (normal ordered according to the creation and destruction operators constructed from x and p) such that

$$V'(x_i,p_i)|_{p_i}=V(x_i-x_j,y_i-y_j)\,, \tag{8}$$

where evaluation is at

$$p_i=\frac{\bar{\rho}}{2}q_jy_j-\sigma_{xy}\sum_{k\neq j}q_jq_k\frac{\partial}{\partial y_j}G(\vec{r}_{jk})\,.$$

Although the relation between p_j and y_j is complicated, the Heisenberg equation of motion for $\vec{r}_i$ is simple and is given by $\bar{\rho}\dot{\vec{r}}_j=\sigma_{xy}^2(\hat{z}\times\vec{\partial}_j)\sum_{k\neq j}q_kV(\vec{r}_{jk})$, the same as the Euler-Lagrangian equation derived from (6). This equation of motion enables a direct analysis of the behavior of the structure factor in the $\vec{q}\to 0$ limit. For small $|\vec{q}|$, the vortex density operator $\rho_v(\vec{q})=\sum_jq_je^{i\vec{q}\cdot\vec{r}_j}$ can be expanded in powers of $\vec{q}$. The density-density correlation function after this expansion is given by

$$\langle\rho_v(\vec{q},t)\rho_v(-\vec{q},0)\rangle=\langle Q(t)Q(0)\rangle+\tfrac{1}{3}|\vec{q}|^2\langle\vec{D}(t)\cdot\vec{D}(0)\rangle+\sum_{\alpha,\beta}\frac{(q_\alpha q_\beta)^2}{4}\langle Q^{\alpha,\beta}(t)Q^{\alpha,\beta}(0)\rangle+\cdots\,, \tag{9}$$

where $Q\equiv\sum_iq_i$ is the net vorticity, $\vec{D}\equiv\sum_iq_i\vec{r}_i$ is the total dipole moment, and $Q^{\alpha,\beta}\equiv\sum_iq_ir_i^\alpha r_i^\beta$ is the total quadrupole moment ($\alpha,\beta=x,y$ refer to the space indices). Since the Heisenberg equations of motion explicitly conserve both Q and $\vec{D}$, the first two contributions in (9) vanish identically. This result establishes the fact that the vortex contributions to the dynamical structure factor vanish like $|\vec{q}|^4$ in the $\vec{q}\to 0$ limit.

Finally, we should like to address the issue associated with the dispersion of the SMA mode. First we consider the case of large momentum. It is useful to gain some intuition by looking at the behavior predicted by the classical equation of motion. For simplicity, let us consider a vortex-antivortex pair. Since the vortex and antivortex attract each other, their classical behavior is to form a dipole with size R that drifts with a center-of-mass velocity chosen so that the Lorentz force balances the Coulomb force. For such a pair, small quantum fluctuations do not involve processes in which one particle goes around another. (The same thing cannot be said for two vortices of the same sign. In that case the classical behavior involves one particle going around the other, and fluctuation in the interparticle distance results in changes in the winding number.) Therefore to analyze the pair problem, we drop the Berry-phase term in (6), which leads to a simpler commutation relation $[x_j,(\bar{\rho}/2)q_jy_j]=i$. From this commutation relation and Eq. (7) we see that the components of the relative-position vector $\vec{r}\equiv\vec{r}_1-\vec{r}_2$ commute with each other, and their eigenstates $|r_x,r_y\rangle$ are also eigenstates of the Hamiltonian with eigenvalue $-\sigma_{xy}^2V(r)$. The density operator $\rho_v(\vec{q})\propto e^{i\vec{q}\cdot\vec{R}}$ $[\vec{R}\equiv(\vec{r}_1+\vec{r}_2)/2]$ is off diagonal in this basis set, since $\vec{R}$ and $\vec{r}$ do not commute. In fact, it can be easily checked from the commutation relation that the operator $e^{iq_xR_x}$ connects an eigenstate $|r_x,r_y\rangle$ to the unique eigenstate $|r_x,r_y+l_0^2/\sigma_{xy}q_x\rangle$, where $l_0=1/(2\pi\bar{\rho})^{1/2}$ is the magnetic length. For large q_x, the dipole configuration created by the density operator has an energy given by

$$E(|q_x|)=2\Delta-\sigma_{xy}^2V\left(\frac{l_0^2}{\sigma_{xy}}|q_x|\right)=2\Delta-\frac{\sigma_{xy}^3e^2}{|q_x|l_0^2}, \qquad (10)$$

where we have substituted the long-range part of the static Coulomb interaction for V. Equation (10) agrees exactly with the result of Kallin and Halperin.[8] This is the description of the SMA mode in the region where $q_x>2\pi/l_0$.

For $q_x=0$, we believe that the SMA mode consists of two dipoles, each with a size of roughly l_0, oriented in the $\hat{x}$ and $-\hat{x}$ directions, respectively, and with centers of mass separated by roughly l_0. This configuration has a quadrupole moment, but no net dipole moment, in accordance with our previous analysis for $q=0$. As q_x increases, so does the total dipole moment in the $\hat{y}$ direction, and these two dipoles rotate rigidly in opposite senses as two dumbbells around their individual centers of mass, until they eventually lie in the $\hat{y}$ direction and coalesce into a single dipole when $q_x=2\pi/l_0$. In this process, the electrostatic energy monotonically decreases. As q_x further increases, the size of the remaining dipole further increases in the $\hat{y}$ direction and so does the energy. In this way we can explain the existence of the roton minimum. A disclaimer is in order here. Although we believe that ignoring the Berry-phase term should be a good approximation when $q_x\gg 2\pi/l_0$, we are uncertain about its validity for q_x near or less than $2\pi/l_0$. Finally, this picture also explains why the SMA is not a good approximation for $q_x<2\pi/l_0$. This is because of the relative degrees of freedom involved in the quadrupole configurations.

We acknowledge useful discussions with C. Hanna, S. Kivelson, and N. Read. S.C.Z. is grateful for the hospitality of the theory group at IBM T. J. Watson Research Center where this collaboration was initiated.

---

[1]S. M. Girvin and A. H. MacDonald, Phys. Rev. Lett. **58**, 1252 (1987).

[2]N. Read, Phys. Rev. Lett. **62**, 86 (1989).

[3]R. B. Laughlin, Phys. Rev. Lett. **50**, 1395 (1983).

[4]S. C. Zhang, T. H. Hansson, and S. Kivelson, Phys. Rev. Lett. **62**, 82 (1989).

[5]S. M. Girvin, A. H. MacDonald, and P. M. Platzman, Phys. Rev. Lett. **54**, 581 (1985); Phys. Rev. B **33**, 2481 (1986).

[6]W. Kohn, Phys. Rev. **123**, 1242 (1961).

[7]M. P. A. Fisher and D-H. Lee, Phys. Rev. B **39**, 2756 (1989); D-H. Lee and M. P. A. Fisher, Phys. Rev. Lett. **63**, 903 (1989).

[8]C. Kallin and B. I. Halperin, Phys. Rev. B **30**, 5655 (1984).

[9]This is done in a way similar to that used in C. Kane, S. A. Kivelson, D-H. Lee, and S. C. Zhang, Phys. Rev. B **43**, 3255 (1991).

[10]A. L. Fetter, Phys. Rev. **162**, 143 (1967); F. D. M. Haldane and Y. S. Wu, Phys. Rev. Lett. **55**, 2887 (1985).

PHYSICAL REVIEW B VOLUME 42, NUMBER 1 1 JULY 1990

Superfluid dynamics of the fractional quantum Hall state

Michael Stone
Department of Physics, Loomis Laboratory, University of Illinois at Urbana-Champaign, 1110 West Green Street, Urbana, Illinois 61801
(Received 25 January 1990)

The fractional quantum Hall effect is due to a novel state of matter with properties reminiscent of superfluidity. I derive the Euler equation and the quantum constraint that together determine the dynamics of the superfluid. I then use these equations to explain the incompressibility of the liquid, to describe the magnetophonon excitations, and to study the forces acting on vortices whose pinning is responsible for the plateaus in the Hall resistance.

I. INTRODUCTION

A neutral substance like ^{4}He owes the existence of its superfluid phase to subtle quantum-mechanical effects. Its bulk properties, however, are still those of a classical fluid: it may be poured, shaken, and stirred like any other liquid and its magical superflow properties are simply a consequence of the fact that the superflow is irrotational, i.e., to d'Alembert's paradox. Below the critical velocity for creating phonons dissipation is due to energy absorbed in the manufacture of quantized vortex lines which behave almost classically.

Superconductors behave similarly[1] except the phonons, being turned into plasmons by the electric charge of the condensate, are even harder to create and all resistive dissipation comes from the depinning and motion of Abrikosov vortices.[2] In the Bardeen-Stephens theory,[3] the magnetic field of a moving vortex induces an electric field which drives a normal, dissipative, current through the vortex core. The forces causing the vortex to move are best understood via a fluid-dynamics analysis:[4] The superconducting condensate behaves as charged superfluid and the Abrikosov vortex has an inner core, of radius approximately equal to the coherence length, about which there is quantized circulation. This core vortex is surrounded by a screening antivortex (whose radius is approximately equal to the the magnetic penetration depth) with vorticity proportional to the magnetic field and total circulation equal and opposite to that of the core; there is thus no circulation at large distances. The force felt by a stationary core vortex embedded in a bulk flow is a *Magnus* force due to the interaction of the circulation and the uniform flow. The Magnus pressure on the core is in turn balanced by the Lorentz force on the current through the surrounding antivortex.[4]

The novel state of matter found in a two-dimensional electron gas (2D EG) at low-temperature and high magnetic fields[5] has many properties in common with superfluid phases,[6,7] including long-range order, no dissipation, and vortex soliton excitations.[8] The vortices contain the deviations of the charge density from uniform, rational fraction, filling of the Landau levels and, presumably,[9-11] it is the pinning of the vortices by impurities that yields plateaus in the Hall conductance. Since this "two-fluid picture" of the fractional quantum Hall effect (FQHE) depends crucially on vortex pinning it would be useful to have the same kind of intuitive picture of the forces on a vortex that we have for the other superfluids. It is the intention of this paper to provide that picture. In addition to plateau formation FQHE vortices have another essentially quantum role: when they are free to move they behave as solitonic quasiparticles with fractional charge and, it is believed, fractional statistics. They may in their turn undergo Bose condensation and give rise to the hierarchy of FQHE states.[12-14]

The flow properties of superfluids are determined by classical Euler equations—with an additional constraint on the vorticity which embodies the quantum mechanics. These Euler equations, being an expression of the laws of conservation of mass and momentum, have greater validity than any particular model used to derive them and can be applied over a wide range of conditions. In the FQHE case we could start from either the microscopic wave function of Laughlin[10] or the more phenomenological Landau-Ginsburg approaches based on Chern-Simons (CS) Lagrangians[6-8] and should arrive at the same equations. In this paper I will take the second route and motivate the Euler equations for the FQHE state from the mean-field Landau-Ginsburg model which I will review in Sec. II. In Sec. III I obtain the Euler equations by the methods used for conventional superfluids and, in Sec. IV, I examine the consequences of the fluid dynamics equations for the properties of the FQHE and more general anyons in a magnetic field.

II. MEAN-FIELD THEORY OF THE FRACTIONAL QUANTUM HALL EFFECT

The need for a Landau-Ginsburg picture of the FQHE was stressed by Girvin in the summary of Ref. 5 and he there indicated the necessary ingredients including a Chern-Simons term in the action. A more complete model, based on a mean-field treatment of the Chern-Simons Lagrangian, was introduced by Zhang, Hansen and Kivelson[6] and by Read.[7] In the approach of Zhang *et al.*[6] one begins with a path integral

$$Z=\int d[\phi]d[\phi^*]d[a_\mu]\exp i \times \int d^3x \left[\phi^* H\phi+\frac{i}{4\Theta_0}\epsilon^{\mu\nu\sigma}a_\mu\partial_\nu a_\sigma\right] \quad (2.1)$$

involving a *commuting* ϕ field representing the electrons, and a "statistics" field a_μ. H is an action for a nonrelativistic particle with effective mass m^*, e.g.,

$$H=\frac{1}{2m^*}(\partial_i-ia_i-iA_i)^2-i(\partial_t-ia_0-iA_0) +\lambda\left[|\phi|^2-\frac{\mu_0}{2\lambda}\right]^2 \quad (2.2)$$

and Θ_0 is the statistics parameter[15-17] which, taking one of the values $(2n+1)\pi$, ensures that the bosonic ϕ field describes particles with Fermi statistics.

In the mean-field approximation to the path integral the CS statistics field is determined by the electron density ρ through

$$2\Theta_0\rho=2\Theta_0|\phi|^2=\nabla\times\mathbf{a}\ . \quad (2.3)$$

If the density is uniform the *curl* of the CS mean field will be constant. When the external magnetic field is also uniform the density may be such that the gauge fields in (2.2) *cancel*. This requires

$$2\Theta_0\rho=2\pi(2n+1)\rho=|B| \quad (2.4)$$

or

$$\rho=\frac{|B|}{2\pi}\frac{1}{2n+1} \quad (2.5)$$

corresponding to a lowest Landau-level filling fraction of $\nu=1/(2n+1)$. In this case the ϕ field has a smooth classical solution, ϕ=const, which will dominate the path integral at low temperature. The Fermi statistics of the electrons has been nullified by the magnetic field, allowing the electrons to Bose condense. The resultant FQHE state is a novel kind of charged superfluid. It is easy to see[6] that the ground state has Hall conductivity

$$\sigma_{xy}=\frac{1}{2n+1}\frac{e^2}{h} \quad (2.6)$$

and has an energy gap resulting in dissipationless flow with $\sigma_{xx}=0$.

The solution appears to depend on the choice of n in the statistics parameter, a choice that should effect no physics, but it is best to regard the picture as being *simplest* for our choice of n rather than depending on it.

This theory is very appealing—but there is some sleight of hand in the derivation implicit in writing down (2.1) as if it were obvious. The "effective mass," m^*, should not be thought of as being the effective mass of the electrons in the 2D EG. As pointed out in Ref. 6, the magnetic field is so large that it has suppressed all the zero-point motion of the electrons and nowhere in the wave functions does any mass m occur. Reference 7 provides a less intuitive derivation of the model but shows clearly that the coefficient of the condensate stiffness, $1/2m^*$, is really a parameter depending on the Coulomb repulsion between the electrons [but see the discussion after Eq. (4.6)].

Including short-wavelength ϕ-field fluctuations in the calculation of the low-energy effective action will induce terms like $(\nabla\times\mathbf{a})^2$ and $\dot{a}^2$. The former has the effect of smearing out the δ function source each particle provides for the statistics field while preserving the essential flux-density relationship. The latter will modify higher frequency motion of the system but should not effect the slower motions that are our primary interests.

In addition to the uniform ground state it is clear by analogy with Abrikosov vortex lines in a conventional superconductor that there will be localized vortex solutions. These will have $\oint(a+A)_\mu dx^\mu=\pm 2\pi$. Since the currents in the thin sample are tiny they will not effect the distribution of the external $|B|$ field and this integral is really $\oint\delta a_\mu dx^\mu$. So, from (2.3), the vortices have a deviation from uniform charge with

$$q=\int\delta\rho d^2x=\pm\frac{1}{2n+1}\ . \quad (2.7)$$

There are two distinct kinds of vortex: the quasihole vortices have $\rho=|\phi|^2$ reduced on average, while the quasiparticle excitations will have it enhanced. In both cases ϕ winds through 2π as we circle the vortex and for finite energy ϕ must vanish at the center: so the hole and particle solutions will have quite distinct charge profiles and cannot be simply related. The Laughlin wave-function picture[4] has excitations with similar properties except there is no vanishing of the charge profile in the center of the quasiparticle.

These excitations cost energy and this energy gap is part of the origin of the stability of the odd-denominator filling-fraction states. Presumably, they are also the origin of plateau formation: any mismatch between the electron density and the magnetic-field strength will lead to the excess or deficit of charge being used to form vortices. If these vortices are pinned by impurities the charge sequestered by them will be immobile and only the "superfluid fraction" will flow and contribute to the observed FQHE. There is therefore a kind of two-fluid model for the FQHE.

Vortices are also anyons with true fractional statistics; the combination of fractional charge and 2π flux gives them a statistics parameter $\Theta_1/\pi=2p_1+\pi/\Theta_0$ where p_1 is an arbitrary integer. These excitations themselves will condense when their density is correctly chosen and there will be vortices in *their* condensate which have statistics parameter $\Theta_2/\pi=2p_2+\pi/\Theta_1$, etc. In this way we see the explanation for the heircharchy of FQHE states with a continued fraction of allowed condensate densities.[12-14]

III. EULER EQUATIONS

In this section the external magnetic field will be uniform unless variations are explicitly introduced. Bold symbols such as **j** will denote two component vectors in the plane of the 2D EG. Vectors in the perpendicular direction such as the B field and the vorticity ω will not be bold—their appearance in vector products such as $\mathbf{v}\times\omega$ should not cause confusion. I will use the symbol

$\mathcal{A}_i$ for the combination, $a_i + A_i$, of the statistics and electromagnetic fields. With this notation the equation of motion for ϕ is

$$i(\partial_t - i\mathcal{A}_0)\phi = -\frac{1}{2m^*}(\partial_i - i\mathcal{A}_i)^2\phi + 2\lambda\phi|\phi|^2 - \mu_0\phi \; . \tag{3.1}$$

The $\mathcal{A}_i$ field is determined by the density, current, and external electromagnetic field by the equations

$$\begin{aligned}(\partial_1\mathcal{A}_2 - \partial_2\mathcal{A}_i) &= -2\Theta_0\rho + B \; , \\ (\partial_2\mathcal{A}_0 - \partial_0\mathcal{A}_2) &= -2\Theta_0 j_1 + E_2 \; , \\ (\partial_0\mathcal{A}_1 - \partial_1\mathcal{A}_0) &= -2\Theta_0 j_2 - E_1 \; .\end{aligned} \tag{3.2}$$

These are consistent with current conservation provided

$$B + \partial_1 E_2 - \partial_2 E_1 = 0 \; , \tag{3.3}$$

i.e., if the Maxwell equation $\dot{B} + \nabla \times \mathbf{E} = 0$ is satisfied.

To obtain the Euler equations we write $\phi = \sqrt{\rho}e^{i\theta}$ and, for the moment, assume ρ to be sufficiently slowly varying that we can ignore its derivatives. Equation (3.1) becomes

$$-(\dot{\theta} - \mathcal{A}_0) = \tfrac{1}{2}m^*(\partial_i\theta - \mathcal{A}_i)^2 - \frac{i}{2m^*}\partial_i(\partial_i\theta - \mathcal{A}_i) + \mu \; . \tag{3.4}$$

The nonlinear terms are bundled into the variable $\mu(x)$. In ordinary hydrodynamics $\mu(x)$ would be the specific enthalpy but here it will be interpreted as a local chemical potential. At $T=0$ the specific enthalpy and the chemical potential coincide, and in the hydrodynamics of superfluids it is the chemical potential that appears in the Bernoulli equation for the superfluid fraction. Defining a flow velocity field by

$$v_i = \frac{1}{m^*}(\partial_i\theta - \mathcal{A}_i) \; , \tag{3.5}$$

we see that the imaginary part of (3.4) asserts that $\nabla \cdot \mathbf{v} = 0$, while taking the gradient of the real part yields

$$\begin{aligned} m^*\partial_t v_1 &= +2\Theta_0\rho v_2 + E_1 - \partial_1\left[\frac{1}{2m^*}v^2 + \mu\right] , \\ m^*\partial_t v_2 &= -2\Theta_0\rho v_2 + E_2 - \partial_2\left[\frac{1}{2m^*}v^2 + \mu\right] . \end{aligned} \tag{3.6}$$

We can recast these equations in a more familiar form by noting that the vorticity is given by

$$\omega = \partial_1 v_2 - \partial_2 v_1 = -\frac{1}{m^*}(\partial_1\mathcal{A}_2 - \partial_2\mathcal{A}_1) = \frac{2\Theta_0}{m^*}\rho - \frac{1}{m^*}B \; , \tag{3.7}$$

i.e.,

$$m^*\omega + B = 2\Theta_0\rho \; . \tag{3.8}$$

While Eq. (3.8) is an almost trivial extension of the $m^*\omega + e^*B = 0$ relation for a superconductor, it is this vorticity relation which is responsible for the incompressibility of the 2D EG: Any local change in density implies a net vorticity, and in two dimensions an isolated vortex has logarithmically divergent kinetic energy.

By writing $\mathbf{j} = \rho\mathbf{v}$, and replacing $2\Theta_0\rho$ by $m^*\omega + B$ we reexpress (3.7) as

$$m^*[\dot{\mathbf{v}} - (\mathbf{v} \times \omega)] = \mathbf{E} + \mathbf{v} \times B - \nabla(\tfrac{1}{2}m^*v^2 + \mu) \; , \tag{3.9}$$

which is a form of the Euler equation for a fluid of particles of mass m^* and unit charge. By comparison with ordinary fluid dynamics we see that μ is related to the pressure P by $\nabla P = \rho\nabla\mu$. Equations (3.8) and (3.9) are the principle results of this section.

The equation of motion (3.9) is compatable with the quantum vorticity condition (3.8) since taking the *curl* of (3.9) and using $\nabla \times \mathbf{E} + \dot{B} = 0$, gives

$$\partial_t(m^*\omega + B) + \nabla \cdot \mathbf{v}(m^*\omega + B) = 0 \tag{3.10}$$

This equation shows that even in the absence of quantum mechanics a perfect two-dimensional charged fluid can only gain or lose vorticity via a change in the magnetic field. The anyonic statistics field ties the combination $(m^*\omega + B)$ to the density ρ and then (3.10) becomes the charge conservation equation

$$2\Theta_0[\dot{\rho} + \nabla \cdot (\rho\mathbf{v})] = 0 \; . \tag{3.11}$$

A consequence of (3.11) is the equations we obtained by assuming ρ essentially constant are actually more general and may be consistently used in situations where the density varies. As is the case of ordinary superfluids the keeping track of density variation terms will add a "quantum pressure" term into $\mu(x)$.[1]

IV. WAVES AND VORTICES

The simplest solutions to Eqs. (3.8) and (3.9) are obviously the steady uniform Hall flows where, e.g., $2\Theta_0\rho v_1 = E_2$. Next easiest to study are the density waves obtained by linearization of the equations of motion. For these we try

$$\begin{aligned} v_1 &= A_1\cos(kx - \Omega t) \; , \\ v_2 &= A_2\sin(kx - \Omega t) \end{aligned} \tag{4.1}$$

corresponding to motion in ellipses, reminiscent of gravity waves in water. With this *ansatz* we find from, $\dot{\rho} + \nabla \cdot \rho\mathbf{v}$, that

$$\rho = \rho_0\left[1 + A_1\frac{k}{\Omega}\cos(kx - \Omega t)\right] . \tag{4.2}$$

This is consistent with the constraint (3.8) provided

$$m^*A_2k\cos(kx - \Omega t) + B = 2\Theta_0\rho\left[1 + A_1\frac{k}{\Omega}\cos(kx - \Omega t)\right] . \tag{4.3}$$

Since $B = 2\Theta_0\rho_0$, this is equivalently

$$m^*\frac{\Omega}{B} = \frac{A_1}{A_2} \; . \tag{4.4}$$

If we now substitute (4.1) into either of equations (3.6) and use $\mu=2\lambda\rho$, we find the dispersion relation

$$(m^*)^2\Omega^2=B^2+2\lambda\rho_0k^2 . \tag{4.5}$$

In this fluid-flow picture of the magnetophonon modes we see that it is the intrinsic vorticity of the 2D EG which, because its long-range influence is governed by the same equations as the 2D Coulomb interaction, plays a role identical to that of the charge in superconducting condensate and opens a gap in the phonon dispersion relation. In this analysis I have ignored the genuine **E** field generated by the nonuniform density—but because it can escape from the plane of the 2D EG it cannot generate a plasmon gap.

The trajectories of the "particles" in the density wave become circular at long wavelength and evolve into conventional longitudinal sound waves as the wavelength decreases. In the $k=0$ mode the whole system of charge moves together in circles and this motion raises some interesting questions. According to Kohn's theorem[18] there is a bulk mode where the particles orbit as a rigid mode at the free particle cyclotron frequency,

$$\Omega_c=B/m , \tag{4.6}$$

independent of interactions. If the present $k=0$ mode is identified with this "Kohn theorem" mode then, the remarks in Sec. II notwithstanding, m^* *is* the effective mass of the individual electrons. This mode, however, is the zero-momentum limit of the *magnetoplasmon* or inter-Landau level branch of excitations. These, in the absence of interactions or at large k, evolve into *excitons* with a particle in the $n=1$ Landau level and a hole in the $n=0$ level. We are interested in the lower energy *intra*-Landau level *magnetophonon* modes and do not wish to make this identification—so we will remain with the interpretation of m^* as a Coulomb derived effective mass and its effects should be regarded only as an "analogue" of inertia.

Because I ignored the quantum pressure in deriving the quasihydrodynamic approximation we do not see any of the larger k phenomena such as the k^4 term, characteristic of the weakly interacting Bose gas model of a superfluid–but then the pressure density relationship in the real system is presumably not the same as in the weakly interacting system. Also, unless we rather artificially take λ negative, we see no sign the magnetoroton dip[19] which occurs at the reciprocal of the mean interparticle spacing and presages the low-density collapse of the FQHE ground state into a Wigner crystal. At the largest momenta the magnetophonon density-wave picture is expected to break down entirely and the excitations are expected, by analogy with the magnetoplasmons, to become vortex antivortex *quasiexciton* pairs.[19] The two vortices in a pair will be separated by a distance proportional to k and will move, each with the flow velocity induced by the other, in parallel rather like a two-dimensional smoke ring. This change of interpretation cannot possibly be described by any quasiclassical model. Despite these deficiencies I think there is some merit in this picture of the phononlike elementary excitation.

To investigate vortex motion we must extend the quantum constraint (3.8) to take into account point defects in the ϕ field. Such point vortices modify (3.8) to

$$m^*\omega(\mathbf{r})+B(\mathbf{r})=2\Theta_0\rho(\mathbf{r})+\sum\kappa_i\delta^2(\mathbf{r}-\mathbf{r}_i) , \tag{3.8a}$$

where $\kappa_i=\pm2\pi$. To produce a low-energy configuration, each of these point vortices is surrounded by an oppositely oriented antivortex and its associated charge density. It is this composite object that is the analog of the quasiparticle and quasihole excitations of the Laughlin ground state.

The forces acting on a vortex held stationary in a steady uniform flow are a combination of electromagnetic forces, hydrodynamic Magnus forces, and the external force provided by the pinning center. To analyze them it is useful to use a version of Bernoulli's theorem that follows most directly from (3.6): assume a steady flow so $\nabla\cdot\rho\mathbf{v}=0$, then we can introduce a stream function χ with

$$\begin{aligned}\rho v_1&=\partial_2\chi ,\\ \rho v_2&=-\partial_1\chi .\end{aligned} \tag{4.7}$$

With $\dot{\mathbf{v}}=0$ and $\mathbf{E}=-\nabla V$, (3.6) can be now be written

$$\nabla(\mu+2\Theta_0\chi+V+\tfrac{1}{2}m^*v^2)=0 \tag{4.8}$$

so the combination

$$\mu_0=\mu+2\Theta_0\chi+V+\tfrac{1}{2}m^*v^2 \tag{4.9}$$

is constant everywhere—not just along streamlines.

The stream function is also useful for evaluating the total Lorentz force $\mathbf{F}_m$ due to a uniform magnetic field acting on an arbitrary steady flow in a region Ω. We can use Stokes theorem and find

$$\mathbf{F}_m=\int_\Omega \mathbf{j}\times B\, d^2x=-B\int_{\partial\Omega}\chi\mathbf{n}dS , \tag{4.10}$$

where $\mathbf{n}$ is the outward normal. The force depends only on asymptotic properties of the flow.

With this information we can evaluate the total force $\mathbf{F}$ on a region Ω, containing a stationary vortex. I will assume that we are in a region where the density takes its asymptotic value $\rho_\infty=B/2\Theta_0$. The force is the sum of the body force and the external pressure

$$\mathbf{F}=\int_\Omega(\mathbf{j}\times B+\rho\mathbf{E})d^2x-\int_{\partial\Omega}P\mathbf{n}dS . \tag{4.11}$$

Using the Bernoulli theorem (4.9) and the relation $\nabla P=\rho\nabla\mu$ to find P along the boundary, we find

$$\mathbf{F}=\int_\Omega \mathbf{E}\rho d^2x+\int_{\partial\Omega}\left[-B\chi+\rho_\infty(2\Theta_0\chi+V+\tfrac{1}{2}m^*v^2)\right]\mathbf{n}dS . \tag{4.12}$$

Since $m^*\omega+B=2\Theta_0\rho$ we see that force reduces to

$$\mathbf{F}=\int_\Omega \mathbf{E}(\rho-\rho_\infty)d^2x+\int_{\partial\Omega}(\rho_\infty\tfrac{1}{2}m^*v^2)\mathbf{n}dS . \tag{4.13}$$

At large distances the flow is uniform and without circulation. The integral of v^2 is zero and so

$$\mathbf{F}_\infty=\int_\Omega \mathbf{E}(\rho-\rho_\infty)d^2x\ . \qquad (4.14)$$

Since there is no net flux of momentum across the boundary the calculated force $\mathbf{F}$ must be balanced by a pinning force on the vortex.

When the $\mathbf{E}$ field is constant the force (4.14) is just the electric buoyancy force on the vortex due to the deviation of the charge density from that of the surrounding fluid. The magnetic forces cancel. The derivation of (4.14) provides a rather convoluted route to the discovery that the force on an isolated stationary vortex is $\mathbf{E}$ times its charge q—but leads naturally into a discussion of *where* on the vortex does the pinning force act. A reasonable assumption is that there are additional conservative forces, due to charged impurities or other inhomogeneities in the interface where the 2D EG resides, acting on the individual electrons. Since such forces can be written as $\mathbf{F}_{\rm imp}=-\rho\nabla W$, I can, without loss of generality, add them into the $\mathbf{E}$ field and, after including them, the integral (4.14) will be zero. I will now argue that such forces *cannot* hold a vortex stationary in a background flow.

Firstly we establish another momentum conservation result. Suppose that we perform the computation of the net force on a region around an *arbitrary* contour but this time use

$$\int_{\partial\Omega} P\mathbf{n}\, dS=\int_{\partial\Omega}\rho\mu\mathbf{n}\, dS-\int_\Omega \mu\nabla\rho d^2x\ , \qquad (4.15)$$

i.e., use the Bernoulli result throughout the region and not just on the boundary. We find after various applications of Eqs. (3.8), (4.9), the identity

$$\mathbf{v}\cdot\nabla\mathbf{v}=\nabla\tfrac{1}{2}v^2-\mathbf{v}\times\omega \qquad (4.16)$$

and $\nabla\cdot\rho\mathbf{v}=0$, etc., that

$$\mathbf{F}=\int_{\partial\Omega} m^*\rho\mathbf{v}(\mathbf{v}\cdot\mathbf{n})dS \qquad (4.17)$$

so that the total force, including pinning forces, on any region between two streamlines must vanish.

Suppose now the velocity field near the vortex comprises three parts: the flow at infinity, $\mathbf{v}_\infty$ equal to the Hall flow $\propto E_\infty/B$, the vortex flow itself, $\mathbf{v}_{\rm rot}$, and the backflow $\mathbf{v}_b$. Close to, but not in, the core of a quasiparticle vortex the streamlines are closed and in a region of slowly varying density so we can write the force on the region within the streamline as

$$\mathbf{F}=\int_\Omega \rho\mathbf{E}d^2x+\int_{\partial\Omega}\tfrac{1}{2}m^*\rho v^2\mathbf{n}dS\ . \qquad (4.18)$$

The integral of v^2 yields the Magnus force equal to the circulation within the contour times the flow past the core (see the Appendix). Because of the relationship between the charge of the vortex q and its circulation given by Eq. (2.7) and because of the relation between the flow $\mathbf{v}_\infty$ and the electric field, this force is $\mathbf{E}q$. By Eq. (4.17) the force is independent of the streamline we use: If we evaluate the integral about a closed streamline further out we would find a smaller Magnus force, because the circulation is reduced by the enclosed part of the antivortex, but the reduction in the Magnus force is exactly compensated by the extra electric body force on the enclosed fluid due to the *uniform* part of the electric field. We see that the whole of the pinning force $\mathbf{E}q$ must be borne by the vortex core. This is exactly what happens in a type-II superconductor[4] where it is quite reasonable that the normal core sustains the force. In the present case the core is *hollow*: There are no electrons there to be acted on by the pinning force—the Magnus force must therefore be zero and the vortex has no recourse but to follow the flow.

Is this a disaster for pinning? Not really—since in listing the flow components near the vortex I have omitted the flow $\mathbf{v}_{\rm imp}$ produced by the pinning force itself. Let us consider the nature of the flow near a charged impurity in the absence of the vortex. As an approximation imagine that the electric field produced by the impurity varies slowly enough that we can use the uniform Hall current formula $\mathbf{v}_{\rm imp}\times\mathbf{B}=-\mathbf{E}_{\rm imp}$. We see that there is a vorticity given by

$$\omega=\frac{1}{B}\nabla\cdot\mathbf{E} \qquad (4.19)$$

implying an induced $\delta\rho$ proportional to $\nabla\cdot\mathbf{E}$. Since the electric field of the impurity is not confined to the plane of the 2D EG, the $\mathbf{E}$ field falls off more rapidly than the solutions of $\nabla^2V=0$ and the in-plane flux of the $\mathbf{E}$ field falls rapidly to zero as we move away. This means that although there will be an induced charge *at* the impurity this charge will come from nearby and there is no *net* charge pulled in by the impurity—this is an example of the incompressibility of the FQHE ground state.

Near the impurity there will be closed streamlines and any particular line will remain closed up to a maximum background current. Since the induced charge is not simply related to the impurity charge it will be energetically favorable for vortices to accumulate near the impurity. Their cores will move with the local flow field but they will orbit the impurity until the background flow is strong enough that their streamline is no longer closed. In particular, there will be a stable point at which the flow velocity is zero until the Hall electric field exceeds the maximum force produced by the impurity potential. A vortex can remain at rest at this point there until this happens. The energetics of this depinning are clearly identical to those of a charge q particle acted on by only the Hall field and the local pinning field. This may give one a slight sense of *nascetur mus* but it must be remembered that the actual situation is more complicated than the conclusion suggests.

V. CONCLUSION

Motivated by the Landau-Ginsburg theory I have made a quasihydrodynamical model of the FQHE. As with the derivation of hydrodynamic equations from the Landau-Ginsburg theory of conventional superfluids it seems reasonable to suppose that such equations will have greater generality than their derivation. A further merit of the classical fluid-flow paradigm is that it enables a direct application of mechanical intuition in any first attempt at understanding new phenomena—although it will in no way substitute for serious quantum-mechanical computation of the relevant parameters and constitutive relations. The fluid picture leads naturally to the effective incompressibility of the FQHE ground state and to the gap in the magneto phonon spectrum. It thus captures the essential physics.

My motivation for the examination of the fluid-flow picture of the FQHE was to seek a model for vortex pinning and unpinning. It is easy to perform a force balance analysis of the FQHE vortices by using slight modifications of the discussion in (Ref. 4), but since there is no "normal core" it seems most likely, as suggested in Sec. IV, that the vortex localization and charge sequestration occur via a slightly different mechanism than pinning in a superconductor.

In studying the vortex dynamics one must be aware that the vortices in the FQHE involve far fewer degrees of freedom and are therefore much more quantum mechanical objects than in some of the other superfluids—but it seems worthwhile to understand the classical behavior before attempting to discuss zero-point fluctuations, quantum delocalization, condensation, and consequent hierarchy of FQHE states at other rational filling factors.

ACKNOWLEDGMENTS

This work was supported by the National Science Foundation Grant Nos. NSF-DMR-88-18713 and NSF-DMR-86-12860. Financial support was also provided by the University of Illinois and by NORDITA. I thank C. J. Pethick and A. Luther for their hospitality at NORDITA. I would also like to thank Henrik Bruus for introducing me to the work on vortex pinning by the group at the H. C. Ørsted Institute and the Danish Institute of Fundamental Metrology and Tony Leggett for discussions about superfluids.

APPENDIX

For completeness I will describe here the theory of the Magnus effect in the case of incompressible, irrotational flow where it is simplest. In such a case the velocity field is described by an analytic stream function $\psi(z)$ such that

$$v_1 - iv_2 = -\frac{d\psi}{dz} . \tag{A1}$$

Suppose there is a cylinder with arbitrary shaped boundary $\partial\Omega$ about which the fluid has circulation. We can use the conventional Bernoulli theorem to write the force on the cylinder as a contour integral round the boundary streamline[20]

$$F_2 + iF_1 = -\tfrac{1}{2}\rho \int_{\partial\Omega} \left[\frac{d\psi}{dz} \right]^2 dz . \tag{A2}$$

Here ρ is the *mass density.* Close to the cylinder there will be backflow and the stream function will be a complicated function of the boundary shape—but at large distance all the complications are irrelevant and the stream function will have the simple form

$$\psi = A + Bz + C\ln z , \tag{A3}$$

where the uniform flow field $\mathbf{U}$ is given by

$$B = -(U_1 - iU_2) \tag{A4}$$

and the circulation is $\kappa = 2\pi iC$. Because of the analyticity the contour integral may be evaluated at infinity where the asymptotic data may be used to express the integrand. The force turns out to be

$$F_1 = -\rho\kappa U_2, \quad F_2 = \rho\kappa U_1 . \tag{A5}$$

There is no drag (d'Alembert's paradox) but only a lift force. In the absence of any outside force to hold the cylinder in place it will have to move with the flow.

---

[1] W. F. Vinen, in *Superconductivity*, edited by R. D. Parks (Decker, New York, 1969).
[2] Y. B. Kim and M. J. Stephen, in *Superconductivity*, edited by R. D. Parks (Decker, New York, 1969).
[3] J. Bardeen and M. J. Stephen, Phys. Rev. **140**, A1197 (1965).
[4] P. Noziers and W. F. Vinen, Philos. Mag. **14**, 667 (1966).
[5] *The Quantum Hall Effect*, edited by R. E. Prange and S. M. Girvin (Springer-Verlag, Berlin, 1987).
[6] S. C. Zhang, T. H. Hansen, and S. Kivelson, Phys. Rev. Lett. **62**, 82 (1989).
[7] N. Read, Phys. Rev. Lett. **62**, 86 (1989).
[8] S. M. Girvin and A. H. MacDonald, Phys. Rev. Lett. **58**, 1252 (1987).
[9] H. Bruus, O. P. Hansen, and E. B. Hansen, J. Phys. C **21**, L375 (1988); Z. Phys. B **73**, 501 (1989); also unpublished work from the H. C. Ørsted Institute.
[10] R. B. Laughlin, Phys. Rev. Lett. **50**, 1395 (1983).
[11] A. M. Chang, M. A. Paalanen, D. C. Tsui, H. C. Stormer, and J. C. M. Hwang, Phys. Rev. B **28**, 6133 (1983).
[12] F. D. M. Haldane, Phys. Rev. Lett. **51**, 605 (1983).
[13] R. B. Laughlin, Phys. Rev. B **23**, 3383 (1983).
[14] B. I. Halperin, Phys. Rev. Lett. **52**, 1583 (1984).
[15] F. Wilczek, Phys. Rev. Lett. **49**, 957 (1982); Y. S. Wu and A. Zee, Phys. Lett. B **147**, 325 (1984); A. S. Goldhaber and R. Mackenzie, *ibid.* **214**, 271 (1984).
[16] F. Wilczek and A. Zee, Phys. Rev. Lett. **51**, 2250 (1983).
[17] D. Arovas, J. R. Schrieffer, F. Wilczek, and A. Zee, Nucl. Phys. B **251**, 117 (1985).
[18] W. Kohn, Phys. Rev. **123**, 1242 (1961).
[19] S. M. Girvin, A. H. MacDonald, and P. M. Platzman, Phys. Rev. B **33**, 2481 (1986); see also the chapter by S. M. Girvin in Ref. 5.
[20] Sir Horace Lamb, *Hydrodynamics* (Dover, New York, 1945).

Chapter 5

EDGE STATES AND CURRENT ALGEBRAS

5.1 Introduction

We have seen that all excitations in the bulk of the quantum Hall fluid have an energy gap. This is not the case at the boundary of the 2DEG. Halperin pointed out long ago [rep.4] that there must be low energy modes in regions where the density of electrons falls to zero. These are the "edge waves", and both classical and quantum instances have been observed [1]. Recently X-G. Wen [rep.39] has provided some fascinating insights into these edge-modes by pointing out that they can be classified by the same Kac-Moody chiral current algebras that appear in conformal field theory. This striking observation provides yet another way to connect the Hall effect with some of the most interesting ideas in contemporary theoretical physics. We reprint in this volume another paper by Wen [rep.40], and two more pedagogical accounts, [rep.41], [rep.42], of my own.

In Section 2 of this Chapter, I give a brief introduction to some of the physics of the edge waves. In Sections 3 and 4, I will discuss some of the properties of Kac-Moody current algebras and their connections with the bulk quasiparticles.

5.2 Edge Waves

Let us begin with a warm-up exercise and consider the simple problem of the motion of a single particle in a two-dimensional harmonic oscillator potential. As always, we have a magnetic field perpendicular to the plane. The classical equations of motion for this system are

$$m\ddot{x} = eB\dot{y} - mx\Omega^2$$

$$m\ddot{y} = -eB\dot{x} - my\Omega^2, \tag{5.2.1}$$

and by writing $z = x + iy = z_0 e^{-i\omega t}$ we find that there are two modes. They have frequencies

$$\omega_{\pm} = \frac{eB}{2m} \pm \sqrt{\left(\frac{eB}{2m}\right)^2 + \Omega^2}. \tag{5.2.2}$$

Inspection of (5.2.2) reveals that as the cyclotron frequency, $\omega_c = eB/m$, becomes large compared to the harmonic oscillator frequency Ω, the ω_+ mode reduces to cyclotron motion at angular frequency

ω_c. The low-frequency mode has motion in the opposite sense, at the much lower frequency $\omega \approx -\Omega^2/\omega_c \propto 1/B$.

Simple as it is, this exercise has consequences for the many-body quantum problem. The center-of-mass motion of even an interacting many-body system obeys (5.2.2), and since the equations are linear, the associated quantum mechanical energy eigenvalues fall into two separate harmonic oscillator families, one containing levels spaced by ω_+, and one with levels spaced by ω_-. For an extended droplet at quantum Hall filling fraction $\nu = 1$, the upper frequency corresponds to the infinite wavelength limit of the *magnetoplasmon* branch of excitations considered in the previous chapter. This mode may be ignored at high magnetic fields. The lower mode is of interest here. It leads to a set of harmonic oscillator energy-levels at

$$E_n \approx \frac{\Omega^2}{\omega_c}(n + \frac{1}{2}), \tag{5.2.3}$$

and these, as we will soon see, correspond to multiple occupation of a family of possible bosonic *edge modes* of the 2DEG.

To understand how this family of modes arises, first consider a straight edged sample with the, by now familiar, wavefunctions

$$\psi_k(x, y) = e^{iky} e^{-\frac{eB}{2}(x-k/eB)^2}. \tag{5.2.4}$$

In the presence of a confining potential, such as the harmonic potential of the warm-up exercise, these states have an energy that depends on k. In particular, if we linearize the potential near the edge of the droplet so that $V(x) = -eEx$, then the energy is

$$\epsilon = -\frac{E}{B}k + const. \tag{5.2.5}$$

By assuming that the negative energy states are occupied and the positive ones empty (a choice of the chemical potential), the Hall droplet forms a physical realization of a "chiral" Dirac sea with the Fermi surface identified with the physical edge of the droplet. Any disturbance in the distribution of electrons will move at the associated group velocity, or "speed of light",

$$c = \frac{\partial \epsilon}{\partial k} = -\frac{E}{B}, \tag{5.2.6}$$

which is simply the Hall drift velocity.

By focussing RF energy onto the edge of a droplet of Hall liquid, it is possible to alternately attract and repel the electrons from some location, and so set up resonant waves around the edge of the droplet. Unlike most resonant modes, these are *traveling* waves whose frequency depends on the number, n, of wavecrests round the edge, and on the length of perimeter of the droplet. For a circular droplet of radius R there will be resonances at frequencies

$$\omega_n = 2\pi\nu_n = 2\pi\frac{E}{B}\frac{n}{2\pi R}. \tag{5.2.7}$$

The edge modes can be distinguished from bulk modes because of the $1/B$ dependence of the resonant frequency.†

---

†There are edge modes for compressible states of the 2DEG, but these will not be as sharply defined because of the presence of low-lying bulk modes.

For the harmonic confining potential considered above, the effective confining force, eE, depends on the radius and is given by $m\Omega^2 R$, so

$$\omega_n = \Omega^2(\frac{m}{eB})n = \frac{\Omega^2}{\omega_c}n. \qquad (5.2.8)$$

We see that the frequency of the lowest edge-mode coincides with that of the lower center-of-mass collective mode. A moment's thought shows that the $n = 1$ mode is simply a displaced, but undeformed, droplet precessing at ω_-, so it *is* the lower center-of-mass mode. Because interactions cancel for the center-of-mass motion, the frequency of the $n = 1$ wave cannot be modified by interactions. This will not be true of the higher edge modes.

The spectrum of edge modes near filling fraction $\nu = 1$ has been calculated by H. W. Wyld and myself [2]. We used a Coulomb interaction

$$V_2(\mathbf{r}_1 - \mathbf{r}_2) = e^2 \frac{1}{|\mathbf{r}_1 - \mathbf{r}_2|}, \qquad (5.2.9)$$

but no confining potential was included because such effects may be added by hand later. In particular, a harmonic force adds a term diagonal in the M basis, and proportional to M.

Motivated by the exactly solvable Luttinger and Tomonaga models [3,4] whose excitations are free bosonic phonons, we compared our numerical spectra with a "non-interacting" boson spectrum whose energies we found by assuming that the energy of a state of angular momentum M_{tot} can be decomposed as

$$E = \sum_M n_M \omega(M) \qquad M_{tot} = \sum_M n_M M, \qquad (5.2.10)$$

where $\omega(M)$ is the energy of the lowest energy state of angular momentum M, and n_M is the occupation number of the boson state of momentum M. The two spectra turn out to be almost identical, so the modes may indeed be best regarded as weakly interacting bosons. The Luttinger model is solved by "bosonization", and a chiral version of applies here, as explained in [rep.42] and [5]. These bosonization techniques are closely related to the Kac-Moody algebras referred to in the introduction.

5.3 A Brief Introduction to Kac-Moody Current Algebras

5.3.1 Historical Background

When we take care in computing the commutators of fermion currents in a relativistic quantum field theory, it often turns out that they have extra terms beyond those anticipated from naive manipulation of the canonical commutation relations. These extra terms are called Schwinger terms, and were first noticed by Jordan [6] in connection with his neutrino theory of light. They formed the essential ingredient in this first example of a fermi-bose equivalence. Some years later, Tomonaga used the Schwinger terms to show that a one-dimensional electron gas is equivalent to a free boson system [3]. I believe that it was in this paper of Tomonaga's that the defining relations of what is now called a Kac-Moody algebra were first written down. Later, Schwinger gave general arguments based on current conservation, the equations of motion, and the existence of a ground state, to show that extra terms were to be expected in all current algebras [7].

A substantial amount of physics has grown out of these early ideas: the Bose-Fermi equivalences in two dimensions [8], the theory of Kac-Moody algebras and their connection with conformally invariant field theories and strings [9], some current algebra in four dimensions, and an enhanced understanding of the origin of anomalies in gauge theories.

We can obtain Schwinger terms in many ways. Jordan and Tomonaga realized that to make their operators well defined they had first to be specific about the states on which the operators were to act, and then to normal-order with respect to these states. A more modern derivation would begin by noting that the short distance current-current correlation function of two right-handed fermions goes as z^{-2}. Using the complex variable machinery developed for conformally invariant theories, this implies that the commutator contains the derivative of a delta function. In the next section I will use more physical arguments to find the Schwinger terms.

5.3.2 Dirac Metals

In condensed matter physics, we find "relativistic" fermions whenever we linearize the electron dispersion relations near a Fermi surface. Consider, for example, electrons moving in one dimension with hamiltonian

$$\hat{H} = \int dx \psi^\dagger \left(-\frac{1}{2m}\partial_x^2\right)\psi. \tag{5.3.1}$$

Here

$$\psi(x) = \sum_k a_k e^{ikx} \tag{5.3.2}$$

is a second-quantized Fermi field. Near the Fermi surface we can approximate this field by

$$\psi = e^{ik_f x}\psi_R(x) + e^{-ik_f x}\psi_L(x) \tag{5.3.3}$$

where, for example,

$$\psi_R(x) = \sum_{k=-\Lambda}^{\Lambda} a_{k-k_f} e^{ikx} \tag{5.3.4}$$

contains only operators annihilating states within a momentum shell $[-\Lambda, \Lambda]$ about the right-hand Fermi surface. The cut-off Λ must be chosen small enough that the electron velocity is essentially constant within the interval, yet large enough that no process is able to change the occupation number of states outside the momentum shell.

Introducing a two-component field $\Psi = (\psi_R, \psi_L)$, and the fermi velocity $v_f = k_f/m$, we have

$$\hat{H} \approx \int dx \Psi^\dagger(x)(-iv_f\sigma_3\partial_x)\Psi(x). \tag{5.3.5}$$

This is a one-dimensional Dirac hamiltonian with the Pauli σ-matrix σ_3 acting on the left/right indices.

The fermion number density and current may be written in terms of densities associated with the right and left-hand Fermi surfaces only:

$$\hat{\rho}(x) = \hat{\rho}_R(x) + \hat{\rho}_L(x) = \psi_R^\dagger\psi_R(x) + \psi_L^\dagger\psi_L(x) = \Psi^\dagger\Psi(x) \tag{5.3.6}$$

$$\frac{1}{v_f}\hat{j}(x) = \hat{\rho}_R(x) - \hat{\rho}_L(x) = \psi_R^\dagger\psi_R(x) - \psi_L^\dagger\psi_L(x) = \Psi^\dagger\sigma_3\Psi(x). \tag{5.3.7}$$

The left and right-handed densities, $\hat{\rho}_L(x)$, and $\hat{\rho}_R(x)$, generate two separate $U(1)$ symmetries, associated respectively with the conservation of left and right-going fermion number. Since $U(1)$ is a commutative group, and since we effectively have an independent $U(1)$ at each point x, we might expect to find

$$[\hat{\rho}_R(x), \hat{\rho}_R(x')] = [\hat{\rho}_L(x), \hat{\rho}_L(x')] = 0. \quad (?) \tag{5.3.8}$$

This is what we would find if we used the canonical commutation relations naively. When we take pains, however, we find instead

$$[\hat{\rho}_R(x), \hat{\rho}_R(x')] = -[\hat{\rho}_L(x), \hat{\rho}_L(x')] = -\frac{i}{2\pi}\partial_x \delta(x - x'). \tag{5.3.8a}$$

The derivative of the delta function is the Schwinger term.

The difficulty in evaluating the commutators directly in the relativistic theory is due to the bottomless Fermi sea. It is easier to return to the original non-relativistic fermions. For these

$$\hat{\rho} = \psi^\dagger \psi \qquad \hat{j} = \frac{-i}{2m}\left(\psi^\dagger \partial_x \psi - (\partial_x \psi^\dagger)\psi\right). \tag{5.3.9}$$

Now, since there is a finite depth to the energy band, we can safely use the canonical commutation relations and find

$$[\hat{\rho}(x), \hat{j}(x')] = \frac{-i}{m}\left(\partial_x \delta(x - x')\psi^\dagger \psi + \delta(x - x')\partial_x \psi^\dagger \psi\right). \tag{5.3.10}$$

Because of the derivative in the "non-relativistic" current operator, there is no surprise at finding the derivative of a delta function in this result. In the "Dirac Limit" where $m \to \infty$, $k_f \to \infty$, with $v_f = k_f/m$ fixed, linearization near the Fermi surface becomes exact and, more importantly, the m in the denominator causes all non-diagonal matrix elements of the operator $\psi^\dagger\psi/m$ to tend to zero. It may be replaced by its expectation value, which is $\langle \psi^\dagger \psi \rangle / m = k_f/\pi m = v_f/\pi$. Thus,

$$[(\hat{\rho}_R + \hat{\rho}_L), v_f(\hat{\rho}_R - \hat{\rho}_L)] \to \frac{-iv_f}{\pi}\partial_x \delta(x - x'). \tag{5.3.11}$$

Finally we argue that the left and right Fermi components must commute with each other, and so recover (5.3.8a). We can confirm this calculation by using a version of the f sum-rule, a non-relativistic analogue of the Schwinger argument.

In the Hall effect, we have one Fermi surface only, either left or right going, and so a *Chiral* fermion field. The commutation relations (5.3.8a) continue to hold, however. An explicit acount of the origin of the Schwinger term and its effect on the group action is given, for the simple $\nu = 1$ case, in [10].

5.3.3 Kac-Moody Algebras

Suppose we now make an $SU(2)$ algebra of currents out of a doublet of "quark" fields transforming under the fundamental representation of $SU(2)$. We think of the $SU(2)$ labels as being "flavour" and also introduce another set of "color" labels running from 1 to k. The $SU(2)$ flavour currents are

$$J^{(a)}(x) = \sum_{i=1}^{k} \psi_i^\dagger(x)\sigma_a \psi_i(x) = \frac{1}{2\pi}\sum_n J_n^{(a)} e^{inx}. \tag{5.3.12}$$

When we take commutators, there will be Schwinger terms only when the commutators involve four fermi fields all with the same indices. We find therefore

$$[J^{(a)}(x), J^{(b)}(y)] = 2i\epsilon_{abc}J^{(c)}(x)\delta(x-y) + \frac{ik}{2\pi}\partial_x\delta(x-y)\mathrm{tr}\,(\sigma_a\sigma_b). \tag{5.3.13}$$

In momentum space, these relations become

$$[J_n^{(a)}, J_m^{(b)}] = 2\epsilon_{abc}J^{(c)}(x)_{m+n} + mk\delta_{m+n,0}\mathrm{tr}\,(\sigma_a\sigma_b), \tag{5.3.14}$$

and this set of infinitely many commutators comprises the defining relations of an $SU(2)$ Kac-Moody current algebra at *level* k.

For a review of Kac-Moody algebras see [11]. For the associated "loop groups" see [12].

5.3.4 Integrable Representations

The representation theory of these infinite dimensional Lie algebras may be developed by analogy to that of finite dimensional Lie groups [11]. The fundamental representations are always infinitely deep Fermi seas, however, so there are naturally a few subtleties. The most important of them leads to the notion of ***integrable representations.***

Consider a representation based on ground state that is annihilated by all the negative fourier component currents, and also annihilated by the step-up charge operator constructed from $\sigma_+ = \frac{1}{2}(\sigma_1 + i\sigma_2)$

$$\begin{aligned} J_n^{(a)}|0\rangle &= 0 \quad n < 0 \\ J_0^{(+)}|0\rangle &= 0. \end{aligned} \tag{5.3.15}$$

There may, however, be a non-vanishing vacuum charge quantum number $j > 0$

$$J_0^{(3)}|0\rangle = j|0\rangle. \tag{5.3.16}$$

In conventional group theory language, j labels the ***greatest weight*** of an $SU(2)$ representation.

Physics demands a positive definite inner-product on our Hilbert space so we must have

$$0 \leq \langle 0|J_{-1}^{(+)}J_{+1}^{(-)}|0\rangle = -j + k. \tag{5.3.17}$$

This means that $j \leq k$, and there is limit as to what quantum numbers we can select for the vacuum charge. For $k = 1$ we have only the fundamental spinor representation where $j = 1$ (notice that my $J^{(a)}$ are normalized to be twice the usual angular momentum $SU(2)$ generators), and the spin-1 vector representation is not allowed until $k = 2$. The allowed representations are said to be ***integrable.*** We briefly alluded to this integrability condition in Chapter 3, where we asserted that the rules for fusing operators was influenced by the level number appearing in the KZ equations.

The physical origin of the restriction is easily seen when thinking of the Fermi sea in terms of Hall effect edge states. If we have k copies of the isospin-up and isospin-down Fermi seas, we can get an isospin j vacuum at the edge of a Hall droplet by inserting some (flavored) flux to pull in the isospin-down particles by $j/2k$ of an orbit, simultaneously pushing out the isospin-up particles by the same amount. The vacuum charge will then change by $k \times j/k = j$. Were we too ambitious, trying to make too great a vacuum charge by pushing too far, we could create the same "vacuum" state by simply taking a particle from the undisturbed isospin-down sea and placing it atop the isospin-up sea. This is an allowed group operation, and we see that the attempt to make a vacuum has actually produced an excited state in another representation. It is no wonder, then, that asserting this state to be a ground-state leads to a violation of positive definiteness.

Insight into the integrability constraint can be also be found by including the vacuum quantum number j in the coherent-state path integral technology used in [rep.41] and [rep.42]. This was done by Iso *et al.* [13]. They found that the usual Wess-Zumino term in the group path integral

$$S_{WZ} = \frac{k}{2\pi}\int \mathrm{tr}\,(g^{-1}\partial_s g\partial_x(g^{-1}\partial_t g))dtdxds \tag{5.3.18}$$

had to be supplemented by an additional term depending on the length L of the x-space loop. This extra term is

$$S_{finite} = -i(\frac{j}{L})\int \mathrm{tr}\,(\sigma_3 g^{-1}\partial_t g)dxdt. \tag{5.3.19}$$

It may seem that we can chose j as we please in (5.3.19), but by twisting the periodic $g(x)$ field by a periodic factor

$$g(x) \to g(x)h(x) = g(x)e^{i\sigma_3\frac{2\pi x}{L}}, \tag{5.3.20}$$

and substituting in (5.3.18), we find

$$S_{WZ} \to S_{WZ} + \frac{k}{2\pi}\int dtdxds\mathrm{tr}\,(\partial_x hh^{-1}[g^{-1}\partial_t g, g^{-1}\partial_s g]). \tag{5.3.21}$$

Now,

$$\partial_s(g^{-1}\partial_t g) - \partial_t(g^{-1}\partial_s g) = [g^{-1}\partial_t g, g^{-1}\partial_s g], \tag{5.3.22}$$

so we can rewrite the effect of the twist on (5.3.18) as

$$S_{WZ} \to S_{WZ} - \frac{ik}{L}\int dxdt\mathrm{tr}\,(\sigma_3 g^{-1}\partial_t g), \tag{5.3.23}$$

and find that S_{WZ} changes by a term of the same form as (5.3.19). The original S_{finite} term itself is unaffected by the twisting. The net effect is to shift $j \to j + k$. Since these twisted configurations are included in the path integral, we see that the vacuum isospin j is defined only modulo k.

For Lie groups with higher rank than $SU(2)$, the weight diagrams are multidimensional and these periodic twists have the effect of translating a weight by k times one of the roots. By combining the twisting operations with the Weyl-group reflections in the planes perpendicular to the roots, we can move any vacuum weight first into a *Weyl chamber* and then into a corner of the chamber close to the origin. This corner is called the *Weyl alcove*. The combined Weyl reflections and twisting operations form a semidirect product group, the *affine Weyl-group*. If the weight lattice of the Lie algebra is denoted by Λ^w, the root lattice by Λ^r, and the Lie algebra Weyl-group by w, the finite dimensional Lie algebra representations are in one-one correspondence with the set Λ^w/w, but the Kac-Moody algebra representations are labeled by points in the smaller set $\Lambda^w/w \times k\Lambda^r$.

5.4 Representations, Wilson Lines, and Quasiholes

In [rep.39] – [rep.38], we see that the edge states in the various quantum Hall effect phases carry representations of Kac-Moody current algebras. The edge-state current algebra has more use, however, than just describing the edge. Imagine inserting some flux to create a small hole in the droplet of electron fluid, and so creating a new "edge". This new edge has its own Hilbert space, its own current algebra, and its own set of excitations corresponding to states that span representations of the algebra. Now we shrink the hole to a point. Only the lowest momentum states survive, and these states form represention of the global charge algebra, a conventional finite-dimensional Lie algebra. The vortex-like quasiholes can be made in this way, and so they inherit topological and group representation properties from the classification of the edge states. This, then, is the origin of the group indices that appeared as labels of the quasiholes in our brief discussion of non-abelian statistics in Chapter 3. The effects described there occur because the Hilbert spaces of different quasi-particles are not independent. As a vortex moves through the electron gas, it influences both the other vortices and the motion of the bulk fluid. The quasi-particle states evolve because of the interaction, producing anyonic statistics, and the bulk motion of the Hall fluid may also be permanently affected by the interchange of two quasiholes, giving rise to the monodromy matrix and to non-abelian statistics.

One way to derive the evolution of the states is to map the Hall effect onto a $2+1$ dimensional Chern-Simons field theory, as done for example in [rep.41]. In this language, our vortex construction is essentially that used by Moore and Seiberg [14] for the insertion of the Wilson loop operators. These are the gauge invariant operators used by Witten to find knot invariants associated with the braiding and monodromy properties of operators in rational conformal field theories. I will sketch here some of the ingredients needed for understanding this.

A Chern-Simons theory coupled to sources

$$S=\int d^3x\,\{\mathrm{tr}\,J_\mu A_\mu\}-\frac{k}{4\pi}\int \mathrm{tr}\,(AdA+\frac{2}{3}A^3) \tag{5.4.1}$$

has as its equation of motion

$$\frac{2\pi}{k}J_\mu=\epsilon_{\mu\nu\sigma}F_{\nu\sigma}. \tag{5.4.2}$$

The Bianchi identity, $dF+[A,F]=0$, or

$$\epsilon_{\mu\nu\sigma}\nabla_\mu F_{\nu\sigma}=0, \tag{5.4.3}$$

then requires covariant conservation of the current J_μ

$$\partial_\mu J_\mu+[A_\mu,J_\mu]=\nabla_\mu J_\mu=0. \tag{5.4.4}$$

The covariant conservation of J is also needed for gauge invariance of (5.4.1) under $A\to g^{-1}Ag+g^{-1}dg$.

In the full quantum theory, covariant conservation is not possible for a c-number source. This is because satisfying (5.4.4) requires equating a term containing a fluctuating A_μ to a non-fluctuating c-number. Suppose, however, we couple A_μ to localized degree of freedom $g(t)$ located at x_0. We do this by replacing the $\int d^3x\,\{\mathrm{tr}\,J_\mu A_\mu\}$ term with

$$S_{x_0}=\frac{k}{2\pi}\int \mathrm{tr}\,(\alpha\{g^{-1}\partial_t g+g^{-1}A_0(x_0,t)\})dt. \tag{5.4.5}$$

Here α is some matrix in the Lie algebra of the gauge group. We have effectively produced a J_0 charge $J_0=g\alpha g^{-1}$.

The g matrix obeys the equation of motion

$$[\alpha, g^{-1}\partial_t g + g^{-1}A_0 g] = 0, \tag{5.4.6}$$

or equivalently

$$g^{-1}\partial_t g + g^{-1}A_0 g + h(t) = 0, \tag{5.4.7}$$

where $h(t)$ is some arbitrary time-dependent Lie-algebra matrix commuting with the fixed α matrix. Because of the $h(t)$, the solutions

$$g(t) = T\exp\left\{-\int dt A_0(t)H(t)\right\} \tag{5.4.8}$$

are not unique, but contain a "gauge" factor $H(t)$ which lies in H, the subgroup of elements of G generated by those Lie algebra generators commuting with α. The solutions are only unique when considered as elements of a coset space G/H. Despite the gauge ambiguity, the solutions do determine J_0 uniquely. The $H(t)$ factors commute through α and cancel. We now see that

$$\partial_t J_0 = -A_0 g\alpha g^{-1} + g\alpha g^{-1}A_0 = -[A_0, J_0], \tag{5.4.9}$$

so this J_0 charge is covariently conserved, and can be used as a source term for the Chern-Simons field.

What we have done here is to re-invent the Wilson line operator. The action S_{x_0} is precisely the classical action whose quantization reproduces the Hilbert space of the group representation determined by α. We know from [rep.41] how to write traces over group representation spaces as a coherent state path integral (see also [15]). The Wilson line can be written

$$\operatorname{tr}\left\{T\exp - \int_0^\beta dt A_0\right\}_\alpha = \int d[g] e^{-\int_0^\beta (\langle g|\partial_t|g\rangle + \langle g|A_0(t)|g\rangle)dt}. \tag{5.4.10}$$

Here $d[g]$ is the Haar measure, and the states $|g\rangle$ are images under the group action of some highest weight state $|\alpha\rangle$:

$$|g\rangle = U(g)|\alpha\rangle. \tag{5.4.11}$$

The highest weight state $|\alpha\rangle$ is an eigenstate of all generators h_i in the Cartan subalgebra,

$$h_i|\alpha\rangle = \alpha_i|\alpha\rangle, \tag{5.4.12}$$

and the α_i are the components of the highest weight vector. The subscript on the trace on the left-hand side of (5.4.10) means that the trace is taken in the representation of G determined by α_i.

Assuming that we have normalized the Cartan algebra generators so that $\operatorname{tr}(h_i h_j) = 2\delta_{ij}$, the exponent of (5.4.10) can be written

$$\begin{aligned}\langle g|\partial_t + A_0|g\rangle &= \langle\alpha|g^{-1}\partial_t g + g^{-1}A_0 g|\alpha\rangle \\ &= \frac{1}{2}\operatorname{tr}\{\sum_i h_i\alpha_i(g^{-1}\partial_t g + g^{-1}A_0 g)\},\end{aligned} \tag{5.4.13}$$

showing the connection between the matrix α appearing in (5.4.5) and the highest weight vector α_i, which determines the group representation of the "quarks" on the Wilson Line. In (5.4.5) it appeared that we could chose any matrix α, but in fact only those matrices that are derivable from allowed weights can be successfully quantized.

It does not take long to realize that the action (5.4.5) for the Wilson lines is exactly what is left of the gauged Chiral boson edge-state action after we shrink the edge to a point and ignore the $\partial_x g(x)$ parts. The Chern-Simons action, and the coupling of different point vortices to each other, occurs because we cannot create the coherent twists of the vortices independently. The quasiholes and their representations are created by inserting flux lines, and other holes feel these fluxes as the A_μ fields in their Wilson lines.

There is much more to this story than I have room for here, especially as a proper quantum mechanical treatment of non-abelian Chern-Simons actions [16] contains many subtle points stemming from the need to fix a gauge.

Reprints for Chapter 5

[rep.39] *Chiral Luttinger liquid and the edge excitations in the fractional quantum Hall states*, X-G. Wen, Phys. Rev. B41 (1990) 12838-12844.

[rep.40] *Edge excitations in the fractional quantum Hall liquids*, D-H. Lee, X. G. Wen, Phys. Rev. Lett. 66 (1991) 1765-1768.

[rep.41] *Edge waves in the quantum Hall effect*, M. Stone, Ann. Phys. 207 (1991) 38-52.

[rep.42] *Schur functions, chiral bosons and the quantum Hall effect edge states*, M. Stone, Phys. Rev. B42 (1990) 8399-8404.

Other References for Chapter 5

[1] D. C. Glattli, E. Y. Andrei, G. Deville, J. Pointrenaud and F. I. B. Williams, Phys. Rev. Lett. 54 (1985) 1710; E. Y. Andrei, D. C Glattli, F. I. B Williams, M. Heiblum, Surface Science 196 (1988) 501; M. Wassermeier, J. Oshinowo, J. P. Kotthaus, A. H. Macdonald, C. T. Foxton, J. J. Harris, Phys. Rev. B41 (1990) 10287; I. Grodensky, D. Heitmann, P. Grambow, K. Ploog, Phys. Rev. Lett. 64 (1990) 788.

[2] M.'Stone, H. W. Wyld, in the proceedings of YKIS'91, *Low Dimensional Field Theories and Condensed Matter Physics*, to appear in Progress of Theoretical Physics (1992).

[3] S. Tomonaga, Prog. Theor. Phys. (Kyoto) 5 (1950) 544.

[4] J. M. Luttinger, J. Math. Phys. 4 (1963) 1154.

[5] M. Stone, Int. Jour. of Mod. Phys. B5 (1991) 509.

[6] P. Jordan, ZS. Phys. 93 (1935) 464.

[7] J. Schwinger, Phys. Rev. Lett. 3 (1959) 296.

[8] S. Coleman, Phys. Rev. D11 (1975) 2088; S. Mandelstam, Phys. Rev. D11 (1975) 3026; M. B. Halpern, Phys. Rev. D12 (1975) 1684; T. Banks, D. Horn and H. Neuberger, Nucl. Phys. B108 (1976) 119; E. Witten, Commun. Math. Phys. 92 (1984) 455.

[9] V. G. Knizhnik and A. B. Zamolodchikov, Nucl. Phys. B247 (1984) 84.

[10] M. Stone, Int. Jour. Mod. Phys. B5 (1991) 509.

[11] P. Goddard, D. Olive, International Journal of Modern Physics A1 (1986) 303.

[12] A. Pressley, G. Segal, *Loop Groups*, Clarendon Press, Oxford 1986.

[13] S. Iso, C. Itoi, H. Mukaida, Phys. Lett. B 244 (1990) 241.

[14] G. Moore, N. Seiberg, Phys. Lett. 220B (1989) 422.

[15] M. Stone, Phys Rev. Lett 63 (1989) 731; Nuc. Phys B 314 (1989) 557.

[16] S. Elitzur, G. Moore, A. Schwimmer, N. Seiberg, Nucl. Phys. B 326 (1989) 108.

PHYSICAL REVIEW B VOLUME 41, NUMBER 18 15 JUNE 1990-II

Chiral Luttinger liquid and the edge excitations in the fractional quantum Hall states

X. G. Wen
School of Natural Sciences, Institute for Advanced Study, Princeton, New Jersey 08540
(Received 25 January 1990)

The low-energy effective theory of the edge excitations in the fractional quantum Hall (FQH) states is derived. The edge excitations are shown to form a new kind of state which is called the chiral Luttinger liquid (χLL). The effective theory is exactly soluble. This enables us to easily calculate all the low-energy properties of the edge excitations. We calculate the electron propagator and the spectral function, which clearly demonstrate the non-Fermi-liquid behaviors of the χLL. We also calculate the interference effects between excitations on different edges. We demonstrate that the properties of the edge excitations are closely related to the properties of the FQH states on compacted spaces. Thus the properties of the edge excitations can be used to characterize the topological orders in the FQH states. We also show that the FQH states with filling fractions $\nu\neq 1/l$ must have at least two branches of edge excitations.

I. INTRODUCTION

In the last few years many people have studied the low-energy dynamical properties of the quantum Hall (QH) states.[1] The experiments clearly observed gapless excitations in finite QH systems. It is generally believed that the gapless excitations are localized at the edges of the systems. This is because the QH states are incompressible and contain no bulk gapless excitations. Using Laughlin's arguments,[2] one can easily prove the existence of the gapless excitations in a finite QH system. The real nontrivial issue is to understand the dynamics of the edge excitations. For integral QH states,[3] the dynamical properties of the edge excitations are shown to be described by one-dimensional (1D) Fermi-liquid theory. While for the FQH states,[4] they are described by the U(1) Kac-Moody (KM) algebras. Some static properties (e.g., dc transport properties) of the edges states in the FQH regime are studied in Ref. 5.

In general, the gapless edge excitations may have many branches.[3-5] The dynamics of the edge excitations is generally described by several U(1) KM algebras in the *low-energy* limit.[4] This is equivalent to say that the charge-zero sector of the edge excitations is described by the charge-zero sector of a Fermi-liquid theory (in the low-energy limit). Such a Fermi-liquid theory contains many branches of fermions. The charges of the fermions are shown to satisfy a sum role[4]

$$\sum_I \frac{v_I}{|v_I|} q_I^2 = \nu e^2 . \tag{1.1}$$

In (1.1) v_I and q_I are the velocities and the charges of the fermions in the Ith branch and ν is the filling fraction of the FQH state. In general, the charges q_I can be irrational numbers.[6] The relation between the edge excitations and the Fermi liquid can be used to calculate many properties of the edge excitations. The responses of the edge states to external electromagnetic fields are calculated in Ref. 6, which lead to a practical way to experimentally measure the charges q_I carried by the fermions.

However, as emphasized in Refs. 4 and 6, although the charge-zero sector of the edge excitations are described by a Fermi-liquid theory, the charged edge states may *not* be described by Fermi-liquid theories. The charges of the charged edge states especially may not be multiples of q_I. Therefore, to be accurate, we will call q_I the optical charges of the edge excitations. This is because q_I are measured only through the current correlation functions and do not correspond to the charges of the charged edge states. Strictly speaking, the edge states in the FQH regime are not Fermi liquids. In this paper, we will derive an effective theory which describes both the charged and the neutral excited edge states. We will concentrate on the non-Fermi-liquid behaviors of the edge excitations. Although the edge states are not Fermi liquids, the effective theory of the edge excitations is still exactly soluble. One can easily obtain all the low-energy properties of the edge excitations from the effective theory.

II. THE EDGE EXCITATIONS ON A DISC

Consider a FQH state on a disc with filling fraction ν. Let us assume that the edge excitations have only one branch. This implies that the charge-zero sector of the edge excitations are described by a single U(1) *KM* algebra:[4,6]

$$\begin{aligned}
&[j_{k'}^+, j_k^+] = e^2 \frac{\nu}{2\pi} k \delta_{k+k'} , \\
&[j_{k'}^+, j_k^-] = [j_{k'}^-, j_k^-] = 0 , \\
&[H, j_k^\pm] = v k j_k^\pm ,
\end{aligned} \tag{2.1}$$

where

$$j_I^\pm = \tfrac{1}{2}\left(j^0 \pm \frac{1}{v} j^\sigma\right) , \quad j_k^\alpha = \int d\sigma \frac{1}{\sqrt{L}} e^{i\sigma k} j^\alpha(\sigma) ,$$

and L is the length of the edge. The (optical) charge of the fermions in the corresponding Fermi-liquid theory is given by $q=\sqrt{\nu}$.

The charged excited states arise from adding (or subtracting) electrons to (from) the edge. Therefore, those charged states are generated by electron creation (or annihilation) operators $\psi^\dagger$ (or ψ). The electron operator ψ carries a unit charge

$$[\psi, Q]=e\psi , \quad (2.2)$$

where $Q=\int d\sigma\, e^{i\sigma k}j^0(\sigma)$. Because all the low-lying excitations have the same velocity v, ψ also satisfies

$$[H, \psi_k]=vk\psi_k . \quad (2.3)$$

If ψ_k had a velocity different from v, the current operator $j_k^\sigma=\sum \psi_{k'}(k+k')\psi_{k+k'}$ would also have that velocity. This would contradict (2.1).

The total Hilbert space of the edge excitations is generated not only by the current operator j^+ but also by the charged operator ψ. Therefore the Hilbert space of the edge excitations forms a representation of the algebra (2.1)–(2.3). To understand the properties of the charged edge states, we first need to find the representation of the algebra (2.1)–(2.3).

The structure of the Hilbert space of the edge excitations can be understood even without any calculations. First, the charge-zero sector of the edge states forms an irreducible representation of the U(1) KM algebra. The Hilbert space of such a representation is denoted $\mathcal{H}_{\rm KM}$. The charge-e excited states are obtained by adding an electron to the system. The system with one more electron is essentially identical to the original system. Thus the charge-e sector also forms the irreducible representation of the KM algebra. A similar result can be obtained for a general charge-Ie sector. From the above discussions, we see that the total Hilbert space of the edge excitations is given by

$$\mathcal{H}_{\rm disc}=\oplus_I \mathcal{H}_{\rm KM}^{(I)}=\mathcal{H}_{\rm KM}\otimes\mathcal{H}_p ,$$

where $\mathcal{H}_p$ is spanned by states $|I\rangle$. The state $|I\rangle$ has a charge Ie and $\mathcal{H}_{\rm KM}^{(I)}=\mathcal{H}_{\rm KM}\otimes\{|I\rangle\}$ corresponds to the charge-Ie sector.

In the following we are going to show that the representation of the algebra (2.1)–(2.3) can be constructed from chiral boson theories. For convenience we will assume that the disc has a unit radius (i.e., $L=2\pi$) and set $e=v=1$. Chiral boson theory is defined by the Lagrangian[7]

$$\mathcal{L}=\frac{1}{8\pi}[(\partial_0\phi)^2-(\partial_\sigma\phi)^2] , \quad (2.4)$$

where the real scalar field ϕ satisfies the "chiral" constraint

$$(\partial_0-\partial_\sigma)\phi=0 . \quad (2.5)$$

In the following, we will follow Ref. 8 to quantize the chiral boson theory (2.4) and (2.5). The operator ϕ satisfies the equation of motion

$$(\partial_0-\partial_\sigma)(\partial_0+\partial_\sigma)\phi=0 . \quad (2.6)$$

The solutions of (2.6) take the form

$$\phi=\phi_0+\bar{\phi}_0+p_\phi(t+\sigma)+\bar{p}_\phi(t-\sigma) +i\sum_{n(\neq 0)}\frac{1}{n}(\alpha_n e^{-in(t+\sigma)}+\bar{\alpha}_n e^{-in(t-\sigma)}) . \quad (2.7)$$

The canonical momentum of ϕ is given by $\pi=(1/4\pi)\partial_0\phi$:

$$\pi=\frac{1}{4\pi}(p_\phi+\bar{p}_\phi)+\frac{1}{4\pi}\sum_{n(\neq 0)}(\alpha_n e^{-in(t+\sigma)}+\bar{\alpha}_n e^{-in(t-\sigma)}) , \quad (2.8)$$

p_ϕ and α_n describe the left-moving excitations, while $\bar{p}_\phi$ and $\bar{\alpha}_n$ describe the right-moving ones. From the commutator between ϕ and π we find that $\bar{\phi}_0$, $\bar{p}_\phi$, and $\bar{\alpha}_n$ satisfy the algebra

$$[\bar{\alpha}_n, \bar{\alpha}_m]=n\delta_{n+m} ,$$
$$[\bar{\phi}_0, \bar{p}_\phi]=i , \quad (2.9)$$
$$\text{others}=0 ,$$

and ϕ_0, p_ϕ, and α_n satisfy

$$[\alpha_n, \alpha_m]=n\delta_{n+m} ,$$
$$[\phi_0, p_\phi]=i , \quad (2.10)$$
$$\text{others}=0 .$$

At this stage we may impose the constraint (2.5) by dropping $\bar{\phi}_0$, $\bar{p}_\phi$, and $\bar{\alpha}_n$. A more systematic and careful treatment of the chiral boson theory can be found in Ref. 7. Notice that algebra (2.10) just describes many independent oscillators. The Hilbert space of the chiral boson theory is defined as the Fock space of the oscillator algebra (2.10). The operators α_n generate the irreducible representation of the KM algebra $\mathcal{H}_{\rm KM}$. The space generated by the "zero modes" ϕ_0 and p_ϕ needs more careful treatment and will be discussed later. The Hamiltonian of the chiral boson theory is given by

$$H=\tfrac{1}{2}p_\phi^2+\sum_{n(>0)}\alpha_n\alpha_{-n} . \quad (2.11)$$

The electrical current in the chiral boson theory is identified as

$$j^\alpha=\frac{\sqrt{\nu}}{2\pi}\epsilon^{\alpha\beta}\partial_\beta\phi_L , \quad (2.12)$$

where

$$\phi_L(t,\sigma)=\phi_0+p_\phi(t+\sigma)+i\sum_{n(\neq 0)}\frac{1}{n}\alpha_n e^{-in(t+\sigma)} . \quad (2.13)$$

The total charge operator is

$$Q=\sqrt{\nu}p_\phi=-\sqrt{\nu}i\partial_{\phi_0} . \quad (2.14)$$

Using (2.10) one can explicitly check that the current in (2.12) satisfies the KM algebra (2.1). Therefore the Hilbert space of the chiral boson theory forms a representation of the KM algebra (2.1).

The charged operators in the chiral boson theory have a form $\Psi=:e^{i\gamma\phi_L}:$. Because we want to identify the charged operator Ψ as an electron operator, Ψ must satisfy the anticommutation relation

$$\Psi(\sigma)\Psi(\sigma')=-\Psi(\sigma')\Psi(\sigma), \quad \sigma'\neq\sigma . \tag{2.15}$$

Using the formula

$$e^{A}e^{B}=e^{[A,B]}e^{B}e^{A} , \tag{2.16}$$

we find that[9]

$$\Psi(\sigma)\Psi(\sigma')=e^{i\gamma^2(\sigma+\sigma')/2}(e^{-i\sigma}-e^{-i\sigma'})^{\gamma^2} \times :e^{i\gamma[\phi_L(\sigma)+\phi_L(\sigma')]}: . \tag{2.17}$$

Therefore Ψ is a fermionic operator if $\gamma^2\equiv l$ is an odd integer.

In order to identify the operator Ψ as an electron operator, we not only require Ψ to be a fermionic operator, we also require Ψ to carry a unit charge. From (2.14) we see that the charge of Ψ is given by $\gamma\sqrt{\nu}$. This implies that only when the filling fraction satisfies $\nu=1/l$, can the operator Ψ be identified as an electron operator:

$$\psi^{\dagger}(t,\sigma)=\eta\Psi=\eta e^{i\sqrt{l}\phi_L(t,\sigma)} , \tag{2.18}$$

where η is a constant which may depend on the cutoff. As a direct consequence of the above result, a FQH state with filling fraction $\nu\neq 1/l$ must have more than one branch of edge excitations.

The physical Hilbert space of the chiral boson theory (2.4) and (2.5) is generated by the operators j_k^+ and ψ, or equivalently by α_n and $e^{i\sqrt{l}\phi_0}$. The operator $e^{i\sqrt{l}\phi_0}$ generates the charged excited states. Because α_n and $e^{i\sqrt{l}\phi_0}$ commute, the Hilbert space can be written as

$$\mathcal{H}_{\rm KM}\otimes\mathcal{H}_p , \tag{2.19}$$

where $\mathcal{H}_p$ is spanned by the states $|I\rangle$. The state $|I\rangle$ carries a charge $Q=Ie$ and $p_\phi=I\sqrt{l}$. Thus the Hilbert space of the chiral boson theory is identical to the Hilbert space $\mathcal{H}_{\rm disc}$ that we obtained before.

The commutation relation (2.3) can easily be derived. First, notice that

$$[H,\psi(t,\sigma)]=-i\partial_0\psi(t,\sigma) . \tag{2.20}$$

From (2.13) we see that $\psi(t,\sigma)$ depend on t and σ only through the combination $t+\sigma$. This implies that

$$\partial_0\psi(t,\sigma)=\partial_\sigma\psi(t,\sigma) ; \tag{2.21}$$

(2.3) can be easily obtained from (2.20) and (2.21). We find that the chiral boson theory (2.4) and (2.5) together with the quantization condition

$$p_\phi=\sqrt{l}\times \text{integers} \tag{2.22}$$

form a representation of the algebra (2.1)–(2.3)

Let us summarize our results. Consider a FQH state on a disc with a filling fraction ν. Assume that the edge excitations only have one branch and assume that it costs infinitely small energy to add a single electron to the FQH state. Under those assumptions we show that the edge excitations of such a FQH state are described by the chiral boson theory (2.4) and (2.5) and (2.22). The chiral boson theory contains the charge-e fermion operator only when $\nu=1/l$, where l is an odd integer. Such an operator is identified as the electron operator. Therefore, a FQH state with filling fraction $\nu\neq 1/l$ must have more than one branch of edge excitations. For the $\nu=1/l$ FQH states, Haldane suggested that the edge excitations have only one branch.[10] In this case, the edge excitations are described by the chiral boson theory (2.4) and (2.5) and (2.22). The Hilbert space of the chiral boson theory is generated by the operators α_n and ψ.

The electron Green's function can be calculated using (2.16). We find that

$$\langle\psi^{\dagger}(t,\sigma)\psi(0,0)\rangle=\eta^2 e^{-il(t+\sigma)/2}(e^{-i(t+\sigma)}-1)^{-l} . \tag{2.23}$$

In the thermodynamic limit, $\sigma\ll L=2\pi$ and $t\ll L/v=2\pi$, and we have

$$\langle\psi^{\dagger}(t,\sigma)\psi(0,0)\rangle=\eta^2\left[\frac{i}{t+\sigma}\right]^{l} . \tag{2.24}$$

In the momentum space (2.24) becomes

$$\langle\psi_k^{\dagger}\psi_k\rangle\propto\frac{(\omega-k)^{l-1}}{\omega+k-i\delta} . \tag{2.25}$$

If electrons are described by the Fermi-liquid theory then the Green's function should be

$$\langle\psi^{\dagger}(t,\sigma)\psi(0,0)\rangle=\frac{i}{t+\sigma} . \tag{2.26}$$

The anomalous exponent in the propagator (2.24) implies that the electrons on the edge of the FQH states do not form a Fermi liquid. The electrons are strongly correlated and form new kind of states. Those states resemble the Luttinger liquid,[11] in which the electron propagator also has an anomalous exponent. Because the excitations in our states only move in one direction, we will call such states chiral Luttinger liquids (χLL).

The Luttinger liquids contain both right-moving and left-moving excitations, while the χLL contain only left (or right) moving excitations. Because the chiral property of the χLL, the exponent in the electron propagator is expected to be a topological invariant. In the chiral boson theory considered here, the exponent is given by $1/\nu$, which is quantized as an odd integer. The exponent remains unchanged no matter how we perturb the Hamiltonian. In contrast, the exponent of the electron propagator in the Luttinger liquid can take arbitrary real values. The exponent depends on the interactions between electrons and is not a topological invariant. From the above discussion we see that the χLL and the Luttinger liquid have some fundamental distinctions.

However, as pointed out in Refs. 4 and 6, the edge states of FQH systems are closely related to a Fermi liquid of *charge-*$q=(1/\sqrt{l})$ *fermions*. Or more precisely, the charge-zero sector of the χLL is described by the charge-zero sector of the charge-$q=(1/\sqrt{l})$ Fermi-liquid theory. If we were only interested in the processes that conserve the total charge of the system, then the χLL could be regarded as a Fermi liquid. But the charged excited states are not described by the charge-q Fermi-liquid theory. In particular the charges of the edge states are quantized as integers instead of as multiplets of $1/\sqrt{l}$.

As we increase the size of the system, the constraint on the total charge becomes less and less important. In fact, the χLL and the charge-q Fermi liquid have the same thermodynamic properties. We may effectively treat the edge states as a charge-$q=(1/\sqrt{l})$ Fermi liquid if we were only interested in those thermodynamic properties.

To further demonstrate the similarity between the χLL and the Fermi liquid, we would like to calculate the "edge capacity" of the FQH states. Assume that a $\nu=1/l$ FQH state is in equilibrium with a charge reservoir of voltage V. The total charge of the FQH state is a function of V, $Q=Q(V)$. The edge capacity is defined as

$$C=\frac{dQ}{dV} . \tag{2.27}$$

The edge capacity can also be obtained from

$$\frac{1}{C}=\frac{d^2E(Q)}{d^2Q} , \tag{2.28}$$

where $E(Q)$ is the total energy of the FQH state. Comparing (2.28) and (2.11) and (2.14), we find that the capacity of the χLL is given by

$$C=\nu=\frac{1}{l}\frac{e^2L}{2\pi v} . \tag{2.29}$$

The capacity of a charge-q Fermi liquid is given by q^2N_0, where $N_0=1$ is the density of states of the Fermi liquid. We find that (2.29) is also the capacity of the charge-$q=(1/\sqrt{l})$ Fermi liquid Therefore, despite the charge, quantization conditions are different, the capacity of the χLL and the capacity of the charge-$q=(1/\sqrt{l})$ Fermi liquid are identical.

To observe the non-Fermi-liquid behaviors of the χLL we need to use the processes that change the total charge of the system and probe the microscopic structures in the states. Electron tunneling and photoemission are two such experiments. Those experiments measure the electron spectral function

$$n_{\omega,k}=\sum_n |\langle n|\psi_k|0\rangle|^2\delta(\omega-\omega_n) , \tag{2.30}$$

where ω_n is the energy of the state $|n\rangle$. From the electron propagator we find that the spectral function and the electron "density of states" in the χLL are given by

$$n_{\omega,k}\propto\omega^{l-1}\delta(\omega+k)\theta(-\omega) ,$$
$$N(\omega)=\int\frac{dk}{2\pi}n_{\omega,k}\propto\omega^{l-1}\theta(-\omega) . \tag{2.31}$$

Measuring the spectral function allows us to determine the anomalous exponent.

The χLL are characterized by the following properties:

(1) The χLL contain a conserved current which forms a U(1) KM algebra with a central charge $q^2/2\pi$. q is called the optical charge of the χLL.

(2) The χLL contain a local charged operator. The U(1) charge of the operator is given by q_0, which may not be equal to q.

(3) All the excited states have the same velocity $v=\varepsilon/k$, where ε and k are the total energy and the total momentum of the excited states.

When $q=q_0$ the χLL is just a charge-q Fermi liquid. When $q\neq q_0$ the χLL is different from Fermi liquid. But the charge-zero sector of the χLL is still described by the charge-zero sector of the charge-q Fermi liquid.

Before ending this section we would like to discuss the relation between the chiral boson theory and the microscopic theory of the FQH states.[10] Consider a FQH system confined in a circular potential well. The filling fraction of the FQH states is $\nu=1/l$. The ground state has an angular momentum M_0 and is given by the Laughlin wave function

$$\Phi_0(z_i)=\left[\prod_{i<j}(z_i-z_j)^l\right]\exp\left[-\frac{1}{4}\sum_i|z_i|^2\right] . \tag{2.32}$$

For such simple FQH states, we may assume that the total energy of the system is a single-valued smooth function of the total angular momentum

$$E=E(M) . \tag{2.33}$$

Haldane[10] pointed out that the charge-zero edge excitations in such a system are generated by multiplying a symmetric polynomial to the ground-state wave function

$$|n_1,n_2,\ldots\rangle=P_{(n_1,n_2,\ldots)}(z_i)\Phi_0(z_i) , \tag{2.34}$$

where

$$P_{(n_1,n_2,\ldots)}(z_i)=\sum_{\{i_1,\ldots,i_{n_1};j_1,\ldots\}}(z_{i_1}\cdots z_{i_{n_1}})(z_{j_1}^2\cdots z_{j_{n_2}}^2)\cdots . \tag{2.35}$$

The excited state $|n_1,n_2,\ldots\rangle$ has angular momentum $K+M_0$ and energy $K\,dE/dM$ [assume $E(M_0)=0$], where $K=\sum jn_j$. Such a state is called the Kth-level excited state. The number of states at the Kth level is given by

$$N_K=\sum_{n_1,n_2,\ldots}\delta\left[\sum jn_j-K\right] . \tag{2.36}$$

In the chiral boson theory the excited states are generated by the operators α_n, $n>0$:

$$|n_1,n_2,\ldots\rangle=\alpha_1^{n_1}\alpha_2^{n_2}\cdots|0\rangle . \tag{2.37}$$

The energy of the state $|n_1,n_2,\ldots\rangle$ is given by $Kv\times(2\pi/L)$, where $K=\sum jn_j$. We will again call such a state the Kth-level state. The number of the Kth-level states in the chiral boson theory is given by the same formula (2.36). Because dE/dM is the angular velocity of the edge excitations $dE/dM=v(2\pi/L)$, the Kth-level states in the FQH states and the Kth-level states in the chiral boson theory have the same energy. Adding m electrons to the system increases the angular momentum by $lm(m+1)/2$ and the energy by

$$\Delta E=\frac{m(m+1)}{2}l\frac{dE}{dM}=m\mu ,$$

where μ is the electron chemical potential. If we choose $\mu=(l/2)(dE/dM)$, we find that $\Delta E=(m^2/2)lv(2\pi/L)$.

This again agrees with the result in the chiral boson theory that the operator $e^{im\sqrt{l}\phi_0}$ creates m electrons and increases the total energy by $\frac{1}{2}m^2lv(2\pi/L)$ [See (2.11)]. From the above discussions we find that the microscopic FQH theory and the chiral boson theory give rise to the same Hilbert space and the same Hamiltonian for the low-lying edge excitations. In this way we show that the chiral boson theory (2.4) and (2.5) and (2.22) describe all of the dynamical properties of the edge excitations in the $v=1/l$ FQH states.

We would like to remark that comparing to the microscopic theory discussed above, the chiral boson theory is more general. The chiral boson theory remains to be valid even when the edge potential is not a smooth function and when the electron interaction is modified near the edge. This is because the KM algebra (2.1) is a consequence of the gauge symmetry. The validity of the KM algebra is independent of the detailed structures of the edge configuration. The chiral boson theory also applies to the hierarchy FQH states, which have many branches of edge excitations. The chiral boson theory in this case contains several boson fields, one boson field for each branch. For the $\nu\neq 1/l$ FQH states, the total energy is not a smooth single-valued function of M. In this case the symmetric polynomials do not generate all the low-lying excitations.

III. EDGE EXCITATIONS ON A CYLINDER

In this section we will discuss the edge excitations on a cylinder. We will assume the FQH state on a cylinder to have a filling fraction $\nu=1/l$. Since the cylinder has two edges, one may naively expect that the Hilbert space of the edge excitations on the cylinder is a direct product of the Hilbert spaces on the edges of two discs, $\mathcal{H}_{\rm disc}\otimes\mathcal{H}_{\rm disc}$, where $\mathcal{H}_{\rm disc}=\mathcal{H}_{\rm KM}\otimes\mathcal{H}_p$ is constructed in the last section. However, this naive expectation is incorrect. There are new kinds of excitations on the edges of the cylinder. Those excitations are not contained in $\mathcal{H}_{\rm disc}\otimes\mathcal{H}_{\rm disc}$. Such new excitations transfer multiples of a *fractional* charge e/l from one edge to the other. The new excitations can be induced by adiabatic turning on the unit flux going through the cylinder (Fig. 1). This adiabatic operation transfer e/l charge from one edge to the other. Because the charges in $\mathcal{H}_{\rm disc}\otimes\mathcal{H}_{\rm disc}$ are quantized as integers, such an excited state is not in $\mathcal{H}_{\rm disc}\otimes\mathcal{H}_{\rm disc}$. Turning on l unit flux transfers one electron between the edges. This excitation adds an electron to one edge and subtracts an electron from the other edge. Such an excitation lies within the Hilbert space $\mathcal{H}_{\rm disc}\otimes\mathcal{H}_{\rm disc}$. From the above considerations we conclude that the edge excitations on the cylinder contain l sectors. Each sector is given by $\mathcal{H}_{\rm disc}\otimes\mathcal{H}_{\rm disc}$, since different sectors are related by adiabatic turning on the unit flux. The total Hilbert space of the edge excitations on the cylinder is given by

$$\mathcal{H}_{\rm cyl}=\bigoplus_{M=1}^{l}(\mathcal{H}_{\rm disc}\otimes\mathcal{H}_{\rm disc})^{(M)}$$
$$=\mathcal{H}_{\rm disc}\otimes\mathcal{H}_{\rm disc}\otimes\mathcal{H}_{\rm glo}\,, \tag{3.1}$$

where $\mathcal{H}_{\rm glo}$ contains l states $|M\rangle$, $M=1,2,\ldots,l$. Those states are generated by an operator T, which transfers the e/l charge from one edge to the other:

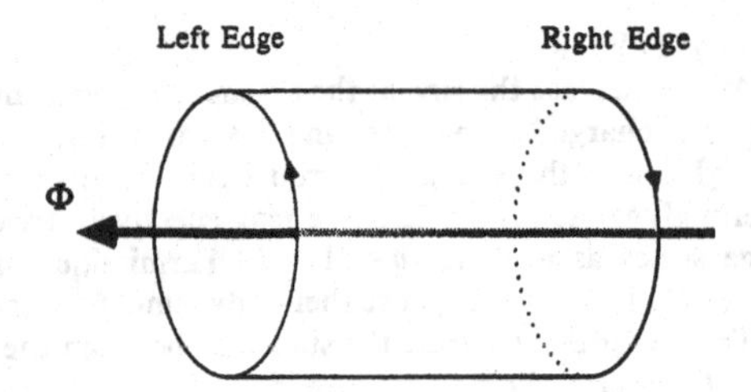

FIG. 1. A FQH state on a cylinder with magnetic flux Φ.

$$|M+1\rangle=T^M|0\rangle\,. \tag{3.2}$$

The operator T is induced by adiabatic turning on the unit flux. Such an operation induces transitions between sectors. But turning on l unit flux does not change the sectors. From the above discussions, we also see that the allowed values of the charges on the two edges are labeled by three integers I_1, I_2, and $M=1,2,\ldots,l$:

$$Q^{(R)}=\left[I_1+\frac{1}{m}M\right],\quad Q^{(L)}=\left[I_2-\frac{1}{m}M\right], \tag{3.3}$$

where the superscript R and L denote the right edge and the left edge (Fig. 1).

In the following we are going to show that the Hilbert space $\mathcal{H}_{\rm cyl}$ can be constructed from a (nonchiral) boson theory

$$\mathcal{L}=\frac{1}{8\pi}[(\partial_0\phi)^2-(\partial_\sigma\phi)^2]\,. \tag{3.4}$$

Repeating the discussions in Sec. II, we may write the ϕ field as

$$\phi(t,\sigma)=\phi_0+\bar\phi_0+p_\phi(t+\sigma)+\bar p_\phi(t-\sigma)$$
$$+i\sum_{n\,(\neq 0)}\frac{1}{n}(\alpha_n e^{-in(t+\sigma)}+\bar\alpha_n e^{-in(t-\sigma)})\,.$$

The operators ϕ_0, $\bar\phi_0$, p_ϕ, $\bar p_\phi$, α_n, and $\bar\alpha_n$ satisfy the following algebra:

$$[\alpha_n,\alpha_m]=n\delta_{n+m}\,,$$
$$[\bar\alpha_n,\bar\alpha_m]=n\delta_{n+m}\,,$$
$$[\phi_0,p_\phi]=[\bar\phi_0,\bar p_\phi]=i\,,$$
$$\text{others}=0\,. \tag{3.5}$$

The operators ϕ_0, p_ϕ, and α_n describe the excitations on the right edge, while $\bar\phi_0$, $\bar p_\phi$, and $\bar\alpha_n$ describe the excitations on the left edge (see Fig. 1). From the preceding section we see that the total charge operators on the two edges are given by

$$Q^{(R)}=\frac{p_\phi}{\sqrt{l}},\quad Q^{(L)}=-\frac{\bar p_\phi}{\sqrt{l}}\,. \tag{3.6}$$

Equations (3.3) and (3.6) imply that p_ϕ and $\bar p_\phi$ are quantized as

$$p_\phi=\left[I_1+\frac{1}{l}M\right]\sqrt{l},\ \bar{p}_\phi=\left[-I_2+\frac{1}{l}M\right]\sqrt{l}\ . \quad (3.7)$$

The charged operators in the boson theory have a form $:e^{i\gamma\phi_L(t+\sigma)+i\bar{\gamma}\phi_R(t-\sigma)}:$. In order for the charged operators to consist of the quantization condition (3.7), γ and $\bar{\gamma}$ must be quantized. The allowed values of γ and $\bar{\gamma}$ are given by

$$\gamma=\left[n_1+\frac{1}{l}n_3\right]\sqrt{l},\ \bar{\gamma}=\left[-n_2+\frac{1}{l}n_3\right]\sqrt{l}\ , \quad (3.8)$$

where n_i, $i=1,2,3$ are three integers. The operator $:e^{i\sqrt{l}\phi_L(t+\sigma)}:$ $(:e^{i\sqrt{l}\phi_R(t-\sigma)}:)$ create an electron on the left (right) edge, while the operator $:e^{i\phi(t,\sigma)/\sqrt{l}}:$ transfers $1/l$ charge from one edge to the other.

The Hilbert space that satisfies the quantization condition (3.7) is generated by the operators α_n, $\bar{\alpha}_n$, $e^{i\sqrt{l}\phi_0}$, $e^{i\sqrt{l}\bar{\phi}_0}$, and $e^{i(\phi_0+\bar{\phi}_0)/\sqrt{l}}$. The operators α_n and $\bar{\alpha}_n$ generate the irreducible representation of the two KM algebras, $\mathcal{H}^{(R)}_{\rm KM}\otimes\mathcal{H}^{(L)}_{\rm KM}$. The total Hilbert space is given by $\mathcal{H}^{(R)}_{\rm KM}\otimes\mathcal{H}^{(L)}_{\rm KM}\otimes\mathcal{H}_{p\bar{p}}$, where $\mathcal{H}_{p\bar{p}}$ is spanned by the states $|I_1,I_2,M\rangle$, with I_1, I_2 denoting the integers and $M=1,2,\ldots,l$. The charges of the state $|I_1,I_2,M\rangle$ are given in (3.3). At this stage it is not difficult to see that the Hilbert space constructed above is identical to $\mathcal{H}_{\rm cyl}=\mathcal{H}_{\rm disc}\otimes\mathcal{H}_{\rm disc}\otimes\mathcal{H}_{\rm glo}$. We conclude that the edge excitations on the two edges of the cylinder are described by the Lagrangian (3.4) and the quantization condition (3.7).

Using the effective theory (3.4), we are able to study the quantum-interference effects when the two edges are brought together to form a torus. Notice that only electrons can tunnel between two edges. The operators that transfer an electron from one edge to the other are given by $e^{\pm i\sqrt{l}\phi}$. After including the electron tunneling between two edges, the system is described by the following low-energy effective Lagrangian:

$$\mathcal{L}=\frac{1}{8\pi}[(\partial_0\phi)^2-(\partial_\sigma\phi)^2+g\cos(\sqrt{l}\,\phi)]\ , \quad (3.9)$$

where g measures the strength of the electron tunneling. Equation (3.9) is the standard sine-Gordon theory (or the clock model).[12] The charged operator $e^{i\gamma\phi}$ has a dimension γ^2. Therefore, the operator $\cos(\gamma\phi)$ is relevant if $\gamma<\sqrt{2}$ and irrelevant if $\gamma>\sqrt{2}$. The operator $\cos(\sqrt{l}\,\phi)$ is relevant only when $l=1$ (i.e., for $\nu=1$ quantum Hall states). In this case, an arbitrary small electron tunneling will open a finite-energy gap to edge excitations. But for $l\geq 3$ the operator $\cos(\sqrt{l}\,\phi)$ is irrelevant. It can open an energy gap only when g is greater than a finite critical value. This is a very nontrivial result. It would be interesting to observe this gap-opening phase transition in the type of experiments discussed in Ref. 13.

The system (3.9) has many degenerate ground states $|m\rangle$ after the energy gap is opened. Those ground states are characterized by $\langle m|e^{i\phi/\sqrt{l}}|m\rangle=e^{i2\pi(m/l)}$, which correspond to the minimums of the potential $\cos(\sqrt{l}\,\phi)$. Different ground states are related by the operator $U=e^{i2\pi p_\phi/\sqrt{l}}=e^{i2\pi\bar{p}_\phi/\sqrt{l}}$ [see (3.7)]:

$$U|m\rangle=|m+1\rangle\ .$$

From the quantization condition (3.7) we find that $|m\rangle=|m+l\rangle$. Therefore the ground states in our system are l-fold degenerate. This is consistent with the well-known result that the $\nu=1/l$ FQH state has l degenerate ground states on a torus.[14,15] The solitons (kinks) in the sine-Gordan theory (2.9) carry charge e/l and correspond to the quasiparticles in the FQH states. As a soliton propagates all the way around the circle, it will transform the ground state $|M\rangle$ into $|M+1\rangle$. This also agrees with the results in Ref. 15.

From the above example we see that the properties of the edge excitations and the properties of the FQH states on compacted space are closely related. We emphasize that this relation is very important. We know that the properties of the FQH states on compacted spaces can be used to characterize the hierarchy structures, or more precisely, the topological orders in the FQH states.[16,15] Because of the above relation, the properties of the edge excitations can also be used to characterize the topological orders in the FQH states. The dynamical properties of the edge excitations provide a practical way to experimentally measure the topological orders in the FQH states. Using this relation, one should also be able to determine the properties of the edge excitations from the properties of the FQH states on compacted spaces.

IV. CONCLUSIONS

In this paper we derive the effective theory of the edge excitations in the FQH states. In particular, we discuss the properties of the charged excited states. The edge excitations are shown to form new kinds of states which are not described by Fermi-liquid theories. Such new states are called chiral Luttinger liquids. The χLL are closely related to Fermi liquids. Actually it can be shown that the charge zero sector of the χLL is identical to the charge zero sector of a charge-q Fermi liquid. Here q is the optical charge of the χLL.

The χLL is described by the chiral boson theory (2.4) and (2.5) and (2.22). The chiral boson theory is exactly soluble. Using this effective theory we can easily obtain all the low-energy properties of the edge excitations. We calculated the electron propagator and the spectral function. The electron tunneling and the photoemission experiments can be used to demonstrate the non-Fermi-liquid behaviors of the χLL.

Using the effective theory, we studied the interference effects between excitations on different edges. We demonstrated that the properties of the edge excitations are closely related to the properties of the FQH states in compacted spaces. The properties of the edge excitations can be used to characterize the hierarchy structures, or the topological orders in the FQH states. Using this relation we can also derive the properties of the edge excitations from the properties of the FQH states in compacted space.

Another nontrivial result obtained from the chiral boson theory is that the FQH states with $\nu \neq 1/l$ must contain more than one branch of edge excitations. This result is very general. It is independent of the edge potentials, electron interactions, etc.

The dynamical properties of edge excitations contain very rich structures which reflect the rich topological orders in the FQH states. Experimental and theoretical studies of edge excitations may lead to a much deeper understanding of the FQH states and may open a new era in the FQH theory.

Note added in proof. Recently the propagator (2.23) at equal time has been confirmed by numerical calculation[17] for the $\nu=\frac{1}{2}$ Laughlin state.

ACKNOWLEDGMENTS

I would like to thank F. D. M. Haldane, D. Tsui, and Y. S. Wu for many discussions. Especially, I would like to thank Y. S. Wu for his careful reading of the manuscript and many helpful comments. This work was supported in part by U. S. Department of Energy Grant No. DE-AC02-76ER02220.

[1]S. J. Allen, Jr., H. L. Störmer, and J. C. M. Hwang, Phys. Rev. B **28**, 4875 (1983); D. B. Mast, A. J. Dahm, and A. L. Fetter, Phys. Rev. Lett. **54**, 1706 (1985); D. C. Glattli *et al.*, *ibid.* **54**, 1710 (1985); S. A. Govorkov *et al.*, Pis'ma Zh. Eksp. Teor. Fiz. **45**, 252 (1987) [JETP Lett. **45**, 316 (1987)]; **94**, 226 (1988) [**67**, 342 (1988)]; I. E. Batove *et al.* (unpublished).

[2]R. Laughlin, Phys. Rev. B **23**, 5632 (1981); B. I. Halperin, *ibid.* **25**, 2185 (1982); R. Tao and Y. S. Wu, *ibid.* **30**, 1097 (1984).

[3]B. I. Halperin, Phys. Rev. B **25**, 2185 (1982).

[4]X. G. Wen (unpublished).

[5]C. W. J. Beenakker, Phys. Rev. Lett. **64**, 216 (1990); A. H. MacDonald, *ibid.* **64**, 220 (1990).

[6]X. G. Wen, Phys. Rev. Lett. **64**, 2206, 1990.

[7]See, for example, D. Gross *et al.*, Nucl. Phys. B **256**, 253 (1985); R. Floreanini and R. Jackiw, Phys. Rev. Lett. **59**, 1873 (1988).

[8]See, for example, M. Green, J. Schwartz, and E. Witten, *Superstring Theory Vol. 1* (Cambridge University Press, Cambridge, 1986), p. 323; D. Gross *et al.*, Phys. Rev. Lett. **54**, 502 (1985).

[9]M. Green, J. Schwartz, and E. Witten, *Superstring Theory Vol. 1* (Cambridge University Press, Cambridge, 1986), p. 332.

[10]F. D. M. Haldane, Lecture in Fractional Statistics and High-T_c Superconductivity Workshop, University of Minnesota, 1989 (unpublished).

[11]J. M. Luttinger, J. Math. Phys. **15**, 609 (1963); D. C. Mattis and E. H. Lieb, *ibid.* **6**, 304 (1965); F. D. M. Haldane, J. Phys. C **14**, 2585 (1981).

[12]See, for example, S. Coleman, Phys. Rev. D **11**, 2088 (1985); J. V. José, L. P. Kadanoff, S. Kirkpatrick, and D. R. Nelson, Phys. Rev. B **16**, 12 (1977); L. P. Kadanoff, Ann. Phys. (N.Y.) **120**, 39 (1979).

[13]R. J. Haug *et al.*, Phys. Rev. Lett. **61**, 2797 (1988); S. Washburn *et al.*, *ibid.* **61**, 2801 (1988).

[14]P. A. Maksym, J. Phys. C **18**, L433 (1985); F. D. M. Haldane, Phys. Rev. Lett. **55**, 2095 (1985); Q. Niu, D. J. Thouless, and Y. S. Wu, Phys. Rev. B **31**, 3372 (1985).

[15]X. G. Wen and Q. Niu, Phys. Rev. B **41**, 9377 (1990).

[16]X. G. Wen, Phys. Rev. B **40**, 7387 (1989); X. G. Wen, Int. J. Mod. Phys. B **4**, 239 (1990).

[17]X. G. Wen (unpublished).

Edge Excitations in the Fractional-Quantum-Hall Liquids

Dung-Hai Lee
IBM Research Division, T. J. Watson Research Center, Yorktown Heights, New York 10598

Xiao-Gang Wen
School of Natural Sciences, Institute for Advanced Study, Princeton, New Jersey 08540
(Received 14 January 1991)

We present an alternative derivation for the macroscopic theory of edge excitations of the fractional-quantum-Hall liquid. The predictions of the macroscopic theory are compared with numerical results for finite systems.

PACS numbers: 73.20.Dx, 72.15.Gd, 73.50.Jt

The incompressible quantum-Hall liquid (the Laughlin liquid) discovered in two-dimensional electron systems represents a new state of matter.[1,2] The precise meaning of the term "new state of matter" can be obtained by studying the effective gauge action generated by a Laughlin liquid when the electronic degrees of freedom are integrated over. To be specific, if we couple a bulk incompressible Laughlin liquid to an external electromagnetic gauge field A_μ, $\mu=0,1,2$ (A_μ is defined not to include the constant vector potential that generates the uniform magnetic field) and integrate out the electronic degrees of freedom, we generate the following bulk gauge action (in Euclidean space-time):

$$L_B=\int d^2r\,\mathcal{L}_B\,,$$
$$\mathcal{L}_B=-i\mathbf{J}_0\cdot\mathbf{A}+\tfrac{1}{2}\sigma_{xy}^2(\bar\rho/m)|(\nabla\times\mathbf{A})_\perp|^2+\tfrac{1}{2}\sigma_{xy}^2(\nabla\times\mathbf{A})_0V(\nabla\times\mathbf{A})_0-\tfrac{1}{2}i\sigma_{xy}\mathbf{A}\cdot\nabla\times\mathbf{A}\,. \tag{1}$$

Here and throughout the rest of the paper, $\mathbf{A}$ and $\vec{A}$ will denote a three-vector and two-vector, respectively. Moreover, in Eq. (1) $\sigma_{xy}=1/k$ (k being an odd integer), m is the effective mass, $\mathbf{J}_0=(\bar\rho,\vec{0})$ ($\bar\rho$ being the average particle density), V is the two-body interaction between electrons, and $(\nabla\times\mathbf{A})_0$ and $(\nabla\times\mathbf{A})_\perp$ denote the time and space components of the three-vector $\nabla\times\mathbf{A}$. This action summarizes the electromagnetic responses of the Laughlin liquid. For example, by calculating $\delta L_B/\delta A_0$ in the limit $\mathbf{A}=0$ we obtain the average density $\bar\rho$. Taking the second derivative with respect to A_0, one can show that the finite-$\vec{q}$ compressibility $K(\omega=0,\vec{q})=m|\vec{q}|^2/\bar\rho k^2$ hence vanishes in the $\vec{q}\to 0$ limit. By differentiating L_B with respect to $\vec{A}$ we obtain the induced current. From the components proportional to $\vec{E}=\partial_0\vec{A}-\vec{\partial}A_0$ and $\hat{z}\times\vec{E}$ we can deduce the longitudinal and Hall conductivities, respectively. The results are as expected, i.e., $\sigma_{xx}\to 0$ and $\sigma_{xy}\to 1/k$ in the dc limit. The Laughlin liquid represents a new state of matter in the sense that the gauge action given in Eq. (1) has a form that is totally unfamiliar.

An interesting question arises when one asks what is the effective gauge action for a droplet of Laughlin liquid. In order to produce such a droplet, we can imagine a system of a finite number of electrons moving in a smooth, spatially confining potential. Naively, we would simply write

$$L_D=\int d^2r\,S(\vec{r},t)\mathcal{L}_B\,, \tag{2}$$

where the "support function" $S(\vec{r},t)=1$ or 0 depending on whether at time t the spatial point $\vec{r}$ is inside or outside the liquid droplet. Upon further investigation, we notice that L_D is in general not gauge invariant.[3] In particular, it is easy to show that if we perform a gauge transformation via $\mathbf{A}\to\mathbf{A}+\nabla\theta$, the droplet action remains invariant only if the support function S satisfies the following equation:

$$\nabla S\cdot(\mathbf{J}_0+\tfrac{1}{2}\sigma_{xy}\nabla\times\mathbf{A})=0\,. \tag{3}$$

Here, due to the spatially confining potential $W(\vec{r})$, $\mathbf{J}_0=(\bar\rho,-i\bar\rho\vec{v})$ (in Euclidean space-time) with $\vec{v}=-(\sigma_{xy}/\bar\rho)\hat{z}\times\vec{\partial}W$. Equation (3) serves as a dynamical constraint on the allowed support function S. If $\nabla\times\mathbf{A}=0$, Eq. (3) allows static solution S_0 where $\vec{\partial}S_0\cdot\vec{v}=0$. Therefore the boundary of the static support traces out an equipotential contour of $W(\vec{r})$. If we define $u(\xi,t)$ (ξ is the coordinate measured along the boundary of S_0) as the normal displacement of the droplet boundary from that of S_0, u must satisfy

$$(\partial_0-iv\partial_\xi)u(\xi,t)=\frac{\sigma_{xy}}{2\bar\rho}\hat{n}(\xi)\cdot\nabla\times\mathbf{A}(\vec{r}_\xi,t)\,, \tag{4}$$

where $\hat{n}(\xi)$ is the unit vector normal to the boundary of S_0 at the point $\vec{r}_\xi$. When $\nabla\times\mathbf{A}=0$, Eq. (4) gives the dispersion of the edge phonon as $\omega_q=vq$. A word of caution is in order here. Since the Laughlin liquid is incompressible, the edge displacement must preserve the droplet area. Consequently, a nonzero $u_{q=0}$ [where $u_q\equiv\int d\xi\,e^{-iq\xi}u(\xi,t)$] should be excluded.

In the presence of $\nabla\times\mathbf{A}$, we use Eq. (4) to calculate the induced displacement. To first order in $\nabla\times\mathbf{A}$, we have

$$u(\xi,t)=\frac{\sigma_{xy}}{2\bar{\rho}}\int d\xi' dt' G(\xi-\xi',t-t') \times\hat{n}(\xi',t')\cdot\nabla\times\mathbf{A}(\vec{r}_{\xi'},t'), \quad (5)$$

where G satisfies

$$(\partial_0-iv\partial_\xi)G(\xi-\xi',t-t')=\delta(\xi-\xi')\delta(t-t'). \quad (6)$$

Substituting Eq. (5) into Eq. (2) and keeping terms to second order in $\nabla\times\mathbf{A}$, we have

$$L_D=\int d^2r\, S_0\mathcal{L}_B+L_E,$$

$$L_E=-\tfrac{1}{2}i\sigma_{xy}\int d\xi\, d\xi' dt' \times[A_0(\vec{r}_\xi,t)-ivA_\xi(\vec{r}_\xi,t)]G(\xi-\xi',t-t') \times[\partial_\xi A_0(\vec{r}_{\xi'},t')-\partial_0 A_\xi(\vec{r}_{\xi'},t')], \quad (7)$$

where A_ξ is the tangential component of $\vec{A}$. This action is in agreement with that obtained by Wen[3] for a single branch of edge excitations.

We will now derive the Lagrangian governing the edge displacement by working backwards from the edge gauge action. In order to make contact with the chiral edge phonon[3-5] description of the edge states, we must first subtract from L_E the contributions due to the vacuum anomaly, namely,

$$L_{E,a}=i\frac{\sigma_{xy}}{2}\int d\xi[A_0(\vec{r}_\xi,t)-ivA_\xi(\vec{r}_\xi,t)]A_\xi(\vec{r}_\xi,t). \quad (8)$$

By differentiating $L_{E,a}+\int d^2r\, S_0\mathcal{L}_B$ with respect to $\mathbf{A}$ we obtain the following edge current:

$$\mathbf{J}_{E,a}=-(\sigma_{xy}/\bar{\rho})\delta(\vec{r}-\vec{r}_\xi)\mathbf{J}_0(\vec{r}_\xi,t)A_\xi(\vec{r}_\xi,t). \quad (9)$$

By letting $A_\xi\rightarrow A_\xi-\partial_\xi\theta$, one can convince oneself that Eq. (8) exhibits the appropriate anomaly. The result of this subtraction gives

$$L'_E=-i\frac{\sigma_{xy}}{2}\int d\xi\, d\xi' dt' \times[A_0(\vec{r}_\xi,t)-ivA_\xi(\vec{r}_\xi,t)]G'(\xi-\xi',t-t') \times[A_0(\vec{r}_{\xi'},t')-ivA_\xi(\vec{r}_{\xi'},t')], \quad (10)$$

where G' satisfies

$$(\partial_0-iv\partial_\xi)G'(\xi-\xi',t-t')=\partial_\xi\delta(\xi-\xi')\delta(t-t'). \quad (11)$$

Here we note that Eq. (8) is a polynomial in A_μ; therefore L_E and L'_E have the same set of poles and hence the same low-energy spectrum.

The goal now is to construct an action L_u for the displacement field, so that after we integrate over u, the action in Eq. (10) is generated. Since the coupling between the edge displacement and $\mathbf{A}$ is given by

$$L_{u-A}=-i\bar{\rho}\int d\xi\, u(\xi,t)[A_0(\vec{r}_\xi,t)-ivA_\xi(\vec{r}_\xi,t)], \quad (12)$$

the action L_u must have the following form:

$$L_u=\frac{i}{2}\frac{\bar{\rho}^2}{\sigma_{xy}}\int d\xi\, d\xi' dt' u(\xi,t)G'^{-1}(\xi-\xi',t-t') \times u(\xi',t'), \quad (13)$$

where

$$\int d\xi' dt' G'(\xi-\xi',t-t')G'^{-1}(\xi'-\xi'',t'-t'')=\delta(\xi-\xi'')\delta(t-t'').$$

If we define a Bose field $a_q(t)$ via

$$a_q(t)\equiv\frac{\bar{\rho}}{(2\sigma_{xy}q)^{1/2}}u_q(t), \quad \bar{a}_q(t)\equiv\frac{\bar{\rho}}{(2\sigma_{xy}q)^{1/2}}u_{-q}(t), \quad (14)$$

where $q>0$, Eq. (14) reduces to

$$L_a=\int\frac{d\omega}{2\pi}\sum_{q>0}\bar{a}_{q,\omega}(i\omega+vq)a_{q,\omega}. \quad (15)$$

Equation (15) is precisely the Lagrangian in the coherent-state path-integral representation of the following Bose Hamiltonian:

$$\mathcal{H}=\sum_{q>0}vqa_q^\dagger a_q, \quad (16)$$

where the operators a and $a^\dagger$ satisfy the canonical Bose commutation relation $[a_q,a_{q'}^\dagger]=\delta_{q,q'}$. Equation (16) is precisely the edge-phonon Hamiltonian obtained when we start from the quantum description of the edge particle-hole excitations.[3-5]

Another useful representation of the edge dynamics can be obtained if we replace $\mathbf{A}$ in Eq. (10) by a pure gauge, i.e., $A_0\rightarrow\partial_0\theta$, $A_\xi\rightarrow\partial_\xi\theta$. The action we obtain is

$$L_\theta=-\tfrac{1}{2}i\sigma_{xy}\int d\xi\,\partial_\xi\theta(\partial_0\theta+v\partial_\xi\theta). \quad (17)$$

Equation (17) is the so-called chiral boson representation of the edge excitations.[5,6] Equation (17) has the property that by replacing $\partial_0\theta\rightarrow\partial_0\theta-iA_0$ and $\partial_\xi\theta\rightarrow\partial_\xi\theta-iA_\xi$ and integrating out θ, we generate L'_E. We may now use Eq. (17) to compute the correlation function

$$\langle\theta(\xi,t)\theta(0,0)\rangle=-\sigma_{xy}^{-1}\ln(vt-i\xi) \quad (18)$$

which can be used to show that

$$\langle e^{i\theta(\xi,t)}e^{-i\theta(0,0)}\rangle\propto(\xi+ivt)^{-1/\sigma_{xy}}. \quad (19)$$

Under the gauge transformation $A_\mu\rightarrow A_\mu+\partial_\mu\phi$, $e^{i\theta}$ transforms like $e^{i\theta}\rightarrow e^{i\phi}e^{i\theta}$; therefore, Eq. (19) is actually the electron propagator along the edge.[5,6]

We can calculate the edge contributions to the three-current by differentiating L_D with respect to $\mathbf{A}$. The result is

$$\mathbf{J}_{\text{tot}} = S_0[\mathbf{J}_0 + \sigma_{xy}\nabla\times\mathbf{A}] + \mathbf{J}_E,$$
$$\mathbf{J}_E = \frac{\sigma_{xy}}{2\bar{\rho}}\delta(\vec{r}-\vec{r}_\xi)\left[\int d\xi' dt' \mathbf{J}_0(\vec{r}_\xi,t)G(\xi-\xi',t-t') \times [\hat{n}(\xi')\cdot\nabla\times\mathbf{A}(\vec{r}_{\xi'},t)]\right]. \quad (20)$$

The time and spatial components of Eq. (20) are

$$\rho_E = \frac{\sigma_{xy}}{2}\delta(\vec{r}-\vec{r}_\xi)\int d\xi' dt' G(\xi-\xi',t-t')E_\xi(\vec{r}_{\xi'},t'),$$
$$J_E = \frac{\sigma_{xy}}{2}\delta(\vec{r}-\vec{r}_\xi)\int d\xi' dt' v\hat{\xi}G(\xi-\xi',t-t')E_\xi(\vec{r}_{\xi'},t'), \quad (21)$$

where E_ξ is the component of the electric field along the boundaries of the droplets.

Above we have presented a macroscopic description of the edge excitations in an incompressible quantum-Hall liquid. We now look at the edge excitations from a microscopic point of view by comparing the macroscopic predictions with numerical results.

To be specific, let us consider spinless electrons moving in an external magnetic field at filling factor $\nu = \frac{1}{3}$. In the limit of $m \to 0$ (m being the effective mass of the electrons), all electrons lie in the first Landau level. For the interaction between electrons we choose $V(\vec{r}) = \alpha\partial^2\delta(\vec{r})$, where α is a parameter. Because (1) $V(\vec{r})$ is positive definite and (2) V has zero expectation value when all pairs of electrons have relative angular momentum ≥ 3, the $\nu = \frac{1}{3}$ Laughlin state[2]

$$\Phi_0(\{z_i\}) = \prod_{i<j}(z_i-z_j)^3\prod_k e^{-|z_k|^2/4} \quad (22)$$

has zero energy and hence is an exact ground state [in writing Eq. (22) we have chosen the symmetric gauge, and define z_j as $z_j \equiv (x_j+iy_j)/l_H$, with l_H being the magnetic length]. There are, however, many other states with zero energy. In fact, the wave function $\Phi(\{z_i\})$ lying solely in the first Landau level has zero energy as long as Φ vanishes at least as fast as $|z_i-z_j|^3$ when particles i and j are brought together. However, since Φ_0 has the lowest angular momentum among all zero-energy wave functions Φ, we can always write

$$\Phi(\{z_i\}) = P(\{z_i\})\Phi_0(\{z_i\}), \quad (23)$$

where $P(\{z_i\})$ is a symmetric polynomial. Setting $P=1$ produces the lowest angular momentum wave function —the Laughlin wave function. This wave function describes a liquid droplet of the smallest size. All the other Φ's have higher angular momenta and describe deformed and/or inflated liquid droplets.[4]

In the presence of a spatially confining potential $W(\vec{r})$, the degeneracy between Φ_0 and all other Φ's is split. Since the Laughlin state represents the droplet with the smallest radius, it has the lowest potential energy. To obtain the low-energy spectrum, we diagonalize $W(\vec{r})$ using as basis linearly independent wave functions of the form given by Eq. (23). In the case of the integer-quantum-Hall effect, all the low-energy wave functions have been explicitly written down by Stone.[5] He shows that the low-energy state with edge-phonon occupation numbers $n_1,n_2,\ldots$ corresponds to the wave functions in Eq. (23) with $P = \prod_k(\sum_i z_i^k)^{n_k}$;

$$|n_1,n_2,\ldots\rangle \equiv \prod_m (a_m^\dagger)^{n_m}|0\rangle \to \prod_m\left[\sum_i z_i^m\right]^{n_m}\Phi_0, \quad (24)$$

where $a_m^\dagger$ is the edge-phonon creation operator in Eq. (16) with $q = 2\pi m/L$ (L being the length of the droplet perimeter). Moreover, for the integer-quantum-Hall effect, one can show that Eq. (23) (or the states generated by the edge-phonon operators) exhaust all low-energy excitations.

For the fractional-quantum-Hall effect, it is not clear whether Eq. (23) (and hence the states generated by the edge-phonon operators) exhausts all low-energy excitations. However, in the following we shall provide indirect evidence for the correctness of the macroscopic description by comparing the degeneracy of the low-energy states obtained from numerical diagonalizations with that predicted by the macroscopic edge-phonon theory. Let M_0 be the angular momentum of Φ_0, and ΔM be the increment in angular momentum of a particular low-energy excited state. Since ΔM is proportional to $\sum_m q_m n_m$, the total linear momentum of the edge phonons, the degeneracy associated with ΔM is

$$\Delta M\text{: } 0,1,2,3,4,5\,, \qquad \text{degeneracy: } 1,1,2,3,5,7\,. \quad (25)$$

Now we compare this prediction with the results of direct diagonalization of a six-particle Hamiltonian. The first 100 energy levels for each value of the total angular momentum are plotted in Fig. 1. The degeneracy of the

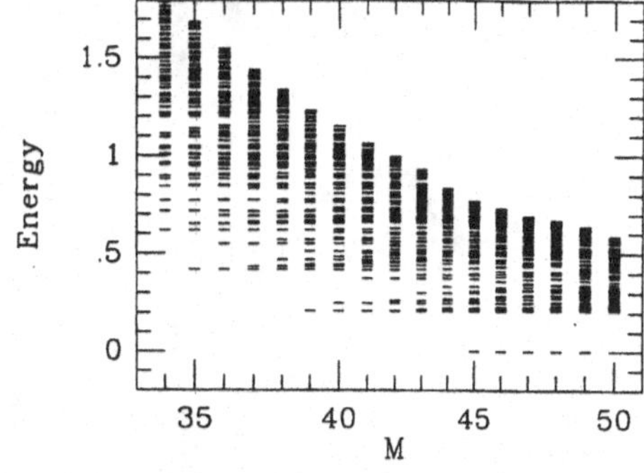

FIG. 1. First 100 energy levels for each total angular momentum. The Hamiltonian contains six particles on the first 20 cyclotron orbitals (i.e., the orbits with angular momenta $0 \to 19$).

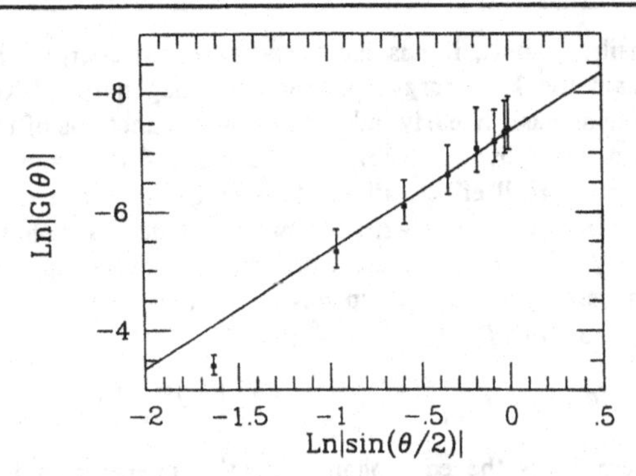

FIG. 2. A Monte Carlo result for the electron propagator $G(\theta)=G(R,Re^{i\theta})$. $R=2\sqrt{N}$ on the edge of the $\nu=\frac{1}{2}$ Laughlin state with $N=36$ particles. The solid line is a fit $|G(\theta)| \propto |\sin(\theta/2)|^{-2}$.

zero-energy states are 1,1,2,3,5,6, for $M=45,\ldots,50$ ($M_0=45$). This exactly agrees with Eq. (25) except at $M=50$. The discrepancy at $\Delta M=5$ is due to the finite-size effect. This result indicates that the edge excitations described by the edge phonons are the only low-energy excitations of the Laughlin states. This result also makes manifest the incompressibility of the Laughlin liquid in a finite system—to reduce the total angular momentum from M_0 to M_0-1 (and to shrink the size of the droplet), a finite-energy gap has to be overcome.

Another important prediction of the macroscopic theory is Eq. (19)—the anomalous exponent in the electron propagator along the edge. To check this prediction, we have calculated the equal-time electron propagator along the edge for a $\nu=\frac{1}{2}$ boson Laughlin state with 36 particles (see Fig. 2). The straight line is the prediction from the macroscopic theory. This anomalous exponent has highly nontrivial implications for the microscopic wave functions. This is because according to the macroscopic theory the exponent of the electron propagator should be exactly equal to $1/\nu$ for all incompressible states that are in the same universality class with the Laughlin states. For a generic two-body interaction, the true ground-state wave functions of the incompressible liquid contain components in higher Landau levels and are in general quite complicated. However, according to the macroscopic theory,[6] all these complications should not change the exponent of the electron propagator along the edge. Hence the exponent of the edge electron propagator is a universal quantity that characterizes the universality classes of the fractional-quantum-Hall states. From this point of view, this exponent plays a role similar to the ground-state degeneracy of the fractional-quantum-Hall states on high genus Riemann surfaces.[7] However, the former can be measured in realistic tunneling experiments.[6]

X.G.W. would like to thank IBM T. J. Watson Research Center for hospitality. X.G.W. is supported in part by DOE Grant NO. DE-FG02-90ER40542.

[1]D. C. Tsui, H. L. Stormer, and A. C. Gossard, Phys. Rev. Lett. **48**, 1599 (1982).

[2]R. B. Laughlin, Phys. Rev. Lett. **50**, 1395 (1983).

[3]X. G. Wen, Institute for Theoretical Physics Report No. NSF-ITP-89-157, 1989 (to be published); X. G. Wen, Phys. Rev. Lett. **64**, 2206 (1990).

[4]F. D. M. Haldane, Bull. Am. Phys. Soc. **35**, 254 (1990); (private communication).

[5]M. Stone (to be published); University of Illinois Report No. IL-PH-90-32, 1990 (to be published).

[6]X. G. Wen, Phys. Rev. B **41**, 12838 (1990).

[7]X. G. Wen and Q. Niu, Phys. Rev. B **41**, 9377 (1990).

Reprinted from ANNALS OF PHYSICS
All Rights Reserved by Academic Press, New York and London

Vol. 207, No. 1, April 1991
Printed in Belgium

Edge Waves in the Quantum Hall Effect

MICHAEL STONE

University of Illinois at Urbana Champaign, Department of Physics,
1110 West Green Street, Urbana, Illinois 61801

Received March 30, 1990; revised August 14, 1990

I show that the bosonized form of the low energy edge excitations at the surface of a droplet of two-dimensional quantum Hall liquid can be understood in terms of coherent "ripplon" deformations in the shape of the droplet. When interacting with an electromagnetic field the resulting chiral boson theory has anomalies that are cancelled, via the Callan–Harvey mechanism, by contributions from the bulk. I also consider non-Abelian generalization of the bosonization that may be useful for chiral spin states. © 1991 Academic Press, Inc.

1. INTRODUCTION

The flow of a globule of incompressible and irrotational fluid is determined entirely by the changing shape of its free surface. Considering the fluid as a dynamical system we would say that all the degrees of freedom reside on the boundary. A Chern–Simons field theory defined in a simply connected two-dimensinal region shares this property [1, 2] and this is no coincidence: a droplet of two-dimensional electron gas (2DEG) in the quantum Hall regime is effectively incompressible and its motion irrotational at energies below the magnetophonon gap [4, 5], and the Hall liquid responds to electromagnetic fields via a Chern–Simons effective action containing these (non-dynamical) external fields. Recent work by Wen [6–8] uses this observation to deduce that the only low-energy excitations, the edge states, form representations of chiral Kac–Moody algebras [9]. In this paper I expand on the results of Wen and show that these algebras may be represented by chiral boson fields [10] describing "ripplon" fluctuations in the shape of the 2DEG droplet. The ripplons may be regarded as the edge degrees of freedom of a Chern–Simons action with *dynamical* fields [2]. The connection between the edge states and the bulk 2DEG provides an example of the Callan–Harvey anomaly cancellation mechanism [11].

The picture that emerges should be of interest to both solid state and particle physicists. In the quantum Hall system many of the mathematical constructions of string theory are available for experimental investigation: The central charges of the Kac–Moody algebras are related to the electric charge of the edge excitations [8], the central charges of the intertwining Virasoro algebras determine the 2DEG

38

0003-4916/91 $7.50

contributions to the specific heat [6], while the uni-directional motion of the edge waves provides a concrete example of the key concept in the heterotic string. It is also possible that the technology developed by string theorists may have other applications in these systems.

In the fractional quantum Hall effect (FQHE) there are many branches of gapless edge excitations [12] but in this paper I will, for simplicity, concentrate on the the case of a single branch. This means that most of my results apply, in the first instance, only to the integer effect or to the lowest level of the FQHE hierarchy. I hope, however, that the technique and language will be useful in the more general cases.

In Section 2 we will examine in detail the case of the integer quantum Hall effect. We will see that there is an essentially exact description of the dynamics of this state in terms of chiral edge states and that these states can be interpreted as bosonic "ripplon" fluctuations in the droplet shape. In Section 3 we will review the bosonization of general chiral excitations. In Section 4 we examine the "anomalies" that appear in the edge system when it is coupled to external electromagnetic fields and how they are removed when we consider the system as a whole. In an appendix we will discuss the analogy with the Callan–Harvey effect.

2. A Chiral Fermi Surface

In an interesting series of papers [6–8] Wen has focused attention on the *edges* of the quantum hall sample and, under very general assumptions, has shown that these edges support operators with the structure of Kac–Moody algebras. Since these algebras are rather abstract objects, it is worth considering the simplest case in some detail and trying to extract the physical content from the mathematics. The simplest example is provided by the edge of the lowest Landau-level in the $n = \nu = 1$ integer Hall effect state.

The application of a large magnetic field supresses many of the degrees of freedom of a quantum mechanical particle. In two dimensions restricting the motion to the lowest Landau level reduces the original four-dimensional phase-space to one with only two canonically conjugate variables, the coordinates x, y. In addition, for fermions, the only low energy degrees of freedom are transitions from an occupied state to an unoccupied state in the same level, excitations to the next level costing energy $\omega_c = eB/m$. In a quantum Hall device the fermions form a two-dimensional droplet confined by a potential well and, for the $\nu = 1$ QHE state, they have uniform density $B/2\pi$. Moving an electron from within the droplet to an unoccupied state outside is stongly reminscent of exciting an electron–hole pair from a Dirac, or Fermi, sea. The purpose of this section is to argue that this analogy is, in essence, *exact*: the $\nu = 1$ droplet *is* a Fermi sea and its surface *is* its Fermi surface.

We will take the particles to have mass $m = 1$ and charge $e = 1$. Using the linear gauge, $A_y = Bx$, and keeping only states from the lowest Landau level, the second-quantizied field operator is

$$\Psi(x, y) = \sqrt{\frac{B}{\pi L_y}} \sum_{n=-\infty}^{+\infty} a_n e^{ik_n y} e^{-B(x-k_n/B)^2/2}. \tag{2.1}$$

For convenience in counting densities of states we have taken the system to be periodic in the y direction so the allowed momenta are $k_n = 2\pi n/L_y$. The wave functions have been normalised in this volume and the annihilation and creation operators obey $\{a_n, a_m^\dagger\} = \delta_{nm}$. Since the physical droplet has a boundary with the topology of a circle these boundary conditions are sensible—indeed, since the edge excitations will turn out to be chiral, there will be no suitable self-adjoint definition for the edge state Hamiltonian unless we take periodic boundary conditions [13].

In the absence of an electric field the states created by $a_n^\dagger$ all have the same energy and carry no net current—but once an electrochemical potential gradient $V(x) = Ex$ is added to the Hamiltonian, the energy of the state with canonical momentum k becomes $\varepsilon(k) = +Ek/B$ and each state carries a Hall curent. The group and phase velocities in the y direction coincide with the Hall current drift velocity E/B. Such a potential occurs natually at the edge of any Hall effect device [14].

Since we are measuring the energy from the Fermi surface, the negative energy states will be occupied by electrons. With $V = Ex$ the physical surface of the 2DEG will then be at $x = 0$, the states to the left being full and those to the right empty. This physical edge of the droplet in x space can be identified, through the relation $x_n = k_n/B$ between the location of the centres of the Gaussian wavefunctions and their momentum k_n, with the *Fermi surface* at $k = 0$, the "speed of light" or "Fermi velocity," v_f, being the drift velocity E/B. Since all the states move in the same direction the droplet is the Dirac sea of a *chiral*, or Weyl, fermion.

To probe the excitations near the Fermi surface we define the y dependent "surface charge" operator, $j(y)$, as the charge density integrated over an x interval $[-\Lambda, \Lambda]$ containing the edge.

$$j(y) = \int_{-\Lambda}^{\Lambda} \Psi^\dagger(x, y)\, \Psi(x, y)\, dx. \tag{2.2}$$

We must take the cutoff Λ larger than the amplitude of any expected ripple in the surface but the Fourier components with non-zero momentum will be independent of the particular choice of Λ.

We can write $j(y) = \sum e^{-ik_n y} j_n$, where $k_n = 2\pi n/L_y$, and

$$j_n = \sum_{m=-\infty}^{+\infty} a_{m+n}^\dagger a_m e^{-k_n^2/4B}. \tag{2.3}$$

By focusing on excitations whose wavelengths are long compared to the magnetic length, the exponential cutoff can be ignored and j_n is precisely the density operator of a "relativistic" right-going chiral fermion with hamiltonian,

$$\hat{H}=\int \psi_R^\dagger(y)(-iv_f\partial_y)\,\psi_R(y)\,dy. \tag{2.4}$$

The total charge operator $\hat{Q}=j_0$ has an unimportant, but Λ dependent, c-number contribution from states far inside the droplet. This is removed in the usual manner by normal ordering the operator with regard to the $k=0$ Fermi surface. It has been shown by many authors, begining with Jordan [15] that the commutator of the j_n operators has a Schwinger term, or central charge, whose magnitude in the case of a normal Fermi sea is determined by the f sum-rule [16]. (A particularly clear demonstration is provided by Manton [17].) The commutator is

$$[j_n, j_m]=-n\delta_{n+m,0} \tag{2.5}$$

and the j_n form the simplest example of a Kac–Moody algebra, i.e., $U(1)$ at level one [9]. It is worth stressing that the Schwinger term is a Berry phase [18] and is thus a property of the states to which the operator is applied, rather than a property of the operator stemming from the normal ordering (which only affects j_0): if the positive k states were filled, rather than the negative k ones, the central charge has opposite sign [19]. The commutator (2.5) is the key ingredient in the solution of the Tomonaga [20] and Luttinger [21] models as well as the physics of a number of other one-dimensional electron systems [16].

In terms of the coordinate along the edge of the droplet the commutator reads

$$[j(y), j(y')]=\frac{-i}{2\pi}\,\partial_y\delta(y-y'). \tag{2.6}$$

The representation of this algebra by bosonic operators has been described in one language by Wen in Ref. [7]. Here I will give a somewhat different view of the same physics. Denote the state of the undisturbed droplet by $|0\rangle$. We can produce coherent excitations of the droplet by using these $j(y)$ operators: define the family of states

$$|\theta(y)\rangle=e^{i\int\theta(y)j(y)\,dy}\,|0\rangle, \tag{2.7}$$

remember that $\langle 0|\,j(y)|0\rangle=0$, and use

$$\begin{aligned} e^{-i\int\theta(y')j(y')\,dy'}\,j(y)\,e^{+i\int\theta(y')j(y')\,dy'} &= j(y)+\frac{1}{2\pi}\int dy'\,\theta(y')\,\partial_y\delta(y'-y)\\ &= j(y)+\frac{1}{2\pi}\,\partial_y\theta(y) \end{aligned} \tag{2.8}$$

to see that the state $|\theta\rangle$ has charge

$$\langle\theta|j(y)|\theta\rangle = +\frac{1}{2\pi}\partial_y\theta(y). \tag{2.9}$$

These states are seen to correspond to ripples in the surface density. For long wavelength the effect of the excitation operator is to change the local value of k to $k+\partial_x\theta$ in the Landau-level eigenstates. Because the new state is a still a linear superposition of lowest Landau-level states, and because for these states there is a connection between k and the location of the state along the x axis, one knows that the surface density fluctuations correspond to changes in the droplet shape—as they should since the 2DEG is incompressible. Further, because all states near the edge propagate at the same speed, the ripples should move at this speed as well. To see this formally we commute the Hamiltonian $\hat{H}$ through $\exp(i\int\theta(y')j(y')\,dy'j(y))$. The Hamiltonian $\hat{H}$ comes from the potential gradient and the dependence of the position of the state on its momentum;

$$\hat{H}=\frac{E}{B}\sum_{n=-\infty}^{+\infty} k_n :a_n^\dagger a_n:, \tag{2.10}$$

so we find

$$e^{-i\hat{H}t}e^{i\int\theta(y)j(y)\,dy}e^{+i\hat{H}t} = e^{i\int\theta(y-v_ft)j(y)\,dy} \tag{2.11}$$

and, as claimed, the ripples move along at the drift velocity with their shape unchanged:

$$e^{-i\hat{H}t}\,|\theta(y)\rangle = |\theta(y-v_ft)\rangle. \tag{2.12}$$

The description of the fermionic edge states in terms of coherent boson states is an example of chiral bosonization. It is by no means obvious that the bosonic $|\theta(x)\rangle$ states exhaust all possible excitations—but the plausibility of this assertion may be demonstrated by examining the partition function: For simplicity move the origin from which we measure the energy so that it lies midway between two levels and the allowed energies become

$$\varepsilon_n=\frac{E}{B}\frac{\pi}{L_y}(2n-1). \tag{2.13}$$

Write $q=\exp(-\beta(E/B)(\pi/L_y))$. Then, remembering that we must count both particle states and hole states, the partition function is

$$\mathscr{Z}=\mathrm{Tr}\{e^{i\theta\hat{Q}-\beta\hat{H}}\}=\prod_{n=1}^{\infty}(1+e^{i\theta}q^{2n-1})(1+e^{-i\theta}q^{2n-1}). \tag{2.14}$$

(Strictly speaking, we should include a Casimir energy factor $e^{+2\pi\beta/12L_y}$. This factor plays a vital role when one is interested in the the modular transformation properties of $\mathscr{Z}$—but it plays no role in the present, purely combinatorial, argument.) The Jacobi triple-product formula [22]

$$\prod_{n=1}^{\infty} (1-q^{2n})(1+e^{i\theta}q^{2n-1})(1+e^{-i\theta}q^{2n-1}) = \sum_{n=-\infty}^{\infty} e^{i\theta}q^{n^2} \tag{2.15}$$

then shows that $\mathscr{Z}$ can be written as

$$\mathscr{Z} = \frac{1}{\prod_{n=1}^{\infty}(1-q^{2n})} \sum_{-\infty}^{\infty} e^{i\theta}q^{n^2}, \tag{2.16}$$

which is the partition function for neutral bosonic excitations (the ripplons) with energies

$$\varepsilon_n^{(b)} = \frac{E}{B}\frac{2\pi}{L_y} n, \tag{2.17}$$

together with a sum over charged sectors, each created by first inserting (or removing) n extra fermions in the lowest available states above the sea, and then creating the ripplons as distortions of the new state. The total energy of these n extra fermions is proportional to $1^2+3^2+5^2+\cdots+2n-1=n^2$ and accounts for the q^{n^2}. In the bosonized language the charged states are composed of solitons where the $U(1)$ field winds once for each extra unit of charge. (This is another way of saying that the $U(1)$ group is disconnected and an irreducible representation of the whole group contains many copies of the representation of the Lie algebra.)

In the next sections we will generalize these coherent excitations and examine their connection to the bulk flow of the Hall liquid.

3. Chiral Bosons

In the previous section we have discussed, in operator language, the edge states and their bosonization. It useful to have a path-integral description of the chiral bosons but the chiral constraint serves to complicate matters [23–25]. It is actually easiest to understand the physics by using a non-Abelian generalization of the chiral bosons invented by Sonnenschein [26] and then specialising to the Abelian case after understanding the broader setting. This general machinary of chiral bosonisation is useful in its own right when we have more than one "flavour" of fermions—as occurs when we wish to include spin excitations or, for example, in the chiral spin states of Wen, Wilczek, and Zee [27, 6].

The simplest route to chiral bosonization [10] is *via* the quantum mechanical concept of coherent-state path-integrals [28]. These begin with an irreducible representation $D(g)$ of some continuous group G and define a collection of generalized coherent states [29], labeled by elements $g \in G$, analogous to the labeling of the states $|\theta\rangle$ by the function $\theta(x)$,

$$|\theta\rangle = D(g)|0\rangle. \tag{3.1}$$

Because of the irreducibility, Schur's lemma shows that these states satisfy an over-completeness relation

$$1 = \text{const} \int d[g] |g\rangle\langle g|, \tag{3.2}$$

where $d[g]$ is the invariant measure on G.

We can use the $|g\rangle$ to give a path-integral representation for the vacuum persistence amplitude,

$$\mathscr{Z} = \mathrm{Tr}\{e^{-it\hat{H}}\}. \tag{3.3}$$

(The trace is taken only over the states in the representation $D(g)$.) By repeatedly inserting the resolution of the identity, (3.2), into the trace, we obtain the path integral

$$\mathscr{Z} = \int d[g]\, e^{-\int_0^t (\langle g|\dot{g}\rangle + i\langle g|\hat{H}|g\rangle)\,dt}, \tag{3.4}$$

where $d[g]$ is the path measure made from taking the invariant measure at each time step and the integral is over all paths in G taking time t.

This construction applies in a natural way to the *loop group* [30] generated by the Kac–Moody algebra [8],

$$[J_i(x), J_j(y)] = if_{ij}^k J_k(x)\, \delta(x-y) - \frac{ik}{2\pi} \mathrm{tr}(\lambda_i \lambda_j)\, \partial_x \delta(x-y), \tag{3.5}$$

for $k=1$. This is discussed for $k=1$ in [10] and for $k>1$ in [31].

The loop group, LG, is the "gauge" group made by taking an element of an ordinary Lie group, G, at each point of the one-dimensional space. Multiplication is done point-by-point and seems trivial—but all non-trivial representations are *projective*, i.e. representations up to a phase. They are spanned (for level $k=1$) by coherent deformations, made by application of group elements $\hat{g}(x,t)$ which are exponentials of integrals of the generating currents $J_i(x)$, of a chiral Dirac sea. These coherent states are labeled by the G-valued fields $|g(x,t)\rangle = \hat{g}(x,t)|0\rangle$.

Setting $v_f = 1$, we calculate the ingredients for (3.4) and find that the action contains a Wess–Zumino term. To write it in a simple form we have to extend the functions $g(x,t)$ to $g(x,t,\tau)$ defined in the interior of the region bounded by the two-dimensional space time. In terms of these extended fields the action in the path integral is found to be [10]

$$S(g) = -\frac{1}{4\pi} \int dx\, dt\, \mathrm{tr}(g^{-1}\partial_x g)^2 + \frac{1}{2\pi} \int dx\, dt\, d\tau\, \mathrm{tr}(g^{-1}\partial_t g \partial_x (g^{-1}\partial_\tau g)). \tag{3.6}$$

Despite the extended domain of definition for the fields, the variation of the action depends only on values of g at physical points. We find

$$\delta S(g)=\frac{1}{2\pi}\int dx\,dt\,\mathrm{Tr}(g^{-1}\delta g\,\partial_x(g^{-1}\partial_x g+g^{-1}\partial_t g)), \tag{3.7}$$

so the clasical equation of motion is

$$\partial_x(g^{-1}\partial_x g+g^{-1}\partial_t g)=0 \tag{3.8}$$

with solutions

$$g(x,t)=g_1(x-t)\,g_2(t), \tag{3.9}$$

where $g_2(t)$ is arbitrary x independent factor.

The solutions look like right-going waves—but, in addition, there is a hidden gauge invariance which shows itself in the factor $g_2(t)$. The reason for the gauge invariance is that any x independent right factor in the $\hat{g}$ leaves the Dirac sea invariant up to a phase so the coherent states are really elements of the coset space LG/G [10, 30].

The physical currents have bosonized form

$$\psi^{\dagger}(x)\,\lambda_i\psi(x)=\frac{1}{2\pi i}\mathrm{Tr}(\lambda_i\partial_x gg^{-1}) \tag{3.10}$$

which is insensitive to this arbitrary factor. (There are also "unphysical" currents which are sensitive to $g_2(t)$ and they can be used for sewing two chiral bosons together to make one Wess–Zumino–Witten bosonized Dirac fermion [32, 33]. These may be useful when two edges are in proximity [8].)

The simplest Hall effect case considered in the previous section uses the Abelian, $U(1)$, group where $g(x,t)=e^{i\theta(x,t)}$. In this case the action becomes

$$S=-\frac{1}{4\pi}\int d^2x((\partial_x\theta)^2+\partial_x\theta\partial_t\theta) \tag{3.11}$$

with equation of motion

$$\partial_x(\partial_x+\partial_t)\,\theta=0 \tag{3.12}$$

and solution

$$\theta(x,t)=\theta_1(x-t)+\theta_2(t) \tag{3.13}$$

exhibiting the undetermined gauge degree of freedom in $\theta_2(t)$.

The momentum field conjugate to $\theta(x, t)$ is $\pi(x, t) = -(1/2\pi)\,\partial_x\theta$ and the commutation relations for the $\pi(x)$ field are

$$[\pi(x), \pi(x')] = -\frac{i}{2\pi}\partial_x\delta(x-x'). \tag{3.14}$$

Writing

$$\theta(x) = q_0 - \frac{2\pi}{L}p_0 x + i\sum_{n\neq 0}\frac{a_n}{n}e^{2\pi inx/L} \tag{3.15}$$

and thus

$$\pi(x) = \frac{1}{L}\left(p_0 + \sum_{n\neq 0} a_n e^{2\pi inx/L}\right). \tag{3.16}$$

We obtain (3.14) by taking

$$[a_n, a_m] = n\delta_{n+m,0}, \qquad [q_0, p_0] = i. \tag{3.17}$$

If the θ field contains a classical soliton twist, where $\theta \to \theta + 2\pi n$ as we circle the boundary, we would expect the eigenvalue of p_0 to equal $-n$. This looks innocuous until we realise that q_0 contains the unphysical gauge degree of freedom and the gauge transformations are generated by p_0. Demanding that the wave function be independent of the q_0 requires $p_0 = 0$ and thus there are no solitions. Although it is customary to require the wave function to be gauge invariant, it is is not strictly necessary: the wave function may be a *section* of fibre bundle over the configuration space of the system; i.e., it changes by a prescribed phase as the unphysical degrees of freedom vary. The phase must, however, form a representation of the global $U(1)$ charge group and this selects the allowed eigenvalues of p_0 to be integers—consistent with the soliton requirement. These phases are natural here as the twisted states have charges and therefore change by a phase under global rotations of the Dirac sea [10].

The Hamiltonian for the chiral boson is[1] a sum of free oscillators and one rotator

$$\hat{H} = \frac{2\pi}{L}\left(\frac{1}{2}p_0^2 + \sum_{n=1}^{\infty} a_{-n}a_n\right) \tag{3.18}$$

from which we may read off the allowed energies and find the partition function to be identical to that found in (2.16).

Wen has also discussed [8] the bosonization of the $\nu = 1/(2n+1)$ FQHE states. These are still level-one algebras (although with a differential normalization of the generators) and the only significant effect is on the compactification radius of the $U(1)$ group where we now have to wind $2n+1$ times to create a state with charge $e = 1$.

[1] I am ignoring the extra gauge generator term which yields $\theta_2(t)$.

4. Attaching the Surface to the Bulk

Suppose the two-dimensional electron liquid occupies a region Ω. We know that the bulk of the liquid responds to an external electromagnetic field by producing currents whose maginitude and direction are precisely quantified by the integer Hall effect. If the filling factor is $\nu = k$ this means that the effective action for the bulk is a Chern–Simons action

$$S(A) = \frac{k}{4\pi} \int_{\Omega \times R} d^3x \varepsilon^{\mu\nu\sigma} A_\mu \partial_\nu A_\sigma, \tag{4.1}$$

where A_μ is the electromagnetic potential. The current derived from this action, $J_\mu = \delta S/\delta A_\mu = (k/2\pi)\, \varepsilon^{\mu\nu\sigma} F_{\mu\nu}$, is conserved *within* the region, but not at the edge, $\partial\Omega$. To make a gauge invariant theory we need to add terms describing the motion of the edge, or, if the walls present expansion, the accumulation of charge there. I have already described the edge motion of a single Landau level in terms of a chiral boson field and it seems reasonable to do the same for the higher (integer) ν cases. We need to see how to consistently sew together the bulk action and a chiral boson field.

So as to have a concrete picture in mind while examining the general case, let me describe the physical origin of the gauge anomaly in the edge states: A chiral fermion interacting with an external electromagnetic field has Hamiltonian

$$\hat{H} = \int_0^L dx\, \psi^\dagger(-i\partial_x - A_x - A_t)\, \psi. \tag{4.2}$$

I have taken the particle to live on a circle of circumference L. The fields occur in the combination $A_x + A_t$ because $\psi^\dagger\psi$ is both the charge density and the current density. Temporarily set $A_t = 0$ and look at the single particle eigenstates at fixed A_x. They are simply

$$\psi_n(x) = e^{2\pi i n x/L}, \qquad \varepsilon_n = \frac{2\pi}{L} n - A_x. \tag{4.3}$$

When A_x is made vary with time as $A_x = -Et$, the particles gain energy from the electric field as they should—but this means that states are crossing the fermi surface at rate $E/2\pi$ per unit length, and, counting from the fermi surface, the particle density is changing. The naive conservation law $(\partial_t + \partial_x)\, \psi^\dagger\psi = 0$ must be replaced by the anomalous version

$$(\partial_t + \partial_x)\, \psi^\dagger\psi = \frac{1}{2\pi} E. \tag{4.4}$$

If these fermions were the whole story the anomaly would prevent us from producing a consistent, gauge invariant, theory—as is the case in the the Standard

Model of Weak Interactions where, complete quark–lepton "generations" of chirally coupled particles are needed for satisfactory physics. In the Hall droplet system an everywhere tangential E field must come from a changing magnetic flux, and the anomaly is simply the expansion or contraction of the droplet as the B field is varied. It is another statement of the Hall effect. The non-conservation of charge implied by the bulk CS effective action will match, and cancel, the non-conservation of charge for the edge states.

For demonstrating the general structure, and, in fact, for case of calculation (the non-abelian group structure provides a structure that guides the calculation), it is again easiest to consider the non-Abelian extension of the problem. In the case of a non-Abelian group the (2 + 1)-dimensional Chern–Simons term is, in the compact notation of differential forms,

$$S_{CS}(A)=\frac{k}{4\pi}\int_{\Omega\times R}\mathrm{Tr}\left(AdA+\frac{2}{3}A^3\right). \tag{4.5}$$

For the non-Abelian case, k must be an integer. Under a variation $A\to A+\delta A$, we have

$$\delta S_{CS}=\frac{k}{2\pi}\int_{\Omega\times R}\mathrm{Tr}(\delta AF)+\frac{k}{4\pi}\int_{\partial\Omega\times R}\mathrm{Tr}(\delta AA). \tag{4.6}$$

F is the field strength 2-form $F=dA+A$. Under a gauge transformation

$$A\to hAh^{-1}-dhh^{-1}, \tag{4.7}$$

the field strength changes by $F\to hFh^{-1}$ and, for infinitessimal $h=1+\varepsilon$,

$$\delta_\varepsilon A=[\varepsilon,A]-d\varepsilon. \tag{4.8}$$

Inserting (4.8) into (4.6) gives

$$\delta_\varepsilon S_{CS}=-\frac{k}{4\pi}\int_{\partial\Omega\times R}\mathrm{Tr}(\varepsilon(F-A^2))=-\frac{k}{4\pi}\int_{\partial\Omega\times R}\mathrm{Tr}(\varepsilon dA), \tag{4.9}$$

and the CS action is not gauge invariant on its own.

At the edge of the droplet we attach a chiral boson for the level k algebra,

$$S_0(g)=-\frac{k}{4\pi}\int_{\partial\Omega\times R}\mathrm{Tr}(g^{-1}\partial_x g)^2+\frac{k}{2\pi}\int_{\Omega\times R}d^3x\,\mathrm{Tr}(g^{-1}\partial_t g\partial_x(g^{-1}\partial_\tau g)) \tag{4.10}$$

whose variation we have already found,

$$\begin{aligned}\delta S_0&=\frac{k}{2\pi}\int_{\partial\Omega\times R}dx\,dt\,\mathrm{Tr}(g^{-1}\delta g\,\partial_x(g^{-1}\partial_x g+g^{-1}\partial_t g))\\&=\frac{k}{2\pi}\int_{\partial\Omega\times R}dx\,dt\,\mathrm{Tr}(\delta g g^{-1}\partial_+(\partial_x gg^{-1})).\end{aligned} \tag{4.11}$$

Here $\partial_+ = \partial_x + \partial_t$. We wish to couple the external gauge fields A_x, A_t to the boson so the resulting theory is gauge invariant under the combined variation $\varepsilon = \delta g g^{-1}$, $A \to A + \delta_\varepsilon A$. Since, as mentioned earlier, the $\partial_x g g^{-1}$ curent represents both the charge density *and* the current density we expect it to couple to the combination $A_+ = A_x + A_t$.

After some exploration one sees that

$$S(g, A) = S_0 - \frac{k}{2\pi} \int_{\partial\Omega \times R} dx\, dt\, \mathrm{Tr}((A_x + A_t)\, \partial_x g g^{-1})$$

$$- \frac{k}{4\pi} \int_{\partial\Omega \times R} dx\, dt\, \mathrm{Tr}((A_x + A_t)\, A_x)) \tag{4.12}$$

has as its gauge variation,

$$\delta_\varepsilon S = \frac{k}{4\pi} \int_{\partial\Omega \times R} \mathrm{Tr}(\varepsilon(F - A^2)), \tag{4.13}$$

which is equal and opposite to the variation of the bulk Chern–Simons effective action. (The actual form of the variation is that of the so-called "consistent anomaly.") The combined action $S_{\mathrm{CS}}(A) + S(g, A)$ is gauge invariant and describes the entire system in a unified manner. This cancellation of a gauge variation in one system by contributions from a system of different dimensionality is an example of the "Callan–Harvey" effect [11] which is reviewed in the Appendix. Notice that there is ome mixing between the A_+ and A_t that might not have been expected. This mixing is familiar from string theory where it is known as the "holomorphy" anomaly.

Discussion

Following the ideas of Wen [6–8] I have shown that the edge of a droplet of quantum Hall liquid can be described in terms of either the fermionic excitation of a "Fermi surface" or as bosonic ripples. In the present paper I have considered only the case of a single branch of excitations where the mapping is straightforward. I have also considered a non-Abelian generalization, which needs additional degrees of freedom, such as unpolarised spins, to be physically realised. These degrees of freedom do occur in chiral spin states so the extension is not entirely accademic.

The fractional Hall effect has been discussed by Wen but the representations were infered from general principles and not constructed explicitly from the underlying electron wavefunction. A concrete construction of the fractional edge states in terms of deformed Laughlin states would be very interesting. It does not seem possible to have non-Abelian analogues of these states as the level of the non-Abelian algebras would be equal to ν but have to be integers [9].

The chiral boson formalism presented here is not as unified as it might be since the component of the electric field providing the "speed of light" for the edge states is treated differently from the other component and from the magnetic field. It may be that using the technology of [2] to express (4.12) entirely in terms of Chern–Simons fields will help here. In addition an alternative formulation of the QHE droplet should exist: one in which the geometry of the droplet shape enters the formalism explicitly—instead of appearing implicitly in the surface charge. Such a formalism would be be able to describe large deformations in the droplet shape and would provide a physical application for the machinary of string theory.

Appendix

In Ref. [11] Callan and Harvey demonstrated the cancellation of anomalies between two systems of differing dimensionality. This cancellation has had previous application to condensed matter systems [34–36]. The original Callan–Harvey effect was driven by a slightly different mechanism than the Hall effect but one that has many features in common. As an illustration of their machinary, consider a $(2+1)$-dimensional massive Fermi system with Dirac–Hamiltonian

$$H = -i\sigma_3 \partial_x - i\sigma_2 \partial_y + m(x)\,\sigma_1. \tag{A1}$$

The mass $m(x)$ is taken to be asymptotically constant, is independent of y, and changes sign as we pass through $x=0$. The vanishing of m creates a one-dimensional "string" or domain boundary. We would expect there to be states near $x=0$ which are bound because they have energies in the asymptotic mass gap—but one of these states is different in that it *must* exist for topological reasons: we look for eigenstates of H of the form

$$\psi(x, y) = e^{ik_y y}\psi_0(x); \tag{A2}$$

then

$$\psi_0 = u_0 \exp\left(\varepsilon \int_0^x m(x')\,dx'\right), \qquad \sigma_2 u_0 = \varepsilon u_0 \tag{A3}$$

is a normalizable eigenstate with

$$H\psi = \varepsilon k_y \psi, \tag{A4}$$

provided

$$\varepsilon = -\tfrac{1}{2}[\mathrm{sgn}(m(\infty)) - \mathrm{sgn}(m(-\infty))]. \tag{A5}$$

These states move in only one direction despite their originating from fermions with both chiralities. A person made out of these particles would think of herself as living in a one-dimensional universe with a net anomaly: On applying an electric field E_y along the string, the k_y momenta are replaced by $k_y + eE_y t$. There is a steady flow of states across $E=0$ at a rate $\varepsilon eE_y/2\pi$ per unit length and since these states are charged, the one-dimensional observer sees charge being created out of the vacuum at a rate of $\varepsilon e^2 E_y/2\pi$, violating charge conservation in accord with the $(1+1)$-dimensional anomalous conservation law for particles with chirality ε:

$$\partial_\mu j^\mu = \varepsilon \frac{e^2}{4\pi} \varepsilon_{\sigma\tau} F^{\sigma\tau}. \tag{A6}$$

Being able to see the larger picture, we know that the extra charge is not really coming out of the depths of an infinite one-dimensional Dirac sea but is actually flowing in from outside the universe: Each momentum k_y looks like a one-dimensional fermion theory and can easily be seen to contribute an amount

$$\frac{e}{2\pi} \partial_t \tan^{-1}\left(\frac{k_y + eE_y t}{m}\right) = \frac{e^2 E_y}{2\pi m} \frac{1}{1 + k^2/m^2} \tag{A7}$$

to the current leading to a total flux of

$$\begin{aligned} j_x(x) &= \frac{e^2 E_y}{2\pi m} \int \frac{dk}{2\pi} \frac{1}{1 + k^2/m^2} \\ &= \frac{1}{4\pi} \operatorname{sgn}(m)\, e^2 E_y. \end{aligned} \tag{A8}$$

This flux is exactly what is needed to account for the extra charge seen by our anomalous observer.

For a general applied field, electric or magnetic, we would have

$$j_\mu = \frac{e^2}{8\pi} \operatorname{sgn}(m)\, \varepsilon_{\mu\nu\sigma} F^{\nu\sigma}. \tag{A9}$$

The motion of the charge perpendicular to the applied field described by (A9) is analogous to the Hall effect but distinct from it. The flow has to be dissipationless as there is a gap, but the gap is created by a conventional mass term rather than by a magnetic field. It is known as the parity "anomaly" because, at first sight, the Levi–Civita symbol suggests that we have a parity-even expression equal to a parity-odd expression. The resolution of this paradoxical dependence on the choice of what to call left and right needs the observation that m, and hence $\operatorname{sgn}(m)$, is parity-odd in an even number of space dimensions: In $2+1$ dimensions, parity is defined by $\psi(x, y) \to \sigma_2 \psi(x, -y)$ ($\psi(x, y) \to \psi(-x, -y)$ is merely a rotation) and $m\bar{\psi}\psi$ changes sign under this transformation.

Acknowledgments

This work was suported by NSF–DMR–88–18713 and by NSF–DMR–86–12860. I thank Eduardo Fradkin for drawing my attention to Refs. [6–8] and for much useful discussion.

REFERENCES

1. E. WITTEN, *Comm. Math. Phys.* **121** (1989), 351.
2. G. MOORE AND N. SEIBERG, *Phys. Lett. B* **220** (1990), 422.
3. Deleted in proof.
4. For a review see "The Quantum Hall Effect" (R. E. PRANGE AND S. M. GIRVIN, Eds.), Springer-Verlag, New York/Berlin, 1987.
5. M. STONE, "The Superfluid Dynamics of the Fractional Quantum Hall States," *Phys. Rev. B* **42** (1990), 212.
6. X. G. WEN, "Gapless Boundary Excitations in the Quantum Hall States and the Chiral Spin States," NSF-ITP-89-157, ITP Santa Barbara preprint.
7. X. G. WEN "Chiral Luttinger Liquid and the Edge Excitation in the Fractional Quantum Hall Effect," IAS Princeton preprint.
8. X. G. WEN, "Electrodynamical Properties of Gapless Edge Excitations in the Fractional Quantum Hall State," IAS Princeton preprint.
9. For a review see P. GODDARD AND D. OLIVE, *Int. J. Mod. Phys. A* **1** (1986), 303.
10. M. STONE, *Phys. Rev. Lett.* **63** (1989), 731; *Nucl. Phys. B* **327** (1989), 399.
11. C. G. CALLAN, JR. AND J. A. HARVEY, *Nucl. Phys. B* **250** (1985), 427.
12. C. W. J. BENNAKKER, *Phys. Rev. Lett.* **64** (1990), 216; A. H. MCDONALD, *Phys. Rev. Lett.* **64** (1990), 220.
13. R. D. RICHTMEYER, "Principles of Advanced Mathematical Physics," p. 15, "Springer-Verlag, 1978.
14. B. I. HALPERIN, *Phys. Rev. B* **25** (1982), 2185.
15. P. JORDAN, *Z. Phys.* **93** (1935), 464.
16. G. D. MAHAN, "Many Particle Physics," Plenum, New York, 1981.
17. N. S. MANTON, *Ann. Phys. (N.Y.)* **159** (1985), 220.
18. M. V. BERRY, *Proc. Roy. Soc. London A* **392** (1984), 45; B. SIMON, *Phys. Rev. Lett.* **51** (1983), 2167.
19. For example, M. STONE AND W. GOFF, *Nucl. Phys. B* **295** [FS21] (1988), 234.
20. S. TOMONAGA, *Prog. Theor. Phys.* **5** (1950), 544.
21. D. C. MATTIS AND E. H. LIEB, *J. Math. Phys.* **6** (1965), 304.
22. E. T. WHITTAKER AND G. N. WATSON, "A Course of Modern Analysis," Chap. XXI, Cambridge Univ. Press, Cambridge, UK, 1927.
23. W. SIEGAL, *Nucl. Phys. B* **238** (1984), 307.
24. C. IMBIMBO AND A SCHWIMER, *Phys. Lett. B* **193** (1987), 455.
25. R. FLOREANINI AND R. JACKIW, *Phys. Lett.* **59** (1987), 1873.
26. J. SONNENSCHEIN, *Nucl. Phys. B* **309** (1988), 752.
27. X. G. WEN, F. WILCZEK, AND A. ZEE, *Phys. Rev. B* **39** (1989), 11413.
28. J. R. KLAUDER AND B. S. SKAGERSTAM (EDS.), "Coherent States," World Scientific, Singapore, 1985.
29. A. M. PERELOMOV, *Commun. Math. Phys.* **26** (1972), 222; reprinted in [28].
30. A. PRESSLEY AND G. SEGAL, "Loop Groups," Clarendon Press, Oxford, 1986.
31. S. ISO AND H. MUKAIDA, "Necessity of Finite Size Term in the WZW Mode," University of Tokyo Preprint UT-555, TIT/HEP-154.
32. M. STONE, "How to Make a Bosonized Dirac Fermion Out of Two Bosonized Weyl Fermions," Preprint NSF-ITP-89-97.
33. K. HARADA, "Equivalence between the Wess-Zumino-Witten Model and Two Chiral Bosons," University of Heidelberg Preprint HD-THEP-90-9.
34. M. STONE AND F. GAITAN, Topolgical charge and chiral anomalies in Fermi superfluids, *Ann. Phys. (N.Y.)* **178** (1987), 89.
35. D. BOYANOVSKI, E. DAGOTTO, AND E. FRADKIN, *Phys. Rev. Lett* **57** (1986), 2967; *Nucl. Phys. B* **285** (1987), 340.
36. A. M. J. SCHAKEL AND P. BATENBURG, *Ann. Phys. (N.Y.)* **195** (1990), 25.

Printed by Catherine Press, Ltd.. Tempelhof 41, B-8000 Brugge, Belgium

PHYSICAL REVIEW B VOLUME 42, NUMBER 13 1 NOVEMBER 1990

Schur functions, chiral bosons, and the quantum-Hall-effect edge states

Michael Stone
Department of Physics, University of Illinois at Urbana-Champaign, 1110 West Green Street, Urbana, Illinois 61801
(Received 27 April 1990)

I demonstrate how the many-body wave function may be used to describe the bosonization of the edge excitations of a droplet of $\nu=1$ quantum-Hall liquid. In particular, I exhibit an isomorphism between the charge-neutral edge-state excitations of the droplet and the space of universal symmetric polynomials. There are two natural bases for this space; the first, the Schur functions, correspond to the fermion picture; the second, generated by the power sums, yields the Bose picture and the Kac-Moody algebra. I also show explicitly how the loop group LU(1) acts to create the coherent-state deformations of the droplet shape used in path-integral bosonization and in the quantization of chiral bosons.

I. INTRODUCTION

The two-dimensional electron gas (2DEG) in a strong magnetic field is a Fermi system that can be mapped into an equivalent Bose system in two distinct, but related, ways. The bosonization of the bulk degrees of freedom by means of a Chern-Simons statistics field[1] leads to the approximate mean-field theory of the fractional quantum-Hall effect (FQHE) which was further developed by Zhang, Hansen, and Kivelson[2] and by Read.[3] The second bosonization concentrates on the edge states, the only low-energy states when the bulk of the two-dimensional electron gas exhibits the quantum-Hall effect (QHE).[4] These states have been shown by Wen[5] to span representations of Kac-Moody current algebras.[6] The fundamental representations of these algebras are unique up to unitary equivalence but their generators may be written in terms of either Bose or Fermi operators and this dual description extends to the dynamics: the fermion operators serve to promote individual fermions to excited states while the chiral boson fields create coherent-edge waves or deformations in the shape of the droplet of 2DEG.[7]

The bosonic description of the edge degrees of freedom is a chiral variant of one-dimensional bosonization, a by now classic idea which goes back to Jordan[8] and Tomonaga[9] and whose modern formulation is due to Luther and Peschel[10] and to Coleman and Mandelstam.[11] The formalism invoked is rather abstract: Commuting variables are introduced via the language of field theory using Schwinger terms, vertex operators, operator product expansions, and renormalization. At no point are the wave functions of the individual electrons mentioned. In the QHE the many-body states are most familiarly exhibited as wave functions[12] and it is from the wave-function picture that much of our physical understanding of the effect is derived.

An alternative approach to bosonization used coherent states and coherent-state path integrals.[13] This formalism is the geometric counterpart of the algebraic representation theory of the Kac-Moody algebras, and focuses on the groups obtained by exponentiating the Lie algebras.[14] Since the geometry of infinite-dimensional Lie groups can seem complicated, it is valuable to have an explicit example of the resulting many-body coherent states. The Hall edge states provide a physical realization of these constructions and this paper is intended to provide an exegesis of the ideas behind them.

In the second section I will describe the physical basis for the Fermi-Bose mapping in the case of filling-fraction $\nu=1$, an observation due to Haldane,[15] and show how the Schur functions map the states of the QHE onto the ring of symmetric polynomials. The third section reviews some of the algebraic properties of this ring. The fourth section will show how it becomes a Hilbert space, and how two natural bases form the Bose and Fermi pictures of the states. The fifth section will use the constructions from the earlier sections to produce the explicit coherent-state wave functions generated by the loop group. I will show that they describe deformations of the edges of the QHE sample. In the discussion section I will briefly discuss the FQHE states.

I should make it clear that none of the mathematics in this paper is original, only the physical application is believed to be new. In particular the key constructions rely heavily on Secs. 10.3 and 10.4 of Ref. 14.

II. SLATER DETERMINANTS AND SCHUR FUNCTIONS

In the symmetric gauge, and with a choice of length scale, the lowest Landau-level single-particle wave functions can be identified with the space of analytic functions, $f(z)\,z=x+iy$, subject to the restriction that

$$\psi(z)=f(z)e^{-(1/2)|z|^2} \tag{2.1}$$

be square integrable. A normalized basis for this space is given by

$$\psi_n(z)=\frac{1}{\sqrt{\pi n!}}z^n e^{-(1/2)|z|^2}\,. \tag{2.2}$$

The simplest N-body state, a homogeneous circular droplet of electron liquid with filling-fraction $\nu=1$, has an antisymmetric Slater determinant of the basis functions as its wave function

$$\Psi_0(z_1,z_2,\ldots,z_N)=\begin{vmatrix} z_1^{N-1} & z_1^{N-2} & \cdots & 1 \\ z_2^{N-1} & z_2^{N-2} & \cdots & 1 \\ \vdots & \vdots & \ddots & \vdots \\ z_N^{N-1} & z_N^{N-2} & \cdots & 1 \end{vmatrix} \times \exp\left[-\tfrac{1}{2}\sum_i |z_i|^2\right] . \tag{2.3}$$

Expectation values in this state are well known[16] to be calculable as statistical averages in a two-dimensional one-component Coulomb gas with the Gaussian factor providing the neutralizing background charge.

In the absence of a perturbing potential all N-body states with energy below the cyclotron energy gap, $\omega_c=eB/m$, are degenerate with this one. They are created by moving some electrons to states outside the droplet. A typical wave function is

$$\Psi_{|\lambda|}(z_1,z_2,\ldots,z_N)=\begin{vmatrix} z_1^{\lambda_1+N-1} & z_1^{\lambda_2+N-2} & \cdots & z_1^{\lambda_N} \\ z_2^{\lambda_1+N-1} & z_2^{\lambda_2+N-2} & \cdots & z_2^{\lambda_N} \\ \vdots & \vdots & \ddots & \vdots \\ z_N^{\lambda_1+N-1} & z_N^{\lambda_2+N-2} & \cdots & z_N^{\lambda_N} \end{vmatrix} \times \exp\left[-\tfrac{1}{2}\sum_i |z_i|^2\right] \tag{2.4}$$

with $\lambda_1 \geq \lambda_2 \geq \lambda_3$, etc.

The wave function $\Psi_{|\lambda|}$ still describes an N-electron state but has $\lambda_1+\lambda_2+\cdots+\lambda_n=M$ extra powers of z. Each λ_i means that an electron has been moved from its position in the "sea" and raised from, say, the z^m state to the $z^{m+\lambda_i}$ state. If the droplet were confined in a potential well, the states near the edge would see a potential gradient and, since the single-particle z^m states are localized near a circle whose radius depends on m, they will have energy $E_m=Cm$. The new many-body state has energy $E=CM$. The Hall droplet becomes a physical realization of a "chiral" Dirac sea where all the states near the Fermi surface, identified with the physical edge of the droplet, move in the direction of the Hall current induced by the gradient.[7]

The set of distinct λ_i with $\sum_i \lambda_i = M$ are in one-to-one correspondence with the partitions of M, which I will label by the collective symbol $\{\lambda\}$,

$$\{\lambda\}\equiv\{\lambda_1,\lambda_2,\ldots,\lambda_N\} . \tag{2.5}$$

The number $p(m)$ of distinct partitions of m is given by the number theory partition function

$$Z=\frac{1}{\prod_{n>0}(1-x^n)}=\sum_m p(m)x^m . \tag{2.6}$$

Put $x=e^{-\beta C}$

$$Z=\frac{1}{\prod_{n>0}(1-e^{-\beta Cn})}=\sum_m p(m)e^{-\beta Cm} \tag{2.7}$$

and the similarity to a statistical-mechanical partition function already suggests a description of the charge-neutral excitations in terms of bosonic oscillators with energy Cn.

To explore what this means for the wave functions we will concentrate on the determinant part of the wave function. Define

$$D_{|\lambda|}(z)=\begin{vmatrix} z_1^{\lambda_1+N-1} & z_1^{\lambda_2+N-2} & \cdots & z_1^{\lambda_N} \\ z_2^{\lambda_1+N-1} & z_2^{\lambda_2+N-2} & \cdots & z_2^{\lambda_N} \\ \vdots & \vdots & \ddots & \vdots \\ z_N^{\lambda_1+N-1} & z_N^{\lambda_2+N-2} & \cdots & z_N^{\lambda_N} \end{vmatrix} \tag{2.8}$$

or, in compact but, I hope, self-explanatory notation

$$D_{|\lambda|}(z)=\det|z_s^{\lambda_t+n-t}| . \tag{2.9}$$

The empty partition corresponds to the Vandermonde determinant

$$D(z)=\begin{vmatrix} z_1^{N-1} & z_1^{N-2} & \cdots & 1 \\ z_2^{N-1} & z_2^{N-2} & \cdots & 1 \\ \vdots & \vdots & \ddots & \vdots \\ z_N^{N-1} & z_N^{N-2} & \cdots & 1 \end{vmatrix}=\prod_{i<j}(z_i-z_j) \tag{2.10}$$

which is a factor of all the $D_{|\lambda|}(z)$—so the quotient

$$\Phi_{|\lambda|}(z)=D_{|\lambda|}(z)/D(z) \tag{2.11}$$

is a *symmetric* polynomial in the z_i. Clearly all charge-neutral "excited" states $\Psi_{|\lambda|}$ are obtained by multiplying the "ground state" Ψ_0 by one of these symmetric polynomials[15] and the quantum Hilbert space of neutral excitations is isomorphic to the linear space spanned by them.

The polynomial $\Phi_{|\lambda|}(z)$ is the *Schur function* associated with the partition $\{\lambda\}$ of M. These functions are familiar in physics as the characters of the groups $GL(n)$ or $U(n)$.[17] The partitions are usually displayed as *Young tableaux*—diagrams with λ_i boxes in the ith row. For example,

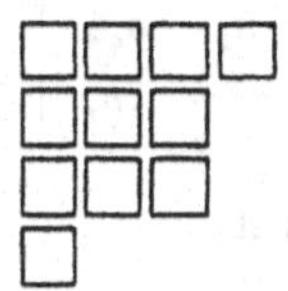

represents the partition $\{4,3,3,1\}$. Schur functions are also important as they form a linear basis for the space of symmetric polynomials and it is in this role, rather than that of group characters, that we will use them.

III. SYMMETRIC FUNCTIONS

In this section I will review those parts of the theory of symmetric functions that we need in the sequel. The material is standard and can be found in many books, e.g., Ref. 18.

Given a set of symbols α_i, $i=1,N$ (for example, complex variables like our z_i) form the ring $\mathcal{S}(\alpha)$ of polynomials in the α_i which are invariant under arbitrary permutations of the α_i.

Elements of this ring include the *elementary symmetric functions*, a_n, $n=1,N$ defined by

$$\prod_i (1-\alpha_i x)=1-a_1 x+a_2 x^2+\cdots \pm a_N x^N , \quad (3.1)$$

and the *homogeneous product sums* h_i, $i=1,\infty$, defined by

$$\frac{1}{\prod_i (1-\alpha_i x)}=1+h_1 x+h_2 x^2+h_3 x^3+\cdots . \quad (3.2)$$

Inverting the power-series definitions, we can express the a_n as polynomials in the h_n and vice versa. Remarkably the coefficients are integers in both directions. Sums and products of either the a_n or the h_n generate the ring $\mathcal{S}(\alpha)$.

The *power sums* S_k, $k=1,\infty$ are defined as

$$S_k(\alpha)=\sum_i \alpha_i^k \quad (3.3)$$

and using the relation

$$\exp\left[\sum_k \frac{1}{k}x^k S^k(\alpha)\right]=1+h_1 x+h_2 x^2+h_3 x^3+\cdots \quad (3.4)$$

we can express the S_n as polynomials in the h_n (with integer coefficients), and the h_n as polynomials (with rational coefficients) in the S_k, so sums of products of the S_n also generate $\mathcal{S}(\alpha)$.

The Schur functions can be written in terms of the h_n as

$$\Phi_{\{\lambda\}}(\alpha)=\det|h_{\lambda_s-s+t}| . \quad (3.5)$$

For example,

$$\Phi_{\{p\}}=h_p \quad (3.6)$$

and

$$\Phi_{\{p,q,r\}}=\begin{vmatrix} h_p & h_{p+1} & h_{p+2} \\ h_{q-1} & h_q & h_{q+1} \\ h_{r-2} & h_{r-1} & h_r \end{vmatrix} . \quad (3.7)$$

We can easily evaluate

$$\Phi_{\{1^N\}}(z)=z_1 z_2\cdots z_N=a_N , \quad (3.8)$$

and this is an example of the expression for $\Phi_{\{\lambda\}}$ in terms of the a_n: We use the *conjugate partition*, the Young tableau with the rows and columns interchanged, in the same manner of determinant as the h expression, e.g.,

$$\Phi_{\{322\}}=\begin{vmatrix} h_3 & h_4 & h_5 \\ h_1 & h_2 & h_3 \\ 1 & h_1 & h_2 \end{vmatrix}=\begin{vmatrix} a_3 & a_4 & a_5 \\ a_2 & a_3 & a_4 \\ 0 & 1 & a_1 \end{vmatrix} , \quad (3.9)$$

where $\{3,2,2\}$

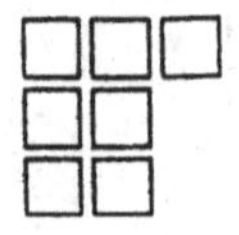

is the partition conjugate to $\{3,3,1\}$

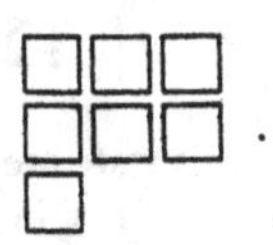

None of the relations between the h_n, a_n, S_n, and $\Phi_{\{\lambda\}}$ depend explicitly on N, but when N is finite the a_i for $i>N$ are zero, as are the Schur functions with more than N rows in the partition $\{\lambda\}$. Similarly only the first N of the h_n and S_n's are functionally independant. It is convenient to regard N as infinite and then we speak of the ring of *universal* symmetric functions.

The Schur functions form a *linear basis* for the ring $\mathcal{S}(\alpha)$: Any element of $\mathcal{S}(\alpha)$ is a linear combination of $\Phi_{\{\lambda\}}$'s. To see this, observe that any symmetric function can be converted to an antisymmetric one by multiplying by $D(\alpha)$, and that antisymmetrizing each monomial in the resulting expression converts it to a $D_{\{\lambda\}}(\alpha)$. Finally dividing out the catalytic $D_{\{\lambda\}}$ returns the symmetric function as a sum of $\Phi_{\{\lambda\}}$.

There is a key identity connecting the Φ's and the $\mathcal{S}_k$'s: Take two sets of indeterminates, α_i and β_i, then

$$\exp\left[\sum_k \frac{1}{k}S_k(\alpha)S_k(\beta)\right]=\sum_{\{\lambda\}}\Phi_{\{\lambda\}}(\alpha)\Phi_{\{\lambda\}}(\beta) . \quad (3.10)$$

It is straightforward to prove (3.10) from Cauchy's determinant identity[19]

$$\det\left|\frac{1}{1-\alpha_s\beta_t}\right|=\frac{D(\alpha)D(\beta)}{\prod_{ij}(1-\alpha_i\beta_j)} . \quad (3.11)$$

Once we have the relation (3.10) we can take all but one, say $\beta_1=x$, of the β_i to be zero. We see that (3.4) is a special case of (3.10).

IV. $\mathcal{S}(z)$ AS A HILBERT SPACE

As mentioned in the Introduction, a general reference for this section is Secs. 10.3 and 10.4 of Ref. 14.

The symmetric algebra $\mathcal{S}(z)$ is both a vector space spanned by the $\Phi_{\{\lambda\}}$ and a ring generated by the $S_n(z)$. We now define an inner product on $\mathcal{S}(z)$ which serves to make it into a Hilbert space. I wish the generators $S_k(z)$ to be independent so we take the number N of the z_i to infinity and this means taking the thermodynamic limit of the 2DEG.

We define the inner product of two polynomial functions of the S_k to be

$$\langle f(S)|g(S)\rangle=\int\prod_k\left[\frac{d^2S_k}{\pi k}\right]f^*(S)g(S)\times\exp\left[-\sum_k\frac{1}{k}|S_k|^2\right]. \quad (4.1)$$

In particular the product of the S_n's themselves is

$$\langle S_k|S_{k'}\rangle=k\delta_{k,k'}. \quad (4.2)$$

A general element in the ring is a sum of products of the S_n such as

$$S_{(l)}(z)=S_1^{l_1}S_2^{l_2}\cdots S_n^{l_n}, \quad (4.3)$$

and their inner products are

$$\langle S_{(l)}|S_{(l')}\rangle=(1^{l_1}l_1!2^{l_2}l_2!\cdots n^{l_n}l_n!)\delta_{(l)(l')}. \quad (4.4)$$

The inner product is essentially that of a Bargman-Fock space where creation operators are represented by multiplication by S_k and annihilation operators by their adjoint, $k\partial_{S_k}$. Our Hilbert space is therefore isomorphic to the space created from a cyclic vector $|0\rangle$ by application of bosonic $a_k^\dagger$'s whose commutation relations are

$$[a_k,a_{k'}^\dagger]=k\delta_{k,k'},\quad k>0. \quad (4.5)$$

If we define $a_{-k}=a_k^\dagger$ we can write

$$[a_k,a_{k'}]=k\delta_{k+k',0}. \quad (4.6)$$

They are the commutation relations of a level-one Abelian Kac-Moody algebra.

The remarkable property of this inner product is that the $\Phi_{\{\lambda\}}$, which we already know to form a linear basis for the Hilbert space $\mathcal{S}(z)$, are *orthonormal* with respect to it. To prove this use the reproducing kernel identity

$$\int\left[\prod_k\frac{d^2S_k(z)}{\pi k}\right]F(S_k(z))\exp\left[-\sum_k\frac{1}{k}|S_k(z)|^2\right]\times\exp\left[\sum_k\frac{1}{k}S_k(\bar z)S_k(z')\right]=F(S_k(z')) \quad (4.7)$$

and (3.9)

$$\exp\left[\sum_k\frac{1}{k}S_k(\bar z)S_k(z')\right]=\sum_{\{\lambda\}}\Phi_{\{\lambda\}}(\bar z)\Phi_{\{\lambda\}}(z') \quad (3.10')$$

with the choice $F(S_k(z))=\Phi_{\{\mu\}}(z)$.

Since the $\Phi_{\{\lambda\}}(z)$ are linearly independent [they are independent on the torus of U(n) where the $z_i=e^{i\theta_i}$, so *a fortiori* independent on the larger space of all z_i] we must have

$$\langle\Phi_{\{\lambda\}}|\Phi_{\{\lambda'\}}\rangle=\int\left[\prod_k\frac{d^2S_k}{\pi k}\right]\Phi^*_{\{\lambda\}}\Phi_{\{\lambda'\}}\times\exp\left[-\sum_k\frac{1}{k}|S_k|^2\right]=\delta_{\{\lambda\}\{\lambda'\}}. \quad (4.8)$$

An alternative demonstration is via Frobenius's famous reciprocity formula connecting the characters of the permutation group with the characters of GL_n.[20] Frobenius's formula asserts that

$$S_{(l)}=\sum_{\{\lambda\}}\chi_{(l)}^{\{\lambda\}}\Phi_{\{\lambda\}}(z), \quad (4.9)$$

where the $\chi_{(l)}^{\{\lambda\}}$ are the characters of the representation $\{\lambda\}$ of a permutation group. The conjugacy classes of the group are labeled by (l). As group characters the $\chi_{(l)}^{\{\lambda\}}$ obey the orthogonality conditions

$$\frac{1}{g}\sum_{(l)}g_{(l)}\chi_{(l)}^{\{\lambda\}}\chi_{(l)}^{\{\lambda'\}}=\delta_{\{\lambda\}\{\lambda'\}}, \quad (4.10)$$

where $(l_1+2l_2+3l_3+\cdots)!=g$ is the order of the permutation group and

$$g_{(l)}=\frac{g}{1^{l_1}2^{l_2}\cdots l_1!l_2!\cdots l_n!} \quad (4.11)$$

is the number of permutation group elements in the conjugacy class. Equation (4.9) can now be inverted to give $\Phi_{\{\lambda\}}$ in terms of the S_k,

$$\Phi_{\{\lambda\}}(z)=\frac{1}{g}\sum_{(l)}g_{(l)}\chi_{(l)}^{\{\lambda\}}S_{(l)} \quad (4.12)$$

and then (4.4) and (4.10) yield (4.6). This demonstration is not as independent as it seems—the conventional proof of (4.9) depends on (3.9)—but the formulas (4.9) and (4.12) relating the two bases of the Hilbert space are worth exhibiting.

We must now ask the crucial question: Is the inner product we have defined on $\mathcal{S}(z)$ the same as the one given by quantum mechanics? The answer is, in general, *No*: the states (2.4) obtained by multiplying a normalized Ψ_0, Eq. (2.3), by $\Phi_{\{\lambda\}}$ are orthogonal with respect to both products—but when regarded as elements of $\mathcal{S}(z)$ they are normalized to unity, while they are *not* [because we need to include the $\sqrt{n!}$'s from (2.2)] normalized wave functions themselves. This is not, however, a problem in the thermodynamic limit. When the curvature of the droplet edge can be ignored on the length scales of interest, states near the edge reduce to those found in the Landau gauge. For these physically accessible states, the normalization is independent of n and the two products do coincide.

The fruit of our labors is two separate descriptions of the space of edge excitations of the Hall droplet: one in terms of the bosonic a_n's, and one in terms of the elementary fermionic excitations corresponding to the Schur functions. The next step is the identification of the a_n operators in the Kac-Moody algebra (4.6) with the Fourier components of the surface currents used in Ref. 7.

or, equivalently, with the mode expansion of the chiral boson describing the edge waves. This we will do in the next section.

V. COHERENT STATES

In the path-integral route to bosonization of chiral fermions[13] one uses generalized coherent states[21,22] which are obtained by the action of a loop group on the Dirac sea. The loop-group LG acts on each first-quantized, single-particle state in the sea by multiplying the wave function with a common position-dependent element of the group G. For example, the loop-group LU(1) has elements $g(x)=e^{i\varphi(x)}$ and acts by

$$e^{ikx} \mapsto e^{i\varphi(x)}e^{ikx} . \tag{5.1}$$

If one thinks of the states in the sea as spanning a hyperplane (or, more precisely, a closed subspace) in the Hilbert space, this common group action rotates the hyperplane bodily into a new orientation. The resulting many-body wave function is the Slater determinant made from any set of wave functions spanning the new subspace. There will be an *isotropy subgroup* of elements which leave the subspace fixed, merely performing rotations within the hyperplane and within its orthogonal complement. The set of coherent states corresponds to the quotient space of the full group by the isotropy subgroup.

In the second-quantized language the group generators, $j(x)$, form a current algebra with commutation relations

$$[j(x),j(x')]=\frac{-i}{2\pi}\partial_x\delta(x-x') , \tag{5.2}$$

or their non-Abelian generalization

$$[j_a(x),j_b(x')]=if^c_{ab}j_c(x)\delta(x-x') - \frac{i}{2\pi}g_{ab}\partial_x\delta(x-x') . \tag{5.3}$$

In this language the coherent states are made by applying exponentials of the generators to the Dirac sea

$$|\varphi(x)\rangle=\exp\left[i\int dx\, \varphi(x)j(x)\right]|0\rangle . \tag{5.4}$$

Since $\langle 0|j(y)|0\rangle=0$ and

$$\begin{aligned}\exp\left[-i\int \varphi(x')j(x')dx'\right]&j(x)\times\exp\left[+i\int \varphi(x')j(x')dx'\right]\\ &=j(x)-\frac{1}{2\pi}\int dx'\varphi(x')\partial_{x'}\delta(x'-x) \quad (5.5)\\ &=j(x)+\frac{1}{2\pi}\partial_x\varphi(x) \quad (5.6)\end{aligned}$$

we see that the $|\varphi\rangle$ has charge

$$\langle\theta|j(x)|\varphi\rangle=+\frac{1}{2\pi}\partial_x\varphi(x) . \tag{5.7}$$

Assuming that x has period 2π we can Fourier decompose $j(x)$

$$j(x)=\sum_n e^{-inx}j_n \tag{5.8}$$

and read off from (5.1) that

$$[j_n,j_{n'}]=-n\delta_{n+n',0} . \tag{5.9}$$

The j_n, $n>0$ are therefore to be interpreted as creation operators, while the j_n, with negative n annihilate $|0\rangle$ and generate the isotropy group. Quotienting out the isotropy group, we can rewrite (5.4) as

$$|\varphi(x)\rangle \propto \exp\left[i\sum_{n>0}\varphi_n j_n\right]|0\rangle . \tag{5.10}$$

In the case of the Hall droplet it is convenient to adjust the magnetic field so the surface of the droplet coincides with the circle $z=e^{i\theta}$ and then an element of the loop-group LU(1) has a Laurent-Fourier expansion

$$g(z)=\exp\left[\sum_{n=-\infty}^{\infty} i\varphi_n z^n\right] . \tag{5.11}$$

The isotropy group (in this context called B^+) is composed of elements which are boundary values of functions analytic in $z>1$

$$g(z)=\exp\left[\sum_{n=-\infty}^{0} i\varphi_n z^n\right] \tag{5.12}$$

as they will always give determinants with some identical columns after multiplying into Ψ_0. The effective part of $g(z)$ is multiplication by elements of the subgroup N^- consisting of boundary values of functions analytic in $z<1$:

$$g(z)=\exp\left[\sum_{n=1}^{\infty} i\varphi_n z^n\right] . \tag{5.13}$$

Each state z^k in the Slater determinant is changed

$$z^k \mapsto \left[\exp\left[\sum_n i\varphi_n z^n\right]\right]z^k \tag{5.14}$$

and this has the effect of multiplying the whole determinant by the factor

$$\exp\left[\sum_n i\varphi_n S_n(z)\right] . \tag{5.15}$$

Comparing with (5.3) confirms that the operation of multiplying a wave function by S_k is the same as acting on it by the j_k surface current generator of the Kac-Moody algebra. We can identity $a_k^\dagger\equiv j_k$, $k>0$.

To see how these operations deform the charge distribution at the edge of the droplet we can use the Coulomb gas analogy for the wave function. Introduce a distribution of point "charges" of magnitude q_i, at the points $Z_i=e^{i\theta_i}$. Since we are only considering neutral excitations assume $\sum_i q_i=0$. The kth Fourier component of the external charge distribution is then $(1/2\pi)\sum_i q_i\bar{Z}_i^k$. Taking

$$i\varphi_k=\frac{1}{k}\sum_i q_i\bar{Z}_i^k, \quad k\neq 0 \tag{5.16}$$

multiplies the wave function by

$$\exp\left[\sum_k \frac{1}{k}\left[\sum_i q_i \bar{Z}_i^k\right] S_k(z)\right] = \frac{1}{\prod_{i,j}(1-\bar{Z}_i z_j)^{q_i}} \tag{5.17}$$

and, in the statistical average, the internal z_i charges will be attracted or repelled by the external ones. They will screen them over the scale set by the magnetic length (which is infinitesimal as we have taken the thermodynamic limit while keeping droplet radius fixed at unity) so the deformed state $|\varphi\rangle$ has charge at the edge. Because the division by $1/k$ in the Fourier sum is equivalent to integrating the charge distribution, the charge distribution is given by differentiating the configuration space $\varphi(\theta)$

$$\langle j(\theta)\rangle = \frac{1}{2\pi}\partial_\theta \varphi(\theta) , \tag{5.18}$$

thus reproducing the second-quantized operator result.

VI. DISCUSSION

The wave-function picture of chiral bosonization, which seems especially suited to the QHE, provides an alternative to both the field theory and to the coherent-state path-integral formalism. Like all bosonization schemes it provides what is essentially a kinematical, or bookkeeping, tool. But, while the results are derived initially for noninteracting fermions, they may be extended to interacting theories by finding bosonic equivalents of the fermionic interaction terms. The utility of the dual Bose-Fermi pictures lies in the approximation schemes they suggest—effects that are nonperturbative in one language become perturbative in the other.

One can extend these tricks and bosonize the edge states in the FQHE. For the $\nu=1/(2m+1)$ stage in the FQHE hierarchy the wave functions are the Laughlin states. These are simple powers of the $\nu=1$ Slater determinants and the only significant result of this is that the Berry phase, which appears in the coherent-state path integral and determines the level k of the Kac-Moody algebra, is multiplied by $2m+1$. for other hierarchy states the construction seems to be more complicated. Wen[23] has suggested a method which seems very similar in spirit to the Goddard, Kent, and Olive construction[24] of introducing new degrees of freedom and gauging out unwanted ones—indeed this is what is being done for the $\nu=2m+1$ states: We decompose the electron into $2m+1$ fermionic partons and use these to make one U(1) current of level $k=2m+1$ and one set of level $k=1$ SU$(2m+1)$ currents. The ground state is then the product of $2m+1$ independent Slater determinants, one for each kind of parton. The unwanted SU$(2m+1)$ degrees of freedom are then frozen out by forcing the each of the $2m+1$ partons to be at the same location as $2m$ others. The individual determinants then coincide and give the Laughlin state.

ACKNOWLEDGMENTS

This work was supported by the National Science Foundation (NSF) Grant Nos. DMR-88-18713 and DMR-86-12860. I would like to thank Eduardo Fradkin for drawing my attention to Refs. 5 and for much useful discussion. I would also like to thank X. G. Wen for pointing out an error in the first version of this paper and for sending me Ref. 23 before publication.

[1] S. M. Girvin and A. H. MacDonald, Phys. Rev. Lett. **58**, 1252 (1987).
[2] S. C. Zhang, T. H. Hansen, and S. Kivelson, Phys. Rev. Lett. **62**, 82 (1989).
[3] N. Read, Phys. Rev. Lett. **62**, 86 (1989).
[4] C. W. J. Bennakker, Phys. Rev. Lett. **64**, 216 (1990); A. H. McDonald, *ibid.* **64**, 220 (1990).
[5] X. G. Wen (unpublished).
[6] For a review, see P. Goddard and D. Olive, Int. J. Mod. Phys. A **1**, 303 (1986); also, see V. G. Kac, *Infinite Dimensional Lie Algebras* (Cambridge University Press, Cambridge, England, 1985).
[7] M. Stone, University of Illinois Report No. IL-Th-90-No. 8 [Ann. Phys. (N.Y.) (to be published)].
[8] P. Jordan, Z. Phys. **93**, 464 (1935).
[9] S. Tomonaga, Prog. Theor. Phys. **5**, 544 (1950).
[10] A. Luther and I. Peschel, Phys. Rev. B **9**, 2911 (1974).
[11] S. Coleman, Phys. Rev. D **11**, 2088 (1975); S. Mandelstam, *ibid.* **11**, 3026 (1975).
[12] For a review, see *The Quantum Hall Effect*, edited by R. E. Prange and S. M. Girvin (Springer-Verlag, Berlin, 1987).
[13] M. Stone, Phys. Rev. Lett. **63**, 731 (1989); Nucl. Phys. **B327**, 399 (1989).
[14] A. Pressley and G. Segal, *Loop Groups* (Clarendon, Oxford, 1986).
[15] F. D. M. Haldane, University of Minnesota report, 1989 (unpublished).
[16] R. B. Laughlin, in *The Quantum Hall Effect* (Ref. 12), Chap. 7.
[17] M. Hamermesh, *Group Theory and Its Application to Physical Problems* (Addison-Wesley, Reading, MA, 1962).
[18] D. E. Littlewood, *University Algebra* (Heinemann, London, 1950).
[19] M. Hamermesh, *Group Theory and Its Application to Physical Problems* (Ref. 17), p. 195.
[20] M. Hamermesh, *Group Theory and Its Application to Physical Problems* (Ref. 17), p. 197; H. Weyl, *Group Theory and Quantum Mechanics* (Dover, New York, 1950), p. 331.
[21] *Coherent States*, edited by J. R. Klauder and B-S. Skagerstam (World Scientific, Singapore, 1985).
[22] A. M. Perelomov, Commun. Math. Phys. **26**, 222 (1972); reprinted in *Coherent States* (Ref. 21).
[23] X. G. Wen (unpublished).
[24] P. Goddard, A. Kent, and D. Olive, Phys. Lett. **152B**, 88 (1985); Commun. Math. Phys. **103**, 105 (1986).